Physiology and Biotechnology Integration for Plant Breeding

BOOKS IN SOILS, PLANTS, AND THE ENVIRONMENT

Physiology and Biotechnology Integration for Plant Breeding

edited by

Henry T. Nguyen
University of Missouri–Columbia
Columbia, Missouri, U.S.A

Abraham Blum
Agricultural Research Organization
Volcani Center
Israel Ministry of Agriculture
Bet Dagan, Israel

CRC Press
Taylor & Francis Group
Boca Raton London New York

CRC Press is an imprint of the
Taylor & Francis Group, an **informa** business

CRC Press
Taylor & Francis Group
6000 Broken Sound Parkway NW, Suite 300
Boca Raton, FL 33487-2742

First issued in paperback 2019

© 2004 by Taylor & Francis Group, LLC
CRC Press is an imprint of Taylor & Francis Group, an Informa business

No claim to original U.S. Government works

ISBN-13: 978-0-8247-4802-9 (hbk)
ISBN-13: 978-0-367-39462-2 (pbk)

Library of Congress Cataloging-in-Publication Data
A catalog record for this book is available from the Library of Congress.

Visit the Taylor & Francis Web site at
http://www.taylorandfrancis.com

and the CRC Press Web site at
http://www.crcpress.com

Preface

The changing economies, the change in weather, the increasing population in developing countries, the desertification of certain parts of the globe and the degrading genetic diversity of our major crop plants all exert an increasing pressure on agriculture. The continuing economic and biological function and survival of agriculture demands increasingly greater physical, economical and intellectual resources. An important means for sustaining productive agriculture, especially in vulnerable ecosystems, is plant breeding. For a variety of reasons, most of which are derived from the above statements, there is an increasing demand for new plant cultivars that either have the potential for higher yield or have the capacity to perform in a stable and dependable manner under abiotic (environmental) and biotic constraints.

This increasing demand for higher yielding and environmentally stable crop cultivars comes exactly at the time when a new era in plant biotechnology is emerging, with the quickly developing disciplines of molecular biology and genomics. There are today great expectations that these new technologies will provide superior crop cultivars. Indeed, the molecular approach to plant breeding is proving to be an important component in the development of new cultivars having improved disease and pest resistance or higher product quality. Progress made in these instances by molecular biology was largely determined by the fact that the plant traits involved were under comparatively simple genetic control and therefore readily amenable to molecular manipulations and genetic transformations.

However, progress in raising the yield potential and in enhancing plant resistance to environmental stress by the use of the molecular approach has been relatively slow. Progress in this area does not only require proficiency in applying molecular technology. The nature of the traits involved in the yield potential and environmental adaptation is very complex physiologically and biochemically. Therefore, progress in improving yield and yield stability under stress environments cannot be achieved without understanding the physiology of yield and plant responses to the environment. Plant breeders and geneticists have long strived to have the capacity to develop crop cultivars by design. This goal has yet to be achieved. The biotechnological and genetic tools for enabling breeding by design are at hand or very nearly so. The missing component is our ability to form a phenotypic plant ideotype in terms of its functional physiology and genetics.

A link must be established between crop physiology and molecular biology in order for plant biotechnology to be effectively applied to the breeding of high yielding and environmentally stable crop plant cultivars. After the successful completion and annotation of model plant genomes, functional genomics has been central to research to monitor gene expression and gene functions. The comprehensive analysis of gene functions using microarray and proteomics tools accelerated investigations of cellular metabolism in specialized tissues or whole organisms responding to environmental changes. The multidimensional approach of system biology to discover and understand biological properties that emerges from the interactions of many system elements facilitates the collection of comprehensive data sets on a wide variety of plant responses.

This book is offered as a form of a dialog between the two disciplines— crop and plant physiology on one hand and plant biotechnology on the other—in order to fuse a better understanding and coordination toward the increasing needs of agriculture. The book offers the most updated information and views on crop physiology in relation to potential yield and environmental adaptation while on the other hand it summarizes the current status of genome mapping, functional genomics and proteomics tools to identify the gene functions leading to the application of molecular techniques for the improvement of crop yield and environmental adaptation. It is therefore intended for scientists and students interested in applying plant biotechnology and molecular biology to the improvement of crop yield and its resistance to environmental stress. This book is not a manual for the breeder or the molecular biologist, but it offers the most current discussion of the options and avenues available to those who are interested in augmenting crop yield and its stability.

Because the topic is very contemporary, there are varying views. The organization of the discussion on crop physiology into various chapters on the various crops by different experts allows us to illuminate the issue from two different perspectives. The first perspective is the plant species and the agroecological niche it occupies and the second perspective is the personal one, reflecting the views and the extensive experience of each of these leading experts. The application of biotechnology to the issue at hand is discussed as a more integrated approach, using the available knowledge in physiology and biochemistry and the options for building upon this knowledge in order to achieve the required genetic modification.

We would like to thank Russell Dekker for his support and production editors Michael Deters and Joe Cacciottoli for their help on the completion of this book. We thank our colleagues for their contribution. Special thanks go to the US-Israel Binational Agriculture Research and Development Fund (BARD) and the Rockefeller Foundation, whose funding support brought us together on research collaborations and interactions that initiated this book project. Finally, we thank our spouses, Devora Blum and Jenny Lam, for their constant support of our scientific journey.

Henry T. Nguyen, Columbia, Missouri, USA
Abraham Blum, Tel Aviv, Israel

Contents

Contributors

José L. Araus, Ph.D. Catedrático de Fisiologia Vegetal, Facultad de Biología, Universidad de Barcelona, Barcelona, Spain

Marianne Bänziger, Ph.D. CIMMYT, Zimbabwe, Harare, Zimbabwe

Francis R. Bidinger, Ph.D. International Crops Research Institute for the Semi-Arid Tropics (ICRISAT), Patancheru, Andhra Pradesh, India

Abraham Blum, Ph.D. Department of Field Crops, The Volcani Center, Bet Dagan, Israel

Brigitte Courtois, Ph.D. CIRAD-Biotrop, Montpellier, France

Gregory O. Edmeades, Ph.D. Pioneer Hi-Bred International, Inc., Waimea, Hawaii, U.S.A.

Anthony E. Hall, Ph.D. Department of Botany and Plant Sciences, University of California, Riverside, Riverside, California, U.S.A.

C. Thomas Hash, Ph.D. International Crops Research Institute for the Semi-Arid Tropics (ICRISAT), Andhra Pradesh, India

Tuan-hua David Ho, Ph.D.   Department of Biology, Washington University, St. Louis, Missouri, U.S.A. and Institute of Botany Academia Sinica, Taipei, Taiwan, Republic of China

David Hoisington, Ph.D.   Applied Biotechnology Laboratory, CIMMYT, Mexico City, Mexico

Abdelbagi M. Ismail, Ph.D.   Crop, Soil and Water Sciences Division, International Rice Research Institute (IRRI), Manila, Philippines

Elizabeth A. Lee, Ph.D.   Department of Plant Agriculture, University of Guelph, Guelph, Ontario, Canada

Henry T. Nguyen, Ph.D.   University of Missouri–Columbia, Columbia, Missouri, U.S.A.

Derrick M. Oosterhuis, Ph.D.   Department of Crop, Soil, and Environmental Sciences, University of Arkansas, Fayetteville, Arkansas, U.S.A.

Ilan Paran, Ph.D.   Department of Plant Genetics and Breeding, The Volcani Center, Bet Dagan, Israel

Andrew H. Paterson, Ph.D.   Center for Applied Genetic Technologies, Department of Crop and Soil Sciences, University of Georgia, Athens, Georgia, U.S.A.

M. S. Pathan, Ph.D.   Plant Sciences Unit, Department of Agronomy, University of Missouri—Columbia, Columbia, Missouri, U.S.A.

Shaobing Peng, Ph.D.   Crop, Soil and Water Sciences Division, International Rice Research Institute (IRRI), Manila, Philippines

Matthew P. Reynolds, Ph.D.   Wheat Program, International Maize and Wheat Improvement Centre (CIMMYT), Mexico City, Mexico

Jean-Marcel Ribaut, Ph.D.   CIMMYT, Mexico City, Mexico

Conxita Royo, Ph.D.   Field Crops Department, Centre UdL-IRTA, Lleida, Spain

Yehoshua Saranga, Ph.D. Department of Field Crops, Vegetables and Genetics, Faculty of Agricultural, Food and Environmental Quality Sciences, Hebrew University of Jerusalem, Rehovot, Israel

Tim L. Setter, Ph.D. Department of Crop and Soil Science, Cornell University, Ithaca, New York, U.S.A.

Gustavo A. Slafer, Ph.D. Departamento de Produccion Vegetal e IFEVA, Facultad de Agronomia, Universidad de Buenos Aires, Buenos Aires, Argentina

James E. Specht, Ph.D. Department of Agronomy and Horticulture, University of Nebraska, Lincoln, Nebraska, U.S.A.

James McD. Stewart, Ph.D. Department of Crop, Soil, and Environmental Sciences, University of Arkansas, Fayetteville, Arkansas, U.S.A.

Charles W. Stuber, Ph.D. USDA-ARS, Department of Genetics, North Carolina State University, Raleigh, North Carolina, U.S.A.

Prasanta K. Subudhi, Ph.D. Department of Agronomy, Agricultural Center, Louisiana State University, Baton Rouge, Louisiana, U.S.A.

Matthijs Tollenaar, Ph.D. Department of Plant Agriculture, University of Guelph, Guelph, Ontario, Canada

Tara T. VanToai, Ph.D. USDA-ARS Soil Drainage Research Unit, Columbus, Ohio, U.S.A.

Ray Wu, Ph.D. Department of Molecular Biology and Genetics, Cornell University, Ithaca, New York, U.S.A.

1

Physiology of Yield and Adaptation in Wheat and Barley Breeding

José L. Araus
Universitat de Barcelona
Barcelona, Spain

Gustavo A. Slafer
Universidad de Buenos Aires
Buenos Aires, Argentina

Matthew P. Reynolds
International Maize and Wheat Improvement Centre (CIMMYT)
Mexico City, Mexico

Conxita Royo
Centre UdL-IRTA
Lleida, Spain

1 INTRODUCTION

1.1 Sources for Improved Production Since the Beginning of Agriculture

Agriculture in the Old World started about 10,000 years ago, coinciding with the beginning of the Holocene. From this time up to the present, C3 cereals, such as bread (*Triticum aestivum* L.) and durum wheat (*Triticum turgidum* L. var. *durum*), as well as barley (*Hordeum vulgare* L.) have remained the outstanding crops in terms of area and food source (e.g., Evans, 1998). Relatively high cereal yields are suggested at the beginning of agriculture (Amir and Sinclair, 1994; Araus et al., 1999, 2001b), so that the yields believed to be attained then (the equivalent to ca. 1 Mg ha^{-1}; see, for instance, Araus et al., 2001b and references therein) were quite similar to the averaged yields attained globally at the beginning of the twentieth century (Calderini and Slafer, 1998; Slafer and Satorre, 1999). This means that the increased demands produced by the growing population since the Neolithic (some 4–10 million people; Minc and Vandermeer, 1990; Evans, 1998) to 1900 (more than 1 billon people) were chiefly satisfied by the enlargement of the cultivated area. During the twentieth century, when Mendel's laws were rediscovered and breeding started its period of scientifically based selection, increases in average yield were still not evident until around the 1950s (Calderini and Slafer, 1998; Slafer and Satorre, 1999). There was still a large increase in growing area as a response to the increased demand during the first half of the twentieth century (Slafer and Satorre, 1999). Since then, average yields increased dramatically in only a few decades. This change, known as the Green Revolution, was due to the introduction of semidwarf varieties with improved harvest index (HI) and, consequently, higher yield potential* (Calderini et al., 1999a; Abeledo et al., 2001), which in turn were more responsive to management improvement (Calderini and Slafer, 1999). It allowed the interaction between genetic and management improvement to express a relative increase in average yields even greater than that in population during the second half of the twentieth century (Slafer and Satorre, 1999). Before the improvement in yield potential during the intermediate decades of the twentieth century, responsiveness to environmental amelioration has been limited. For instance, the increase in atmospheric CO_2 concentrations from ca. 270 ppm before the beginning of the industrial revolution to the levels observed in 1950s (ca. 350 ppm) appears to have affected only marginally the yield levels of cereals (Slafer and Satorre, 1999).

Therefore, the extraordinary increase in average yields (which was not counterbalanced by a reduction in yield stability; see Smale and McBride, 1996; Calderini and Slafer, 1998) during the last few decades was due to the contribution of both genetic and management improvement. Although it is highly likely that the interaction between both factors was important, several

---

*In this paper, the term yield potential is used to define the productivity of adapted, high-yielding cultivars achieved in the absence of yield reductions due to either the presence of diseases, weeds, and insects or insufficient availability of water and major nutrients (see, for more details, Evans and Fischer, 1999).

analyses estimated relative breeding contributions to total yield increases obtained by farmers ranged from ca. 30% in Mexico (Bell et al., 1995) to ca. 50% in most other countries (e.g., Slafer et al., 1994a). The remaining 50% of the improvement came from changes in agronomic practices such as increased use of N fertilizer, P fertilizer applications, fitting of sowing density and crop phenology, use of herbicides for weed control, irrigation, and mechanization.

1.2 Sources for Improved Production in the Future

Global demand for wheat is expected to rise by approximately 1.3% $year^{-1}$ (and by approximately 1.8% $year^{-1}$ in developing countries) over the next 20 years (Rosegrant et al., 1995). Meeting these demands by increasing wheat production through increased land use is not very likely. Cultivated areas of wheat in developing countries are expected to rise by only 0.14% $year^{-1}$ through 2020 (Rosegrant et al., 1995). Thus, most of the needed increase in production must come from increases in average yields. Both economic and environmental analyses suggest that we can only marginally depend on improved crop management to provide the increase in yield needed to keep pace with population increase. Yield gaps (potential minus actual) in high production environments are rather small: Actual yields account for 70% to 80% of maximum yields* obtained on experiment stations (Byerlee, 1992; Pingali and Heisey, 1999); in those environments, there is little to expect from adoption of improved management. In other lower-yielding environments, the gap between potential and actual yields is large, which may be taken as an indication of the potential contribution that management improvement might make in these cases. It is believed that half of the large contribution made by management improvement to raise average yields in the last decades came solely from the increase in the use of N fertilizers. However, the widespread belief that intensive agriculture should be discouraged because of its detrimental effect on the environment will lead to more sustainable agricultural systems, therefore limiting in the future the rate of resource application (e.g., Austin, 1999). On the other hand, it is expected that a steady increase in atmospheric CO_2 concentrations will take place during the next few decades, which should promote yield. However, although a positive effect on yield is likely in some high-yielding conditions (e.g., Marderscheid and Weigel, 1995; Amthor, 2001), in most other cases, the environmental constraints accompanying CO_2 rise may make these responses negligible or even negative. This is the case when the increased CO_2 is expected to be accompanied by warmer or drier conditions or insufficient levels of N fertilization. Even in some high-yielding environments, the expected increases in temperature associated with

---

* Also termed potential yield, the highest possible yields obtainable with ideal management and nutrient under specific soil and weather conditions.

the global increase in greenhouse gases can counteract the effect of increasing atmospheric CO_2. Detrimental effects of high temperatures on both grain number per unit land area (ca. Fischer, 1985) and individual grain weight (ca. Wardlaw and Moncur, 1995) are well known.

All in all, it seems that we will strongly rely on genetic gains in yield of wheat and barley (and other cereals) as the exclusive source for the required increased production in the near future (Slafer et al., 1999; Araus et al., 2002).

1.3 Objectives

In this review, we summarize the major physiological changes that took place due to the breeding process of the twentieth century and conclude on the most promising traits to be considered in the future of cereal-breeding programs. Readers interested in more basic physiological ecological and molecular aspects of yield determination in wheat and barley may consult some of the recent books on these issues (Satorre and Slafer, 1999; Smith and Hamel, 1999; Slafer et al., 2001b) as well as the renown classic books by Evans (1975, 1993), Smith and Banta (1983), Rasmusson (1985), Heyne (1987), and Evans and Peacock (1981).

This chapter focuses mainly on analytical approaches directly linked to our ecophysiological understanding of the crop through:

(i)   The classical view of yield (GY) as a function of its components ($GY = Spikes/m^{-2} \times Grains/spikes \times GWt$; i.e., the number of spikes per unit land area, the grains per spike, and the averaged grain weight);

(ii) The carbon–economy-based relationship [$GY = RAD \times \%RI \times RUE \times HI$], where RAD is the total quantity of incident solar radiation throughout the growing period; %RI is the fraction of RAD intercepted by the canopy averaged across the crop cycle; RUE is the radiation-use efficiency, which is the overall photosynthetic efficiency of the crop; and HI is the harvest index or the fraction of the total dry matter harvested as yield; or

(iii) The water-use-based relationship [$GY = W \times WUE \times HI$], where W is the water transpired by the crop plus direct evaporation from the soil; WUE is the water-use efficiency, i.e., the ability of the crop to produce biomass per unit of water evapotranspired; and HI is the harvest index.

2 PHYSIOLOGICAL CHANGES ASSOCIATED WITH GENETIC GAINS IN GRAIN YIELD

One popular approach to identify physiological attributes contributing to increased yield potential consists in determining the physiological bases of the differences in yield between cultivars released at different eras (e.g., Austin 1999; Austin et al., 1980a; Feil, 1992; Loss and Siddique, 1994; Slafer et al., 1994a; Calderini et al., 1999a; Abeledo et al., 2001).

Most of these studies have been carried out under different field conditions, which seems to be adequate to produce reliable figures of genetic gains in yield and its determinants (Calderini et al., 1999a). These studies have been often performed on wheat (reviewed by Slafer and Andrade, 1991; Feil, 1992; Loss and Siddique, 1994; Slafer et al., 1994a; Calderini et al., 1999a) rather than barley (Abeledo et al., 2001). Most of the traits identified in the retrospective analyses have been shown to be constitutive.

2.1 Yield Gains from Wheat and Barley Breeding

In both cereals, genetic improvements in yield seemed to have been small during the first part of the twentieth century and much faster later on (see extended discussions on this fact in reviews by Slafer et al., 1994a; Calderini et al., 1999a; Abeledo et al., 2001). This trend has been discussed regarding actual farmer yields (see above) and provides indirect support to the idea that increasing yield potential is essential in improving the broad tolerance to mild and moderate stresses (Slafer and Araus, 1998). Although the positive interaction between genetic and management improvements on measured yield gains may overestimate the contribution frequently assigned to improvements in yield potential, results from conventional breeding also supports this idea. For instance, the wheat program at International Maize and Wheat Improvement Center (CIMMYT) focuses on raising yield potential (e.g., Reynolds et al., 1996); some high-yielding cultivars were also tolerant to a number of environmental stresses (Rajaram and van Ginkel, 1996).

Absolute gains in yield from breeding differed greatly among countries, but much of the difference was related to the environmental conditions under which the breeding process took place. In general, the better the agronomic conditions, the faster is the rate of genetic gain, as expected from theory (Richards, 1996b; Slafer et al., 1999). For instance, genetic gains in wheat yield from the 1860s to the 1980s have been far larger in the United Kingdom (23.3 kg ha^{-1} yr^{-1}; Austin et al., 1989) than in Australia (5.3 kg ha^{-1} yr^{-1}; Perry and D'Antuono, 1989), which agrees with differences between both countries in environmental growing conditions (Calderini and Slafer, 1998). However, if yields of cultivars released in different eras are expressed in relative terms of the average yield of the study (as an indirect indicator of the environmental condition), yield gains are similar (Fig. 1). Thus, when considering the gains in relative terms (% yr^{-1}), most differences among breeding programs of countries contrasting in genetic gains disappeared (Calderini et al., 1999a; Abeledo et al., 2001).

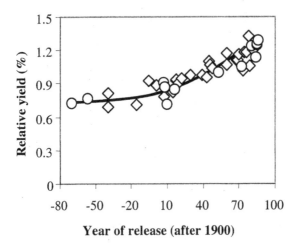

Figure 1 Grain yield of cultivars released at different eras when grown simultaneously in the field in favorable (U.K., O) and stressed (Australia, ◇) conditions. Yields are given in relative values (the yield of each cultivar as a percentage of the average yield of the corresponding experiment). (From Calderini and Slafer, 1999 and Araus et al., 2001b.)

2.2 Physiological Changes Produced During Genetic Improvement of Yield

2.2.1 Crop Phenology

Phenology largely determines the adaptability of a cereal crop to a certain range of environmental conditions and may also strongly affect yield potential (Slafer et al., 2001a). However, the length of the period between seedling emergence and heading was only occasionally modified in a systematic way by breeders (Slafer et al., 1994a; Abeledo et al., 2001). For instance, it has been found in several studies with wheat cultivars released at different eras that time to anthesis was not (or was only slightly) modified by breeding (e.g., Deckerd et al., 1985; Waddington et al., 1986; Hucl and Baker, 1987; Cox et al., 1988; Austin et al., 1989; Slafer and Andrade, 1989; Calderini et al., 1997). However, developmental traits in cereals are particularly significant in water-limiting conditions of Mediterranean environments, where available water becomes increasingly scarce toward the end of the growing season (Richards, 1991; Slafer et al., 1994a; Royo et al., 1995; Passioura, 1996; Bort et al., 1998; González et al., 1999). The better adapted cultivars develop faster, reaching anthesis earlier, thus minimizing the exposure to terminal drought stress.

Therefore, selection for yield under such conditions led to a systematic short-ening of the duration from sowing to anthesis such as in Australian wheat (Perry and D'Antuono, 1989; Siddique et al., 1989a,b; Yunusa et al., 1993; Loss and Siddique, 1994).

In the few studies available in barley, there were no clear trends for changes in time to anthesis. In general, the relationship between time from seedling emergence to heading and the year of release of the cultivars was not significant (e.g., Wych and Rasmusson, 1983; Boukerrou and Rasmusson, 1990; Bulman et al., 1993; Muñoz et al., 1998), or the slope of such relation-ship was relatively small (Martiniello et al., 1987; Riggs et al., 1981, Jedel and Helm, 1994b). It is speculated that as in wheat (Perry and D'Antuono, 1989; Siddique et al., 1989a), barley selection for yield in these regions may have given preference to early flowering genotypes (Loss and Siddique, 1994).

2.2.2 Plant Height

The stature of the stem is important in the determination of cereal produc-tivity as it affects both yield potential and lodging. Semidwarf wheats are able to receive higher rates of fertilizer application because they are generally resistant to lodging. There is a general concensus that breeding programs throughout the world consistently reduced stem height in modern cultivars as compared to older ones in both wheat (e.g., Austin et al., 1980a,b, 1989; Cox et al., 1988; Perry and D'Antuono, 1989; Siddique et al., 1989a; Slafer and Andrade, 1989; Canevara et al., 1994; Calderini et al., 1995) and barley (e.g., Riggs et al., 1981; Jedel and Helm, 1994b; Martiniello et al., 1987). While in wheat, the reduction in plant height has largely been a continuous process throughout the twentieth century (Calderini et al., 1999a), in barley, stronger reductions are seen at the beginning of the century and smaller changes thereafter (Abeledo et al., 2001). The optimum height of barley heights in terms of productivity was attained earlier than in wheat, where height has been optimized with the introduction of *Rht* dwarfing genes.

Lodging is an important problem in wheat and barley, often reducing yields by 30–40% (Stapper and Fischer, 1990). Lodging resistance has been an important trait in plant-breeding programs (Abeledo et al., 2001). Lodging is largely independent of stem height within a range of relatively short plants (stem diameter is then more critical). Lodging is positively related to stem height as plant height increase beyond a certain threshold. It has been em-pirically demonstrated in a wide range of environmental conditions that optimum height ranges between 70 and 100 cm (e.g., Fischer and Quail, 1990; Richards, 1992a; Miralles and Slafer, 1995b; Austin, 1999). Average height of modern wheats and barleys is already within the range that optimizes yield in virtually all countries analyzed (Calderini et al., 1999a; Abeledo et al., 2001), implying that breeders would not further reduce plant height in the future.

2.2.3 Dry Matter Production and its Partitioning

In general terms, the large increase in yield produced with the release of newer cultivars was far more associated with the improved partitioning to yield than with total dry matter production. In wheat, the increases in grain yield were almost entirely independent of modifications in biomass (see reviews by Slafer and Andrade, 1991; Feil, 1992; Evans, 1993; Loss and Siddique, 1994; Slafer et al., 1994a; Calderini et al., 1999a); however, some exceptions can be found (Waddington et al., 1986; Hucl and Baker, 1987). The general scenario is similar for barley, in which biomass either (i) did not show any trend with the year of release of the cultivars (Martiniello et al., 1987; Bulman et al., 1993; Jedel and Helm, 1994a); or (ii) was increased with the year of release of the cultivars, but at a slower pace than that for yield (Riggs et al., 1981; Wych and Rasmusson, 1983; Boukerrou and Rasmusson, 1990).

The lack of systematic historical changes in biomass production with breeding indicates that little has been done to change the ability of the crop to either intercept more radiation or use it more efficiently. With exceptions, cultivars released at different eras generally did not differ in their leaf area index (Austin et al., 1980a,b; Deckerd et al., 1985; Feil and Geisler, 1988; Canevara et al., 1994; Calderini et al., 1997). Consequently, neither radiation interception (Deckerd et al., 1985; Slafer et al., 1990; Calderini et al., 1997) nor radiation-use efficiency (RUE) at the canopy level has been modified substantially, whenever the length of the crop cycle remained unmodified (Slafer et al., 1990; Calderini et al., 1997). However, breeding increased total biomass accumulation such as in the case of some durum wheats, which was associated with a shift toward later maturity (Pfeiffer et al., 2000).

Single-leaf photosynthesis has not been related to historical yield increase by breeding (Austin et al., 1982; Austin, 1989). Although maximum photosynthesis appeared to be related to the year of release of CIMMYT's cultivars (Fischer et al., 1998), it was mainly due to increase in stomatal conductance (Richards, 2000), without an associated increases in biomass production.

Thus, apart from a few examples in which breeders may have increased crop biomass as the main way to achieve yield gains, the yield increase in wheat and barley due to breeding has been strongly dependent on biomass partitioning, as reflected by higher harvest index (HI) at maturity (Fig. 2).

In fact, retrospective analyses shows that HI was recently increased not only under stress-free conditions but also under drought (Calderini et al., 1999a) and low-fertility (Ortiz-Monasterio et al., 1997) conditions (Fig. 2). Associated with the higher HI in the modern cultivars is a higher N-use efficiency (Reynolds et al., 1999) and a shorter plant stature (Byerlee and Moya, 1993; Austin, 1999; Slafer et al., 1999). However, short stature may

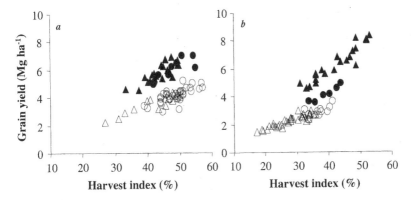

Figure 2 Yield barley (*a*) and wheat (*b*) cultivars released at different eras in each country vs. harvest index. Countries in this example (United Kingdom, ▲; Italy, ●; United States, △; and Canada, ○) are characterized by substantially different environmental conditions for cereal growing (responsible for the large differences in average yield among them). (Original data: From Wych and Rasmusson, 1983; Boukerrou and Rasmusson, 1990; Martiniello et al., 1987; Bulman et al., 1993; Jedel and Helm, 1994a. Adapted from Abeledo et al., 2001.)

have negative consequences under drought-prone environments, where kernel growth may be sustained by reserve assimilates accumulated in the upper internodes before anthesis (Loss and Siddique, 1994). This could explain the lack of (negative) relationship in durum wheat between plant stature and yield under moderate–severe drought (e.g., Villegas et al., 2000).

Because large genetic gains in yield achieved during the second half of the twentieth century were heavily associated with improvements in partitioning of assimilates, modern cultivars in most cereal regions reached already high values of HI (Fig. 2). Thus, further genetic improvement of wheat and barley yield cannot depend on improving HI.

2.2.4 Yield Components

Yield variation of cereals is more often related to differences in number of grains per square meter than to diffenerces in individual grain weight [see, for example, differences in yield due to locations and years (Magrin et al., 1993); or to timing and doses of fertilizer application (Fischer, 1993a)]. Breeding higher-yielding cultivars also had a more pronounced effect on number of grains per unit land area than on the average individual grain weight in both wheat (Feil, 1992; Loss and Siddique, 1994; Slafer et al., 1994a; Calderini et al., 1999a) and barley (Abeledo et al., 2001); however, few exceptions may be found. Modern wheat cultivars were found to exhibit a superior ability to sustain the development of floret primordia just before anthesis thus allowing a higher survival rate, thereby increasing the number of fertile florets at anthesis (Siddique et al., 1989b, Slafer and Andrade, 1993; Slafer et al., 1994b). Moreover, the greater number of florets per spike at anthesis in semidwarf than in standard height genotypes has been attributed to the reduced mortality of floret primordia between flag leaf appearance and anthesis rather than to differences in the maximum number of floret primordia initiated (Kirby 1988). An example may be given for durum wheat changes through time in Spanish genotypes because the number of spikelets per main spike has been reduced to 3%, comparing old genotypes (from the 1920s) and recently released varieties, whereas the number of grains per spike has increased to more than 30% (Royo et al., unpublished data). This change is critical because only a small proportion of the florets initiated in each spike is able to produce fertile florets at anthesis (Kirby, 1988); that was the result of an improved growth of the spikes immediately before anthesis in association with a reduction in stem height (Siddique et al., 1989b; Slafer and Andrade, 1993). The clearest example comes from the impact of the major dwarfing genes in wheat. *Rht1* and *Rht2* genes, which reduce cell size in most aboveground organs, also inhibit the elongation of the internodes, thus releasing a relatively large amount of assimilates that would have been used in stem growth otherwise (Miralles et al., 1998). This results in an increased spike growth and improved floret fertility (Fischer and Stockman, 1980; Brooking and Kirby, 1981; Slafer and Miralles, 1993; Richards, 1996b) associated with a remarkable reduction in floret abortion (Youssefian et al., 1992; Miralles and Slafer, 1995b).

The relative importance of grains per square meter than the individual grain weight in the historical improvement of yield can be seen from experiments with shading, thinning, degraining, or defoliations after the grain set phase (a week or so after anthesis). Under such treatments, growth of the grains is not, or is only slightly, affected (e.g., Martinez-Carrasco and Thorne, 1979; Borghi et al., 1986; Koshkin and Tararina, 1989; Savin and Slafer, 1991; Slafer and Miralles, 1992; Bulman and Smith, 1993; Nicolas and Turner, 1993; Slafer and Savin, 1994; Dreccer et al., 1997; Kruk et al., 1997).

Even within semidwarf germplasm, where plant height has been optimized, progress in yield of irrigated wheat has been most strongly associated with improved HI and grain number (Kulshrestha and Jain, 1982; Waddington et al., 1986, 1987; Calderini et al., 1995; Sayre et al., 1997; Calderini and Slafer, 1998).

Tillering capacity is a major determinant of grain yield in barley, has been associated to yield stability (Simmons et al., 1982), and is a critical trait for the crop recovery from early season drought (Blum et al., 1990a,b; El

Hafid et al., 1998). Semidwarf genotypes have greater tillering rates than normal genotypes (see references in McMaster, 1997), and the recessive gene *Tin* (Richards, 1988) inhibits tillering in wheat. Reduced tillering may contribute to a higher HI both in the presence and the absence of drought (Richards et al., 2002).

2.3 Genetic Basis of Yield Improvement

CIMMYT Wheat Improvement Program has expanded the genetic base of modern wheat lines using conventional breeding as well as cytogenetic approaches (Table 1, see also Evans and Fischer, 1999). The Veery lines of bread wheat produced in the early 1980s (Rajaram et al., 1990) resulted from a cross to a winter wheat parent containing the 1B/R translocation from rye. The Veerys show outstanding yield potential as well as wide adaptation and other physiological characteristics. For example, the variety Seri 82 shows high leaf photosynthetic rate, high stomatal conductance, and greater leaf greenness relative to a set of hallmark varieties developed both before and after its release (Fischer et al., 1998). Broad adaptation and superior performance may also be related to increased stress tolerance (Villareal et al., 1997a; Slafer and Araus, 1998).

The development of synthetic hexaploid wheat, the product of wide crossing between durum wheat and wild (D genome) diploids, has enabled new genetic diversity to be introduced into the bread wheat gene pool. Although initial objectives centered on improved disease resistance (Villareal et al., 1995; CIMMYT, 1996), many synthetic lines also have good yield characteristics (Villareal et al., 1997b; Calderini and Reynolds, 2000; Calder-

Table 1 Examples of the Successful Application of Wide Crossing to Introduce Alien Genes into the Hexaploid Wheat Genome

Alien genes	Effect	Reference
1B/R translocation from rye	Present in >300 cultivars	Rajaram et al., 1990
1B/R translocation	Increased stress tolerance	Villareal et al., 1997a
Triticum tauschii × *T. turgidum*	Good yield characteristics	Villareal et al., 1997b
Synthetic hexaploids	Used in approximately 15% of all CIMMMYT bread wheat crosses	van Ginkel, 1997, personal communication
Lr19 (*A. elongatum*)	Significant increase in yield and biomass over check lines	Singh et al., 1998; Reynolds et al., 2001

(From Reynolds et al., 1999.)

ini et al., 2001). Synthetic hexaploid lines are now used in approximately 15% of all crosses made by CIMMYT's Wheat Improvement Program.

Recently, a chromosome translocation containing the *Lr19* gene from *Agropyron elongatum* has been shown to be associated with a significant increase in yield and biomass when introduced into already high-yielding backgrounds (Singh et al., 1998; Reynolds et al., 2001a).

The semidwarf photoperiod-insensitive variety Jori 69, developed by the Durum Program at CIMMYT, was bred primarily for irrigated subtropical areas. With Jori 69 and other early semidwarfs, the project internationalized and globalized with varieties such as Cocorit 71, Mexicali 75, and Yavaros 79. All varieties were developed for irrigated conditions, but showed some adaptation to moisture stress situations. In particular, Yavaros 79 showed high additional adaptation to high rainfall conditions in addition to a more dramatic increase in yield potential (Pfeiffer et al., 2000). After that, Altar 84 and Aconchi 89 were obtained from ideotype breeding for erectophile leaf canopy. Later on, more emphasis was put in the development of varieties with higher yield potential, which allowed the recent release of Atil 2000, a high water-use and input efficiency variety (Pfeiffer et al., 2001).

2.4 Some Pointers for the Future

The theoretical limit to HI was estimated by Austin et al. (1980a) to be 60%, and even a figure closer to 0.50 appears realistic. Because this value is not much higher than that of modern cultivars in many cereal regions worldwide (e.g., Fig. 2) and because plant heights have already been optimized (Richards, 1992a; Miralles and Slafer, 1995a; Austin, 1989), it would seem that continued increase in yield potential must come about in the future by exploiting different physiological traits from those successfully used in the past. It is clear that future yield gains will far more strongly depend upon biomass increases (as indicated by the above CIMMYT case) than on further increase in HI. Regarding yield components, further increases in number of grains per unit land area would be required. A stronger sink than source limitation for yield during grain filling in both wheat and barley have been usually reported (Slafer and Savin, 1994; Miralles and Slafer, 1995b; Miralles et al., 1996; Richards, 1996b; Dreccer et al., 1997; Kruk et al., 1997; Voltas et al., 1998).

However, recent results raise again the dilemma of sink vs. source limitation. In the case of six high biomass lines from CIMMYT containing chromosome substitution associated with the *Lr19* gene, improved yield (average 13%) and biomass (average 10%) were shown to be associated with an improved source–sink balance manifested by increased investment of biomass in spike at anthesis and up to 25% higher rates of flag leaf photo-

Table 2   Performance Recent CIMMYT Bread Wheat Lines in Obregon, NW Mexico, 1997–1998

Cultivars	Yield (Mg ha^{-1})	Biomass at maturity (M ha^{-1})	Harvest index	Grain number (m^{-2})	Grain weight (mg)	Days to maturity (day)	Plant height (cm)
Seri-82	9.2	18.7	0.43	18,250	44.5	132	95
Bacanora-88	9.4	20.8	0.40	24,000	34.3	129	90
Baviacora-92	9.8	20.2	0.43	17,200	50.0	132	110
LSD (0.05)	664	2030	0.426	1804	2.4	5.1	6

(From Reynolds et al., 1999).

synthesis during grain filling (Reynolds et al., 2001a). This data suggest that genetic increases in biomass can be achieved, as seen in Table 2.

Nevertheless, lines with high yield associated with improved biomass do not exhibit a unique physiological pathway to yield, as can be seen when comparing, for example, yield components and other characteristics of recent CIMMYT lines (Table 2). Semierect leaves are apparent only in Bacanora-88, while Seri-82 has high chlorophyll and stay-green characteristics, and Baviacora-92 is a tall robust-looking plant type with large kernel size. Nonetheless, the idea initially proposed by Donald (1968) that yield potential could be improved by reducing individual plant competitiveness with a population is supported by data from CIMMYT wheat lines (Reynolds et al., 1994a) as well as in modern maize hybrids (Duvick, 1992; Otegui and Slafer, 2000).

3 PHYSIOLOGICAL AVENUES TO INCREASE YIELD POTENTIAL

3.1 Biomass Increase

When considering avenues for increasing yield, it is important to assess the biological limitations on yield potential. Theoretical wheat yield potential has been calculated for the Yaqui Valley, NW Mexico (Reynolds et al., 2000), based on the potential range of RUE for wheat as estimated by Loomis and Amthor (1996). Accumulated solar radiation for this environment averages 2250 MJ m^{-2} for the growth cycle, with canopy-intercepted active absorption being 1750 MJ m^{-2}. At quantum requirements ranging from 15 to 24 mol of photons per mol of CO_2 fixed as CH_2O (Fischer, 1983; McCree, 1971), RUE varies from 1.5 to 2.6 g CH_2O MJ^{-1} solar radiation (Loomis and Amthor, 1996). Using the value of 1750 MJ m^{-2} of absorbed radiation, biomass would vary between 2600 and 4550 g m^{-2}. Irrigated wheat in this environment currently approaches an aboveground biomass of 2100 g m^{-2} (Reynolds et al., 1999). Assuming that the cost of root growth and maintenance under irrigated conditions is not substantially higher than previously estimated (i.e., 10% of aboveground biomass, Weir et al., 1984), there would seem to be room for improving wheat biomass and, therefore, yield potential.

An increase in biomass must be achieved without any negative interaction that may cause yield reduction such as lodging or reduced HI. Although the benefits for biomass production may be found in increasing the length of the stems (it would improve light distribution within the canopy and then RUE), it would most likely reduce rather than increase cereal yields due to both reductions in HI expected from an increased stem growth (Siddique et al., 1989b; Slafer and Andrade, 1993) and increased lodging. Therefore, biomass must be increased, while maintaining the height within the optimum range, by increasing the amount of radiation intercepted and/or the efficiency of the photosynthetic tissues to use the intercepted radiation. This may be achieved through modifications in photosynthetic metabolism, canopy architecture, and source–sink balance.

3.1.1 Radiation Interception

It has been long recognized that biomass is largely linearly related to the amount of accumulated intercepted (actually absorbed) photosynthetically active radiation (e.g., Puckridge and Donald, 1967; Sibma, 1970; Monteith, 1977; Biscoe and Gallagher, 1978; Fischer, 1983). More intercepted radiation may be accumulated by the crop by (i) increasing the length of the growing cycle so that the canopy is exposed by a longer time duration larger to incoming radiation, or (ii) increasing total daily interception by the canopy.

Although lengthening the cycle of the crop would be easily manipulated genetically through manipulation of photoperiod and vernalization genes (Slafer and Rawson, 1994; Snape, 1996, 2001; Slafer and Whitechurch, 2001), this is mostly unlikely to be practical. In fact, time to anthesis is the primary trait improved when the crop is bred for a new region because it determines the adaptation of the crop to the growing conditions and has therefore been optimized in most regions making unlikely any further change, in traditional cereal areas at least (Slafer et al., 1996).

Alternatively, increasing the ability of the crop to intercept incoming radiation may be proposed. The yield of cereals is quite dependent to the rate of growth of the crop (Fig. 3), specially during the period of stem elongation (e.g., Fischer, 1985; Thorne and Wood, 1987; Savin and Slafer, 1991), and where the growing cycle is not markedly short, cereal crops normally achieve maximum radiation interception before the onset of stem elongation. In these conditions, it may be only possible to increase the ability of the canopy to

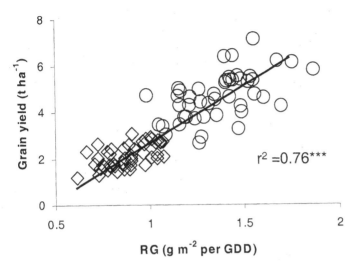

Figure 3 Relationship between the mean rate of growth and the grain yield of 25 durum wheat genotypes grown under irrigated (O) and rainfed (◇) environments during two crop seasons. Determination coefficients for the relationships across the 25 genotypes within rainfed and irrigated environments were 0.33 and 0.32 ($P<0.001$), respectively. (From Villegas et al., 2001.)

increase interception during early phases of development when the crop yield is not responsive and, therefore, mostly meaningless. However, for environments with very short growing seasons or subjected to moderate–severe water stresses, and because the short crop cycle required, fast early coverage of the soil may be instrumental to guarantee maximum interception at the start of the critical stem elongation phase. In these environments, vigorous genotypes are likely to yield more grain and biomass than less vigorous types (Richards et al., 2002). Genetic variation in early vigor is important when the target environments are characterized by conditions preventing full coverage by the onset of stem elongation, frequently where the season is markedly short, such as in Nordic countries, or under severe stress conditions (Richards, 2000). Rapid canopy establishment may be beneficial for increasing radiation interception and reducing evaporation of soil moisture if conditions are favorable in the early part of the cycle (Richards, 1996a) or for water-limited environments, where most water comes as rain on the crop. Traits contributing to early vigor include minimizing the time from sowing to seedling emergence, involving temperature response (Slafer and Rawson, 1994), thin leaves (with large leaf area-to-weight ratio), or the presence of coleoptile tillers (Liang and Richards, 1994; Richards, 1996a).

However, except for very short growing seasons and/or severe stress, it is unlikely to improve yield through improvements of radiation interception (as it has been already maximized during the critical stem elongation phase). The realistic alternative for improving biomass by breeding may be to attempt increasing RUE, particularly during the critical period of stem elongation.

3.1.2 Radiation-Use Efficiency

This is a rather complex character that was little modified by breeding during the twentieth century (Siddique et al., 1989b; Slafer et al., 1990; Calderini et al., 1997). However, the values recorded in these and other studies with modern cereals leave room for at least some substantial improvements: In fact, the values frequently reported (ca. 1.2 g MJ^{-1} on a solar radiation basis; Calderini et al., 1999a) are quite low, less than one-half of the potential RUE calculated for C3 cereals (Loomis and Amthor, 1996). Such expectation is particularly evident for an improvement in postanthesis, when estimations show RUE to often be even lower, which may be brought about as a consequence of the increased demand of a larger sink (Calderini et al., 1997; Miralles and Slafer, 1997). A larger sink strength during postanthesis may be associated with either a further improved number of grains per square meters or an increased potential size of the individual grains (see below). Apart from altering the source–sink balance, other potential alternatives to increase RUE include the manipulation of leaf photosynthesis, photorespiration, and radiation distribution.

Leaf Photosynthesis. The relationship between grain yield and leaf photosynthetic rate per unit leaf area at light saturation (photosynthetic capacity, A_{max}) is complicated. Firstly, yield differences in wheat are generally better associated with HI than with RUE (Slafer et al., 1994a; Calderini et al., 1995, 1999a). Even when yield is source-limited, an association between A_{max} and yield cannot necessarily be anticipated. In addition, instantaneous readings of leaf A_{max} cannot be extrapolated to predict canopy photosynthesis throughout the day or the crop cycle. Moreover, leaf thickness or nitrogen content per unit leaf area (leaf characteristics determining A_{max}) is often negatively related with total leaf blade area (see references in Araus et al., 1989). As a consequence, there is negative or no relationships between leaf photosynthesis and yield (Evans and Dunstone, 1970; Austin et al., 1982; Johnson et al., 1987; Carver et al., 1989; Evans, 1993) so that the selection for higher photosynthesis failed in producing yield gains (Austin, 1989) and vice versa. However, in a recent study under hot irrigated conditions, where yield increases of cultivars were related to greater biomass, association between A_{max}, biomass, and yield was reported (Reynolds et al., 1994b; Gutiérrez-Rodríguez et al., 2000). The latter study demonstrated genetic gains in yield in response to selection for flag leaf photosynthetic rate and stomatal conductance in F_5 sister lines (Table 3).

Another recent work has reported an association between yield and photosynthetic rate under temperate conditions (Fischer et al., 1998). In these studies at CIMMYT, the increases in A_{max} were associated with larger stomatal conductance (g_s); in fact, this was the photosynthetic trait most strongly related to yield, even when mesophyll conductance also seemed to increase (Fischer et al., 1998; see also Richards, 2000). This relationship across different wheat genotypes between stomatal conductance and yield has already been observed for the emerging "green revolution" semidwarf wheat cultivars (Shimshi and Ephrat, 1975). A higher g_s may reflect decreased stomatal sensitivity to VPD or to subtle water stress, excessive leaf cooling particularly at warmer temperatures, or increased sink strength (Reynolds et al., 1994b; Fischer et al., 1998, Richards, 2000).

Leaf nitrogen concentration may also affect A_{max} (Evans, 1983, 1989; Araus and Tapia 1987; Sinclair and Horie, 1989; Dreccer et al., 2000) and thus RUE (e.g., Sinclair and Muchow, 1999). An improved distribution of the absorbed nitrogen within the canopy, with more N being partitioned to the top layer of leaves and less N remaining in the darker layers, would theoretically benefit photosynthetic rates in the more illuminated leaves (Field, 1983; Hirose and Werger, 1987, Dreccer et al., 1997).

Reducing Photorespiration. Another theoretical approach to increase RUE might be to reduce photorespiration, perhaps by increasing the affinity of Rubisco for CO_2, thereby decreasing its oxygenase activity. Modest variation for CO_2 specificity has been found in land plants (Parry et al., 1989; Delgado et al., 1995), with wheat having among the highest values for crop species. However, much higher values are reported in marine algae (Read and Tabita, 1994). Molecular techniques may offer the possibility of genetically

Table 3 Correlation Between Photosynthetic Traits Measured of 16 Individual F_5 Plants and Performance in $F_{5:7}$ Yield Plots, 1995–1996

Individual F_5 plants (grain filling)	$F_{5:7}$ yield plots		
	Yield	Biomass	A_{max}
A_{max}	0.66[a]	0.67[a]	0.68[a]
Stomatal conductance	0.65[a]	0.68[a]	0.68[a]

[a] Denotes statistical significance at the 1% probability level. (From Gutiérrez-Rodríguez et al., 2000).

transforming wheat Rubisco from its current specificity of 95 to potential values of 195, corresponding to thermophilic alga *Galderia partita*. However, Rubisco also plays a protective role in dissipating excess energy, with O_2 uptake in the light playing a significant role in preventing chronic photo-inhibition under field conditions (Osmond and Grace, 1995). In which case, it may not make sense to alter Rubisco's oxygenase specificity especially in crops subjected to abiotic (such as drought or temperature) stresses.

Distribution of the Incoming Radiation. Plant height may affect the distribution of radiation within the canopy, but it has been already discussed that there is little, if any, margin to manipulate stem length in most situations. Alternatively, another theoretical approach to improve RUE is in improving the distribution of the incoming radiation through the crop by changing the optical properties of the canopy, mainly reducing its coefficient of light attenuation (Slafer et al., 1999 and references cited therein). Given the difficulties associated with measuring or accurately simulating canopy photosynthesis, traits which could potentially modify RUE at the canopy level must be tested empirically by measuring their effect in near-isogenic material. Simulation is, however, a good way to test leaf size and angle effects on canopy photosynthesis as a function of latitude, date, and cloudiness. Several lines of evidence suggest that more erectophile canopies may have higher RUE (Duncan, 1971; Angus et al., 1972; Innes and Blackwell, 1983). Smaller and more erect upper-culm leaves may allow the incoming radiation to be more evenly distributed through the canopy (Richards, 1996b). Moreover, genetic manipulation of leaf angle is not complex, probably involving only two to three genes (Carvalho and Qualset, 1978). However, more erect leaves may prove to be inefficient in improving yields perhaps due to the pleitropic negative effects not only on flag leaf area but more important on yield components such as the number of kernels per spike (Araus et al., 1993). Nevertheless, more erect leaf canopy types are characteristic of many of CIMMYT's best-yielding wheat lines and temperate maize hybrids of the 1970 and 1980s (Fischer, 1996; Duvick, 1992).

Source–Sink Balance. Slafer and Savin (1994) in their analysis of data from 15 studies, where source–sink balance was manipulated at or after anthesis, showed that while source and sink can both limit yield, sinks are generally more limiting during grain filling. Increased wheat yield and biomass at CIMMYT has been shown in one case to be associated with a chromosomal translocation containing the *Lr19* gene from *A. elongatum* (Singh et al., 1998). In a study with six near-isogenic pairs, improved source–sink balance was shown to be associated with increased biomass (Reynolds et al., 2001a). This was reflected by a higher partitioning of assimilates to the developing spike at anthesis in some backgrounds, indicating that higher photosynthetic rates observed during grain filling in these lines were a response to increased sink strength. However, in other backgrounds, larger spike size was associated with more total biomass at flowering as well as higher photosynthetic rates in grain filling (Reynolds, unpublished data). The isogenic lines were compared with respect to their phenology and, in particular, duration of spike growth phase because it has been suggested that this may influence source–sink balance (Slafer et al., 1996). However, in the case of *Lr19*, although source–sink balance was improved, it was not associated with changes in the relative duration of spike growth phase or any other phenological stage (Reynolds et al., 2001a).

3.2 Sink Strength: Opportunities to Increase Number of Grains and Final Grain Weight

The number of grains per square meter is the final result of a process in which a large number of primordia are generated during the initial phases of development followed by an abortion of most of these structures (Fischer, 1983; Kirby, 1988; Miralles et al., 1998). This major yield component, which is the combination of number of spikes per square meter plus the number of grains per spike, is thus formed throughout the whole period from sowing to anthesis (Slafer and Rawson, 1994). Clearly, the phase coinciding with the mortality (or survival) of floret primordial is far more important for yield formation than the initiation phase (e.g., Fischer, 1985; Thorne and Wood, 1987; Kirby 1988; Savin and Slafer, 1991; Slafer and Savin, 1994). This is why yield is generally well related to the growth of the spikes during stem elongation (Fischer, 1985; Siddique et al., 1989b; Slafer and Andrade, 1993; Slafer, 1995; Miralles et al., 1998). An increase the number of grains per square meter will require to produce more growth of the spikes during the spike growth period (approximately the last half of the stem elongation phase), occurring normally during the 2–3 weeks immediately before anthesis (Kirby, 1988; Siddique et al., 1989b; Slafer and Andrade, 1993). In fact, this has been the main mode of yield increased affected by the dwarfing genes (see above). Due to the limit in HI, in the future, there seems to be little room for further improvements in biomass partitioning (e.g., Slafer et al., 1999 and references therein) through the genetic manipulation of phasic development (Miralles et al., 2000), carpel size (Calderini et al., 1999c), sterile tiller reduction (Richards, 1988), and grain set (Fischer et al., 1998). Thus, the possibility of enhancing the growth of the spikes is rather restricted to increasing the growth of the whole canopy while maintaining the actual, high values of biomass partitioning. This might be attained by increasing RUE during this phase (see above) or by lengthening the phase of spike growth so that growth during the critical phase is greater (e.g., Slafer et al., 2001a).

The length of the critical period for the determination of grain number per square meter may be extended by manipulations of genes responsible for the sensitivity to photoperiod or for earliness per se (Snape et al., 2001a,b), provided the sensitivity to photoperiod during this late reproductive phase (Slafer and Rawson, 1997, Miralles and Richards, 2000) and the variability for earliness per se are large (Slafer, 1996). For instance, stem elongation (and spike growth) phase of both wheat and barley was lengthened when exposed to shorter photoperiods and vice versa (Miralles and Richards, 2000; Miralles et al., 2000; Slafer et al., 2001a). The extra time for spike growth would result in larger spikes with more fertile florets due to reduced abortion and more grains (Slafer et al., 2001a); however, photoperiodic effects beyond those mediated through an extended duration of spike growth should not be overlooked (González et al., 2003).

In addition, grain-setting efficiency (i.e., the number of grains set per unit of spike dry matter) of the cultivars may also be an avenue increasing the number of grains per unit land area. Abbate et al. (1998) have identified genetic variation for the trait, which might be related to the dry matter partitioning within the spike between the spikelets and the rachis (Slafer and Andrade, 1993).

The average grain weight might also be improved. It has been demonstrated that under cool conditions, durum wheat kernel weight is the main determinant of yield, whereas the number of spikes per square meter plays a major role in yield formation in warmer environments (García del Moral et al., 2003). As many of the source–sink studies during grain filling mentioned above have shown, grain growth is hardly affected by the source strength. In other words, the photosynthetic capacity of the crop during grain filling together with the translocatable reserves in the vegetative organs are in most nonstress situations adequate or in excess compared with the grain demand (Savin and Slafer, 1991; Slafer and Miralles, 1992; Richards, 1996b). In this context, any increase in the potential size of the grains may be translated into higher yields (Slafer et al., 1999). However, we need to learn first which are the physiological determinants of the potential size of the grains. As it seems that the vascular system would unlikely be responsible of the final weight of the grains (Evans, 1993), candidates are structural features imposing a limit to the grains for further growth. For instance, Calderini et al. (1999b) and Calderini and Reynolds (2000) convincingly showed that the growth of the carpels immediately before anthesis is critical in the determination of an upper limit to grain growth, suggesting that the improved growth of the carpels may lead to greater average grain weight by raising the final weight of the smaller grains within the spikes (Calderini et al., 2001). This may also impact some grain utilization consideration such as in malting barley.

3.3 Water Use and Its Efficiency

Improved crop performance in environmental conditions characterized by water shortages may be achieved through improvements in water use, water-use efficiency (WUE), and HI (Passioura, 1977, 1996). Increasing water use is relevant where there is still soil water available at maturity or when deep-rooted genotypes access deep soil moisture. WUE and HI become more significant when all available water is normally used up by the end of the crop cycle. The most important attribute determining performance under water stress in Mediterranean-type conditions has been phenology (matching crop development and seasonal rainfall pattern). Phenology may affect either water use, WUE, or HI. Early sowings may drastically increase WUE and biomass, but the final grain yield is often unchanged due to the low harvest index of early-sown crops (Gomez-Macpherson and Richards, 1995). However, varieties with modified dry matter partition and allocation should be developed to overcome this problem (Richards et al., 2002). Other developmental traits, such as deeper root systems and early vigor, may also help the crop to use more water. Greater early vigor and/or early flowering reduces the evaporative loss of water from the soil surface, on one hand, while also ensuring that more growth and transpiration occur when the vapor pressure deficit is small. However, the value of early vigor and earliness as drought adaptive traits will strongly rely on the profile of drought stress.

If water is a major limitation to yield, then maximized growth during periods of cool weather and relatively low vapor pressure deficits will raise WUE and biomass production (Richards, 1991; Gomez-MacPherson and Richards, 1995). Increasing the relative duration of the spike growth phase, discussed above as a strategy to further raise the number of grains per unit land area in potential conditions, might theoretically contribute to increase WUE, as an increased proportion of the transpired water would be used during the critical period for yield determination.

It must be taken into account, before defining a selection strategy for increased water use or WUE, that both characteristics are not independent. Besides being mathematically dependent, they are frequently negatively related as the greater the amount of water captured and transpired by the crop, the greater will be transpiration and, consequently, the lower the WUE.

3.4 Target Environments and Breeding Strategy

The design of a breeding program depends on its target environment. The fact that uniform criteria cannot be easily applied is clearly illustrated by the controversy between breeding strategies of CIMMYT and International Rice Research Institute (IRRI), on one hand, and ICARDA, on the other hand.

CIMMYT coined the term "megaenvironment", where wheat is selected mainly under stress-free conditions (Reynolds et al., 1996) with further testing for yield in a wide range of environmental conditions, ranging from stress-free to mild and moderate stress environments. The same can be said for IRRI on rice. ICARDA breeding efforts on barley are more devoted to improving performance under severe stress; thus, yield stability is weighed more important than yield potential. Selection is performed in drought-prone, poor environments characterized by yields frequently below 1.0 Mg ha^{-1} of barley and durum wheat (Ceccarelli and Grando, 1996). Both strategies fit well with the model of Finlay and Wilkinson (1963), whereby the yield of particular genotypes is evaluated against the environmental index, which is constructed from the average yield of all participating genotypes tested over a range of environments. If the range of environmental (E) index and genotype (G) variation is large enough, a crossover $G \times E$ interaction might be seen (see Araus et al., 2002 for details). This crossover would indicate that genotypes selected for low yield conditions will probably perform better than those released fo. high-yielding environments when grown under very poor environments. Thus, crop selection performed in nurseries with good growing conditions is frequently translated to cultivars with higher water use and larger productivity in a wide range of growing conditions (considered as a whole as a megaenvironment), from nonlimiting (e.g., with yields beyond 7.0 Mg ha^{-1}) to mild stress (ca. 4.5–7.0 Mg ha^{-1}) and moderate stress (ca. 2.0–4.5 Mg ha^{-1}) environments. However, in more stressed environments, the situation may reverse, with the genotypes selected in good environments performing worse than those selected under the poor conditions of the target environment. Normally, such crossover is placed at low-yield levels and appears to be slightly higher for barley (2.0–2.5 Mg ha^{-1}; Ceccarelli and Grando, 1991) than for wheat (1.0–1.5 Mg ha^{-1}; Laing and Fischer, 1977; Fischer, 1993b).

One of the most difficult problems that have to be faced by breeders is the large genotype × environment (both site and/or year) interactions that usually occur when the performance of elite lines is evaluated in a range of target environments. These interactions mask the expression of the differences between genotypes and slow genetic advance.

From an ecophysiological perspective, traits associated with abiotic stresses, such as drought, may be divided into two distinct categories: those conferring the ability to survive under extreme stress, and those permitting agronomically acceptable levels of productivity under a relatively large range of less stressful conditions (by avoiding stress). The expression of the former may have penalties in yield under less severe conditions because these traits, conferring stress tolerance (and usually related with a low growth), are constitutive and expressed independently of the degree of stress (Blum, 1996a). By contrast, the latter tend to be facultative in their response (e.g., deeper rooting under drought or evaporative cooling under heat stress) which confer broader adaptation. The fact that selection for higher-yielding performance has frequently resulted in higher yields in a wide range of environments (see examples in Calderini and Slafer, 1999; Araus et al., 2002) is probably largely a consequence of facultative stress avoidance traits. In these examples, modern cultivars have consistently outyielded older cultivars even in the lowest-yielding conditions of each particular study (Austin et al., 1980a; Perry and D'Antuono, 1989; Slafer and Andrade, 1989, 1993; Calderini et al., 1995). Shorter crop duration is the most conspicuous strategy of drought escape. It is a constitutive trait which confers genotypes' better performance (in terms of yield and stability) in severe-to-moderate drought environments, particularly when the drought is terminal such as in Mediterranean regions (Loss and Siddique, 1994). However, this trait may also have negative implications in yield potential if the reduction in cycle compromises light interception during the critical phase of stem growth.

4   USING PHYSIOLOGICAL TOOLS TO COMPLEMENT YIELD SELECTION: A HYPOTHESIS

By combining information on the physiological basis of yield with new physiological selection tools, the probability of accelerating the rate of genetic progress through plant breeding should be significantly increased. Parents can be selected for improved physiological traits and can be crossed to high-yielding agronomically elite materials. Identification of the progeny phenotypes with the favorable interactions among genes permitting the expression of higher yield can be enhanced by: (i) eliminating inferior agronomic phenotypes visually in early generations; (ii) selecting superior physiological phenotypes using rapid detection techniques in intermediate generations; and (iii) selecting for higher performance in yield trials in advanced generations (Reynolds et al., 2000). Examples of physiological approaches that have been used in wheat breeding and that have had impact over the years are the introgression of erect leaf angle (Fischer 1996), the carbon isotope discrimination in Australia (Richards et al., 2002; Condon et al., 2002), or the canopy temperatures for selection in drought- and heat-stressed nurseries (Blum et al., 1982; Reynolds et al., 1998).

However, in the past, the use of yield-related physiological selection traits in cereal breeding has fallen short of expectations. Many possible reasons have been stated to explain this lack of success. One is the great integrating power of empirical selection for characteristics influenced by many processes and genes as yield potential and yield stability (Evans, 1993). Another is the difficulty of understanding—in spite of the enormous amount of information that has been accumulated about physiological processes—what causes low grain yields and how putative traits may enhance drought resistance and contribute to grain yield in water-limited environments (Ludlow and Muchow, 1990). Other causes are related to the inadequate methodology used in the investigations, which sometimes lead to incorrect conclusions, or the differences in the approaches to dealing with genotype by environment interactions in the disciplines of crop physiology and plant breeding (Jackson et al., 1996).

From a breeding perspective, there are, of course, specific requirements that any physiological selection criterion should fulfill before being included in a breeding program. Namely, it must exhibit enough genetic variability, a high genetic correlation with yield and a higher heritability than yield itself in genetic populations representative of those being evaluated (Jackson et al., 1996). Moreover, evaluation of these traits must be fast, easy, and cheap (Araus, 1996; Slafer and Araus, 1998; Araus et al., 2001a, 2002).

Any trait to be considered must be directly related to yield. The literature is filled with proposed traits at the lower levels of organization (i.e., molecular, biochemical), which frequently show only poor and inconsistent relationships with crop yield in the field (Richards, 1996a; Araus et al., 2001a, 2002). In this category, we can include metabolic traits such as enzyme activities (e.g., rubisco or nitrate reductase), levels of substrates (e.g., proline or sugars), and growth regulators such as abscisic acid (ABA; an updated review on these mechanisms for barley is provided in Araus, 2002). Yield is an integrated trait, from molecular level to the canopy. Therefore, any trait consistently related to yield should also be integrative, either in time (by being determined through part, if not all, of the crop cycle), in level of organization (by representing a level close, if not identical, to that of yield), or both (Araus, 1996; Slafer and Araus, 1998; Araus et al., 2001a, 2002). In fact, even yield per plant is not well related with crop yield. Given that wheat yield has been shown to be strongly associated with adaptation to high plant density (Reynolds et al., 1994a), it is not surprising that individual plant yield is not well associated with plot yield.

Besides the difficulty of finding physiological traits simpler than yield itself, but unequivocally linked to it, a major obstacle in practice is the slow methods of their measurement which is unsuitable for work in large breeding populations. Recently, surrogates for estimating physiological traits instead of the direct slow methods were proposed. A comprehensive description can be found in Araus (1996), Slafer et al. (1999), Reynolds et al. (2001b), and Araus et al. (2002). In what follows, we will focus on some of the promising techniques for evaluating traits addressed to improving yield under favorably low to moderate stress situations (as covered in 3.4). These surrogate techniques may be grouped into (i) remote sensing (of radiation reflected or temperature) of crop canopies and (ii) measuring the contribution of stable isotopes in the biomass produced. This is not intended to be an exhaustive review.

4.1 Spectroradiometrical Indices

Spectral reflectance of the canopy is extremely useful for estimating the structure of the canopy in a fast way. Due to the differential spectral absorption and reflection of the soil and plant organs, it is possible to estimate, for example, leaf area index, radiation interception, canopy chlorophyll content, etc. by measuring the spectral properties of the canopy (Field et al., 1994). As the reflectance between the near-infrared and red is quite similar in soils, but different for leaves, the proportion of red wavelength is (i) reduced with increasing the fraction of radiation intercepted by the canopy and (ii) increased with increasing the fraction of radiation intercepted by the soil. The most widely used indexes are the normalized difference vegetation index [NDVI; developed by National Aeronautics and Space Administration (NASA) research] and the simple ratio (SR; see references in Araus, 1996; Araus et al., 2001a). As the relationship between light interception and leaf area index is curvilinear, the accuracy of the estimates for discriminating among genotypes in a breeding program should be maximum during early developmental phases. The speed and low cost make the use of spectroradiometry well-suited for screening purposes in breeding programs (Elliott and Regan, 1993; Bellairs et al., 1996; Peñuelas et al., 1997; Peñuelas and Filella, 1998; Aparicio et al., 2000; Araus et al., 2001a). Besides the usefulness of spectroradiometry in assessing canopy structure and thus biomass, there is the theoretical possibility of estimates RUE as well. For this, the useful wavelengths are different (530 vs. 570 nm), and they should reflect the activity of the xanthophyll cycle which is negatively associated with RUE (Filella et al., 1996). The proposed index is the photochemical reflectance index (PRI; Araus et al., 2001a).

4.2 Using Stomatal Aperture-Related Traits to Select for Yield

Leaf photosynthesis cannot be measured rapidly in the field; hence, the trait does not lend itself to large-scale screening of segregants in a breeding program. However, certain traits that are related to photosynthesis, namely, stomatal conductance (g_s) and transpiration-driven canopy temperature depression (CTD) can be measured within a few seconds on a leaf and plot basis, respectively (Blum et al., 1982; Rebetzke et al., 1996, Amani et al., 1996). Both traits, g_s and CTD, have been shown to be associated with performance of irrigated wheat under high radiation levels (Araus et al., 1993; Reynolds et al., 1994b, 1999; Fischer et al., 1998). CTD is measured with the infrared (IR) thermometer and g_s can be measured with viscous flow porometer, taking about 10–20 sec per leaf in irrigated wheat (Richards et al. 2001). In the same way, stable carbon isotope discrimination ($\Delta^{13}C$) has also proved itself a potentially powerful approach for wheat and barley (Farquhar and Richards, 1984; Acevedo, 1993; Araus et al., 1998), with the advantage of integrating not only the functioning of the crop at its highest level of organization (the canopy level, as CTD does) but also through at least part of the plant's growing cycle. The possible advantage of using CTD and $\Delta^{13}C$ is highlighted by studies that suggested that g_s seems to be a better indicator of the plant water status effect on photosynthesis than, for example, water potential or the relative water content (Sharkey, 1990; Flexas et al., 2000).

4.2.1 Canopy Temperature

A major function of transpiration in plants is leaf cooling. When plant water status is reduced, the stomata close leaf temperature rises due to the lack of transpirational cooling. Hence, canopy temperature can serve as an indirect probe of plant transpiration and plant water status. This probe has been widely developed and used in agronomy and plant breeding with the development of the infrared thermometer that can sense canopy temperature remotely and speedily. Thus, canopy temperature has been used to develop a crop water stress index (CWSI) in wheat and other crops (see http://www.plantstress.com/articles/drought_i/drought_i_files/CWSI_phoenix.pdf) leading to its use in irrigation scheduling.

Canopy temperature is also being used as a screening tool for drought resistance (avoidance). On a relative basis, genotypes having lower canopy temperature at midday have relatively better water status and are taken as drought avoidant (Blum et al., 1982; Garrity and O'Toole 1995). A significant positive relationship was found across a large number of wheat breeding materials between midday canopy temperature under stress and yield stability under stress (Blum et al., 1990b), even when less supportive reports have also been published (see, for example, Villegas et al., 2000; Royo et al., 2002 for durum wheat). The main consideration in using canopy temperature for the selection for drought avoidance in wheat is that plants must be under adequate stress, typically around −2.0 to −2.5 MPa of leaf water potential at midday and a full soil covering by the canopy. Additional guidelines for using the method are given at http://www.plantstress.com/admin/files/IRT_protocol.htm. However, the relationship between canopy temperature depression [canopy minus air temperature (CTD)] and yield is best expressed under well-watered conditions and high evaporative demand. High evaporative demand develops under high air temperature and high vapor pressure deficit. It was shown that under such conditions, CTD can be used to select for high yield potential wheat materials (Reynolds et al., 1994b, 1998; Amani et al., 1996).

The potential of aerial IR imagery for screening purposes has also been demonstrated; a positive association was found between yield of recombinant bread wheat inbred lines and their plot temperatures when sensed from a height of 800 m (Reynolds et al., 1999).

4.2.2 Carbon Isotope Discrimination

The techniques of remote sensing (spectroradiometry and CTD) discussed above need a canopy structure and are not useful for selection with isolate plants (e.g., in early generations). Moreover, these techniques provide instantaneous readings; consequently, to reflect a time-integrative behavior, a plot must be measured at different stages. By contrast, the proportion of different stable isotopes in the dry matter compared with the natural proportion of them in the environmental source, although requiring the sampling and analyses of part of the material, may, in many cases, integrate the behavior of the genotype during at least part of the growing season and allows the estimation to be made in early generations. As the abundance of ^{13}C in C3 plants (including small grain cereals) is commonly less than that in air CO_2 to be photosynthesized, plants, such as wheat and barley, discriminate against ^{13}C in the photosynthetic process (Farquhar et al., 1982, 1989a; Hall, 1990). When measured in dry matter, the carbon isotope discrimination ($\Delta^{13}C$) indicates the reduction in the proportion of ^{13}C relative to ^{12}C (both stable isotopes of C in the air) experimented during plant growth. Whereas the main discrimination against ^{13}C takes place during carboxylation (Farquhar et al., 1989a), $\Delta^{13}C$ is associated with the ratio of the intracellular to air concentrations of CO_2 ($C_i:C_a$ ratio; e.g., Hall, 1990; Farquhar et al., 1989b), and then $\Delta^{13}C$ increases as C_i remains high. Therefore, and providing water pressure deficit is steady, it can be assumed that a higher $\Delta^{13}C$ is an indicator of a lower photosynthesis-to-transpiration ratio (namely, provided that evaporation is steady, water-use efficiency). Plants discriminating heavily against ^{13}C, and thus having a low water-use efficiency, are those able to keep a relatively high C_i, either due to a higher g_s or, less commonly, a low internal photosynthetic capacity, or both together (Richards, 2000). There is abundant literature supporting a positive correlation between grain yield and $\Delta^{13}C$ for bread wheat (e.g., Condon et al., 1987; Araus et al., 1993; Sayre et al., 1995, 1997; Fischer et al., 1998), durum wheat (Araus et al. 1998; Villegas et al., 2000; Royo et al., 2002), and barley (Romagosa and Araus, 1991; Araus et al., 1999; Voltas et al., 1999) among other C3 crops. These positive relationships are found particularly in the range of moderately drought-stressed to full-irrigated environments. Provided that no differences in phenology exist among genotypes, the positive correlation between $\Delta^{13}C$ and yield may be explained by the fact that plants with higher $\Delta^{13}C$ are those having either better status during crop cycle and thus probably maintaining higher g_s (Richards 1996a; Condon et al., 2002; Araus et al., 2003b) or exhibiting a constitutive higher g_s (Farquhar and Richards, 1984; Fischer et al., 1998). Therefore, high values of $\Delta^{13}C$ are frequently indicating lines with higher efficiency for converting the intercepted radiation into new dry matter (Slafer et al., 1999). However, the relationship between $\Delta^{13}C$ and yield tend to shift to negative when the environmental conditions are characterized by a relatively severe drought and then reduced g_s is related to higher WUE and yield (see discussion in Farquhar and Richards, 1984; Hubick and Farquhar, 1989; Acevedo, 1993; Slafer and Araus, 1998). Negative relationships between $\Delta^{13}C$ and yield have also been reported for cereals in water-limited environments, where crop growth is most dependent on soil moisture stored from rain that falls outside the main crop growth phase (Richards et al., 2002).

To conclude, a large degree of variation among cereals genotypes in $\Delta^{13}C$ has been reported and this trait seemed to be highly heritable (Farquhar et al., 1989a; Condon et al., 1990; Richards and Condon, 1993; Sayre et al., 1995; Araus et al. 1998), evidencing the likelihood of using $\Delta^{13}C$ in realistic breeding programs for the selection of either improved ability to capture more water (and then keep transpiration flux less restricted) or greater water-use efficiency (Richards, 2000; Araus et al., 2003b).

5 POTENTIAL USE OF BIOTECHNOLOGY TO RAISE CEREAL YIELDS

The above discussion underlined the need to look for ways to complement conventional breeding done by selection for yield per se. Integration of novel techniques and methodologies into conventional programs is needed to facilitate the identification, the characterization, and the manipulation of genetic variation for continued and accelerated progress (Sorrells and Wilson, 1997). Biotechnology offers two new ways for improving wheat and barley: one through the development and application of molecular markers, and the other through genetic engineering. However, the different size of wheat (16 × 10^9 bp for bread wheat, *T. aestivum*, 10 × 10^9 bp for durum wheat *T. turgidum* L. var. *durum*) and barley (5 × 10^9 bp) genomes and their different structure (genomes ABD in bread wheat, AB in durum wheat, but only H in barley) makes much more complex the use of molecular markers for breeding and selection in wheat than in barley. Moreover, bread and durum wheat have a lower level of polymorphism than barley or other cereals (Chao et al., 1989; Devos et al., 1995), and the level of polymorfism is not consistent across genomes or crosses (Langridge et al., 2001).

Molecular markers technology offers a novel approaches to improve the efficiency of selection. Marker-mediated genetical analyses are useful in the prediction and tracking of valuable alleles, the analysis of the genetic control of specific traits, and the analysis of the whole genome. Molecular markers have become a critical tool in studies of diversity in cereals because they offer an easily quantifiable measure of genetic variation within a given species. Their use is enlarging the variability available for breeding programs and allowing the detection of the origin of specific alleles introduced on recent varieties.

The development of comprehensive genetic maps based on molecular markers has enormously improved the power of genetic analysis (Snape et al., 2001). In addition, the use of marker-assisted selection (MAS) will enhance selection efficiency not only for Mendelian traits for which individual phenotypes provide large information about the underlying genotypes but also for most of the complex traits of agronomic interest for which phenotypes are less informative about the underlying genotypes (McCouch, 2001; Tuberosa et al., 2002). The detection and location of quantitative trait loci (QTLs) enables the use of MAS for attributes difficult to manage by conventional breeding approaches, leading to a potentially more reliable, quick, and efficient selection. Traits that in the past were recalcitrant to analysis, such as abiotic stress responses, are now amenable, and individual major genes and QTL mediating the variation can be identified (Snape et al., 2001).

Molecular markers linked to traits of economic importance have been identified in wheat (see examples in Langridge et al., 2001 and a broad catalogue in *http://wheat.pw.usda.gov*) and barley (Barr et al., 2000). This has allowed the development of strategies for manipulating the phenology of genotypes or introducing genes that enable the plant to tolerate stress (Snape et al., 2001). For instance, genes for vernalization response and cold tolerance have been located in chromosomes of homeologous group 5 of wheat (Snape et al., 2001). It is potentionally possible to tailor wheat varieties with different growth cycle length and, thus, change yield potential by adjusting the allele at the *Vrn-A1* locus, one of the five loci that control vernalization requirements in wheat (Snape et al., 2001). Group 5 also carries genes controlling a range of stress responses such as tolerance to freezing, drought, osmotic stress, and high temperatures (Snape et al., 2001).

While some workers report that they have identified QTLs for yield, WUE, and other complex quantitative characteristics (examples in Slinkard 1998 and Yin et al., 1999), QTL identification is much more complex than for simple traits and the $G \times E$ interactions on the expression of these QTLs are frequently large (e.g., Kjaer and Jensen, 1996). It is therefore important to understand of the causes underlying that $G \times E$ interaction through a collaboration between crop physiologists, plant breeders, and molecular biologists. However, to date, the impact of marker-based QTL analysis on the development of new varieties with enhanced quantitative traits has been less than expected, partially due to the detachment between QTL studies and variety development (Tanksley and Nelson, 1996). QTL analyses have been recently carried out for some yield-related traits, such as grain weight (Varshney et al., 2000), ear compactness (Sourdille et al., 2000b), lodging resistance (Keller et al., 1999), heading time (Sourdille et al., 2000a), or abiotic stresses such as responses to drought (Quarrie et al., 1995), salt tolerance (Mano and Takeda, 1997), or manganese efficiency (Pallotta et al., 2000). However, many of these traits are poorly understood at the physiological or biochemical level (Langridge et al., 2001). Physiological markers are likely to be more useful at least until denser molecular maps can be developed. These promising techniques may then be put to wider use to increase yield potential and tolerance to complex and largely unpredictable stresses (Slafer and Otegui, 2000). The role of markers in screening breeding populations for yield and adaptation will increase when these complex traits will be tagged with molecular markers (Langridge et al., 2001).

Examples of drought-related traits studied by QTL analysis include leaf ABA content in wheat (Quarrie et al., 1994) and water status, water-soluble carbohydrate, osmotic adjustment, plant architecture, growth habit, and chlorophyll content in barley (Teulat et al., 1997, 1998, 2001). A problem highlighted by QTL analysis is the association of abiotic stress traits with genetic loci of agronomic importance (Forster et al., 2000b). For instance, genetic linkage between salt tolerance at germination and ABA response has been found from QTL mapping in barley (Mano and Takeda, 1997). Important genes for adaptation to target environments, such as genes responsive to vernalization and photoperiod, and semidwarf genes frequently show pleiotropic effects on stress tolerance. In addition, QTLs for stress responses can coincide with yield and quality QTLs or other important genomic regions (Forster et al., 2000a).

Genetic transformation technology opens up opportunities to raise cereal yields. Apart from the direct inclusion of specific foreign genes in a cereal genome, transformation techniques open the possibility of drastically increasing the genetic variability available for plant breeding. There are two approaches to the transformation of cereals. One consists in modifying and reinserting native genes or promoters or suppressors back into the same species in order to increase the expression of these genes, modify their products, or to switch them off. The second approach consists in generating novel germplasm through the introduction of genes from alien sources such as virus, bacteria, plants, or animals (Snape, 1998). Progress in cereal transformation have been faster in rice and maize (e.g., Lazzeri and Shewry, 1993; Tanksley and McCouch, 1997; Sheehy et al., 2000) than in wheat and barley (Barceló et al., 1998).

Wheat or barley of genetically modified varieties are not cultivated at present in part due to the obstacles derived from ethical, environmental, and political aspects. Moreover, the reluctance of the industry and the consumers to accept transgenic varieties will restrict the application of the technology. However, the expectation is that many genes will be amenable to manipulation via genetic engineering. Field trials of genetically engineered wheat (Langridge et al., 2001) and barley (Lörz et al., 2000) are now well advanced in many countries. A major difficulty at present in transformation is derived from the restricted number of genotypes which can be handled and regenerated successfully in vitro, but other problems that have to be addressed relate to the efficiency of transformation, the number of integrated transgenes, the unpredictable variation in regenerated plants and progeny, and the stability of transgene expression (Lörz et al., 2000).

At present, the work is focused in the acquisition of genes that code for agronomically and qualitatively interesting characteristics. Traits, such as photoperiod and dwarfing genes, that contributed significantly to increase yields in the past are often considered to be useful for future gains in yield potential (Worland et al., 1998; Sears, 1998; Slafer et al., 2001a), but interest should also be focused on producing cultivars with faster growth rates and greater biomass during grain filling (Austin, 1999; Villegas et al., 2001) even when these are much more complex traits. Transgenic wheat plants with higher water-use efficiency and improved total biomass, root and shoot dry weights have been obtained introducing the ABA-responsive barley gene *HVA1* (Sivamani et al., 2000). Increased yield in transgenic varieties may also be accomplished indirectly through the introduction of resistance to herbicides, resistance to viral and some fungal pathogens that are the main targets considered so far (Langridge et al., 2001).

Functional genomics has the potential to reveal the genetic basis phenotypic response to the environment and, hence, opens up the possibility for genetic improvement by transformation. However, as outlined above, adaptation to stress at the whole plant level involves the interaction of many genes which are expressed at multiple levels (i.e., different environmental condition, plant tissue, phenological stage, etc.). At present, there is a massive database resource to explore (e.g., *http://wheat.pw.usda.gov/genome*). The challenge now is to sift through these databases to identify genes associated with QTLs for drought and other stress tolerances (Forster et al., 2000b). Considerable investment will be required before genetic engineering as a means of improving cereal cultivars for different target environments becomes routine. Furthermore, molecular research often addresses processes related to dehydration tolerance and recovery (Cushman and Bohnert, 2000) which are not important factors for crop production under drought stress such as, for example, dehydration avoidance and WUE (Richards et al., 2001).

Some functional genomics studies of wheat stress responses may develop into useful application. Late-embryogenesis-abundant (LEA) proteins appear when drying initiates in developing seeds and disappear after imbibition (Roberts et al., 1993). The genes are similar to those expressed in drought-stressed vegetative tissue of wheat (Curry et al., 1991); ABA can induce expression of these proteins. Sugar synthesis also seems to play a role in drought, providing compatible solutes for osmotic adjustment (Bohnert et al., 1995), or through various protective roles including protection of membranes (Crowe et al., 1992). Antioxidants, such as superoxide dismutase and ascorbate peroxidase, increase in response to drought stress (Mittler and Zilinskas, 1994) and probably play a role in tolerance because excess radiation and increased photorespiration associated with stress can result in accumulation of active oxygen species. Some other relatively simple biochemical processes involved in drought which may lend themselves to genetic transformation include osmotic adjustment, repair and degradation of proteins, and structural adjustment, for example, of the cell wall (Ingram and Bartels, 1996). However, a deeper crop ecophysiological understanding is required to take full advantage of biotechnology on cereal breeding (Araus et al., 2003a; Slafer, 2003).

ACKNOWLEDGMENTS

This study was supported by the research CICYT projects AGF 99-0611-C03 and AGL2002-04285-C03 (Spain). We are also grateful to the R. Thalmann Program (UBA, Argentina) and to the Generalitat de Catalunya (Spain) for their financial support.

REFERENCES

Abbate PE, Andrade FH, Lázaro L, Barifi JH, Berardocco HG, Inza VH, Marturano F. Grain yield increase in recent Argentina wheat cultivars. Crop Science 1998; 38:1203–1209.

Abeledo LG, Calderini DF, Slafer GA. Physiological changes associated with breeding progress. In: Slafer GA, Molina-Cano JL, Savin R, Araus JL, Romagosa I, eds. Barley Science: Recent Advances from Molecular Biology to Agronomy of Yield and Quality. New York: Food Product Press, 2002:361–386.

Acevedo E. Potential of carbon isotope discrimination as a selection criterion in barley breeding. In: Ehleringer JR, Hall AE, Farquhar GD, eds. Stable Isotopes and Plant Carbon/Water Relations. New York: Academic Press, 1993:399–417.

Amani I, Fischer RA, Reynolds MP. Canopy temperature depression association with ·

yield of irrigated spring wheat cultivars in hot climate. J Agron Crop Sci 1996; 176:119–129.

Amir J, Sinclair TR. Cereal grain yield: biblical aspirations and modern experience in the Middle East. Agron J 1994; 86:362–364.

Amthor JS. Effects of atmospheric CO_2 concentration on wheat yield: review of results from experiments using various approaches to control CO_2 concentration. Field Crops Res 2001; 73:1–34.

Angus JF, Jones R, Wilson JH. A comparison of barley cultivars with different leaf inclinations. Aust J Agric Res 1972; 23:945–957.

Aparicio N, Villegas D, Casadesús J, Araus JL, Royo CA. Spectral reflectance indices for assessing durum wheat biomass, green area, and yield under Mediterranean conditions. Agron J 2000; 92:83–91.

Araus JL. Integrative physiological criteria associated with yield potential. In: Reynolds MP, Rajaram S, McNab A, eds. Increasing Yield Potential in Wheat: Breaking the Barriers. Mexico, DF: CIMMYT, 1996:150–167.

Araus JL. Physiological basis of the processes determining barley yield under potential and stress conditions: current research trends on carbon assimilation. In: Slafer GA, Molina-Cano JL, Savin R, Araus JL, Romagosa I, eds. Barley Science: Recent Advances from Molecular Biology to Agronomy of Yield and Quality. New York: Food Product Press, 2002:269–306.

Araus JL, Tapia L. Photosynthetic gas exchange characteristics of wheat flag leaf blades and sheaths during grain filling. The case of a spring crop grown under Mediterranean climate conditions. Plant Physiol 1987; 85:667–673.

Araus JL, Tapia L, Alegre L. The effect of changing sowing date on leaf structure and gas exchange characteristics of wheat flag leaves grown under Mediterranean conditions. J Exp Bot 1989; 40:639–646.

Araus JL, Reynolds MP, Acevedo E. Leaf posture, grain yield, growth, leaf structure and carbon isotope discrimination in wheat. Crop Sci 1993; 33:1273–1279.

Araus JL, Amaro T, Casadesús J, Asbati A, Nachit MM. Relationships between ash content, carbon isotope discrimination and yield in durum wheat. Aust J Plant Physiol 1998; 25:835–842.

Araus JL, Slafer GA, Romagosa I. Durum wheat and barley yields in antiquity estimated from ^{13}C discrimination of archaeological grains: a case study from the Western Mediterranean Basin. Aust J Plant Physiol 1999; 26:345–352.

Araus JL, Casadesús J, Bort J. Recent tools for the screening of physiological traits determining yield. In: Reynolds MP, Ortiz-Monasterio JI, McNab A, eds. Application of Physiology in Wheat Breeding. Mexico, DF: CIMMYT, 2001a:59–77.

Araus JL, Slafer GA, Romagosa I, Molist M. Wheat yields during the emergence of agriculture estimated from the carbon isotope discrimination of grains: evidence from a tenth millennium BP site on the Euphrates. J Archeol Sci 2001b; 28:341–350.

Araus JL, Slafer GA, Reynolds MP, Royo C. Plant breeding and drought in C3 cereals: what to breed for? Ann Bot 2002; 89:925–940.

Araus JL, Bort J, Steduto P, Villegas D, Royo C. Breeding cereals for Mediterranean conditions: ecophysiological clues for biotechnology application. Ann Appl Biol 2003a; 142:129–141.

Araus JL, Villegas D, Aparicio N, García del Moral LF, El Hani S, Rharrabti Y, Ferrio JP, Royo C. Evironmental Factors Determining Carbon Isotope Discrimination and Yield in Durum Wheat Under Mediterranean Conditions. Crop Sci 2003b; 43(1).

Austin RB. Genetic variation in photosynthesis. J Agric Sci (Cambridge) 1989; 112: 287–294.

Austin RB. Augmenting yield-based selection. In: Hayward MD, Bosemark NO, Romagosa I, eds. Plant Breeding: Principles and Prospects. London: Chapman & Hall, 1993:391–405.

Austin RB. Yield of wheat in the United Kingdom: Recent advances and prospects. Crop Sci 1999;391604–1610.

Austin RB, Bingham J, Blackwell RD, Evans LT, Ford MA, Morgan CL, Taylor M. Genetic improvement in winter wheat yields since 1900 and associated physiological changes. J Agric Sci (Cambridge) 1980a; 94:675–689.

Austin RB, Morgan CL, Ford MA, Blackwell RD. Contributions to grain yield from preanthesis assimilation in tall and dwarf barley phenotypes in two contrasting seasons. Ann Bot 1980b; 45:309–319.

Austin RB, Morgan CL, Ford MA, Bhagwat SCA. Flag leaf photosynthesis of *Triticum aestivum* and related diploid and tetraploid species. Ann Bot 1982; 49:177–189.

Austin RB, Ford MA, Morgan CL. Genetic improvement in the yield of winter wheat: a further evaluation. J Agric Sci (Cambridge) 1989; 112:295–301.

Barceló P, Rasco-Gaunt S, Sparks C, Cannell M, Salgueiro S, Rooke L, He GY, Lamacchia C, DelaViña G, Shewry PR. Transformation of wheat: State of the technology and examples of application. In: Slinkard AE, ed. Proceedings of the 9th International Wheat Genetics Symposium. Vol. 1. Saskatoon: University of Saskatchewan Extension Press, 1998:143–147.

Barr AR, Jefferies SP, Warner P, Moody DB, Chalmers KJ, Langridge P. Marker assisted selection in theory and practice. In: Logue S, ed. Proceedings of the 8th International Barley Genetics Symposium. Vol. 1. Adelaide: Adelaide University, 2000:167–178.

Bell MA, Fischer RA, Byerlee D, Sayre K. Genetic and agronomic contributions to yield gains: A case study for wheat. Field Crops Res 1995; 44:55–65.

Bellairs SM, Turner NC, Hick PT, Smith RCG. Plant and soil influences on estimating biomass of wheat in plant-breeding plots using field spectral radiometers. Aust J Agric Res 1996; 47:1017–1034.

Biscoe PV, Gallagher JN. Weather, dry matter production and yield. In: Landsberg J, Cutting C, eds. Environmental Effects on Crop Physiology. London: Academic Press, 1978.

Blum A. Constitutive traits affecting plant performance under stress. In: Edmeades GO, Bänziger M, Mickelson HR, Peña-Valdivia CB, eds. Developing Drought- and Low Nitrogen-Tolerant Maize. Mexico, DF: CIMMYT, 1996a:131–135.

Blum A. Yield potential and drought resistance: are they mutually exclusive? In: Reynolds MP, Rajaram S, McNab A, eds. Increasing Yield Potential in Wheat: Breaking the Barriers. Mexico, DF: CIMMYT, 1996b:90–100.

Blum A. www.plantstress.com. Web site dedicated to plant environmental stress in agriculture and biology, 2000.

Blum A, Mayer J, Golan G. Infrared thermal sensing of plant canopies as a screening technique for dehydration avoidance in wheat. Field Crops Res 1982; 5:137–146.

Blum A, Ramaiah S, Kanemasu ET, Paulsen GM. Wheat recovery from drought stress at the tillering stage of development. Field Crops Res 1990; 24:67–85.

Blum A, Shpiler L, Golan G, Mayer J. Yield stability and canopy temperature of wheat genotypes under drought stress. Field Crops Res 1990; 22:289–296.

Bohnert HJ, Nelson DE, Jensen RG. Adaptations to environmental stresses. Plant Cell 1995; 7:1099–1111.

Borghi B, Corbellini M, Cattaneo M, Fornasari MA, Zucchelli L. Modification of the sink/source relationship in bread wheat and its influence on grain yield and protein content. J Agron Crop Sci 1986; 157:245–254.

Bort J, Araus JL, Hazzam H, Grando S, Ceccarelli S. Relationships between early vigour, grain yield, leaf structure and stable isotope composition in field grown barley. Plant Physiol Biochem 1998; 36:889–897.

Boukerrou L, Rasmusson DD. Breeding for high biomass yield in spring barley. Crop Sci 1990; 30:31–35.

Brooking IR, Kirby EJM. Interrelationships between stem and ear development in winter wheat: the effects of a Norin 10 dwarfing gene *Gai/Rht2*. J Agric Sci (Cambridge) 1981; 97:373–381.

Bulman P, Smith DL. Grain protein response of spring barley to high rates and post-anthesis application of fertilizer nitrogen. Agron J 1993; 85:1109–1113.

Bulman P, Mather DE, Smith DL. Genetic improvement of spring barley cultivars grown in eastern Canada from 1910 to 1988. Euphytica 1993; 71:35–48.

Byerlee D. Technical change, productivity, and sustainability in irrigated cropping systems of South Asia: Emerging issues in the post-green revolution era. J Int Dev 1992; 4:477–496.

Byerlee D, Moya P. Impact of international wheat breeding research in the developing world, 1966–1990. Mexico, DF: CIMMYT, 1993.

Calderini DF, Reynolds MP. Changes in grain weight as a consequence of de-graining treatments at pre- and post-anthesis in synthetic hexaploid lines of wheat (*Triticum durum* × *T. tauschii*). Aust J Plant Physiol 2000; 27:183–191.

Calderini DF, Slafer GA. Changes in yield and yield stability in wheat during the 20th century. Field Crops Res 1998; 57:335–347.

Calderini DF, Slafer GA. Has yield stability changed with genetic improvement of wheat yield? Euphytica 1999; 107:453–460.

Calderini DF, Dreccer MF, Slafer GA. Genetic improvement in wheat yield and associated traits. A re-examination of previous results and the latest trends. Plant Breed 1995; 114:108–112.

Calderini DF, Dreccer MF, Slafer GA. Consequences of breeding on biomass radiation interception and radiation use efficiency in wheat. Field Crops Res 1997; 52:271–281.

Calderini DF, Reynolds MP, Slafer GA. Genetic gains in wheat yield and main physiological changes associated with them during the 20th century. In: Satorre EH,

Slafer GA, eds. Wheat: Ecology and Physiology of Yield Determination. New York: Food Product Press, 1999a:351–377.

Calderini DF, Abeledo LG, Savin R, Slafer GA. Final grain weight in wheat as affected by short periods of high temperature during pre- and post-anthesis under field conditions. Aust J Plant Physiol 1999b; 26:453–458.

Calderini DF, Abeledo LG, Savin R, Slafer GA. Effect of temperature and carpel size during pre-anthesis on potential grain weight in wheat. J Agric Sci 1999c; 132:453–459.

Calderini DF, Savin R, Abeledo LG, Reynolds MP, Slafer GA. The importance of the immediately preceding anthesis period for grain weight determination in wheat. Euphytica 2001; 119:199–204.

Canevara MG, Romani M, Corbellini M, Perenzin M, Borghi B. Evolutionary trends in morphological, physiological, agronomical and qualitative traits of *Triticum aestivum* L. cultivars bred in Italy since 1900. Eur J Agron 1994; 3:175–185.

Carvalho FIF, Qualset CO. Genetic variation for canopy architecture and its use in wheat breeding. Crop Sci 1978; 18:561–567.

Carver BF, Johnson RC, Rayburn AL. Genetic analysis of photosynthetic diversity in hexaploid and tetraploid wheat and their interspecific hybrids. Photosynth Res 1989; 20:105–118.

Ceccarelli S, Grando S. Selection environment and environmental sensitivity in barley. Euphytica 1991; 57:157–167.

Ceccarelli S, Grando S. Drought as a challenge for the plant breeder. Plant Growth Regul 1996; 20:149–155.

Chao S, Sharp PJ, Worland AJ, Warham EJ, Koebner MD, Gale MD. RFLP-based genetic linkage maps of wheat homeologous group 7 chromosomes. Theor Appl Genet 1989; 78:495–504.

CIMMYT. CIMMYT 1995/96 World Wheat Facts and Trends: Understanding Global Trends in the Use of Wheat Diversity and International Flows of Wheat Genetic Resources. Mexico, DF: CIMMYT, 1996.

Condon AG, Richards RA, Farquhar GD. Carbon isotope discrimination is positively correlated with grain yield and dry matter production in field grown wheat. Crop Sci 1987; 27:996–1001.

Condon AG, Farquhar GD, Richards RA. Genotypic variation in carbon isotope discrimination and transpiration efficiency in wheat. Leaf gas exchange and whole plant studies. Aust J Plant Physiol 1990; 17:9–22.

Condon AG, Richards RA, Rebetzke GJ, Farquhar GD. Improving intrinsic water-use efficiency and crop yield. Crop Sci 2002; 42:122–131.

Cooper M, Stucker RE, DeLacy IH, Harch BD. Wheat breeding nurseries, target environments, and indirect selection for grain yield. Crop Sci 1997; 37:1168–1176.

Cox TS, Shroyer JP, Ben-Hui L, Sears RG, Martin TJ. Genetic improvement in agronomic traits of hard red winter wheat cultivars from 1919 to 1987. Crop Sci 1988; 28:756–760.

Crowe JH, Hoerkstra FA, Crowe LM. Anhydrobiosis. Annu Rev Plant Physiol 1992; 54:579–599.

Curry J, Morris CF, Walker-Simmons MK. Sequence analysis of a cDNA encoding a group 3 LEA mRNA inducible by ABA or dehydration stress in wheat. Plant Mol Biol 1991; 16:1073–1076.

Cushman JC, Bohnert HJ. Genomic approaches to plant stress tolerance. Curr Opin Plant Biol 2000; 3:117–124.

Deckerd EL, Busch RH, Kofoid KD. Physiological aspects of spring wheat improvement. In: Harper JE, Schrader LE, Howell RW, eds. Exploitation of Physiological and Genetic Variability to Enhance Crop Productivity. Rockland: American Society of Plant Physiology, 1985:45–54.

Delgado E, Medrano H, Keys AJ, Parry MAJ. Species variation in Rubisco specificity factor. J Exp Bot 1995; 292:1775–1777.

Devos KM, Dubcovsky J, Dvorak J, Chinoy CN, Gale MD. Structural evolution of wheat chromosomes 4A, 5A, and 7B and its impact on recombination. Theor Appl Genet 1995; 91:282–288.

Devos KM, Gale MD. Comparative genetics in the grasses. Plant Mol Biol 1997; 35:3–15.

Donald CM. The breeding of crop ideotypes. Euphytica 1968; 17:385–403.

Dreccer MF, Grashoff C, Rabbinge R. Source–sink ratio in barley (*Hordeum vulgare* L.) during grain filling: effects on senescence on grain nitrogen concentration. Field Crops Res 1997; 49:269–277.

Dreccer MF, Schapendonk AHCM, Slafer GA, Rabbinge R. Comparative response of wheat and oilseed rape to nitrogen supply: absorption and utilisation efficiency of radiation and nitrogen during the reproductive stages determining yield. Plant and Soil 2000; 220:189–205.

Duncan WG. Leaf angles, leaf area, and canopy photosynthesis. Crop Sci 1971; 11:482–485.

Duvick DN. Genetic contributions to advances in yield of US maize. Maydica 1992; 37:69–79.

El Hafid R, Smith DH, Karrou M, Samir K. Morphological attributes associated with early-season drought tolerance in spring durum wheat in a Mediterranean environment. Euphytica 1998; 101:273–282.

Elliott GA, Regan KL. Use of reflectance measurements to estimate early cereal biomass production on sandplain soils. Aust J Exp Agric 1993; 33:179–183.

Evans JR. Nitrogen and photosynthesis in the flag leaf of wheat (*Triticum aestivum* L.). Plant Physiol 1983; 72:297–302.

Evans JR. Photosynthesis and nitrogen relationships in leaves of C3 plants. Oecologia 1989; 78:9–19.

Evans LT. Crop Physiology. Cambridge: Cambridge University Press, 1975.

Evans LT. Crop Evolution, Adaptation and Yield. Cambridge: Cambridge University Press, 1993.

Evans LT. Adapting and improving crops: the endless task. Philos Trans R Soc Lond B 1997; 352:901–906.

Evans LT. Feeding the Ten Billion: plants and population growth. Cambridge: Cambridge University Press, 1998.

Evans LT, Dunstone RL. Some physiological aspects of evolution in wheat. Aust J Biol Sci 1970; 23:725–741.

Evans LT, Fischer RA. Yield potential: its definition, measurement, and significance. Crop Sci 1999; 39:1544–1551.

Evans LT, Peacock WJ. Wheat Science Today and Tomorrow. Cambridge: Cambridge University Press, 1981.

Farquhar GD, Richards RA. Isotopic composition of plant carbon correlates with water-use-efficiency of wheat genotypes. Aust J Plant Physiol 1984; 11:539–552.

Farquhar GD, Ball MC, von Caemmerer S, Roksandic Z. Effects of salinity and humidity on ^{13}C value of halophytes-evidence for diffusional isotope fractionation determined by the ratio of intracellular/atmospheric partial pressure of CO_2 under different environmental conditions. Oecologia 1982; 52:121–124.

Farquhar GD, Ehleringer JR, Hubick KT. Carbon isotope discrimination and photosynthesis. Annu Rev Plant Physiol Plant Mol Biol 1989a; 40:503–537.

Farquhar GD, Wong SC, Evans JR, Hubick KT. Photosynthesis and gas exchange. In: Jones HG, Flowers TJ, Jones MB, eds. Plants Under Stress: Biochemistry, Physiology and Ecology and their Application to Plant Improvement. Cambridge: Cambridge University Press, 1989b.

Feil B. Breeding progress in small grain cereals—A comparison of old and modern cultivars. Plant Breed 1992; 108:1–11.

Feil B, Geisler G. Untersuchungen zur Bildung und Verteilung der Biomasse bei alten und neuen deutschen Sommerweizensorten. J Agron Crop Sci 1988; 161:148–156.

Field CA. Allocating leaf nitrogen for the maximisation of carbon gain: leaf age as a control on the allocation program. Oecologia 1983; 56:341–347.

Field CB, Gamon JA, Peñuelas J. Remote sensing of terrestrial photosynthesis. In: Schulze D, Caldwell MM, eds. Ecophysiology of Photosynthesis. Berlin: Springer-Verlag, 1994.

Filella I, Amaro T, Araus JL, Peñuelas J. Relationship between photosynthetic radiation-use efficiency of barley canopies and the photochemical reflectance index. Physiol Plant 1996; 96:211–216.

Finlay KW, Wilkinson GN. The analysis of adaptation in a plant-breeding programme. Aust J Agric Res 1963; 14:342–354.

Fischer RA. Wheat. In: Smith WH, Banta JJ, eds. Potential Productivity of Field Crops under Different Environments. Los Baños: IRRI, 1983:129–154.

Fischer RA. Number of kernels in wheat crops and the influence of solar radiation and temperature. J Agric Sci (Cambridge) 1985; 105:447–461.

Fischer RA. Irrigated spring wheat and timing and amount of nitrogen fertilizer. II. Physiology of grain yield response. Field Crops Res 1993a; 33:57–80.

Fischer RA. Cereal breeding in developing countries: progress and prospects. In: Buxton DR, Shibles R, Forsberg RA, Blad BL, Asay KH, Paulsen GM, Wilson RF, eds. International Crop Science I. Madison: Crop Science Society of America, 1993b:201–209.

Fischer RA. Wheat physiology at CIMMYT and raising the yield plateau. In: Reynolds MP, Rajaram S, McNab A, eds. Increasing Yield Potential in Wheat: Breaking the Barriers. Mexico: CIMMYT, 1996:195–203.

Fischer RA, Quail KJ. The effect of major dwarfing genes on yield potential in spring wheats. Euphytica 1990; 46:51–56.

Fischer RA, Stockman YM. Kernel number per spike in wheat (*Triticum aestivum* L.): Responses to preanthesis shading. Aust J Plant Physiol 1980; 7:169–180.

Fischer RA, Rees D, Sayre KD, Lu Z-M, Condon AG, Larqué-Saavedra A. Wheat yield progress associated with higher stomatal conductance and photosynthetic rate, and cooler canopies. Crop Sci 1998; 38:1467–1475.

Flexas J, Briantais JM, Cerovic Z, Medrano H, Moya I. Steady-state and maximum chlorophyll fluorescence responses to water stress in grapevine leaves: a new remote sensing system. Remote Sens Environ 2000; 73:283–297.

Forster BP, Ellis RP, Thomas WTB, Newton AC, Tuberosa R, This D, El-Enein RA, Bahri MH, Ben Salem M. The development and application of molecular markers for abiotic stress tolerance in barley. J Exp Bot 2000a; 51:19–27.

Forster BP, Ellis RP, Thomas WTB, Waugh R, Ivandic V, Tuberosa R, Talame V, This D, Teulat-Merah B, El-Enein RA, Bahri H, Ben Salem M. Research Developments in genetics of drought tolerance in barley. In: Logue S, ed. Proceedings of the 8th International Barley Genetics Symposium. Vol. 1. Adelaide: Adelaide University, 2000b: 233–237.

García del Moral LF, Rharrabti Y, Villegas D, Royo C. Evaluation of grain yield and its components in durum wheat under Mediterranean conditions: an ontogenic approach. Agron J 2003; 95:266–274.

Garrity DP, O'Toole JCA. Selection for reproductive stage drought avoidance in rice, using infrared thermometry. Agron J 1995; 87:773–779.

Gomez-Macpherson H, Richards RA. Effect of sowing time on yield and agronomic characteristics of wheat in south-eastern Australia. Aust J Agric Res 1995; 46:1381–1399.

González A, Martin I, Ayerbe L. Barley yield in water stress conditions. The influence of precocity, osmotic adjustment and stomatal conductance. Field Crops Res 1999; 62:23–34.

González FG, Slafer GA, Miralles DJ. Floret development and spike growth as affected by photoperiod during stem elongation in wheat. Field Crops Res 2003; 81:29–38.

Gutiérrez-Rodríguez M, Reynolds MP, Larqué-Saavedra A. Photosynthesis of wheat in a warm, irrigated environment. II: Traits associated with genetic gains in yield. Field Crops Res 2000; 66:51–62.

Hall AE. Physiological ecology of crops in relation to light, water and temperature. In: Carroll CR, Vandermeer JH, Rosset PM, eds. Agroecology. New York: Mc Graw-Hill Publishing Company, 1990.

Hanson PR, Riggs TJ, Klose SJ, Austin RB. High biomass genotypes in spring barley. J Agric Sci (Cambridge) 1985; 105:73–78.

Hay RKM. Harvest index: a review of its use in plant breeding and crop physiology. Ann Appl Biol 1995; 126:197–216.

Heyne EG. Wheat and Wheat Improvement. 2d ed. Madison: American Society of Agronomy, 1987.

Hirose T, Werger MJA. Maximizing daily canopy photosynthesis with respect to the leaf nitrogen allocation pattern in the canopy. Oecologia 1987; 72:520–526.

Hubick KT, Farquhar GD. Carbon isotope discrimination and the ratio of carbon gained to water lost in barley cultivars. Plant Cell Environ 1989; 12:795–804.

Hucl R, Baker RJ. A study of ancestral and modern Canadian spring wheats. Cananian Journal of Plant Science 1987; 67:87–97.

Ingram J, Bartels D. The molecular basis of dehydration tolerance in plants. Annu Rev Plant Physiol Plant Mol Biol 1996; 47:377–403.

Innes P, Blackwell RD. Some effects of leaf posture on the yield and water economy of winter wheat. J Agric Sci (Cambridge) 1983; 101:367–376.

Jackson P, Robertson M, Cooper M, Hammer G. The role of physiological understanding in plant breeding; from a breeding perspective. Field Crops Res 1996; 49:11–39.

Jedel P, Helm JH. Assessment of western Canadian barleys of historical interest: I. Yield and agronomic traits. Crop Sci 1994a; 34:922–927.

Jedel P, Helm JH. Assesment of western Canadian barleys of historical interest: II. Morphology and Phenology. Crop Sci 1994b; 34:927–932.

Johnson RC, Kebede H, Mornhinweg DW, Carver BF, Rayburn AL, Nguyen HT. Photosynthetic differences among *Triticum* accessions at tillering. Crop Sci 1987; 27:1046.

Keller M, Karutz C, Schmid JE, Stamp P, Winzeler M, Keller B, Messmer MM. Quantitative trait loci for lodging resistance in a segregating wheat × spelt population. Theor Appl Genet 1999; 98:1171–1182.

Kirby EJM. Analysis of leaf, stem and ear growth in wheat from terminal spikelet stage to anthesis. Field Crops Res 1988; 18:127–140.

Kjaer B, Jensen J. Quantitative trait loci for grain yield and yield components in a cross between a 6-rowed and a 2-rowed barley. Euphytica 1996; 90:39–48.

Koshkin EI, Tararina VV. Yield and source/sink relations of spring wheat cultivars. Field Crops Res 1989; 22:297–306.

Kruk B, Calderini DF, Slafer GA. Source–sink ratios in modern and old wheat cultivars. J Agric Sci (Cambridge) 1997; 128:273–281.

Kulshrestha VP, Jain HK. Eighty years of wheat breeding in India: Past selection pressures and future prospects. Zeitschrift für Pflanzenzüchtg 1982; 89:19–30.

Laing DR, Fischer RA. Adaptation of semidwarf wheat cultivars to rainfed conditions. Euphytica 1977; 26:129–139.

Langridge P, Lagudah ES, Holton TA, Appels R, Sharp PJ, Chalmers KJ. Trends in genetic and genome analyses in wheat: a review. Aust J Agric Res 2001; 52:1043–1077.

Lazzeri PA, Shewry PR. Cereal biotechnology. Biotechnol Genet Eng Rev 1993; 11:79–146.

Liang YL, Richards RA. Coleoptile tiller development is associated with fast early vigour in wheat. Euphytica 1994; 80:119–124.

Loomis RS, Amthor JS. Limits to yield revisited. In: Reynolds MP, Rajaram S, McNab A, eds. Increasing Yield Potential in Wheat: Breaking the Barriers. Mexico: CIMMYT, 1996:76–89.

Lörz H, Serazetdinova L, Leckband G, Lütticke S. Transgenic barley—A journey with obstacles and milestones. In: Logue S, ed. Proceedings of the 8th International Barley Genetics Symposium. Vol. 1. Adelaide: Adelaide University, 2000:189–193.

Loss SP, Siddique KHM. Morphological and physiological traits associated with wheat yield increases in Mediterranean environments. Adv Agron 1994; 52:229–276.

Ludlow MM, Muchow RC. Critical evaluation of traits for improving crop yields in water-limited environments. Adv Agron 1990; 43:107–153.

Lungu D, Kaltsikes PJ, Larter EN. Honeycomb selection for yield in early generations of spring wheat. Euphytica 1987; 36:831–839.

Magrin GO, Hall AJ, Baldy C, Grondona MO. Spatial and interannual variations in the phototemal quotient: implications for the potential kernel number of wheat crops in Argentina. Agric For Meteorol 1993; 67:29–41.

Mano Y, Takeda K. Mapping quantitative trait loci for salt tolerance at germination and the seedling stage in barley (*Hordeum vulgare* L.). Euphytica 1997; 94:263–272.

Marderscheid R, Weigel HJ. Do increasing atmospheric CO_2 concentrations contribute to yield increases of German crops? J Agron Crop Sci 1995; 175:73–82.

Martinez-Carrasco R, Thorne GN. Physiological factors limiting grain size in wheat. J Exp Bot 1979; 30:669–679.

Martiniello P, Delogu G, Oboardi M, Boggini G, Stanca AM. Breeding progress in grain yield and selected agronomic characters of winter barley (*Hordeum vulgare* L.) over de last quarter of a century. Plant Breed 1987; 99:289–294.

McCree KJ. The action spectrum, absorptance and quantum yield of photosynthesis in crop plants. Agric Meteorol 1971; 9:191–216.

McCouch SR. Genomics and Synteny. Plant Physiol 2001; 125:152–155.

McMaster GS. Phenology development and growth of the wheat (*Triticum aestivum* L.) shoot apex: a review. Adv Agron 1997; 59:63–118.

Minc LD, Vandermeer JH. The origin and spread of agriculture. In: Carroll CR, Vandermeer JH, Rosset P, eds. Agroecology. New York: McGraw-Hill Publishing Company, 1990:65–111.

Miralles DJ, Richards RA. Response of leaf and tiller appearance and primordia development to interchanged photoperiod in wheat and barley. Ann Bot 2000; 85:655–663.

Miralles DJ, Slafer GA. Yield, biomass and yield components in dwarf, semidwarf and tall isogenic liens of spring wheat under recommended and late sowing dates. Plant Breed 1995a; 114:392–396.

Miralles DJ, Slafer GA. Individual grain weight responses to genetic reduction in culm length in wheat as affected by source–sink manipulations. Field Crops Res 1995b; 43:55–66.

Miralles DJ, Slafer GA. Radiation interception and radiation use efficiency of near isogenic wheat lines with different height. Euphytica 1997; 97:201–208.

Miralles DJ, Dominguez C, Slafer GA. Relationship between grain growth and postanthesis leaf area duration in dwarf and semidwarf isogenic lines of wheat. J Agron Crop Sci 1996; 177:115–122.

Miralles DF, Katz SD, Colloca A, Slafer GA. Floret development in near isogenic wheat lines differing in plant height. Field Crops Res 1998; 59:21–30.

Miralles DF, Richards RA, Slafer GA. Duration of the stem elongation period influences the number of fertile florets in wheat and barley. Aust J Plant Physiol 2000; 27:931–940.

Mitchell JW, Baker RJ, Knott DR. Evaluation of honeycomb selection for single plant yield in durum. Crop Sci 1982; 22:840–843.

Mittler R, Zilinskas BA. Regulation of pea cytosolic ascorbate peroxidase and other antioxidant enzymes during the progression of drought stress and following recovery from drought. Plant J 1994; 5:397–405.

Monteith JL. Climate and the efficiency of crop production in Britain. Philos Trans R Soc Lond, B 1977; 281:277–297.

Moore G, Devos K, Wang Z, Gale MD. Grasses, line up and form a circle. Curr Biol 1995; 5:737–739.

Muñoz P, Voltas J, Araus JL, Igartua E, Romagosa I. Changes over time in the adaptation of barley releases in north-eastern Spain. Plan Breed 1998; 117:531–535.

Nicolas ME, Turner NCA. Use of chemicals desiccants and senescing agents to select wheat lines maintaining stable grain size during post-anthesis drought. Field Crops Res 1993; 31:155–171.

Ortiz-Monasterio JI, Sayre KD, Rajaram S, McMahon M. Genetic progress in wheat yield and nitrogen use efficiency under four nitrogen rates. Crop Sci 1997; 37:898–904.

Osmond CB, Grace SCA. Perspectives on photoinhibition and photo-respiration in the field. Quintessential inefficiencies of the light and dark reactions of photosynthesis? J Exp Bot 1995; 46:1351–1362.

Otegui ME, Slafer GA. Physiological Bases for Maize Improvement. New York: Food Product Press, 2000.

Pallotta MA, Graham RD, Langridge P, Sparrow DHB, Barker SJ. RFLP mapping of manganese efficiency in barley. Theor Appl Genet 2000; 101:1100–1108.

Parry MAJ, Keys AJ, Gutteridge S. Variation in the specificity factor of C3 higher plant Rubiscos determined by the total consumption of ribulose-P2. J Exp Bot 1989; 40:317–320.

Passioura JB. Grain yield, harvest index and water use of wheat. J Aust Inst Agric Sci 1977; 43:117–120.

Passioura JB. Drought and drought tolerance. Plant Growth Regul 1996; 20:79–83.

Peñuelas J, Filella I. Visible and near-infrared reflectance techniques for diagnosing plant physiological status. Trends Plant Sci 1998; 3:151–156.

Peñuelas J, Isla R, Filella I, Araus JL. Visible and near-infrared reflectance assessment of salinity effects on barley. Crop Sci 1997; 37:198–202.

Perry MW, D'Antuono MF. Yield improvement and associated characteristics of some Australian spring wheat cultivars introduced between 1860 and 1982. Aust J Agric Res 1989; 40:457–472.

Pfeiffer WH, Sayre KD, Reynolds MP. Enhancing genetic grain yield potential and yield stability in durum wheat. Durum Wheat Improvement in the Mediterranean Region: New Challenges. Proceedings of the Seminar. Options Mediterraneennes, Serie A, n° 40, 12–14 April, 2000. Zaragoza, Spain: CIHEAM-IRTA, 2000: 83–93.

Pfeiffer WH, Sayre KD, Reynolds MP, Payne TS. Increasing yield potential and yield stability in durum wheat. Wheat in a Global Environment, International Wheat

Conference, 6th, 5–9 June, 2000. Budapest, Hungary. Developments in Plant Breeding. Vol. 9. Dordrecht: Kluwer Academic Publishers, 2001:569–577.

Pingali PL, Heisey PW. Cereal Crop Productivity in Developing Countries. CIMMYT Economics Paper 99-03. Mexico, DF: CIMMYT, 1999.

Puckridge DW, Donald CM. Competition among wheat plants sown at a wide range of densities. Aust J Agric Res 1967; 18:193–211.

Quarrie SA, Galiba G, Sutka J, Snape JW. Association of a major vernalization gene of wheat with stress-induced abscisic acid production. In: Dorffling K, Brettschneider B, Tatau H, Pitham K, eds. Crop Adaptation to Cool Climates. Proc. COST 814 Workshop, Hamburg, Oct. 1994. Brussels: European Comission, 1995:403–414.

Quarrie SA, Gulli M, Calestani C, Steed A, Marmiroli N. Location of a gene regulating drought-induced abscisic acid production on the long arm of chromosome 5A of wheat. Theor Appl Genet 1994; 89:794–800.

Rajaram S, van Ginkel M. Yield potential debate: Germplasm vs. methodology, or both. In: Reynolds MP, Rajaram S, McNab A, eds. Increasing Yield Potential in Wheat: Breaking the Barriers. Mexico, DF: CIMMYT, 1996:11–18.

Rajaram S, Villareal R, Mujeeb-Kazi A. The global impact of 1B/1R spring wheats. Agronomy Abstracts. San Antonio: American Society of Agronomy, 1990:105.

Rasmusson DCA. Barley. Madison: American Society of Agronomy, 1985.

Read BA, Tabita FR. High substrate specificity factor ribulose bisphosphate carboxylase/oxygenase from eukaryotic marine algae and properties of recombinant cyanobacterial Rubisco containing "Algal" residue modifications. Arch Biochem Biophys 1994; 312:210–218.

Rebetzke GJ, Condon AG, Richards RA. Rapid screening of leaf conductance in segregating wheat populations. In: Richards RA, Wrigley CW, Rawson HM, Davidson JL, Brettell RIS, eds. Proceedings Eighth Assembly, Wheat Breeding Society of Australia, 1996:130–134.

Reynolds MP, Acevedo E, Sayre KD, Fischer RA. Yield potential in modern wheat varieties: its association with a less competitive ideotype. Field Crops Res 1994a; 37:49–160.

Reynolds MP, Balota M, Delgado MIB, Amani I, Fischer RA. Physiological and morphological traits associated with spring wheat yield under hot, irrigated conditions. Aust J Plant Physiol 1994b; 21:717–730.

Reynolds MP, van Beem J, van Ginkel M, Hoisington D. Breaking the yield barriers in wheat: a brief summary of the outcomes of an international consultation. In: Reynolds MP, Rajaram S, McNab A, eds. Increasing Yield Potential in Wheat: Breaking the Barriers. Mexico, DF: CIMMYT, 1996:1–10.

Reynolds MP, Singh RP, Ibrahim A, Ageeb OA, Larqué-Saavedra A, Quick JS. Evaluating physiological traits to compliment empirical selection for wheat in warm environments. Euphytica 1998; 100:85–94.

Reynolds MP, Rajaram S, Sayre KD. Physiological and genetic changes of irrigated wheat in the post-green revolution period and approaches for meeting projected global demand. Crop Sci 1999; 39:1611–1621.

Reynolds MP, van Ginkel M, Ribaut J-M. Avenues for genetic modification of radiation use efficiency in wheat. J Exp Bot 2000; 51:459–473.

Reynolds MP, Calderini DF, Condon AG, Rajaram S. Physiological basis of yield gains in wheat associated with the LR19 translocation from *A. elongatum*. Euphytica 2001a; 119:139–144.

Reynolds MP, Ortiz-Monasterio JI, McNab A, eds. Application of Physiology in Wheat Breeding. Mexico, DF: CIMMYT, 2001b:240.

Richards RA. A tiller inhibition gene in wheat and its effect on plant growth. Aust J Agric Res 1988; 39:749–757.

Richards RA. Crop improvement for temperate Australia: Future opportunities. Field Crops Res 1991; 26:141–169.

Richards RA. The effect of dwarfing genes in spring wheat in dry environments I. Agronomic characteristica. Aust J Agric Res 1992; 43:517–522.

Richards RA. The effect of dwarfing genes in spring wheat in dry environments II. Growth, water use and water use efficiency. Aust J Agric Res 1992; 43:529–539.

Richards RA. Defining selection criteria to improve yield under drought. Plant Growth Regul 1996a; 20:57–166.

Richards RA. Increasing yield potential in wheat—source and sink limitations. In: Reynolds MP, Rajaram S, McNab A, eds. Increasing Yield Potential in Wheat: Breaking the Barriers. Mexico, DF: CIMMYT, 1996b:134–149.

Richards RA. Selectable traits to increase crop photosynthesis and yield of grain crops. J Exp Bot 2000; 51:447–458.

Richards RA, Condon AG. Challengers ahead in using carbon isotope discrimination in plant breeding programs. In: Ehleringer JR, Hall AE, Farquhar GD, eds. Stable Isotopes and Plant Carbon–Water Relations. San Diego: Academic Press, 1993: 451–462.

Richards RA, Condon AC, Rebetzke GJ. Traits to improve yield in dry environments. In: Reynolds MP, Ortiz-Monasterio JI, McNab A, eds. Application of Physiology in Wheat Breeding. Mexico, DF: CIMMYT, 2001.

Richards RA, Rebetzke GJ, Condon AG, van Herwaarden AF. Breeding opportunities for increasing the efficiency of water use and crop yield in temperate Cereals. Crop Sci 2002; 42:111–121.

Riggs TJ, Hanson PR, Start ND, Miles DM, Morgan CL, Ford MA. Comparison of spring barley varieties grown in England and Wales between 1880 and 1980. J Agric Sci (Cambridge) 1981; 97:599–610.

Roberts JK, DeSimone NA, Lingle WL, Dure L III. Cellular concentrations and uniformity of cell-type accumulation of two Lea proteins in cotton embryos. Plant Cell 1993; 5:769–780.

Romagosa I, Araus JL. Genotype-environment interaction for grain yield and ^{13}C discrimination in barley. Barley Genet 1991; VI:563–567.

Rosegrant MW, Agcaoili-Sombilla M, Perez ND. Global Food Projections to 2020: Implications for Investment. Washington: IFPRI, 1995.

Royo C, Soler C, Romagosa I. Agronomical and morphological differentiation among winter and spring triticales. Plant Breed 1995; 114:413–416.

Royo C, Voltas J, Romagosa I. Remobilization of pre-anthesis assimilates to the grain for grain only and dual-purpose (forage and grain) triticale. Agron J 1999; 91:312–316.

Royo R, Villegas D, García del Moral LF, El Hani S, Aparicio N, Rharrabti Y, Araus JL. Comparative performance of carbon isotope discrimination and canopy temperature depression as predictors of genotype differences in durum wheat yield in Spain. Aust J Agric Res 2002; 53:561–569.

Satorre EH, Slafer GA. Wheat: Ecology and Physiology of Yield Determination. New York: Food Product Press, 1999.

Savin R, Slafer GA. Shading effects on the yield of an Argentinian wheat cultivar. J Agric Sci (Cambridge) 1991; 116:1–7.

Sayre KD. The role of crop management research in CIMMYT in addressing bread wheat yield potential issues. In: Reynolds MP, Rajaram S, McNab A, eds. Increasing Yield Potential in Wheat: Breaking the Barriers. Mexico, DF: CIMMYT, 1996: 203–208.

Sayre KD, Acevedo E, Austin RB. Carbon isotope discrimination and grain yield for three bread wheat germplasm groups grown at different levels of water stress. Field Crops Res 1995; 41:45–54.

Sayre KD, Rajaram S, Fischer RA. Yield potential progress in short bread wheats in northwest Mexico. Crop Sci 1997; 37:36–42.

Sears RG. Strategies for improving wheat grain yield. In: Braun HJ, Altay F, Kronstad WE, Beniwal SPS, McNab A, eds. Wheat: Prospects for Global Improvement. Dordrecht: Kluwer Academic Publishers, 1998:17–22.

Sharkey TD. Water stress effects on photosynthesis. Photosynthetica 1990; 24:651.

Sheehy JE, Mitchell PL, Hardy B. Redesigning Rice Photosynthesis to Increase Yield (Studies in Plant Science, 7). Los Baños: IRRI and Amsterdam: Elsevier Science B.V., 2000.

Shimshi D, Ephrat J. Stomatal behavior of wheat cultivars in relation to their transpiration, photosynthesis and yield. Agron J 1975; 67:326–331.

Sibma L. Relation between total radiation and yield of some field crops in the Netherlands. Neth J Agric Sci 1970; 18:125–131.

Siddique KHM, Belford RK, Perry MW, Tennant D. Growth, development and light interception of old and modern wheat cultivars in a Mediterranean type environment. Aust J Agric Res 1989a; 40:473–487.

Siddique KHM, Kirby EJM, Perry MW. Ear-to-stem ratio in old and modern wheats; relationship with improvement in number of grains per ear and yield. Field Crops Res 1989b; 21:59–78.

Simmons SR, Rasmusson DC, Wiersma JV. Tillering in Barley: Genotype, row spacing, and seeding rate effects. Crop Sci 1982; 22:801–805.

Sinclair TR, Horie T. Leaf nitrogen, photosynthesis and crop radiation use efficiency: A review. Crop Sci 1989; 29:90–98.

Sinclair TR, Muchow R. Radiation use efficiency. Adv Agron 1999; 65:215–265.

Singh RP, Huerta-Espino J, Rajaram S, Crossa J. Agronomic effects from chromosome translocations 7DL.7Ag and 1BL.1RS in spring wheat. Crop Sci 1998; 38:27–33.

Sivamani E, Bahieldin A, Wraith JM, Al-Niemi T, Dyer WE, Ho THD, Rongda Q. Improved biomass productivity and water use efficiency under water deficit conditions in transgenic wheat constitutively expressing the barley $HVA1$ gene. Plant Sci 2000; 155:1–9.

Slafer GA. Genetic Improvement of Field Crops. New York: Marcel Dekker Inc., 1994.

Slafer GA. Wheat development as affected by radiation at two temperatures. J Agron Crop Sci 1995; 175:249–263.

Slafer GA. Differences in phasic development rate amongst wheat cultivars independent of responses to photoperiod and vernalization. A viewpoint of the intrinsic earliness hypothesis. J Agric Sci (Cambridge) 1996; 126:403–419.

Slafer GA. Genetic basis of yield as viewed from a crop physiologist's perspective. Ann Appl Biol 2003; 142:117–128.

Slafer GA, Andrade FH. Genetic improvement in bread wheat (*Triticum aestivum*) yield in Argentina. Field Crops Res 1989; 21:289–296.

Slafer GA, Andrade FH. Changes in physiological attributes of the dry matter economy of bread wheat (*Triticum aestivum*) through genetic improvement of grain yield potential at different regions of the world. A review. Euphytica 1991; 58:37–49.

Slafer GA, Andrade FH. Physiological attributes related to the generation of grain yield in bread wheat cultivars released at different eras. Field Crops Res 1993; 31:351–367.

Slafer GA, Araus JL. Keynote address: Improving wheat responses to abiotic stresses. In: Slinkard AE, ed. Proceedings of the 9th International Wheat Genetics Symposium. Vol. 1. (Keynote address). Saskatoon: University of Saskatchewan Extension Press, 1998.

Slafer GA, Miralles DJ. Green area duration during the grain filling period of wheat as affected by sowing date, temperature and sink strength. J Agron Crop Sci 1992; 168:191–200.

Slafer GA, Miralles DJ. Fruiting efficiency in three bread wheat (*Triticum aestivum* L) cultivars released at different eras and their number of grains per spike. J Agron Crop Sci 1993; 170:251–260.

Slafer GA, Otegui ME. Is there a niche for physiology in future genetic improvement of maize yields? In: Otegui ME, Slafer GA, eds. Physiological Bases for Maize Improvement. New York: Food Product Press, 2000:1–13.

Slafer GA, Peltonen-Sainio P. Yield trends of temperate cereals in high latitude countries from 1940 to 1998. Agric Food Sci Finl 2001; 10:121–131.

Slafer GA, Rawson HM. Sensitivity of wheat phasic development to major environmental factors: a re-examination of some assumptions made by physiologists and modellers. Aust J Plant Physiol 1994; 21:393–426.

Slafer GA, Rawson HM. Phyllochron in wheat as affected by photoperiod under two temperature regimes. Aust J Plant Physiol 1997; 24:151–158.

Slafer GA, Satorre EH. An introduction to the physiological–ecological analysis of wheat yield. In: Satorre EH, Slafer GA, eds. Wheat: Ecology and Physiology of Yield Determination. New York: Food Product Press, 1999:3–12.

Slafer GA, Savin R. Sink-source relationships and grain mass at different positions within the spike in wheat. Field Crops Res 1994; 37:39–49.

Slafer GA, Whitechurch EM. Manipulating wheat development to improve adaptation and to search for alternative opportunities to increase yield potential. In: Reynolds MP, Ortiz-Monasterio JI, McNab A, eds. Application of Physiology in Wheat Breeding. Mexico, DF: CIMMYT, 2001:160–170.

Slafer GA, Andrade FH, Satorre EH. Genetic-improvement effects on pre-anthesis physiological attributes related to wheat grain yield. Field Crops Res 1990; 23:255–263.

Slafer GA, Satorre EH, Andrade FH. Increases in grain yield in bread wheat from breeding and associated physiological changes. In: Slafer GA, ed. Genetic Improvement of Field Crops. New York: Marcel Dekker Inca, 1994a:1–68.

Slafer GA, Calderini DF, Miralles DJ, Dreccer MF. Preanthesis shading effects on the number of grains of three bread wheat cultivars of different potential number of grains. Field Crops Res 1994b; 36:31–39.

Slafer GA, Calderini DF, Miralles DJ. Yield components and compensation in wheat: opportunities for further increasing yield potential. In: Reynolds MP, Rajaram S, McNab A, eds. Increasing Yield Potential in Wheat: Breaking the Barriers. Mexico, DF: CIMMYT, 1996:101–134.

Slafer G, Araus JL, Richards RA. Promising traits for future breeding to increase wheat yield. In: Satorre EH, Slafer GA, eds. Wheat: Ecology and Physiology of Yield Determination. New York: Food Product Press, 1999:379–415.

Slafer GA, Abeledo LG, Miralles DJ, Gonzalez FG, Whitechurch EM. Photoperiod sensitivity during stem elongation as an avenue to rise potential yield in wheat. Euphytica 2001a; 119:191–197.

Slafer GA, Molina-Cano JL, Savin R, Araus JL, Romagosa I. Barley Science: Recent Advances from Molecular Biology to Agronomy of Yield and Quality. New York: Food Product Press, 2001b.

Slinkard AE. Proceedings of the 9th International Wheat Genetics Symposium. Vol. 1, Section 7 Abiotic Stresses. Saskatoon: University of Saskatchewan Extension Press, 1998.

Smale M, McBride T. Understanding global trends in the use of wheat diversity and international flows of wheat genetic resources. Part 1: CIMMYT 1995/96 World Wheat Facts and Trends. México, DF: CIMMYT, 1996.

Smith WH, Banta JJ. Potential Productivity of Field Crops under Different Environments. Los Baños: IRRI, 1983.

Smith DL, Hamel CA. Crop Yield, Physiology and Processes. Berlin: Springer-Verlag, 1999.

Snape JW. The use of doubled haploids in plant breeding. Induced Variability in Plant Breeding. International Symposium of the Section Mutations and Polyploidy of Eucarpia. 1981. Wageningen: Centre for Agriculture Publishing and Documentation, 1982:52–58.

Snape JW. The contribution of new biotechnologies to wheat breeding. In: Reynolds MP, Rajaram S, McNab A, eds. Increasing Yield Potential in Wheat: Breaking the Barriers. Mexico, DF: CIMMYT, 1996:167–179.

Snape JW. Golden calves or white elephants? Biotechnologies for wheat improvement. Euphytica 1998; 100:207–217.

Snape JW, Butterworth K, Whitechurch E, Worland AJ. Waiting for fine times: genetics of flowering time in wheat. Euphytica 2001; 119:185–190.

Snape JW, Sarma R, Quarrie SA, Fish L, Galiba G, Sutka J. Mapping genes for flowering time and frost tolerance in cereals using precise genetic stocks. Euphytica 2001; 120:309–315.

Sorrells ME, Wilson WA. Direct classification and selection of superior alleles for crop improvement. Crop Sci 1997; 37:691–697.

Sourdille P, Snape JW, Cadalen T, Charmet G, Nakata N, Bernard S, Bernard M. Detection of QTLs for heading time and photoperiod response in wheat using a doubled-haploid population. Genome 2000a; 43:487–494.

Sourdille P, Tixier MH, Charmet G, Gay G, Cadalen T, Bernard S, Bernard M. Location of genes involved in ear compactness in wheat (*Triticum aestivum*) by means of molecular markers. Mol Breed 2000b; 6:247–255.

Stapper M, Fischer RA. Genotype, sowing date and planting spacing influence on high-yielding irrigated wheat in southern New South Wales. I Phasic development, canopy growth and spike production. Aust J Agric Res 1990; 41:997–1019.

Stuber CW, Polacco M, Lynn Senior M. Synergy of empirical breeding, marker-assisted selection, and genomics to increase crop yield potential. Crop Sci 1999; 39:1571–1583.

Tanksley SD, Nelson JCA. Advanced backcross QTL analysis: a method for the simultaneous discovery and transfer of valuable QTLs from unadapted germplasm into elite breeding lines. Theor Appl Genet 1996; 92:191–203.

Tanksley SR, McCouch SR. Seed banks and molecular maps: Unlocking genetic potential for the wild. Science 1997; 277:1063–1066.

Teulat B, Borries C, This D. New QTLs identified for plant water status, water-soluble carbohydrate and osmotic adjustment in a barley population grown in a growth-chamber under two water regimes. Theor Appl Genet 2001; 103:161–170.

Teulat B, Monneveux P, Wery J, Borries C, Souyris I, Charrier A, This D. Relationships between relative water content and growth parameters under water stress in barley: a QTL study. New Phytol 1997; 137:99–107.

Teulat B, This D, Khairallah M, Borries C, Ragot C, Sourdille P, Leroy P, Monneveux P, Charrier A. Several QTLs involved in osmotic-adjustment trait variation in barley (*Hordeum vulgare* L.). Theor Appl Genet 1998; 96:688–698.

Thorne GN, Wood DW. Effects of radiation and temperature on tiller survival, grain number and grain yield in winter wheat. Ann Bot 1987; 59:413–426.

Tuberosa R, Salvi S, Sanguineti MC, Landi P, Maccaferri M, Conti S. Mapping QTLs regulating morpho-physiological traits and yield in drought stressed maize: case studies, shortcomings and perspectives. Ann Bot 2002; 89:941–963.

Varshney RK, Prasad M, Roy JK, Harjit-Singh NK, Dhaliwal HS, Balyan HS, Gupta PK. Identification of eight chromosomes and a microsatellite marker on 1AS associated with QTL for grain weight in bread wheat. Theor Appl Genet 2000; 100:1290–1294.

Villareal RL, Fuentes-Davila G, Mujeeb-Kazi A. Synthetic hexaploids × *Triticum aestivum* advanced derivatives resistant to Karnal Bunt (*Tilletia indica* Mitra). Cereal Res Commun 1995; 27:127–132.

Villareal RL, Bañuelos O, Mujeeb-Kazi A. Agronomic performance of related durum wheat (*Triticum turgidum* L.) stocks possessing the chromosome substitution T1BL.1RS. Crop Sci 1997a; 37:1735–1740.

Villareal RL, Bañuelos O, Borja J, Mujeeb-Kazi A, Rajaram S. Agronomic perform-ance of some advanced derivatives of synthetic hexaploids (*T. turgidum* × *T. tauschii*). Annu Wheat Newslett 1997b; 43:175–176.

Villegas D, Aparicio N, Nachit MM, Araus JL, Royo CA. Photosynthetic and developmental traits associated with genotypic differences in durum wheat yield across the Mediterranean basin. Aust J Agric Res 2000; 51:891–901.

Villegas D, Aparicio N, Blanco R, Royo CA. Biomass accumulation and main stem elongation of durum wheat grown under Mediterranean conditions. Ann Bot 2001; 88:617–627.

Voltas J, Romagosa I, Araus JL. Growth and final weight of central and lateral barley grains under Mediterranean conditions as influenced by sink strength. Crop Sci 1998; 38:84–89.

Voltas J, Romagosa I, Lafarga A, Armesto AP, Sombrero A, Araus JL. Genotype by environment interaction for grain yield and carbon isotope discrimination of barley in Mediterranean Spain. Aust J Agric Res 1999; 50:1263–1271.

Waddington SR, Ransom JK, Osmanzai M, Saunders DA. Improvement in the yield potential of bread wheat adapted to northwest Mexico. Crop Sci 1986; 26:698–703.

Waddington SR, Osmanzai M, Yoshida M, Ransom JK. The yield of durum wheats released in Mexico between 1960 and 1984. J Agric Sci (Cambridge) 1987; 108:469–477.

Wardlaw IF, Moncur L. The response of wheat to high temperature following anthesis. I. The rate and duration of kernel filling. Aust J Plant Physiol 1995; 22:391–397.

Weir AH, Bragg PL, Porter JR, Rayner JH. A winter wheat crop simulation model without water or nutrient limitations. J Agric Sci (Cambridge) 1984; 102:371–382.

Worland AJ, Börner A, Koyrun V, Li WM, Petrovic S, Sayers EJ. The influence of photoperiod genes on the adaptability of European winter wheats. Euphytica 1998; 100:385–394.

Wych RD, Rasmusson DCA. Genetic improvement in malting barley cultivars since 1920. Crop Sci 1983; 23:1037–1040.

Yin X, Stam P, Dourleijn CJ, Kropff MJ. AFLP mapping of quantitative trait loci for yield-determining physiological characters in spring barley. Theor Appl Genet 1999; 99:244–253.

Youssefian S, Kirby EJM, Gale MD. Pleiotropic effects of the G.A. insensitive *Rht* dwarfing gene in wheat. 2. Effects on leaf, stem and ear growth. Field Crops Res 1992; 28:191–210.

Yunusa IAM, Siddique KHM, Belford RK, Karimi MM. Effect of canopy structure on efficiency of radiation interception and use in spring wheat cultivars during the pre-anthesis period in a Mediterranean-type environment. Field Crops Res 1993; 35:113–122.

2

Genetic Yield Improvement and Stress Tolerance in Maize

Matthijs Tollenaar and Elizabeth A. Lee
University of Guelph
Guelph, Ontario, Canada

1 YIELD IMPROVEMENT IN MAIZE AND THE PHYSIOLOGICAL BASIS OF GENETIC YIELD IMPROVEMENT

1.1 Yield Improvement in the United States and Canada

Average U.S. farm maize (*Zea mays* L.) grain yields have increased from about 1 Mg/ha prior to the introduction of commercial corn hybrids in the early 1930s to about 7 Mg/ha in the late 1990s (Fig. 1), and, similarly, average maize yields in Ontario, Canada, increased from about 2 Mg/ha in the early 1940s, when commercial corn hybrids were introduced, to about 7 Mg/ha in the late 1990s (Tollenaar and Wu, 1999). Yield improvement during the hybrid era has been attributed both to adoption of improved agronomic practices and genetic gains made through plant breeding. Results of studies with Corn Belt varieties and hybrids from the 1920s to the 1980s in side-by-side studies (Duvick, 1984, 1992; Russell, 1991) have indicated that 40% to 60% of the yield improvement in Corn Belt hybrids was attributable to genetic improvement, and Cardwell (1982) concluded that 58% of the corn

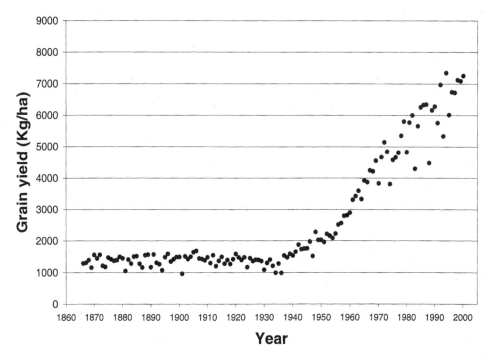

Figure 1 Average U.S. maize yield from 1865 to 2001. Data compiled by the USDA.

yield improvement in Minnesota from 1930 to 1980 could be attributed to genetic improvement. When U.S. yield improvement is examined in terms of the amount of applied nitrogen and increased plant population density since 1964, and precipitation and temperatures for the 1950 to 1995 period, 63% of the gain since 1950 was attributable to genetics (Smith, 1998). However, the trend of the genetic gain was quadratic and the results suggested that the rate of gain in 2000 was only 38% of what it was in 1950 (Smith, 1998).

The "two-component" concept for grain-yield improvement from the 1930s to the 1990s is illustrated in Fig. 2. The difference in yield between maize grown in the 1930s and 1990s is the sum of genetic improvement (CB) and improvement in agronomic practices (DC). In contrast, we contend that the yield improvement during the hybrid era is predominantly the result of the interaction between the two components. Any estimate of the relative contribution of either genetic or management to the yield improvement in maize is difficult as the increase in grain yield due to improved genetics is directly associated with a change in crop management (e.g., plant density, fertilizer amendments), and the increase in grain yield due to improved

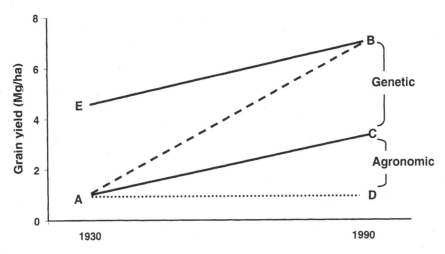

Figure 2 Mean grain yield of maize grown in producer's fields in the U.S. from the 1930s to the 1990s (AB) and the hypothetical contribution of the genetic (CB) and agronomic management (DC) components to the yield improvement during this period. (From Tollenaar and Lee, 2002.)

management is directly linked with the capacity of the maize hybrids to utilize or tolerate the change in crop management (Tollenaar and Lee, 2002). The "interaction" concept for grain-yield improvement from the 1930s to the 1990s is supported by data reported by Duvick (1997). Duvick (1997) showed that grain yield of the 1990s' hybrids did not differ from that of the 1930s' hybrids when grown at 10,000 plants/ha and that grain yields of the 1930s' hybrids did not differ significantly when grown at either 10,000 or 79,000 plants/ha. If we make the assumption that the agronomic contribution to the difference in grain yield between the 1990s' and 1930s' hybrids is represented by increased plant density, then: (i) when the agronomic contribution to grain-yield improvement is eliminated by comparing hybrids at 10,000 plants/ha, Duvick's results show that hybrids did not differ in grain yield, which implies that the genetic yield potential per plant has not changed (i.e., CB = AE = 0 in Fig. 2). (ii) Yield of the 1930s' hybrids did not change with increasing plant density, which implies that the agronomic component per se did not contribute to yield improvement (i.e., DC = 0 in Fig. 2). In other words, there was no simple effect of either the "genetic" or the "agronomic" component of grain-yield improvement in U.S. Corn Belt maize hybrids. Hence, these results imply that all yield improvement is attributable to hybrid × environment interactions. We contend that this genotype × environment is an expression of the improvement in general stress tolerance, i.e., the ability to mitigate the impact of stresses on the physiological processes involved in resource capture and utilization (Tollenaar and Lee, 2002).

1.2 Physiological Basis of Genetic Improvement

Yield increases can be achieved by either increasing total dry matter accumulation during the life cycle, increasing harvest index (i.e., the proportion of aboveground dry matter at maturity that is allocated to the grain), or increasing both dry matter accumulation and harvest index.

Harvest index. Genetic improvement in maize yield is not associated with an increase in harvest index under most environmental conditions. This is in contrast to reports in the literature before 1990 on genetic yield improvement in a range of crop species (e.g., Gifford, 1986) that showed that harvest index was the factor that had contributed most to yield improvement. Harvest index of commercial maize hybrids grown at their optimum plant density for grain yield has remained relatively stable during the past seven decades (Tollenaar et al., 1994). Differences in harvest index among hybrids representing different eras have been reported (Russell, 1985), but differences were apparent only when hybrids were grown at high plant population densities. We compared two older hybrids and two newer hybrids grown at plant densities ranging from 0.5 to 24 plants/m^2 (Tollenaar, 1992) and found that harvest index was approximately 50% for all hybrids grown at plant densities of 8 plants/m^2 or lower. Harvest index in the Tollenaar (1992) study started to decline at 12 plants/m^2 and continued to decline with an increase in plant density to 24 plants/m^2, with the decline being steeper for the older than for the newer hybrids. These results show that harvest index does not differ among older and newer North American maize hybrids unless the comparisons are made at very high plant densities. Consequently, except for these extreme stressful conditions, genetic improvement in maize yield must be attributable to increased dry matter accumulation during the life cycle.

Dry matter accumulation. Increased dry matter accumulation can result from increased light interception by the crop canopy because of a greater leaf area index (LAI), a longer duration of green leaf area (that is, greater resource capture), and a higher canopy photosynthetic rate per unit absorbed irradiance (that is, a greater resource utilization). Higher canopy photosynthetic rate throughout the life cycle may be attributable to either a more even distribution of the intercepted radiation by the canopy, a greater leaf carbon exchange rate (CER) per unit leaf area, a greater tolerance of leaf CER to abiotic stresses, or a combination of these three factors.

Canopy light interception. Light absorption by a canopy (I_A) is a function of incident radiation (I_O), leaf area index (LAI), and the light extinction coefficient of the canopy (k) as:

$$I_A = I_O[1 - \exp(-k \times \text{LAI})] \tag{1}$$

Maximum leaf area per plant of Corn Belt hybrids has remained fairly stable from the 1930s to 1980s (Crosbie, 1982; Duvick 1997), whereas leaf area per plant of short-season hybrids in Ontario increased from the 1950s to the 1980s (Tollenaar, 1991). Maximum LAI has increased in both United States and Ontario hybrids because newer hybrids are grown at higher plant densities than older hybrids, but the impact of the higher LAI on light interception by the crop canopy has been generally small. For instance, when LAI increases from 3 to 4 or 33% and the canopy light extinction coefficient (k) is 0.65 [Eq. (1)], absorptance (I_A/I_O) will increase only by 8% (i.e., from 0.86 to 0.93). The increase in LAI with year of release in U.S. Corn Belt hybrids is associated with a more erect leaf-angle distribution (Duvick, 1997), and, consequently, the increase in light interception resulting from a greater LAI is mitigated, in part, by a reduction in the light extinction coefficient k [see Eq. (1)]. The characteristic most frequently associated with genetic yield improvement in maize is delayed leaf senescence or "stay green" (Crosbie, 1982; Tollenaar, 1991; Duvick, 1997). Delayed leaf senescence increases the interception of incident solar radiation by green leaf area by extending the period during which the maize canopy intercepts incident solar radiation. We distinguish between changes in visible leaf senescence (e.g., increased "stay green") and changes in functional leaf senescence (e.g., increased leaf photosynthesis), as the former may or may not result in increased dry matter accumulation.

Canopy architecture. A more even distribution of incident solar radiation across a crop canopy can increase canopy photosynthesis (Tollenaar and Dwyer, 1999). The light extinction coefficient [i.e., k in Eq. (1)] is a function of the leaf-angle distribution of the crop canopy, and as the leaf angle increases, k becomes smaller and the incident solar radiation is distributed more evenly across the crop LAI. Leaf angle in Corn Belt hybrids has increased during the hybrid era (Duvick 1997). It has been shown in a theoretical analysis that an increase in canopy leaf angle from 30° to 60° could result in a potential increase in rate of maize dry matter accumulation of between 15% and 30% after complete leaf-area expansion (Tollenaar and Dwyer, 1999).

Potential leaf CER. Results of our studies with older and newer maize hybrids from Ontario (Ying et al., 2000, 2002) and with Duvick's (1997) Corn Belt hybrids from the 1930s to the 1990s (Tollenaar et al., 2000) have shown that potential leaf CER is not associated with genetic yield improvement. Maximum leaf CER in maize is attained after the leaf blade is fully expanded (Thiagarajah et al., 1981). Consequently, potential leaf CER is defined here as leaf CER of young, completely expanded leaves measured at a photosynthetic photon flux density of 2000 μmol m^{-2} sec^{-1}, for plants grown under apparent optimal conditions. Older and newer hybrids generally did not differ in potential leaf CER around silking, although sometimes the leaf CER was higher in the older than in the newer Ontario hybrids used in our studies (Ying et al., 2000, 2002). Leaf CER declined during the grain-filling period, and the decline in leaf CER was greater in the older than in the newer maize hybrids (Fig. 3).

Leaf CER under suboptimal conditions. In contrast to potential leaf CER, leaf CER when stress is imposed on a plant is higher in newer than in older hybrids. In earlier studies, we focused on an older hybrid Pride 5, released in Ontario in 1959, and a newer hybrid Pioneer 3902, released in Ontario in 1988. Results of these studies showed that leaf CER of the newer hybrid was more tolerant than that of the older hybrid when field-grown plants were exposed to cool night temperatures during the grain-filling period (Dwyer and Tollenaar, 1989), high plant density (Dwyer et al., 1991), the herbicide bromoxynil (Tollenaar and Mihajlovic, 1991), water stress (Dwyer et al., 1992), and low N supply (McCullough et al., 1994). In addition, when plants were exposed to a water-deficit stress by withholding water supply until net canopy photosynthesis (i.e., whole-plant CER) declined to zero, canopy photosynthesis declined to zero faster in the older than in the newer hybrid (Nissanka et al., 1997). Cumulative canopy photosynthesis and transpiration during the drying cycle were reduced more in the older than in the newer hybrid, but integrated stem water potential from the beginning of the drying cycle until rehydration was more negative in the newer than in the older hybrid. When plants were subsequently rewatered, canopy photosynthesis recovered slower in the older hybrid than in the newer hybrid (Nissanka et al., 1997). More recently, we have examined the effect of a low night temperature exposure ($4°C$) during the grain-filling period on leaf CER of plants grown hydroponically in the field (Tollenaar et al., 2000; Ying et al., 2000, 2002). Results of these studies showed that leaf CER was reduced by low night temperature exposure and the decline was two times greater in the older hybrid Pride 5 than in the newer hybrid Cargill 1877 (Fig. 3). A comparison of Duvick's (1997) U.S. Corn Belt hybrids showed that the decline in leaf CER owing to low night temperature was also significantly greater in the older than in the newer U.S. hybrids (Tollenaar et al., 2000).

Figure 3 Leaf CER and reduction in leaf CER of maize hybrids Pride 5 (released in 1959), Pioneer 3902 (released in 1988), and Cargill 1877 (released in 1993) after exposure to a single night at $4°C$ from tassel emergence to 6 weeks after silking. Data shown are measurements taken at 12:00 on the day following the night of the cold exposure. Vertical bars are standard errors of the mean. (From Ying et al., 2000.)

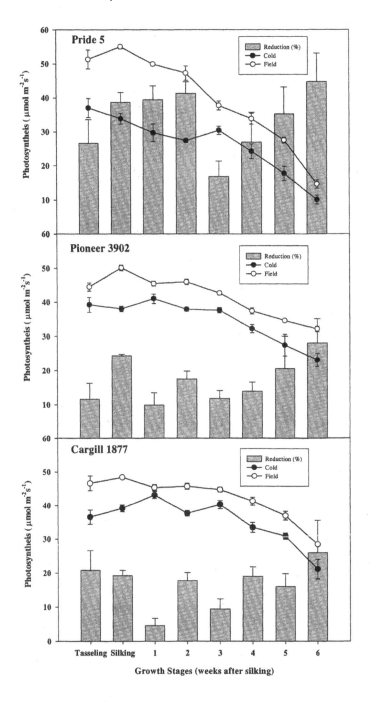

Analysis of the physiological basis of grain-yield improvement shows that potential rates of several important physiological processes involved in maize yield formation have not changed and that yield improvement has resulted predominantly from increased stress tolerance, in particular, during the grain-filling period. This conclusion is consistent with reports showing that (i) yield differences between newer and older hybrids are greatest at high plant densities (e.g., Tollenaar, 1991; Duvick, 1997), (ii) differences in dry matter accumulation between an old and a new Ontario hybrid are small prior to silking and differences become large after silking (Fig. 4), and (iii) tropical maize populations selected for mid-season drought stress tolerance exhibited a concomitant increased tolerance to N stress (Lafitte and Edmeades, 1995; Bänziger et al., 1999). Quantifying the effect of stress across a range of environments on final grain yield can be difficult as timing, duration, and intensity of the stress can all independently influence the outcome (e.g., Bruce et al., 2002). High plant-density stress is the stress most commonly associated with genetic improvement in maize. The grain-yield response to high plant density

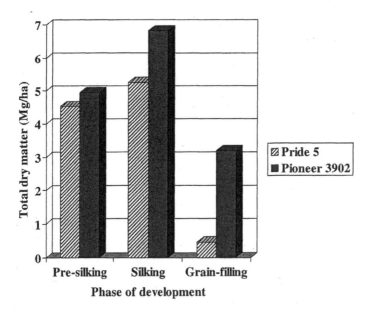

Figure 4   Aboveground dry matter accumulation of an older hybrid (Pride 5) and a newer hybrid (Pioneer 3902) from seeding to 1 week before silking ("Pre-silking"), from 1 week before silking to 3 weeks after silking ("Silking"), and from 3 weeks after silking to maturity ("Grain filling"). Means across four location/years and four plant densities. (From Tollenaar, 1991.)

is the result of the integration of a crop's responses to numerous, relatively mild, above- and belowground stresses during its whole life cycle. A simplified framework of the response of newer and older hybrids to three hypothetical plant-density stress environments in terms of physiological parameters discussed above can be summarized as the following:

(a) *Low stress.* When maize is grown at a very low plant density, conditions for plant growth are close to optimal, and, consequently, leaf CER of young leaves will not differ between older and newer hybrids under these conditions. Leaf angle of the canopy will be greater in newer than in older hybrids, which will result in lower light interception by the newer hybrids at a very low plant density. The reduction in light interception by the newer hybrids is counteracted, in part, by greater "stay green" and higher leaf CER of the newer hybrids during the second half of the grain-filling period. Indeed, Duvick's (1997) results show that yield of the 1990s' hybrids did not differ from that of the 1930s' hybrids when grown at 10,000 plants/ha. Maize grown at 10,000 plants/ha in the U.S. Corn Belt under otherwise standard agronomic practices is exposed to minimum stress as, in general, both aboveground and belowground resources are abundant.

(b) *Medium stress.* Maize grown at plant densities optimum for grain yield, or at plant densities slightly higher than those for optimum grain yield, is exposed to medium stress conditions. Under these conditions, light interception of newer hybrids will be fairly similar to that of older hybrids at silking, as the effect of higher LAI is offset, in part, by a lower light extinction coefficient (k) in newer hybrids. Rates of dry matter accumulation will become greater in newer than in older hybrids as development progresses towards maturity (i.e., stay green). Differences in leaf CER between newer and older hybrids will also increase when plants are approaching maturity.

(c) *High stress.* A high-stress environment for maize can either be a plant density that is much greater than the optimum plant density for grain yield or exposure to severe abiotic stresses such as soil-moisture deficit or nutrient deficiency during the life cycle. Under these conditions, light interception may be similar for newer and older hybrids, but leaf CER could be much lower in the older than the newer hybrids when the high-stress conditions prevail. Differences in total dry matter accumulation between newer and older hybrids will be greatest under high-stress conditions. In addition, differences in leaf CER between newer and older hybrids during the silking period may cause differences in harvest index between the hybrids. Low leaf CER during the silking period will reduce crop growth rate, which can result in a disproportional decline in kernel number per plant (Tollenaar et al., 1992).

In conclusion, genetic yield improvement in maize does not appear to be associated with changes in potential rates of production processes such as light interception, maximum leaf photosynthesis, and assimilate partitioning to the grain (i.e., harvest index). Genetic yield improvement in maize appears to be predominantly associated with stress tolerance. Although differences in grain yield between newer and older hybrids are associated with higher tolerance to a range of abiotic stresses such as water, N, low temperature, and herbicide stress, the most commonly observed phenomenon is the association between yield improvement and tolerance to high plant density. Most stresses that occur during one or more phases of the life cycle of maize will result in increased leaf senescence (i.e., visual senescence) and reduced rates of leaf photosynthesis (i.e., functional senescence) during the grain-filling period.

1.3 General Stress Tolerance in Maize

Yield improvement in maize has been associated with increased stress tolerance and, in particular, tolerance to high plant density (Duvick, 1977, 1984, 1992, 1997; Tollenaar, 1989; Tollenaar et al., 1994, 1997, 2000; Bruce et al., 2002). The physiological mechanisms that confer increased stress tolerance to newer maize hybrids are not known nor is it always clear which abiotic stresses are involved in yield depression. Improved stress tolerance of newer maize hybrids is a result of maize breeding, but it is not known which part or parts of the breeding process have led to increased stress tolerance (i.e., selection in maize breeding programs has focused on grain yield, and, to the best of our knowledge, stress tolerance per se was not a selection criterion). Hence, genetic yield improvement is associated with increased stress tolerance, but very little is known about either the nature of the stresses or the mechanisms by which maize confers tolerance to these stresses. We define general stress tolerance as the ability of a genotype to maintain its grain-yield level under unfavorable conditions for rate of dry matter accumulation per plant, relative to that of one or more other genotypes.

High plant-density stress is the stress most commonly associated with yield improvement in maize. High plant density imposes a stress because available resources, such as incident solar radiation and soil nutrients, will have to be shared by an increasing number of plants, resulting in lower resource availability per plant. High plant-density tolerance is therefore a result of either or both increased resource-capture efficiency (e.g., N uptake relative to soil N content) or resource-use efficiency (e.g., dry matter accumulation or grain yield per unit N uptake). High plant-density stress is associated with a number of relatively mild stresses operating in concert as a consequence of competition for resources like incident solar radiation, soil moisture, and soil nutrients during the entire life cycle of the crop. Little is known about the physiological mechanisms that influence the relationship between plant density and grain yield (e.g., Duncan, 1984), but the general nature of the high plant-density stress (i.e., multiple small stresses across the life cycle) makes it an ideal method for evaluating general stress tolerance.

High plant density is a relatively simple management factor that can be utilized to quantify the relative stress tolerance of maize hybrids. Duvick (1997) showed that differences in grain yield between newer and older U.S. Corn Belt hybrids continued to decline when plant density at which the hybrids were grown was decreased from 7.9 to 2 plants/m^2, at which plant density the difference between the newer and older hybrids became nonsignificant. In contrast, in a comparison of two older and two newer short-season maize hybrids grown at plant densities ranging from 0.5 to 24 plants/m^2, the grain-yield ratio of newer and older hybrids declined from 1.49 at 20 plants/m^2 to 1.15 at 4 plants/m^2, and then increased again to 1.60 at 0.5 plants/m^2 (Tollenaar, 1992). It has been speculated that the increase in the grain-yield ratio between older and newer hybrids at extremely low plant densities is associated with wind damage and/or photoinhibition, but no evidence to that effect has been reported. In addition to high plant-density stress, newer hybrids appear to be also more tolerant to weed competition (Tollenaar et al., 1997). To what extent the greater tolerance of improved maize genotypes to high plant density and weed interference as a result of improved tolerance of soil moisture deficit (Lafitte and Edmeades, 1995; Nissanka et al., 1997; Bruce et al., 2002) and soil N (Bänziger et al., 1999) is not known.

Numerous, relatively small abiotic stress effects throughout the growing season may reduce grain yield in maize under field conditions, and grain yield could be increased if genotypes would be more tolerant to these abiotic stresses. A number of lines of circumstantial evidence indicate that the rate of dry matter accumulation of maize can be much greater under apparent "no-stress" conditions than under apparent agronomical optimal conditions. (i) The record maize yield, obtained in a nonirrigated field, is three times greater than the average grain yield in the United States (Tollenaar and Lee, 2002). The difference between the record and average yield cannot be attributed to either of the major stresses such as soil water and N deficits. (ii) Rates of dry matter accumulation of maize grown under controlled-environment conditions are much greater than that expected from analyses of studies conducted under field conditions. We have reported results of a 4-year study in which we carefully monitored rates of dry matter accumulation, grain yield, interception of incident radiation, and leaf photosynthesis of maize canopies grown under controlled-environment conditions (Tollenaar and Migus, 1984). Results showed that the rates of dry matter accumulation of maize canopies intercepting >95% of incident radiation were 50–100% greater under environment-controlled conditions than under field conditions at double incident irradiance (Tollenaar and Migus, 1984). (iii) Both theoretical analyses and experimental evidence have shown that there is a linear relationship between intercepted solar radiation and accumulated crop dry matter accumulation, and the slope of the relationship is called radiation use efficiency (Sinclair and Muchow, 1999). However, the relationship between crop growth rates and incident solar radiation in the range 14 to 22 MJ m^{-2} day^{-1} has been shown to be weak in detailed studies with maize canopies intercepting >90% of incident solar radiation from the end of July to early September (Fig. 5). Maize in these studies was grown under experimental conditions in which stresses were neither imposed nor stress effects were visually apparent, and the variability in the response of crop growth rate to incident solar radiation is likely the result of a number of small abiotic stress effects. If the difference between potential and actual maize yield is a result of many small abiotic stresses, grain yield can be improved by enhancing general stress tolerance.

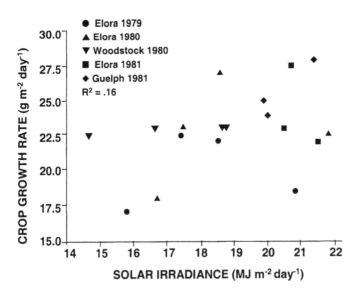

Figure 5 Relationship between crop growth rate of maize grown in southern Ontario in 1979, 1980, and 1981 and incident solar irradiance during periods of complete leaf area expansion. Crop growth rates were measured over 2-week periods, and each date point is the mean of 24 to 48 well-bordered plots each measuring 2.4 to 3.2 m^2. (From Tollenaar, 1986.)

Increased stress tolerance in newer hybrids has been the result of selection by maize breeders. One of the factors that may have contributed to the increase in stress tolerance is a shift in hybrid evaluation philosophy that occurred in the early 1980s in North America. Instead of emphasizing relatively high precision per location at a few locations, the evaluation procedure emphasized relatively low precision at a large number of locations (Bradley et al., 1988). This shift encompassed an increase in the number of locations and years and a change in the type of location. The type of environments shifted from high-yielding environments to environments that are most likely to occur in commercial maize production, including stress environments. This shift occurred because maize-breeding companies were interested in identifying widely adapted maize hybrids (Bradley et al., 1988), thereby increasing yield stability. We have shown that phenotypic yield stability, quantified using a regression analysis as described by Finlay and Wilkenson (1963), of some very high-yielding commercial U.S. Corn Belt maize hybrids was as high or higher than that of a basket of commercial contemporary hybrids (Tollenaar and Lee, 2002). Another reason for the increase in stress tolerance, which may have resulted in the increase in yield stability, is that hybrid evaluation has occurred under conditions that reflected the changes in management practices, in particular, an increase in plant density. Duvick (cf., Tollenaar and Lee, 2002) has suggested another important reason for the continuing increase in stability and stress tolerance. Maize breeding has depended heavily on recycling successful inbred lines through pedigree breeding, with breeders using inbred lines from the most popular hybrids as parents of the next generation. The most popular hybrids were those that were high yielding and dependable in producers' fields. Consequently, maize producers were doing yield testing on a very large scale of parents that the breeders used to make the next generation of inbred lines.

In conclusion, the ability of a genotype to maintain its grain yield under unfavorable conditions for dry matter accumulation relative to that of other genotypes (i.e., general stress tolerance) is the factor associated most with genetic yield improvement in maize. Evaluation of and selection for general stress tolerance are difficult because physiological mechanisms underlying the tolerance are not known and probably involve multiple processes, controlled by multiple genes.

2 EVALUATION OF GENERAL STRESS TOLERANCE AT THE CANOPY LEVEL OF ORGANIZATION

The high plant-density response can be used to evaluate general stress tolerance, but this methodology has a number of drawbacks and we are presenting a methodology that evaluates a similar response but without the drawbacks of high plant density. First, the plant-density response is confounded with leaf area per plant (Major et al., 1972). The effect of an increase in plant density on plant-to-plant competition for incident solar radiation is positively associated with the leaf area per plant, which will probably result in a leaf area per plant × plant-density interaction when leaf area varies among the genotypes that are evaluated. Second, high random variability is associated with increased plant density and using high plant density, as a selection criterion for stress tolerance will require a large plot size. Third, although the methodology of selection under high plant density is simple in principle, the methodology is impractical when evaluating either large numbers of lines and/or when seed number per line is limited, like in a quantitative trait locus (QTL) analysis for stress tolerance. For instance, the seed requirement for the high plant-density treatment at one location in an experiment consisting of four replications and a plot size of six 0.76-m-wide rows 8.5 m long, sown at two plants per hill to obtain a plant density after thinning at the seedling phase of 11 plants/m^2, is about 3400 seeds (note that two border rows at each side of the sample area are the absolute minimum when evaluating yield at a high plant density because of the likelihood of plant lodging).The objective of selection under high plant density is to quantify the grain-yield response to stress, i.e., assuming no difference in yield at low plant density (low stress), yield differences at high plant density are indicative of stress tolerance. We have shown that the grain-yield response of maize to a nonuniform plant stand vs. a uniform plant stand was similar to that of high plant-density vs. low plant-density stress (cf., Tollenaar and Wu, 1999). The nonuniform plant stand is created by seeding the entries in an emerging stand of maize, that is, maize at the two-leaf stage. The nonuniform plant stand as a methodology to evaluate stress tolerance does not have the drawbacks associated with the high plant-density stress, i.e., stature of the entry is irrelevant because the entries are always smaller than plants surrounding the entries, no border rows are required, and only a small number of seeds are required in this procedure.

2.1 Tolerance to High Plant Density and a Nonuniform Plant Stand

The response of three short-season maize hybrids to an increase from a suboptimal plant density (i.e., 3.5 plants/m^2) to a super-optimal plant density (i.e., 11 plants/m^2) was examined in three short-season maize hybrids grown in a uniform and a nonuniform plant stand across three location/years (Table 1). Stress tolerance of Hybrid A relative to that of Hybrid B is defined as:

$$RST_{AB} = [Yield_A/Yield_B](\text{high plant density})H[Yield_A/Yield_B](\text{low plant density}) \quad (2)$$

where RST_{AB} is the relative stress tolerance of Hybrid A vs. Hybrid B and $Yield_A$ and $Yield_B$ are grain yields of Hybrid A and Hybrid B, respectively. In

Table 1 Grain Yield Across 3 Location/Years[a] of Three Maize Hybrids Grown at Two Plant Densities, in Either a Uniform Plant Stand or a Nonuniform Plant Stand

Hybrids	Uniform stand		Nonuniform stand[b]	
	3.5 plants/m^2	11 plants/m^2	3.5 plants/m^2	11 plants/m^2
	Yield (kg/ha)			
Pride 5	4430 (100)[c]	3240 (100)	3920 (100)	1540 (100)
Pioneer 3902	5610 (127)	6390 (197)	5170 (132)	3900 (253)
Pioneer 3893	6340 (143)	7120 (220)	5560 (142)	3800 (247)
Mean	5460 (123)	5580 (172)	4880 (124)	3080 (200)
LSD (0.05)	260 (6)	790 (24)	440 (11)	690 (45)
CV (%)	5.5	16.5	10.6	26.2

[a] Experiments were carried out at the Elora Research Station, Ontario, in 1999 and 2000, and at the Cambridge Research Station, Ontario, in 1999. The experimental design was a split-plot RCB with four replications, with plant density as main factor and plant-stand uniformity as subfactor; sample area was 6.84 m^2. Sowing date was the same for the uniform plant stand and the border plants in the nonuniform plant stand.
[b] The sample area in nonuniform plant stand consisted of the following: (i) one-third of plants were sown at the same sowing date as plants in the border rows of this treatment; (ii) one-third of plants were sown 2 thermal leaf units (Tollenaar et al., 1979) before plants in the border rows; (iii) one-third of plants were sown 2 thermal units after plants in the border rows. (Note that the duration from seeding to plant emergence is approximately two leaf stages, and, consequently, plants that were sown 2 thermal leaf units after the sowing date of the border plants were sown when plants in the border rows emerged.) Each plant in the sample area of the nonuniform plant stand was bordered within the row by plants belonging to the other two sowing-date treatments. Grain yield depicted in the nonuniform stand represent yield of plants sown 2 thermal units after the sowing date of the plants in the uniform stand.
[c] Numbers in brackets indicate grain yield as a percentage of Pride 5.

a uniform stand, stress tolerance of Pioneer 3902 (a single-cross hybrid released in the late 1980s) relative to that of Pride 5 (a double-cross hybrid released in the late 1950s) was 1.55 (197/127), and the stress tolerance of Pioneer 3983 (a single-cross hybrid released in the early 1990s) was 1.54 (220/143). Results for the comparison of Pioneer 3902 and Pride 5 in this study are similar to the results that were reported by Tollenaar (1989) from studies carried out during four location/years, i.e., stress tolerance of Pioneer 3902 relative to Pride 5 when grown at 4 and 13 plants/m^2 was 1.66 (i.e., [(9.5/4.54)H(8.78/6.97)]).

A hybrid × plant–density response was also apparent in the nonuniform plant stand. Grain yield in the 3.5-plants/m^2 treatment was reduced by about

10% when the sowing date was delayed by two leaf stages relative to that of the uniform plant stand, but the response of the three hybrids was similar in the uniform and nonuniform plant stand at 3.5 plants/m^2 (Table 1). Grain yield of plants that emerged two leaf stages later than neighboring plants in the 11-plants/m^2 treatment was almost 50% lower than that in the uniform plant stand, and reduction in yield was greater for Pride 5 than for the other two hybrids (Table 1). Relative stress tolerance based on the response to delayed sowing in the high plant-density treatment was 1.99 (253/127) for Pioneer 3902 and 1.73 (247/143) for Pioneer 3893. Results are similar when yield in the low plant density of the uniform stand were used as the comparison for relative stress tolerance, i.e., values were 192 (253/132) for Pioneer 3902 and 174 (247/142) for Pioneer 3893.

The CV for grain yield was close to three times greater in the high plant density than in the low plant density in both the uniform stand and the nonuniform stand (Table 1). These results confirm the contention that plant-to-plant variability for grain yield is positively associated with the stress level that plants have been exposed to (Tollenaar and Wu, 1999). As the mean yield for the two plant densities in the uniform stand were similar, the comparison of variability between the high and low plant densities in this treatment is particularly relevant in this respect. The CV for grain yield was almost two times greater in the uniform stand than in the nonuniform stand (Table 1). Differences in variability between the uniform and the nonuniform stand may be attributable, in part, to the differences in sample area (i.e., the number of plants sampled in the nonuniform stand was one-third of those sampled in the uniform stand). This contention is supported by the results showing that mean grain yield and differences in grain yield among the hybrids were similar between the uniform stand and nonuniform stand at 3.5 plants/m^2.

2.2 Yield Potential and Stress Tolerance

The difference in yield between two genotypes is attributable to either differences in yield potential, stress tolerance, or both. Although genetic improvement of commercial U.S. Corn Belt hybrids did not appear to be related to potential yield per plant (Duvick, 1997), yield potential may vary among any set of maize genotypes. In addition, yield in a low-stress environment (i.e., potential yield) is a necessary input in the calculation of relative stress tolerance, as relative stress tolerance is an expression of the genotype × environment interaction [cf., Eq. (2)]. A measure of relative yield potential can be obtained by comparing maize genotypes grown at low plant density. Relative stress tolerance of maize hybrids can be evaluated by comparing the response of the hybrids to low-stress conditions relative to the response to high-stress conditions. High-stress conditions can be created by imposing a delayed-sowing treatment (i.e., previous section). The utility of these variables in quantifying differences in relative stress tolerance is illustrated below.

In order to evaluate eight short-season hybrids for yield potential and relative stress tolerance, a study was carried out that included low plant-density and delayed-sowing treatments across four location/years. Grain yield of the hybrid Pride 5 sown in a stand of maize at emergence was 29% of the yield of Pride 5 grown in a uniform stand of 3.5 plants/m^2 (Table 2). Grain yield of Pride 5 varied among the four location/years. A low yield in 2000 in both treatments was attributable, in part, to below-average temperatures during the 2000 growing season and poor conditions for maize growth during the first 3 weeks following seeding. The low yield in the stress treatment in 2001 was attributable, in part, to a delay of 3 thermal leaf units of the hybrids relative to the stand of maize in the stress treatment (compared to a delay of 2 thermal leaf units in the other location/years). The effect of the increased delay in sowing thereby creating a variable stand was greater in Pride 5 than in the other hybrids (Table 2).

Differences among the hybrids were greater in the high-stress than in the low-stress treatments. The mean yield of the seven hybrids in the low-stress treatment was 19% greater than that of Pride 5, and relative yield among the hybrids varied from 109 to 128. Grain yields in the low-stress treatment were associated with relative maturity rating of the hybrids: the two hybrids with the highest relative maturity (Pioneer 3893 and Pioneer 39K38) also recorded the highest grain yield, and yield of the two hybrids with the lowest relative maturity (Pickseed 2459 and Maizex MZ128) was lower than the mean yield. Yield under high-stress conditions was 69% greater than that of Pride 5, and relative yield among the seven hybrids ranged from 122 to 209. Results show that the differences in apparent yield potential between more recent hybrids and the older hybrid Pride 5 were small in 3 out of the 4 years and yield differences between the seven hybrids and Pride 5 were much greater under high-stress conditions.

Mean relative stress tolerance was 44% greater in the seven hybrids than in Pride 5, and relative stress tolerance ranged from 113 to 169. Among the seven hybrids, two hybrids can be classified as high stress-tolerant (Pioneer 39K38 and Pioneer 3902), three hybrids can be classified as average stress-tolerant (NK N17-R3, Maizex MZ128, and Pioneer 3893), and one hybrid each can be classified as low stress-tolerant (Cargill 1877) and very low stress-tolerant (Pickseed 2459). Although substantial variation in relative stress tolerance occurred among the four location/years (Table 2), the consistency in the relative stress tolerance of the hybrid Pioneer 3902 relative to the hybrid Pride 5 is high in four studies across different location/years and using different methodologies to quantify relative stress tolerance: 1.64 in this study, 1.55 in the plant-density study shown in Table 1, 1.99 for the delayed-sowing

Table 2 Grain Yield of Pride 5 and Relative Yield of Seven Other Maize Hybrids Grown Under Low Stress (i.e., 3.5 plants/m²) and High Stress (i.e., delayed sowing) in Four Location/Years[a]

	Low stress					High stress					Relative stress tolerance[c]				
	Elora			Camb		Elora			Camb		Elora			Camb	
Hybrid[b]	1999	2000	2001	1999	Mean	1999	2000	2001	1999	Mean	1999	2000	2001	1999	Mean
	Grain yield per plant (g/plant)												%		
Pride 5	178	52.4	151	178	142	55.6	19.4	31.7	51.6	41	100	100	100	100	100
	Relative Yield ([Yield Hybrid H / Yield Pride 5] × 100%)														
Pioneer 3902	104	158	112	110	121	150	220	234	175	195	145	141	210	159	164
Pioneer 3893	119	143	123	128	128	153	160	249	152	178	129	114	206	121	143
Cargill 1877	103	155	129	99	122	107	186	186	145	156	104	120	144	148	129
Pickseed 2459	111	108	107	110	109	103	131	153	101	122	93	121	145	91	113
Pioneer 39K38	115	141	118	120	124	180	231	248	176	209	156	162	210	147	169
NK N17-R3	108	140	102	110	115	139	222	187	130	170	129	158	187	119	149
Maizex MZ128	107	155	90	96	112	140	183	163	128	154	131	119	199	133	145
Mean[d]	110	143	112	110	119	139	190	203	144	169	127	134	186	131	144
LSD (0.05)					7					14					13

[a] Experiments were carried out at the Elora Research Station, Ontario, in 1999, 2000, and 2001, and at the Cambridge Research Station, Ontario, in 1999.

[b] Short-season maize hybrids used were (relative maturity rating is indicated in brackets): Pride 5 (2600), Pioneer 3902 (2650), Pioneer 3893 (2700), Pickseed 2459 (2450), Cargill 1877 (2600), Pioneer 39K38 (2700), NK N17-R3 (2600), and Maizex MZ128 (2450). Pride 5 was released in the late 1950s, Pioneer 3902 was released in the late 1980s, Pioneer 3893 was released in the early 1990s, and the other hybrids were released in the middle to late 1990s. The experimental design was a slit-plot RCB with four replications, with stress level as main factor and hybrids as subfactor. Plot size was four 0.76-m-wide and 8.5-m-long rows; sample area was a 4.6-m² area in the center of the four-row plot. Plots for the high-stress treatment were established in an emerging stand of the hybrid Pioneer 3905 sown at 7 plants/m². Hybrids were sown at a plant density of 3.5 plants/m² in both the high- and low-stress treatments at the same date at two seeds per hill and thinned at the seedling stage to one plant per hill. A plot in the low-stress treatment consisted of four 8.5-m rows and a plot in the high-stress treatment consisted of two 8.5-m rows sown 5 to 10 cm off the center of the two center rows of the emerging stand of Pioneer 3905.

[c] Relative stress tolerance of Hybrid A was defined as the ratio of the yield of Hybrid A vs. the yield of Pride 5 grown under high-stress conditions and the yield of Hybrid A vs. the yield of Pride 5 grown under low-stress conditions, i.e., Eq. (2).

[d] Value comprised of means of seven hybrids (i.e., all hybrids except Pride 5).

treatment in the study shown in Table 1, and 1.66 in the study reported by Tollenaar (1989).

2.3 Stress Tolerance and Hybrid Vigor

We contend that stress tolerance is attributable, in part, to heterosis and, in part, to additive genetic effects. We have attempted to quantify the relative contribution of heterosis on stress tolerance in short-season maize hybrids. The traditional approach of comparing inbred parent mean performance to hybrid performance when examining heterosis is problematic. Phenotypically, hybrids and inbred lines are very different in stature and the planting density that is appropriate for a hybrid may not be appropriate for an inbred. As an alternative to this approach, we chose to look at inbreeding depression or loss of heterosis and how that impacts stress tolerance. To do this, we compared F1s and their respective F2s, which are theoretically 50% more homozygous than the F1. In addition, F2s did not differ from F1s in stature. The three measures of grain yield in this study distinguish between effects during the grain-filling period (i.e., yield per grain-bearing ear), effects during the silking period and grain-filling period (i.e., yield per plant), and effects during the entire life cycle (i.e., yield per plot). Inbreeding depression (IBD) was estimated as

$$IBD = [(Yield\ F1 - Yield\ F2)\ H\ Yield\ F1] \qquad (3)$$

where Yield F1 and Yield F2 are grain yield per plant or per unit area for a maize hybrid (F1) and its F2, respectively.

A difference in relative stress tolerance between an F1 hybrid and its F2 is indicative of the contribution of heterosis to stress tolerance in that hybrid. Also, the smaller the difference in stress tolerance between an F1 and its F2, the smaller the contribution of heterosis to stress tolerance and, by implication, the greater the contribution of additive gene effects to stress tolerance. The application of these concepts is illustrated in a study in which commercial maize hybrids (F1s) and their F2s were grown at low-stress and high-stress conditions. Details of the experimental conditions and results are depicted in Table 3.

Differences in yield relative to the yield of Pride 5 among newer hybrids (F1s) and among the F2s of newer hybrids were much greater under high-stress than under low-stress conditions. On average, newer hybrids yielded 12% more than Pride 5 under low stress and 54% more than Pride 5 under high stress for yield of grain-bearing ears and 103% more than Pride 5 under high stress for yield per plant. Note that the yield difference between the older hybrid and the newer hybrids did not differ when the comparison was based on either yield per plant or yield per plot, indicating that the impact of stress on the difference in yield between Pride 5 and the newer hybrids was confined

Table 3 Grain Yield of Hybrid Seed (F1) of Pride 5 and Grain Yield of the Seed of Selfed Hybrid (F2) of Pride 5, Relative Yield and Relative Stress Tolerance of the F1s and F2s of Seven Other Maize Hybrids, and Inbreeding Depression of Maize Grown Under Low Stress and High Stress in 2001[a]

Hybrid[b]	Low stress F1 (g/plant)	Low stress F2 (g/plant)	High stress (F1) Ear (g/ear)	High stress (F1) Plant (g/plant)	High stress (F1) Area (g/m²)	High stress (F2) Ear (g/ear)	High stress (F2) Plant (g/plant)	High stress (F2) Area (g/m²)	Relative stress tolerance[c] F1 (%)	Relative stress tolerance[c] F2 (%)	Inbreeding depression[d] Low stress (%)	Inbreeding depression[d] High stress (%)
Pride 5	151	100	44.9	31.7	426	14.6	5.62	80.6	100	100	34	81
		Relative yield ([Yield Hybrid H/Yield Pride 5] × 100%)										
Pioneer 3902	112	122	171	234	255	216	301	301	208	255	27	77
Pioneer 3893	123	160	176	249	293	281	552	508	206	344	14	61
Cargill 1877	129	114	142	186	205	203	366	371	144	326	41	63
Pickseed 2459	107	107	109	153	158	150	244	200	145	227	33	71
Pioneer 39K38	118	133	183	248	260	231	470	470	210	365	25	66
NK N17-R3	102	132	132	187	212	176	343	362	187	263	14	68
Maizex MZ128	90	114	163	163	163	166	242	253	199	213	24	73
Mean[e]	112	125	154	203	221	203	360	353	186	285	25	68
LSD (0.05)	16	15	33	50	59	73	141	132	23	31	-	12
CV (%)	27	27	14	17	18	24	26	25	65	130	-	12

[a] The study was carried out at the Elora Research Station, Elora, Ontario, in 2001.

[b] Genotypes that were evaluated in the study were the same eight hybrids described in Table 2 (henceforth termed F1s) and genotypes that were derived from selfing the eight hybrids during the 2000 growing season (henceforth termed F2s). Methodology was the same to that described in Table 2, with the exception that the experimental design was a split, split-plot design with four replications, with F1 and F2 as main treatments, high- and low-stress treatment as subtreatments, and genotypes as sub-subtreatments. In the high-stress treatment, yield was determined as (i) total yield per plot of 16 plants, (ii) yield per plant, i.e., total yield per plot divided by number of plants per plot at maturity, and (iii) yield per grain-bearing ear.

[c] Relative stress tolerance was estimated from yield per plant in the low- and high-stress treatments. Relative stress tolerance of Hybrid A was defined as the ratio of the relative yield of Hybrid A vs. the yield of Pride 5 grown under high-stress conditions and the relative yield of Hybrid A vs. the yield of Pride 5 grown under low-stress conditions, i.e., Eq. (2).

[d] Inbreeding depression was estimated using Eq. (3).

[e] Value comprised of means of seven hybrids (i.e., all hybrids except Pride 5).

to the silking and grain-filling periods. Results indicate that, on average, the proportion of the yield difference that was due to the grain-filling period per se was approximately 50% in the F1s (i.e., 54/106) and 60% in the F2s (i.e., 103/260). The relative difference between F2 derivatives of newer hybrids and that of Pride 5 was also much greater under high-stress than under low-stress conditions. On average, F2s of newer hybrids yielded 26% more than the F2 of Pride 5 under low stress, 103% more than the F2 of Pride 5 under high stress for yield of grain-bearing ears, and 260% more under high stress for yield per plant.

Inbreeding depression of maize hybrids in this study was influenced greatly by the relative stress level the genotypes were exposed to. Inbreeding depression of Pride 5 increased from 34% under low-stress conditions to 81% under high-stress conditions. Inbreeding depression of the newer hybrids tended to be lower than that of Pride 5, and inbreeding depression under high-stress conditions was negatively associated with relative stress tolerance of the F2s of the seven newer hybrids ($R = 0.74$).

The relative stress tolerance of the newer hybrids (i.e., stress tolerance relative to that of Pride 5) was greater for the F2s than for the F1s (Table 3). The mean stress tolerance of the newer hybrids was 1.86 times that of Pride 5, whereas the mean stress tolerance of the F2s of newer hybrids was 2.85 times that of the F2 of Pride 5. The change in relative stress tolerance differed among the hybrids; the increase in relative stress tolerance from the F1s to the F2s was 84% for hybrids Pioneer 3893, Cargill 1877, and Pioneer 39K38 (i.e., from 187 to 334), and the increase was 30% for the other four hybrids (i.e., from 185 to 240). The increase in relative stress tolerance from the F1s to the F2s indicates that a greater proportion of the stress tolerance of the hybrid is attributable to additive gene effects, or, alternatively, less is attributable to heterosis.

In conclusion, (a) the magnitude of inbreeding depression is a function of the conditions under which the study has been conducted; inbreeding depression is greater when plants are exposed to stress. (b) Hybrid vigor or heterosis confers stress tolerance. Relative stress tolerance of hybrids (F1s) was always greater than that of their F2s. (c) The difference in relative stress tolerance between the hybrid (F1) and its F2 is smaller in newer, more stress-tolerant hybrids than in older hybrids, which may indicate that more of the stress tolerance in newer hybrids is fixed in additive genes. Duvick (1999) showed that the contribution of heterosis to grain yield has not changed in U.S. Corn Belt hybrids from the 1930s to the 1990s, and, consequently, the relative contribution of heterosis to grain yield has declined as yields increased from the 1930s to 1990s. Whether the increase in the relative contribution of additive gene effects is associated with increased stress tolerance awaits further investigation.

3 MOLECULAR APPROACHES FOR UNDERSTANDING STRESS TOLERANCE

The previous sections in this chapter show that the biological basis of general stress tolerance is poorly understood, the environmental parameters creating the stress are poorly defined, and stress tolerance is a complex trait that behaves as a typical quantitative trait. Yet stress tolerance has had an important impact of yield improvement in maize. Traditional molecular approaches to understanding stress tolerance have taken reductionist approaches, focusing on one gene at a time (e.g., *sos1* and *sas1*), working with a very specific stress (e.g., salt), and treating the response to stress as a qualitative trait (e.g., survival vs. death) (Shi et al., 2000, Nublat et al., 2001). These rather simplistic approaches are not very useful when the trait of interest is difficult to define at the biological level, the trait is influenced by environmental factors that are not tangible, and the trait is quantitative in its expression. A holistic approach is required to understand and manipulate general stress tolerance, and recent advances in molecular biology could potentially be useful in this approach. In this section, we are attempting to introduce the reader to some of the approaches currently available for understanding the biological/molecular mechanisms underlying stress tolerance. Substantial technological advances have been made within the past decade in the area of comprehensively surveying genomes and in dissecting biological mechanisms at the molecular level. Information gathered from the approaches and methodology discussed below can be conceptually viewed as antagonistic and protagonistic interactions between and among various molecular components of a cell and the environment culminating in a whole plant phenotype (Roberts, 2002) (Fig. 6). The whole plant phenotype can be considered a quantitative trait that is controlled by the expression of many genes, and the environment influences the genes. The molecular components represented by proteome, transcriptome, and metabolome can be defined on varying levels of biological complexity, cell compartments, cell, tissues, or whole plants. Each of the molecular components is also considered a quantitative trait since high-throughput methods have been developed that can accurately measure quantitative changes in them, and they, like the whole plant phenotype, are controlled by many genes and are influenced by the environment (e.g., Consoli et al., 2002).

3.1 Whole Plant Phenotype

"Whole plant phenotype" is considered a quantitatively inherited trait. It is controlled by many genes, influenced by the environment, and is generally quantified using a continuous scale. Quantitative trait locus (QTL) analysis is used for identifying chromosome regions that influence the expression of a

Phenotype

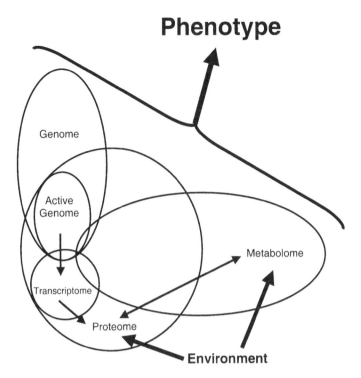

Figure 6 Conceptual relationship of the molecular components represented by genome, proteome, transcriptome, and metabolome and how the environment influences them and ultimately results in a phenotype. Only a subset of the genome encodes genes and only a subset of those genes are transcriptionally active in a given tissue at a given time point in development, represented by the "active genome." The represented overlap between the various molecular components represents physical interactions, proteins interacting with DNA, metabolites interacting with proteins, and metabolites interacting with DNA. The environment interacts directly with the molecular components through either the proteome or the metabolome, which then interacts with the genome via signaling mechanisms, proteins, or metabolites. Phenotype is then the result of the genetics of the plant represented by the genome, the environment, and the interaction between genetics and the environment that occurs via proteins and metabolites. (From Roberts, 2002.)

quantitatively inherited trait (for overview of methodologies, see Liu, 1998). In maize, QTL analysis has identified regions of the maize genome that influence many diverse traits: resistance to European corn borer (e.g., Cardinal et al., 2001; Jampatong et al., 2002), maysin accumulation in silk tissues and the corresponding resistance to corn earworm (e.g., Lee et al., 1998), pollen germination and pollen tube growth under heat stress (Frova and Sari-Gorla, 1994), dry milling properties (Séne et al., 2001), anthesis–silk interval and other flowering parameters under drought stress conditions (Ribaut et al., 1996), ABA concentration in drought-stressed leaves (Tuberosa et al., 1998), root characteristics and grain yield under different watering regimes (Tuberosa et al., 2002), and adaptation to highland vs. lowland growing conditions (Jiang et al., 1999). While by themselves this information is interesting and potentially useful for marker-based selections, the power of using QTL analysis comes from the prospect of being able to associate actual genes, proteins, or metabolites with the regions identified as containing the QTL.

3.2 Genomics

The "genome" of any plant consists of the entire DNA content of a cell, that is, the plastid, mitochondrial, and nuclear DNA components of a cell. Not all of the DNA sequence in a genome represents genes. For some species, a considerable fraction of the genome is actually comprised of repetitive, noncoding DNA (e.g., SanMiguel et al., 1996). The "active genome" simply refers to that portion of the genome that contains genes. Sequencing of two "model" plant genomes is completed or nearing completion. Genomic sequence of the dicot arabidopsis was released in December 2000 (The Arabidopsis Genome Initiative, 2000), and the sequence of the monocot rice is nearing completion (Burr, 2002; Chen et al., 2002; Wu et al., 2002). Once the genomes of the "model" plants, arabidopsis and rice, were sequenced, another approach to identifying genes involved in expression of a trait became possible—the use of synteny. Synteny or genome colinearity is the conservation of ancestral linkage blocks over wide evolutionary distances. The less related the two species are, the smaller or rarer the syntenic regions. By using genomic and QTL information gathered from a related species, it is possible to build upon that information in another related species (e.g., Fatokun et al., 1992; Lin et al., 1995). Unfortunately for the monocots such as maize, the arabidopsis genome does not appear to be all that useful since the usable colinearity in gene order between rice and arabidopsis is relatively rare (Liu et al., 2001, van Buuren et al., 2002). However, the degree of colinearity between maize and rice is considerable, making the genomic sequence of rice useful for gene discovery in maize (Ahn and Tanksley, 1993; Ahn et al., 1993;

Benetzen and Freeling, 1997). So we can now start examining QTL containing chromosomal regions for candidate genes corresponding to those QTLs, and we can also start to take advantage of QTL information and genomic sequence from related species for gene discovery.

3.3 Expression Arrays

"Transcriptome" is the entire mRNA content of a cell or tissue at an instant in time. Several analytical methodologies have been developed to examine the transcriptome. One of those approaches, microarray technology, is essentially a reverse Northern, where transcript levels of thousands of unique coding regions are simultaneously examined from one or two RNA sources (for review, see Ahroni and Vorst, 2001). This results in an expression profile that is unique for that genotype and/or environmental condition. Microarrays can be used to analyze any kind of variability in gene transcriptional levels between given samples. When coupled with QTL analysis or by using "selected" genotypes, expression profiles can reveal clues of the underlying biology of a quantitative trait (e.g., Consoli et al., 2002; Bruce et al., 2001; Zinselmeier et al., 2002).

3.4 Proteomics

Proteomics is the identification and characterization of all the proteins present in a "body" (Wilkens et al., 1996; Roberts, 2002). "Proteome" is a snapshot of all the proteins present in a cell or tissue at an instant in time (Washburn et al., 2001; Prime et al., 2000; Peltier et al., 2000). High-throughput approaches for efficiently cataloging and quantifying the proteins have been developed (for review, see Kersten et al., 2002; Roberts, 2002). The protein "snapshot" is far more complex than the RNA "snapshot" because it integrates posttranscriptional, translational, and posttranslational events that influence quantity, stability, localization, and functionality of the final product. Proteomics can also be partnered with QTL analysis or "selected" genotype approaches to reveal the underlying biology of the trait of interest (e.g., Consoli et al., 2002).

3.5 Metabolic Profiling

Metabolic profiling captures "snapshots" of the levels of metabolites in a cell or tissue at an instant in time and examines how those levels and compositions change under different conditions (Teusink et al., 1998; Fiehn, 2002). "Metabolome" simply refers to the set of metabolites synthesized by an organism (Oliver et al., 1998). High-throughput methodologies have been developed in plants permitting the simultaneous analysis of metabolites using gas chromatography (GC) and GC–mass spectrometry coupled with multi-variate data mining tools (Roessner et al., 2000, 2001a,b). The approach enables automatic identification and quantification of large numbers of distinct compounds. Using these profiles, a metabolic phenotype that is unique for that genotype and/or environmental condition can be established (Fiehn et al., 2000; Roessner et al., 2001a,b). Like transcriptome and proteome analysis, the methodologies for metabolome analysis are amenable for coupling with QTL analysis and "selected" genotype approaches.

4 CONCLUSIONS

Although the genetic improvement of grain yield in maize during the past seven decades has been large, the understanding of the physiological basis of the improvement is only starting to emerge now. Evidence indicates that higher yield of newer vs. older maize hybrids is not attributable to a higher yield potential of the former but to the capacity of newer hybrids to tolerate abiotic stresses, resulting in higher rates of crop dry matter accumulation. The genetic yield improvement of maize during the past decades has been the result of empirical selection for grain yield by maize breeders. An understanding of the physiological mechanisms that have led to increased yield of newer hybrids can help in the formulation of more precise and effective selection procedures if the physiological mechanisms at the canopy level of organization can be linked to biological mechanisms at the molecular level and molecular genetics. Substantial technological advances have been made during the past decades in the area of molecular genetics and in dissecting biological mechanisms at the molecular level.

However, the challenges involved in successfully linking processes at the canopy level with molecular genetics are large. First, the nature of the abiotic stresses that reduce maize yield under apparent low-stress growing conditions is poorly understood. Mechanisms involved in conferring tolerance to major stresses such as soil-moisture deficit, N deficiency, and low temperature may or may not confer tolerance to a whole range of minor stresses (e.g., the difference in yield between maize grown under apparent agronomical optimal conditions and maize grown under record-yield conditions). Procedures that invoke a general stress response, such as the response to high plant density and the response to the nonuniform plant stand discussed in this chapter, may shed some light on this issue. Second, the general stress-tolerance trait is most likely quantitative in its expression, as the continuous improvement during the past seven decades in stress tolerance has been gradual and slow. Traditional molecular approaches to understanding specific stress tolerance using reductionist approaches are not appropriate for highly complex traits such as general stress tolerance. Clearly, a better understanding of canopy-level processes, molecular genetics, and the interface between the two will be a prerequisite for utilization of molecular techniques in grain-yield improvement.

REFERENCES

Ahn S, Tanksley SD. Comparative linkage maps of the rice and maize genomes. Proc Natl Acad Sci 1993; 90:7980–7984.

Ahn S, Anderson JA, Sorrels ME, Tanksley SD. Homoeologous relationships of rice, wheat and maize chromosomes. Mol Gen Genet 1993; 241:483–490.

Ahroni A, Vorst O. DNA microarrays for functional plant genomics. Plant Mol Biol 2001; 48:99–118.

Bänziger M, Edmeades GO, Lafitte HR. Selection for drought tolerance increases maize yields across a range of nitrogen levels. Crop Sci 1999; 39:1035–1040.

Bennetzen JL, Freeling M. The unified grass genome: synergy in synteny. Genome Res 1997; 7:301–306.

Bradley JP, Knittle KH, Troyer AF. Statistical methods in seed corn product selection. J Prod Agric 1988; 1:34–38.

Bruce WL, Edmeades GO, Barker TC. Molecular and physiological approaches to maize improvement for drought tolerance. J Exp Bot 2002; 53:13–25.

Bruce WL, Desbons P, Crasta O, Folkerts O. Gene expression profiling of two related maize inbred lines with contrasting root-lodging traits. J Exp Bot 2001; 52:459–468.

Burr B. Mapping and sequencing the rice genome. Plant Cell 2002; 14:521–523.

Cardinal AJ, Lee M, Sharopova N, Woodman-Clikeman WL, Long MJ. Genetic mapping and analysis of quantitative trait loci for resistance to stalk tunneling by the European corn borer in maize. Crop Sci 2001; 41:835–845.

Cardwell VW. Fifty years of Minnesota corn production: sources of yield increase. Agron J 1982; 74:984–990.

Chen M, Presting G, Barbazuk WB, Goicoechea JL, Blackmon B, Fang G, Kim H, Frisch D, Yu Y, Sun S, Higingbottom S, Phimphilai J, Phimphilai D, Thurmond S, Gaudette B, Li P, Liu J, Hatfield J, Main D, Farrar K, Henderson C, Barnett L, Costa R, Williams B, Walser S, Atkins M, Hall C, Budiman MA, Tomkins JP, Luo M, Bancroft I, Salse J, Regad F, Mohapatra T, Singh NK, Tyagi AK, Soderlund C, Dean RA, Wing RA. An integrated physical and genetic map of the rice genome. Plant Cell 2002; 14:537–545.

Consoli L, Lefèvre A, Zivy M, de Vienne D, Damerval C. QTL analysis of proteome and transcriptome variations for dissecting the genetic architecture of complex traits in maize. Plant Mol Biol 2002; 48:575–581.

Crosbie TM. Changes in physiological traits associated with long-term efforts to improve grain yield of maize. In: Loden HD, Wilkinson D, eds. Proc 37th Corn Sorghum Res Conf. Washington, DC: Am Seed Trade Assn, 1982:206–233.

Duncan WG. A theory to explain the relationship between corn population and grain yield. Crop Sci 1984; 24:1141–1145.

Duvick DN. Genetic rates of gain in hybrid maize during the past 40 years. Maydica 1977; 22:187–196.

Duvick DN. Genetic contributions to yield gains of US hybrid maize, 1930 to 1980. In: Fehr WR, ed. Genetic Contributions to Yield Gains of Five Major Crop Plants. Madison, WI: CSSA Spec Publ 7, ASA/CSSA/SSSA, 1984:1–47.

Duvick DN. Genetic contributions to advances in yield of US maize. Maydica 1992; 37:69–79.

Duvick DN. What is yield? In: Edmeades GO, Bänziger B, Mickelson HR, Pena-Valdivia CB, eds. Developing Drought and Low N-Tolerant Maize. El Batan, Mexico: CIMMYT, 1997:332–335.

Duvick DN. Heterosis: feeding people and protecting resources. In: Coors JG, Pandey S, eds. The Genetics and Exploitation of Heterosis in Crops. Madison, WI: ASSA/CSSA/SSA, 1999:19–29.

Dwyer LM, Tollenaar M. Genetic improvement in photosynthetic response of hybrid maize cultivars, 1959 to 1988. Can J Plant Sci 1989; 69:81–91.

Dwyer LM, Tollenaar M, Stewart DW. Changes in plant density dependence of leaf photosynthesis of maize hybrids, 1959 to 1988. Can J Plant Sci 1991; 71:1–11.

Dwyer LM, Stewart DW, Tollenaar M. Analysis of maize leaf photosynthesis under drought. Can J Plant Sci 1992; 72:477–481.

Fatokun CA, Menancio-Hautea DI, Danesh D, Young ND. Evidence of orthologous seed weight genes in cowpea and mung bean based on RFLP mapping. Genetics 1992; 132:841–846.

Fiehn O. Metabolomics—the link between genotypes and phenotypes. Plant Mol Biol 2002; 48:155–171.

Fiehn O, Kopka J, Dörmann P, Altmann T, Trethewey RN, Willmitzer L. Metabolite profiling for plant functional genomics. Nat Biotechnol 2000; 11:1157–1161.

Finlay KW, Wilkenson GN. The analysis of adaptation in a plant breeding programme. Ast J Agric Res 1963; 14:742–754.

Frova C, Sari-Gorla M. Quantitative trait loci (QTLs) for pollen thermotolerance detected in maize. Mol Gen Genet 1994; 245:424–430.

Gifford RM. Partitioning of photoassimilate in the development of crop yield. In: Cronshaw J, Lucas WJ, Giaquinta RT, eds. Phloem Transport. New York: Alan R Liss Inc, 1986:535–549.

Jampatong C, McMullen MD, Barry BD, Darrah LL, Byrne PF, Kross H. Quantitative trait loci for first- and second-generation European corn borer resistance derived from the maize inbred Mo47. Crop Sci 2002; 42:584–593.

Jiang C, Edmeades GO, Armstead IP, Lafitte HR, Hayward MD, Hoisington D. Genetic analysis of adaptation differences between highland and lowland tropical maize using molecular markers. Theor Appl Genet 1999; 99:1106–1119.

Kersten B, Burkle L, Kuhn EJ, Giavalisco P, Konthur Z, Lueking A, Walter G, Eickhoff H, Schneider U. Large-scale plant proteomics. Plant Mol Biol 2002; 48:133–141.

Lafitte HR, Edmeades GO. Stress tolerance in tropical maize is linked to constitutive changes in ear growth characteristics. Crop Sci 1995; 35:820–826.

Lee EA, Byrne PF, McMullen MD, Snook ME, Wiseman BR, Widstrom NW, Coe EH. Genetic mechanisms underlying apimaysin and maysin synthesis, and corn earworm antibiosis in maize (*Zea mays* L). Genetics 1998; 149:1997–2006.

Lin Y-R, Schertz KF, Patterson AH. Comparative analysis of QTLs affecting plant height and maturity across the Poaceae, in reference to an interspecific sorghum population. Genetics 1995; 141:391–411.

Liu BH. Statistical Genomics—Linkage, Mapping, and QTL Analysis. Boca, FL: CRC Press, 1998.

Lui H, Sachidanandam R, Stein L. Comparative genomics between rice and arabidopsis shows scant collinearity in gene order. Genome Res 2001; 11:2020–2026.

Major DJ, Hunter RB, Kannenberg LW, Daynard TB, Tanner JW. Comparison of inbred and hybrid corn grain yield measured at equal leaf area index. Can J Plant Sci 1972; 52:315–319.

McCullough DE, Aguilera A, Tollenaar M. N uptake, N partitioning, and photosynthetic N-use efficiency of an old and a new maize hybrid. Can J Plant Sci 1994; 74:479–484.

Nissanka SP, Dixon MA, Tollenaar M. Canopy gas exchange response to moisture stress in old and new maize hybrid. Crop Sci 1997; 37:172–181.

Nublat A, Desplans J, Casse F, Berthomieu P. Sas1, an *Arabidopsis* mutant over-accumulating sodium in the shoot, shows deficiency in the control of the root radical transport of sodium. Plant Cell 2001; 13:125–137.

Oliver SG, Winson MK, Kell DB, Baganz R. Systematic functional analysis of the yeast genome. Trends Biotechnol 1998; 16:373–378.

Peltier JB, Friso G, Kalume DE, Roepstorff P, Nilsson F, Adamska I, van Wijk KJ. Proteomics of the chloroplast: systematic identification and targeting analysis of lumenal and pericperal thylakoid proteins. Plant Cell 2000; 12:319–341.

Prime TA, Sherrier DJ, Mahon P, Packman LC, Dupree P. A proteomic analysis of organelles from *Arabidopsis thaliana*. Electrophoresis 2000; 21:3488–3499.

Ribaut J-M, Hoisington D, Deutsch J, Jiang C, Gonzalez-de-Leon D. Identification of quantitative trait loci under drought conditions in tropical maize. 1. Flowering parameters and the anthesis-silking interval. Theor Appl Genet 1996; 92:905–914.

Roberts JKM. Proteomics and a future generation of plant molecular biologists. Plant Mol Biol 2002; 48:143–154.

Roessner U, Willmitzer L, Fernie AR. High-resolution metabolic phenotyping of genetically and environmentally diverse potato tuber systems. Identification of phenocopies. Plant Physiol 2001; 127:749–764.

Roessner U, Wagner C, Kopka J, Trethewey RN, Willmitzer L. Simultaneous analysis of metabolites in potato tuber by gas chromatography–mass spectrometry. Plant J 2000; 23:131–142.

Roessner U, Luedemann A, Brust D, Fiehn O, Linke T, Willmitzer L, Fernie AR. Metabolic profiling allows comprehensive phenotyping of genetically or environmentally modified plant systems. Plant Cell 2001; 13:11–29.

Russell WA. Evaluations for plant, ear, and grain traits of maize cultivars representing different eras of breeding. Maydica 1985; 30:85–96.

Russell WA. Genetic improvement of maize yields. Adv Agron 1991; 46:245–298.

SanMiguel P, Tikhonov A, Jin Y-K, Motchoulskaia N, Zakharova D, Melake Berhan A, Springer PS, Edwards KJ, Avramova Z, Bennetzen JL. Nested retrotransposons in the intergenic regions of the maize genome. Science 1996; 274:765–768.

Séne M, Thevenot C, Hoffmann D, Benetrix F, Causse M, Prioul J. QTLs for grain dry milling properties, composition and vitreousness in maize recombinant inbred lines. Theor Appl Genet 2001; 102:591–599.

Shi H, Ishitani M, Kim C, Z J-K. The *Arabidopsis thaliana* salt tolerance gene SOS1 encodes a putative Na+/H+ antiporter. Proc Natl Acad Sci 2000; 97:6896–6901.

Sinclair TR, Muchow RC. Radiation use efficiency. Adv Agron 1999; 65:215–265.

Smith OS. Trend analysis of US maize yields from 1950–1994 regression model based on agronomic inputs, weather, and genetic trend. Proc 53rd Corn Sorghum Res Con. Washington, DC: Am Seed Trade Assn, 1998:170–179.

Teusink B, Baganz F, Westerhoff HV, Oliver SG. Metabolic control analysis as a tool in the elucidation of the function of novel genes. Methods Microbiol 1998; 26:297–336.

The Arabidopsis Genome Initiative. Analysis of the genome sequence of the flowering plant *Arabidopsis thaliana*. Nature 2000; 408:796–815.

Thiagarajah MR, Hunt LA, Mahon JD. Effects of position and age on leaf photosynthesis in corn (*Zea mays*). Can J Bot 1981; 59:28–33.

Tollenaar M. Effect of assimilate partitioning during the grain filling period of maize on rate of dry matter accumulation. In: Cronshaw J, Lucas WJ, Giaquinta RT, eds. Phloem Transport. New York: Alan R Liss Inc, 1986:551–556.

Tollenaar M. Genetic improvement of grain yield of commercial corn hybrids grown in Ontario from 1959 to 1988. Crop Sci 1989; 29:1365–1371.

Tollenaar M. Physiological basis of genetic improvement of maize hybrids in Ontario from 1959 to 1988. Crop Sci 1991; 31:119–124.

Tollenaar M. Is low plant density a stress in maize? Maydica 1992; 37:305–311.

Tollenaar M, Migus W. Dry matter accumulation of maize grown hydroponically under controlled-environment and field conditions. Can J Plant Sci 1984; 64:475–485.

Tollenaar M, Mihajlovic M. Bromoxynil tolerance during the seedling phase is associated with genetic grain yield improvement. Can J Plant Sci 1991; 71:1021–1027.

Tollenaar M, Dwyer LM. Physiology of maize. In: Smith DL, Hamel C, eds. Crop Yield, Physiology and Processes. Berlin: Springer Verlag, 1999:169–204.

Tollenaar M, Wu J. Yield improvement in temperate maize is attributable to greater stress tolerance. Crop Sci 1999; 39:1597–1604.

Tollenaar M, Lee EA. Yield potential, yield stability and stress tolerance in maize. Field Crops Res 2002; 75:161–170.

Tollenaar M, Daynard TB, Hunter RB. The effect of temperature on rate of leaf appearance and flowering date in maize. Crop Sci 1979; 19:363–366.

Tollenaar M, Dwyer LM, Stewart DW. Ear and kernel formation in maize hybrids representing three decades of grain yield improvement in Ontario. Crop Sci 1992; 32:432–438.

Tollenaar M, McCullough DE, Dwyer LM. Physiological basis of the genetic improvement of corn. In: Slafer GA, ed. Genetic Improvement of Field Crops. New York: Marcel Dekker, 1994:183–236.

Tollenaar M, Aguilera A, Nissanka SP. Grain yield is reduced more by weed interference in an old than in a new maize hybrid. Agron J 1997; 89:239–246.

Tollenaar M, Ying J, Duvick DN. Genetic gain in corn hybrids. Proc 55th Corn Sorghum Res Conf. Washington, DC: Am Seed Trade Assn, 2000:53–62.

Tuberosa R, Sanguineti MC, Landi P, Salvi S, Casarini E, Conti S. RFLP mapping of quantitative trait loci controlling abscisic acid concentration in leaves of drought-stressed maize (*Zea mays* L). Theor Appl Genet 1998; 97:744–755.

Tuberosa R, Sanguineti MC, Landi P, Giuliani MM, Salvi S, Conti S. Identification of QTLs for root characteristics in maize grown in hydroponics and analysis of their overlap with QTLs for grain yield in the field at two water regimes. Plant Mol Biol 2002; 48:697–712.

van Buuren LM, Salvi S, Morgante M, Serhani B, Tuberosa R. Comparative genomic mapping between a 745 kb region flanking DREB1A in *Arabidopsis thaliana* and maize. Plant Mol Biol 2002; 48:741–750.

Washburn MP, Wolters D, Yates JR. Large-scale analysis of the yeast proteome by multidimensional protein identification technology. Nat Biotechnol 2001; 19:242–247.

Wilkens MR, Pasquali C, Appel RD, Ou K, Golaz O, Sanchez J-C, Yan JX, Gooley AA, Hugues G, Humphery-Smith L, Williams KL, Hochstrasser DF. From proteins to proteomes: large-scale protein identification by two-dimensional electrophoresis and amino acid analysis. Biotechnol 1996; 14:61–65.

Wu J, Maehara T, Shimokawa T, Yamamoto S, Harada C, Takazaki Y, Ono N, Mukai Y, Koike K, Yazaki J, Fujii F, Shomura A, Ando T, Kono I, Waki K, Yamamoto K, Yano M, Matsumoto T, Sasaki T. A comprehensive rice transcript map containing 6591 expressed sequence tag sites. Plant Cell 2002; 14:525–535.

Ying J, Lee EA, Tollenaar M. Response of maize leaf photosynthesis to low temperature during the grain-filling period. Field Crops Res 2000; 68:87–96.

Ying J, Lee EA, Tollenaar M. Response of leaf photosynthesis during the grain-filling period of maize to duration of cold exposure, acclimation and incident PPFD. Crop Sci 2002; 42:1164–1172.

Zinselmeier C, Sun Y, Helentjaris T, Beatty M, Yang S, Smith H, Habben J. The use of gene expression profiling to dissect the stress sensitivity of reproductive development in maize. Field Crops Res 2002; 75:111–122.

3

Physiological Basis of Yield and Environmental Adaptation in Rice

Shaobing Peng and Abdelbagi M. Ismail
International Rice Research Institute (IRRI)
Manila, Philippines

1 INTRODUCTION

Rice is the most important food crop in the world. Rice accounts for more than 40% of caloric intake in tropical Asia, reaching more than 60% in many countries and for many poor people. During the past 35 years, world rice production doubled as a result of the adoption of modern varieties and improved cultural practices. It is estimated that world rice production has to increase by 1% annually in the next 20 years to meet the growing demand for food (Rosegrant et al., 1995). It is a big challenge to continuously increase world rice production at such pace since the production environment will be very different from those in the past. Rice productivity is showing signs of decline, expansion of area is limiting, investments in irrigation have virtually ceased, high fertilizer use is causing concern, and good rice lands are being lost to other purposes. The only option available is to raise rice yield potential in favorable environments and to enhance adaptability of rice cultivars in less favorable environments through genetic improvement.

2 PHYSIOLOGICAL BASIS OF RICE YIELD POTENTIAL

Irrigated rice land contributes more than 75% of total rice production, although it accounts for about 50% of total rice area. Irrigated rice planting area is about 80 million hectares worldwide. Majority of these areas are considered favorable for rice production: water and nutrients are not the major limiting factors for rice growth. In tropical Asia, the yield potential of current high-yielding cultivars grown under favorable environments is about 7 t ha^{-1} during the wet season and about 10 t ha^{-1} during the dry season. Potential yield has been estimated at 9.5 and 15.9 t ha^{-1} in this region during the wet and dry seasons, respectively, based on the level of solar radiation (Yoshida, 1981). The challenge is to narrow the gap between yield potential and potential yield through genetic improvement.

Grain yield is determined by biomass production and harvest index (HI). Biomass production is a function of the rate and duration of photosynthesis and respiration rate. Optimum canopy architecture for maximum crop photosynthesis, increased photosynthetic capacity of individual leaves, and delayed leaf senescence for longer photosynthetic duration are means to increase biomass production. Harvest index is affected by sink size (spikelets per square meter), canopy photosynthetic rate during ripening phase, assimilate partitioning, and grain-filling percentage. Increased number of grains per panicle and reduced partitioning of dry matter to unproductive tillers will result in an improved HI. Understanding the physiological processes governing yield potential and identifying plant traits associated with high yield are essential for crop improvement through plant breeding and molecular technology.

2.1 Single-Leaf Photosynthesis

Crop physiologists have tried selecting for high single-leaf net photosynthetic rate under light saturation (P_n) in several crop species, but no cultivar has been released from these selection programs (Nelson, 1988). Direct selection for high P_n sometimes resulted in lower yield (Evans, 1990). Although genetic variation in P_n has been reported in rice, the relationships between photosynthetic capacity and biomass production were poor (McDonald et al., 1974). In spite of these problems, the hypothesis that higher P_n is necessary for increased yields is still popular (Elmore, 1980). Zelitch (1982) stated that the lack of a strong positive relationship is due to artifact of P_n measurement rather than biological reasons. Austin (1993) believes that genotype × ontogeny and genotype × environment interactions for P_n cause poor correlation. Traits that are pleiotropically and negatively related to P_n may offset any gains from higher P_n (Austin, 1993).

Yield enhancement in rice by conventional breeding has mainly resulted from improvement in plant type, which has increased canopy net photosynthesis, especially during the grain-filling period, and HI. Austin (1993) argued that no substantial improvement in biomass production could be obtained by selecting for modified canopy morphology since the canopy architecture of modern varieties is close to optimum. This suggests that increasing single-leaf photosynthesis could be a more effective way to substantially enhance rice yield potential. With a better understanding of the limiting processes in photosynthesis, advances in measurement methodology, and the advent of biotechnology, which enables the modification of content or activity of individual enzymes, the possibility of enhancing biomass production by improving single-leaf photosynthesis should be re-examined. Single-leaf photosynthetic rate is affected by morphological and physiological traits of leaves in addition to environmental factors.

2.1.1 Stomatal Density and Aperture

Stomata are found in the leaf blade, leaf sheath, rachis, rachis branch, and the lemma and palea of the spikelet. Hence leaf sheath, rachis, and spikelets of rice plants are active in photosynthesis. Maeda (1972) found a small number of poorly developed and uncompleted stomata in the glume of the spikelet. The stomata of rice plants are smaller than other plant species, although rice has a relatively high stomatal density. The stomatal density of rice is about seven times greater than that of wheat (Chen et al., 1990; Teare et al., 1971), ranging from 150 to 650 mm^{-2}, depending on leaf positions on the stem, cultivar, and growing conditions (Matsuo et al., 1995). The stomatal density of the indica type and indica/japonica hybrids is higher than that of the japonica type (Maruyama and Tajima, 1986). The distribution of the stomata on the adaxial and abaxial sides of the leaf depends on leaf position and genotype.

The stomatal aperture of rice is much smaller than other species. The maximum aperture of rice stomata is 1.5 μm, while the average stomatal aperture of several other species is as high as 6 μm (Meidner and Mansfield, 1968). Ishihara et al. (1971) observed rice stomatal aperture to vary from 0.5 to 1.2 μm. Stomatal aperture and conductance in leaves increase with higher leaf N concentration. Stomatal aperture decreases under high solar radiation, high temperature, and low air humidity as a result of the lowered leaf water potential caused by increased transpiration (Ishihara and Saito, 1983). A close relationship between stomatal aperture and stomatal conductance was observed in rice leaves with different N concentrations.

2.1.2 Leaf Thickness and Leaf Shape

Specific leaf weight (g leaf dry matter m^{-2} leaf area) is a measure of leaf thickness. Thick leaves usually have high chlorophyll content and high content of photosynthetic enzymes per unit leaf area. Although a positive association between leaf thickness and yield potential has not been documented in rice, leaf thickness is positively correlated with leaf photosynthetic rate (Murata, 1961). A thick leaf does not tend to expand horizontally and therefore tends to be narrow and erect. Thicker leaves are thought to be desirable (Yoshida, 1972), and erect leaf trait provides a visual selection criterion for the new plant type breeding (Peng et al., 1994). Tsunoda and Kishitani (1976) claimed that, at a given leaf area, a narrow leaf can increase P_n by as much as 30% because of reduced boundary layer resistance on the leaf surface.

2.1.3 Leaf Nitrogen Concentration

Yoshida and Coronel (1976) and Makino et al. (1988) reported a linear relationship between P_n and leaf N concentration per unit leaf area (N_a) in rice when N_a ranged from 0.5 to 2.1 g m^{-2}. This relationship holds whether differences in N_a are due to environment or genotype (Tsunoda, 1972). On the other hand, a curvilinear relationship between light-saturated P_n and N_a was observed by Takano and Tsunoda (1971) and Cook and Evans (1983): a linear relationship when N_a was below 1.6 g N m^{-2}, but P_n leveled off above 1.6 g N m^{-2}. This curvilinear relationship might have resulted from growth limitations in some of the primitive genotypes caused by factors other than leaf N. Wheat usually has higher N_a than rice, which may explain the difference in radiation-use efficiency between the two species (Mitchell et al., 1998).

During the vegetative and reproductive stages, P_n can be increased by increasing leaf N concentration. Because of the high N-absorbing capacity of the rice root system, modern rice varieties respond quickly to N application by increasing leaf N concentration (Peng and Cassman, 1998). High leaf N concentration results in an increase in tiller production and leaf area expansion, which cause mutual shading and an actual reduction in biomass production and grain yield. If increased leaf N concentration and P_n do not lead to excessive leaf area index (LAI) and more unproductive tillers, high P_n should contribute to increased biomass and grain yield. The concern is how to increase leaf N concentration without significant increases in tiller number and leaf area. For the cultivars with moderate tillering capacity and semitall stature, leaf area and tiller production are not very sensitive to increased leaf N concentration compared to the cultivars with high tillering capacity and semidwarf stature. Genotypic variation in the sensitivity of leaf area and tiller production to leaf N concentration that is independent of plant height may exist in rice.

2.1.4 Ribulose 1,5-Bisphosphate Carboxylase/Oxygenase (Rubisco) Content

About 50% of total soluble protein and 25% of total N are associated with Rubisco protein in rice leaves (Makino et al., 1984). During leaf senescence, specific carboxylase activity of Rubisco did not change and a decline in activity was caused by the reduction in the amount of Rubisco protein (Makino et al., 1983). Rice has similar K_m (CO_2) value and ratio of Rubisco to total soluble protein as other C_3 plants, but its V_{max} may be much lower (Makino et al., 1985). The V_{max} of Rubisco in rice is 45% lower than that in wheat. Among species of *Oryza*, the differences in enzymatic properties of Rubisco are small (Makino et al., 1987). Within *Oryza sativa*, the differences in K_m (CO_2), V_{max}, and the ratio of Rubisco to total soluble protein are small among indica, temperate japonica, and intermediate types (Makino et al., 1987). The tropical japonica type has a slightly higher K_m (CO_2) and V_{max} than other subspecies. The differences are also small among the cultivars within a subspecies. These results suggest that it is difficult to improve the enzymatic characteristics of Rubisco by crossbreeding since the genetic variation in enzymatic properties is small even when the comparison was expanded to the other species of *Oryza*.

Rubisco is the most abundant protein and is also the most inefficient enzyme in terms of carboxylase reaction. This is because of its slow reaction turnover rate, low affinity for CO_2, and unavoidable oxygenase reaction (Yokota et al., 1998). Increasing the specificity of Rubisco for CO_2 fixation has been a target of genetic engineering (Gutteridge et al., 1995). A Rubisco with a strong specificity for CO_2 fixation (2.5-fold that of higher plant Rubisco) has been found in thermophilic red algae (Uemura et al., 1997). It has been proposed to improve Rubisco of crop plants by replacing the present enzyme with the more efficient red algae form (Mann, 1999). Yokota et al. (1998) proposed to create a super Rubisco in C_3 crop plant with *Galdieria* Rubisco to increase relative specificity and affinity for CO_2. Mae (1997) stated that Rubisco is a major N source, which is used as remobilized N for grain filling. Therefore we should improve Rubisco efficiency for carboxylase reaction without reducing the amount because of its function as N storage for the growth of grains.

2.1.5 Chlorophyll Content

Rabinowitch (1956) stated that chlorophyll content seldom limits P_n under sufficient photosynthetically active radiation (PAR) because more chlorophyll is contained in an ordinary leaf than necessary. Under low PAR, however, chlorophyll content may limit P_n since the rate of light reaction may limit the overall process of photosynthesis (Murata, 1965). Paulsen (1972) reported that P_n and chlorophyll content were closely correlated in six rice cultivars under low PAR but not under high PAR. In rice, chlorophyll content is closely correlated with N concentration, so an apparent close relationship exists between chlorophyll content and P_n. However, if the variation in chlorophyll content is caused by different genotypes or by other nutrients such as phosphorus or potassium, chlorophyll content is no longer correlated with P_n (Yoshida et al., 1970). During leaf senescence, chlorophyll remains relatively more stable than Rubisco (Makino et al., 1983). Under excessive solar radiation, high chlorophyll content may be disadvantageous due to its susceptibility to oxidative stress (Bennett, personal communication, 1997).

2.1.6 Stomatal Conductance

High stomatal density compensates for the small aperture so that the stomatal conductance of a rice leaf is comparable to or even greater than that of other species. It is not easy to compare reported values of conductance because measurements were made under different conditions and using different gas exchange measurement systems. In addition, different boundary layer conductance might have been used to calculate stomatal conductance in various studies.

The 20% to 30% decline in P_n in the afternoon despite sufficient PAR is accompanied by a reduction in stomatal conductance (Ishihara and Saito, 1987). Murata (1961) observed midday depression in P_n of a rice leaf. The main cause of midday P_n depression in the rice leaf is also attributed to stomatal closure and a reduction in the CO_2 supply to mesophyll cells. O'Toole and Tomer (1982) have shown that rice leaves can suffer water deficits even when the plants are kept flooded.

Irrigated rice has much higher stomatal density and stomatal conductance than wheat (Teare et al., 1971; Dai et al., 1995), suggesting that P_n in irrigated rice plants is unlikely to be limited by stomatal conductance. Small differences in carbon isotope discrimination among varieties and over a wide range of N input levels (Laza, personal communication, 1996) also suggest that there is little chance to improve P_n by increasing stomatal conductance in irrigated rice. However, recent studies provide strong evidence that grain yield may be improved by increasing stomatal conductance. Ku et al. (2000) reported that transgenic rice plants overexpressing maize C_4-specific phosphoenolpyruvate carboxylase (PEPC) and pyruvate, orthophosphate dikinase (PPDK) enhanced P_n by increasing stomatal conductance. Horie (2001) observed strong correlation between grain yield and canopy conductance across rice subspecies. The yield potential of wheat varieties released by International Maize and Wheat Improvement Center (CIMMYT) has increased by 0.83% per year over the last 30 years. This increase was mainly attributed to increased stomatal conductance and canopy temperature depression (Fischer et al., 1998).

2.1.7 Photorespiration and Respiration

About 30% to 40% of the fixed carbon is consumed by photorespiration in C_3 plants. Yeo et al. (1994) reported that photorespiration reduced CO_2 fixation in rice and two other *Oryza* species by about 30%. Therefore rice does not seem to be markedly different from other C_3 plants in terms of photorespiration. Yeo et al. (1994) found that *Oryza rufipogon* had significantly lower photorespiration than *O. sativa*. Varietal differences in photorespiration rate have been observed by Kawamitsu et al. (1989). However, no significant difference in photorespiration rate was observed by Akita et al. (1975) among cultivars or across the species of *Oryza*.

Net photosynthesis is gross photosynthesis minus concurrent respiration. The respiration rate is about 10% of P_n within the optimum range of temperature for net photosynthesis. It is generally believed that the rate of respiration is the same in the light as in the dark (Matsuo et al., 1995). Respiratory rate among leaves does not vary as much as photosynthetic rate (Tanaka et al., 1966). Respiration can be separated into maintenance and growth terms. The maintenance respiration of rice plants is 0.003 to 0.005 g CH_2O g^{-1} DM day^{-1} at 20°C (Mitchell et al., 1998). The growth respiration of rice plants is 0.55 to 0.65 g DM g^{-1} CH_2O (Mitchell et al., 1998).

Scientists proposed to increase P_n by suppressing photorespiration and reducing maintenance respiration (Penning de Vries, 1991). There is little evidence that photorespiration can be suppressed in C_3 plants under current photosynthetic pathway, and although there is evidence of genetic variation in maintenance respiration, the magnitude of such differences is small (Gifford et al., 1984).

2.1.8 Photoinhibition

High sunlight induces photoinhibition of photosynthesis and even photodamage of the photosynthetic apparatus when photosynthetic carbon assimilation is affected by severe environmental conditions such as drought and temperature extremes (Xu, 2001). Horton and Ruban (1992) believe that operational photosynthesis in the field do not actually reach the intrinsic maximum photosynthetic rate. During the course of the day and the entire growing season, photosynthesis operates at maximum level over a very short period of time. Internal (feedback inhibition or sink limitation) and external (photo-oxidative stresses or photoinhibition) factors limit attainment of the full potential of photosynthesis. Preliminary studies indicate that alternative dissipative electron transfer pathways, such as the xanthophyll cycle, and free radical-scavenger enzymes, such as superoxide dismutase, catalase, and ascorbate peroxidase, give plants overall tolerance for photo-oxidative stresses. The capacity of photoprotection is variable between species (Johnson et al., 1993). Tu et al. (1995) reported genotypic variation in photoinhibition and midday photosynthetic depression under high light-induced conditions, suggesting scope for improvement by breeding.

Photodamage was not observed in field-grown rice plants in irrigated system (Horton, 2000). Senescing and N-deficient leaves are more susceptible to photoinhibition under excessive solar radiation (Murchie et al., 1999; Chen et al., 2003). Murchie et al. (1999) reported that erect leaves reduced the level of intercepted irradiance at each leaf surface at solar noon when incident irradiance level is maximal. Therefore erect leaves are less susceptible to photoinhibition than horizontally positioned leaves.

2.1.9 Leaf Senescence

Increasing photosynthetic duration is often achieved by delaying leaf senescence. Senescence is associated with the degradation of Rubisco and chlorophyll (Makino et al., 1983). Gan and Amasino (1995) introduced a DNA fragment from Agrobacterium encoding isopentenyl transferase (*ipt*), an enzyme that catalyzes the rate-limiting step in cytokinin biosynthesis, into tobacco. The transgenic tobacco plants exhibited a significant delay in leaf senescence and an increase in growth rate. The *ipt* gene was also successfully introduced into rice plants (Zhang and Bennett, personal communication, 2000). Studies are underway to determine if the transgenic rice plants demonstrate delayed leaf senescence and increased grain yield. Increased late-season N application protects Rubisco from degradation, which delays flag leaf senescence and increases photosynthetic duration. However, delaying senescence of the flag leaf does not always result in greater yield if the sink is limiting. Moreover, delaying senescence of the flag leaf results in a reduction in nutrient translocation from the flag leaf to grain, although the quantitative effects remain to be determined.

Zhang et al. (2003) studied N, chlorophyll, and Rubisco contents in the top three leaves of field-grown rice plants during natural senescence. The patterns of leaf senescence of the top three leaves were not very different, and the senescence of rice leaves was a function of leaf age (e.g., sequential senescence). Rubisco content declined earlier than N and chlorophyll during the senescence of all the top three leaves. The rate of decline was much faster in Rubisco than in N and chlorophyll. Camp et al. (1982) found that Rubisco content decreased at a much faster rate than P_n during leaf senescence. Field data indicated that photosynthetic capacity remained significant when leaf Rubisco content approached zero (Murchie, personal communication, 2000). Tsunoda (1972) stated that leaf N content correlated better with P_n than with Rubisco content. It was speculated that this could be due to the fact that leaf N can be measured more accurately than Rubisco content (Peng, 2000). Makino et al. (1983) reported that loss of chlorophyll during leaf senescence did not necessarily cause the decrease in photosynthetic activity. Kura-Hotta et al. (1987) found that photosynthetic capacity decreased more rapidly than with chlorophyll content during leaf senescence. Close correlation between leaf N content and P_n has been reported in many studies, regardless of plant age or leaf age (Yoshida and Coronel, 1976; Makino et al., 1988; Peng et al.,

1995). The determination of leaf N and chlorophyll is relatively simple compared with Rubisco measurement. Therefore leaf N content could be a more suitable parameter to quantify leaf senescence than Rubisco or chlorophyll contents if leaf senescence will be used as a selection criteria in the breeding program.

2.1.10 C_4 Rice

Transforming C_3 rice plant into C_4 rice plant by genetic engineering of photosynthetic enzymes and required anatomic structure aims to enhance photosynthetic rate. High-level expression of maize PEPC, PPDK, and NADP-malic enzyme (NADP-ME) in transgenic rice plants has been achieved (Agarie et al., 1998; Ku et al., 1999). Ku et al. (2000) reported that PEPC and PPDK transgenic rice plants had up to 30% to 35% higher photosynthetic rate than untransformed plants. This increased photosynthetic rate was associated with an enhanced stomatal conductance and a higher internal CO_2 concentration, which was as high as 275 ppm vs. 235 ppm in untransformed plants. In addition, PEPC transgenic plants had higher light saturation point and were less susceptible to photoinhibition than untransformed plants. As a result, grain yield was 10% to 30% higher in PEPC and 30% to 35% higher in PPDK transgenic rice plants compared with control plants. The mechanism of maintaining a higher stomatal conductance by the transgenic plants is unknown (Ku et al., 2000). Stomatal density and aperture between PEPC and PPDK transgenic and untransformed plants were not compared. Achieving C_4 photosynthesis without Kranz leaf anatomy is possible as evidenced in the primitive aquatic angiosperm *Hydrilla verticillata* (Bowes and Salvucci, 1989). Voznesenskaya et al. (2001) observed C_4 photosynthesis without Kranz anatomy in *Borszczowia aralocaspica* and concluded that Kranz anatomy is not essential for terrestrial C_4 plant photosynthesis.

2.2   Canopy Photosynthesis

There is little doubt that canopy net photosynthesis rate (CP_n) correlates with biomass production and grain yield. Yield enhancement in modern rice cultivars has mainly resulted from improvement in canopy structure, which has increased CP_n, especially during the grain-filling period. Canopy photosynthesis is difficult to measure because of CO_2 fluxes in the aerenchyma of the plant and from the soil and water. On a seasonal basis, CP_n reaches a maximum between panicle initiation and the booting stage when leaf area index (LAI) is the highest (Tanaka, 1972). CP_n is affected by environmental factors such as PAR, temperature, and ambient CO_2 concentration. Plant morphophysiological traits such as LAI, leaf orientation, P_n, and crop respiration also control CP_n. Among these morphophysiological traits, LAI is the most important determinant of light interception and therefore of CP_n, with canopy architecture being of secondary importance.

2.2.1 Leaf Area Index

Peak LAI value of 6 to 10 is generally observed around the heading stage. At heading, flag leaves contribute 19% of LAI, second leaves contribute 28%, and third leaves contribute 27% (Yoshida et al., 1972). Crop management practices such as fertilizer application aim to optimize CP_n and yield, mainly by controlling LAI. Optimal LAI is defined as the LAI at which CP_n and crop growth rate (CGR) are maximum, and beyond which CP_n and CGR will decrease. This is because canopy gross photosynthesis (CP_g) increases with an increase in LAI up to a point where CP_g reaches its maximum, whereas canopy respiration increases proportionally to the increase in LAI (Monsi and Saeki, 1953). This concept was supported by the study of Murata (1961). The optimal LAI values depend on growth stage, plant type, and PAR (Tanaka et al., 1966). Cultivars with erect leaves have a higher optimum LAI than those with horizontal leaves (Yoshida, 1981). However, Yoshida et al. (1972) questioned the existence of optimal LAI in rice. Cock and Yoshida (1973) reported that canopy respiration increased linearly with CP_g but curvilinearly with LAI. Thus CP_n and CGR only leveled off but did not decline beyond a certain LAI, which was defined as ceiling LAI. Yoshida (1981) reported that CGR reached a maximum at an LAI of about 6 for cv. IR8, etc. and about 4 for Peta, beyond which it remained the same. However, if lodging occurs at high LAI, CP_n and CGR will decrease with an increase in LAI and optimal LAI will appear. Therefore Tanaka (1983) explained that ceiling LAI existed in the high-yielding cultivars with short and erect leaves (such as IR8) and grown in the dry season, whereas optimal LAI appeared in leafy cultivars (such as Peta) grown in the wet season. Crops with ceiling LAI are easier to manage than those with optimal LAI since excessive N application does not have a negative effect on crops with the ceiling LAI (Tanaka, 1983). Critical LAI is defined as the LAI when 95% of PAR is intercepted by the crop canopy (Gifford and Jenkins, 1982).

2.2.2 Canopy Structure

Monsi and Saeki (1953) stated that rice cultivars with erect, short, and thick leaves have a small light extinction coefficient (K). The value of K is related to the leaf spread, which is determined by the leaf angle and curvature of the leaf blade (Tanaka et al., 1966). Tanaka et al. (1969) studied the effect of leaf inclination on CP_n. The droopy-leaf canopy was created with weights hung on the leaf tips. Light saturation for CP_n was observed in the droopy-leaf plot but not in the control plot with erect leaves. The maximum CP_n of the droopy-leaf plot was 68% of that of the control. Sinclair and Sheehy (1999) argued about the benefit of erect leaves in light interception for a rice crop with LAI greater than 4.2. They stated that the major benefit of erect leaves is to sustain a high LAI for N storage for a high-yielding rice crop. The effect of erect leaves on the proportion of diffused light in the low portion of the canopy and the benefit of erect leaves for reducing photoinhibition under high light intensity (Murchie et al., 1999) deserve further investigation.

Cultivars with erect leaves give a higher CP_n only when LAI is larger than 5, whereas cultivars with droopy leaves give a higher CP_n when LAI is less than 3. Therefore an ideal variety should have a droopy-leaf canopy in the very early vegetative stage to intercept PAR effectively. As the crop grows, a plant community with vertically oriented leaves has better light penetration and a higher CP_n at high LAI. The beneficial effect of erect leaves is pronounced when light intensity is high (Yoshida, 1981). Erect leaves have a considerable advantage in rice, some advantage in wheat, but little advantage in maize (Paulsen, 1972). However, Duvick and Cassman (1999) reported that newly released maize hybrids had more erect leaves than the old ones. Erect leaf between panicle initiation and flowering is one of the major morphological traits that rice breeders have been selecting for. It was reported recently that V-shaped leaf blades reduce mutual shading and increase canopy photosynthesis as do erect leaves (Sasahara et al., 1992).

2.2.3 Plant and Panicle Height

The semidwarf plant type reduces susceptibility to lodging at high N inputs and increases HI (Tsunoda, 1962). Recent studies, however, claim that the height of semidwarf rice and wheat may limit canopy photosynthesis and biomass production (Kuroda et al., 1989; Gent, 1995). Under a given LAI, a taller canopy has better ventilation and therefore higher CO_2 concentration inside the canopy than a shorter canopy (Kuroda et al., 1989). Similarly, light penetrates better in the tall canopy than in the short one (Kuroda et al., 1989). If stem strength can be improved, the height of modern rice varieties can be increased to improve biomass production.

Panicles that droop below the flag leaves (lower panicles) increase the interception of PAR by leaves and consequently increase canopy photosynthesis (Setter et al., 1995a,b). However, the adverse effects of lowering the panicles on panicle exsertion and panicle diseases need to be investigated. In addition, the panicle contributes to photosynthesis. On the basis of projected area, the P_n of the panicle is 20% of that of the flag leaf and the gross photosynthetic rate of the panicle is 30% of that of the flag leaf. On the basis of chlorophyll, the photosynthetic capability of spikelets is similar to that of the flag leaf (Imaizumi et al., 1990). It was estimated that panicle photosynthesis contributed 20% to 30% of the dry matter in grain (Imaizumi et al., 1990). However, the superhybrid rice that was developed by crossing Pei-ai

64S and E32 using two-line system has longer flag leaves and lower panicle height compared with common plant type (Normile, 1999). This hybrid yielded 17 t ha^{-1} in Yunnan, P.R. China (Yuan, 2001).

2.2.4 Nitrogen Distribution Within a Canopy

Increased leaf N concentration results in increases in LAI and P_n, and therefore enhancing CP_n. Under favorable rice-growing conditions, CP_n is limited by LAI during the vegetative growth stage and by foliage N concentration during the reproductive stage (Schnier et al., 1990). The foliage N compensation point (N concentration at which CP_n is zero) increases with LAI (Dingkuhn et al., 1990). CP_n correlates better with foliage N concentration on a dry-weight basis than on a leaf-area basis (Dingkuhn et al., 1990). Fertilizer-N application increases CP_g and canopy respiration. Under excessive N application, CP_n may be reduced due to lodging and pest damage. Simulation modeling suggests that a steeper slope of the vertical N concentration gradient in the leaf canopy with more N present in the uppermost stratum enhances canopy photosynthesis (Dingkuhn et al., 1991).

2.2.5 Tillering Capacity

Tillering capacity plays an important role in determining rice grain yield since it is closely related to panicle number per unit ground area. Too few tillers result in too few panicles; but excessive tillers cause high tiller abortion, small panicles, poor grain filling, and thus reduced grain yield (Peng et al., 1994). Leaf area index and plant N status are two major factors that influence tiller production in rice crop (Zhong et al., 2002). Leaf area index probably affects tillering by attenuation of light intensity and/or by influencing light quality at the base of the canopy where tiller buds and young tillers are located. High light intensity at the base of the canopy stimulates tillering (Yoshida, 1981; Graf et al., 1990).

Modern semidwarf indica rice varieties tiller profusely. Although each rice hill includes 3–5 plants and produces 30–40 tillers under favorable growing conditions, only 15–16 produce panicles. Unproductive tillers compete with productive tillers for assimilates, solar energy, and mineral nutrients—particularly nitrogen. Elimination of the unproductive tillers could direct more nutrients to grain production, but the magnitude of the potential contribution to yield by eliminating unproductive tillers has not been quantified. Furthermore, the dense canopy that results from excessive tiller production creates a humid microenvironment favorable for diseases, especially endogenous pathogens like sheath blight (*Rhizoctonia solani*) and stem rot (*Sclerotium oryzae*) that thrive in N-rich canopies (Mew, 1991).

Reduced tillering is thought to facilitate synchronous flowering and maturity, more uniform panicle size, and efficient use of horizontal space

(Janoria, 1989). Low-tillering genotypes are also reported to have a larger proportion of high-density grains (Padmaja Rao, 1987). High-density grains are those that remain submerged in a solution of specific gravity greater than 1.2. Ise (1992) found that a single semidominant gene controlled the low tillering trait, and that this gene had pleiotropic effects on culm length and thickness and panicle size. Therefore the low tillering trait was hypothesized to be associated with larger panicle size, and it became a target trait for International Rice Research Institute's (IRRI) new plant type breeding.

2.2.6 Radiation Use Efficiency

Crop growth and yield depend on solar radiation through photosynthesis. Radiation use efficiency (RUE) is commonly defined as grams of above-ground dry matter produced per unit intercepted PAR in MJ. Radiation use efficiency is an empirical quantity, which depends on canopy photosynthesis and respiration. Mitchell et al. (1998) compared RUE of rice, wheat, maize, and soybean using published figures in the literature. The average RUE for maize, wheat, rice, and soybean is 3.3, 2.7, 2.2, and 1.9 g DM MJ^{-1}, respectively. The reason why wheat has higher RUE than rice is not very clear. One speculation is that wheat usually contains a higher amount of N per unit leaf area than rice, which may result in higher canopy photosynthetic rate in wheat than rice. Another explanation is that most figures of RUE were determined from tropical rice. Respiration cost of rice in tropics is higher than that of wheat in the subtropical or temperate regions due to temperature difference. This was supported by the fact that the calculated value of RUE for 15 t ha^{-1} rice crop in subtropical Yunnan, P.R. China is about 2.6 g DM MJ^{-1} (Mitchell et al., 1998).

2.3 Assimilate Partitioning

Converting the increased biomass production into economic yield is another challenge for increasing yield potential. In general, HI is negatively correlated with biomass. The strategy of breeding for high yield potential is to increase biomass production and at the same time to maintain HI at the level of 0.50 to 0.55. This can be done by increasing the partitioning of C and N assimilates to grains. Enhancements in sink size, sink strength, grain-filling rate, and grain-filling duration are effective approaches for the maintenance of high HI at increased biomass production.

2.3.1 Sink Size

Sink size (spikelet number per square meter) is determined by spikelet num-ber per panicle and panicle number per square meter. Since a strong compensation mechanism exists between the two yield components, an increase in one component will not necessarily result in an increase in the overall sink size. Sink size could be increased by selecting for large panicles only if the panicle number per square meter is maintained. The way to break the strong negative relationship between the two components is to increase biomass production during the critical phases of development when sink size is determined. Slafer et al. (1996) stated that breeders should select for a greater growth during the time when grain number is determined rather than select for panicle size or number. The critical period that determines sink size was reported to be 20–30 days before flowering in wheat (Fischer, 1985). In rice, spikelet number per square meter was highly related to dry matter accumulation from panicle initiation to flowering (Kropff et al., 1994). High light intensity and CO_2 enrichment enhanced the number of differentiated spikelets (Yoshida and Parao, 1976). Wada and Matsushima (1962) also reported that spikelet formation is strongly affected by both N uptake and availability of carbohydrates during panicle initiation to flowering. Akita (1989) stated that there is genotypic variation in spikelet formation efficiency (the number of spikelets produced per unit of growth from panicle initiation to flowering). To increase sink size, one should select for higher spikelet formation efficiency.

Fischer (1985) reported that accelerating development during the period of active spike growth through increases in air temperature reduced the final number of grains in wheat. Slafer et al. (1996) proposed to extend the stem elongation phase (from terminal spikelet initiation to flowering) in order to increase biomass accumulation in the same phase and final spikelet number. Temperature and photoperiod are the main environmental factors that affect the rate of development. Slafer and Rawson (1994) showed varietal differences in degree of sensitivity to temperature during stem elongation in wheat. Sheehy (1995, personal communication) observed that a large proportion of primordia were aborted in the tropical rice plant, probably due to fast development rate caused by high temperature or shortage in N uptake. Yoshida (1973) proved that the number of spikelets per panicle was reduced under high temperature. Several other approaches were suggested to increase sink size. Richards (1996) proposed to increase carbon supply to the developing panicles by reducing the size of the competing sinks. This could be achieved by reducing the length of peduncle (the internode between the uppermost leaf node and the panicle) and reducing the unproductive tillers.

2.3.2 Grain Filling

Grain filling has larger influence on yield potential as sink size increases. Spikelets can be fully filled, partially filled, or empty. Since the grain size is rigidly controlled by hull size, the weight of fully filled spikelet is relatively constant for a given variety (Yoshida, 1981). Breeders rarely select for heavy grain weight because of the negative linkage between grain weight and grain number. This does not mean that there is no opportunity to increase rice yield potential by selecting for heavy grains. However, the major efforts should be directed to reduce the proportion of partially filled and empty spikelets by improving grain filling.

Filled spikelet percentage is determined by the source activity relative to sink size, the ability of spikelets to accept carbohydrates, and the translocation of assimilates from leaves to spikelets (Yoshida, 1981). These factors determine the rate of grain filling. Akita (1989) reported a close relationship between crop growth rate at heading and filled spikelet percentage. Carbon dioxide enrichment during the ripening phase increased crop growth rate, filled spikelet percentage from 74% to 86%, and grain yield from 9.0 to 10.9 t ha^{-1} (Yoshida and Parao, 1976). Increasing late-season N application led to increased leaf N concentration, photosynthetic rate, and grain yield (Kropff et al., 1994). The ability of spikelets to accept carbohydrates is often referred to as sink strength. Starch is reported to be a critical determinant of sink strength (Kishore, 1994). Starch levels in a developing sink organ can be increased by increasing the activity of ADP glucose pyrophosphorylase (Stark et al., 1992). Zhang et al. (1996) introduced a gene from *Escherichia coli* encoding for ADP-glucose pyrophosphorylase into rice with the goal of increasing starch biosynthesis during seed development and increasing the sink strength of developing grains. Plant hormones such as cytokinins that regulate cell division and differentiation in the early stage of seed development also affect sink strength (Quatrano, 1987). Yang et al. (2000) reported that grain-filling percentage was significantly correlated with cytokinin contents in the grains and roots at the early and middle grain-filling stages of rice plants. Indole-3-acetic acid and gibberellin contents in the grains and roots did not correlate with grain-filling percentage. Application of cytokinin at and after flowering improved grain filling and yield of rice plants, probably through increased sink strength and/or delayed leaf senescence (Singh et al., 1984). The capacity of transporting assimilates from source to sink could also limit grain filling (Ashraf et al., 1994). Indica rice has more vascular bundles in the peduncle relative to the number of primary branches of panicle than japonica rice (Huang, 1988). It is not clear if the number of vascular bundles is more important than their size in terms of assimilate transport.

Simulation modeling suggests that prolonging grain-filling duration will result in an increase in grain yield (Kropff et al., 1994). Varietal differences in grain-filling duration were reported by Senadhira and Li (1989), but only main culm panicles were monitored in this study. It is unknown if grain-filling duration differs among varieties within subspecies when the entire population of panicles is considered. Grain-filling duration is controlled mainly by temperature. Slafer et al. (1996) proposed to increase grain-filling duration through the manipulation of the responses to temperature. Hunt et al. (1991) reported genotypic variation in sensitivity to temperature during grain filling in wheat. Such variation in grain-filling duration in response to temperature has not been reported in rice.

High-density grains tended to occur on the primary branches of the panicle, while the spikelets of the secondary branches had low grain weight (Ahn, 1986). Padmaja Rao (1987) reported that the top of the panicle (superior spikelet positions) has more high-density grains than the lower portion of the panicle (inferior spikelet positions). Varietal differences in the number of high-density grains per panicle were reported, and this trait appeared to be heritable (Venkateswarlu et al., 1986). It was suggested that rice grain yield could be increased by 30% if all the spikelets of an 8 t ha^{-1} crop were high-density grains (Venkateswarlu et al., 1986). However, source limitation and regulation of assimilate allocation within the panicle make it difficult to achieve. Iwasaki et al. (1992) found that superior spikelets are the first to accumulate dry matter and N during grain filling, while inferior spikelets do not begin to fill until the dry weight accumulation in superior spikelets is nearly finished. This apical dominance within the panicle was immediately altered upon the removal of superior spikelets. It is unknown if overall grain filling can be improved by weakening this apical dominance.

2.3.3 Harvest Index

Comparisons between semidwarf and traditional rice cultivars attributed improvement in yield potential to the increase in HI rather than to biomass production (Takeda et al., 1983; Evans et al., 1984). When comparisons were made among the improved semidwarf cultivars, however, high yield was achieved by increasing biomass production (Akita, 1989; Amano et al., 1993). Hybrid rices have about 15% higher yield than inbreds mainly due to an increase in biomass production rather than in HI (Song et al., 1990). This suggests that further improvement in rice yield potential might come from increased biomass production rather than increased HI.

Harvest index decreases with increased growth duration (Vergara et al., 1966). Akita (1989) reported that HI decreased from 55% to 35% as growth duration increased from 95 to 135 days. For a specific environment, there is an optimum growth duration giving high grain yields and high HI. Varieties with an HI of 0.55 or more have growth duration between 100 and 130 days (Vergara and Visperas, 1977). Chandler (1969) reported that the mean HI of N-responsive varieties was 0.53, while varieties that respond poorly to added N had HI values of 0.36. Year-round monthly planting experiments with different varieties demonstrate that HI is higher in the dry season and lower in the wet season with a range from 0.44 to 0.58 for improved varieties and 0.12 to 0.48 for traditional varieties (Vergara and Visperas, 1977). Clearly, there are limits to how far HI can be further increased in improved varieties. Austin et al. (1980) estimated that the maximum possible HI in wheat is 0.63. If the same was true for rice, the present yield potential of 10 t ha^{-1} with an HI of 0.53 would increase to 12 t ha^{-1} assuming an HI of 0.63 for a new rice plant type (Evans, 1990).

2.4 Lodging Resistance

It is impossible to further increase yield potential of irrigated rice without improving its lodging resistance. The types of lodging are bending or breakage of the shoot and root upheaval (Setter et al., 1994a,b). Lodging reduces grain yield by reducing canopy photosynthesis, increasing respiration, reducing translocation of nutrients and carbon for grain filling, and increasing susceptibility to pests and diseases (Hitaka, 1969). The magnitude of damage from lodging depends on the degree of lodging and the time when it occurs. Lodging results from the interaction and the balance of three forces: straw strength, environmental factors affecting straw strength, and the impact of external forces such as wind and rain (Setter et al., 1994a,b). Excessive supply of N, deficiencies of K, Si, and Ca, low solar radiation, and diseases affecting the leaves, sheaths, and culm reduce straw strength (Chang and Loresto, 1985). Leaf sheath wrapping, basal internode length, and the cross-sectional area of the culm are the major plant traits that determine straw strength (Chang and Vergara, 1972). Before the start of internode elongation, the leaf sheaths support the whole plant. Even after the completion of internode elongation, the leaf sheaths contribute to the breaking strength of the shoot by 30–60% (Chang, 1964). Therefore the sheath biomass and extent of wrapping will always be an important trait for selection against lodging at all developmental stages (Setter et al., 1994a,b). Ookawa and Ishihara (1992) reported that the breaking strength of the basal internode was doubled due to leaf sheath covering and was tripled due to the large area of the basal internode cross section.

Terashima et al. (1995) found that greater root mass and root number distributed in the subsoil (where soil bulk density is high) were associated with increased resistance to root lodging in direct-seeded rice. Further reduction in stem height of present semidwarf varieties is not a good approach to increase lodging resistance because it will cause a reduction in biomass production. Lowering the height of the panicle could have a profound effect on increasing lodging tolerance because of the reduction in the height of the center of gravity of the shoot (Setter et al., 1995a,b). Ookawa et al. (1993) found that the densities of lignin, glucose, and xylose in the cell wall materials of the fifth internode of different rice varieties grown under different conditions were associated with stem strength.

3 IMPROVING YIELD STABILITY IN LESS FAVORABLE ENVIRONMENTS

Rainfed rice-growing environments constitute virtually half of the total rice-growing areas of the world. Based on hydrology, rainfed ecosystems are classified into rainfed lowland, upland, deepwater, and tidal wetland ecosystems (IRRI, 1984). Growing conditions of these environments are considered less favorable compared to irrigated ecosystem due to the complexity of biotic and abiotic constraints prevailing in these environments and the uncertainties associated with rainfall patterns. Farmers often apply minimum levels of inputs to reduce risks encountered from crop failures with the effect of lower yields. Yield gains associated with growing improved varieties are much less notable compared to favorable environments and improved varieties are adopted at much slower rate. As a consequence, average yield is only about 2 t ha^{-1} compared to more than 5 t ha^{-1} under irrigated lowland environments.

Superior traits that allow traditional cultivars to survive and produce well under such extreme conditions need to be incorporated into modern cultivars. This will require a methodical understanding of the genetics and physiology of such traits together with careful evaluation of the target environments to select for pertinent traits. Research over the past 25 years uncovered many potentially useful traits among cultivated and wild rice germ plasm, making genetic improvement a viable strategy. We will attempt to highlight some of the progress made on major abiotic stresses common to rice production environments.

3.1 Drought

Drought has long been recognized as the primary constraint and a serious yield-limiting factor for production and yield stability of rice under rainfed conditions. Erratic rainfall and high evaporative demands supersede the benefit of the high annual rainfall totals common to most of these areas. Adaptation of rice to drought conditions is complex and unique compared to most other crops. This complexity arises from the requirements to adapt to extremely different hydrological conditions ranging from submergence to drought to vacillation between the two extremes during the growing season (Nguyen et al., 1997). Improving drought tolerance in rice requires a meticulous understanding of its physiology and careful characterization of target environments. Recent research efforts are reviewed in greater details elsewhere (Fukai and Cooper, 1995; Mackill et al., 1996; Nguyen et al., 1997; Ito et al., 1999).

Drought stress affects rice plant directly and indirectly. The direct effects include reduction of growth rate and tiller number, delayed flowering, and even whole plant death when drought is severe. The major indirect effects include reduction in nutrient uptake, changes in nutrient balance, and weed competition. As soil dries, it becomes more aerobic, which will affect crop performance through changes in nutrient availability due to disruption of the aqueous pathway of ionic movement as well as changes in soil redox conditions. Changes in soil redox potential are known to alter the availability of some nutrients such as phosphorus and silicon and directly or indirectly affect availability of others (Ponnamperuma, 1977). When water is scarce, many weed species become more competitive than rice due to ability of some weeds to maintain higher water potential under drought and grow faster than rice, which will hasten the depletion of stored soil moisture (O'Toole and Chang, 1978; Mackill et al., 1996).

Drought stress is less damaging during the seedling stage than the reproductive stage. Young seedlings can recover much better upon relief of stress (O'Toole and Chang, 1979), although reduction in yield may be anticipated if leaf area and tiller numbers are drastically reduced. Drought during vegetative stage could alter rice phenology by delaying panicle initiation and flowering (Turner et al., 1986a; Lilley and Fukai, 1994). During reproductive phase, drought stress causes desiccation of spikelets and anthers, reduction in pollen shedding, inhibition of panicle exsertion, and increase in sterility (O'Toole and Namuco, 1983; Ekanayake et al., 1989).

The diversity of adapted germ plasm to different rice-growing environments supports the notion that numerous adaptive mechanisms may be found in rice that enable the crop to cope with water deficit under these different climatic conditions (O'Toole and Chang, 1978). The pertinence of these mechanisms will depend on the different environmental conditions unique to each area.

Broadly, mechanisms of drought adaptation can be classified as those that permit plants to escape drought and those that help in resisting drought. Drought resistance mechanisms are those that enable cultivars to produce greater economic yield when subjected to soil or atmospheric drought (Hall, 2000). They could further be subdivided into mechanisms that aid plants to avoid dehydration and those that are involved in dehydration tolerance. These mechanisms will be discussed with reference to rice.

3.1.1 Drought Escape

Hall (2000) defined drought escape as "where drought-sensitive stages of plant development are completed during part of the season when drought is not present." For rice, drought escape is probably one of the most effective adaptive mechanisms to ensure productivity. In areas where rainy season is short, plants that have short growth duration may complete their reproductive stage before the onset of severe drought stress. One commonly used method to select for drought escape is to opt for varieties with life cycle short

enough to fit into duration of sufficient rainfall. One major problem anticipated with this approach is the low yield potential and the inadequate
plasticity to cope with mild or intermittent drought.

Photoperiod sensitivity is another method of drought escape especially
for lowland rice, when flowering occurs on certain calendar dates regardless of
the sowing or transplanting date. The sensitive reproductive stages are photo-
periodically controlled to coincide with the period of ample rainfall, allowing
the crop to complete its grain filling before the onset of severe water stress.
This is particularly useful in locations where rainfall distribution follows a
bimodal pattern or where the monsoon rainy season ends sharply (Vergara
and Chang, 1976; O'Toole and Chang, 1978). Most traditional rainfed
lowland rice cultivars are sensitive to photoperiod (Mackill et al., 1996).

3.1.2 Dehydration Avoidance

Dehydration avoidance mechanisms enable plants to maintain high water
potential and avoid the damaging effect of stress. By and large, these
mechanisms entail the maximization of water uptake from the soil and control
of water loss to the atmosphere. The adaptive mechanisms that enhance water
uptake under drought are related to rooting characteristics that maximize
root–soil interface. Rooting characteristics that are adaptive to a particular
habitat will depend on the physical characteristics of the soil, such as depth
and hydraulic conductivity, as well as the availability of water at depth.

Rooting characteristics are particularly important under upland conditions where rice is grown aerobically in deep soils with extreme diversity and
topography. Water stress development is generally more severe than under
lowland conditions. Under upland conditions, differences in rooting characteristics are related to drought adaptation with cultivars having longer,
thicker, or denser roots being more tolerant to drought (Ekanayake et al.,
1985b). Upland rice cultivars generally outdo lowland cultivars in root traits
associated with drought avoidance, but they are not adapted to transplanted,
anaerobic lowland conditions.

Lowland rice varieties are especially vulnerable to drought stress
because of their shallow root system, although drought stress usually develops more slowly. Most roots are confined to the top layer of the soil in areas
where compacted soil or hardpan developed as a result of soil puddling.
Hardpans improve retention of surface water but hinder root penetration to
reach moisture in deeper soil. Improving root penetration through hardpans
or compacted layers may substantially improve drought adaptation of lowland rice. Genotypic variation in root penetration ability has been found
among rice cultivars (Yu et al., 1995), which could further be exploited in
breeding. The ability of rice roots to reach deep soil or to penetrate hard
compacted soil layers is associated with the ability to develop few thick and

long roots (Yoshida and Hasegawa, 1982; Yu et al., 1995; Nguyen et al., 1997).

Deep rooting characteristics as measured by ratio of deep roots to shoot biomass showed a good association with drought resistance in upland rice (O'Toole and Chang, 1978), which suggests that cultivars with higher deep root to shoot ratio are more capable of extracting water stored deep in the soil. The association between deep root/shoot ratio and drought tolerance may suggest that field screening for deeper roots may successfully and indirectly be carried out based on visual scoring for drought tolerance. Thicker roots tend to have larger diameter xylem vessels (Yambao et al., 1992), which are expected to have higher hydraulic conductivity and allow more water to be delivered to shoots. Depth of the root system is associated with tiller number and not with plant height, with early tillers having longer and thicker roots than late tillers (Yoshida and Hasegawa, 1982). Thus deeper roots can be combined with the short stature of the modern rice varieties (Mackill et al., 1996). Plants with larger number of tillers tend to have more late tillers and shallower root system.

Root-pulling force (RPF) has been used to study variations in rooting depth under lowland conditions (O'Toole and Soemartono, 1981). Genotypes with high RPF are characterized by larger, thicker, and denser root system. The RPF is also dependent on the root length density of the portion of the root system that remains in the soil (Ekanayake et al., 1986) and is positively correlated with drought scores under upland conditions (Ekanayake et al., 1985a). This suggests that high RPF is related to the ability of plants to develop deeper and thicker roots with greater penetration ability.

Root length, thickness, dry weight, and root length density are polygenic traits that are moderately to highly heritable (Armenta-Soto et al., 1983; Ekanayake et al., 1985b); hence selection for these traits in early segregating generations may be successful. Xylem vessels cross-sectional area seems to be controlled by few genes with additive effects and with moderate broad-sense heritability. However, the heritability of root-pulling force was relatively low (Ekanayake et al., 1985a).

Considerable potential for improving drought resistance in rice is present, based on selection for root traits. The greater limitation to this approach is the laborious nature of the screening methods, which makes it extremely difficult to measure any specific root phenotype for selection in field nurseries. An alternative approach is the use of molecular tagging of the relevant traits. Many of these traits had been recently mapped using populations developed from crosses between japonica × indica lines, and the association of most of them with drought tolerance has been confirmed (Champoux et al., 1995; Ray et al., 1996; Yadav et al., 1997). Tagging of these traits may speed breeding for drought tolerance using marker-assisted selection approach.

Adaptive mechanisms of the shoot also play an important role in dehydration avoidance in rice such as increased stomatal sensitivity, well-developed cuticle, and leaf rolling. Stomatal conductance has been investigated as a tool for determining drought resistance of rice cultivars. Comparison between upland and lowland rice varieties showed that upland varieties generally had higher stomatal resistance than lowland varieties under drought (IRRI, 1975). However, stomatal behavior is more complex due to its dynamic nature and the intricacy of environmental factors that contribute to drought which make the development of repeatable screening procedures more difficult.

Predawn leaf water potential is a good indicator of the capability of genotypes to rehydrate during the dark period because leaf-water potential rises to its highest level at this time. This may also reflect genetic variation in root system development since higher leaf water potential may develop only in genotypes with roots in contact with soil moisture. Using visual scoring, genotypic variation in drought tolerance was found to correlate well with predawn leaf-water potential (O'Toole and Chang, 1978, 1979).

High cuticular resistance is often associated with drought adaptation in many plant species, and it relates to the amount of wax on the leaf surface. Under severe drought, epicuticular transpiration accounts for a greater proportion of total leaf water loss than does stomatal transpiration. In rice, genotypic variation in cuticular resistance was found and with relatively higher values for upland cultivars (Yoshida and de los Reyes, 1976; O'Toole et al., 1979). Transfer of this trait from upland varieties could therefore improve drought tolerance of lowland cultivars. Unfortunately, rice has relatively much lower epicuticular wax compared to other upland crops as sorghum, corn, and wheat (Yoshida and de los Reyes, 1976; O'Toole et al., 1979). This trait is also difficult to measure and quantitatively inherited with moderate heritability (Haque et al., 1992), making its use in breeding for drought tolerance more challenging.

Leaf rolling is the most obvious symptom of drought and has been frequently identified both as a symptom of drought stress as well as an adaptive response to drought (Mackill et al., 1996). Leaf rolling occurs when turgor or leaf pressure potential decreased. It reduces the area of leaf surface exposed to the atmosphere, reducing heat load and increasing resistance to water loss. In this respect, leaf rolling could be adaptive if it occurs in response to soil or atmospheric drought and with ability to unfold when drought stress is partially relieved as in cooler evenings. Diurnal expression of leaf rolling may successfully be used to select for rooting characteristics because cultivars with deeper roots may be capable of rehydrating overnight (Chang et al., 1974). O'Toole and Chang (1978) observed an association between leaf rolling scores and leaf water potential for 16 upland cultivars, but with some genotypic variation, and suggested that leaf rolling is a good indicator of leaf water potential. Leaf rolling can reduce leaf gas exchange and photosynthesis, but improve water-use efficiency, the ratio of carbon assimilation to transpiration. Drought-resistant cultivars are found to be better able to maintain their water uptake and avoid leaf rolling for longer intervals than susceptible cultivars, and screening tests showed delayed leaf rolling to be associated with drought resistance (Singh and Mackill, 1991). Leaves of some upland cultivars were found to roll at higher water potentials and higher turgor pressures, at midday, than leaves of lowland cultivars (Turner et al., 1986a), suggesting variation in threshold turgor potential at which leaf rolling occurs and support the notion that leaf rolling may be adaptive if it occurs diurnally to reduce water loss under severe atmospheric drought. However, it should be noted that plants that permanently roll their leaves at higher leaf water potential might have reduced yields during periods of moderate water stress (Dingkuhn et al., 1989). It is probably clear that caution must be taken when using leaf rolling as an index for drought adaptation.

3.1.3 Dehydration Tolerance

Dehydration tolerance refers to the extent to which plants maintain their metabolic function when leaf water potential is low. Mechanisms of dehydration tolerance in plants are poorly understood (Hall, 2000). Osmotic adjustment is the trait that is mostly studied and considered associated with dehydration tolerance. Other traits such as extent of assimilate translocation and accumulation of protective metabolites as sugars and proteins during drought stress have also been suggested, and further studies are needed to confirm their involvement in dehydration tolerance in rice.

Osmotic adjustment involves accumulation of solutes in response to water stress, leading to the maintenance of turgor potential and continued functioning (Morgan, 1984). It may also enhance water uptake and enable root expansion into deeper moist soil profiles. Several compatible solutes are known to accumulate in response to water deficit in different plant species such as sugars, proteins, organic acids, amino acids, sugar alcohols, or ions, most commonly K^+. Maintenance of turgor pressure at relatively high level, despite reduction in leaf water potential, results in continuance of metabolic processes that are sensitive to reduction in cell turgor such as cellular expansion.

Osmotic adjustment is probably more important for lowland varieties where short intermittent drought periods are encountered. Genetic variation in osmotic adjustment was observed in rice. Turner et al. (1986b) observed greater variation in osmotic adjustment among lowland cultivars than in upland cultivars and with maximum variation of about 0.5 MPa. Lilley and Ludlow (1996) showed a larger genotypic variation in osmotic adjustment among 61 rice cultivars with a range of 0.4 to 1.5 MPa. They concluded that there is a good potential for improving dehydration tolerance and osmotic adjustment of the current rice cultivars. Indica lines have greater osmotic adjustment under drought stress than lines with japonica background.

Osmotic adjustment was reported to delay leaf rolling and leaf death in rice (Hsiao et al., 1984), but to date, there is no evidence for the association between genetic variation in osmotic adjustment with growth and grain yield. However, the positive role of osmotic adjustment on growth and yield of some other cereals such as sorghum (Tangpremsri et al., 1995) and wheat (Morgan and Condon, 1986) has been established. Research is needed to scrutinize the usefulness of osmotic adjustment in rice under field conditions before it can successfully be used as selection criteria for breeding. Osmotic adjustment in rice develops quickly and the maximum adjustment is maintained during drought periods; thus it may be effective in buffering against deleterious effects of mild, intermittent water stress (Fukai and Cooper, 1995).

A single quantitative trait locus (QTL) with major effect on osmotic adjustment was observed in a rice recombinant inbred population developed from a cross between an upland japonica×lowland indica cultivars (Lilley et al., 1996). The authors postulated that this major QTL might be homologous with a single recessive gene previously identified for the same trait in wheat by Morgan (1991). The effect of this QTL was negatively associated with rooting characteristics related to dehydration avoidance. Linkage between these traits may need to be broken in order to combine high osmotic adjustment with extensive root system (Lilley et al., 1996).

3.2 Salinity

Rice is a salt-sensitive crop with tolerance threshold of about 3 dS m^{-1}; beyond this threshold, yield start to decline at a rate of about 12% per unit increase in salinity (Maas, 1986). However, due to its ability to grow well under flooded conditions, rice is recommended as a desalinization crop because the standing water in rice fields can help leach the salts from the topsoil to a level low enough for subsequent crops (Bhumbla and Abrol, 1978). Despite its high sensitivity to salinity, considerable variation in salinity resistance is found in rice (Akbar et al., 1972; Flowers and Yeo, 1981).

Sensitivity to salinity in rice varies with the stage of development being tolerant during germination, active tillering, and maturity and sensitive during early seedling stage, panicle initiation, pollination, and fertilization (Pearson and Bernstein, 1959; Akbar et al., 1972). Moreover, environmental factors such as light intensity, humidity, and temperature can impose a dramatic effect on rice susceptibility to salinity. Damage to plants could result from water deficit due to low osmotic potential imposed externally, or internally, when uptake exceeds the need for osmotic adjustment, resulting in high inorganic salts in the intercellular spaces of leaves. Ion toxicity could result from undue ion entry in excess of appropriate compartmentation. Salinity can also affect nutrient balance of the soil; for example, displacement of potassium, an essential element, with sodium, which is chemically similar but physiologically functionless, can cause nutritional stress. Fageria (1985) observed a decline in concentrations of P and K in the shoot of two rice cultivars with increasing salinity. Application of K improved photosynthesis and yield and reduced the concentration of sodium in the straw (Bohra and Doerffling, 1993). Moderate levels of P were also found to improve both grain and straw yield of rice in salt-affected soils but with P toxicity occurring at lower P levels than under nonsaline conditions (Aslam et al., 1996). It is therefore discernible that there is likelihood for correcting nutritional imbalances encountered from soil salinity. Further research is needed to uncover such imbalances and enhance its management. Genetic variability in responses to these nutrient imbalances is also observed (Aslam et al., 1996) and could be explored.

Numerous abnormalities were noted in rice due to salt injury as stunted growth, rolling of leaves, white leaf tips, white blotches in leaves, drying of older leaves, poor root growth, reduced survival, and spikelet sterility (Ponnamperuma and Bandyopadhya, 1980). Selection against one or few of these traits may not be effective in breeding. Yeo and Flowers (1986) reported similar reduction in relative growth rate of both tolerant and intolerant cultivars under salinity and concluded that reduction in relative growth rate is not related to subsequent mortality, neither is it a reliable indicator of resistance.

Salinity tolerance in rice is conferred by the sum of a number of contributing traits, the most important of which are seedling vigor, salt exclusion, preferential compartmentation of Na^+ in older leaves, and tissue tolerance (Yeo and Flowers, 1986).

3.2.1 Seedling Vigor

Rice cultivars differ substantially in their growth rate with the most vigorous lines being the traditional varieties. Dwarfing genes were incorporated in most of the modern varieties and breeding lines to increase harvest index and reduce lodging. Naturally occurring salt-resistant varieties invariably belong to these traditional tall varieties. The high vigor of landraces may enable them to tolerate growth reduction. Vigorous growth may also have a dilution effect; one tall, salt-tolerant landrace had the same net transport of Na^+ through its roots as a semidwarf susceptible line, but had much lower shoot Na^+ concentration (Yeo and Flowers, 1984a, 1986). Differences in vigor among rice cultivars accounted for much of the variation in their survival of salinity (Yeo et al., 1990). However, vigorous growth is a notorious trait and has been selected against in developing modern varieties because of the low yield potential associated with tallness and susceptibility to lodging with the intensive input system currently practiced under irrigated conditions. Therefore overall vigor may have limited usefulness as a physiological trait for improving salinity tolerance in rice. However, early seedling vigor is desirable due to high sensitivity during this stage coupled with the high salinity levels normally encountered at the beginning of the season.

3.2.2 Salt Exclusion

Numerous reports suggested that rice is relatively ineffective in controlling influx of sodium and chloride ions to the shoot at high external ion concentrations and has a high tissue salt concentration even at moderate external salinities (Flowers and Yeo, 1981; Yeo and Flowers, 1984a). Yeo and Flowers (1986) concluded that sodium uptake is not regulated in rice and seems to occur passively rather than carrier-mediated. This is because the concentration of sodium in leaves tends to increase with time, saturating only at leaf death. Root membrane selectivity was not improved by calcium normally associated with membrane stability (Yeo and Flowers, 1985). The authors further hypothesized that the passive leakage could be via lipid bilayer of the plasma membrane, through transitory pores or ion channels, or could be via apoplastic contact to the stele at root apices before the formation of the casparian strip and where lateral roots cause local disruption of the endosperms, causing a proportion of the transpiration stream to reach the xylem without crossing a membrane (Yeo and Flowers, 1984a). On the other hand, Akita and Cabuslay (1990) reported that sodium uptake into rice plant is dependent both on passive processes, determined by leaf area, and active processes determined by root selectivity.

The specific properties of root membranes that permit such variability in net Na^+ transport are still unknown. Addition of low concentration of polyethylene glycole (PEG-1540) was found to improve survival and decrease Na^+ flux to the shoot (Yeo and Flowers, 1984b). The authors postulated that this is probably due to multiple attachments between PEG and charged sites on plasma membrane which would stabilize it in a manner analogous to divalent metal ions, reducing passive leakage via membranes. PEG also increases the osmotic potential of the external solution and may influence the overall transpiration rate and, consequently, salt accumulation in plant tissue. The role of the elaborate aerenchyma developed in rice roots in this passive bypass flow of ions is still unknown and worth further investigation. Transport of sodium in rice under salinity is most likely controlled by genes affecting root anatomy and development, rather than specific carriers or ion channels.

Substantial genetic variability in the rate of sodium uptake by rice roots was reported (e.g., Yeo and Flowers, 1983, 1986; Yeo et al., 1990), signifying a sizable potential for genetic improvement. For a set of rice cultivars with diverse origin, a strong negative relationship between leaf sodium concentration and survival was found (Yeo and Flowers, 1983), but when a larger number of genotypes were investigated, sodium uptake was found to account for only a small proportion of the total variation in survival (Yeo et al., 1990), suggesting the importance of other traits in salinity tolerance besides ion uptake.

3.2.3 Leaf-to-Leaf Compartmentation

Under saline conditions, rice plants maintain a gradient in sodium concentration from leaf to leaf with higher concentrations of salt in older leaves. Sodium content rose rapidly in older leaves upon exposure to salinity with a distinct lag before it rose in younger leaves. This leaf-to-leaf gradient did not develop with time but arose rapidly and maintained with time with no evident retranslocation. Older leaves reach lethal ion concentrations and were lost, while new ones are initiated (Yeo and Flowers, 1982). The ability of rice cultivars to compartmentalize ions in older leaves could crucially affect plant survival. Maintaining younger leaves at low salt concentrations probably contributes to the ability of certain varieties to survive saline conditions if they maintain their rates of leaf initiation at least equal to rates of leaf death. Rice cultivars differ in their ability to maintain the younger leaves at low sodium concentrations, and variability in leaf-to-leaf compartmentation of up to fivefold was reported (Yeo et al., 1990). Selection of plants on the bases of shoot appearance and/or whole shoot sodium content may not always reflect resistance in genotypes where this mechanism plays a major role in their salt tolerance. Mechanism by which this compartmentation is accomplished and sustained is unknown.

3.2.4 Tissue Tolerance

A cellular component of salt tolerance has been discovered in rice. Similar concentration of salt in leaves is found to cause different degrees of toxicity in different varieties (Yeo and Flowers, 1983, 1986; Yeo et al., 1990) and is termed tissue tolerance. Tissue tolerance is commonly measured as the concentration of Na^+ in leaf tissue that causes 50% loss of chlorophyll. Substantial differences were observed between rice cultivars with sodium concentration causing 50% loss of chlorophyll ranging from 135 to 500 mol m^{-3} and is interpreted as reflecting differences in apoplastic salt load (Yeo and Flowers, 1983; Yeo et al., 1990). Varieties that had high tissue tolerance also maintain the ultrastructure of their cells and had higher net photosynthesis at higher levels of tissue salt concentration than sensitive ones (Flowers et al., 1985), providing further evidence for a cellular component of salt tolerance in rice. Nonetheless, varieties with the greatest tolerance to sodium in their leaf tissues were not necessarily those with the greatest salinity tolerance as measured by survival rate. No correlation between tissue tolerance and survival was found for a large set of diverse rice cultivars and with inverse association between tissue tolerance and vigor. The tall traditional varieties have leaf tissue that is sensitive, while the most tissue-tolerant lines are from the dwarf modern cultivars (Yeo et al., 1990). Based on overall survival normally followed in mass screening, a variety could be rejected because its tissue tolerance alone is not enough to dominate its overall performance under saline conditions, while it could be an important source of genes for tissue tolerance. Contribution to the tolerance of these lines from such trait may be masked by their poor response in other characteristics such as excess salt entry and may require independent assessment (Yeo and Flowers, 1983).

To date, physiological traits associated with salinity tolerance have not been found favorably combined in any exact genotype, and all the known salt-tolerant cultivars have either one or few of these traits. Useful genetic variation is present for each character (Yeo et al., 1990); thus salt tolerance of rice can be improved beyond the present phenotypic range by use of physiological criteria to select independently for individual contributing traits, which may subsequently be combined.

3.2.5 Physiological Approach for Improving Tolerance to Salinity

Physiological bases of salinity tolerance are possibly the most studied and well understood of all abiotic stresses affecting rice and may provide a classic example for breeding based on physiological criteria. Yeo and Flowers (1984a) summarized the value of physiological approaches to breeding for salinity tolerance in three major points. First, the use of physiological criteria may simplify mass screening by virtue of being more rapid and objective. Second, physiological parameters may provide more information about the system. Third, the major role of physiology comes in the ability to combine several traits (building blocks), which do not, on their own, lead to improved survival in saline conditions. Combining physiological traits causative to overall salinity resistance in rice is therefore a logically desirable long-term ambition. Physiological methods can be used to identify potential donors for various traits, which can then be combined by conventional breeding.

The use of physiological criteria in breeding, however, is restricted by some important constraints. Many individuals are needed to obtain a single assessment, making selection in early generations virtually impossible. Second, screening methods can logically be used to select parental lines but cannot generally be used to select individuals from a large number of plants as the case in early segregating populations. Third, the destructive nature of many of the physiological assays made them impractical for field screening especially in situations where the whole plant needs to be sampled. A final

constraint is that in order to differentiate between stress-tolerant and susceptible phenotypes, it is often necessary to expose plants to salt concentrations
that are lethal to the sensitive individuals, as the expression of the resistant
genotype requires high salt concentration. This procedure may not be
applicable when individual plants are to be chosen and also may result in
selecting individuals with only one or few mechanisms with strong expression
(Flowers et al., 2000). Evaluation and selection may be feasible only in
advanced generations.

Developing strategies that use molecular markers to trace putative
quantitative trait loci (QTL) presumed to underlie physiological phenotype
might surmount these problems. Although the initial mapping of QTLs is a
prolonged process, once markers are generated they can be used swiftly and
plausibly on various populations using marker aided or assisted selection.
Molecular markers have the potential to indicate unequivocally the genotype
of a single plant, and the procedure is nondestructive because only small
amount of tissue is needed. Besides the information sought is genotypic, and it
is not important to expose the plant to stress, as is the case in assessing the
physiological phenotype. The use of DNA-based marker technology is
becoming routine and capable of dealing with large numbers of samples.

Few attempts have been carried out to identify QTLs associated with
salinity tolerance in rice. Using a double haploid population derived from a
cross between a moderately salt-tolerant indica cultivar IR64, and a susceptible japonica cultivar, Azucena, Prasad et al. (2000) identified seven QTLs for
seedling traits associated with salt stress and were mapped to five different
chromosomes. Zhang et al. (1995) mapped a major gene for salt tolerance on
chromosome 7 using F_2 population derived from a cross between a salt-
tolerant japonica rice mutant, M-20, and the sensitive original variety 77-170.
Using the same population, a RAPD marker was also identified that links to
the salt tolerance gene (Ding et al., 1998). Zhang et al. (1999) demonstrated
that allelic variation in one copy of a small family of H^+ ATPase genes from
variety 77-170 correlated with a QTL locus for salt tolerance located on
chromosome 12. Transcripts of this gene were found to accumulate in roots of
the salt-tolerant mutant; M-20 and the authors interpreted this as indicative
of its active role in restricting salt uptake into roots.

Koyama et al. (2001) identified and mapped QTLs associated with
different mechanisms of salinity tolerance in rice. They were able to describe
genetic determinants of the net quantity of ions transported to the shoot by
clearly distinguishing between QTLs associated with ion uptake, which affect
the quantity of ions in the shoot, from those that affect the overall vigor and
hence the concentration of ions in the shoot. The QTLs identified independently govern the uptake of Na and K and Na:K selectivity. Quantitative trait
loci for Na and K uptake are on different chromosomes supporting the earlier

reports of their independent inheritance (Garcia et al., 1997) and the mechanistically different pathways suggested before (Yeo and Flowers, 1986; Yadav et al., 1996). In a concomitant study using related populations of recombinant inbred lines (Flowers et al., 2000), none of the markers associated with the said QTLs showed any association with similar traits, which caution against any expectations of a general applicability of markers for physiological traits. The authors concluded that direct knowledge of the genes involved is needed. Evidently, more studies are needed to help tag the major component traits contributing to salinity tolerance to facilitate pyramiding the traits involved and shorten the breeding cycle.

3.3 Submergence

Rice is the only crop plant adapted to aquatic environments and can grow well under waterlogged conditions. This adaptation arises from the well-developed aerenchyma tissues that facilitate oxygen diffusion through continuous air spaces from shoot to root and avoid anoxia development in roots. Although rice is well adapted to waterlogged conditions, complete submergence due to frequent flooding can adversely affect plant growth and yield. More than 16% of rice lands of the world in lowland and deepwater rice areas are unfavorably affected by flooding due to complete submergence (Khush, 1984).

Two types of flooding cause damages to rice. The first type is flash flooding, which results in rapid ascending of water levels with complete submergence for 7 to 14 days. Plants adapt to these environments essentially through maintenance processes that provide necessary energy supply for survival and minimize energy losses. Shoot elongation under flash flood conditions is disadvantageous because of energy loss, and the taller plants tend to lodge once the water level recedes.

The second type is deepwater and floating rice where water depth could exceed 100 cm and stagnate for several months. Plants may become completely submerged for short periods if flooding is severe. Elongation ability of leaves and internodes under these conditions is essential to keep pace with the increasing water levels and escape complete submergence. This will ensure O_2 supply to roots via the continuum of aerenchyma tissue to avoid anoxia and gain access to CO_2 and light and maintain energy supply (Setter et al., 1997). Traditional varieties adapted to these environments are low-yielding due to their low-tillering ability, long droopy leaves, susceptibility to lodging, and poor grain quality (Mallik et al., 1995). Improved varieties are needed that combine yield attributes with submergence tolerance and elongation ability.

Elongation ability of the coleoptiles of germinating seeds is also considered a desirable trait especially with direct seeding to effect emergence above anaerobic waterlogged soils. Variability in the ability of coleoptiles to elongate under anoxia was observed in rice and was related to the rate of alcoholic fermentation (Setter et al., 1994a,b).

3.3.1 Effects of the Floodwater Environment

Characterization of floodwater environments in most rice-growing areas pointed to gas diffusion as the most important limiting factor under complete submergence (Setter et al., 1997; Ram et al., 2002). This is because gas diffusion is 10^4-fold less in water than in air (Armstrong, 1979). The importance of gas diffusion during submergence is clearly demonstrated in many experiments. When submerged rice is flushed with air at high partial pressure of CO_2, plants can survive for up to 3 months of complete submergence compared with only 1 to 2 weeks when submergence water is in equilibrium with air (Setter et al., 1989). Oxygen levels in floodwater vary with location and time of day, usually below air saturation during the night but may become supersaturated during the day. Anoxia for 24 hr can kill sensitive rice varieties (Crawford, 1989) presumably because of the need for O_2 for respiration to maintain survival and elongation growth processes.

One important component of survival during submergence is maintenance of carbon assimilation to supply needed energy for maintenance and growth process. Carbon assimilation is influenced by several factors during submergence, including capacity for underwater photosynthesis, CO_2 supply, irradiance, and temperature of the floodwater. CO_2 supply may be limiting due to both its lower level especially in stagnant water and the large boundary layer effects (Smith and Walker, 1980). Carbon assimilation of submerged rice plants could also be reduced by low irradiance due to water turbidity or growth of surface algal flocks (Setter et al., 1995a,b, 1997). Reduction of photosynthesis due to these constraints may impede carbohydrate supply needed for respiration or alcoholic fermentation.

3.3.2 Mechanisms of Submergence Tolerance

Under flash flooding, few characters were identified as playing a key role in submergence tolerance in rice, the most critical of which are the following: maintenance of high carbohydrate concentration, high rates of alcoholic fermentation, and energy conservation by restraining elongation growth during submergence (Setter et al., 1997).

Stem Carbohydrates. Carbohydrate concentration before and during submergence has long been recognized as an important factor in submergence tolerance in rice. For example, using 30-day-old seedlings, Mallik et al. (1995) found a strong positive correlation between carbohydrate concentration prior to submergence and tolerance to submergence ($R = 0.95$). Studies using techniques that alter the concentration of carbohydrates before or during submergence support the significant role carbohydrates play during submergence, e.g., (a) reduction of light intensity by shading, increasing water turbidity or water depth reduced survival (Palada and Vergara, 1972), (b) time of day submergence which affects the diurnal cycle of carbohydrate concentration (Ram et al., 2002), (c) CO_2 supply which affects underwater photosynthesis (Setter et al., 1989), (d) seed size; high correlation observed between carbohydrate level and anoxia tolerance of rice seeds (IRRI, 1996), and (e) seedling age as older seedlings have higher carbohydrate levels and better survival (Chaturvedi et al., 1995). Studies are needed to test for genetic variability in underwater carbon assimilation, as this will have a great impact on maintenance of carbohydrates in shoots and plant survival. Genetic variability in stem carbohydrate content is present among rice cultivars but is greatly influenced by growth conditions before submergence. Selection of genotypes with high stem carbohydrates could provide an increment of submergence tolerance.

Alcoholic Fermentation.   Alcoholic fermentation (AF) is one of the major metabolic adaptations that plants assume when they are submerged or faced with lack of oxygen. Under anaerobic conditions, aerobic respiration shifts to a less-efficient anaerobic fermentation to provide energy and sustain plant life. Although the amount of ATP produced by this process is very small ($\sim 5\%$) compared to ATP produced through aerobic respiration, it is still vital for survival. Efficiency of AF pathway depends on the supply of substrate carbohydrates and the activity of two key enzymes: pyruvate decarboxylase (PDC), which decarboxylates pyruvate to acetaldehyde, and alcohol dehydrogenase (ADH), which then reduces acetaldehyde to ethanol. Acetaldehyde is very toxic to plants and its reduction to ethanol by ADH regenerates NAD^+ needed to maintain glycolysis and substrate level phosphorylation under anaerobic conditions (Davies, 1980).

For rice, this pathway is probably important both under anoxic as well as aerated floodwater where boundary layer limits diffusion of O_2 in water and may result in anoxic conditions within tissues. The importance of increased rates of AF during anoxia for plant survival has been demonstrated by several experimental observations (Setter et al., 1997; Ram et al., 2002). Examples of these are the following: (a) enzymes of AF often increase under flooding, (b) hypoxia pretreatment increased tolerance to anoxia and presumably the induction of AF enzymes, (c) mutants lacking ADH die more quickly under anoxia, (d) rates of AF are related to the tolerance of several species to flooding or water logging, and (e) high sugar supply improved survival, presumably due to continued functioning of AF. Besides, recent studies with rice plants overexpressing PDC gene (*pdc1*) further confirmed the importance of AF during submergence. Transgenic lines showed higher activities of PDC enzyme, higher ethanol production, and better survival after submergence (Quimio et al., 2000).

The role of AF end products in plant injury during anoxia (especially ethanol) has been debated since the hypothesis of metabolic injury during anoxia was published (Crawford, 1978). In rice, recent studies using photo-acoustic spectroscopy for online measurements of ethanol, acetaldehyde, and CO_2 revealed that both submergence-tolerant (FR13A) and susceptible (CT6241) rice cultivars had almost similar rates of AF on exposure to 4 hr of anoxia. However, on switching to postanoxia, the intolerant cultivar produced greater amount of acetaldehyde and CO_2, twice as large as that of the tolerant cultivar (Zuckermann et al., 1997; Ram et al., 2002).

Stem Elongation. Under flash flooding, stem elongation is not enviable and limited stem elongation growth is found to be associated with cultivars' ability to survive flash flooding. This is probably due to energy conservation during flooding for maintenance and survival processes. A strong negative correlation between percent survival and elongation growth of 14-day-old seedlings was observed (IRRI, 1996). Submergence tolerance of sensitive cultivars was substantially improved when underwater elongation was minimized by the application of a gibberellin biosynthesis inhibitor, paclobutrazol. Addition of GA increased elongation and reduced survival of even submergence-tolerant lines. Similar observations were also made with a GA-deficient rice mutant (IRRI, 1996).

Aerenchyma Formation. Aerenchyma comprises gas-filled spaces within plant tissue and is considered an essential anatomical adaptive trait for survival under flooded conditions (Justin and Armstrong, 1987). Sufficient substantiation of its role in submergence tolerance was presented by several experiments in providing a diffusion path of low resistance for the transport of oxygen from shoot to roots in waterlogged soils as well as diffusion of volatile compounds produced in anaerobic soils and plant tissue during flooding (Vartapetian and Jackson, 1997; Visser et al., 1997). In rice, the formation of aerenchyma occurs both in roots and shoots to provide a continuum from root to shoot (Vartapetian and Jackson, 1997).

3.3.3 Postsubmergence Events

When the water level recedes after complete submergence, plants are subjected to both high light intensity and higher oxygen levels. Visual symptoms of injury are normally not evident immediately after desubmergence but develop progressively during postsubmergence (Gutteridge and Halliwell, 1990; Crawford, 1992). This postanoxic injury is caused by generation of reactive oxygen species and toxic oxidative products as acetaldehyde (Monk et al., 1989; Crawford, 1992). Oxygen is one possible source of active oxygen species. When O_2 gets reduced, one electron leaks out from the electron transfer system converting it into superoxide anion (O_2^-), which in turn produces more active O_2 species as hydrogen peroxide (H_2O_2) and hydroxyl radical (OH). These highly reactive oxygen species can oxidize unsaturated fatty acids in cellular membranes and intercellular organelles (Scandalios, 1993).

In a recent study with one tolerant and one intolerant rice cultivars (Kawano et al., 2002), levels of H_2O_2 and malondialdehyde (a product of lipid peroxidation) were lower in the tolerant cultivar both during and after submergence. This was associated with higher levels of reduced ascorbate antioxidant and activities of antioxidant enzymes superoxide dismutase and glutathione reductase, suggesting the involvement of active oxygen scavenging species in lowering the harmful effects associated with oxygen re-entry.

3.3.4   Genetics of Submergence Tolerance

Research results suggested both simple and quantitative inheritance for submergence tolerance. In one study, Suprihatno and Coffman (1981) reported the involvement of at least three genes and low to moderate broad sense heritability. In another study, analysis of segregating F_2 and backcross populations made between tolerant and intolerant lines suggested the involvement of one major dominant gene in the submergence tolerance of three tolerant lines, FR13A, BKNFR(76106-16-0-1-0), and Kurkaruppan. Crosses made between the three lines did not produce any susceptible genotypes suggesting that the three lines possess the same submergence tolerance gene at the same locus. However, the segregation pattern of populations derived from crosses between the above tolerant cultivars with another tolerant line, Goda Heenati, showed that this line did not have the same gene in the same locus relative to the other tolerant lines (Setter et al., 1997). Thus this line may offer some hope for pooling genes to improve submergence tolerance beyond the current level, but whether Goda Heenati possesses a new mechanism of tolerance warrants further investigation. In a set of double haploid population developed between a susceptible and a tolerant line and screened in Thailand under high irradiance in the field, a strong bimodal distribution was observed with equal number of individuals in the tolerant and intolerant categories suggesting the involvement of a single gene in submergence tolerance and agreed with the model discussed above. However, when the same population was screened under low irradiance in the Philippines, the results suggest multiple genes or a more complex inheritance (IRRI, 1995). These differences in response may be due to differences in environmental conditions. When plants are grown at high irradiance, carbohydrate levels would have been high and possibly eliminating its effect as a limiting factor.

Using a cross between an indica submergence-tolerant line (IR40931-26) and a susceptible japonica line (PI543851), Xu and Mackill (1996) mapped a major QTL associated with submergence tolerance, designated *Sub1*, to chromosome 9. This QTL accounted for 70% of the phenotypic variance of submergence tolerance in the population studied, which is extremely high for a QTL and again suggests the involvement of a major gene. The donor line for this QTL was derived from the most submergence-tolerant line, FR13A, and with similar level of tolerance (Mackill et al., 1993). In a subsequent study (Nandi et al., 1997), the importance of *Sub1* in submergence tolerance was confirmed and four additional QTLs were identified on four different chromosomes. Moreover, Xu et al. (2000) fine-mapped *Sub1* locus using a high-resolution map, paving the way for its positional cloning and use in marker-assisted selection. The map position of *Sub1* does not correspond to that of mapped enzymes associated with alcohol fermentation, such as the three forms of PDC identified in rice, which had been map to chromosomes 3, 5, and 7 (Huq et al., 1999), and ADH gene, which had been mapped to chromosome 11 by trisomic analysis (Ranjhan et al., 1988). It is noteworthy that one of the four QTLs identified by Nandi et al. (1997) also maps to chromosome 11 and in the vicinity of the genes *Adh1* and *Adh2* for ADH. Due to similarity in function and chromosomal position, the authors further postulated that this QTL might correspond to the *Adh* genes.

The dramatic effect of *Sub1* locus on what is essentially a quantitative trait implies that this locus is more likely a regulatory locus rather than being associated with a specific enzyme. It could either be a transcription factor or is involved in signal transduction in response to submergence stress (Setter et al., 1997; Xu et al., 2000). An explicit answer to these arguments will likely be evident after cloning the putative gene.

Under flash flooding, physiological findings of the enviable role of carbohydrates and alcoholic fermentation and the adverse effect of elongation growth offer a good opportunity for breeding to enhance submergence tolerance in rice. Improving our understanding of the physiological and molecular aspects of these mechanisms will help in improving submergence tolerance of modern rice cultivars.

3.4 Low Temperature

Low-temperature stress is a major problem for rice production particularly in temperate zones. Many countries are affected including P.R. China, Nepal, Korea, Japan, European countries, Australia, Iran, Bangladesh, and United States (California). Even in tropical and subtropical areas, low temperature is a major constraint to crop production for crops grown at high elevations or for dry-season crops grown at higher latitudes (Mackill et al., 1996). Both cool weather and cold irrigation water are damaging to rice, and losses of more than 50% in grain yield has been reported (Takita, 1994). The low-temperature threshold for rice is relatively high, and damage to reproductive structures has been frequently reported at temperatures as high as 18°C to 20°C.

Responses of rice to low temperature vary with the temperature pattern of the locality and also with the stage of development. Low temperature is particularly damaging to rice during germination and seedling establishment, active tillering, and during reproductive development. Damage by low temperature can occur even during maturity causing lower yield and reducing grain quality. During reproductive development, low-temperature injury is particularly noticeable for long growth duration varieties especially at higher latitudes.

3.4.1 Effect of Cold Temperature During Vegetative Stage

Chilling temperature at sowing reduces germination, delays emergence, decreases seedling vigor, and causes discoloration of leaves of the young seedlings. Stunted growth, reduced tiller number, and leaf discoloration are common symptoms when low temperature prevails during active tillering (Kaneda and Beachell, 1974; Chung et al., 1983). In northern Japan, low-temperature injury during the vegetative stage was minimized by using early maturing varieties to avoid effect on delayed growth and by establishing seedlings in warm enclosures before transplanting (Takita, 1994). Increasing water temperature (Satake, 1986) or depth of water (Koike, 1991) in the paddy fields a few days before and during the critical stages also reduces injury from cold weather.

3.4.2 Effects of Cold Temperature During Reproductive Stage

Rice is more sensitive to low temperature during reproductive development. The young microspore stage, which occurs 10 to 14 days before panicle initiation, and the booting stage are particularly more sensitive. Low temperature during microsporogenesis causes immature pollen and anther indehiscence (Satake and Hayase, 1970), and it also reduces the length of the anther and pollen number per anther (Satake, 1986) leading to male sterility. Rice varieties that have high proportion of viable pollen grains are likely to be more tolerant to cold stress and have higher fertility (Satake, 1986). At low temperature, anther length was strongly related to the number of viable pollen grains and thus tolerance to low temperature (Satake, 1986; Huaiyi et al., 1988). Varieties that develop long anthers are more tolerant to low temperature, and this could be used as a selection criterion in breeding. Besides reducing pollen shedding, low temperature at booting also reduces germination ability of pollen grain (Ito et al., 1970). The low germination percentage of pollen grains is probably due to the abnormal development of pollen grains when low temperature occurs at the booting stage (Koike, 1991).

Rice is also susceptible to low temperature during flowering (Satake and Koike, 1983; Khan et al., 1986). Low temperature during this stage reduces pollination by decreasing viability of pollen grains and reducing pollen germination on the stigma. Pollinating flowers developed at low temperature with pollen produced at optimum temperatures restore the fertility (Satake and Koike, 1983), suggesting that male sterility caused by cold temperature is the sole cause of reduced fertility and with no effect on receptivity of the stigma or viability of the ovary. Low temperature at flowering also inhibits complete exsertion of the panicle from the flag leaf sheath, which is known to increase sterility. Partial panicle exsertion is related to suppressed elongation of the internodes, which also relate to delayed heading and stunting (Kaneda and Beachell, 1974). During grain filling, low temperature causes irregular maturity, reduces grain weight, and increases the percentage of immature grains or grains of poor quality. The use of early maturing varieties in areas where cool temperatures are frequent late in the season may help escape the damage.

Low temperature at the end of the dark period is more injurious than at the end of the light period (Koike, 1989). This may be elucidated either by a protective role played by high soluble carbohydrates accumulating at the end of the light period or certain metabolic processes are more sensitive to low temperature early in the morning. Excessive nitrogen fertilizer before the critical stages was found to increase low-temperature injury in rice, however, excessive phosphorus addition reduced the injury when added before the booting stage (Koike, 1991). Together with controlling water depth in the field, these management options may minimize losses in grain yield especially in areas of sporadic incidences of low temperature.

High heritabilities for germination rate, plumule greening, and seedling vigor at low temperatures were reported (Sthapit and Witcombe, 1998). Based on leaf yellowing, a single dominant gene was found to control chilling tolerance at seedling stage (Kwak et al., 1984; Nagamine and Nakagahra, 1991). Four or more genes were found to control seedling vigor at low temperature with moderate to high heritabilities (Li and Rutger, 1980). Inheritance of cold tolerance at the reproductive stage is polygenic and with moderate to high heritability (Khan et al., 1986). Resistance to cold damage at booting stage does not seem to correlate with resistance at flowering (Koike, 1991), suggesting the involvement of different mechanisms.

Achieving tolerance to low temperature in rice is clearly a complex endeavor. Tolerance to low temperature at one stage may not correlate with tolerance at another. One more difficulty is that most of the sources of cold tolerance are from japonica background, and breeding efforts to incorporate cold tolerance from japonicas to indica types have encountered some major difficulties (Kwon, 1985). Studies in the past were conducted to elucidate types of injuries caused by cold temperatures. Physiological and molecular studies are needed to divulge individual mechanisms involved in cold tolerance to allow their pyramiding in order to achieve higher levels of tolerance. Molecular tagging of the traits involved will speed the breeding process. Isolation of genes directly involved in chilling tolerance (Binh and Oono, 1992) will help in understanding the mechanisms involved and in designing efficient screening methods.

3.5 Phosphorus Deficiency

Low level of available P in soils is one of the major constraints for rice production in the world. This is particularly apparent under upland conditions commonly characterized by poorly fertile, erodible, badly leached, highly acidic, and P-fixing soils, normally with little or no fertilizer applied (IRRI, 1997). Even under lowland conditions, P deficiency is identified as a main factor limiting the performance of modern rice varieties to approach their optimum yields. Under favorable conditions, P nutrition of rice has received little attention due to the markedly less response to P compared to nitrogen. Phosphorus deficiency is likely to be an increasingly important constraint as P is removed from soils under intensive rice production. Rate of plant nutrient removal from the soil by modern rice varieties is about three times that by traditional varieties (De Datta and Biswas, 1990).

Application of P fertilizers is a quick remedy for P deficiency in rice soils. However, nonorganic fertilizers are not always available to a large sector of poor rice farmers. Besides, some rice soils that are low in available P can also fix it into a highly less soluble mineral. Dobermann et al. (1998) estimated that more than 90% of the added fertilizer P may rapidly be transformed to P forms that are not easily available to plants. An attractive and cost-effective alternative to the use of fertilizers is the development of rice cultivars capable of extracting higher proportion of the fixed P. This could offer a more suitable and sustainable long-term solution than relying on fertilizer application alone.

Large variability among lowland (Wissuwa and Ae, 2001) and upland (Fageria et al., 1988) rice cultivars in their ability to exploit soil and fertilizer P was observed. In a recent study with a set of 30 rice cultivars and landraces tested on a P-deficient soil, large variation for P uptake was found ranging from 0.6 to 12.9 mg P plant^{-1} and with the traditional landraces being superior to modern varieties (Wissuwa and Ae, 2001). Hence genetic variation in tolerance to P deficiency could effectively be exploited for rice improvement.

Largely, two types of mechanisms confer tolerance to P deficiency: internal mechanisms associated with the efficient use of P by plant tissue and external mechanisms that allow greater P uptake by plant roots. The main external mechanisms could probably be summarized as follows: (a) ability to develop long, fine hairy roots in soil zones containing available P, (b) ability of rice cultivars to solubilize P through pH changes or the release of chelating agents, (c) ability to utilize soil organic P through release of phosphate enzymes, and (d) ability to associate with mycorrhizal fungi (Kirk et al., 1993; Hedley et al., 1994). However, mycorrhizae are less important for fine-rooted crops such as rice especially in flooded soils.

Variability in internal P-use efficiency, as measured by shoot dry weight per unit total P uptake, was reported under low soil P; however, this superior internal efficiency was associated with low P uptake (Wissuwa and Ae, 1999), and the authors concluded that this was due to an indirect effect of P uptake on P-use efficiency because most lines with high efficiency had very low uptake and dry weight and apparently experiencing extreme P deficiency stress.

Genetic variation in external efficiency is probably the most important mechanism underlying tolerance to P deficiency in rice (Hedley et al., 1994; Wissuwa and Ae, 1999). However, mechanisms responsible for this efficiency still await further investigation. Due to the slow mobility of P in soils, morphological characteristics of plant roots such as root length, surface area, fitness, and intensity of root hairs are found to influence P uptake in many crop species (Otani and Ae, 1996; Kirk and Du, 1997). In one study with rice cultivars of different origins, tolerance to P deficiency was entirely dependent on genotypic variation in P uptake, which is dependent on root size and root efficiency but with stronger association with root size, and genetic variability for both traits was observed (Wissuwa and Ae, 2001). Large root system may therefore be adaptive and may provide a more reliable criterion to identify genotypes with tolerance to P deficiency.

The ability of rice cultivars to solubilize P fixed in the soil has been suggested in many studies (Hedley et al., 1994; Saleque and Kirk, 1995; Kirk et al., 1999). Under flooded conditions, rice roots can acidify soils in their immediate vicinity and changes of more than 2 pH units had been reported (Saleque and Kirk, 1995). This rhizosphere acidification arises from the release of H^+, both from roots, to balance excess intake of cations over anions, and from oxidation of Fe^{2+} by root-released O_2. Mechanisms of P solubilization are different under aerobic soils and mainly involve the secretion of low molecular weight organic acids such as citrate that increase P solubilization through the formation of soluble metal-citrate chelates (Kirk et al., 1999).

Involvement of hydrolytic agents such as phosphatase enzymes in solubilizing soil P was ruled out (Kirk et al., 1993; Hedley et al., 1994).

However, chelating agents such as organic acids may help solubilize P in the soil via chelation of Al and Fe in the solution. This will result in dissolution of Al and Fe solid phases on which P is held. High rates of release of P-solubilizing organic acid anions from roots in response to P deficiency have been reported (Kirk et al., 1993).

Numerous studies showed that the high yields of the modern wheat varieties are mainly due to their high harvest index and the efficient use of P for grain production owing to efficient remobilization of P from vegetative to reproductive tissue. In rice, although genotypic differences in P deficiency tolerance were reported long ago, efforts were limited to screening available cultivars and advanced breeding lines for superior performance under P-deficient soils rather than developing new cultivars especially adapted to P-deficient soil (Fageria et al., 1988; Hedley et al., 1994). The fact that traditional varieties were more superior to modern varieties (Wissuwa and Ae, 2001) may suggest the need for such breeding programs to incorporate P-deficiency tolerance into modern cultivars.

Tolerance to P-deficiency is quantitatively inherited with both additive and dominant genetic effects (Chaubey et al., 1994). Attempts were made to detect putative QTLs controlling P-deficiency tolerance in rice, and four QTLs were identified for P uptake. One major QTL was located on chromosome 12 and controlling most of the variation in P-deficiency tolerance (Wissuwa et al., 1998; Wissuwa and Ae, 1999). The major QTL increased P uptake by threefolds under P-deficient soils and with no apparent effect when P is not limiting (Wissuwa and Ae, in preparation). Studies are underway to clone and characterize the genes involved and also to test for the possibility of using this QTL in marker-assisted selection to improve P uptake under P-deficient conditions.

3.6 Zinc Deficiency

Zinc (Zn) deficiency is one of the most widespread soil constraints to rice production with as much as 50% of all lowland rice soils prone to Zn deficiency (Yoshida et al., 1973; White and Zasoski, 1999). Zn deficiency is normally associated with perennial soil wetness and occurs particularly in alkaline, organic, and poorly drained soils (Yoshida et al., 1973; Forno et al., 1975). High rates of Zn depletion by intensive cultivation of modern rice cultivars provoked the problem of Zn deficiency over the past several years.

Zinc deficiency in rice occurs in the first few weeks after soil flooding. Surviving plants can then recover spontaneously within 6–8 weeks, although the vegetative stage might be prolonged (Forno et al., 1975). The mechanism of Zn solubilization in soils low in Zn was similar to the mechanism of P solubilization under flooded soils. Kirk and Bajita (1995) reported that rice roots solubilize Zn through acidification of the rhizosphere in the vicinity of roots. This acidification occurs as a result of H^+ released from the roots to balance excess intake of cations over anions and H^+ generated in the oxidation of iron by O_2 released by roots.

Since Zn fertilizers are by far not available, the use of cultivars that can tolerate Zn deficiency is probably the most prudent solution. Noticeable differences between rice cultivars in ability to grow under low Zn conditions were observed (Yang et al., 1994), and this variability can be exploited in breeding. In a recent survey of rice germ plasm screened for Zn deficiency at the International Rice Research Institute in the Philippines, Quijano-Guerta et al. (2002) reported that there is no yield cost associated with Zn-deficiency tolerance and that tolerant genotypes often showed tolerance to both salinity and P deficiency, but the reason behind this cross-tolerance awaits further investigation.

Despite useful variation among rice cultivars in their ability to extract Zn, no serious breeding programs has yet been initiated to incorporate this trait into elite breeding lines and varieties. Studies of the inheritance of this trait and the unraveling of the physiological mechanisms underlying the observed genetic variation are prerequisites for a successful breeding program.

4 CONCLUSIONS

Further increase in world rice production relies both on enhancing yield potential of favorable environments and on improving yield stability in less favorable environments. In the irrigated ecosystem, which is considered favorable for rice production, yield potential will be increased primarily by enhancing biomass production. Enhancing photosynthesis at both single-leaf and canopy levels offers great opportunities for improving biomass production than reducing photorespiration and respiration. There is certain scope to increase canopy photosynthesis by fine-tuning the plant type and optimum crop management. However, substantial improvement in photosynthesis relies on the modification of physiological processes through the molecular approach. Successfully transforming C_3 rice plant into C_4 rice plant proves that such approach is feasible. In addition to increasing maximum photosynthetic rate, improving operational photosynthetic rate by reducing photoinhibition and prolonging photosynthetic duration by delaying leaf senescence also contribute to greater biomass production. Further improvement in HI seems difficult. At the increased level of biomass production, maintaining HI at the level of current cultivars is not an easy task. Sink size, sink strength, and grain filling have to be improved in order to avoid the decline in HI at higher biomass. Lodging resistance deserves more attention at the increased level of biomass production.

Rainfed ecosystem that is considered less favorable for rice production constitutes 50% of the area devoted to rice production in the world but provides only about 25% of the total production. Yields are very low and highly unstable compared to irrigated or favorable conditions. This is due to the lack of suitable cultivars adapted to the complex biotic and abiotic stresses affecting rice production in these environments and often forcing farmers to adopt a risk-aversion strategy that minimizes input costs but also results in low yields. The potential for increasing rice production in these areas is immense because of the existing enormous yield gaps. Germ plasm improvement is probably the most sound and sustainable solution to most of these problems. Huge gaps in our knowledge of the physiology of adaptive traits still exist especially for drought, temperature extremes, and nutrient deficiencies and toxicities. Extensive efforts are required to advance our knowledge of the physiology and genetics of adaptive traits together with the implementation of recent tools for genetic evaluation.

Anticipated climatic changes in the near future due to global warming are expected to provoke problems facing rice production especially in tropical and subtropical regions. More severe droughts, higher temperatures, greater problems with flooding, and salinity due to rising sea levels and storms are expected. The impact of these changes on rice production in both favorable and unfavorable environments needs to be appraised, and more efficient breeding programs are needed to cope with these changes.

REFERENCES

Agarie S, Tsuchida H, Ku MSB, Nomura M, Matsuoka M, Miyao-Tokutomi M. High level expression of C4 enzymes in transgenic rice plants. In: Garab G, ed. Photosynthesis: Mechanisms and Effects. Vol. V. Dordrecht, Netherlands: Kluwer Academic Publishers, 1998:3423–3426.

Ahn JK. Physiological factors affecting grain filling in rice. Ph.D. dissertation, University of the Philippines, Los Baños, Laguna, Philippines, 1986.

Akbar M, Yabuno T, Nakao S. Breeding for saline resistant varieties of rice I. Variability for salt tolerance among some rice varieties. Jpn J Breed 1972; 22:277–284.

Akita S. Improving yield potential in tropical rice. Progress in Irrigated Rice Research. Manila, Philippines: International Rice Research Institute, 1989:41–73.

Akita S, Cabuslay GS. Physiological basis of differential response to salinity in rice cultivars. Plant Soil 1990; 123:277–294.

Akita S, Tanaka I, Noma F. Studies on the mechanism of differences in photosynthesis among species. 5. An analysis of the factors changing photorespiration. Proc Crop Sci Soc Jpn 1975; 44(Suppl 1):151–152.

Amano T, Zhu Q, Wang Y, Inoue N, Tanaka H. Case studies on high yields of paddy rice in Jiangsu Province, China. I. Characteristics of grain production. Jpn J Crop Sci 1993; 62(2):267–274.

Armenta-Soto J, Chang TT, Loresto GC, O'Toole JC. Genetic analysis of root characters in rice. SABRAO J 1983; 15:103–116.

Armstrong W. Aeration in higher plants. In: Woolhouse HWW, ed. Advances in Biochemical Research. London: Academic Press, 1979:26–232.

Ashraf M, Akbar M, Salim M. Genetic improvement in physiological traits of rice yield. In: Slafer GA, ed. Genetic Improvement of Field Crops. New York: Marcel Dekker, 1994:413–455.

Aslam M, Flowers TJ, Qureshi RH, Yeo AR. Interaction of phosphate and salinity on the growth and yield of rice (*Oryza sativa* L.). J Agron Crop Sci 1996; 176:249–258.

Austin RB. Crop photosynthesis: can we improve on nature? International Crop Science I. Madison, Wis., USA: Crop Science Society of America, 1993:697–701.

Austin RB, Bingham J, Blackwell RD, Evans LT, Ford MA, Morgan CL, Taylor M. Genetic improvements in winter wheat yields since 1900 and associated physiological changes. J Agric Sci (Cambridge) 1980; 94:675–689.

Bhumbla DR, Abrol IP. Saline and sodic soils. Soils and Rice. Manila, Philippines: International Rice Research Institute, 1978:719–738.

Binh LT, Oono K. Molecular cloning and characterization of genes related to chilling tolerance in rice. Plant Physiol 1992; 99:1146–1150.

Bohra JS, Doerffling K. Potassium nutrition of rice (*Oryza sativa* L.) varieties under NaCl salinity. Plant and Soil 1993; 299–303.

Bowes G, Salvucci ME. Plasticity in the photosynthetic carbon metabolism of submersed aquatic macrophytes. Aquat Bot 1993; 34:233–249.

Camp PJ, Huber SC, Burke JJ, Moreland DE. Biochemical changes that occur during senescence of wheat leaves. Plant Physiol 1982; 70:1641–1646.

Champoux MC, Wang G, Sarkarung S, Mackill DJ, O'Toole JC, Haung N, McCouch SR. Locating genes associated with root morphology and drought avoidance in rice via linkage to molecular markers. Theor Appl Genet 1995; 90:969–981.

Chandler RF Jr. Plant morphology and stand geometry in relation to nitrogen. In: Eastin JD, Haskins FA, Sullivan CY, van Bavel CHM, eds. Physiological Aspects of Crop Yield. Madison, Wisconsin: ASA, 1969:265–285.

Chang TT. Varietal differences in lodging resistance. Int Rice Comm Newsl 1964; 13(4):1–11.

Chang TT, Loresto GC. Lodging in rice. Rice Production and Extension, Management Specialist Training Course. Manila, Philippines: International Rice Research Institute, 1985.

Chang TT, Loresto GC, Tagumpay O. Screening rice germplasm for drought resistance. SABRAO J 1974; 6:9–16.

Chang TT, Vergara BS. Ecological and genetic information on adaptability and yielding ability in tropical varieties. Rice Breeding. Manila, Philippines: International Rice Research Institute, 1972:431–453.

Chaturvedi GS, Misra CH, Singh ON, Pandey CB, Yadav VP, Singh AK, Dwivedi JL, Singh BB, Singh RK. Physiological basis and screening for tolerance for flash flooding. In: Ingram KT, ed. Rainfed Lowland Rice: Agricultural Research for High-Risk Environments. Manila: International Rice Research Institute, 1995:79–96.

Chaubey CN, Senadhira D, Gregorio GB. Genetic analysis of tolerance for phosphorus deficiency in rice (*Oryza sativa* L.). Theor Appl Genet 1994; 89:313–317.

Chen YZ, Murchie EH, Hubbart S, Horton P, Peng S. Effects of season-dependent irradiance levels and nitrogen-deficiency on photosynthesis and photoinhibition in field-grown rice (*Oryza sativa* L.). Physiol Plant 2003; 117:343–351.

Chen W, Xu Z, Zhang L, Yang S. Comparative studies on stomatal density and its relationships to gas diffusive resistance and net photosynthetic rate in rice leaf. Chin J Rice Sci 1990; 4(4):163–168.

Chung GS, Vergara BS, Heu MH. Breeding strategies for development of cold tolerant rice varieties. International Rice Research, Project reviews no. 14. International Rice Research Institute, Los Baños, Laguna, Philippines, 18–22 Apr 1983.

Cock JH, Yoshida S. Photosynthesis, crop growth, and respiration of tall and short rice varieties. Soil Sci Plant Nutr 1973; 19:53–59.

Cook MG, Evans LT. Some physiological aspects of the domestication and improvement of rice (*Oryza* spp.). Field Crops Res 1983; 6:219–238.

Crawford RMM. Metabolic adaptation to anoxia. In: Hook DD, Crawford RMM, eds. Plant Life in Anaerobic Environment. Ann Arbor, Michigan, Ann Arbor Science, 1978:119–136.

Crawford RMM. Studies in Plant Survival: Ecological Case Histories of Plant Adaptations to Adversity. Oxford: Blackwell Scientific, 1989.

Crawford RMM. Oxygen availability as an ecological limit to plant distribution. Adv Ecol Res 1992; 93–185.

Dai Q, Peng S, Chavez AQ, Vergara BS. Effects of UVB radiation on stomatal density and opening in rice (*Oryza sativa* L.). Ann Bot 1995; 76:65–70.

Davies DD. Anaerobic metabolism and the production of organic acids. In: Davies DD, ed. The Biochemistry of Plants. Vol. 2. New York: Academic Press, 1980:581–611.

De Datta SK, Biswas TK, Charoenchamratcheep C. Phosphorus requirements and management for lowland rice. Phosphorus Requirements for Sustainable Agriculture in Asia and Oceania. Manila: International Rice Research Institute, 1990: 307–323.

Ding HY, Zhang GY, Guo Y, Chen SL, Chen SY. RAPD tagging of a salt-tolerant gene in rice. Chin Sci Bull 1998; 43:330–332.

Dingkuhn M, Cruz RT, O'Toole JC, Doerffling K. Net photosynthesis, water use efficiency, leaf water potential, and leaf rolling as affected by water deficit in tropical upland rice. Aust J Agric Resour 1989; 40:1171–1181.

Dingkuhn M, Penning de Vries FWT, De Datta SK, van Laar HH. Concepts for a new plant type for direct seeded flooded tropical rice. Direct Seeded Flooded Rice in the Tropics. Manila, Philippines: International Rice Research Institute, 1991:17–38.

Dingkuhn M, Schnier HF, De Datta SK, Doerffling K, Javellana C, Pamplona R. Nitrogen fertilization of direct-seeded flooded vs. transplanted rice. II. Interactions among canopy properties. Crop Sci 1990; 30:1284–1292.

Dobermann A, Cassman KG, Mamaril CP, Sheehy JE. Management of phosphorus, potassium, and sulfur in intensive irrigated lowland rice. Field Crops Res 1998; 56:113–138.

Duvick DN, Cassman KG. Post-green revolution trends in yield potential of temperate maize in the north-central United States. Crop Sci 1999; 39:1622–1630.

Ekanayake IJ, De Datta SK, Steponkus PL. Spikelet sterility and flowering response of rice to water stress at anthesis. Ann Bot 1989; 63:257–264.

Ekanayake IJ, Garrity DP, Masajo TM, O'Toole JC. Root pulling resistance in rice: inheritance and association with drought resistance. Euphytica 1985a; 34: 905–913.

Ekanayake IJ, Garrity DP, O'Toole JC. Influence of deep root density on root pulling resistance in rice. Crop Sci 1986; 26:1181–1186.

Ekanayake IJ, O'Toole JC, Garrity DP, Masajo TM. Inheritance of root characters and their relations to drought resistance in rice. Crop Sci 1985b; 25:927–933.

Elmore CD. The paradox of no correlation between leaf photosynthetic rates and crop yields. In: Hesketh JD, Jones JW, eds. Predicting Photosynthesis for Ecosystem Models. Vol. II. Boca Raton, Fla., USA: CRC Press, 1980:155–167.

Evans LT. Raising the ceiling to yield: the key role of synergisms between agronomy and plant breeding. In: Muralidharan K, Siddiq EA, eds. New Frontiers in Rice Research. Hyderabad, India: Directorate of Rice Research, 1990:13–107.

Evans LT, Visperas RM, Vergara BS. Morphological and physiological changes among rice varieties used in the Philippines over the last seventy years. Field Crops Res 1984; 8:105–124.

Fageria NK. Salt tolerance of rice cultivars. Plant Soil 1985; 88:237–243.

Fageria NK, Morais OP, Baligar VC, Wright RJ. Response of rice cultivars to phosphorus supply on an oxisol. Fertil Res 1988; 16:195–206.

Fischer RA. Number of kernels in wheat crops and the influence of solar radiation and temperature. J Agric Sci 1985; 100:447–461.

Fischer RA, Rees D, Sayre KD, Lu ZM, Condon AG, Saavedra AL. Wheat yield progress associated with higher stomatal conductance and photosynthetic rate, and cooler canopies. Crop Sci 1998; 38:1467–1475.

Flowers TJ, Duque E, Hajibagheri MA, McGonigle TP, Yeo AR. The effect of salinity on the ultrastructure and net photosynthesis of two varieties of rice: further evidence for a cellular component of salt resistance. New Phytol 1985; 100:37–43.

Flowers TJ, Koyama ML, Flowers SA, Sudhakar C, Singh KP, Yeo AR. QTL: their place in engineering tolerance of rice to salinity. J Exp Bot 2000; 51:99–106.

Flowers TJ, Yeo AR. Variability in the resistance of sodium chloride salinity within rice (*Oryza sativa* L.) varieties. New Phytol 1981; 88:363–373.

Forno DA, Yoshida S, Asher CJ. Zinc deficiency in rice. I. Soil factors associated with the deficiency. Plant Soil 1975; 42:537–550.

Fukai S, Cooper M. Development of drought-resistant cultivars using physio-morphological traits in rice. Field Crops Res 1995; 40:67–86.

Gan S, Amasino RM. Inhibition of leaf senescence by autoregulated production of cytokinin. Science 1995; 270:1986–1988.

Garcia A, Rizzo CA, Ud-din J, Bartos SL, Senadhira D, Flowers TJ, Yeo AR.

Sodium and potassium transport to the xylem are inherited independently in rice, and the mechanism of sodium:potassium selectivity differs between rice and wheat. Plant Cell Environ 1997; 20:1167–1174.

Gent MPN. Canopy light interception, gas exchange, and biomass in reduced height isolines of winter wheat. Crop Sci 1995; 35:1636–1642.

Gifford RM, Jenkins CLD. Prospects of applying knowledge of photosynthesis toward improving crop production. In: Govindjee, ed. Photosynthesis: Development, Carbon Metabolism, and Plant Productivity. Vol. II. New York, USA: Academic Press Inc., 1982:419–457.

Gifford RM, Thorne JH, Hitz WD, Gianquinta RT. Crop productivity and photoassimilate partitioning. Science 1984; 225:801–808.

Graf B, Rakotobe O, Zahner P, Delucchi V, Gutierrez AP. A simulation model for the dynamics of rice growth and development. I. The carbon balance. Agric Syst 1990; 32:341–365.

Gutteridge JMC, Halliwell B. Reoxigination injury and antioxidant protection: a tale of two paradoxes. Arch Biochem Biophys 1990; 283:223–226.

Gutteridge S, Newman J, Herrmann C, Rhoades D. The crystal structures of Rubisco and opportunities for manipulating photosynthesis. J Exp Bot 1995; 46:1261–1267.

Hall AE. Crop Responses to Environment. New York: CRC Press LLC, 2000.

Haque MM, Mackill DJ, Ingram KT. Inheritance of leaf epicuticular wax content in rice. Crop Sci 1992; 32:865–868.

Hedley MJ, Kirk GJD, Santos MB. Phosphorus efficiency and the forms of soil phosphorus utilized by upland rice cultivars. Plant Soil 1994; 158:53–62.

Hitaka H. Studies on the lodging of rice plants. JARQ 1969; 4(3):1–6.

Horie T. Increasing yield potential in irrigated rice: breaking the yield barrier. In: Peng S, Hardy B, eds. Proceedings of the International Rice Research Conference: Rice Research for Food Security and Poverty Alleviation. Manila, Philippines: International Rice Research Institute, 2001:3–25.

Horton P. Prospects for crop improvement through the genetic manipulation of photosynthesis: morphological and biochemical aspects of light capture. J Exp Bot 2000; 51:475–485.

Horton P, Ruban AV. Regulation of photosystem II. Photosynth Res 1992; 34:375–385.

Hsiao TC, O'Toole JC, Yambao EB, Turner NC. Influence of osmotic adjustment on leaf rolling and tissue death in rice. Plant Physiol 1984; 75:338–341.

Huaiyi W, Jianhua X, Sizhu Z, Kunihiro Y, Horisue N. Anther length and cold tolerance in rice in relation to breeding. International Symposium on Rice Breeding Through the Utilization of Unexploited Genetic Resources Proceedings. Proceedings of a Symposium on Tropical Agricultural Research. Tsukuba, Japan, Tropical Agriculture Research Center, Series no. 21, 1988:93–104.

Huang H. Japonica and indica differences in large vascular bundles in culm. Intl Rice Res Newsl 1988; 13(1):77.

Hunt LA, van der Poorten G, Pararajasingham S. Post-anthesis temperature effects on duration and rate of grain filling in some winter and spring wheats. Can J Plant Sci 1991; 71:609–617.

Huq E, Harrington S, Hossain MA, Wen F, McCouch SR, Hodges TK. Molecular characterization of *pdc2* and mapping of three *pdc* genes from rice. Theor Appl Genet 1999; 98:815–824.

Imaizumi N, Usuda H, Nakamoto H, Ishihara K. Changes in the rate of photosynthesis during grain filling and the enzymatic activities associated with the photosynthetic carbon metabolism in rice panicles. Plant Cell Physiol 1990; 31(6): 835–843.

IRRI. Annual Report for 1974. Manila, Philippines, International Rice Research Institute, 1975.

IRRI. Terminology for Rice-Growing Environments. Manila, Philippines, International Rice Research Institute, 1984.

IRRI. Program Report for 1994. Manila, International Rice Research Institute, 1995.

IRRI. Program Report for 1995. Manila, International Rice Research Institute, 1996.

IRRI. Rice Almanac. 2d ed. Manila, Philippines: International Rice Research Institute, 1997.

Ise K. Inheritance of a low-tillering plant type in rice. Intl Rice Res Newsl 1992; 17(4):5–6.

Ishihara K, Ishida Y, Ogura T. The relationship between environmental factors and behaviour of stomata in rice plants. 2. On the diurnal movement of the stomata. Proc Crop Sci Soc Jpn 1971; 40:497–504.

Ishihara K, Saito K. Relationship between leaf water potential and photosynthesis in rice plants. JARQ 1983; 17:81–86.

Ishihara K, Saito K. Diurnal changes in photosynthesis, respiration, and diffusive conductance in the single-leaf of rice plants grown in the paddy field under submerged conditions. Jpn J Crop Sci 1987; 56:8–17.

Ito N, Hayase H, Satake T, Nishiyama I. Male sterility caused by cooling temperature at the meiotic stage in rice plants. III. Male abnormalities at anthesis. Proc Crop Sci Soc Jpn 1970; 39:60–64.

Ito O, O'Toole J, Hardy B. Genetic Improvement of Rice for Water-Limited Environments. Manila: International Rice Research Institute, 1999.

Iwasaki Y, Mae T, Makino A, Ohira K, Ojima K. Nitrogen accumulation in the inferior spikelet of rice ear during ripening. Soil Sci Plant Nutr 1992; 38(3):517–525.

Janoria MP. A basic plant ideotype for rice. Intl Rice Res Newsl 1989; 14(3):12–13.

Johnson GN, Young AJ, Scholes JD, Horton P. The dissipation of excess excitation energy in British plant species. Plant Cell Environ 1993; 16:673–679.

Justin SHFW, Armstrong W. The anatomical characteristics of roots, and plant response to soil flooding. New Phytol 1987; 106:465–495.

Kaneda C, Beachell HM. Response of indica-japonica rice hybrids to low temperatures. SABRAO J 1974; 6:17–32.

Kawamitsu Y, Cheng WJ, Katayama T, Agata W. Varietal differences in photorespiration rate in rice plants. Sci Bull Fac Agric Kyushu Univ 1989; 43:135–144.

Kawano N, Ella E, Ito O, Yamauchi Y, Tanaka K. Metabolic changes in rice seedlings with different submergence tolerance after desubmergence. Environ Experimt Bot 2002; 47:195–203.

Khan DR, Mackill DJ, Vergara BS. Selection for tolerance to low temperature-induced spikelet sterility at anthesis in rice. Crop Sci 1986; 26:694–698.

Khush GS. Terminology for Rice Growing Environments. Manila: International Rice Research Institute, 1984.

Kirk GJD, Bajita JB. Root-induced iron oxidation, pH changes and zinc solubilization in the rhizosphere of lowland rice. New Phytol 1995; 131:129–137.

Kirk GJD, Du LV. Changes in rice root architecture, porosity, and oxygen and proton release under phosphorus deficiency. New Phytol 1997; 135:191–200.

Kirk GJD, Hedley MJ, Bouldin DR. Phosphorus efficiency in upland rice cultivars. In: IBSRAM Reports and Papers on the Management of Acid Soils (IBSRAM/ASIALAND). Network Document No. 6. Bangkok: IBSRAM, 1993:279–295.

Kirk GJD, Santos EE, Santos MB. Phosphate solubilization by organic anion excretion from rice growing in aerobic soil: rates of excretion and decomposition, effects on rhizosphere pH and effects on phosphate solubility and uptake. New Phytol 1999; 142:185–200.

Kishore GM. Starch biosynthesis in plants: identification of ADP glucose pyrophosphorylase as a rate-limiting step. In: Cassman KG, ed. Breaking the Yield Barrier. Manila, Philippines: International Rice Research Institute, 1994:117–119.

Koike S. Diurnal variation in the chilling sensitivity of rice seedlings. In: Iyama S, Takeda G, eds. Breeding Research: The Key to the Survival of the Earth. Proceedings of the 6th International Congress of SABRAO, Tsukuba, Japan, 1989:253–256.

Koike S. Rice low temperature tolerance research in Japan. Paper presented at the International Rice Research Conference: Focus on Irrigated Rice, 27–31 Aug 1990, Seoul, Korea. IRRI, Los Baños, Philippines, 1991.

Koyama M, Levesley A, Koebner RMD, Flowers TJ, Yeo AR. Quantitative trait loci for component physiological traits determining salt tolerance in rice. Plant Physiol 2001; 125:406–422.

Kropff MJ, Cassman KG, Peng S, Matthews RB, Setter TL. Quantitative understanding of yield potential. In: Cassman KG, ed. Breaking the Yield Barrier. Manila, Philippines: International Rice Research Institute, 1994:21–38.

Ku MSB, Agarie S, Nomura M, Fukayama H, Tsuchida H, Ono K, Hirose S, Toki S, Miyao M, Matsuoka M. High-level expression of maize phosphoenolpyruvate carboxylase in transgenic rice plants. Nat Biotechnol 1999; 17:76–80.

Ku MSB, Cho D, Ranade U, Hsu TP, Li X, Jiao DM, Ehleringer J, Miyao M, Matsuoka M. Photosynthetic performance of transgenic rice plants overexpressing maize C4 photosynthesis enzymes. In: Sheehy JE, Mitchell PL, Hardy B, eds. Redesigning Rice Photosynthesis to Increase Yield. Amsterdam: Elsevier Science, 2000:193–204.

Kura-Hotta M, Satoh K, Katoh S. Relationship between photosynthesis and chlorophyll content during leaf senescence of rice seedlings. Plant Cell Physiol 1987; 28:1321–1329.

Kuroda E, Ookawa T, Ishihara K. Analysis on difference of dry matter production between rice cultivars with different plant height in relation to gas diffusion inside stands. Jpn J Crop Sci 1989; 58(3):374–382.

Kwak TS, Vergara BS, Nanda JS, Coffman WR. Inheritance of seedling cold tolerance in rice. SABRAO J 1984; 16(2):83–86.

Kwon YW. Fundamental physiological constraints in breeding for a higher yielding cold tolerant rice variety. Potential Productivity and Yield Constraints of Rice in East Asia, Proceedings of the International Crop Science Symposium. Fukuoka, Japan, Crop Science Society of Japan, Kyushu, Japan, 1985:19–38.

Li CC, Rutger JN. Inheritance of cool-temperature seedling vigor in rice and its relationship with other agronomic characters. Crop Sci 1980; 20:295–298.

Lilley JM, Fukai S. Effect of timing and severity of water deficit on four diverse rice cultivars. III. Phenological development, crop growth and grain yield. Field Crops Res 1994; 37:225–234.

Lilley JM, Ludlow MM. Expression of osmotic adjustment and dehydration tolerance in diverse rice lines. Field Crops Res 1996; 48:185–197.

Lilley JM, Ludlow MM, McCouch SR, O'Toole JC. Locating QTL for osmotic adjustment and dehydration tolerance in rice. J Exp Bot 1996; 47:1427–1436.

Maas EV. Salt tolerance of plants. Appl Agric Res 1986; 1:12–26.

Mackill DJ, Amante MM, Vergara BS, Sarkarung S. Improved semidwarf rice lines with tolerance to submergence of seedlings. Crop Sci 1993; 33:749–753.

Mackill DJ, Coffman WR, Garrity DP. Rainfed Lowland Rice Improvement. Manila: International Rice Research Institute, 1996.

Mae T. Physiological nitrogen efficiency in rice: Nitrogen utilization, photosynthesis, and yield potential. Plant Soil 1997; 196:201–210.

Maeda E. Observations on the surface structure of unhulled rice grains through a scanning electron microscope. Proc Crop Sci Soc Jpn 1972; 41:459–471.

Makino A, Mae T, Ohira K. Photosynthesis and ribulose-1,5-bisphosphate carboxylase in rice leaves: changes in photosynthesis and enzymes involved in carbon assimilation from leaf development through senescence. Plant Physiol 1983; 73:1002–1007.

Makino A, Mae T, Ohira K. Relation between nitrogen and ribulose-1,5-bisphosphate carboxylase in rice leaves from emergence through senescence. Plant Cell Physiol 1984; 25(3):429–437.

Makino A, Mae T, Ohira K. Enzymatic properties of ribulose-1,5-bisphosphate carboxylase/oxygenase purified from rice leaves. Plant Physiol 1985; 79:57–61.

Makino A, Mae T, Ohira K. Variations in the contents and kinetic properties of ribulose-1,5-bisphosphate carboxylase among rice species. Plant Cell Physiol 1987; 28:799–804.

Makino A, Mae T, Ohira K. Differences between wheat and rice in the enzymatic properties of ribulose-1,5-bisphosphate carboxylase/oxygenase and the relationship to photosynthetic gas exchange. Planta 1988; 174:30–38.

Mallik S, Kundu C, Banerji C, Nayak DK, Chatterjee SD, Nada K, Ingram KT, Setter TL. Rice germplasm evaluation and improvement for stagnant flooding. In: Ingram KT, ed. Rainfed Lowland Rice: Agricultural Research for High Risk Environments. Manila: International Rice Research Institute, 1995:97–109.

Mann CC. Genetic engineers aim to soup up crop photosynthesis. Science 1999; 283:314–316.

Maruyama S, Tajima K. Factors causing the difference of leaf diffusive resistance between japonica and indica rice. Jpn J Crop Sci 1986; 55(Suppl 1):228–229.

Matsuo T, Kumazawa K, Ishii R, Ishihara K, Hirata H. Science of the Rice Plant. Physiology. Vol. II. Tokyo, Japan: Food and Agriculture Policy Research Center, 1995:1240.

McDonald DJ, Stansel JW, Gilmore EC. Breeding for high photosynthetic rate in rice. Indian J Genet 1974; 34A:1067–1073.

Meidner H, Mansfield TA. Physiology of Stomata. London: McGraw-Hill, 1968:52.

Mew T. Disease management in rice. In: Pimentel D, ed. CRC Handbook of Pest Management in Agriculture. Vol. III. 2d ed. Boston: CRC Press Inc., 1991:279–299.

Mitchell PL, Sheehy JE, Woodward FI. Potential yields and the efficiency of radiation use in rice. IRRI Discussion Paper Ser No. 32. Manila, Philippines: International Rice Research Institute, 1998:62.

Monk LS, Fagerstedt KV, Crawford RMM. Oxygen toxicity and superoxide dismutase as an antioxidant in physiological stress. Physiol Plant 1989; 76:456–459.

Monsi M, Saeki T. Über den Lichtfaktor in den Pflanzengesellschaften und seine Bedeutung für die Stoffproduktion. Jpn J Bot 1953; 14:22–52.

Morgan JM. Osmoregulation and water stress in higher plants. Annu Rev Plant Physiol 1984; 35:299–329.

Morgan JM. A gene controlling differences in osmoregulation in wheat. Aust J Plant Physiol 1991; 18:249–257.

Morgan JM, Condon AG. Water use, grain yield and osmoregulation in wheat. Aust J Plant Physiol 1986; 13:523–532.

Murata Y. Studies on the photosynthesis of rice plants and cultural significance. Bull Natl Inst Agric Sci Jpn Ser D 1961; 9:1–169.

Murata Y. Photosynthesis, respiration, and nitrogen response. The Mineral Nutrition of the Rice Plant. Baltimore, Md, USA: Johns-Hopkins Press, 1965:385–400.

Murchie EH, Chen Y, Hubbart S, Peng S, Horton P. Interactions between senescence and leaf orientation determine in situ patterns of photosynthesis and photoinhibition in field-grown rice. Plant Physiol 1999; 119:553–563.

Nagamine T, Nakagahra M. Genetic control of chilling injury in rice seedlings detected by low-temperature treatment. Rice Genetics II, Proceedings of the 2nd International Rice Genetic Symposium. Manila: International Rice Research Institute, 1991:737–739.

Nandi S, Subudhi PK, Senadhira D, Manigbas NL, Sen-Mandi S, Huang N. Mapping QTLs for submergence tolerance in rice by AFLP analysis and selective genotyping. Mol Gen Genet 1997; 255:1–8.

Nelson CJ. Genetic associations between photosynthetic characteristics and yield: review of the evidence. Plant Physiol Biochem 1988; 26:543–554.

Nguyen HT, Babu RC, Blum A. Breeding for drought resistance in rice: physiology and molecular genetics considerations. Crop Sci 1997; 37:1426–1434.

Normile D. Crossing rice strains to keep Asia's rice bowls brimming. Science 1999; 283:313.

Ookawa T, Ishihara K. Varietal difference of physical characteristics of the culm related to lodging resistance in paddy rice. Jpn J Crop Sci 1992; 61(3):419–425.

Ookawa T, Todokoro Y, Ishihara K. Changes in physical and chemical characteristics of culm associated with lodging resistance in paddy rice under different growth conditions and varietal difference of their changes. Jpn J Crop Sci 1993; 62(4):525–533.

Otani T, Ae N. Sensitivity of phosphorus uptake to changes in root length and soil volume. Agron J 1996; 88:371–375.

O'Toole JC, Chang TT. Drought and Rice Improvement in Perspective. IRRI Res Paper No. 14. Manila: International Rice Research Institute, 1978.

O'Toole JC, Chang TT. Drought resistance in cereals-rice: a case study. In: Mussel H, Staples RC, eds. Stress Physiology in Crop Plants. New York: John Wiley and Sons, 1979:373–405.

O'Toole JC, Cruz RT, Sieber JN. Epicuticular wax and cuticular resistance in rice. Physiol Plant 1979; 47:239–244.

O'Toole JC, Namuco OS. Role of panicle exsertion in water stress induced sterility. Crop Sci 1983; 23:1093–1097.

O'Toole JC, Soemartono. Evaluation of a simple technique for characterizing rice root systems in relation to drought resistance. Euphytica 1981; 30:283–290.

O'Toole JC, Tomer VS. Transpiration, leaf temperature and water potential of rice and barnyard grass in flooded fields. Agric Meteorol 1982; 26:285–296.

Padmaja Rao S. High density grain among primary and secondary tillers of short- and long-duration rices. Intl Rice Res Newsl 1987; 12(4):12.

Palada MC, Vergara BS. Environmental effect on the resistance of rice seedlings to complete submergence. Crop Sci 1972; 12:209–212.

Paulsen GM. Relationship Between Photosynthesis Rates and Other Physiological Traits in Rice. Manila, Philippines: International Rice Research Institute. IRRI Saturday Seminar, 1972.

Pearson GA, Bernstein L. Salinity effects at several growth stages of rice. Agron J 1959; 51:654–657.

Peng S. Single-leaf and canopy photosynthesis of rice. In: Sheehy JE, Mitchell PL, Hardy B, eds. Redesigning Rice Photosynthesis to Increase Yield. Amsterdam: Elsevier Science, 2000:213–228.

Peng S, Cassman KG. Upper thresholds of nitrogen uptake rates and associated nitrogen fertilizer efficiencies in irrigated rice. Agron J 1998; 90:178–185.

Peng S, Cassman KG, Kropff MJ. Relationship between leaf photosynthesis and nitrogen content of field-grown rice in the tropics. Crop Sci 1995; 35:1627–1630.

Peng S, Khush GS, Cassman KG. Evolution of the new plant ideotype for increased yield potential. In: Cassman KG, ed. Breaking the Yield Barrier. Manila, Philippines: International Rice Research Institute, 1994:5–20.

Penning de Vries FWT. Improving yields: designing and testing VHYVs. Systems Simulation at IRRI. IRRI Research Paper Series 5. Manila, Philippines: International Rice Research Institute, 1991:13–19.

Ponnamperuma FN. Physiochemical Properties of Submerged Soils in Relation to

Fertility. IRRI Res Paper Ser No. 5. Manila: International Rice Research Institute, 1977.

Ponnamperuma FN, Bandyopadhya AK. Soil salinity as a constraint on food production in the humid tropics. Priorities for Alleviating Soil-related Constraints to Food Production in the Tropics. Los Baños, Philippines: IRRI, 1980:203–216.

Prasad SR, Bagali PG, Hittalmani S, Shashidhar HE. Molecular mapping of quantitative trait loci associated with seedling tolerance to salt stress in rice (*Oryza sativa* L.). Curr Sci 2000; 78:162–164.

Quatrano RS. The role of hormones during seed development. In: Davies PJ, ed. Plant Hormones and their Role in Plant Growth and Development. Dordrecht, Netherlands: Kluwer Academic Publishers, 1987:494–514.

Quijano-Guerta C, Kirk GJD, Portugal AM, Bartolome VI, Mclaren GC. Tolerance of rice germplasm to zinc deficiency. Field Crops Res 2002; 76:123–130.

Quimio CA, Torrizo LB, Setter TL, Ellis M, Grover A, Abrigo EM, Oliva NP, Ella ES, Carpena AL, Ito O, Peacock WJ, Dennis E, Datta S. Enhancement of submergence tolerance in transgenic rice overproducing pyruvate decarboxylase. J Plant Physiol 2000; 156:516–521.

Rabinowitch EI. Photosynthesis and Related Processes. New York: Interscience Publishers, Inc., 1956.

Ram PC, Singh BB, Singh AK, Ram P, Singh PN, Singh HP, Boamfa I, Harren F, Santosa E, Jackson MB, Setter TI, Reuss J, Wade LJ, Singh VP, Singh RK. Submergence tolerance in rainfed lowland rice: physiological basis and prospects for cultivar improvement through marker-aided breeding. Field Crops Res 2002; 76:131–152.

Ranjhan S, Glaszmann JC, Ramirez DA, Khush GS. Chromosomal localization of four isozyme loci by trisomic analysis in rice (*Oryza sativa* L.). Theor Appl Genet 1988; 75:541–545.

Ray JD, Yu L, McCouch SR, Champoux MC, Wang G, Nguyen HT. Mapping quantitative trait loci associated with root penetration ability in rice (*Oryza sativa* L.). Theor Appl Genet 1996; 92:627–636.

Richards RA. Increasing yield potential: source and sink strength. In: Reynolds MP, Rajaram S, McNab A, eds. Increasing Yield Potential in Wheat: Breaking the Barriers. Mexico: International Maize and Wheat Improvement Center, 1996:134–149.

Rosegrant MW, Sombilla MA, Perez N. Global food projections to 2020: implications for investment. Food, Agriculture and the Environment Discussion Paper No. 5. Washington, D.C.: IFPRI, 1995.

Saleque MA, Kirk GJD. Root-induced solubilization of phosphate in the rhizosphere of lowland rice. New Phytol 1995; 129:325–336.

Sasahara T, Takahashi T, Kayaba T, Tsunoda S. A new strategy for increasing plant productivity and yield in rice. Int Rice Comm Newsl 1992; 41:1–4.

Satake T. Anther length as indicator to estimate chilling tolerance at booting stage in rice plants. In: Napomapeth B, Subhandrabandhu S, eds. New Frontiers in Breeding Researches. Proceedings of the 5th International Congress of SABRAO, Bangkok, Thailand, 1986:221–228.

Satake T, Hayase H. Male sterility caused by cooling treatment at the young microspore stage in rice. V. Estimation of pollen developmental stage and the most sensitive stage to coolness. Proc Crop Sci Soc Jpn 1970; 39:468–473.

Satake T, Koike S. Sterility caused by cooling treatment at the flowering stage in rice plants. I. The stage and organ susceptible to cool temperature. Jpn J Crop Sci 1983; 52:207–214.

Scandalios JG. Oxygen stress and superoxide dismutases. Plant Physiol 1993; 101:7–12.

Schnier HF, Dingkuhn M, De Datta SK, Mengel K, Wijangco E, Javellana C. Nitrogen economy and canopy carbon dioxide assimilation of tropical lowland rice. Agron J 1990; 82:451–459.

Senadhira D, Li GF. Variability in rice grain-filling duration. Intl Rice Res Newsl 1989; 14(1):8–9.

Setter TL, Conocono EA, Egdane JA, Kropff MJ. Possibility of increasing yield potential of rice by reducing panicle height in the canopy. I. Effects of panicle on light interception and canopy photosynthesis. Aust J Plant Physiol 1995a; 22:441–451.

Setter TL, Ella ES, Valdez AP. Relationship between coleoptile elongation and alcoholic fermentation in rice exposed to anoxia. II. Cultivar differences. Ann Bot 1994a; 74:273–279.

Setter TL, Ellis M, Laureles EV, Ella ES, Senadhira D, Mishra SB, Sarkarung S, Datta S. Physiology and genetics of submergence tolerance in rice. Ann Bot 1997; 79:67–77.

Setter TL, Peng S, Kirk GJD, Virmani SS, Kropff MJ, Cassman KG. Physiological considerations and hybrid rice. In: Cassman KG, ed. Breaking the Yield Barrier. Manila, Philippines: International Rice Research Institute, 1994b:39–62.

Setter TL, Ramakrishnayya G, Ram PC, Singh BB. Environmental characteristics of floodwater in Eastern India: relevance to flooding tolerance of rice. Indian J Plant Physiol 1995b; 38:34–40.

Setter TL, Waters I, Wallace I, Wiengweera A, Bhekasut P, Greenway H. Submergence of rice. I. Growth and photosynthetic response to CO_2 enrichment of floodwater. Aust J Plant Physiol 1989; 16:251–263.

Sinclair TR, Sheehy JE. Erect leaves and photosynthesis in rice. Science 1999; 283:1456–1457.

Singh BN, Mackill DJ. Genetics of leaf rolling under drought stress. Rice Genetics II. Manila: International Rice Research Institute, 1991:159–166.

Singh G, Singh S, Gurung SB. Effect of growth regulators on rice productivity. Trop Agric 1984; 61:106–108.

Slafer GA, Calderini DF, Miralles DJ. Generation of yield components and compensation in wheat: opportunities for further increasing yield potential. In: Reynolds MP, Rajaram S, McNab A, eds. Increasing Yield Potential in Wheat: Breaking the Barriers. Mexico: International Maize and Wheat Improvement Center, 1996:101–133.

Slafer GA, Rawson HM. Sensitivity of wheat phasic development to major environmental factors: a re-examination of some assumptions made by physiologists and modellers. Aust J Plant Physiol 1994; 21:393–426.

Smith FA, Walker NA. Photosynthesis by aquatic plants. Effects of unstirred layers in relation to assimilation of CO_2 and HCO_3^- and to carbon isotope discrimination. New Phytol 1980; 86:245–259.

Song XF, Agata W, Kawamitsu Y. Studies on dry matter and grain production of F_1 hybrid rice in China. I. Characteristic of dry matter production. Jpn J Crop Sci 1990; 59:19–28.

Stark DM, Timmerman KP, Barry GF, Preiss J, Kishore GM. Regulation of the amount of starch in plant tissues by ADP glucose pyrophosphorylase. Science 1992; 258:287–292.

Sthapit BR, Witcombe JR. Inheritance of tolerance to chilling stress in rice during germination and plumule greening. Crop Sci 1998; 38:660–665.

Suprihatno B, Coffman WR. Inheritance of submergence tolerance in rice (*Oryza sativa* L.). SABRAO J 1981; 13(2):98–108.

Takano Y, Tsunoda S. Curvilinear regression of the leaf photosynthetic rate on leaf nitrogen content among strains of *Oryza* species. Jpn J Breed 1971; 21:69–76.

Takeda T, Oka M, Agata W. Studies on the dry matter and grain production of rice cultivars in the warm area of Japan. I. Comparison of the dry matter production between old and new types of rice cultivars. Jpn J Crop Sci 1983; 52:299–306.

Takita T. Rice breeding in Japan with emphasis on high yield and cold tolerance. In: Humphreys E, Murray EA, Clampett WS, Lewin LG, eds. Temperate Rice—Achievements and Potential. Proceedings of the Temperate Rice Conference. Vol. I. Australia: Yanco, NSW, 1994:35–41.

Tanaka A. Physiological aspects of productivity in field crops. Potential Productivity of Field Crops under Different Environments. Manila, Philippines: International Rice Research Institute, 1983:61–80.

Tanaka A, Kawano K, Yamaguchi J. Photosynthesis, respiration, and plant type of the tropical rice plant. IRRI Tech Bull 1966; 7:1–46.

Tanaka T. Studies on the light-curves of carbon assimilation of rice plants. Bull Natl Inst Agric Sci Jpn Ser A 1972; 19:88–100.

Tanaka T, Matsushima S, Kojo S, Nitta H. Analysis of yield-determining process and the application to yield-prediction and culture improvement of lowland rice. 90. Relationships between the structure of a plant community and the light-carbon assimilation curve. Proc Crop Sci Soc Jpn 1969; 38:287–293.

Tangpremsri T, Fukai S, Fischer KS. Growth and yield of sorghum lines extracted from a population for differences in osmotic adjustment. Aust J Agric Res 1995; 46:61–74.

Teare ID, Peterson CJ, Law AG. Size and frequency of leaf stomata in cultivars of *Triticum aestivum* and other *Triticum* species. Crop Sci 1971; 11:496–498.

Terashima K, Akita S, Sakai N. Physiological characteristics related with lodging tolerance of rice in direct sowing cultivation. III. Relationship between the characteristics of root distribution in the soil and lodging tolerance. Jpn J Crop Sci 1995; 64:243–250.

Tsunoda S. A developmental analysis of yielding ability in varieties of field crops. IV. Quantitative and spatial development of the stem-system. Jpn J Breed 1962; 12:49–56.

Tsunoda S. Photosynthetic efficiency in rice and wheat. Rice Breeding. Manila, Philippines: International Rice Research Institute, 1972:471–482.

Tsunoda S, Kishitani S. Photosynthesis, transpiration and leaf temperature of hosoba (narrow-leaf) lines in rice. Jpn J Breed 1976; 26(Suppl 2):215–216.

Tu ZP, Lin XZ, Cai WJ, Yu ZY. Reprobing into rice breeding for high photosynthetic efficiency. Acta Bot Sin 1995; 37(8):641–651.

Turner NC, O'Toole JC, Cruz RT, Namuco OS, Ahmed S. Responses of seven diverse rice cultivars to water deficits. I. Stress development, canopy temperature, leaf rolling and growth. Field Crops Res 1986a; 13:257–271.

Turner NC, O'Toole JC, Cruz RT, Yambao EB, Ahmed S, Namuco OS, Dingkuhn M. Responses of seven diverse rice cultivars to water deficits. II. Osmotic adjustment, leaf elasticity, leaf extension, leaf death, stomatal conductance and photosynthesis. Field Crops Res 1986b; 13:273–286.

Uemura K, Anwaruzzaman, Miyachi S, Yokota A. Ribulose-1,5-bisphosphate carboxylase/oxygenase from thermophilic red algae with a strong specificity for CO_2 fixation. Biochem Biophys Res Commun 1997; 233:568–571.

Vartapetian BB, Jackson MB. Plant adaptation to anaerobic stress. Ann Bot 1997; 79(Suppl A):3–20.

Venkateswarlu B, Vergara BS, Parao FT, Visperas RM. Enhanced grain yield potentials in rice by increasing the number of high density grains. Philipp J Crop Sci 1986; 11:145–152.

Vergara BS, Chang TT. The flowering response of the rice plant to photoperiod: a review of literature. IRRI Tech Bull 1976; 8:1–75.

Vergara BS, Tanaka A, Lilis R, Puranabhavung S. Relationship between growth duration and grain yield of rice plants. Soil Sci Plant Nutr 1966; 12:31–39.

Vergara BS, Visperas RM. Harvest Index: Criterion for Selecting Rice Plants with High Yielding Ability. IRRI Saturday Seminar. Manila, Philippines: International Rice Research Institute, 1977.

Visser EJW, Nabben RHM, Blom CWPM, Voesenek LACJ. Elongation by primary lateral roots and adventitious roots during conditions of hypoxia and high ethylene concentrations. Plant Cell Environ 1997; 20:647–653.

Voznesenskaya EV, Franceschi VR, Kiirats O, Freitag H, Edwards GE. Kranz anatomy is not essential for terrestrial C_4 plant photosynthesis. Nature 2001; 414: 543–546.

Wada G, Matsushima S. Analysis of yield determining processes and its application to yield prediction and culture improvement of lowland rice. Proc Crop Sci Soc Jpn 1962; 31:15–18.

White JG, Zasoski RJ. Mapping soil micronutrients. Field Crops Res 1999; 60:11–26.

Wissuwa M, Ae N. Molecular markers associated with phosphorus uptake and internal phosphorus-use efficiency in rice. In: Gissel-Nielsen G, Jensen A, eds. Plant Nutrition—Molecular Biology and Genetics, Kluwer, The Netherlands, 1999:433–439.

Wissuwa M, Ae N. Genotypic variation for tolerance to phosphorus deficiency in rice and the potential for its exploitation in rice improvement. Plant Breed 2001; 120:43–48.

Wissuwa M, Yano M, Ae N. Mapping of QTLs for phosphorus-deficiency tolerance in rice (*Oryza sativa* L.). Theor Appl Genet 1998; 97:777–783.

Xu DQ. Progress in photosynthesis research: From molecular mechanisms to green revolution. Acta Phytophysiol Sin 2001; 27(2):97–108.

Xu K, Mackill DJ. A major locus for submergence tolerance mapped on rice chromosome 9. Mol Breed 1996; 2:219–224.

Xu K, Xu X, Ronald PC, Mackill DJ. A high-resolution linkage map of the vicinity of the rice submergence tolerance locus *Sub1*. Mol Gen Genet 2000; 263:681–689.

Yadav R, Courtois B, Huang N, Mclaren G. Mapping genes controlling root morphology and root distribution in a double-haploid population of rice. Theor Appl Genet 1997; 94:619–632.

Yadav R, Flowers TJ, Yeo AR. The involvement of the transpirational bypass flow in sodium uptake by high- and -low-sodium-transporting lines of rice developed through intravarietal selection. Plant Cell Environ 1996; 19:329–336.

Yambao EB, Ingram KT, Real JG. Root xylem influence on the water relations and drought resistance in rice. J Exp Bot 1992; 43:925–932.

Yang J, Peng S, Visperas RM, Sanico AL, Zhu Q, Gu S. Grain filling pattern and cytokinin content in the grains and roots of rice plants. Plant Growth Regul 2000; 30(3):261–270.

Yang X, Romheld V, Marschner H. Uptake of iron, zinc, manganese, and copper by seedlings of hybrid and traditional rice cultivars from different soil types. J Plant Nutr 1994; 17:319–331.

Yeo AR, Flowers TJ. Accumulation and localization of sodium ions within the shoot of rice (*Oryza sativa*) varieties differing in salinity resistance. Physiol Plant 1982; 56:343–348.

Yeo AR, Flowers TJ. Varietal differences in the toxicity of sodium ions in rice leaves. Physiol Plant 1983; 59:189–195.

Yeo AR, Flowers TJ. Mechanisms of salinity resistance in rice and their role as physiological criteria in plant breeding. In: Staples RC, Toenniessen GA, eds. Salinity Tolerance in Plants: Strategies for Crop Improvement. New York: Wiley, 1984a:151–170.

Yeo AR, Flowers TJ. Nonosmotic effects of polyethylene glycols upon sodium transport and sodium–potassium selectivity by rice roots. Plant Physiol 1984b; 75:298–303.

Yeo AR, Flowers TJ. The absence of an effect of the Na/Ca ratio on sodium chloride uptake by rice (*Oryza sativa* L.). New Phytol 1985; 99:81–90.

Yeo AR, Flowers TJ. Salinity resistance in rice (*Oryza sativa* L.) and a pyramiding approach to breeding varieties for saline soils. Aust J Plant Physiol 1986; 13:161–173.

Yeo AR, Yeo ME, Flowers SA, Flowers TJ. Screening of rice (*Oryza sativa* L.) genotypes for physiological characters contributing to salinity resistance, and their relationship to overall performance. Theor Appl Gen 1990; 79:377–384.

Yeo ME, Yeo AR, Flowers TJ. Photosynthesis and photorespiration in the genus *Oryza*. J Exp Bot 1994; 45:553–560.

Yokota A, Okada S, Miyake C, Sugawara H, Inoue T, Kai Y. Super-RuBisCo for improving photosynthesis. In: Watanabe K, Komamine A, eds. Proceedings of the 12th Toyota Conference: Challenge of Plant and Agricultural Sciences to the Crisis of Biosphere on the Earth in the 21st Century. Austin, TX: Eurekah.com, 1998:183 190.

Yoshida S. Physiological aspects of grain yield. Annu Rev Plant Physiol 1972; 23:437 464.

Yoshida S. Effects of temperature on growth of the rice plant (*Oryza sativa* L.) in a controlled environment. Soil Sci Plant Nutr 1973; 19:299–310.

Yoshida S. Fundamentals of Rice Crop Science. Manila, Philippines: International Rice Research Institute, 1981:269.

Yoshida S, Ahn JS, Forno DA. Occurrence, diagnosis and correction of zinc deficiency of lowland rice. Soil Sci Plant Nutr 1973; 19:83–93.

Yoshida S, Cock JH, Parao FT. Physiological aspects of high yields. Rice Breeding. Manila, Philippines: International Rice Research Institute, 1972:455–514.

Yoshida S, Coronel V. Nitrogen nutrition, leaf resistance, and leaf photosynthetic rate of the rice plant. Soil Sci Plant Nutr 1976; 22(2):207–211.

Yoshida S, de los Reyes E. Leaf cuticular resistance of rice varieties. Soil Sci Plant Nutr 1976; 22:95–98.

Yoshida S, Hasegawa S. The rice root system: its development and function. Drought Resistance of Crops with Emphasis on Rice. Manila: International Rice Research Institute, 1982:97–114.

Yoshida S, Nakabayashi K, Perez PH. Photosynthesis of the rice plant. Manila, Philippines: International Rice Research Institute. IRRI Saturday Seminar, 1970.

Yoshida S, Parao FT. Climatic influence on yield and yield components of lowland rice in the tropics. Climate and Rice. Manila, Philippines: International Rice Research Institute, 1976:471 494.

Yu L, Ray JD, O'Toole JC, Nguyen HT. Use of wax-petrolatum layers for screening rice root penetration. Crop Sci 1995; 35:684 687.

Yuan L. Breeding of super hybrid rice. In: Peng S, Hardy B, eds. Proceedings of the International Rice Research Conference: Rice Research for Food Security and Poverty Alleviation. Manila, Philippines: International Rice Research Institute, 2001:143 149.

Zelitch I. The close relationship between net photosynthesis and crop yield. BioScience 1982; 32:796–802.

Zhang C, Peng S, Laza RC. Senescence of top three leaves in field-grown rice plants. Vol. 26. J Plant Nutr 2003.

Zhang GY, Guo Y, Chen SL, Chen SY. RFLP tagging of a salt-tolerance gene in rice. Plant Sci 1995; 110:227–234.

Zhang JS, Xie C, Li ZY, Chen SY. Expression of the plasma membrane H^+-ATPase gene in response to salt stress in rice salt-tolerant mutant and its original variety. Theor Appl Genet 1999; 99:1006–1011.

Zhang Y, Chantler SE, Gupta S, Zhao Y, Leisy D, Hannah LC, Meyer C, Weston J, Wu MX, Preiss J, Okita TW. Molecular approaches to enhance rice productivity through manipulations of starch metabolism during seed development. In: Khush GS ed. Rice Genetics III. Proceedings of the Third International Rice Genetics

Symposium. Manila, Philippines: International Rice Research Institute, 1996:809–813.

Zhong X, Peng S, Sheehy JE, Visperas RM, Liu H. Relationship between tillering and leaf area index: Quantifying critical leaf area index for tillering in rice. J. Agric Sci 2002; 138:269–279.

Zuckermann H, Harren FJM, Reuss J, Parker DH. Dynamics of acetaldehyde production during anoxia and post-anoxia in red bell pepper studied by photo-acoustic techniques. Plant Physiol 1997; 113:925–932.

4

Sorghum Physiology

Abraham Blum
The Volcani Center, Bet Dagan, Israel

1 INTRODUCTION

By the token of its evolution, domestication, and migration (Doggett, 1988), sorghum developed into an important crop plant serving to sustain people in very diverse and often very harsh environments. Basically, sorghum is a warm-season, daylength-sensitive, C_4-type metabolism plant. However, for almost any feature used to describe the plant, diversity rather than homogeneity is a more fitting characterization. This is linked to the wide environmental adaptation of sorghum. Different sorghum races or cultivars may express adaptation to temperate or tropical climates, high or low altitudes, water logging, or drought stress conditions.

Surprisingly, comprehensive reviews of sorghum physiology are scarce (Wilson and Eastin, 1982) and are often limited to certain topics, such as drought and heat stress responses (e.g., Sullivan and Eastin, 1974) or mineral nutrition (e.g., Clark and Duncan, 1991). Artschwager (1948) and Doggett (1988) provided the basic morphological and anatomical description of sorghum.

In the historical perspective, considerable research in sorghum physiology was done in the Western developed countries with genetic materials adapted to temperate climates. Such sorghum represents a very small fraction of the available genetic diversity of cultivated sorghum. Since the early work of J. R. Quinby in Plainview, Texas, through the work of Hugh Doggett in Africa, the sorghum conversion program in Texas, and the sorghum physiology work at the International Crops Research Institute for the Semi-Arid Tropics (ICRISAT), India, we are constantly reminded of our limited knowledge of the physiology of sorghums outside the temperate region.

Much of the early sorghum physiology research followed the development of the crop in the United States, especially after its wider acceptance upon the development of hybrids. Topics were wide-ranging, from germination physiology to the crop energy balance. A strong affiliation between sorghum physiology research and applied work in breeding and agronomy was very eminent.

John Martin and his associates at the USDA were among the first to address sorghum physiology in the context of problem solving in farming. Roy Quinby pioneered the investigation into the genetics of sorghum physiology and phenology, offering original lines of thought. Much of the following sorghum physiology research in Texas was affected by Quinby's work and thinking. This was true also for sorghum physiology work in Oklahoma, Kansas, Arizona, New Mexico, Indiana, and other states, in the sense that a large share of sorghum physiology research has been taken up by sorghum breeders. The establishment in the late 1960s and early 1970s of a sorghum physiology research group at the University of Nebraska, Lincoln, had a significant and a continuous impact on the discipline in terms of developing knowledge and techniques and in educating many sorghum scientists from all parts of the world. That program actually established sorghum physiology as an important discipline of agricultural research.

The inception of ICRISAT in India recognized the importance of physiology research in solving problems of sorghum in the developing countries of the semiarid tropics, on both national and international levels. An ongoing contribution of sorghum physiology research to plant breeding and cropping systems research and development at ICRISAT is well documented. ICRISAT is now effectively extending physiology research into environmental stress problems in Africa.

International Sorghum and Millet Research (INTSORMIL) program invested a large share of its resources in physiological research of sorghum as means to advance sorghum breeding and management in Africa and other parts of the developing world. INTSORMIL program has been an important vehicle of training newcomers into the area of sorghum physiology, agronomy, and breeding. Important physiological work has been done also by national programs in various institutions in Africa such as in Uganda, Nigeria, and at various Centers in Western Africa, such as the one in Bambey,

Senegal, and the various stations cooperating with the French Institute de Recherche de Agronomie Tropicale (IRAT).

The introduction of sorghum into new lands in Australia was accompanied by an effective crop physiology research program, which was strongly related to problem solving and was executed mostly in the field. That research may be taken as an excellent example of the contribution of crop physiology to agricultural development.

The development of sorghum growth models has been an ongoing endeavor in several centers around the world, from Texas, Kansas, and Florida to Australia (e.g., Brar et al. 1992; Carberry et al., 1993a,b; Hammer et al., 1993; Heininger et al., 1997; Muchow and Craberry, 1990; Robertson et al., 1993a,b; Sinclair et al., 1997; Stockle and Kiniry, 1990). Much of the past and present sorghum physiology research is being continuously incorporated into various crop growth models that are having an important feedback effect on the direction of research in sorghum agronomy, physiology, and breeding. However, a comprehensive discussion of the available models and submodels is beyond the scope of this review.

2 SEED GERMINATION AND SEEDLING ESTABLISHMENT

Only about 33% to 51% of sorghum seed planted in the field in various parts of Australia resulted in established plants (Radford and Henzell, 1990). Israeli sorghum growers routinely planted 22 to 23 viable and chemically protected seed in order to establish 14 to 18 seedlings per 1 m of row under dryland conditions. The discrepancy between laboratory germination rate and the final number of established seedlings in the field is explained by the nature of germination, emergence, and seedling establishment of sorghum in relation to the seedbed environment.

2.1 Dormancy

Although certain sorghums can germinate as early as 8 to 15 days after pollination (Wall and Ross, 1970), seed dormancy, which is expressed in delayed germination of freshly harvested seed, is not uncommon. Clark et al. (1968) confirmed earlier work that the seed pericarp and testa may contain germination inhibitors. Working with segregants of white (Kafir) and brown (Shallu) seed of sorghum, they demonstrated that seed with white pericarp and testa germinated readily within about a week, whereas brown seed took more than 3 weeks to germinate. The importance of seed dormancy is in the resistance to preharvest sprouting in the field and other phenomena of seed deterioration and molding as associated with damp weather conditions during maturation. Various tannins and other phenolic compounds may be present in the seed testa, especially of brown sorghums, which are relatively resistant to molds. However, not all brown sorghums are mold resistant. The exact biochemical factors in the sorghum seed testa, which are involved with dormancy and resistance to molds, were not fully understood (e.g., Jambunathan et al., 1986). Furthermore, mold resistance in sorghum grains may also be conditioned by certain proteins found in the endosperm (e.g., Kumari et al., 1992). Several sorghum seed antifungal proteins, including sormatin, chitinase, glucanase, and a ribosome-inhibiting protein, were extracted from sorghum seed and were found to inhibit spore germination of *Fusarium moniliforme*, *Curvularia lunata*, and *Aspergillus flavus* (Seetharaman et al., 1997).

Abscisic acid (ABA), which accumulates in the seed during its development, is a known inhibitor of seed sprouting. Its concentration is normally reduced toward maturity. Seed sprouting of different sorghum cultivars is likely affected by embryonic sensitivity to seed ABA (Steinbach et al., 1995, 1997) and possibly also gibberellic acid (Steinbach et al., 1997). However, the absolute concentration of ABA in the maturing seed and seed sensitivity to ABA may vary with environmental conditions during seed development with a variable effect on seed sprouting (e.g., Benech-Arnold et al., 1991). Abscisic acid is not an important control of seed dormancy after full maturation.

2.2 Germination

The definition of germination in various reports dealing with sorghum is not necessarily the same. It ranges from the appearance or the projection of the radicle from the seed to the development of a normal radicle and plumule. In this discussion germination is referred to by the latter definition.

The normal moisture content of sorghum seed is around 8% to 12%, depending on the conditions during seed maturation and storage. When seed are imbibed at 25°C their moisture content rises to a maximum of about 30% within 20 to 24 h (Meyers et al., 1984b).

Information on the exact temperature response of sorghum germination is scattered because rarely do investigations include a full range of relevant temperatures. Generally, the maximum rate of sorghum germination has been reported to be around 30° to 35°C, with little genetic variation in this respect. Genetic diversity in germination rate is seen at the nonoptimal temperatures, and more frequently at the suboptimal range (e.g., Radford and Henzell, 1990). A comprehensive study with nine sorghum genotypes (Harris et al., 1987) indicated that the base temperature for germination (T_b) was 8.5° to 11.9°C; optimum temperature for germination was 33.2° to 37.5°C; and maximum temperature for germination was 46.8° to 49.2°C, depending on genotype. T_b was found to vary from 6.9° to 13.4°C among different genotypes (Thomas and Miller, 1979), and it was further shown (Mann et al., 1985) that in tropically adapted sorghum T_b was about 7°C whereas in temperate sorghum it was about 10°C. The reason for this consistent divergence is not clear and there are various speculations based on evolutionary and environmental considerations. Lawlor et al. (1990) found that the difference in T_b between tropical and temperate sorghums held only for germination but not for seedling root or shoot elongation. The base temperature is apparently affected by the temperature sensitivity of germination immediately after imbibition. At 10°C sorghum seed imbibed normally but did not germinate (Meyers et al., 1984b).

The genetic variation in cold-temperature germination of sorghum is apparently common (e.g., Tiryaki and Andrews, 2001). Furthermore, there is heterosis ("hybrid vigor") for cold germination of sorghum (e.g., Blum, 1969), which allowed taking advantage of earlier planting in certain dryland environments (e.g., Pinthus and Rosenblum, 1961; Blum, 1972). This is further discussed below. There is significant general combining ability for cold germination. Cold tolerance in the female (seed) parent would be more important than in the male parent of hybrids (Tiryaki and Andrews, 2001).

The effect of the seed production environment on the temperature response of sorghum seed germination was not studied very extensively. Data obtained by Lawlor et al. (1990) for three locations in Midwestern United States allow the conclusion that if such an effect exists it is limited to the germination stage only, while subsequent stages of seedling root and shoot elongation are unaffected. In another study (Harris et al., 1987), different panicle temperatures during seed development did not have any effect on the temperature response of seed germination.

2.3 Emergence

Critical data on factors involved with the physiological and environmental control of sorghum emergence are lacking. The few studies that addressed emergence often confound variations in seed germination with final rates of emergence. Clearly, conditions affecting germination are not necessarily the same as those affecting emergence (e.g., Lawlor et al., 1990).

Not all seed that germinate will emerge from the seedbed. A host of biotic and abiotic seed and soil-related factors could drastically reduce emergence of germinating seed. Already in 1935 Martin et al. demonstrated that both seedbed temperature and sowing depth affect the rate and time of emergence, independently of germination. Emergence was decreased appreciably as temperature was reduced from 20° to 15°C, especially at a sowing depth of 3 cm or more.

Brar et al. (1992) investigated the emergence of sorghum cv. Richardson-9112 in a Pullman clay loam at a bulk density of 1.33, as affected by soil moisture and temperature. Sorghum rate of emergence (about 80%) was not reduced at temperatures between 20.5° and 30.2°C and at soil water potential between −0.03 and −0.1 MPa. Cooler temperature of 15.9°C did not reduce emergence as long as soil water potential was high (−0.03 MPa). Thus, field conditions, which combine high temperature and high vapor pressure deficit (VPD), are expected to reduce emergence of germinated seed, and the effect is accentuated when seedbed water potential is below −0.1 MPa.

In many sorghum-growing environments sorghum is planted into wet soil and emergence is totally dependent on stored water in the soil. This is a situation where the germinating seed and the emerging seedling must capture moisture from a slowly drying seedbed. In most cases this entails deep planting that in itself reduces emergence (Martin et al., 1935). Strong compaction of the seedbed after planting by heavy compaction wheels serve to press the seed into the wet soil while reducing its distance from the surface.

The germinating sorghum seedling emerges by the elongation of the mesocotyl and the coleoptile (Fig. 1), as in maize, rice, and oats and unlike in wheat and barley where the mesocotyl does not elongate (Hoshikawa, 1969). Because wheat mesocotyl does not elongate, there is a strong positive correlation between the potential (genotypic) coleoptile's length and wheat emergence from deep sowing. Sorghum emergence depends on the extension of both the mesocotyl and the hypocotyl. In rice the total potential length of the mesocotyl and the hypocotyl enables the prediction of rice emergence from different soil depths (Turner et al., 1982). This is not evident in sorghum, because the temperature response of sorghum coleoptile elongation is somewhat different from that of mesocotyl elongation (Radford and Henzell, 1990). Significant genetic variation exists in sorghum emergence at different temperatures and from deep sowing (Harris et al., 1987; Soman, 1990; Radford and Henzell, 1990).

2.4 Seedling Establishment

The emerging seedling will develop into a growing plant only after it is established. The longevity of the seminal root of sorghum is limited to about a month (Blum et al., 1977a). This is supported by the observation that the experimental excision of the crown roots in hydroponically grown sorghum seedlings caused a sharp reduction in shoot growth between 15 and 35 days after sowing (Jesko, 1972). The mesocotyl is often observed to deteriorate, whether spontaneously or under the infection of soil-inhabiting pathogens. The reasons for the inherent limited viability of the seminal root and meso-

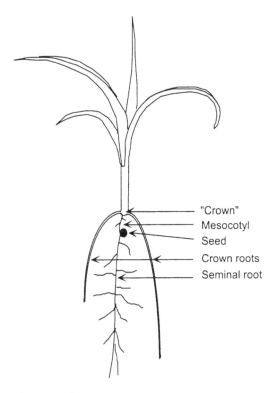

"Crown"
Mesocotyl
Seed
Crown roots
Seminal root

Figure 1 Schematic morphology of a sorghum seedling during its establishment. (From Martin et al., 1935, and Blum, 1988.)

cotyl of sorghum are not fully understood, but it can be seen (Fig. 1) that the deterioration of either can bring about the collapse of the seedling, unless the crown roots (adventitious roots) develop and grow into the wet soil. Thus, sorghum seedling establishment is determined by the successful shift from dependence on seminal root to dependence on crown roots. Crown roots begin to initiate spontaneously when the first two to three leaves are fully expanded. Thermal time from emergence to the appearance of the first crown root is about 160°C days (base temperature of 10°C) (Soman and Seethar-ama, 1992). Crown roots are initiated in weekly cycles, and two to four roots are initiated in each cycle (Blum et al., 1977a,b). The first cycle roots will determine the seedling's establishment. Whether the initiated crown roots will grow into the soil depends on the soil and the atmospheric environment. It has been found in sudan grass and maize (Wenzel et al., 1989) that late metaxylem differentiation in the crown root proceeds only after the root reached a length of about 30 cm. Because the axial hydraulic resistance of the root is high as long as late metaxylem is not developed, sorghum seedlings with only young crown roots may be relatively vulnerable to water stress.

The topsoil becomes hard as it dries. Crown roots will not penetrate hard drying topsoil. The exact relationship between crown root penetration and topsoil strength or topsoil moisture status (independently of strength) is not known. The role of the shoot in affecting crown root growth is clearer. Crown root growth rate, especially in small seedlings, is determined by the amount and rate of carbon partitioned to the root (see further discussion in Section 4). In this respect, vigorous seedlings, such as those of hybrids, have a clear advantage (Blum et al., 1977b). The superior seedling establishment of sorghum hybrids was an important factor in their rapid acceptance in dryland farming.

High topsoil temperatures are a major impediment to achieving a proper stand in some semiarid tropical environments, even when the soil is wet. Emerged sorghum seedlings die at high soil temperatures (>45°C) (Peacock et al., 1990, and the review therein). The cause for heat killing of emerged seedlings was ascribed to reduced carbon partitioning to the seminal root, caused by high temperatures and by phloem restriction of transport to the root ("heat girdling").

Sorghum seedlings are capable of accumulating heat shock proteins (HSPs) during imbibition and coleoptile growth (Ougham and Stoddart, 1986; Howarth, 1990a) and afterward (Jorgensen et al., 1993). Their accumulation seems to be positively correlated with acquired seedling thermotolerance (Ougham and Stoddart, 1986; Sivaramakrishnan et al., 1990; Howarth, 1990a,b). Many HSPs are produced constitutively and of these some are synthesized to large amounts under heat stress (Jorgensen et al., 1993). The synthesis of HSPs and the induction of seedling thermotolerance were found to be rapid, reversible, and reinducible (Howarth and Skot, 1994). Maximal thermotolerance was obtained after treatments that induced the full complement of HSPs. Subsequent treatments that repressed HSP synthesis, also abolished thermotolerance. The presence of HSPs before heat stress, however, was not sufficient for the tissue to be in a thermotolerant state and the results suggested that either their de novo synthesis, or some other factor, is required for the induction of thermotolerance.

Translatable RNAs encoding HSPs were found to be present in sorghum seedlings and also in quiescent embryos (Howarth, 1990a). The presence of these RNAs in dry seed led to suggest that RNAs stored in the maturing seed in the field provides for early HSP production upon germination. This would explain, at least partly, some of the known effects of the environment during seed maturation on its environmental response during germination. It would also require standardizing experiments for the seed maturation environment when seedling response to environmental stress is tested.

The effect of temperature on crown root development and seedling establishment has not been studied in detail. Martin et al. (1935) noted that when sorghum seed were planted at a depth of 2.5 cm and subjected to soil temperatures of 15° and 25°C, the crown developed in the soil at 15°C and above the soil at 25°C. This corresponds very well with data of Radford and Henzell (1990), who found that mesocotyl elongation was accelerated by high temperatures between 15° and 30°C. Crown development above the soil may cause failure to establish crown roots.

Undoubtedly, the final stand established under the dryland field conditions is a function of different stress perturbations during seed imbibition, germination, emergence and establishment. Although variations in heat shock proteins may be associated with germination (Howarth, 1990) or emergence of certain genotypes under heat stress (Sivaramakrishnan et al., 1990), more research is required to understand the complete chain of events leading to satisfactory sorghum seedling establishment under environmental stress.

3 SHOOT ONTOGENY, GROWTH, AND DEVELOPMENT

During the early 1970s, Quinby put forward his hypothesis that very few major genes, which were basically those controlling plant height and flowering, controlled sorghum growth and development. This hypothesis is especially compelling today after the fact that only few quantitative trait loci (QTLs) were found to be involved in modifying wild species into well-adapted domesticated crops, such as sorghum (Paterson et al., 1995). Quinby's hypothesis went further to propose that these genes affected growth by way of involving the balance between different endogenous growth hormones. Although this hypothesis has not been fully rejected or accepted yet (e.g., Wright et al., 1983a,b; Morgan and Quinby, 1987; Beall et al., 1991), there is no doubt that few major genes exercise a major control on how the sorghum plant grows and develops.

3.1 Photoperiod Response and the Maturity Genes

Sorghum is a quantitative short-day plant. It will flower earlier as the days become shorter, depending on the maturity genotype (Ma). Sorghum is receptive to the photoperiod signal from 4 to 9 days after emergence to about the onset of panicle initiation (Ellis et al., 1997). After sensing the photoperiod stimuli, the inductive effect persisted for 4–14 days in short days (SD), and for

15–33 days in long days (LD) depending on genotype (Alagarswamy et al., 1997).

Tropical cultivars, which require very short days for flowering and will not flower in the long days of the summer in the temperate region, are basically dominant for all four loci that control the time to flowering (Quinby, 1974). The substitution of one locus from dominant Ma_1 to recessive ma_1 have converted the tropical sorghum to a temperate one that will flower in high latitudes (Quinby, 1974; Major et al., 1990). However, such converted sorghums still varied in their date of flowering according to the effect of alleles in the other three maturity loci. The earliest genotype in Quinby's (1974) study was the recessive in all four maturity loci. The recessive $ma_3{}^R$ allele from "Ryer" milo eliminates photoperiod sensitivity altogether (Pao and Morgan, 1986b; Major et al., 1990).

Miller et al. (1968) investigated the effect of photoperiod on sorghum development under the tropical conditions of Puerto Rico, where variations in seasonal temperatures were low as 2°C. He established the fact that different sorghum maturity genotypes have different critical photoperiod requirements, but all will flower if exposed to the very short days of December in Puerto Rico. They also established five groups of genotypes according to their critical photoperiodic requirement. In the most sensitive group 11.1- to 11.2-h long days will delay flowering, while the least sensitive group required 12.0–12.6 h to delay flowering. The temperate sorghums (as developed in the United States) belonged off course to the least sensitive or the totally insensitive group.

Most of the effect of the maturity genes on time of flowering is mediated through their effect on the duration of the period from seedling emergence to panicle initiation (GS1) (Quinby, 1974). Because leaves are being formed in the shoot apex as long as it does not differentiate into a panicle, later flowering evidently entails the formation of proportionally more leaves (Fig. 2) and often, taller stems. The maturity genes cause additional morphological and developmental variation in sorghum. Some of this variation may result from a possible effect of the maturity genes also on the duration of the period from panicle initiation to flowering (GS2). Quinby's (1974) work and that of others (e.g., Sorrells and Meyers, 1982) indicated that the effect exists, although it is small in magnitude as compared with the effect on GS1. On the other hand, the effect of photoperiod on duration of GS2 was not seen in a number of early-to-medium maturity commercial U.S. and Australian hybrids (Hammer et al., 1989).

Pao and Morgan (1986) found that most of the variation in plant developmental characteristics (other than flowering time) caused by the maturity genes could be assigned mainly to the presence and effect of the $ma_3{}^R$ allele. However, an effect of maturity alleles independent of $ma_3{}^R$ ($Ma2Ma3$) on

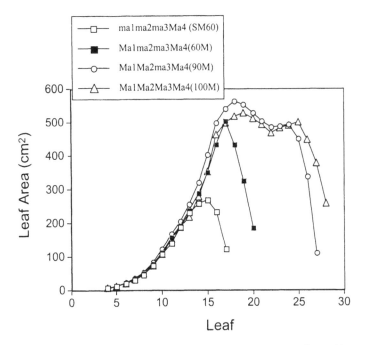

Figure 2 The maximal area of successive leaves as affected by maturity genes, in Plainview, Texas. (Adapted from Quinby, 1974.)

various root developmental characteristics was observed (Blum et al., 1977a,b), as discussed below. Effects of the maturity genes on plant developmental traits other than the time to flowering are not strictly pleiotropic. In controlling the number of leaves and their area (Fig. 2), the maturity genes mediate processes and events which involve carbon assimilation, partitioning and source–sink interactions—which may carry far fetching effects beyond those of flowering time.

The premature decline in area of late developed leaves in the two late genotypes depicted in Fig. 2 was most probably the consequence of stress conditions in the field where they where grown. Maas et al. (1987) established very well the unimodal distribution of leaf size along the sorghum stem. As can be seen in Fig. 2, the onset of the decline in the maximum area of successive leaves takes place at higher leaf insertion, as panicle initiation is delayed. In fact, progressively larger leaf areas along the stem were produced when panicle initiation was delayed indefinitely under sufficiently long days (McCree, 1983). Maas et al. (1987) reasonably suggested that the onset of the decline in the size of successive leaves results from intraplant competition. Based on the above cited results and results for other determinate plant species it may be concluded that this competition is driven mainly by the growing stem and panicle within the plant.

Major et al. (1990) proposed the basic concepts for predicting leaf number of sorghum according to its photoperiodic response, as employed already in several crop growth models such as the CERES maize model. They proposed that:

$$LN = BVP + PS(Phtpd - MOP) \tag{1}$$

where LN is leaf number, BVP is the basic vegetative phase (in leaf number) that is independent of photoperiod, PS is photoperiod sensitivity (in increased leaf number per hour delay of photoperiod), Phtpd is photoperiod in hours, and MOP is the minimum optimal photoperiod (in hours). For the very different maturity genotypes and a Phtpd of 11 to 15 h as used by Major et al. (1990), the following were the appropriate value ranges: LN 9 to > 35; BVP 9.1 to 12.7 leaves; PS 0 to 4.84 leaves per hour; and MOP 0 to 12.85 h. BVP may be largely affected by the number of embryonic leaves in the seed, which was four in the cultivated grain sorghums inspected by Clark (1970).

While modeling the photoperiodic response and developing the capability to predict phasic development is important, the control of flowering time in sorghum is not quite resolved. Some of the questions were dealt with in a series of papers on the nature of the photoperiod insensitive ma_3^R allele. Apart from its effect toward photoperiod insensitivity, this early maturity inducing allele generates plants that are taller, tiller less, have longer and narrower leaves and exhibit earlier rapid shoot elongation immediately after floral initiation (Pao and Morgan, 1986b; Beall et al., 1991). The effect of this genotype could be mimicked in photoperiod-sensitive later flowering isogenic lines by exogenous GA_3 application (Pao and Morgan, 1986b). This mimicking suggested that the maturity genes affect early flowering by regulation of gibberellin production in the plant. Indeed, it was later seen (Beall et al., 1991) that induced chemical inhibition of endogenous gibberellin synthesis in ma_3^R created a phenocopy of the late and photoperiod-sensitive genotypes. It was further shown that the ma_3^R genotype was responsible for a two- to threefold increase in GA_1 in various plant parts, as compared with later flowering and photoperiod-sensitive genotypes.

However, Morgan and Quinby (1987) showed that while exogenous application of GA_3 hastened floral initiation in late-flowering and photoperiod-sensitive genotypes, it did not affect the time of flowering. It was partly the result of a de-differentiation of the panicle primordium into leaves, in the absence of the photoperiodic induction. This may indicate that very possibly floral initiation and floral differentiation are genetically independent.

At least two phytochromes were detected in sorghum (Childs et al., 1992): a light-labile 126-kDa phytochrome that predominates in etiolated tissue and a 123-kDa phytochrome that predominates in green tissue. This 123-kDa phytochrome was not detected in the photoperiod-insensitive $Ma_1Ma_1Ma_2Ma_2ma_3^R$ genotype but was abundant in photoperiod-sensitive $Ma_1Ma_1Ma_2Ma_2ma_3ma_3$ genotype. It was later established that the Ma_3 locus in sorghum is a PHYB gene that encodes a 123-kDa phytochrome (Childs et al., 1995, 1997).

It was suggested that photoperiod control of flowering in sorghum was correlated with the presence of the 123-kDa phytochrome in green tissues (Childs et al., 1992). Thus, the ma_3^R allele, which ascribes photoperiod insensitivity in sorghum, could be associated with overproduction of gibberellin and the deficiency in the 123-kDa phytochrome. This assumption was later overruled (Foster and Morgan, 1995) by experimental evidence showing that the absence of the 123-kDa phytochrome in ma_3^R ma_3^R disrupted the diurnal regulation of the GA19 to GA20 step rather then simply causing the overproduction of GA20 and GA1. After studying the diurnal fluctuations in GA20 and GA1 it was further suggested that the diurnal rhythm of GA levels play a role in floral initiation (Lee et al., 1998). More recently, Finlayson et al. (1999) found that a sorghum mutant deficient in functional phytochrome B and exhibiting reduced photoperiodic sensitivity expressed high-amplitude ethylene rhythms.

There is a very possible photoperiod-by-temperature interaction with respect to flowering and leaf number. Caddel and Weibel (1971) and Quinby et al. (1973) were the first to note that under long days the number of leaves increased with temperature in certain genotypes but not in others. However, in most genotypes leaf number increased as night temperature rose from $23°$ to $29°C$ (Quinby et al., 1973). Sorrells and Meyers (1982) suggested that the night/day temperature differential was responsible for the interaction with photoperiod. Major et al. (1990) identified the Ma_2 locus as mediator of the interaction, but its exact role was not clear. In a later study, Craufurd et al. (1998) published mean values for base ($8.5°C$) and optimum ($27°C$) temperatures for the time duration from planting to panicle initiation when different cultivars were grown at photoperiods of 11.0 to 12.3 h. The interaction between temperature and photoperiod was not resolved.

Morgan et al. (1987) revealed that photoperiod-sensitive genotypes carrying the Ma_1Ma_2 alleles, responded to asynchrony of thermoperiod and photoperiod. For example, when the 12-h thermoperiod was shifted 0.5 or 2.5 h forward (asynchronous state) floral initiation was hastened as compared with the synchronized state. The ma_3^R genotype was insensitive in this respect. These investigators suggest that the phenomenon may have evolved in sorghum in its near-equatorial region of origin, where photoperiod sensitivity alone would not allow sufficiently wide genetic adaptation to seasonal change.

Hammer et al. (1989) presented a model for estimating sorghum phenology by photoperiod and temperature over a given set of cultivated hybrids. The model is largely based on principles discussed earlier. From their data collected in the greenhouse and the field they described the temperature sensitivity of sorghum development as curvilinear, with the fastest rate of development at 30°C. They noted that newer sorghum hybrids tended to have slower rates of development than older hybrids, especially close to 30°C. The reduced physiological sensitivity to temperature variations, in this case at the higher range of temperatures, is most likely an expression of heterosis in sorghum (Blum, 1989), which is expected to be greater in newer hybrids.

Recently, Rooney and Aydin (1999) identified two additional photosensitivity-controlling loci, designated as Ma_5 and Ma_6.

3.2 Plant Height

Stem length is controlled by four height (Dw_1 through Dw_4) gene loci, which affect stem internode length. The recessive allele in each locus decreases stem length, and there are three major height groups, ranging from the shortest (4-dw) to the tallest (1-dw) (Quinby, 1974). Schertz et al. (1971) found that the dw_3 allele as compared with Dw_3 specified a high level of peroxidase production in stem internodes and they suggested it inhibits growth-promoting substance activities in the stem. In other cereals, such as wheat, the effect of the dwarfing genes was mediated largely by gibberellin metabolism, to the extent that the height genotype could be identified by the phenotypic response to exogenous gibberellin application (e.g., Gale and Youssefian, 1985). This may be expected also for sorghum (Morgan et al., 1977; Wright et al., 1983b).

The exact phenotypic height depends also on the environmental conditions and possible modifying genes. Other implications of plant height on sorghum growth and productivity are further discussed below.

3.3 Tillering

The information on sorghum tillering is very lacking. Sorghum tillering has not received as much attention as tillering in the small grains, for the reason perhaps that a much smaller proportion of grain yield is normally produced by tillers in sorghum, as compared with the small grains.

The tillering capacity of sorghum is the main reason why this plant is often referred to as "perennial." Sorghum can regrow from basal tillers after the crop is harvested at grain maturity. This regrowth may result in subsequent crop(s) defined as "ratoon." The number of ratoon crops is limited mainly by the season and other environmental and biotic factors. The regrowth capacity of sorghum from basal tillers is off course the grounds for using sudan grass and other forage sorghums for multiple harvests and grazing.

The tillering capacity at the juvenile plant growth stage is an important component of sorghum plasticity and ability to compensate for varying environmental resources in time and space. Many of the general issues concerning the physiology of tillering in crop plants may be relevant for sorghum, but most were never verified for this crop plant. For example, the effect of light quality on tillering was established for wheat (Casal, 1988) and it may or may not occur in sorghum.

Tillering of sorghum is based on buds located in each stem internode (Artschwager, 1948), which under certain conditions may differentiate. Whether such tillers would grow or degenerate depends on a set of conditions. Sorghum may potentially develop axial tillers from any stem internode above the crown if the appropriate conditions prevail, such as panicle removal in certain genotypes.

The basic conditions for initiating tillers are determined by photoperiod and temperature. Perhaps, as in other plant species, these effects are mediated by the carbon concentration in the region of initiation and plant hormone involvement in apical dominance. The establishment of growing tillers is largely dependent on soil conditions and the light regime.

There is a clear role for plant hormones in controlling tiller initiation. Exogenous GA_3 applications were effective in eliciting apical dominance and in inhibiting tillering (Isbell and Morgan, 1982). Tiller outgrowth was promoted by ancymidol treatment, which inhibits endogenous GA. The photoperiod-insensitive $ma_3{}^R$ allele reduces tillering as a result of abundant endogenous GA_1 production (Beall et al., 1991).

Apart from the specific effect of this allele, tillering tends to increase under conditions that delay panicle initiation, namely, longer days with photoperiod-sensitive genotypes (Downes, 1968; Escalada and Plucknet, 1975a,b). Using "Combine Kafir" (a temperate cultivar) under short days, Downes (1972) found that tillering required that the seedling be exposed to temperature below 18°C. Plants were most responsive to this temperature when they had about 4 to 6 expanded leaves. Downes suggested that more developed plants did not tiller in response to low temperature probably because panicles initiated and apical dominance prevented tillering. He further argued that because other tropical grasses tiller under much higher temperatures, the requirement for low temperature may have been specifically bred into the temperate sorghum during its adaptation to the conditions of the Midwestern United States.

Using "Pride 550 Br" sorghum hybrid, which was defined as "somewhat photoperiod insensitive," Escalada and Plucknet (1975b) found that when photoperiod requirement was supplied, tillering increased as temperature rose from 23.9°/15.5° to 32.2°/23.9°C (day/night) temperatures. The disagreement between this study and that of Downes (1972) may stem from a temperature by genotype interaction for tillering in sorghum. Secondly, Escalada and Plucknet (1975b) dealt with tillering from mature stubble (ratoon) while Downes (1972) worked with juvenile plants prior to flowering induction. Thirdly, genotypes may vary in their basic tillering capacity irrespective of their photoperiod response.

Beyond the effect of genotype, photoperiod, and temperature, additional factors may influence tiller initiation. These may involve plant density, irradiance, and mineral nutrition, mainly nitrogen. Youngquist and Maranville (1992) concluded that an increase in the number of productive tillers was the main mechanism by which grain yield increased under high nitrogen fertilization.

Tiller establishment, defined as the proportion of initiated tillers that reach maturity, is even less understood. Data by Escalada and Plucknet (1975a) are perplexing in that they show in pot-grown plants that sometimes the first tiller(s) may die and in other cases, the later tillers may die, without any apparent pattern or reason. It may be assumed that the probability for tiller establishment is high if it developed early before being shaded by the developing canopy. Tillers will establish better if they establish roots successfully. The probability for effective rooting is expected to decrease as tillers are initiated from higher (epigeal) internodes. Such tillers are also the later ones. A dry and hard soil surface is an important impediment to penetration by crown roots (Blum and Ritchie, 1984), and therefore it may impel tiller establishment.

Depending on the environmental conditions, the tiller is generally smaller, it may flower up to a week later, and it may have about four leaves less than the main stem when grown in a temperate climate (Gerik and Neely, 1987). Undoubtedly, the tiller grows under a totally different environment from that under which the main stem grows, both spatially and temporally. Under most sorghum growing conditions, the tiller is significantly smaller than the main stem. From reviewing the many plant density experiments performed with grain sorghum it may be generally concluded that under the normal and modern management conditions, up to three tillers may reach maturity and that their relative contribution to total yield is far less than in the small grains. Tillers did not contribute to sorghum yield when plant density was sufficient (≥ 12.5 plant m^{-2}) (Gerik and Neely, 1987). One reason for the moderate importance of sorghum tillering under normal growing conditions is the large developmental plasticity of the panicle (see further below).

3.4 The Leaf and Its Function

Sorghum leaf growth and expansion follows the general pattern typical of the Gramineae and it will not be discussed here. Leaf–water relations are discussed later.

Sorghum leaf longevity is affected by ontogeny, the environment, and the genotype. Leaf age, water stress, and the deficiency in or export of certain nutrients such as nitrogen are all very important in affecting leaf senescence. Because most of these factors involve the breakdown of leaf proteins, leaf chlorophyll content is a convenient measure of leaf senescence in sorghum (e.g., Duncan et al., 1981; Khanna-Chopra and Sinha, 1988). Plant senescence in relation to yield and other whole plant considerations is discussed in a following section.

3.4.1 Photosynthesis

Sorghum possesses the typical C_4 pathway of photosynthetic carbon assimilation, with the associated leaf anatomical traits and environmental consequences. The reader is referred to papers by Pearcy and Ehleringer (1984) and by Brown and Hattersley (1989) for a general discussion of C_4-type metabolism. Being a C_4 plant sorghum is characterized by relatively high rate of photosynthesis, no photorespiration, high water-use efficiency (or transpiration ratio), light saturation of photosynthesis at high irradiance, and adaptation of plant metabolism to warm climate.

During the 1970s and 1980s a continuous debate had been taking place in the literature on the relative importance of stomatal and nonstomatal (photochemical) factors in controlling leaf photosynthetic response to environmental conditions such as irradiance, ambient CO_2 concentration, and plant water deficit. Farquhar and Sharkey (1982) proposed a new approach to understanding the relative interplay of stomata and chloroplast biochemistry in controlling photosynthesis. Their model depicted that direct stomatal limitation to photosynthesis is relatively small. Most of the effect of the environment on photosynthesis is mediated by affecting photosystem II (PSII) biochemistry, mainly ribulose biphosphate (RuBP) regeneration. When photosystem II activity is reduced by the environment, leaf internal CO_2 concentration increases (relative to external CO_2 concentration) and stomata gradually respond by closure. When stomata gradually close internal CO_2 is reduced, depending on chloroplast activity. Thus, the stomata operate to maintain a constant ("optimum") ratio of internal to external CO_2 concentration. This ratio determines the transpiration ratio of the leaf.

Krieg and Hutmacher (1986) confirmed that the primary cause for variation in photosynthetic carbon fixation in sorghum under the effect of leaf age, irradiance, and water deficit is photosystem activity rather than stomatal activity. They determined that the ratio of leaf internal to external CO_2 concentration in sorghum was 0.58 to 0.60 in field-grown plants under semiarid conditions. This ratio, which is inversely proportional to transpiration ratio, varies genetically and it can be well estimated in C_3 plants by measuring the plant stable carbon isotope discrimination (Δ) (e.g., Farquhar et al., 1989).

Carbon isotope discrimination was theoretically related to the ratio of internal to external CO_2 concentration, and therefore it was considered as an estimator of transpiration ratio of leaves or the water-use efficiency of whole plants. A wealth of evidence in several C_3 plants confirmed this theory. Subsequently, carbon isotope discrimination and hence water-use efficiency were found to vary also among different sorghum genotypes (Hubick et al., 1990).

Genetic variation in leaf carbon assimilation, transpiration, stomatal conductance, and transpiration ratio has been long demonstrated in sorghum (e.g., Blum and Sullivan, 1972; Peng and Krieg, 1992). However, in the light of the model of Farquhar and Sharkey (1982) it is now realized that the important component of this genetic variation is the photosynthetic capacity (or photosystem capacity to reduce leaf internal CO_2 concentration). For a C_4 plant such as sorghum, Hubick et al. (1990) suggest that meaningful genetic variation in photosynthetic capacity and/or water-use efficiency may result from variable "leakiness" of the bundle-sheath cells or from variable ratio of assimilation rate to stomatal conductance. Thus, for example, genetic variation and even heterosis exists in sorghum for the ratio of carbon exchange rate (CER) to stomatal conductance (Blum, 1989) and the increase in this ratio expressed very well the effect of heat hardening on the photochemical component of sorghum assimilation under very high temperatures.

The importance of the photochemical component in mediating genetic and environmental effects on photosynthesis was partly responsible for the increasing popularity of chlorophyll fluorescence as a probe of photosystem integrity and activity in various plants (e.g., Baker, 1991) including sorghum (Ludlow and Powels, 1988; Havaux, 1989).

While the model of Farquhar and Sharkey (1982) describes the effect of drought stress on leaf photosynthesis by way of reducing the photosynthetic capacity, upon which stomata follow suit, new evidence recently indicates that a direct effect of soil moisture stress on stomatal closure is possible. Such an effect, defined as a "nonhydraulic root signal," is discussed further later.

There is not a very good agreement in the literature on the temperature response of photosynthesis in sorghum, probably because it varies so much with environmental preconditioning and genotype. Experimental data tend to vary according to the thermal preconditioning of leaves prior to their measurement. Furthermore, in some past experiments data on photosynthesis were related to air temperature rather than to leaf temperature.

El-Sharkawy and Hesketh (1964) found that net photosynthesis increased with the rise of leaf temperature from 30° to about 44°C. Loreto et al. (1995) found photosynthesis to be maximized around 37°C. Photosynthesis clearly declined above 40°C. When leaf CER was measured after a realistic protocol of hardening (Blum, 1989), CER increased steadily with leaf temperature above 32°C and it was maximized between 37° and 40°C in different genotypes. However, transpiration ratio reduced steadily with temperature between 34° and 44°C, depending on genotype. Genetic variation in maximum temperature for photosynthesis (in the range of 40° and 43°C) was also reported by Sullivan and Ross (1979).

The response of sorghum photosynthesis to low (chilling) temperatures is not well documented. Photosynthesis still proceeds at a low rate at 15°C (Downes, 1970). Havaux (1989) found significant genetic variation among sweet sorghum and sudan grass cultivars in the reduction of the photochemical quenching of chlorophyll fluorescence in intact leaves at 3°C.

It is the consensus that single leaf photosynthesis is not representative of canopy assimilation and that single leaf measurements of photosynthesis have a low predictive value for estimating variations in crop growth or yield. However, an exception to the consensus is being noted for several crops including sorghum (Peng et al., 1990), especially when genetic variation in leaf photosynthesis is considered within a group of genotypes of similar phenology and canopy architecture. Genetic variation in the biochemistry of photosynthesis is now being revisited with the emerging options for genetic engineering. Thus, single leaf photosynthesis, quantum yield, and chlorophyll fluorescence are becoming important criteria for assessing the genetic potential of the photosystem in relations to plant breeding.

3.4.2   The Leaf Surface

Sorghum leaves are heavily covered with epicuticular wax, visually recognized as the "waxy bloom" on the abaxial surfaces of the leaf lamina and the leaf sheath. On a microscopic scale the depositions take an amorphic, a flaky, or a "starlike" shape on the leaf lamina while they form a dense, thin fiberlike mass on the leaf sheath (e.g., Blum, 1975; Tarumoto et al., 1981; Maiti et al., 1984; Traore et al., 1989). These depositions are genetically controlled by the *Bm* gene locus, where the recessive allele(s) (*bm*) reduces leaf lamina epicuticular wax load to about a quarter of that in the wild type (Blum, 1975; Ebercon et al., 1977) and may also reduce cuticle thickness and weight (Jenks et al., 1994). There are several *bm* (bloomless) mutants in sorghum.

Leaf sheath cuticular waxes on *Bm* sorghum were approximately 96% free fatty acids, with the C28 and C30 acids being 77% and 20% of these acids, respectively. In 12 *bm* mutants the reduction in the amount of C28 and C30 acids accounted for essentially all of the reduction in total wax load relative to the *Bm* genotype (Jenks et al., 2000).

Epicuticular wax load on sorghum leaf laminae, as measured colorimetrically (Ebercon et al., 1977), varied with environmental conditions and genotype (Jordan et al., 1983) between about 0.6 and 2.3 mg dm^{-2}. Water deficit was found to increase epicuticular wax load (Jordan et al., 1983; Premachandra et al., 1992). The environmental factors that promote epicuticular wax deposition on leaves of plants are generally those that lead to an increase in transpirational demand plant water deficit, namely, high irradiance, high temperature, high vapor pressure deficit, and soil moisture deficit. Certain conditions may also modify the physical state of the depositions, as seen in other plant species. The specific effects of these environmental variables on epicuticular wax deposition in sorghum leaves were not well investigated. For other plant species the chemical composition of the wax determines the shape of its deposits.

Indian sorghum breeders have long noticed that shoot-fly (*Atherigona varia soccata*)-resistant sorghums were characterized by a light green glossy appearance of seedling leaves (e.g., Maiti et al., 1984), as in cultivar M-35-1. The glossy leaf appeared to be related to nonpreference for oviposition by the shoot fly (Maiti et al., 1984). The glossy leaf trait is controlled by a single recessive gene and can be easily identified by the adherence of sprayed water to the leaf (Tarumoto, 1980; 1981). The frequency of the glossy trait in a large sample of the world sorghum collection was about 2.8% (Maiti et al., 1984). Glossiness was associated with reduced wax depositions, increased cuticular transpiration and leaf wetness, trichome appearance at early stages in young leaves (Traore et al., 1989; Tarumoto et al., 1981; Maiti et al., 1984; Sree et al., 1994). While claims were made for a greater drought resistance in glossy genotypes, this cannot be reconciled with their absence of or reduced epicuticular wax and increased cuticular conductance. The specific effect of the glossy leaf gene on sorghum water relations warrants further investigation.

The general implications of epicuticular wax load on the gas exchange and the spectral characteristics of leaves are discussed elsewhere (Blum, 1988). In sorghum, high epicuticular wax load increase leaf surface reflectance (Blum, 1975; Grant et al., 1995), reduce net radiation by about 3% to 5% at midday (Blum, unpublished data) and reduce cuticular transpiration (Blum, 1975; Chatterton et al., 1975; Jordan et al., 1984; Traore et al., 1989; Premachandra et al., 1995) irrespectively whether stomata are open or closed (Blum, 1988). The effect of epicuticular wax toward reduced transpiration is expressed also in increased leaf water-use efficiency (transpiration ratio) (Premachandra et al., 1995). However, the effects of epicuticular wax are finite (Jordan et al., 1984) and it is reasonable to assume that for many normal sorghums, epicuticular wax load is already optimized. Any further increase in load such as above $0.7 \, mg \, dm^{-2}$ (Jordan et al., 1984) or $1.5 \, mg \, dm^{-2}$ (Blum, unpublished data) would not reduce transpiration any further. This threshold value may very well be affected by the composition of the wax.

Reduced epicuticular wax load, as achieved by the bloomless genotype, significantly improved the estimated forage digestibility by ruminants (Cummins and Dobson, 1972).

3.4.3 Hydrocyanic Acid (HCN) Potential

The cyanogenic glucoside dhurrin [D-glucopyranosyl-oxy-(*S*)-*p*-hydroxy-mandelonitrile] is synthesized within cells of sorghum shoots and roots

(Adewuai, 1990) but not seed (Halkier and Lindberg, 1989, and the review therein). It was not synthesized in roots of etiolated seedlings but it was present in roots of green seedlings (Adewuai, 1990). Biosynthesis takes place in etiolated seedlings or green plants, at higher rate in the light than in the dark (Halkier and Lindberg, 1989; Wheeler et al., 1990).

Sorghum seedlings synthesize dhurrin from L-tyrosine. Intermediates in the pathway are N-hydroxytyrosine, p-hydroxyphenylacetaldoxime, p-hydroxyphenylacetonitrile, and p-hydroxymandelonitrile. The latter compound is converted to dhurrin by specific UDP-glucose glucosyltransferase (Halkier and Moller, 1989). Dhurrin metabolic turnover in young sorghum seedlings is high. Although the rate of biosynthesis was high, 27% and 34% of the synthesized dhurrin was broken down, in the shoot and root, respectively (Adewuai, 1990). Still, dhurrin accumulation in young sorghum plants may even reach 5% of total dry matter (cited in Kojima et al., 1979) and its cyanogenic capacity may peak at about 2500 mg kg^{-1} dry matter of hydrocyanic acid (HCN) (Wheeler et al., 1990).

An additional cyanogenic glucoside, dhurrin-6-glucoside, has been recently identified in sorghum leaves (Selmar et al., 1996). It is a relatively minor cyanogenic glucoside, which occurs only in low concentrations but may be present in significant amounts in guttation droplets of young sorghum seedlings.

It has been established for white clover that dhurrin has no vital physiological importance for the growing plant. However, in view of its high turnover rate, it has been suggested by several authors that dhurrin may provide carbon atoms for other biosynthetic pathways of physiological significance. Evolutionary, the cyanogenic capacity of plants may have a role in deterring predators. In agriculture, the cyanogenic capacity of sorghum is hazardous to animals feeding on young sorghum forage, under certain circumstances.

In green leaf blades of young sorghum seedlings dhurrin is located entirely in the epidermal layers (Kojima et al., 1979). The two enzymes responsible for its catabolism, namely, dhurrin beta-glucosidase and hydroxynitrile lyase, reside almost exclusively in the mesophyll cells. There are two isozymes (genes) of the cyanogenic beta-glucosidase dhurrinase: dhurrinase-1 (Dhr1) and dhurrinase-2 (Dhr2), with the expression of the former being organ specific (Cicek and Esen, 1998). The compartmentation of dhurrin and its catabolic enzymes in different tissues prevents its large-scale hydrolysis under normal physiological conditions. However, any condition that disrupts the tissues and allows mixture of substrate and enzyme would cause cynogenesis.

HCN potential increases with nitrogen fertilization and water stress. However, the effect of water stress in this respect was not always repeatable (Wheeler et al., 1979).

HCN potential of sorghum develops already in the emerging first leaf during germination (Halkier and Moller, 1989) and it increases to maximum in the young plant. It then reduces curvilinearly with age (Wheeler et al., 1990). The reduction in HCN potential may depend on genotype, whereas in some cultivars HCN potential hardly reduces with age. Large variations exist in HCN potential among genotypes of sorghum and sudan grass, and the most practical approach to reducing the hazard has been by breeding for low HCN potential.

3.4.4   The Brown Midrib Trait

Similar to previously identified mutants in maize, the brown midrib mutant (*bmr*) was also isolated in sorghum (Porter et al., 1978). This mutant is recognized phenotypically by the brown pigmentation of the midrib and is more pronounced on the abaxial surface. When leaves are senescing, the mutant is easily recognized in its stem pith color, which is brownish yellow to dark brown as compared with light green or white pith of the normal type. The most prominent effect of the *bmr* mutation was found to be the decrease in lignin content of up to 51% in mature stems and 25% in leaves (Porter et al., 1978). However, it was later found that there were no significant differences in total lignin contents between *bmr* and normal lines as determined by the acetyl bromide procedure or the sum of the acid-insoluble lignin and acid-soluble lignin. It was suggested that the mutant was characterized by higher amounts of lignin with a lower degree of polymerization than the normal genotype (Lam et al., 1996). The *bmr* sorghum had also a lower *trans-p*-coumaric acid concentration and a *p*-coumaric acid to ferulic acid ratio (Fritz et al., 1990). These modifications, and especially those associated with lignin structure and content, were associated with the improved dry matter and cell wall digestibility by rumen animals of the brown midrib genotype, as discussed elsewhere (e.g., Porter et al., 1978; Cherney et al., 1986; Fritz et al., 1990; Thorstensson et al., 1992).

4   ROOT GROWTH AND FUNCTION

The crown roots of sorghum constitute a fairly organized system. Crown roots are initiated from buds in the basal stem internodes. The initiation rate is distinctly cyclic at about weekly amplitude, reaching a rate of up to 1.5 roots day^{-1} (Blum et al., 1977a,b).

Typical crown root axis growth rate is about 3 cm day^{-1}. Undoubtedly, root growth rate varies with environmental and genetic factors, but the most prominent factor in affecting growth of initiated roots must be the amount of carbon partitioned to the root. Indirect evidence to that effect is available from various sources. For example, when some of the root axes or the seminal root are excised, the remaining axes immediately branch profusely (Blum et al., 1977a). When young tillers in field grown sorghum were excised, root length density increased (Fukai et al., 1986). Presumably, young tillers compete with the root for assimilates from the main shoot.

Wilson (1988) concluded that under the effect of the most common environmental factors, such as water, light, major nutrients, and ambient CO_2 concentration, root growth can be explained reasonably well by its relationship to shoot growth in terms of the relative concentrations, fluxes, and partitioning of carbon and nitrogen pools. Still exceptions are noted, such as the case when shading of the canopy in the field did not reduce root length density (Robertson et al., 1993b). Hole et al. (1984) argued for a stronger dependence of root growth models on the developmental and anatomical features of the specific plant, which most certainly is a strong case also for sorghum.

An example of the relevance of both developmental aspects and the partitioning of carbon in the control of root growth is seen in the results of Blum and Ritchie (1984) (Fig. 3). When sorghum was grown under conditions of continuous soil wetting, it established the full potential of crown root initiation and penetration into the soil and the root system comprised of many short crown root axes. This created the typical root distribution of irrigated sorghum where root density (or dry matter distribution) is large at shallow soil and it decreases sharply with soil depth (e.g., Merrill and Rawlings, 1979; Meyers et al., 1984a; Kaigama et al., 1977). When sorghum was grown in drying topsoil, while water was ample at deeper soil, the newly initiated crown roots did not penetrate and therefore did not grow into the soil. The remaining roots at deeper soil layers were the only root sinks to receive assimilates and therefore they continued to grow. The root system was then composed of few but long and deep crown roots. This created the typical root distribution of dryland sorghum where root density is distributed more evenly along the soil profile and maximum root depth is often greater than under irrigation (e.g., Merrill and Rawlings, 1979; Meyers et al., 1984a; Kaigama et al., 1977).

Similar control of root growth distribution by the soil environment is seen also in the case of waterlogging. Waterlogged sorghum plants were characterized by promoted crown root initiation while existing roots were not growing (Pardales et al., 1991). In this case the newly initiated crown roots supported the recovery of plant growth after soil drainage.

The advantage of a small number of root axes in root penetration to deeper soil as represented by Fig. 3 is supported by more recent results (Salih et al., 1999) indicating deeper soil moisture extraction by a sorghum cultivar having relatively a smaller number of root axes.

As in many other crop plants, soil coring, soil moisture depletion, root observations in various containers and root media, and in situ root video-

TOP-SOIL MOISTURE CONTENT

21.6%   2.4%

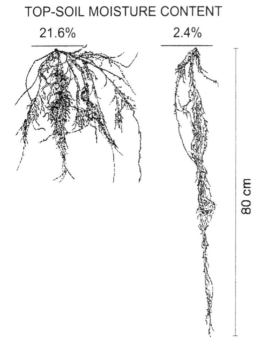

Figure 3 Computer-enhanced drawing of photographed root systems of 24-day-old sorghum plants grown in soil in deep containers over a water table at a depth of 100 cm. Plant on left received also frequent watering from the top while plant on right received water only from water table while topsoil became dry and hard.

graphy in minirhizotrones were used to study sorghum roots. The information collected under all these different situations is very difficult to compare. However, it has been indicated that for the most part, soil moisture extraction occurs where the roots are and therefore patterns of soil moisture extraction describe reasonably well where the active roots are (Meyers et al., 1984a; Robertson et al., 1993b).

Robertson et al. (1993b) found that soil moisture extraction front coincided with the root front under conditions of continuous soil drying. This relationship was evidently complicated after soil rewetting (Blum and Arkin, 1984) when moisture extraction occurred below or where existing roots branched in response to watering. Root growth, as depicted by the progress of the root front and the extraction front, is relatively rapid. Roots may reach a depth of roughly 90 to 180 cm by the boot stage and they can efficiently extract water to a lateral distance of 160 cm from the plant (Blum and Naveh, 1976). However, as discussed earlier, these values depend on irrigation (or rainfall) and its frequency, as well as on the medium in which roots grow. Henceforth, descriptive studies showing a difference between crops in root depth cannot be conclusive when performed with one genotype on a specific soil (Stone et al., 2001).

Root growth and proliferation in relation to soil moisture status is not perfectly clear. Robertson et al. (1993b) concluded that root proliferation in a given soil layer appeared to halt when soil moisture at that layer was reduced to around 20% to 40% of the extractable soil moisture content. Blum and Arkin (1984) suggested that root growth continued even when all extractable moisture was taken up, but suspected that roots tended to proliferate in a layer above where water was available. Irrespective of the response of root growth to soil moisture status, roots are active and extract soil moisture even below a soil water potential of −1.5 MPa (Blum and Arkin, 1984; Hundal and De Datta, 1984).

The relationship between the large structural and functional changes occurring in water-stressed sorghum root and the associated reductions in root conductance (Cruz et al., 1992) is not well understood. It may even be possible that water taken up by one part of the root (where root turgor is high) may reach and rehydrate another part of the root (where turgor is lower). This can be hypothesized based on the findings that water may flow from shoot to root (Blum and Johnson, 1992) or from root to soil (Blum and Johnson, 1992; Xu and Bland, 1993b).

While root lignification, suberization, and reduced conductance (Cruz et al., 1992) are serious consequences of water stress, sorghum still has an impressive capacity to resume soil moisture extraction when rewatered after severe water stress (Sanchez-Diaz and Kramer, 1973; Xu and Bland, 1993a).

Root growth, in terms of total dry matter weight or root length density, terminate at about the flowering growth stage (Zartman and Woyewodzic, 1979; Robertson et al., 1993b). The reduction in root mass or length density after flowering was taken as an expression of root senescence, which was more prominent in a "senescent" than in a "nonsenescent"-type hybrid (Zartman and Woyewodzic, 1979). Calculations based on reported data indicate that root length density of sorghum can be reduced by about 30% to 50% between heading and maturity, depending on soil depth and genotype. However, root mortality also occurs at high rates before heading, but this is masked by the growth of new roots when instantaneous observations of root length density are performed. Sorghum roots were traced as they appeared on a 60- by 180-cm glass panel of a large root observation/rhizotron installation (Blum and Arkin, 1984). It was found that of the total cumulative root length traced from emergence, more than two thirds was not present at heading. Evidently, there is an extensive turnover of roots during plant growth, which is not well studied in any crop. This turnover involves natural aging as affected by the plant hormone status (Ambler et al., 1992) as well as the effect of soil physical, chemical, and biotic factors.

Robertson et al. (1993a) proposed a simple simulation model to describe sorghum root growth, based on the approach used in the CERES crop growth models. The model has five components: (1) daily accumulation of root length is proportional to aboveground biomass growth, (2) the root front descends at a constant rate from sowing until early grain filling, (3) daily accumulation of root length in water nonlimiting conditions is partitioned among the occupied soil layers in an exponential pattern with depth, (4) proliferation of root length is restricted in any layer if the extractable soil water in that layer declines below a threshold, and (5) a fixed proportion of existing root length is lost due to senescence each day. The parameter values for the relationships were derived from data collected on sorghum grown in soil with no physical or chemical restrictions to root growth in the subhumid subtropics of Australia.

Cultivar and genetic variation in different parameters of root growth of sorghum have long been observed (e.g., McClure and Harvey, 1962). Jordan et al. (1979) demonstrated that for several parameters of root growth there was no genetic diversity among common U.S. breeding lines of the period, while variation was large among various exotic sorghum introductions. It is quite possible that selection pressure for a rather narrow range of plant morphology and phenology as required for adaptation to a given environment (e.g., the Midwestern United States) have also narrowed the genetic variation in root development and morphology.

The height genes were often suspected by sorghum breeders to have an effect on root development. However, when root data were normalized for leaf area per plant, no differences were found in root numbers between 3-dwarf and 1-dwarf isogenic lines (Jordan et al., 1979). For other crop plants, different studies on the effect of dwarfing genes on root development very often yielded conflicting results, probably because of involvement of additional phenotypic and genetic factors in the materials studied. In sorghum, the rate of tillering may affect root dry matter, and reduced tillering appears to promote root growth (Fukai et al., 1986).

Jordan and Miller (1980) discussed the importance of selecting for deep root development in order to maximize soil moisture extraction in sorghum grown in the southern Midwestern United States, which would be supported by the model of Robertson et al. (1993a). On the other hand, Blum (1974) and Blum and Naveh (1976) argued for a moderated initial shoot and root development when sorghum is grown on limited stored soil moisture.

Evidently, before genetic optimization of the root system is pursued in breeding programs, we need a better understanding of root growth and function within the whole system. It is important to understand the basic physiological and developmental parameters that are involved in the genetic control of "large," "small," "shallow," or "deep" roots. It has already been pointed out above that shoot development has a critical effect on root development. Thus, when genotypes are to be evaluated for their root attributes independently of leaf area and phenology, data must be normalized for these shoot traits (e.g., Jordan et al., 1979; Blum and Arkin, 1984).

Root exudates of sorghum consist primarily of a dihydroquinone that is quickly oxidized to a p-benzoquinone named sorgoleone. Ten to 125 micromolar concentration of sorgoleone inhibit the growth of various plants in vitro, indicating strong allelopathic effect to sorghum (Einhellig and Souza, 1992). Sorgoleone was found to be a potent inhibitor of state 3 and state 4 respiration rates in both soybean and corn (Rasmussen et al., 1992). In various test plants sorgoleone inhibited photosynthetic electron transport as effectively as DCMU (diuron) [N'-(3,4-dichlorophenyl)-N,N-dimethylurea] (Nimbal et al., 1996).

Witchweed (*Striga* spp.) parasitism of sorghum is conditioned by the exudation of witchweed germination stimulants exuded from sorghum roots. Sorgoleone is a major witchweed germination stimulant (Oliver and Leroux, 1992). Witchweed resistance in sorghum is at least partially conditioned by the rate of sorgoleone production in roots, although other witchweed germination stimulants are exuded by sorghum roots (Hess et al., 1992). Strigol is the major witchweed seed germination stimulant in maize and proso millet root exudates but only a minor component of the total activity in sorghum root exudates (Siame et al., 1993).

5 FROM PLANT TO CANOPY

Plant productivity is measured by the aboveground amount of dry matter produced by the crop throughout its life cycle. Other definitions may be used for specific situations or crop plants. Crop dry matter accumulation can also be defined by the amount of radiation intercepted by the canopy and its efficiency of conversion into dry matter. Partitioning of dry matter to different organs will determine the final amount of dry matter invested in certain plant parts of economic interest, such as root or grain. It is therefore important to understand how sorghum canopy develops.

Sorghum canopy structure is determined mainly by leaf number per plant, leaf size, leaf aspect, leaf senescence, and plant density and arrangement in the field. Leaf area index (LAI) is a major determinant of radiation interception, assimilation per unit land area, and crop growth. The green portion of the LAI is the assimilating component. Knowledge of senescence rates allows accounting for the difference between total LAI and green LAI.

To predict the green leaf area for a given genotype it is necessary to (1) predict leaf appearance as a function of thermal time, (2) predict total leaf number as a function of photoperiod and thermal time, (3) predict total leaf area per plant based on leaf number, and (4) predict the proportion of senesced leaf area and discount it from the total to give the remaining green leaf area. In many respects the work of Muchow and Craberry (1990) validated previous knowledge and served well to organize it into several mathematical functions useful for the simulation of green leaf area in a given genotype, probably a nontillering one. They confirmed that thermal time from emergence to panicle initiation and leaf number decreased with shorter daylength in a photoperiod-sensitive ("tropical") genotype. Leaves were initiated at the rate of 41 °C-days per leaf. They described the area of leaves as Quinby did in 1974 (Fig. 2), only that sowing dates replaced in their work the different maturity genes in expressing the role of photoperiod. Consequently they calculated that the appearance rate of fully expanded leaves was 69 °C-days per leaf. The area per leaf was a function of its insertion and dependent on planting date (or photoperiod response). Leaf senescence was better related to calendar time than to thermal time, which indicates an effect on senescence of factors other then plant age alone. They proposed nitrogen depletion from leaves as a possible major influence on leaf senescence. However, another factor in delaying senescence is the accumulation of solutes in the leaves, which is under genetic control (Sowder et al., 1997). In later modeling exercises they were able to simulate leaf area per plant as a function of thermal time over a certain base temperature and within the bounds of a given maximal leaf area (Hammer et al., 1993; Carberry et al., 1993a). The rate of leaf senescence after heading was calculated in their model (Carberry et al., 1993b) also as a function of thermal time and assuming a constant rate of individual leaf senescence.

The root may have a role in affecting shoot senescence (Jackson, 1993). Soil conditions may be sensed by the shoot via three major endogenous plant hormones, which are produced in the root: the cytokinins, abscisic acid, and ethylene. Certain soil conditions, such as dryness or salinity may reduce cytokinin delivery from the root to the shoot, causing accelerated shoot senescence. Soil dryness or hardness may increase ABA delivery, and root anoxia may increase ethylene delivery from the root to the shoot, causing accelerated senescence. Genotypic effects can be involved in root-to-shoot communications or in conditioning plant hormonal balance, which affects plant senescence. This may be the case for the nonsenescence trait in sorghum (e.g., Duncan et al., 1981), thought to be conditioned by cytokinins produced in the root (Ambler et al., 1992).

Maximum LAI of sorghum is normally around 4 to 6 $m^2 m^{-2}$, but it may reach even 10 $m^2 m^{-2}$ in sorghum planted under conditions of high fertility and high plant density (Blum and Feigenbaum, 1969; Fischer and Wilson, 1975c). Intercepted radiation by the canopy is commonly measured by placing radiation sensors above and below the canopy, such that the difference between the two sensors is the amount intercepted. Interception increases with LAI in an exponential manner as described by the expression:

$$T = \exp(-k \times \text{LAI})$$

where T is the fraction of the transmitted photosynthetically active radiation (PAR) of the total incoming PAR and k is the extinction coefficient. From data collected for two sorghum hybrids over several planting dates at Temple, Texas, Rosenthal et al. (1993) calculated that k was 0.51, as compared with a slightly lower value (0.49) previously observed by others in Texas (Arkin et al., 1976). However, k changes drastically with plant density and it may reach values approaching those of grasses ($k = 0.29$) when LAI is closer to 10 m^2 m^{-2} (Fischer and Wilson, 1975c). Goldsworthy (1970b) demonstrated very well the effect of tall and late genotypes on the variations in light interception by the canopy as the season progressed. His data strongly imply that k may be relatively predictable in temperate short sorghums (e.g., close to 0.50 at typical plant density) and far more difficult to predict in tall and late-flowering genotypes.

Under favorable growing conditions and with short productive sorghums, high-density sorghum will produce more biomass and grain than normal-density sorghum. To a large extent this can be ascribed to the relatively higher LAI and the better distribution of light into the canopy due to the more erect and smaller leaves at high density, which result in greater total canopy photosynthesis (Fischer and Wilson, 1976a,b). The canopy at high densities was also found to be more productive per unit increment of LAI above 2 m^2 m^{-2} (Fischer and Wilson, 1975c). The proportional contributions to total canopy photosynthesis of different parts of the canopy were generally consistent. It was found to be 21%, 24%, 21%, and 13% for flag leaf and leaves 2, 3, and 4, respectively, and 14% for the panicle (Fischer and Wilson, 1976b). Growth and dry matter accumulation curves for different crops of sorghum in temperate and tropical environments can be found in Goldsworthy, (1970a), Vanderlip (1972), and Fischer and Wilson (1975b).

The slope of the linear regression of the accumulated biomass on the accumulated PAR is defined as radiation-use efficiency (RUE) (see example in Fig. 3 of Rosenthal et al., 1993). If RUE is given a unique value over a wide range of conditions or certain limits of conditions, biomass production can be predicted from measurements or calculations of intercepted radiation for those conditions (e.g., Arkin et al., 1976; Huda et al., 1984; Rosenthal et al., 1989). However, the utility of using RUE in crop growth simulation models is controversial (Demetriades-Shah et al., 1992; Arkebauer et al., 1994). RUE varies extensively. Published values of RUE for sorghum range from 1.2 to 4.9 g MJ^{-1}. (e.g., Huda et al., 1984; Rosenthal et al., 1989, 1993; Hammer et al., 1989; Stockle and Kiniry, 1990; Muchow and Sinclair, 1994). RUE values reported for a wide range of temperate, high-yielding sorghums in the United States were between 2.3 and 4.9 g MJ^{-1}, while values obtained with locally bred sorghums in ICRISAT India were only between 1.2 and 2.8 g MJ^{-1} for nonirrigated and frequently irrigated conditions, respectively (Huda et al., 1984).

As would be expected, sorghum RUE varies with plant age, crop management practices (which affect plant water and nutrient status), temperature, and atmospheric vapor pressure deficit. Further research is most likely to reveal additional sources of variation in RUE, perhaps even the optical properties of leaves. These variations are the essence of crop growth simulation and their nature should be understood. For example, carbon assimilation by the sorghum panicle (estimated at about 14% of total crop assimilation) (Fischer and Wilson, 1976b) and even the effect of panicle morphology in this respect (Eastin and Sullivan, 1969) can introduce an appreciable error into RUE as it is being calculated on the basis of LAI.

On the other hand, this does not mean that RUE is a conceptually useless parameter. Within its own limitations it is a valid tool for analyzing the complex relations between plant assimilation and plant productivity. RUE is useful, to the extent that models are, for merging basic plant physiology, biochemistry, and genetics with a total systems approach. An example can be seen in the work of Hammer et al. (1989), who made good use of RUE to explore genotype by temperature interactions affecting sorghum productivity.

6 THE FORMATION OF YIELD

Depending on its management and utilization, sorghum may be used for the production of different commodities such as feed grain, food grain, forage, broom, syrup, sugar, and alcohol. The following discussion is limited to grain yield.

As in most field crops, the foundation of yield is the production of aboveground total biomass, part of which constitutes the grain, which develops during the latter quarter or third part of the plant life. The maximization of biomass and/or harvest index (HI), either genetically or culturally, will increase grain yield (e.g., Howell, 1990). The partitioning between total biomass and grain mass is a major consideration in understanding yield formation. Partitioning is often discussed in terms of source–sink relationships, with the understanding that source and sink interact in a complex manner. The source and its "strength" are determined by transient assimilation of carbon and its partitioning to the sinks, as well as the partitioning of stored preanthesis assimilates from different plant parts. Sink size, and thus its potential for importing assimilates, is determined by the development of the different yield components (namely, the number of panicles per unit area, the number of kernels per panicle, and kernel weight) as controlled by processes of initiation, differentiation, cell enlargement, and intraplant competition for assimilates. "Sink strength" is sometimes an elusive and a controversial term (Farrar, 1993), which may involve not only sink size but also some controls over partitioning. Such may be the case for hormonal control of assimilate partitioning to the inflorescence, possibly emanated from or regulated by the inflorescence (e.g., Khanna-Chopra and Sinha, 1988). It is generally accepted that both sink and source or their interaction or balance may affect yield, and the relative importance of each may vary with the genotype or the environment.

6.1 Biomass Accumulation

Sorghum displays the typical sigmoid growth curve and very generally total plant growth plateau soon after flowering time. Leaf area index is generally maximized (normally at 4 to 6 $m^2 m^{-2}$) shortly before or at heading. The rate of reduction in LAI after its peak is a function of genotype (inherent rate of plant senescence) and the level of biotic and abiotic stresses occurring after heading. Stem dry weight may continue to increase after heading and even after grain maturity (Goldsworthy, 1970a; Vietor et al., 1990; Vietor and Miller, 1990). This is especially noted (although not exclusively) for sugar accumulation in sweet sorghum stems and for stem dry matter accumulation in nonsenescent sorghum genotypes (Duncan et al., 1981; Vietor et al., 1990). Depending on the dynamics and the balance of growth among the different plant parts, crop growth rate is maximized toward the end of the exponential phase of the plant's sigmoid growth curve when LAI is close to maximum (Fischer and Wilson, 1975c). Then it reduces to minimum around heading and again increases steadily as the kernels grow (Goldsworthy, 1970a). It is important to remember that in sorghum the stem constitutes 50% or more of the total aboveground biomass at maturity, especially in the taller genotypes.

It is evident from the earlier discussion that sorghum biomass and yield potential generally increase with the duration of growth, if the environment is optimal. Normalized for phenology, mean daily biomass production calculated over the whole growth duration of the crop may vary with the environment and the genotype. For example, under potential conditions sorghum hybrids tend to have greater mean daily biomass production than open-pollinated cultivars, but not necessarily so under stress conditions (Blum et al., 1992). The use of biomass in the analysis of crop growth and productivity is therefore more meaningful if data are normalized for variations in crop phenology.

Biomass production is controlled by the gross input of carbon from photosynthesis, the efficiency of synthesis of new biomass from carbon input, and the maintenance energy requirement of existing biomass. In sorghum, the synthesis efficiency increased from 0.70 to 0.78 gC gC^{-1}, while maintenance requirement decreased from 24 to 6 mgC gC^{-1} day^{-1}, in young and mature plants, respectively (Stahl and McCree, 1988).

Photosynthesis has already been discussed above. On a canopy basis it has been estimated that the normal mean maximum rate of productivity in sorghum is about 3 g of dry matter per megajoule of PAR absorbed (Huda et al., 1984). The seasonal interception of PAR and thus total crop assimilation can be manipulated to an extent by cultural and genetic means. Fine-tuning of the canopy radiation balance is possible by modifying planting dates, plant arrangement, plant architecture, and leaf surface properties. Regretfully, these controls are not sufficiently flexible to fit the changing and unpredictable field environment, especially with regard to water supply. Most crop management decisions are made at planting, when the coming season is largely unknown.

6.2 Harvest Index

It has been shown for various cereal crops, including sorghum (Blum et al., 1991), that the genetic improvement of yield during the period of scientific agriculture was achieved mainly by increasing HI rather than by increasing total biomass. The highest HI values of about 50% were recorded in the modern, high-yielding, semidwarf wheat. HI in sorghum may vary from about 6% in tall and late-maturing African landraces (Blum et al., 1991) to about 50% in modern temperate hybrids (Prihar and Stewart, 1991). Plant breeding increased HI in sorghum mainly by reducing plant height and growth duration. The effect of growth duration on HI is prominent, whether in open-pollinated cultivars or in hybrids (e.g., Blum et al., 1992). Because biomass increases while HI decreases with growth duration, the first approach to the improvement of yield involves an optimization of the balance among phenology, biomass, and HI.

Although generally, harvest index was not found to vary within a narrow range of environmental conditions (Howell, 1990), it does change in sorghum with the environment (e.g., Prihar and Stewart, 1991). As in other cereals, HI is reduced appreciably under any situation that is conducive to vegetative growth before flowering and detrimental to reproductive growth during the latter part of the plant cycle, such as the case is for late-season drought stress.

6.3 Panicle Differentiation and Growth

The basic structure of the mature panicle consists of whorls of primary branches emerging from each internode of the panicle rachis. Each primary branch is divided into secondary branches that carry the spikelets. Thus, the number of grains per panicle is determined by the number of branch whorls, the number of primary branches per whorl, and the number of grains per primary branch. The number of grains per branch is normally the highest in the basal whorls and it decreases acropetaly (Blum, 1967, 1970a).

Lee et al. (1974) performed a thorough study of panicle differentiation in sorghum. They found that primary branches were differentiated acropetaly along the panicle rachis while the differentiation of spikelets was basipetal (from the tip to the base). They suggested that the total number of primary branches in the panicle was affected by the size of the apical dome. They noted that the apical dome was larger when the duration of the vegetative phase was longer, giving rise to larger panicles. Secondly, they speculated that a delay in the onset of the basipetal spikelet initiation would allow more time for panicle branches and secondary branches to be initiated. This, according to their opinion, should increase the spikelet-carrying capacity of the basic structure of the panicle.

The extent to which the full potential of grain number per panicle is realized depends on the conditions before and during panicle differentiation and growth. Shading and plant-thinning experiments (e.g., Fischer and Wilson, 1975a) demonstrated that the availability of assimilates during panicle development determines the number of grains per panicle. Most of this effect is seen in the basal panicle branches. Generally, most of the exposed parts of the panicle contain chlorophyll and are capable of photosynthesis.

Pollen formation, pollination, male sterility, anthesis, and fertilization are obviously of a major interest to plant breeders and geneticists. Ample information on the subject may be found in breeding manuals (e.g., House, 1983). The progress of flowering and pollination within the individual panicle is basipetal. The individual panicle completes flowering in about 8 days and most of the panicle flowers within 3 days (Pendleton et al., 1994). A uniform sorghum field may flower for about 2 weeks.

Poor seed set reduce grain number. Severe drought stress before heading may result in poor seed set (Fig. 4).

Temperature extremes are an established cause for male sterility. Sorghum pollen is killed at a temperature of $\geq 42\,^\circ C$ (Stephens and Quinby, 1933). Chilling night temperatures of $10\,^\circ C$ (Downes, 1971; Brookings, 1976) to $13\,^\circ C$ (Downes, 1971) cause failure of pollen mother cells during meiosis, without affecting female fertility. The period of chilling sensitivity is extended from flag leaf ligule emergence until the flag leaf sheath has elongated to about

Figure 4 Poor seed set in a sorghum panicle subjected to drought stress during pollen differentiation. (Original photograph by A. Blum.)

20 cm, a period of about 6 to 7 days at moderate temperatures (Brookings, 1976). As seen also in other species (Hong-Qi and Croes, 1983), high proline content of the pollen ascribes better viability to sorghum pollen under temperature stress (Brookings, 1976, Lansac et al., 1996), to the extent that proline content of the pollen was suggested as an assay for pollen viability.

Grain number and weight per grain determine the total grain mass per panicle. A negative association between grain number per panicle and weight per grain is common in sorghum (e.g., Blum, 1970a, 1973; Fischer and Wilson, 1975a) especially when grain number per panicle is high. The effect of intrapanicle competition on potential grain size is determined within 1 week after anthesis (Fischer and Wilson, 1975a). This competitive association seems to indicate a potentially limited source, especially when the sink is very large due to genetic or environmental effects. However, as will be seen in the following sections, sorghum productivity does not seem to be uniquely source limited.

6.4 Grain Growth

Information on the physiology of sorghum grain growth and development is not as developed as in the small grains.

The important component for grain mass development and yield is the endosperm and the accumulation of starch in endosperm cells. Endosperm cell division is completed at about 10 to 12 days after anthesis. The number of endosperm cells may vary with the environmental conditions during cell division. For example, the reduction in irradiance during cell division decreases endosperm cell number (Kiniry and Mausser, 1988). However, reduction in endosperm cell number does not necessarily limit starch accumulation. All caryopsis tissues are thought to be associated with carbohydrate influx to the endosperm (vascular tissue, chalazal tissue, remnant nucellar tissue, the placental sac, and aleurone transfer cells) (Maness and McBee, 1986). The placental sac was identified as an intermediate apoplastic sink for assimilate accumulation to be imported as hexose by endosperm transfer cells.

The principal period for starch accumulation and grain mass increase is after endosperm cell division is completed and it is defined as the exponential growth phase of the grain. Growth rate plateaus toward physiological maturity, when maximal seed dry matter weight is attained. This is preceded by visible signs of the degeneration of endosperm tissues in the carbohydrate import route (Maness and McBee, 1986). Physiological maturity is conveniently recognized in sorghum by the appearance of a dark layer on the placental area of the grain (Eastin et al., 1973). The timing of the appearance of the dark layer was found to coincide with the cessation of assimilate import into the grain (Weibel et al., 1982).

The biochemistry of endosperm starch accumulation is reasonably understood for the small grains (Jenner et al., 1991; Lopes and Larkins, 1993). Briefly, starch is synthesized from sucrose imported into the grain. Sucrose is converted (via its cleavage to hexoses) to glucose-1-phosphate, which is incorporated into ADP-glucose. The latter serves as substrate for polymerization of sucrose into amylose and amylopectin by action of starch synthase and branching enzymes, presumably via a primer. In sorghum sucrose enters into the base of the kernel as such and is then hydrolyzed (Singh et al., 1991). Some of the sucrose may be hydrolyzed just before entering the kernel. Soluble acid invertase was the predominant sucrose-cleaving enzyme. Before being converted to starch, sucrose seemed to be reconstituted from reducing sugars within the endosperm.

The effect of plant and environmental factors on starch synthesis and accumulation in the endosperm is well understood. Starch synthesis is sensitive to temperature, and soluble starch synthase activity may be the major limiting step in this respect in wheat and maize (Singletary et al., 1994; Keeling et al., 1994). The heat sensitivity of this enzyme is most probably the main limitation to starch accumulation at high temperatures, above 34°C. No specific information is available in this respect for sorghum.

Environmental conditions other than temperature may affect starch accumulation by affecting source activity and the availability to the grain of carbon stored in the shoot. It has been suggested that some of the starch-synthesizing enzymes may be sucrose inducible and that sugar-responsive genes may control starch synthesis in the endosperm (Lopes and Larkins, 1993). This can be one way to explain source–sink interaction in affecting grain filling. Any plant stress that would limit sucrose availability to the grain may induce a modification in starch synthesis. On the other hand, Jenner et al. (1991) suggest that wheat starch synthesis in the grain is quite insensitive to variations in sucrose supply within the "normal" range. They conclude that the control over the rate and duration of starch accumulation most probably resides within or at close proximity to the grain.

Temperature has a major influence on sorghum grain filling, as in other cereals. Under temperate conditions the duration of grain filling is about 400 to 600°C-days, with a significant genetic variation in this respect (Heiniger et al., 1993). Generally, sorghum grain growth duration was reduced (linearly) and grain growth rate increased as temperature was raised from 15° to 30°C in the growth chamber (Kiniry and Mausser, 1988). Muchow (1990a) confirmed the increase in grain filling rate with the rise in mean temperature from about 25° to 30°C in the field. However, the relationship between mean temperature and the duration of grain filling was somewhat scattered ($R^2 = 0.65$) in his field experiments. Maximum sorghum grain weights were attained around 15° to 25°C (Kiniry and Mausser, 1988), roughly about 5°C higher than for wheat. More data is needed in this respect, especially over the higher temperature range to which sorghum grain filling is normally subjected in the tropics and semiarid tropics.

Apart from temperature, the rate of grain filling and final kernel size is strongly influences by intrapanicle interactions. For example, within the sorghum panicle kernel growth rate and final weight decrease basipetally (e.g., Blum, 1967) in correspondence with the delay in pollination (Heiniger et al., 1993). Grain filling rates vary within the panicle as a function of differences in assimilate supply. Earlier pollinated florets will produce larger kernels because they have a competitive advantage as growing sinks over later formed kernels within the panicle (Heiniger et al., 1993). The number of developing kernels in the panicle affects intrapanicle competition. It is therefore not an uncommon observation that when growing conditions or the plant genetic potential are favorable for developing a large number of kernels per panicles, final mean kernel weight may be reduced, as compared with smaller panicles (e.g., Blum et al., 1992).

Assimilate supply for grain growth may come from current photosynthesis by leaves and the panicle and from preflowering assimilates stored in the plant. About 75% to 85% of panicle yield was attributed to current photosynthesis by the upper four leaves, while the rest was attributed to panicle photosynthesis (Fischer and Wilson, 1971b, 1976b). The relative contribution of the upper three leaves was nearly similar, although efficiency (contribution per unit leaf area) was greatest in the flag leaf. The contribution by the panicle was equally divided between the assimilation of atmospheric CO_2 and the assimilation of CO_2 released be respiration from the grain (Fischer and Wilson, 1971b). The role of the panicle in supplying assimilate to the grain may be underestimated. While about half of the assimilation by the panicle is based on CO_2 evolved by respiration, it has been noted that dark respiration in the panicle is appreciably less thermosensitive than in the leaf (Gerik and Eastin, 1985). This may imply a relative advantage to panicle assimilation over leaf assimilation when sorghum is subjected to high temperatures during grain filling.

CO_2 fixation by the panicle tends to decrease sharply as grains develop and panicle organs loose chlorophyll upon senescence (Eastin and Sullivan, 1969). It is not perfectly clear what is the relative role of actual senescence and of the increasing panicle density as grains develop with time in reducing carbon assimilation by the maturing panicle. Loose panicle architecture was found to extend the duration of panicle assimilation until the soft dough stage of grain development, as compared with a compact architecture (Eastin and Sullivan, 1969). Thus, the increasing panicle density by the enlarging kernels may in itself reduce carbon exchange by the panicle.

An important source of assimilates during grain filling of small grain cereals are preflowering assimilates stored in the plant, mainly in stems (Schnyder, 1993). The relative importance of these reserves for grain filling is greater when stress reduces transient photosynthesis during grain filling. The contribution of stored stem reserves to grain filling depends on the size of the storage pool, the availability of current photosynthate for grain filling, the demand by the grain and the rate of reserve remobilization to the grain in relations to grain growth rate (Blum, 1997). Depending on these factors, remobilized stem reserves may sometimes account for more than half of the final grain mass in wheat. Data on sorghum in this respect are not abundant. In a study of nine sorghum hybrids grown under three levels of water supply, Borrell et al. (2000b) found that stem reserves did not contribute to grain yield under fully irrigated conditions, yet eight out of nine hybrids mobilized some stem reserves (accounting for up to 15% of yield) during grain filling under a postanthesis water deficit.

Using ^{14}C labeling in the glasshouse and presumably with unstressed sorghum plants, Fischer and Wilson (1971a) estimated that 12% of sorghum grain yield per panicle was attributed to preanthesis stored assimilates. The amount of stored assimilates in sorghum stem at anthesis can be very appreciable, typically between 200 and 300 g/kg of total nonstructural carbohydrates (TNCs) (Vietor and Miller, 1990; Vietor et al., 1990; Kiniry et al., 1992). Fischer and Wilson (1975a) already observed that dry matter production in excess of the demand for grain production was accumulated in the shoot and the root. However, data from different reports or for different conditions or different genotypes show a small reduction, a stable or an increase in stem TNC between anthesis and maturity.

Kiniry et al. (1992) subjected grain sorghum to severe shading (98% shade) during grain filling, which would be expected to reduce current assimilation and therefore promote the use of the stem carbon storage for grain filling. Shading caused a severe reduction in panicle weight. Stem TNC content at anthesis was ample, around 250 g/kg. Stem dry matter reduction during grain filling was generally small and it could be accounted for by the maintenance respiration of the vegetative plant part. Thus, ample assimilates stored in sorghum stems did not contribute to grain filling, even when current photosynthesis was severely inhibited and panicle weight was reduced. In fact, the stem of grain sorghum may perhaps even compete with the developing grain for assimilates during early grain filling stages, especially in nonsenescent types. In a study of nine sorghum hybrids varying in rate of leaf senescence, Borrell et al. (2000b) reported that most (>80%) of the increase in panicle growth during the second half of the grain filling period in an intermediate hybrid could be accounted for by reserves mobilized from the stem, assuming 100% conversion efficiency. However, because stem mass remained relatively constant during the grain-filling period in the "stay-green" and senescent hybrids, it is likely that panicle growth was largely dependent on photo assimilation rather than stem reserves in these hybrids.

The conclusions on the limited role of stem reserves in sorghum grain filling were derived from work with combine height (typically 3-dw) sorghums. Tall (and nonsweet) sorghums are an exception. The 2-dw genotype as compared with the 3-dw genotype was found to ascribe greater stem reserve storage and subsequent greater utilization of storage for grain filling when stress conditions developed during grain filling (Blum et al., 1997).

There are very interesting and important interrelationships among the yield sink development, stem reserve accumulation or depletion, and stem integrity and disposition to stalk rots such as charcoal rot (*Macrophomina phaseolina*). Seetharama et al. (1991) evaluated some of these associations and demonstrated their agronomic importance, especially under drought stress. Generally, the yield sink development draws upon stem reserves and causes their depletion from the stem, which in turn may predispose the stem to stalk rots and lead to lodging. Sorghum defoliation at anthesis promoted lodging, presumably because of carbohydrate depletion from the stem (Rajewski and Francis, 1991). On the other hand, the removal of the yield sink by genetic male sterility causes almost a twofold increase in assimilates storage in stems (Fortmeier and Schubert, 1995), which would be expected to reduce lodging. In this sense, a large yield sink constitutes a physiological load on the shoot, and the impact of this load is aggravated under drought stress when the demand for carbon from stored reserves increases (Khanna-Chopra and Sinha, 1988).

Nonsenescent (stay-green) and sweet sorghums generally tend to accumulate more stem TNC during grain filling than nonsweet or senescent grain sorghums (Vietor and Miller, 1990; Vietor et al., 1990; Kiniry et al., 1992; Mcbee and Miller, 1993). Nonsenescent sorghums also store more assimilates in leaves, probably at the expense of partitioning assimilates to the grain (Sowder et al., 1997). Tarpley et al. (1994) concluded that the decline in soluble sucrose-degrading enzymes appears to be a prerequisite for the accumulation of sucrose in the stems of sorghum. However, this decline cannot account for the difference in stem sucrose content between sweet sorghum and grain sorghum. This concurs with the conclusion of Hoffmannthoma et al. (1996) that none of sugar biosynthetic enzymes that they studied in sweet sorghum stems were fully responsible for sucrose storage and they proposed that other processes, such as transport, should be looked at as an explanation.

Nonsenescent crop cultivars are generally recognized as important for plant production under postflowering drought stress. Nonsenescence can result mainly from delayed onset of senescence or slow rate of senescence (Thomas and Howarth, 2000). In sorghum nonsenescence is also associated with resistance to charcoal rot (*Macrophomina phaseolina*) induced lodging, even though nonsenescence and charcoal rot reaction are genetically independent (Tenkouano et al., 1993).

Assimilate storage in the stem seems to be a dominant tendency in sorghum, which may be related to its basic perennial nature. This is supported by the finding that sugar accumulation in sweet sorghum subjected to drought stress is given preference over growth and photosynthesis (Massacci et al., 1996). The nonsenescent (stay-green) character, which retains assimilates in the stem and foliage, may perhaps be considered to be the "wild type" in this respect. From this standpoint, at least, the recombination of both high-yield potential and the stay-green trait were considered as complex (Tuinstra et al., 1997). Recently, Borrell et al. (2000a,b) produced conclusive evidence showing that nonsenescence in sorghum, whether by way of delayed onset or reduced rate of senescence is very important for supporting biomass and grain yield under postflowering drought stress. Borrell and Hammer (2000) went on to show that specific leaf nitrogen and nitrogen uptake during grain filling were the main cause behind genotypic variations in delayed onset and reduced rate of leaf senescence.

6.5 Heterosis

Despite the immense impact of heterosis (hybrid vigor) in plant breeding, farming, and agribusiness, the phenomenon is not understood in its basic genetics, physiology, or biochemistry. This is especially perplexing because the understanding of heterosis could have been used as a key to unlock the question of the genetic improvement of crop yield potential in general. There has been only limited progress in understanding the physiology of heterosis in crop plants since its review two decades ago (Sinha and Khanna, 1975; Srivastava, 1981) and after considering later reviews by McDaniel (1986), Rhodes et al. (1992), and Tsaftaris (1995).

The rate of heterosis in sorghum depends on the combining ability of the parental lines. For some traits, heterotic effect is general while for other traits it may be specific to the cross. While heterosis has been estimated by comparing the hybrid with the mean value of its parental lines or with a standard open-pollinated cultivar, the critical estimate (as used here) is by comparing the hybrid with its best parent. Thus, percent heterosis is $H = [(F-P)/P] \times 100$, where F and P are values for the F_1 hybrid and best parent, respectively.

The development of F_1 hybrids in sorghum through the use of cytoplasmic–genetic male sterility system in the 1950s (Quinby, 1974) allowed for an average heterosis in grain yield of about 20–25%. A further mean increase of about 10% was achieved by improving the parental lines used in hybrids (Duncan et al., 1991a). The main impact could be ascribed to the use of converted introductions, mainly of the Zerazera group. Haussmann et al. (1999, 2000) showed that heterosis in yield is well expressed under a wide range of stress conditions.

As the case is for other crops, a physiological–biochemical explanation of heterosis in sorghum is not at hand. Rood et al. (1992) and Rood (1995) presented correlative evidence that gibberellins affect heterosis for growth in sorghum. A central role for gibberellins in heterosis would be a very attractive proposition in view of the possible general role of this hormone in affecting growth of sorghum (discussed earlier). However, further evidence on the involvement of gibberellins in controlling heterosis will be required in view of both the distinctive manifestations of heterosis in sorghum, as described in the following.

The most consistent and stable manifestation of heterosis in sorghum is the increase in yield by way of increased kernel number per panicle. This effect is structural and directly related to the number of spikelets (or florets) per panicle and not the rate of seed set. Heterosis in the number of panicles per plant or in kernel weight is relatively small and dependent on the specific cross.

Heterosis in the number of grains per panicle is largely expressed in the basal whorls of the panicle (Blum, 1970a). When panicle differentiation was studied in a heterotic sorghum hybrid and its two parents (Blum, 1977) it was found that the hybrid had a time lapse of 4 days between the termination of the differentiation of panicle branches and the onset of spikelet initiation, while no lapse was seen in both parents. As suggested by Lee et al. (1974), such time lapse must have allowed more spikelets to be formed, especially at the basal panicle branches. Furthermore, the hybrid had a larger apical dome at the onset of panicle initiation, as compared with its parents (Blum, 1977). However, the larger apical dome in the hybrid was not a result of a longer vegetative period. The duration of the vegetative period in the hybrid was even slightly shorter than that in both parents. The larger apical dome could be traced to a larger seedling in the hybrid, indicating that at least part of the potential for panicle size is determined already at the seedling growth stage.

Generally, heterosis in sorghum is expressed also in biomass production (Bartel, 1949) with little or no effect on harvest index (Quinby, 1974; Gibson and Schertz, 1977). Biomass increases in the hybrid with very little change in plant height and with some reduction in growth duration, as compared with the parents. Thus, heterosis is generally expressed in high plant or crop growth rate (Gibson and Schertz, 1977). Leaf area per plant and LAI show initial heterosis during the early growth stages, but this heterosis is not sustained through and after flowering. It appears that the larger developing panicle in the hybrid competes with the growth expansion of the last three to four leaves (Quinby, 1974), which would tend to limit the expansion of leaf area toward panicle exertion. Thus, the greater biomass and dry matter accumulation in the hybrid is at least a compounding result of the initial seedling vigor and its large leaf area, up to about the boot stage. Subsequently, most growth heterosis is invested in the panicle. The initial plant vigor and dry matter accumulation is also expressed in the root, in terms of the growth rate of crown root axes and their branches (Blum et al., 1977b). Seminal root growth and the number of crown roots are unaffected by heterosis. Hence, heterosis in roots is not a structural phenomenon but rather a function of greater assimilates partitioning to the roots, at least in the juvenile plant. The fact that root/shoot ratio does not display heterosis (Arkin and Monk, 1979) indicates a balanced heterosis in root and shoot growth. Ultimately, the effect of heterosis on the root persists or compounds to account for heterosis in total root size at or after flowering (McClure and Harvey, 1962).

General heterosis was not clearly seen in sorghum photosynthesis or leaf CER. It appeared that CER in hybrids was largely dependent on the specific cross (Hofmann et al. 1984) or the specific parental lines (Kidambi et al., 1990a,b; Blum, 1989). However, it was seen that sorghum hybrids tend to maintain higher CER than the parental lines over a wider range of temperatures (Blum, 1989) and extreme environmental conditions (Blum et al., 1990). Sorghum hybrids also tend to perform better than their parental lines or standard open-pollinated cultivars at low (chilling) temperatures during germination and emergence (Pinthus and Rosenblum, 1961; Blum, 1969).

Consequently, it seems that heterosis in grain sorghum productivity is based on unique developmental events in the formation of the panicle structure and most likely involves the stability of some key plant processes, such as photosynthesis, under varying environmental conditions.

7 PLANT–WATER RELATIONS

J. B. Adams, a farmer from El Dorato, Kansas, who had much to do with establishing sorghum ("kafir corn") as a crop in his county, made the following observation in 1911 (quoted by Fields, in Martin, 1930):

> It was a season of unusual severity, the driest, the hottest in Kansas, as it was in Oklahoma since 1901. A pitiless sun burned up the Indian corn and parched the native grass upon the prairies. Throughout this trying ordeal, our unfailing friend, the hardy and indomitable kafir, stood sentinel upon the prairies with the patient fortitude inherent in its nature, born of centuries of hardship upon the desert; it bided its time and silently waited for rain, springing triumphantly into new life with the first downpour from heavens. Our prairie hay turned out less than a third of a crop and our alfalfa only a little better than half a crop. But notwithstanding this accumulation of calami-

ties, we pushed back the impudent face of famine, cheated the hot winds and whipped the drought to a standstill with kafir corn.

Since that time we have been attempting to understand and improve the reputable drought resistance of sorghum. Several reviews were published on the water relations and drought response of sorghum (Sullivan and Ross, 1979; Seetharama et al., 1982; Bidinger and Johansen, 1988). This review deals with some of the main aspects of plant water relations in sorghum and its relative drought resistance. The general advantage of sorghum as a C_4 plant has been discussed above.

7.1 The Development of Plant Water Stress

At ample soil moisture supply the environment largely controls sorghum transpiration. Leaf water potential (LWP) is generally high and it displays the normal diurnal fluctuations, typically between −0.2 and −0.4 MPa during the night and between −1.0 and −1.5 MPa at midday (e.g., Ackerson and Krieg, 1977; Acevedo et al., 1979). Under these conditions atmospheric VPD has an important effect on transpiration and LWP (Grantz, 1990). Crop water use is largely controlled by its developmental stage and leaf area. It increases up to an LAI of about 3.5 m^{-2} m^{-2}, after which it decreases.

Sorghum stomata are relatively less sensitive to VPD as compared with maize (Ackerson and Krieg, 1977). Under well-watered conditions turgor potential displays a diurnal cycle and it can fall especially under very demanding atmospheric conditions at midday. However, midday wilting (zero turgor) under well-watered conditions is rare in sorghum, even at high VPD and windy conditions, because of the ability of the sorghum root system to provide for very high rates of transpiration. This is in contrast to plants, which in the course of meeting excessive transpirational demand may suffer serious midday loss of turgor and a near cessation of transpiration, despite ample soil moisture. Even when all of the roots were exposed to polyethylene-glycol-induced water deficit, sorghum performed better than sunflower, a relatively drought-resistant species, due to the ability of sorghum for maintaining high leaf water status (Zhang and Kirkham, 1995).

When soil moisture cannot supply the transpirational demand the plant enters a state of water deficit. Sorghum whole-plant transpiration or evapotranspiration (ET) or the ratio of ET to potential evapotranspiration (E_0) has been shown to decline when soil moisture was reduced to around 20% to 40% of the total available or "extractable" (Ritchie et al., 1972; Blum and Arkin, 1984). Specifically in sorghum this reduction in canopy transpiration results initially from the reduction in leaf area per plant due to the death of the lower leaves, while younger growing leaves maintain normal or near-normal transpiration rates (Blum and Arkin, 1984). Garrity et al. (1984) found that on a seasonal basis the reduction in leaf area under drought stress was more important than stomatal conductance in controlling transpiration.

On a single-leaf basis, sorghum E_0 as a fraction of R_n (net radiation) began to decrease when leaf relative water content (RWC) was reduced to about 90% (Ritchie and Jordan, 1972). However, this value is very dynamic, depending on leaf physiology as well as its environment. Sorghum displays the general sigmoid relationship between LWP and RWC (Acevedo et al., 1979). This relationship is strongly dependent on osmotic adjustment, such that a higher RWC is maintained for a given LWP when osmotic adjustment is high. This is most likely the reason why sorghum maintains a higher RWC than maize at given LWP (Sanchez-Diaz and Kramer, 1973).

The relationship between transpiration and RWC in a sorghum leaf depends on osmotic adjustment, as seen in Fig. 5. In that experiment, stomatal closure and leaf rolling (as a symptom of turgor loss) occurred at lower RWC in cv. C-63 than in cv. C-192, because the former cultivar had greater osmotic adjustment than the latter. This was supported by the strong and negative linear relationship between osmotic adjustment and RWC at stomatal closure, across sorghum and millet cultivars (Fig. 5, inset).

Because leaf rolling in different sorghum cultivars consistently occurred at RWC very close to that where stomata close (Fig. 5) it emerges that leaf rolling in sorghum is a simple indicator of turgor loss. In the field, it is not rare to observe a marked difference in leaf rolling between different sorghum cultivars subjected to same soil moisture stress. This difference may result from a respective difference in LWP or in osmotic adjustment, both of which determine the leaf turgor potential at any given time in the field. Thus, delayed leaf rolling in the field would indicate a better capacity to maintain higher LWP or better osmotic adjustment. This does not mean that leaf rolling is a negative trait per se. It has evolved in the Gramineae as a means for protecting the leaf from excessive radiation load when this radiation cannot be dissipated by transpiration, namely, when turgor is lost. If the leaves do not transpire and remain unrolled their temperature will rise to the killing point. Older sorghum leaves seem to lose their capacity to roll and this could be one reason why older leaves desiccate first when stress develops (Blum and Arkin, 1984). Matthews et al. (1990b) suggested that under very harsh conditions wheat leaf rolling allows stomatal conductance and gas exchange to be sustained with relatively little leaf water loss. However, it appears that in sorghum stomatal closure precedes leaf rolling (Fig. 5).

From the foregoing discussion it is seen that osmotic adjustment (OA) plays a major role in the development and the implications of plant water stress. OA is strongly dependent on the rate of development and the extent of plant water deficit, as measured by LWP (Jones and Rawson, 1979; Chapman

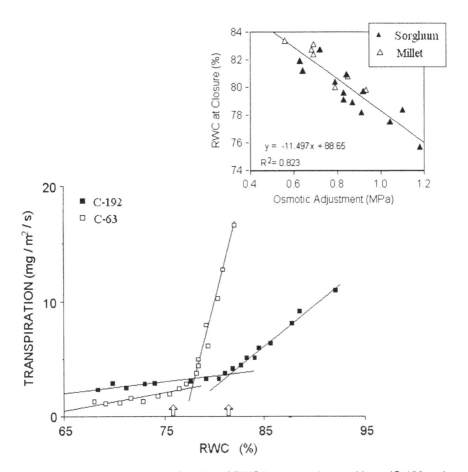

Figure 5 Transpiration as a function of RWC in two sorghum cultivars (C-192 and C-63). Lines are fitted linear regressions at R^2 = 0.79. For each cultivar, the intercept between the two regression lines (which represent stomatal transpiration and cuticular transpiration, respectively) marks the RWC at stomatal closure. Arrows mark the RWC when leaf rolling occurred in the two cultivars. Inset: The relationship between osmotic adjustment and the RWC at stomatal closure over 13 sorghum and 7 millet cultivars. (Unpublished data by Blum, 1982.)

and Fischer, 1988; Blum et al., 1989; Tangpremsri et al., 1991). Osmotic adjustment in three sorghum cultivars increased as the rate of dehydration decreased from 0.153 to 0.093 MPa day^{-1} (Basnayake et al., 1996). Water deficit must develop to a certain level for a sufficiently long time in order to induce significant solute accumulation in a responsive cultivar. It follows that the assessment of genotypic variations in OA requires a testing protocol, which would standardize the rate and extent of plant water deficit across all genotypes or account for variations in the extent of deficit.

When water stress develops, cell division and expansion are inhibited before carbon assimilation is (e.g., McCree, 1986). As a result it was found in sorghum that carbon storage was greater in a water-stressed than in a non-stressed plant. Part of this storage was used for OA. Upon rewatering, OA was lost within 3 days and the stored carbon was released for regrowth (McCree et al., 1984; Richardson and McCree, 1985).

Osmotic adjustment may vary with plant phenology. For a group of sorghum cultivars it was found (Ackerson et al., 1980) that OA developed when stress occurred after flowering but not before flowering. However, others have found that OA occurred both before and after flowering (Ludlow et al., 1990; Santamaria et al., 1990; Girma and Krieg, 1992a). Preflowering OA may be less or nonexistent in early-flowering genotypes because they tend to escape water deficit before flowering, as compared with late ones (Blum 1988). This does not imply that early genotypes lack the ability to accumulate solutes when leaf tissues are desiccating. Sorghum has the constitutive ability to accumulate solutes in vegetative plant parts, including nonexpanding leaves after heading. This may constitute a basis for effective OA after heading and it may well explain why stomata at this stage require a very large reductions in LWP to affect their closure (Ackerson et al., 1980).

There is large and consistent genotypic variation in sorghum for OA (Ackerson et al., 1980; Blum and Sullivan, 1986; Blum et al., 1989; Ludlow et al., 1990; Santamaria et al., 1990; Tangpremsri et al.,1991; Basnayake et al., 1993, 1995, 1996; Tangpremsri et al., 1995). In a study of OA in crosses between lines of high OA (Tx2813 and TAM422) and of low OA (QL27) it was found that one major gene exercised main control over OA in each of the two high-OA lines, with additional minor effects (Basnayake et al., 1995). The level of OA commonly encountered in different sorghum genotypes is between 0.4 and 1.2 MPa, although higher values are not rare (e.g., Blum and Sullivan, 1986; Basnayake et al., 1993).

The main solutes accounting for OA in sorghum are sugars and potassium (Premachandra et al., 1992, 1995a,b). The contribution of sugars (glucose and sucrose) and inorganic ions (potassium and chloride) to OA of fully expanded leaves was about equal (Jones et al., 1980). Differences in OA between cultivars were ascribed mainly to variations in potassium and sugar accumulation (Premachandra et al., 1995). The amino acids proline (Blum and Ebercon, 1977; Bhaskaran et al., 1985), alanine, and aspartic acid were found to accumulate in sorghum leaves subjected to water deficit, but their contribution to OA was found to be insignificant or small. Glycinebetaine is an important solute accumulating in barley and other cereals under drought and salinity stress. Sorghum accumulated glycinebetaine under salinity stress but the concentration in leaves was far less than that in barley (Ishitani et al.,

1993). When sorghum was subjected to water deficit (-2.3 MPa of leaf water potential) glycinebetaine level increased 26-fold and proline level increased 108-fold. The accumulation of both solutes allowed maximal osmotic adjustment of 0.41 MPa (Wood et al., 1996). It is interesting that exogenous spray application of glycinebetaine was found to increase sorghum yield under drought stress, as compared with nontreated control (Agboma et al., 1997).

Osmotic adjustment could not be an evolutionary coincidence, since it is widely implicated also in plant tolerance to freezing and salinity, both of which involve a component of water deficit (Blum, 1988). Evidence for the positive effect of OA on crop performance under drought stress in the field is difficult to develop, but it is being produced for several crop plants, including sorghum. Osmotically adjusting genotypes were found to yield 15% to 34% more than less adjusting genotypes when stress occurred before flowering (Santamaria et al., 1990). Most of the effect was transcribed through a greater number of kernels per panicle. Osmotic adjustment accounted for a 48% increase in root length, 21% greater transpiration, and 16% more dry matter produced before flowering. While this growth advantage was not reflected in final biomass at harvest, the sustained growth during stress before flowering was invested mainly in the panicle (Santamaria et al., 1990). When stress developed after flowering, OA accounted for a 24% increase in grain yield through a greater number of kernels per panicle and heavier kernels (Ludlow et al., 1990). Again, this yield advantage was achieved with no effect on aboveground biomass at maturity but with a positive effect on root growth during stress. The positive effect on grain yield in osmotically adjusting cultivars was seen when data for different cultivars grown in large containers were normalized for LWP (Tangpremsri et al., 1991). In this case, OA was associated with greater biomass production, root length density, and soil moisture extraction during grain filling. Finally, high-OA lines performed better than low-OA lines under drought stress in the field in terms of leaf area, total dry matter, grain number, and grain yield (Tangpremsri et al., 1995).

The exact physiological role of OA and its transcription into an agronomic advantage under drought stress is not completely understood. The possible protective role of the accumulated solutes on various cellular functions has been often suggested (e.g., Blum, 1988), but evidence for sorghum is unavailable. On the other hand, the maintenance of turgor by OA was found to be responsible for delaying plant death by 10 days, when stress was terminal (Basnayake et al., 1993). As seen for many other plant species, OA in sorghum brings about the maintenance of a higher turgor potential at given leaf water potential (Flower et al., 1990; Girma and Krieg, 1992a; Premachandra et al., 1992). The benefit in sustaining relatively better turgor during a drying cycle is often assumed but it is not always evidenced in experiments. Sustained turgor was believed to be an important factor in maintaining sink activity and kernel number per panicle under drought stress (Santamaria et al., 1990). Sustained root growth as a function of shoot (or root?) OA is strongly implicated in several experiments mentioned earlier, but a link to turgor maintenance has not been established. Stomatal conductance, leaf gas exchange (Flower et al., 1990; Ludlow et al., 1990; Girma and Krieg, 1992b) or sometimes growth rate (Flower et al., 1990; Girma and Krieg, 1992b) were not necessarily better in genotypes of high OA capacity, as compared with genotypes of low OA.

The lack of a unison association between stomatal activity and leaf cellular turgor seems peculiar, in view of the documented positive result of OA on plant performance under stress. A very plausible explanation may be in the involvement of hormonal root signals on stomatal activity and shoot growth, independent of leaf water status and OA (Davies and Zhang, 1991).

Abscisic acid is produced in roots subjected to a drying soil. It is transported to the shoot via the xylem sap inducing stomatal closure and growth retardation, irrespective of tissue leaf water potential, turgor poten-

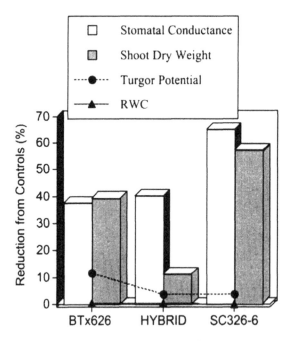

Figure 6 Percent reduction in midday stomatal conductance, shoot dry matter weight, midday leaf turgor potential and midday RWC at 32 days after emergence in a sorghum hybrid and its two parents grown under conditions of a drying topsoil, as compared with frequently irrigated controls. (Unpublished data, Blum and Sullivan, 1992. Seed were provided courtesy of Dr. F. R. Miller, Texas A&M University, College Station, Texas.)

tial, or RWC. Conditions favoring a root signal effect were designed in an experiment with sorghum (Blum and Sullivan, 1992, unpublished data). Plants were grown in soil-filled columns and irrigated from the bottom of the column. This allowed maintaining a very favorable plant water status, while roots in the topsoil were exposed to a drying soil. These columns were compared with columns irrigated normally from the top (controls) (Fig. 6).

Cultivars did not differ significantly in RWC (average of 97.7%) and turgor potential (average of 0.32 MPa) in the controls. Under the conditions of a drying topsoil (assumed to induce a hormonal root signal) RWC did not differ significantly from the control. Turgor potential in the hybrid and SC326-6 was the same under a drying topsoil conditions as in the controls and it relatively decreased slightly by soil drying in BTx626. Although the drying topsoil had no effect on the RWC and turgor potential of SC326-6 and its hybrid, it still caused an appreciable reduction in stomatal conductance and shoot growth in these genotypes. It is assumed that in this case stomatal closure and growth inhibition were induced by a hormonal signal produced in roots under the effect of the drying topsoil. It is interesting to note that SC326-6 was relatively more sensitive than the other two genotypes in this respect.

Ebel et al. (1994) noted that mycorrhiza-infected sorghum plants seemed to be less affected than nonmycorrhizal plants by the growth-reducing effect of an apparent hormonal root signal.

7.2 Major Repercussions of Plant Water Stress

7.2.1 Plant Growth, Development, and Phenology

Cell enlargement is probably the most sensitive physiological process to water deficit, which has far-reaching consequences toward most aspects of plant development and productivity.

Sorghum leaf appearance rate (in terms of thermal time) can be slowed and even inhibited by water stress. The relationship between leaf appearance rate and predawn LWP (PLWP) was linear between about 0.1 and -0.6 MPa of PLWP. Leaf appearance ceased at PLWP of -0.55 MPa (Craufurd et al., 1993). Leaf appearance returned to nonstress rates once plants were rewatered and recovered. Similarly, the rate of panicle development (in terms of thermal time) was reduced with plant stress. Evidently, panicle development is less sensitive to water stress than leaf appearance. The most prominent phenological feature associated with these changes is that the duration of GS1 and GS2 are delayed as water stress increases, until panicles fail to exert altogether (e.g., Matthews et al., 1990a; Craufurd et al., 1993). Heading commenced to delay under drought stress when the ratio of transpiration under stress to transpiration under nonstress was ≤ 0.55 (Donatelli et al., 1992). Failure of panicle exertion (e.g., Santamaria et al., 1990) is very common and discernible symptoms of preflowering water stress in sorghum. Good panicle exertion is used as a simple criterion for drought resistance in field selection (e.g., O'Neill et al., 1983). It has been noted that the deviation from normal plant height under stress is also a sensitive indicator and integrator of drought stress effect during the growing cycle (e.g., Blum et al., 1989; Donatelli et al., 1992).

Manjarrez-Sandoval et al. (1989) imposed drought stress at 10 consecutive growth stages in sorghum grown in pots. Stress was imposed until half of the plants remained wilted at dawn, after which irrigation was resumed. Although plant water status was not measured and the time duration under drought stress was not recorded for each growth stage under stress, the results were very clear at least on a relative scale. The most drought-sensitive growth stage was microsporogenesis. Stress at earlier growth stage (between panicle initiation and pistil primordia appearance) reduced grain number per panicle by about 35% to 50%. Stress during grain filling was most injurious at the milk-ripening growth stage, causing 50% reduction in kernel weight.

The ability of sorghum to recover well after drought stress seems to be a widely observed phenomenon, whether it is expressed in stomatal function and leaf gas exchange (e.g., Glover 1959; Sanchez-Diaz and Kramer, 1973; Blum and Sullivan, 1974), the resumption of root function (Xu and Bland, 1993a), or the return to normal growth and development processes (e.g., Matthews et al., 1990a). From results with wheat (Munns et al., 1979) it may be postulated that the high capacity for osmotic adjustment may be behind the good recovery of sorghum after drought stress. This, however, requires proof.

The developmental flexibility of the yield components in sorghum and the high capability for component compensation has already been discussed earlier. This capability is well expressed under conditions of transient drought stress. When drought stress decreased tillering and the number of panicles per unit area, the number of kernels per panicle increased even though stress persisted (Blum, 1973). Whereas drought stress decreased the number of branch whorls per panicle, the larger number of kernels per panicle under stress was attributed solely to an increase in the number of panicle branches per whorl, mainly in the basal whorls.

7.2.2 Carbon Assimilation

The effect of plant water deficit on photosynthesis is a well-documented topic in plant science. Sullivan and Eastin (1974) were the first to explore the relative roles of stomatal and nonstomatal factors in affecting sorghum leaf photosynthesis under drought stress. They noted that several enzymes of the photosynthetic pathway, and especially the Hill reaction in isolated chloroplast, were very little affected or unaffected when chloroplasts were isolated from sorghum leaves which have been desiccated slowly to around -1.8 to

−2.4 MPa of LWP. They argued that activity of the photosynthetic enzymes was retained to low water potentials if slow hardening (acclimation) of the plant was allowed.

Both drought hardening (Blum and Sullivan, 1974) and heat hardening (Blum, 1989) have a pronounced effect on sorghum photosynthesis and stomatal activity under stress. Stomata of previously stressed sorghum plants were only partly open and hardly responded to a decreasing leaf water deficit in a second cycle of drought stress (Blum and Sullivan, 1974). Similarly, the reduction in both stomatal conductance and net photosynthesis with decreasing LWP was smaller when the rate of reduction in LWP was slower (Jones and Rawson, 1979). These effects of hardening may have been transduced by ABA accumulated under stress (e.g., Larque-Saavedra and Wain, 1976).

Krieg (1983a) and Krieg and Hutmacher (1986) investigated the effect of stomatal and nonstomatal factors in affecting sorghum photosynthesis. They concluded that carbon assimilation was reduced under the effect of water stress (when LWP was less than −1.5 MPa) mainly due to problems related to nonstomatal factors, rather than stomatal conductance. Assimilation rate per unit leaf internal CO_2 concentration and per unit photon flux density was reduced under drought stress. Citing unpublished thesis work by Morrissett, Krieg (1983b) stated that RuBP carboxylase activity explained more of the photosynthetic rate response of sorghum under both drought stress and nonstress conditions than did any of the other biochemical factors evaluated. However, a recent study shows that water deficit causes appreciable reduction in leaf photosynthesis, and the fraction of PSII reaction centers in an open state was very low, although constant (Loreto et al., 1995). The optimum temperature for photosynthesis was reduced with water deficit and that reduction was associated with the lower efficiency of energy capture of PSII reaction centers.

Excess light may cause damage to the primary photosynthetic reactions, when leaves are subjected to environmental stresses that directly affect the photosystem. This damage, defined as photoinhibition, was found to occur also in sorghum (Ludlow and Powels, 1988). When sorghum plants were subjected to drought stress, it was estimated that photoinhibition was about 20%. However, under the conditions of that experiment, the reduction of radiation load on the plants by shading improved growth and yield under stress by way of improving plant water status rather than by way of reducing photoinhibition. It was therefore concluded that photoinhibition induced by drought stress was not of agronomic importance in dryland sorghum.

While nonstomatal factors reduce carbon assimilation in individual water-stressed leaves, carbon assimilation by the whole canopy in the field may behave completely differently from single leaves. When midday LWP of nonirrigated sorghum was −0.1 to −to 0.5 MPa below that of irrigated sorghum, yield was reduced by 36% (Garrity et al., 1984). Under these dry-land conditions, when midday LWP was about -1.8 to -2.0 MPa, seasonal carbon assimilation per unit leaf area was unaffected, but seasonal crop carbon assimilated per unit land area was reduced. The main reason for crop carbon assimilation and yield reduction was the inhibition of leaf expansion and, subsequently, the reduction leaf area index under stress, as compared with conditions of adequate irrigation. Hence, the primary control over a sorghum crop assimilation and transpiration in the field stress environment is achieved by leaf area modification, including arrested leaf expansion and death of older leaves. The photosystem is affected only when water deficit becomes very severe.

7.2.3 Leaf Killing

Sorghum leaves may desiccate and die at leaf water potentials of about -3.0 to -4.0 MPa and RWC of about 35% to 55%, depending on genotype (Sullivan and Eastin, 1974; Basnayake et al., 1993). These values are not very low as compared with other cereal crop plants and they indicate a relative sensitivity to desiccation. Evidently, the ability of sorghum to perform under drought stress depends mainly on avoidance rather than tolerance of desiccation (e.g., Blum, 1979; Matthews et al., 1990a,b). It was found that desiccation tolerance (in terms of RWC at leaf death) was correlated across sorghum cultivars with low capacity for OA (Basnayake et al., 1993). This seemingly peculiar association may have its roots in evolutionary considerations and correlated effects during selection. However, there are substantial differences between sorghum genotypes in LWP at leaf or plant death, which evidently determine survival under very acute water deficit. Furthermore, desiccation tolerance and OA are not necessarily mutually exclusive. Cultivar M-35-1 is an example of a relatively desiccation tolerant "rabi" sorghum cultivar (Sullivan and Eastin, 1974), which also sustains a reasonable level of OA (Shackel et al., 1982).

7.2.4 Water-Use Efficiency

As a C_4 plant, sorghum has a relatively high ratio of leaf carbon assimilation to transpiration (A/T) of about 0.003 (Blum, 1989; Peng and Krieg, 1992). As discussed earlier, carbon assimilation is more sensitive than stomatal conductance to reduction in leaf water potential. Consequently, A/T would tend to decrease with increasing water stress. Indeed, when mean LWP of 10 sorghum cultivars was reduced to -1.8 MPa, A/T was reduced to a mean of 88% of that of the well-watered (control) plants (Blum, unpublished data). However, A/T ranged from 44% to 100% of that of the well-watered plants in the different cultivars.

From studies of carbon isotope discrimination (Hubick et al., 1990) and leaf gas exchange measurements (Kidambi et al., 1990b) it has been indicated that A/T as a measure of the ratio of total above ground biomass or yield to seasonal crop ET (WUE) at the leaf level was correlated across different sorghum genotypes with grain yield or biomass production. This was supported also by work of Peng and Krieg (1992). However, considering the effects of leaf water deficit, temperature, atmospheric vapor pressure deficit, and possibly the effect of a hormonal root signal on A/T, correlations between A/T and plant production require further evaluations on a broader range of conditions. It is still very reasonable that genotypic variations in A/T are basically controlled by variations in carbon assimilation rather than by variation in transpiration (e.g., Peng and Krieg, 1992).

WUE may (Peng and Krieg, 1992) or may not (Blum, unpublished data) correlate with A/T. This discrepancy may stem from the environmental conditions under which WUE and A/T are measured as affected by factors controlling crop evapotranspiration. For example, WUE is affected by the ratio of crop transpiration to crop evaporation. When, for example, soil surface evaporation is reduced, WUE would increase (e.g., Steiner, 1986) without involving A/T. Secondly, it has been shown earlier that sorghum crop transpiration is strongly influenced by leaf area reduction through leaf death, before transpiration per unit leaf area is seriously affected.

Sorghum WUE for grain yield normally ranges between 10 and 20 kg ha^{-1} mm^{-1}. WUE for total above ground biomass is around 40 to 45 kg ha^{-1} mm^{-1} (Virmani and Sivakumar, 1984). These ranges delineate the wide boundaries within which genetic and cultural manipulations have an important role in improving crop performance with limited moisture supply. However, it would be a mistake to assume that higher WUE is always associated with high productivity, especially under stress. The accumulated data on genetic variation in WUE as estimated by carbon isotope discrimination in C_3 plants has led to the conclusion that in a large number of cases better yields under drought stress were associated with lower WUE (Blum, 1992). This was seen also in sorghum (Virmani and Sivakumar, 1984; Donatelli et al., 1992). Hence, the results from selection for higher WUE under drought stress depend on the conditions of stress and the specific plant traits that control yield under stress. Based on a large volume of results for C_3 species (Hall et al., 1994), it can be concluded that a high Genotype × Environment interaction may be expected for WUE—which would determine the association between WUE and yield.

8 MINERAL NUTRITION

The determinants of sorghum mineral nutrition are plant growth and its productivity, the capacity of the plant to acquire minerals from the soil, and the mineral status of the soil. The soil factors are beyond the scope of this review, while factors affecting sorghum growth and production were discussed in previous sections. Here, only sorghum physiological capacity for acquiring and utilizing minerals will be discussed. The reader is referred to several general reviews on the physiology of plant mineral nutrition (Clarkson and Hawkesford, 1993; Jeschke and Hartung, 2000; Liptay and Arevalo, 2000). Visual symptoms of nutritional disorders of sorghum are described and discussed by Grundon et al. (1987).

The net inflow of minerals at the root surface follows the Michaelis–Menten relationship and it depends on the maximum flow at saturated mineral supply and the gradient of concentrations across the root surface. Both depend on plant demand and the rate of ion transport away from the root. The kinetics of ion influx are affected by root growth and aging, by shoot growth, and by plant mineral status and adaptation to mineral stress conditions. Most likely the kinetics of ion influx are not important under potential mineral supply. The modulation of kinetics is the very early response to mineral deficiency before any growth inhibition takes place.

The general importance of the kinetics of ion influx into the root is often questioned when the effects of mineral stress are quantified, especially in view of the overriding demanding effects of shoot and root growth on plant mineral status. Indeed in was seen in sorghum (Roy and Wright, 1974) that when nutrient supply was sufficient, nitrogen, phosphorus, and potassium uptake was highest during periods of active growth.

The first symptom of mineral deficiency stress is growth inhibition, before any of the classical visual symptoms such as leaf discoloration appear. The signaling of shoot growth inhibition by mineral deficiency status at the root or at the shoot level is not understood. The most conspicuous effect of mineral deficiency is the partitioning of more carbon to the root and an increase in root growth relative to shoot growth. There is no information on the control of assimilate partitioning to the root under the influence of mineral deficiency. Whether ABA is involved in mediating whole-plant response to mineral deficiency stress remains under controversy (Coleman and Schneider, 1996).

The concentration of ions at the root surface decreases as a function of absorption, and a depleted "shell" is created around the region of absorption. The continuation of absorption depends partly on mass flow, which can deliver solutes into the shell. Mass flow can deliver the sufficient quantities of the ions, depending on concentrations in soil and diffusivity of the specific ion. For low diffusing nutrients, such as phosphorus and in soils of low nutrient status, root growth becomes a major determinant of the plant's capacity to acquire nutrients.

The importance of root length density in mineral acquisition cannot be overemphasized. While root growth attributes and carbon partitioning to roots are critical in this respect, root length density can be effectively extended by root hairs and by symbiotic associations with mycorrhizas. Root hairs are most effective in increasing absorption of slow diffusing minerals. The most efficient organization of root hairs in terms of carbon invested per increase in ion absorption is high density of short hairs rather than low density of long hairs (Clarkson, 1985). Information on sorghum in this respect is unavailable.

Sorghum as other crops can develop mycorrhizal symbiosis, which affects the mineral nutrient uptake by the plant, especially for minerals of low mobility such as phosphorus. Concise review of symbiotic ectomicorrhiza (ECM), ericoid mycorrhiza (EM), and vesicular–arbuscular mycorrhiza (VAM) involvement with plant mineral nutrition has been published by Marschner and Dell (1994). Raju et al. (1990) have clearly shown that the positive effect of VAM symbiosis on sorghum P status and growth was maintained irrespective of the P status of the soil or the plant. VAM symbiosis increased shoot dry matter, root length, and shoot P content under conditions where P content of the shoot in nonsymbiotic (control) plants ranged between 0.29 and 5.09 g plant^{-1}. At very low soil P status, VAM colonization had very little effect on sorghum growth, although P deficiency symptoms were reduced by VAM (Miranda et al., 1989). As in other plants, a low P status in the soil generally improves root colonization by VAM, although in sorghum small additions of P to a deficient soil improved root colonization and VAM hyphae growth (Marschner and Dell, 1994).

Another root-colonizing endophyte, the nitrogen-fixing bacteria *Azospirillum brasilense*, has been claimed to have a positive effect on sorghum nitrogen status, as the case was argued for other cereals and grasses. These claims were later rejected by evidence from various sources, including work in sorghum (e.g., Pacovsky, 1988). Using ^{15}N natural abundance method, Sarig et al. (1984) provided clear evidence for the absence of biological N fixation in sorghum inoculated with *Azospirillum*. However, the positive effect of *Azospirillum* root inoculation on growth of several grass species is an unwavering observation. Sarig et al. (1984, 1988, 1990) investigated the effect of this endophyte on sorghum growth and yield under variable moisture and nutrient conditions. Typical results from three field experiments under dryland conditions in Israel (Sarig et al., 1988) showed that *Azospirillum* inoculation at planting increased total stover dry matter by 19% and grain yield by 15% to 18%. The major yield component affected was the number of kernels per panicle. Inoculated plants had a better plant water status, higher stomatal conductance, and lower canopy temperature, as compared with noninoculated controls. The better water status of inoculated plants was explained by a greater seasonal water use (15%) due to the deeper soil moisture extraction by the inoculated plants. The ability of the inoculated sorghum roots to meet transpirational demand better than noninoculated roots under conditions of limited moisture supply seems to be a consistent phenomenon also under controlled culture conditions (Sarig et al., 1984). The greater nitrogen content observed in shoots of *Azospirillum*-inoculated sorghum in these experiments evidently resulted from the improved root function in uptake from soil, rather then from biological fixation.

Information on sorghum general response curves to various mineral nutrients is very limited, unless one considers data from fertilizer application experiments without accompanying information on plant mineral status. The quadratic response of irrigated hybrid sorghum yield to soil nitrogen plateaued at about 80 to 100 ppm of nitrate in soil (Blum and Feigenbaum, 1969). Depending on which leaf was measured (Rego et al., 1986), grain yield in an irrigated field (Blum and Feigenbaum, 1969) or total dry matter of potted plants (Rego et al., 1986) increase with the increase in leaf nitrogen concentration to at least 4% (dry matter basis). Sorghum yield response curve to nitrogen fertilizer over 40 test sites in Australia was found to depend on seasonal rainfall and nitrate-N concentration at shallow (15 to 30 cm) soil depth (Holford et al., 1997). Water or osmotic stress constrained the response to soil nitrogen with a concomitant decrease in leaf nitrogen concentration. However, nitrogen concentration in the root was found to be generally high under stress (Rego et al., 1986). Youngquist and Maranville (1992) found that up to 89% of final total plant N had been accumulated in the plant by anthesis, depending upon genotype and soil N level. At that stage, 44% to 57% of plant N was in the stem and 43% to 56% was in the leaves. These responses are generally in agreement with a later study by Utzurrum et al. (1998), which also underlined the role of nitrogen storage in stems as a source for grain N accumulation. Nitrate reductase was found to be a factor limiting N accumulation and use efficiency in grain sorghum (Traore and Maranville, 1999). Nitrogen response of sorghum was modeled by Sinclair et al. (1997), where nitrogen uptake was related to crop thermal units and the calculated soil nitrogen availability.

The critical concentration of P in soils was found to vary with soil type, understandably. In vertisol, 90% relative grain yield of sorghum was obtained at 2.8 mg kg^{-1} Olsen extractable P while in alfisol, 90% relative grain yield was achieved at 5.0 mg P kg^{-1} soil (Sahrawat et al., 1995).

The commonly observed nitrogen by phosphorus interaction in crop fertilization practice is expressed very well also in sorghum. This interaction could be accounted for by the positive effect of phosphorus supply on sorghum nitrogen uptake and leaf nitrogen concentration (Blum and Feigenbaum, 1969). This is a general phenomenon also with respect to other major nutrients (e.g., Roy and Wright, 1974). It can be explained by the promoting effect of accelerated growth on mineral uptake, where growth promotion by one nutrient creates a demand for another, underlining again the importance of plant demand on mineral uptake from the soil.

Generally, mineral uptake continues as long as growth persists and biomass increases (e.g., Roy and Wright, 1974), with exceptions associated with certain soils or special climatic conditions (Muchow, 1990b). The extent to which uptake of major nutrients continues during grain filling is debatable (Muchow, 1990b). It is probably linked to the rate of biomass production after flowering and at maturity (see previous discussion), the total demand for nutrients created by the growing grain and the concentration of nutrients in the shoot.

Borrell and Hammer (2000) observed increased nitrogen uptake in stay-green compared with senescent sorghums under postanthesis drought. They hypothesized that higher nitrogen uptake by stay-green hybrids is a result of greater biomass accumulation during grain filling in response to increased sink demand (higher grain numbers), which, in turn, is the result of increased radiation use efficiency due to higher specific leaf nitrogen (SLN). Delayed leaf senescence resulting from higher SLN should, in turn, allow more carbon and nitrogen to be allocated to the roots of stay-green hybrids during grain filling, thereby maintaining a greater capacity to extract N from the soil compared with senescent hybrids. It is possible that nitrate transporters, for example, may also be associated with enhanced nitrogen uptake in stay-green hybrids (Borrell, personal communication). The critical and minimum SLN for implementation in the N-routines of crop growth simulation models is further discussed by van-Oosterom et al. (2001).

Remobilization of nutrients from the shoot into the growing kernels is seen especially under conditions where uptake from soil was limited due to plant or soil constraints (Roy and Wright, 1974).

Considering that total nitrogen yield in sorghum grain may reach 50 to 100 g m^{-2}, remobilization of nitrogen from the shoot into the growing kernel can be an important determinant of grain development when uptake from soil is limited. Under reasonable crop management conditions, sufficient quantities of nitrogen are available for remobilization from the shoot at flowering (e.g., Blum and Feigenbaum, 1969; Roy and Wright, 1974). Youngquist and Maranville (1992) estimated that during grain filling, stem and leaves lost 45–69% of their N content, with stems having the higher values. The relative contribution of remobilized N to grain N was greater under lower soil N content.

The extent to which sorghum can reduce the mineral content of the soil is an important question not only for maximizing nutrient use by the crop but also for serious environmental and ecological considerations. Israeli agronomists realized some 40 years ago the apparent effectiveness of sorghum in utilizing residual nitrogen left after a wheat crop and it led to the practice of growing successful crops of unfertilized dryland grain sorghum in the second year after well-fertilized wheat. However, it is unclear whether this can be ascribed mainly to the ability of the single root to reduce the mineral content of the soil or the extensive growth of the whole root system of sorghum and its ability to capture nitrogen from deep soil. From discussions of the kinetics of ion influx into the root it may perhaps be hypothesized that the important factor in this respect is deep root growth. However, from the limited literature on the subject in sorghum (Myers and Asher, 1982) it appears that the question is unresolved.

The utilization efficiency of mineral nutrients, especially in relations to its genetic improvement, is becoming an important area of research (e.g., Moll et al., 1982a,b; Glass, 1989; Clark and Duncan, 1991). Information in this respect in sorghum is lacking, so that this topic is discussed here very briefly. There are two main components of mineral-use efficiency: (1) the efficiency of uptake and (2) the efficiency of utilization for the production of the plant material of interest. The efficiency of uptake may be defined as the ratio of the total content of the mineral in the plant at maturity to the total content of the mineral in the soil at germination, with or without accounting for soil mineral losses, such as leaching. If, for example, uptake efficiency involves the capacity of the plant to reduce leaching, then the measurement of efficiency should not be normalized for leaching. The efficiency of utilization may be defined as the ratio of the plant production component of interest to the total content of the mineral at the growth stage when the production component is measured (say, maturity). Other definitions of efficiency are also being addressed in research, such as yield per unit fertilizer applied, but their use does not allow developing the much-needed insight into the mechanisms that control efficiency.

It has been found in maize (Moll et al., 1982a,b) that at low nitrogen supply differences among hybrids in nitrogen-use efficiency (NUE) were due primarily to variations in utilization of accumulated nitrogen, while at high nitrogen supply they were due largely to variations in nitrogen uptake. Ma and Dwyer (1998) have shown that delayed leaf senescence in some maize genotypes is conducive to higher NUE.

The physiological control of efficiency of mineral uptake or utilization is not clear. Much of the work in this area is limited to actual screening for apparent efficiency. Factors, which may control uptake efficiency, can be recognized from the preceding discussion of mineral uptake and the cited reviews therein. It is evident that uptake efficiency may be influenced by demand and therefore is in an interaction with utilization and its efficiency. Utilization efficiency is associated with processes of translocation, assimilation and redistribution (Moll et al., 1982) or with physical and metabolic compartmentation (Wieneke, 1990), such as structural components of the panicle or biochemical components such as enzymes associated with photosynthesis (Lafitte and

Loomis, 1988). Because nitrogen affects leaf size and LAI, it is strongly involved with radiation interception and crop carbon assimilation, especially under nitrogen stress. Nitrogen-use efficiency may therefore be defined also as a low nitrogen threshold for leaf expansion growth.

The depletion of plant stored minerals to assist grain yield formation is an important component of efficiency in sorghum (e.g., Lafitte and Loomis, 1988; Youngquist et al., 1992). In maize hybrids (Moll et al., 1982) the proportion of plant nitrogen remobilized to the grain was important for nitrogen use-efficiency at low nitrogen supply. In sorghum, about 65% of the plant nitrogen and phosphorus were remobilized from the stem and leaves to the grain (Hocking, 1993). Leaves were more important in this respect then stems. Borrell and Hammer (2000) estimated that 41 and 52 kg N ha^{-1} were available for grain filling in stay-green and senescent hybrids via translocation from the leaf, and that a further 27 kg N ha^{-1} was available via translocation from the stems. Of the nitrogen taken up by the grain, 49% was extracted from the soil in senescent hybrids compared with 64% in stay-green hybrids. The importance of nutrient remobilization from the shoot into the developing grain seems to corroborate the conclusion that harvest index is among the best predictors of genotypic variation in grain yield NUE of sorghum (Youngquist et al., 1992).

Nitrogen-use efficiency, as the ratio of total dry matter or grain production to total nitrogen uptake, varied among sorghum genotypes, irrespective of irrigation regime (Zweifel et al., 1987). Nitrogen-use efficiency increased under nitrogen stress as compared with conditions of adequate nitrogen supply. There is a general observation of an apparent increase in mineral-utilization efficiency in terms of production per unit mineral uptake, as mineral uptake or mineral concentration in the plant is reduced.

High NUE for total dry matter production, as tested in nutrient culture for several sorghum cultivars, was found to be associated with thick leaves, low N concentration in leaves, large leaf phloem transectional area, rapid solubilization and remobilization of N from older to younger leaves, and lower dark respiration rates (Gardner et al., 1994). On the other hand, in a field study of 14 sorghum hybrids Kamoshita et al. (1998) found genotypic variation in N uptake, but not in NUE.

Sorghum genotypes differ in phosphorus uptake per unit root weight and in phosphorus-utilization efficiency as dry matter production per unit phosphorus absorbed (Furlani et al., 1984; Wieneke, 1990). It was suggested (Wieneke, 1990) that the better phosphorus-utilization efficiency in efficient sorghum genotypes (such as NB9040) was related to redistribution of phosphorus in different plant parts, which positively affected growth even at low phosphorus concentrations in the plant.

Sorghum is relatively susceptible to iron deficiency chlorosis. Iron deficiency is well recognized visually by the yellowing of the younger leaves, which is a function of their low chlorophyll content under stress (Peterson and Onken, 1992). Resistance to Fe deficiency (also termed as "Fe efficiency") is quite common sorghum (e.g., Bowen and Rodgers, 1987).

In the grasses, Fe acquisition from the soil proceeds by a constitutive capacity for the release of Fe(III) chelating root exudates (phytosiderophores, PS). The rate of Fe(III)PS uptake by roots increase by a factor of about 5 under conditions of Fe stress (Marschner et al., 1986). Sorghum is relatively sensitive to Fe chlorosis because of the low rates of PS release from the root in the seedling stage. Because Fe(III) PS release and uptake was found to be under genetic control in several grass species, sorghum resistance to Fe deficiency may perhaps be developed by directly improving its capacity for PS release and uptake (Romheld and Marschner, 1990). The recorded decrease in iron chlorosis when root temperatures were lowered from 27° to 12°C (Clarke and Reinhard, 1991) may be a function of the higher root/ shoot ratio at low temperatures. At the same time this may also be related to the effect of temperature on PS exudation and its dynamics in the soil.

9 MINERAL TOXICITY

Sorghum is relatively susceptible to soil acidity and the associated mineral deficiencies and toxicities. The general aspects of aluminum and manganese toxicities in crops grown on acid soils are relevant also to sorghum and were reviewed extensively elsewhere (e.g., Delhaize and Ryan, 1995). The effect of aluminum toxicity on sorghum plants is expressed, as in other cereal crops, in severely degenerated roots and the associated repercussions, such as water deficit and mineral deficiencies (e.g., Baligar et al., 1993). The common basis for Al tolerance in several crop plants including sorghum is pointed out by Ishikawa et al. (2001), who indicated the durability of the cellular plasma membrane as a major factor of tolerance.

The relative importance of aluminum-induced magnesium deficiency under relatively moderate soil acidity is pointed out for sorghum (Tan et al., 1992).

Most research on aluminum toxicity in crop plants has been carried out in wheat and it is not known whether sorghum is uniquely different in this respect from wheat. It has been proposed that the major repercussion of soil acidity complex in sorghum is mediated by "inhibition of plant growth regulator (PGR) transport, biosynthesis, and degradation" (Duncan et al., 1991b).

Sorghum accumulates silicon in various tissues including the root (Parry and Kelso, 1975). In a study of sorghum in hydroponics, it was found that silicon in the growth medium enabled plants to overcome Al toxicity symptoms and enhanced shoot and root growth. Silicon counteracted many deleterious effects of Al on nutrient balances because it helped to maintain a higher root mass in the presence of toxic levels of Al (Galvez and Clark, 1991). After measuring Si and Al concentrations in different sorghum root tissues challenged by Al, Hodson and Sangster (1993) suggested that Al was sequestered in the Al–Si deposit in the outer tangential wall of the root epidermis. Cocker et al. (1998) proposed that root cell walls are the main internal sites of aluminosilicate (AS) and/or hydroxyaluminosilicate (HAS) formation and of Al detoxification. Factors promoting AS/HAS formation in this compartment include high apoplastic pH, the presence of organic substances (e.g., malate), and the presence of suitable local concentrations of reactive forms of Al and Si, on or within the surfaces of the wall matrix. This mechanism is specific to sorghum, barley, or soybean but not wheat.

There is very large genetic variation in sorghum for Al tolerance (e.g., Baligar et al., 1993; Boye-Goni and Marcarian, 1985; Gourley et al., 1990; Foy et al., 1993; Shuman et al., 1993) and tolerant germplasm has been continuously identified and registered (e.g., Duncan, 1984; Duncan et al., 1992).

Sorghum is considered as moderate in response to soil alkalinity and salinity. The mean EC_{50} of the saline medium for sorghum grain yield is between 5 and 11 dS m^{-1}, depending on the study, the genotype, and test conditions.

Advisory literature indicated a salinity threshold (no yield reduction) for sorghum of 6.8 dS/m, and 50% yield reduction at 5 to 11 dS/m of soil electrical conductivity (EC), depending on genotype and study. The correct value is probably in the lower range (Igartua et al., 1995). However, such data are normally developed by growing sorghum with saline irrigation water. When sorghum was tested for three years on saline soil under dryland conditions (Daniells et al., 2001), it was found that yield was reduced by 50% at EC levels as low as 2.8 dS/m, indicating greater than expected sensitivity to salt. Soil salinity of EC = 3.6 dS m^{-1} was found to reduce seedling establishment by 77% and grain yield by 84% (Hocking, 1993).

Comparative studies of sorghum and other crop species are very limited. Furthermore, much of the past research on the effect of salinity on sorghum has been performed with short exposures to salinity stress. It is now realized (Amzallag et al., 1990; Munns et al., 1995) that plant adaptation and subsequent expression of plant resistance to salinity requires time, around 3 weeks. Given this time, sorghum can adapt to grow in 300 mol m^{-3} NaCl, while it will be severely inhibited by half that concentration when not given enough time (Amzallag et al., 1990). This adaptation does not only involve osmotic adjustment. Abscisic acid was found to decrease the time duration required for adaptation from 20 days to 8 days (Amzallag et al., 1991). Adaptation is more effective at a juvenile growth stage (Seligmann et al., 1993) and it may strongly involve the root (Amzallag, 1997). Sorghum photosystem II activity was found to be relatively tolerant to salt stress (Lu and Zhang, 1998). On the other hand, salinity stress was found to predispose sorghum photosystem to photoinhibition, resulting in lower quantum efficiency (Sharma and Hall (1991). Again, such conflicting results might evolve from differences in the stress protocol used in experiments.

Sorghum root growth in saline media was inhibited in an exponential relationship to Na accumulation in the root, while wheat root growth was promoted by salinity inasmuch as Na was transported from root to shoot (Devitt et al., 1984). McNeilly's (1990) suggestion that root growth after 2-week exposure to saline media represents the relative salinity tolerance of different sorghum genotypes is therefore reasonable. However, a validation of this conclusion by field tests is not available. It was confirmed, however, that sorghum could exclude Na from the shoot by its relative accumulation in the root (Weimberg et al., 1982; Grieve and Maas, 1988; Igartua et al., 1995). K/Na ratio appears to be a promising indicator of whole-plant salinity tolerance in sorghum (Yang et al., 1990; Igartua et al., 1995), as the case is in several other crop plants. Grieve and Maas (1988) considered calcium acquisition from the saline medium as a factor in sorghum tolerance to sodic soil conditions.

Overall plant response to saline water irrigation could be sensed by infrared canopy temperature measurement, and this technique was suggested as suitable for field screening in selection for salinity resistance (Kluitenberg and Biggar, 1992). Finally, salinity stress was found to ascribe heat tolerance to sorghum photosystem II activity, and this effect was independent of salt concentration in the medium (Lu Zhang, 1998).

REFERENCES

Acevedo E, Fereres E, Hsiao TC, Henderson DW. Diurnal growth trends water potential and osmotic adjustment of maize and sorghum leaves in the field. Plant Physiol 1979; 64:476–480.

Ackerson RC, Krieg DR. Stomatal and nonstomatal regulation of water-use in cotton, corn and sorghum. Plant Physiol 1977; 60:850–853.

Ackerson RC, Krieg DR, Sung FJM. Leaf conductance and osmoregulation of field grown sorghum. Crop Sci 1980; 20:10–14.

Adewuai SRA. Turnover of dhurrin in green sorghum seedlings. Plant Physiol 1990; 94:1219–1224.

Agboma PC, Jones MGK, Peltonensainio P, Rita H, Pehu E. Exogenous glycinebetaine enhances grain yield of maize, sorghum and wheat grown under two supplementary watering regimes. J Agron Crop Sci 1997; 178:29–37.

Alagarswamy G, Reddy DM, Swaminathan G. Durations of the photoperiod-sensitive and -insensitive phases of time to panicle initiation in sorghum. Field Crops Res 1997; 55:1–10.

Ambler JR, Morgan PW, Jordan JW. Amounts of zeatin and zeatin riboside in xylem sap of senescent and nonsenescent sorghum. Crop Sci 1992; 32:411–419.

Amzallag GN. Influence of periodic fluctuations in root environment on adaptation to salinity in *Sorghum bicolor*. Aust J Plant Physiol 1997; 24:579–586.

Amzallag GN, Lerner HR, Poljakoff-Mayber A. Induction of increased salt tolerance in *Sorghum bicolor* by NaCl pretreatment. J Exp Bot 1990; 41:29–34.

Amzallag GN, Lerner HR, Poljakoff Mayber A. Exogenous ABA as a modulator of the response of sorghum to high salinity. J Exp Bot 1991; 41:1529–1534.

Arkebauer TJ, Weiss A, Sinclair RT, Blum A. In defense of radiation use efficiency: a response to Demetriades-Shah et al. 1992. Agric For Meteorol 1994; 68:221–227.

Arkin GF, Monk RL. Seedling photosynthetic efficiency of a grain sorghum hybrid and its parents. Crop Sci 1979; 19:128–130.

Arkin GF, Vanderlip RL, Ritchie JT. A dynamic grain sorghum growth model. Trans Am Soc Agric Eng 1976; 19:622–630.

Artschwager E. Anatomy and Morphology of the Vegetative Organs of *Sorghum vulgare*. USDA Tech Bull No. 957. Washington, DC: USDA, 1948:55.

Baker NR. A possible role for photosystem-II in environmental perturbations of photosynthesis. Physiol Plant 1991; 81:563–570.

Baligar VC, Schaffert RE, Dos Santos HL, Pitta GVE, Bahia Filho AF, de C. Growth and nutrient uptake parameters in sorghum as influenced by aluminum. Agron J 1993; 85:1068–1074.

Barker DJ, Sullivan CY, Moser LE. Water deficit effects on osmotic potential, cell wall elasticity, and proline in 5 forage grasses. Agron 1993; 85:270–275.

Bartel AT. Hybrid vigor in sorghums. Agron J 1949; 41:147–152.

Basnayake J, Cooper M, Henzell RG, Ludlow MM. Influence of rate of development of water deficit on the expression of maximum osmotic adjustment and desiccation tolerance in three grain sorghum lines. Field Crops Res 1996; 49:65–76.

Basnayake J, Cooper M, Ludlow MM, Henzell RG, Snell PJ. Inheritance of osmotic adjustment to water stress in three grain sorghum crosses. TAG 1995; 90:675–682.

Basnayake J, Ludlow M, Cooper M, Henzell RG. Genotypic variation of osmotic adjustment and desiccation tolerance in contrasting sorghum inbred lines. Field Crops Res 1993; 35:51–62.

Beall FR, Morgan PW, Mander LN. Genetic regulation of development in *Sorghum bicolor* v the ma3r allele results in gibberellin enrichment. Plant Physiol 1991; 95:116–125.

Benech-Arnold RLB, Fenner M, Edwards PJ. Changes in germinability, ABA content and ABA embryonic sensitivity in developing seeds of *Sorghum-bicolor* (L.) Moench induced by water stress during grain filling. New Phytol 1991; 118:339–347.

Bhaskaran S, Smith RH, Newton RJ. Physiological changes in cultures sorghum cells in response to induced water stress I. Free proline. Plant Physiol 1985; 79:266–269.

Bidinger FR, Johansen C, eds. Drought Research Priorities for the Dryland Tropics. Patancheru: ICRISAT, 1988:219.

Blum A. Effect of soil fertility and plant competition sorghum panicle morphology and panicle weight components. Agron 1967; 59:400–402.

Blum A. Seedling emergence and establishment of sudangrass varieties and sorghum × sudangrass hybrids under sub-optimal temperatures. Isr J Agric Res 1969; 19:101–104.

Blum A. Nature of heterosis in grain production by the sorghum panicle. Crop Sci 1970a; 10:28–31.

Blum A. Effects of plant density and growth duration on sorghum yield under limited water supply. Agron J 1970b; 62:333–336.

Blum A. Effect of planting date on water-use and its efficiency in dryland grain sorghum. Agron J 1972; 64:775–778.

Blum A. Component analysis of yield response to drought of sorghum hybrids. Expl Agric 1973; 9:155–167.

Blum A. Genotypic responses in sorghum to drought stress I. Response to soil moisture stress. Crop Sci 1974; 14:361–364.

Blum A. Effect of the Bm gene on epicuticular wax deposition and the spectral characteristics of sorghum leaves. SABRAO J Breed Genet 1975; 7:45–52.

Blum A. The basis of heterosis in the differentiating sorghum panicle. Crop, 1977, 880–882.

Blum A. Genetic improvement of drought resistance in crop plants: a case for sorghum. In: Mussell H, Staples RC, eds. Stress Physiology in Crop Plants. New York: Wiley Interscience, 1979:429–445.

Blum A. Plant Breeding for Stress Environments. Boca Raton: CRC Press, 1988: 208.

Blum A. The temperature response of gas exchange in sorghum leaves and the effect of heterosis. J Exp Bot 1989; 40:453–460.

Blum A. Selection for sustainable production in water-deficit environments. Proceedings of the 1st International Crop Science Congress. Dordrecht: Ames, 1992: 343–347.

Blum A. Improving grain filling under stress by stem reserve mobilization. In: Braun HJ, Altay F, Kronstad WE, Beniwal SPS, McNab A, eds. Wheat: Prospects for Global Improvement. Dordrecht: Kluwer Academic Publ, 1997:135–142.

Blum A, Arkin GF. Sorghum root growth and water-use as affected by water supply and growth duration. Field Crops Res 1984; 9:131–142.

Blum A, Ebercon A. Genotypic responses in sorghum to drought stress III. Proline accumulation and drought resistance. Crop Sci 1977; 16:428–431.

Blum A, Feigenbaum S. The effect of soil fertility on hybrid grain sorghum grown under conditions favouring maximum yield. Qual Plant Mater Veg 1969; 17:273–285.

Blum A, Johnson JW. Transfer of water from roots into dry soil and the effect on wheat water relations and growth. Plant Soil 1992; 145:141–146.

Blum A, Naveh M. Improved water-use efficiency by promoted plant competition in dryland sorghum. Agron J 1976; 68:111–116.

Blum A, Ritchie JT. Effect of soil surface water content on sorghum root distribution in the soil. Field Crops Res 1984; 8:169–176.

Blum A, Arkin GF, Jordan WR. Sorghum root morphogenesis and growth I. Effect of maturity genes. Crop Sci 1977a; 17:149–153.

Blum A, Jordan WR, Arkin GF. Sorghum root morphogenesis and growth: II. Manifestation of heterosis. Crop Sci 1977b; 17:153–157.

Blum A, Sullivan CY. Leaf water potential and stomatal activity in sorghum as influenced by soil moisture stress. Isr J Bot 1974; 23:42–47.

Blum A, Sullivan CY. The comparative drought resistance of landraces of sorghum and millet from dry and humid regions. Ann Bot 1986; 57:835–846.

Blum A, Sullivan CY. A laboratory method for monitoring net photosynthesis in leaf segments under controlled water stress: experiments with sorghum. Photosynthetica 1972; 6:18–23.

Blum A, Golan G, Mayer J. Progress achieved by breeding open-pollinated cultivars as compared with landraces of sorghum. J Agric Sci Camb 1991; 117:307–312.

Blum A, Golan G, Mayer J, Sinmena B. The effect of dwarfing genes on sorghum grain filling from remobilized stem reserves under stress. Field Crops Res 1997; 52:43–54.

Blum A, Mayer J, Golan G. Agronomic and physiological assessments genotypic variation for drought resistance in sorghum. Aust J Agric Res 1989; 40:49–61.

Blum A, Ramaiah S, Kanemasu ET, Paulsen GM. The physiology of heterosis in sorghum with respect to environmental stress. Ann Bot 1990; 65:149–158.

Blum A, Golan G, Mayer J, Sinmena B, Obilana T. The comparative productivity and drought response of semi-tropical hybrids and open-pollinated varieties of sorghum. J Agric Sci Camb 1992; 118:29–36.

Borrell AK, Hammer GL. Nitrogen dynamics and the physiological basis of stay-green in sorghum. Crop Sci 2000; 40:1295–1307.

Borrell AK, Hammer GL, Douglas ACL. Does maintaining green leaf area in sorghum improve yield under drought? I. Leaf growth and senescence. Crop Sci 2000a; 40:1026–1037.

Borrell AK, Hammer GL, Henzell RG. Does maintaining green leaf area in sorghum improve yield under drought? II. Dry matter production and yield. Crop Sci 2000b; 40:1037–1048.

Bowen CR, Rodgers DM. Evaluation of a greenhouse screening technique for iron-deficiency chlorosis of sorghum. Crop Sci 1987; 27:1024–1029.

Boye-Goni SR, Marcarian V. Diallel analysis of aluminum tolerance in selected lines of grain sorghum. Crop Sci 1985; 25:749–752.

Brar GS, Steiner JL, Unger PW, Prihar SS. Modeling sorghum seedling establishment from soil wetness and temperature of drying seed zones. Agron J 1992; 84:905–910.

Brookings IR. Male sterility in *Sorghum bicolor* (L) Moench induced by low night temperature I. Timing of the stage of sensitivity. Aust J Plant Physiol 1976; 3:589–596.

Brown RH, Hattersley PW. Leaf anatomy of C3–C4 species as related to evolution of C4 photosynthesis. Plant Physiol 1989; 91:1543–1550.

Caddel JL, Weibel DE. Effect of photoperiod and temperature on the development of sorghum. Agron J 1971; 63:799–803.

Carberry PS, Hammer GL, Muchow RC. Modelling genotypic and environmental control of leaf area dynamics in grain sorghum 3. senescence and prediction of green leaf area. Field Crops Res 1988; 33:329–351.

Carberry PS, Muchow RC, Hammer GL. Modelling genotypic and environmental control of leaf area dynamics in grain sorghum 2. Individual leaf level. Field Crops Res 1993a; 33:311–328.

Carberry PS, Hammer GL, Muchow RC. Modelling genotypic and environmental control of leaf area dynamics in grain sorghum 3. Senescence and prediction of green leaf area. Field Crops Res 1993b; 33:329–351.

Casal JJ. Light quality effects on the appearance of tillers of different order in wheat (*Triticum aestivum*). Ann Appl Biol 1988; 112:167–173.

Chapman SC, Fischer KS. Osmotic adjustment in *Sorghum bicolor* (L) Moench grown under moisture stress in soil and osmotically modified solution cultures. Plant Soil 1988; 107:57–62.

Chatterton NJ, Hanna WW, Powell JB, Lee DR. photosynthesis and transpiration of bloom and bloomless sorghum. Can J Plant Sci 1975; 55:641–643.

Cherney JH, Moore KJ, Volenec JJ, Axtell JD. Rate and extent of digestion of cell wall components of brown-midrib sorghum species. Crop, 1986, 1055–1059.

Childs KL, Cordonnier-Pratt MM, Pratt LH, Morgan PW. Genetic regulation of development in *Sorghum bicolor* VII Ma3R flowering mutant lacks a phytochrome that predominates in green tissue. Plant Physiol 1992; 99:765–770.

Childs KL, Lu JL, Mullet JE, Morgan PW. Genetic regulation of development in *Sorghum bicolor* X. Greatly attenuated photoperiod sensitivity in a phytochrome-deficient sorghum possessing a biological clock but lacking a red light-high irradiance response. Plant Physiol 1995; 108:345–351.

Childs KL, Miller FR, Mullet JE. The sorghum photoperiod sensitivity gene, Ma3, encodes a phytochrome B. Plant Physiol 1997; 113:611–619.

Cicek M, Esen A. Structure and expression of a dhurrinase (beta-glucosidase) from sorghum. Plant Physiol 1998; 116:1469–1478.

Clark LE. Embryonic leaf number in sorghum. Crop Sci 1970; 10:307–309.

Clark RB, Duncan RR. Improvement of plant mineral nutrition through breeding. Field Crops Res 1991; 27:219–240.

Clarkson DT. Factors affecting mineral nutrient acquisition by plants. Ann Rev Plant Physiol 1985; 30:77–115.

Clarkson DT, Hawkesford MJ. Molecular biological approaches to plant mineral nutrition. Plant Soil 1993; 155:21–36.

Clark RB, Reinhard N. Effects of soil temperature on root and shoot growth traits and iron deficiency chlorosis in sorghum genotypes grown on a low iron calcareous soil. Plant Soil 1991; 130:97–103.

Clark LE, Collier JW, Langston R. Dormancy in *Sorghum bicolor* (L) Moench. II. Effect of pericarp and testa. Crop Sci 1968; 8:155–158.

Cocker KM, Evans DE, Hodson MJ. The amelioration of aluminium toxicity by silicon in higher plants: solution chemistry or an in plants mechanism? Physiol Plant 1998; 104:608–614.

Coleman JS, Schneider KM. Evidence that abscisic acid does not regulate a centralized whole-plant response to low soil-resource availability. Oecologia 1996; 106:277–283.

Craufurd PQ, Flower DJ, Peacock JM. Effect of heat and drought stress on sorghum (*Sorghum bicolor*) I. Panicle development and leaf appearance. Exp Agric 1993; 29:61–76.

Craufurd PQ, Qi AM, Ellis RH, Summerfield RJ, Roberts EH, Mahalakshmi V. Effect of temperature on time to panicle initiation and leaf appearance in sorghum. Crop Sci 1998; 38:942–947.

Cruz RT, Jordan WR, Drew MC. Structural changes and associated reduction of hydraulic conductance in roots of *Sorghum bicolor* (L) following exposure to water deficit. Plant Physiol 1992; 99:203–212.

Cummins DG, Dobson JW Jr. Digestibility of bloom and bloomless sorghum leaves as determined by a modified in vitro technique. Agron J 1972; 64:682–683.

Daniells IG, Holland JF, Young RR, Alston CL, Bernardi AL. Relationship between yield of grain sorghum (*Sorghum bicolor*) and soil salinity under field conditions. Aust J Expl Agric 2001; 41:211–217.

Davies WJ, Zhang JH. Root Signals and the regulation of growth and development of plants in drying soil. Annu Rev Plant Physiol Plant Mol Biol 1991; 42:55–76.

Delhaize E, Ryan PR. Aluminum toxicity and tolerance in plants. Plant Physiol 1995; 107:315–321.

Demetriades-Shah TH, Fuchs M, Kanemasu ET, Flitcroft I. A note of caution concerning the relationship between cumulated intercepted solar radiation and crop growth. Agric For Meteorol 1992; 58:193–207.

Devitt D, Stolzy LH, Jarell WM. Response of sorghum and wheat to different K Na ratios at varying osmotic potentials. Agron J 1984; 76:68–688.

Doggett H. Sorghum. Essex: Longman, 1988:512.

Donatelli M, Hammer GL, Vanderlip RL. Genotype and water limitation effects on phenology, growth, and transpiration efficiency in grain sorghum. Crop Sci 1992; 32:781–786.

Downes RW. The effect of temperature on tillering of grain sorghum seedlings. Aust J Agric Res 1968; 219:59–64.

Downes RW. Effect of light intensity and leaf temperature on photosynthesis and transpiration in wheat and sorghum. Aust J Biol Sci 1970; 23:775–882.

Downes RW. Low temperature induced male sterility in *Sorghum bicolor*. Aust J Expl Agric 1971; 11:352–356.

Downes RW. Effect of temperature on the phenology and grain yield of *Sorghum bicolor*. Aust J Agric Res 1972; 23:594–858.

Duncan RR. Registration of acid soil tolerant sorghum germplasm. Crop Sci 1984; 24:1006.

Duncan RR, Bockholt AJ, Miller FR. Descriptive comparison of senescent and non-senescent sorghum genotypes. Agron J 1981; 73, 849–853.

Duncan RR, Bramel-Cox PJ, Miller FR. Contributions of introduced sorghum germplasm to hybrid development in the USA. In: Shands HL, Wiesner L, eds. Use of Plant Introduction in Cultivar Development. Madison: CSSA, 1991a:69 102.

Duncan RR, Wilkinson RE, Shuman LM, Ramseur EL. Acid soil tolerance mechanisms for juvenile stage sorghum (*Sorghum bicolor*). Dev Plant Soil Sci 1991b; 45: 1037–1045.

Duncan RR, Waskom RM, Miller DR, Hanning GE, Timm DA, Nabors MW. Registration of GC103 and GC104 acid-soil tolerant TX430 sorghum. Crop Sci 1992; 6:1520.

Eastin JD, Hultquist JH, Sullivan CY. Physiologic maturity in grain sorghum. Crop Sci 1973; 13:175–178.

Eastin JD, Sullivan CY. Carbon dioxide exchange in compact and semi-open sorghum inflorescence. Crop Sci 1969; 9:165–166.

Ebel RC, Stodola AJW, Duan XR, Auge RM. Non-hydraulic root-to-shoot signaling in mycorrhizal and non-mycorrhizal sorghum exposed to partial soil drying or root severing. New Phytol 1994; 127:495–505.

Ebercon A, Blum A, Jordan WR. A rapid colorimetric method for epicuticular wax content of sorghum leaves. Crop Sci 1977; 17:179–180.

Einhellig FA, Souza IF. Phytotoxicity of sorgoleone found in grain sorghum root exudates. J Chem Ecol 1992; 18:1–11.

Ellis RH, Qi A, Craufurd PQ, Summerfield RJ, Roberts EH. Effects of photoperiod, temperature and asynchrony between thermoperiod and photoperiod on development to panicle initiation in sorghum. Ann Bot 1997; 79:169–178.

El-Sharkawy MA, Hesketh JD. Effects of temperature and water deficit on leaf photosynthesis rate of different species. Crop Sci 1964; 4:514–518.

Escalada RG, Plucknett DL. Ratoon cropping of sorghum I. Origin, time of appearance, and fate of tillers. Agron J 1975a; 67:473–478.

Escalada RG, Plucknett DL. Ratoon cropping of sorghum II. Effect of daylength and temperature on tillering and plant development. Agron J 1975b; 67:479–484.

Farquhar GD, Sharkey TD. Stomatal conductance and photosynthesis. Annu Rev Plant Physiol 1982; 33:317–345.

Farquhar GD, Ehleringer JR, Hubick K. Carbon isotope discrimination and photosynthesis. Annu Rev Plant Physiol Plant Mol Biol 1989; 40:503–537.

Farrar JF. Sink strength—what is it and how do we measure it—introduction. Plant Cell Environ 1993; 16:1045–1046.

Finlayson SA, Lee IJ, Mullet JE, Morgan PW. The mechanism of rhythmic ethylene production in sorghum. The role of phytochrome B and simulated shading. Plant Physiol 1999; 119:1083–1089.

Fischer KS, Wilson GL. Studies of grain production in *Sorghum bicolor* (L) Moench I. The contribution of pre-flowering photosynthesis to grain yield. Aust J Agric Res 1971a; 22:33–37.

Fischer KS, Wilson GL. Studies of grain production in *Sorghum bicolor* (L) Moench II. Sites responsible for grain dry matter production during the post anthesis period. Aust J Agric Res 1971b; 22:39–47.

Fischer KS, Wilson GL. Studies of grain production in *Sorghum bicolor* (L) Moench III. The relative importance of assimilate supply, grain growth capacity and transport system. Aust J Agric Res 1975a; 26:11–23.

Fischer KS, Wilson GL. Studies of grain production in *Sorghum bicolor* (L) Moench IV. Some effects of increasing and decreasing photosynthesis at different stages of the plant's development on the storage capacity of the inflorescence. Aust J Agric Res 1975b; 26:25–30.

Fischer KS, Wilson GL. Studies of grain production in *Sorghum bicolor* (L) Moench V. Effect of plant density on growth and yield. Aust J Agric Res 1975c; 26:31–41.

Fischer KS, Wilson GL. Studies of grain production in *Sorghum bicolor* (L) Moench VI. Profiles of photosynthesis, illuminance and foliage arrangement. Aust J Agric Res 1976a; 27:35–44.

Fischer KS, Wilson GL. Studies of grain production in *Sorghum bicolor* (L) Moench VII. Contribution of plant parts to canopy photosynthesis and grain yield in field stations. Aust J Agric Res 1976b; 27:235–242.

Flower DJ, Rani AU, Peacock JM. Influence of osmotic adjustment on the growth, stomatal conductance and light interception of contrasting sorghum lines in a harsh environment. Aust J Plant Physiol 1990; 17:91–105.

Fortmeier R, Schubert S. Storage of non-structural carbohydrates in sweet sorghum [*Sorghum bicolor* (L) Moench]: comparison of sterile and fertile lines. J Agron Crop Sci 1995; 175:189–193.

Foster KR, Morgan PW. Genetic regulation of development in *Sorghum bicolor* IX. The ma3R allele disrupts diurnal control of gibberellin biosynthesis. Plant Physiol 1995; 108:337–343.

Foy CD, Duncan RR, Waskom RM, Miller DR. Tolerance of sorghum genotypes to an acid, aluminum toxic tatum subsoil. J Plant Nutr 1993; 16:97–127.

Fritz JO, Moore KJ, Jaster EH. Digestion kinetics and cell wall composition of brown midrib sorghum X. Sudangrass morphological components. Crop Sci 1990; 30:213–219.

Fukai S, Liwa CJ, Henderson CWL, Maharjan BB, Hermus RC, Herbert SC. The field performance of an induced uniculm grain sorghum (*Sorghum bicolor*) in South-East Queensland, Australia. Expl Agric 1986; 22:393–403.

Furlani AMC, Clark RB, Maranville GW, Ross WM. Sorghum genotypic differences in phosphorus uptake rate and distribution in plant parts. J Plant Nutr 1984; 7:1113–1127.

Gale MD, Youssefian S. Dwarfing genes of wheat. In: Russell GE, ed. Progress in Plant Breeding. London: Butterworths, 1985:1–35.

Galvez L, Clark RB. Effects of silicon on growth and mineral composition of sorghum (*Sorghum bicolor*) grown with toxic levels of aluminum. Dev Plant Soil Sci 1991; 45:815–823.

Gardner JC, Maranville JW, Paparozzi ET. Nitrogen use efficiency among diverse sorghum cultivars. Crop Sci 1994; 34:728–733.

Garrity DP, Sullivan CY, Watts DG. Changes in grain sorghum stomatal and photosynthetic response to moisture stress across growth stages. Crop Sci 1984; 24:441–446.

Gerik TJ, Eastin JD. Temperature effects on dark respiration among diverse sorghum genotypes. Crop Sci 1985; 25:957–961.

Gerik TJ, Neely CL. Plant density effects on main culm and tiller development of grain sorghum. Crop Sci 1987; 27:1225–1230.

Girma FS, Krieg DR. Osmotic adjustment in sorghum I. Mechanisms of diurnal osmotic potential changes. Plant Physiol 1992a; 99:577–582.

Girma FS, Krieg DR. Osmotic adjustment in sorghum II. Relationship to gas exchange rates. Plant Physiol 1992b; 99:583–588.

Gibson PT, Schertz KF. Growth analysis of sorghum hybrid and its parents. Crop Sci 1977; 17:387–391.

Glass ADM. Physiological mechanisms involved with genotypic differences in ion absorption and utilization. HortScience 1989; 24:559–564.

Glover J. Apparent behavior of maize and sorghum stomata during and after drought. J Agric Sci 1959; 53:412–416.

Goldsworthy PR. The growth and yield of tall and short sorghum in Africa. J Agric Sci 1970a; 75:109–122.

Goldsworthy PR. Canopy structure of tall and short sorghum. J Agric Sci 1970b; 75:123–131.

Gourley LM, Rogers SA, Ruizgomez C, Clark RB. Genetic aspects of aluminum tolerance in sorghum. Plant Soil 1990; 123:211–216.

Grant RH, Jenks MA, Rich PJ, Peters PJ, Ashworth EN. Scattering of ultraviolet and photosynthetically active radiation by *Sorghum bicolor*: influence of epicuticular wax. Agric For Meteorol 1995; 75:263–281.

Grantz DA. Plant response to atmospheric humidity. Plant Cell Environ 1990; 13:667–679.

Grieve CM, Maas EV. Differential effects of sodium/calcium ratio on sorghum genotypes. Crop Sci 1988; 28:659–665.

Grundon NJ, Edwards DG, Takkar PN, Asher CJ, Clark RB. Nutritional disorder of grain sorghum. Cranberra: Australian Disorder Center for International Agricultural Research, 1987:99.

Halkier BA, Lindberg MB. Biosynthesis of the cyanogenic glucoside dhurrin in seedlings of *Sorghum bicolor* (L) Moench and partial purification of the enzyme system involved. Plant Physiol 1989; 90:1552–1559.

Hall AE, Richards RA, Condon AG, Wright GC, Farquhar GD. Carbon isotope discrimination and plant breeding. Plant Breed Rev 1994; 12:81–113.

Hammer GL, Carberry PS, Muchow RC. Modelling genotypic and environmental control of leaf area dynamics in grain sorghum I. Whole plant level. Field Crops Res 1993; 33:293–310.

Hammer GL, Vanderlip RL, Gibson G, Wade LJ, Henzell RG, Younger DR, Warren J, Dale AB. Genotype-by-environment interaction in grain sorghum II. Effects of temperature and photoperiod on ontogeny. Crop Sci 1989; 29:376–384.

Harris D, Hamdi QA, Terry Oda AC. Germination and emergence of *Sorghum bicolor*: genotypic and environmentally induced variation in the response to temperature and depth of sowing. Plant Cell Environ 1987; 10:501–508.

Haussmann BIG, Obilana AB, Ayiecho PO, Blum A, Schipprack W, Geiger HH. Quantitative-genetic parameters of sorghum [*Sorghum bicolor* (L) Moench] grown in semi-arid areas of Kenya. Euphytica 1999; 105:109–118.

Haussmann BIG, Obilana AB, Ayiecho PO, Blum A, Schipprack W, Geiger HH. Yield and yield stability of four population types of grain sorghum in a semi-arid area of Kenya. Crop Sci 2000; 40:319–329.

Havaux M. Fluorimetric determination of the genetic variability existing for chilling tolerance in sweet sorghum and sudan grass. Plant Breed 1989; 102:327–332.

Heiniger RW, Vanderlip RL, Kofoid KD. Caryopsis weight patterns within the sorghum panicle. Crop Sci 1993; 33:543–549.

Heiniger RW, Vanderlip RL, Welch SM. Developing guidelines for replanting grain sorghum. I. Validation and sensitivity analysis of the SORKAM sorghum growth model. Agron J 1997a; 89:75–83.

Heiniger RW, Vanderlip RL, Williams JR, Welch SM. Developing guidelines for replanting grain sorghum III. Using a plant growth model to determine replanting options. Agron J 1997b; 89:93–100.

Hess DE, Ejeta G, Butler LG. Selecting sorghum genotypes expressing a quantitative biosynthetic trait that confers resistance to striga. Phytochemistry 1992; 31:493–497.

Hocking PJ. Distribution and redistribution of mineral nutrients and dry matter in grain sorghum as affected by soil salinity. J Plant Nutr 1993; 16:1753–1774.

Hodson MJ, Sangster AG. The interaction between silicon and aluminum in *Sorghum bicolor* (L) Moench: growth analysis and X-ray microanalysis. Ann Bot 1993; 72: 389–400.

Hoffmannthoma G, Hinkel K, Nicolay P, Willenbrink J. Sucrose accumulation in sweet sorghum stem internodes in relation to growth. Physiol Plant 1996; 97:277–284.

Hofmann WC, O'Neill MK, Dobrenz AK. Physiological responses of sorghum hybrids and parental lines to soil moisture stress. Agron J 1984; 76:223–228.

Hole CC, Thomas TH, Barnes A, Scott PA, Rankin WEF. Dry matter distribution between shoot and storage root of carrot, parsnip, radish and red beet. Ann Bot 1984; 53:625–631.

Holford ICR, Holland JF, Good AJ, Leckie C. Yield and protein responses to nitrogen, and nitrogen fertiliser requirements of grain sorghum, in relation to soil nitrate levels. Aust J Agric Res 1997; 48:1187–1197.

Hong-Qi L, Croes AF. Protection of pollen germination from adverse temperatures: a possible role for proline. Plant Cell Environ 1983; 6:471–474.

Hoshikawa K. Underground organs of the seedlings and the systematics of *Gramineae*. Bot Gaz 1969; 130:192–203.

House LR. A Guide to Sorghum Breeding. Patancheru: International Crops Research Institute for the Semi-Arid Tropics (ICRISAT), 1983:238.

Howarth C. Heat shock proteins in *Sorghum bicolor* and *Pennisetum americanum* II. Stored RNA in sorghum seed and its relationship to heat shock protein synthesis during germination. Plant Cell Environ 1990a; 13:57–64.

Howarth CJ. Heat shock proteins in sorghum and pearl millet—ethanol, sodium arsenite, sodium malonate and the development of thermotolerance. J Exp Bot 1990b; 41:877–883.

Howarth CJ, Skot KP. Detailed characterization of heat shock protein synthesis and induced thermotolerance in seedlings of *Sorghum bicolor* L. J Exp Bot 1994; 45: 1353–1363.

Howell TA. Grain, dry matter yield relationships for winter wheat and grain sorghum Southern High Plains. Agron J 1990; 82:914–918.

Hubick KT, Hammer GL, Farquhar GD, Wade LG, Voncaemmerer S, Henderson

SA. Carbon isotope discrimination varies genetically in C-4 species. Plant Physiol 1990; 90:534–537.

Huda AKS, Sivakumar MVK, Virmani SM, Seetharama N, Singh S, Sekaran JG. Modeling the effect of environmental factors on sorghum growth and development. In: Virmani SM, Sivakumar MVK, eds. Agrometeorology of Sorghum and Millet in the Semi-Arid Tropics. Patancheru: ICRISAT, 1984:277–287.

Hundal SS, De Datta SK. Water table and tillage effects on root distribution, soil water extraction and yield of sorghum grown after wetland rice in a tropical soil. Field Crops 1984; 9:291–303.

Igartua E, Gracia MP, Lasa JM. Field responses of grain sorghum to a salinity gradient. Field Crops Res 1995; 42:15–25.

Isbell VR, Morgan PW. Manipulation of apical dominance in sorghum with growth regulators. Crop Sci 1982; 22:30–35.

Ishikawa S, Wagatsuma T, Takano T, Tawaraya K, Oomata K. The plasma membrane intactness of root-tip cells is a primary factor for Al-tolerance in cultivars of five species. Soil Sci Plant Nutr 2001; 47:489–501.

Ishitani M, Arakawa K, Mizuno K, Kishitani S, Takabe T. Betaine aldehyde dehydrogenase in the *Gramineae*: levels in leaves of both betaine-accumulating and nonaccumulating cereal plants. Plant Cell Physiol 1993; 34:493–495.

Jackson MB. Are plant hormones involved in root to shoot communication? Adv Bot Res 1993; 19:103–187.

Jambunathan R, Butler LG, Bandyopadhyay R, Mughogho LK. Sorghum endosperm that may be involved in resistance to grain moulds of mold-susceptible and mold-resistant sorghum cultivars. J Sci Food Agric 1986; 60:275–282.

Jenks MA, Joly RJ, Peters PJ, Rich PJ, Axtell JD, Ashworth EN. Chemically induced cuticle mutation affecting epidermal conductance to water vapor and disease susceptibility in *Sorghum bicolor* (L) Moench. Plant Physiol 1994; 105:1239–1245.

Jenks MA, Rich PJ, Rhodes D, Ashworth EN, Axtell JD, Ding CK. Leaf sheath cuticular waxes on bloomless and sparse-bloom mutants of *Sorghum bicolor*. Phytochemistry 2000; 54:577–584.

Jenner CF, Ugalde TD, Aspinall D. The physiology of starch and protein deposition in the endosperm of wheat. Aust J Plant Physiol 1991; 18:211–226.

Jeschke WD, Hartung W. Root–shoot interactions in mineral nutrition. Plant Soil 2000; 226:57–69.

Jesko T. Removal of all nodal roots initiating the extension growth in *Sorghum saccharatum* (L) Moench I. Effect on photosynthetic rate and dark respiration. Photosynthetica 1972; 6:51–56.

Jones MM, Osmond CB, Turner NC. Accumulation of solutes in leaves of sorghum and sunflower in response to water deficits. Aust J Plant Physiol 1980; 7:193–205.

Jones MM, Rawson HM. Influence of rate of development of leaf water deficits upon photosynthesis, leaf conductance, water use efficiency and osmotic potential in sorghum. Physiol Plant 1979; 45:103–111.

Jordan WR, Miller FR. Genetic variability in sorghum root systems: implication for drought tolerance. In: Turner NC, Kramer PJ, eds. Adaptation of Plants to Water and High Temperature Stress. New York: John Wiley & Sons, 1980:383–399.

Jordan WR, Miller FR, Morris DE. Genetic variation in root and shoot growth of sorghum in hydroponics. Crop Sci 1979; 19:465–472.

Jordan WR, Monk RL, Miller FR, Rosenow DT, Clark RE, Shouse PJ. Environmental physiology of sorghum I. Environmental and genetic control of epicuticular wax load. Crop Sci 1983; 23:552–555.

Jordan WR, Shouse PJ, Blum A, Miller FR, Monk RL. Environmental physiology of sorghum II. Epicuticular wax load and cuticular transpiration. Crop Sci 1984; 24: 1168–1173.

Jorgensen JA, Rosenow DT, Nguyen HT. Genotypic comparison of heat shock protein synthesis in sorghum. Crop Sci 1993; 33:638–641.

Kaigama BK, Teare ID, Stone RL, Powers WL. Root and top growth of irrigated and nonirrigated grain sorghum. Crop Sci 1977; 17:555–559.

Kamoshita A, Fukai S, Muchow RC, Cooper M. Genotypic variation for grain yield and grain nitrogen concentration among sorghum hybrids under different levels of nitrogen fertiliser and water supply. Aust J Agric Res 1998; 49:737–747.

Keeling PL, Banisadr R, Barone L, Wasserman BP, Singletary GW. Effect of temperature on enzymes in the pathway of starch biosynthesis in developing wheat and maize grain. Aust J Plant Physiol 1994; 21:807–827.

Khanna-Chopra R, Sinha SK. Enhancement of drought induced senescence by the reproductive sink in fertile lines of wheat and sorghum. Ann Bot 1988; 61:649–653.

Kidambi SP, Krieg DR, Nguyen HT. Parental Influences on gas exchange rates in grain sorghum. Euphytica 1990a; 50:139–146.

Kidambi SP, Krieg DR, Rosenow DT. Genetic variation for gas exchange rates in grain sorghum. Plant Physiol 1990b; 92:1211–1214.

Kiniry JR, Mausser RL. Response of kernel weight of sorghum to environment early and late in grain filling. Agron J 1988; 80:606–610.

Kiniry JR, Tischler CR, Rosenthal WD, Gerik TJ. Nonstructural carbohydrate utilization by sorghum and maize shaded during grain growth. Crop Sci 1992; 32:131–137.

Kluitenberg GJ, Biggar JW. Canopy temperature as a measure of salinity stress on sorghum. Irrig Sci 1992; 13:115–121.

Kojima M, Poulton JE, Thayer SS, Conn EE. Tissue distributions of dhurrin and of enzymes involved in its metabolism in leaves of *Sorghum bicolor*. Plant Physiol 1979; 63:1022–1028.

Krieg DR. Whole-plant responses to water deficits: carbon assimilation and utilization. In: Taylor HM, Jordan WR, Sinclair TR, eds. Limitations to Efficient Water Use in Crop Production. Madison: American Society of Agronomy, 1983a: 319–330.

Krieg DR. Sorghum. In: Teare ID, Peet MM, eds. Crop–Water Relations. New York: John Wiley & Sons, 1983b:351–388.

Krieg DR, Hutmacher RB. Photosynthetic rate control in sorghum: stomatal and nonstomatal factors. Crop Sci 1986; 26:112–117.

Kumari SR, Chandrashekar A, Shetty HS. Proteins in developing sorghum endosperm that may be involved in resistance to grain moulds. J Sci Food Agric 1992; 60:275–279.

Lafitte HR, Loomis RS. Growth and composition of grain sorghum with limited nitrogen. Agron J 1988; 80:492–498.

Lam TBT, Iiyama K, Stone BA. Lignin and hydroxycinnamic acids in walls of brown midrib mutants of sorghum, pearl millet and maize stems. J Sci Food Agric 1996; 71:174–178.

Lansac AR, Sullivan CY, Johnson BE. Accumulation of free proline in sorghum (*Sorghum bicolor*) pollen. Can J Bot 1996; 74:40–45.

Larque-Saavedra A, Wain RL. Studies on plant growth regulating substances XLII. Abscisic acid as a genetic character related to drought tolerance. Ann Appl Biol 1976; 83:291–297.

Lawlor DJ, Kanemasu ET, Albrecht WC III, Johnson DE. Seed production environment influence on the base temperature for growth of sorghum genotypes. Agron J 1990; 82:643–647.

Lee IJ, Foster KR, Morgan PW. Photoperiod control of gibberellin levels and flowering in sorghum. Plant Physiol 1998; 116:1003–1011.

Lee K, Lommasson RC, Eastin JD. Developmental studies on the panicle initiation in sorghum. Crop Sci 1974; 14:80–84.

Liptay A, Arevalo AE. Plant mineral accumulation, use and transport during the life cycle of plants: a review. Can J Plant Sci 2000; 80:29–38.

Jambunathan R. Polyphenol concentrations in grain, leaf, and callus tissues of mold-susceptible and mold-resistant sorghum cultivars. J Agric Food Chem 1986; 34:425–429.

Lopes MA, Larkins BA. Endosperm origin, development, and function. Plant Cell 1993; 5:1383–1399.

Loreto F, Tricoli D, Dimarco G. On the relationship between electron transport rate and photosynthesis in leaves of the C4 plant *Sorghum bicolor* exposed to water stress, temperature changes and carbon metabolism inhibition. Aust J Plant Physiol 1995; 22:885–892.

Lu CM, Zhang JH. Thermostability of photosystem II is increased in salt-stressed sorghum. Aust J Plant Physiol 1998; 25:317–324.

Ludlow MM, Powels SB. Effects of photoinhibition induced by water stress on growth and yield of grain sorghum. Aust J Plant Physiol 1988; 15:179–194.

Ludlow MM, Santamaria JM, Fukai S. contribution of osmotic adjustment to grain yield in *Sorghum bicolor* (L) Moench under water-limited conditions 2. Water stress after anthesis. Aust J Agric Res 1990; 41:67–78.

Ma BL, Dwyer LM. Nitrogen uptake and use of two contrasting maize hybrids differing in leaf senescence. Plant Soil 1998; 199:283–291.

Maas SJ, Arkin GF, Rosenthal WD. Relationship between the area of successive leaves on grain sorghum. Agron J 1987; 79:739–745.

Maiti RK, Delarosaibarra M, Sandoval ND. Genotypic variability in glossy sorghum lines for resistance to drought, salinity and temperature stress at the seedling stage. J Plant Physiol 1994; 143:241–244.

Maiti RK, Prasada Rao KE, Raju PS, House LR. The glossy trait in sorghum: its characteristics and significance in crop improvement. Field Crops Res 1984; 9:279–290.

Major DJ, Rood SB, Miller FR. Temperature and photoperiod effects mediated by the sorghum maturity genes. Crop Sci 1990; 30:305–310.

Maness NO, McBee GG. Role of placental sac in endosperm carbohydrate import in sorghum caryopses. Crop Sci 1986; 26:1201–1207.

Manjjarez-Sandoval P, Gonzalez-Hernandez VA, Mendoza-Onofre LE, Engelman EM. Drought stress effects on the grain yield and panicle development of sorghum. Can J Plant Sci 1989; 69:631–641.

Mann JA, Gbur EE, Miller FR. A screening index for adaptation in sorghum cultivars. Crop Sci 1985; 25:593–598.

Marschner H, Dell B. Nutrient uptake in mycorrhizal symbiosis. Plant Soil 1994; 159:89–102.

Marschner H, Römheld V, Kissel M. Different strategies in higher plants in mobilization and uptake of iron. J Plant Nutr 1986; 9:695–713.

Martin JH. The comparative drought resistance of sorghum and corn. Agron J 1930; 22:993–1003.

Martin JH, Taylor JW, Leukel RW. Effect of soil temperature and depth of planting on the emergence and development of sorghum seedlings in the greenhouse. J Am Soc Agron 1935; 27:660–665.

Massacci A, Battistelli A, Loreto F. Effect of drought stress on photosynthetic characteristics, growth and sugar accumulation of field-grown sweet sorghum. Aust J Plant Physiol 1996; 23:331–340.

Matthews RB, Azam-Ali SN, Peacock JM. Response of four sorghum lines to mid-season drought: II. Leaf characteristics. Field Crops Res 1990a; 25:297–308.

Matthews RB, Reddy DM, Rani AU, Azam-Ali SN, Peacock JM. Response of four sorghum lines to mid-season drought I. Growth, water-use and yield. Field Crops Res 1990b; 25:279–296.

Mcbee GG, Miller FR. Stem carbohydrate and lignin concentrations in sorghum hybrids at 7 growth stages. Crop Sci 1993; 33:530–534.

McClure JW, Harvey C. Use of radiophosphorus in measuring root growth in sorghum. Agron J 1962; 54:457–459.

McCree KJ. Carbon balance as a function of plant size in sorghum plants. Crop Sci 1983; 23:1173–1177.

McCree KJ. Whole-plant carbon balance during osmotic adjustment to drought and salinity stress. Aust J Plant Physiol 1986; 13:33–43.

McCree KJ, Kallsen CE, Richardson SG. Carbon balance of sorghum plants during osmotic adjustment to water stress. Plant Physiol 1984; 76:898–902.

McDaniel RG. Biochemical and physiological basis of heterosis. CRC Rev Plant Sci 1986; 4:227–246.

McNeilly T. Selection and breeding for salinity tolerance in crop species—a case for optimism. Acta Oecol 1990; 11:595–610.

Merrill SD, Rawlings SL. Distribution and growth of sorghum roots in response to irrigation frequency. Agron J 1979; 71:738–745.

Meyers RJK, Foale MA, Done AA. Response of grain sorghum to varying irrigation frequency in the Ord irrigation area: II. Evapotranspiration water-use efficiency. Aust J Agric Res 1984a; 35:31–42.

Meyers SP, Nelson CJ, Horrocks RD. Temperature effects on imbibition, germination and respiration of grain sorghum seeds. Field Crops Res 1984b; 8:135–142.

Miller FR, Barnes DK, Cruzado HJ. Effect of tropical photoperiods on the growth of sorghum when grown in 12 monthly plantings. Crop Sci 1968; 8:499–502.

Miranda de JCC, Harris PJ, Wild A. Effects of soil and plant phosphorus concentrations on vesicular–arbuscular mycorrhiza in sorghum plants. New Phytol 1989; 112: 405–410.

Moll RH, Kamprath EJ, Jackson WA. Analysis and interpretation of factors which contribute to efficiency of nitrogen utilization. Agron J 1982a; 74:562–564.

Moll RH, Kamprath EJ, Jackson WA. Analysis and interpretation of factors which contribute to efficiency of nitrogen utilization. Agron J 1982b; 74:562–564.

Morgan PW, Guy LW, Pao CI. Genetic regulation of development in *Sorghum bicolor*: III. Asynchrony of thermoperiods with photoperiods promotes floral initiation. Plant Physiol 1987; 83:448–450.

Morgan PW, Quinby JR. Genetic regulation of development in *Sorghum bicolor*: IV GA3 hastens floral differentiation but not floral development under nonfavorable photoperiods. Plant Physiol 1987; 85:615–620.

Morgan PW, Miller FR, Quinby JR. Manipulation of sorghum growth and development with gibberellic acid. Agron J 1977; 69:789–793.

Muchow RC. Effect of high temperature on the rate and duration of grain growth in field-grown *Sorghum bicolor* (L) Moench. Aust J Agric Res 1990a; 41:329–337.

Muchow RC. Effect of nitrogen on partitioning and yield in grain sorghum under differing environmental conditions in the semi-arid tropics. Field Crops Res 1990b; 25:265–278.

Muchow RC, Craberry PS. Phenology and leaf area development in a tropical grain sorghum. Field Crops Res 1990; 23:221–237.

Muchow RC, Sinclair TR. Nitrogen response of leaf photosynthesis and canopy radiation use efficiency in field-grown maize and sorghum. Crop Sci 1994; 34:721–727.

Munns R, Brady CJ, Barlow WR. Solute accumulation in the apex and leaves of wheat during water stress. Aust J Plant Physiol 1979; 6:379–389.

Munns R, Schachtman DP, Condon AG. The significance of a two-phase growth response to salinity in wheat and barley. Aust J Plant Physiol 1995; 22:561–569.

Myers RJK, Asher CJ. Mineral nutrition of grain sorghum: macronutrients. In: House LR, Mughogho LK, Peacock JM, eds. Sorghum in the Eighties. Pantacheru: ICRISAT, 1982:161–177.

Nimbal CI. Herbicidal activity and site of action of the natural product sorgoleone. Pest Biochem Physiol 1996; 54:73–83.

Oliver A, Leroux GD. Root development and production of a witchweed (*Striga* spp) germination stimulant in sorghum (*Sorghum bicolor*) cultivars. Weed Sci 1992; 40: 542–545.

O'Neill MK, Hofmann W, Dobrenz AK, Marcarian V. Drought response of sorghum hybrids under a sprinkler irrigation gradient system. Agron J 1983; 75:102–107.

Ougham HJ, Stoddart JL. Synthesis of heat shock proteins and acquisition of thermotolerance in high-temperature tolerant and high-temperature susceptible lines of sorghum. Plant Sci 1986; 44:163–167.

Pacovsky RS. Influence of inoculation with *Azospirillum brasilense* and *Glomus fasciculatum* on sorghum nutrition. Plant Soil 1988; 110:283–287.

Pao CI, Morgan PW. Genetic regulation of development in *Sorghum bicolor*: I. Role of the maturity genes. Plant Physiol 1986a; 82:575–580.

Pao CI, Morgan PW. Genetic regulation of development in *Sorghum bicolor*: II. Effect of the ma3R allele mimicked by GA3. Plant Physiol 1986b; 82:584–584.

Pardales JR, Kono Y, Yamauchi A. Response of the different root system components of sorghum to incidence of waterlogging. Environ Exp Bot 1991; 31:107–115.

Parry DW, Kelso M. the distribution of silicon deposits in the root of *Molinia caerulea* (L) Moench and *Sorghum bicolor* (L) Moench. Ann Bot 1975; 39:995–1001.

Paterson AH, Lin YR, Li Z, Schertz KF, Doebley JF, Pinson SRM, Liu SC, Stansel JW, Irvine JE. Convergent domestication of cereal crops by independent mutations at corresponding genetic loci. Science 1995; 269:1714–1718.

Peacock JM, Miller WB, Matsuda K, Robinson DL. Role of heat girdling in early seedling death of sorghum. Crop Sci 1990; 30:138–143.

Pearcy RW, Ehleringer J. Comparative ecophysiology of C3 and C4 plants. Plant Cell Environ 1984; 7:1–13.

Pendleton BB, Teetes GL, Peterson GC. Phenology of sorghum flowering. Crop Sci 1994; 34:1263–1266.

Peng SB, Krieg DR. Gas exchange traits and their relationship to water use efficiency of grain sorghum. Crop Sci 1992; 32:386–391.

Peng SB, Krieg DR, Girma FS. Leaf photosynthetic rate is correlated with biomass and grain production in grain sorghum lines. Photosynth Res 1990; 28:1–7.

Peterson GC, Onken AB. Relationship between chlorophyll concentration and iron chlorosis in grain sorghum. Crop Sci 1992; 32:964–967.

Pinthus MJ, Rosenblum J. Germination and seedling emergence of sorghum at low temperature. Crop Sci 1961; 1:293–295.

Porter KS, Axtell JD, Lechtenberg VL, Colenbrander VF. Phenotype, fiber composition and in vitro dry matter disappearance of chemically induced brown midrib (*bmr*) mutants of sorghum. Crop Sci 1978; 18:205–209.

Premachandra GS, Hahn DT, Axtell JD, Joly RJ. Epicuticular wax load and water use efficiency in bloomless and sparse bloom mutants of *Sorghum bicolor* L. Environ Exp Bot 1995a; 34:293–301.

Premachandra GS, Hahn DT, Rhodes D, Joly RJ. Leaf water relations and solute accumulation in two grain sorghum lines exhibiting contrasting drought tolerance. J Exp Bot 1995b; 46:1833–1841.

Premachandra GS, Saneoka H, Fujita K, Ogata S. Leaf water relations, osmotic adjustment, cell membrane stability, epicuticular wax load and growth as affected by increasing water deficits in sorghum. J Exp Bot 1992; 43:1569–1576.

Prihar SS, Stewart BA. Sorghum harvest index in relation to plant size, environment, and cultivar. Agron J 1991; 83:603–608.

Quinby JR. Sorghum Improvement and the Genetic of Growth. College Station: Texas A&M University Press, 1974:108.

Quinby JR, Hesketh JD, Voigt RL. Influence of temperature and photoperiod on floral initiation and leaf number in sorghum. Crop Sci 1973; 13:243–246.

Radford BJ, Henzell RG. Temperature affects the mesocotyl and coleoptile length of grain sorghum genotypes. Aust J Agric Res 1990; 41:79–87.

Rajewski JF, Francis CA. Defoliation effects on grain fill, stalk rot, and lodging of grain sorghum. Crop Sci 1991; 31:353–359.

Raju PS, Clark RB, Ellis JR, Maranville JW. Mineral uptake and growth of sorghum colonized with VA mycorrhiza at varied soil phosphorus levels. J Plant Nutr 1990; 13:843–859.

Rasmussen JA, Hejl AM, Einhellig FA, Thomas JA. Sorgoleone from root exudate inhibits mitochondrial functions. J Chem Ecol 1992; 18:197–207.

Rego TJ, Grundon NJ, Asher CJ, Edwards DG. Effects of water stress on nitrogen nutrition of grain sorghum. Aust J Plant Physiol 1986; 13:499–508.

Rhodes D, Ju GC, Yang W, Samaras Y. Plant metabolism and heterosis. Plant Breed Rev 1992; 10:53–91.

Richardson SG, McCree KJ. Carbon balance and water relations of sorghum exposed to salt and water stress. Plant Physiol 1985; 79:1015–1020.

Ritchie JT, Burnett E, Henderson RC. Dryland evaporative flux in a subhumid climate: III. Soil water influence. Agron J 1972; 64:168–173.

Ritchie JT, Jordan WR. Dryland evaporative flux in a subhumid climate: IV. Relation to plant water status. Agron J 1972; 64:173–176.

Robertson MJ, Fukai S, Hammer GL, Ludlow MM. Modelling root growth of grain sorghum using the CERES approach. Field Crops Res 1993a; 33:113–130.

Robertson MJ, Fukai S, Ludlow MM, Hammer GL. Water extraction by grain sorghum in a sub-humid environment 2. Extraction in relation to root growth. Field Crops Res 1993b; 33:99 112.

Romheld V, Marschner H. Genotypical differences among graminaceous species in the release of phytosiderophores and uptake of iron phytosiderophores. Plant Soil 1990; 123:147–153.

Rood SB. Heterosis and the metabolism of gibberellin A20 in sorghum. Plant Growth Regul 1995; 16:271–278.

Rood SB, Witbeck JET, Major DJ, Miller FR. Gibberellins and heterosis in sorghum. Crop Sci 1992; 32:713–718.

Rooney WL, Aydin S. Genetic control of a photoperiod-sensitive response in *Sorghum bicolor* (L) Moench. Crop Sci 1999; 39:397–400.

Rosenthal WD, Gerik TJ, Wade LJ. Radiation-use efficiency among grain sorghum cultivars and plant densities. Agron J 1993; 85:703–705.

Rosenthal WD, Vanderlip RL, Jackson BS. SORKAM: a grain sorghum crop growth model. Texas Agricultural Experiment Station College Station, TX, Misc Publ MP-1669, 1989:21.

Roy RN, Wright BC. Sorghum growth and nutrient uptake in relation to soil fertility II. N, P, and K uptake patterns by various plant parts. Agron J 1974; 66:5–10.

Sahrawat KL, Pardhasaradhi G, Rego TJ, Rahman MH. Relationship between extracted phosphorus and sorghum yield in a Vertisol and an Alfisol under rainfed cropping. Fertil Res 1995; 44:23–26.

Salih AA, Ali IA, Lux A, Luxova M, Cohen Y, Sugimoto Y, Inanaga S. Rooting, water uptake, and xylem structure adaptation to drought of two sorghum cultivars. Crop Sci 1999; 39:168–173.

Sanchez-Diaz MF, Kramer PJ. Turgor differences and water stress in maize and sorghum leaves during drought and recovery. J Exp Bot 1973; 24:511 515.

Santamaria JM, Ludlow MM, Fukai S. contribution of osmotic adjustment to grain yield in *Sorghum bicolor* (L) Moench under water-limited conditions 1. Water stress before anthesis. Aust J Agric Res 1990; 41:51–65.

Sarig S, Blum A, Okon Y. The improvement of water status and yield of field-grown grain sorghum (*Sorghum bicolor*) by inoculation with *Azospirillum brasilense*. J Agric Sci Camb 1988; 110:271–277.

Sarig S, Kapulnik Y, Nur I, Okon Y. Response of non-irrigated *Sorghum bicolor* to *Azospirillum inoculation*. Exp Agric 1984; 20:59 66.

Sarig S, Okon Y, Blum A. Promotion of leaf area development and yield in *Sorghum bicolor* inoculated with *Azospirillum brasilense*. Symbiosis 1990; 9:235–245.

Schertz KF, Sumpter NA, Sarkissian IV, Hart GE. Peroxidase regulation by the 3-dwarf height locus in sorghum. J Heredity 1971; 62:235–238.

Schnyder H. The role of carbohydrate storage and redistribution in the source–sink relations of wheat and barley during grain filling—a review. New Phytol 1993; 123: 233–245.

Seetharama N, Subba Reddy BV, Peacock JM, Bidinger FR. Sorghum improvement for drought resistance. Drought Resistance in Crops with Emphasis on Rice. Los Banos: IRRI, 1982:317–338.

Seetharama N, Sachan RC, Huda AKS, Gill KS, Rao KN, Bidinger FR, Reddy DM. Effect of pattern and severity of moisture-deficit stress on stalk-rot incidence in sorghum: 2. Effect of source sink relationships. Field Crops Res 1991; 26:355 374.

Seetharaman K, Whitehead E, Keller NP, Waniska RD, Rooney LW. In vitro activity of sorghum seed antifungal proteins against grain mold pathogens. J Agric Food Chem 1997; 45:3666–3671.

Seligmann H, Amzallag GN, Lerner HR. Perturbed leaf development in *Sorghum bicolor* exposed to salinity — a marker of transition towards adaptation. Aust J Plant Physiol 1993; 20:243–249.

Selmar D, Irandoost Z, Wray V. Dhurrin-6′-glucoside, a cyanogenic diglucoside from *Sorghum bicolor*. Phytochemistry 1996; 43:569 572.

Shackel KA, Foster KW, Hall AE. Genotypic differences in leaf osmotic potential among grain sorghum cultivars grown under irrigation and drought. Crop Sci 1982; 22:1121–1125.

Sharma PK, Hall DO. Interaction of salt stress and photoinhibition on photosynthesis in barley and sorghum. J Plant Physiol 1991; 138:614–619.

Shuman LM, Wilson DO, Duncan RR. Screening wheat and sorghum cultivars for aluminum sensitivity at low aluminum levels. J Plant Nutr 1993; 16:2383–2395.

Siame BA, Weerasuriya Y, Wood K, Ejeta G, Butler LG. Isolation of strigol, a germination stimulant for *Striga asiatica*, from host plants. J Agric Food Chem 1993; 41:1486–1491.

Sinclair TR, Muchow RC, Monteith JL. Model analysis of sorghum response to nitrogen in subtropical and tropical environments. Agron J 1997; 89:201 207.

Singh R, Goyal RK, Bhullar SS, Goyal R. Factors, including enzymes, controlling the import and transformation of sucrose to starch in the developing sorghum caryopsis. Plant Physiol Biochem 1991; 29:177 183.

Singletary GW, Banisadr R, Keeling Pl. Heat stress during grain filling in maize: effects on carbohydrate storage and metabolism. Aust J Plant Physiol 1994; 21:829–841.

Sinha SK, Khanna R. Physiological biochemical and genetic basis of heterosis. Adv Agron 1975; 27:123–175.

Sivaramakrishnan S, Patell VZ, Soman P. Heat shock proteins of sorghum [*Sorghum bicolor* (L) Moench] and pearl millet [*Peninisetum glaucum* (L)] cultivars with differing heat tolerance at seedling establishment stage. J Exp Bot 1990; 41:249–254.

Soman P. Development of a technique to study seedling emergence in response to moisture deficit in the field—the seed bed environment. Ann Appl Biol 1990; 116: 357–364.

Soman P, Seetharama N. Genotypic and environmental variation in nodal root growth of post-rainy season (rabi) sorghum. Exp Agric 1992; 28:331–341.

Sorrells ME, Meyers O. Duration of developmental stages of 10 milo maturity genotypes. Crop Sci 1982; 22:310–314.

Sowder CM, Tarpley L, Vietor DM, Miller FR. Leaf photoassimilation and partitioning in stress-tolerant sorghum. Crop Sci 1997; 37:833–838.

Sree PS, Nwanze KF, Butler DR, Reddy DDR, Reddy YVR. Morphological factors of the central whorl leaf associated with leaf surface wetness and resistance in sorghum to shoot fly, *Atherigona soccata*. Ann Appl Biol 1994; 125:467–476.

Srivastava HK. Inergenomic interaction heterosis and improvement of crop yield. Adv Agron 1981; 34:117–190.

Stahl RS, McCree KJ. Ontogenetic changes in respiration coefficients of grain sorghum. Crop Sci 1988; 28:111–114.

Steinbach HS, Arnold BRL, Kristof G, Sanchez RA, Marcucci-Poltri S. Physiological basis of pre-harvest sprouting resistance in *Sorghum bicolor* (L) Moench. ABA levels and sensitivity in developing embryos of sprouting-resistant and -susceptible varieties. J Exp Bot 1995; 46:701–709.

Steinbach HS, Benech-Arnold RL, Sanchez RA. Hormonal regulation of dormancy in developing sorghum seeds. Plant Physiol 1997; 113:149–154.

Steiner JL. Dryland grain sorghum water use light interception and growth responses to planting geometry. Agron J 1986; 78:720–726.

Stephens JC, Quinby JR. Bulk emasculation of sorghum flowers. J Am Soc Agron 1933; 25:233–234.

Stockle CO, Kiniry JR. Variability in crop radiation-use efficiency associated with vapor-pressure deficit. Field Crops Res 1990; 25:171–181.

Stone LR, Goodrum DE, Jaafar MN, Khan AH. Rooting front and water depletion depths in grain sorghum and sunflower. Agron J 2001; 93:1105–1110.

Sullivan CY, Eastin JD. Plant physiological responses to water stress. Agric Meteorol 1974; 14:113–127.

Sullivan CY, Ross WM. Selecting for drought and heat resistance in grain sorghum. In: Mussell H, Staples RC, eds. Stress Physiology in Crop Plants. New York: Wiley Interscience, 1979:263–282.

Tan KZ, Keltjens WG, Findenegg GR. Acid soil damage in sorghum genotypes—role of magnesium deficiency and root impairment. Plant Soil 1992; 139:149–155.

Tangpremsri T, Fukai S, Fischer KS, Henzell RG. Genotypic variation in osmotic adjustment in grain sorghum 1. Development of variation in osmotic adjustment under water-limited conditions. Aust J Agric Res 1991; 42:747–757.

Tangpremsri T, Fukai S, Fischer KS. Growth and yield of sorghum lines extracted from a population for differences in osmotic adjustment. Aust J Agric Res 1995; 46: 61–74.

Tarpley L, Lingle SE, Vietor DM, Andrews DL, Miller FR. Enzymatic control of nonstructural carbohydrate concentration in stems and panicles of sorghum. Crop Sci 1994; 34:446–452.

Tarumoto I. Inheritance of glossiness of leaf blades in sorghum *Sorghum bicolor* (L) Moench. Jpn J Breed 1980; 30:237–239.

Tarumoto I, Miyazaki M, Matsumura T. Scanning electron microscopic study of the surfaces of glossy and non-glossy leaves in sorghum (*Sorghum bicolor* L Moench). Bull Natl Grassl Res Inst (Japan) 1981; 18:38–44.

Tenkouano A, Miller FR, Frederiksen RA, Rosenow DT. Genetics of nonsenescence and charcoal rot resistance in sorghum. TAG 1993; 85:644–648.

Thomas H, Howarth CJ. Five ways to stay green. J Exp Bot 2000; 51:329–337.

Thomas GL, Miller FR. Base temperature for germination for temperate and tropically adapted sorghum. Proceedings of the 11th Biennial Grain Sorghum Research and Utilization Conference, Wichita, 1979:25–28.

Thorstensson EMG, Buxton DR, Cherney JH. Apparent inhibition to digestion by lignin in normal and brown midrib stems. J Sci Food Agric 1992; 59:183–188.

Traore A, Maranville JW. Nitrate reductase activity of diverse grain sorghum genotypes and its relationship to nitrogen use efficiency. Agron J 1999; 91:863–869.

Traore M, Sullivan CY, Rosowski JR, Lee KW. Comparative leaf surface morphology and the glossy characteristic of sorghum, maize and pearl millet. Ann Bot 1989; 64: 447–453.

Tsaftaris SA. Molecular aspects of heterosis in plants. Physiol Plant 1995; 94:362–370.

Tuinstra MR, Grote EM, Goldsbrough PB, Ejeta G. Genetic analysis of post-flowering drought tolerance and components of grain development in *Sorghum bicolor* (L) Moench. Mol Breed 1997; 3:439–448.

Turner FT, Chen CC, Bollich CN. Coleoptile and mesocotyl lengths in semidwarf rice seedlings. Crop Sci 1982; 22:43–46.

Tiryaki I, Andrews DJ. Germination and seedling cold tolerance in sorghum: II. Parental lines and hybrids. Agron J 2001; 93:1391–1397.

Utzurrum SB, Fukai S, Foale MA. Effect of late nitrogen application on growth and nitrogen balance of two cultivars of water-stressed grain sorghum. Aust J Agric Res 1998; 49:687–694.

Vanderlip RL. How a Sorghum Plant Develops. Kansa State University, Cooperative Extension Service C-447, 1972:19.

van-Oosterom EJ, Carberry PS, Muchow RC. Critical and minimum N contents for development and growth of grain sorghum. Field Crops Res 2001; 70:55–73.

Vietor DM, Miller FR. Assimilation, partitioning, and nonstructural carbohydrates in sweet compared with grain sorghum. Crop Sci 1990; 30:1109–1115.

Vietor DM, Miller FR, Cralle HT. Nonstructural carbohydrates in axillary branches and main stem of senescent and nonsenescent sorghum types. Crop Sci 1990; 30:97–100.

Virmani SM, Sivakumar MVK, eds. Agrometeorology of Sorghum and Millet in The Semi-Arid Tropics. Patancheru: ICRISAT, 1984:322.

Wall JS, Ross WM, eds. Sorghum Production and Utilization. Westport: AVI Pub. Co., 1970:702.

Weibel DE, Sotomayor-Rios A, Pava HM, McNew RW. Relationship of black layer to sorghum kernel moisture content and maximum kernel weight in the tropics. Crop Sci 1982; 22:219–223.

Weimberg R, Lerner HR, Poljakoff-Mayber A. A relationship between potassium and proline accumulation in salt-stressed *Sorghum bicolor*. Physiol Plant 1982; 55:5–10.

Wenzel CL, McCully ME, Canny MJ. Development of water conducting capacity in the root systems of young plants of corn and some other C4 grasses. Plant Physiol 1989; 89:1094–1101.

Wheeler JL, Mulcahy C, Walcott JJ, Rapp GG. Factors affecting the hydrogen cyanide potential of forage sorghum. Aust J Agric Res 1990; 41:1093–1100.

Wieneke J. Phosphorus efficiency and phosphorus remobilization in two sorghum (*Sorghum bicolor* (L) Moench) cultivars. Plant Soil 1990; 123:139–145.

Wilson JB. A review of evidence on the control of shoot:root ratio, in relation to models. Ann Bot 1988; 61:433–449.

Wilson GL, Eastin JD. The plant and its environment. In: House LR, Mughogo LK, Peacock JM, eds. Sorghum in the Eighties. Patancheru: ICRISAT, 1982:101–121.

Wood AJ, Saneoka H, Rhodes D, Joly RJ, Goldsbrough PB. Betaine aldehyde dehydrogenase in sorghum—molecular cloning and expression of two related genes. Plant Physiol 1996; 110:1301–1308.

Wright GG, Smith RCG, Morgan JM. Differences between two sorghum genotypes in adaptation to drought stress: III. Physiological response. Aust J Agric Res 1983a; 34:637–651.

Wright SA, Jordan WR, Morgan PW, Miller FR. Genetic and hormonal control of shoot and root growth in sorghum. Agron J 1983b; 75:682–686.

Xu XD, Bland WL. Resumption of water uptake by sorghum after water stress. Agron J 1993a; 85:697–702.

Xu XD, Bland WL. Reverse water flow in sorghum roots. Agron J 1993b; 85:388–395.

Yang YW, Newton RJ, Miller FR. Salinity tolerance in sorghum 1. Whole plant response to sodium chloride in *S. bicolor* and *S. halepense*. Crop Sci 1990; 30:775–781.

Youngquist JB, Bramel-Cox P, Maranville JW. Evaluation of alternative screening criteria for selecting nitrogen-use efficient genotypes in sorghum. Crop Sci 1992; 32:1310–1313.

Youngquist JB, Maranville JW. Patterns of nitrogen mobilization in grain sorghum hybrids and the relationship to grain and dry matter production. J Plant Nutr 1992; 15:445–455.

Zartman RE, Woyewodzic RT. Root distribution patterns of two hybrid grain sorghums under field conditions. Agron J 1979; 71:325–328.

Zhang J, Kirkham MB. Water relations of water-stressed, split-root C4 (*Sorghum bicolor*; *Poaceae*) and C3 (*Helianthus annuus*; *Asteraceae*) plants. Am J Bot 1995; 82:1220–1229.

Zweifel TR, Maranville JW, Ross WM, Clark RB. Nitrogen fertility and irrigation influence on grain sorghum nitrogen efficiency. Agron J 1987; 79:419–422.

5

Pearl Millet

Francis R. Bidinger and C. Thomas Hash
International Crops Research Institute for the Semi-Arid Tropics (ICRISAT)
Patancheru, Andhara Pradesh, India

1 INTRODUCTION

Pearl millet (*Pennisetum glaucum* (L.) R. Br.) is the sixth most important cereal globally, and the fourth most important of the tropical cereals, after rice, maize, and sorghum. It is grown on approximately 26 million ha annually, 11 million each in south Asia and West Africa, and 2 million each in eastern/southern Africa and in Brazil (Bonamigo, 1999; Harinarayana et al., 1999; ICRISAT and FAO, 1996). It shares with barley (among the temperate cereals) a specific adaptation to the most marginal, driest, and least-fertile end of the spectrum of cereal-growing environments. Its main areas of cultivation are in low-fertility, light-textured soils, receiving less than 500–600 mm of rainfall, where sorghum and (especially) maize are subject to frequent crop failures (Harinarayana et al., 1999). It is primarily a crop of subsistence agriculture, both because of the limited resource base, and/or significant environmental constraints, of the environments in which it is grown. As a result of which, evolution has necessarily favored adaptation and survival over a high level of grain productivity in the crop.

However, this general description masks a very versatile and very unique crop plant. On the one hand, it has been termed "a cereal of the

Sahel" (de Wet et al., 1992), because of its origin and evolution, and its suite of physiological and developmental traits that provide specific adaptation to marginal and arid environments. These include rapid germination, short duration of key developmental periods, nonsynchronous development of tillers, high temperature tolerance, and the production of large seed numbers. However, in addition to its adaptive features, its reproductive characteristics make it the most versatile and most exciting of the major cereals for genetic and plant breeding research. These include the production of 1000–3000 seeds per panicle, the ability to exploit protogyny to cross-pollinate the crop without emasculation, the ability to exploit both population and pure-line plant breeding techniques, and the existence of significant heterosis and multiple cytoplasmic-genetic male-sterility systems. Finally, its molecular genetic traits make it uniquely suitable among the major cereals for the application of molecular techniques of crop improvement. These include its diploid nature and large chromosomes, its moderate haploid DNA content and relatively low recombination rates (resulting in a short genetic map), its high degree of polymorphism at both phenotypic and molecular levels, plus the ease with which both self- and cross-pollinated progenies can be generated. This chapter will attempt to document each of these unique aspects of the crop; and to make the case that pearl millet has much to offer to the improvement of other cereals, especially if the promises of the new molecular methods for crop improvement are realized.

2 THE PEARL MILLET CROP

2.1 Origin, Evolution, and Ecology of the Crop

Pearl millet was domesticated somewhere along the southern fringes of the Sahara desert, at least 3000 years ago, when then current drying period in this area necessitated a change from Mediterranean cereals to other species better adapted to changing rainfall patterns and increasing aridity (Brunken et al., 1977). Its progenitor was almost certainly the wild *Pennisetum fallax–P. violaceum* (Stapf and Hubbard, 1934) complex that is still widely distributed in both the desert margins and the highland areas within the desert itself (de Wet et al., 1992). Brunken et al. (1977) reclassified both progenitors in the (wild) subspecies *monodii* of what is now referred to as *P. glaucum* (de Wet, 1987), based on the fact that both of these species cross readily with the cultivated subspecies. The major differences in the cultivated and wild subspecies reflect the domestication process: the loss of the abscission layer at the base of the spikelet that results in shattering in the wild subspecies, a significant increase in both inflorescence and seed sizes, supported by a greater plant size, and a concomitant reduction in productive tiller numbers (Brunken et al., 1977; Poncet et al., 2000, 2002). However, significantly, most of these differences appear to be because of changes at only a few loci (Poncet et al., 2000, 2002), and cultivated pearl millet retains many of the valuable adaptive features of its wild progenitor, which are responsible for its unique adaptation to environments characterized by variable moisture patterns, high temperatures, and short growing seasons.

The cross-fertility between the cultivated and wild subspecies produces a surviving intermediate hybrid population, which Brunken et al. (1977) classify as subspecies *stenostachyum*. Part of this population mimics the cultivated plant type but retains the shattering character of its wild progenitor and thus cross-pollinates and self-seeds in farmers' fields. These weedy intermediates or "shibras," which exist as stable populations, cannot be distinguished from the cultivated type in the vegetative stage and thus escape from being removed during the weeding of fields, unlike plants of wild relatives and the wild subspecies. However, shibras also survive, and adapt through continued cross-pollination with normal cultivated plants in the farmers' variety. Shibra × cultivated crosses appear to have lower rates of seed set than either shibra × shibra or cultivated × cultivated crosses (Amoukou, 1993), helping to preserve both as distinct subspecies, However, shibras still function as a constant bridge for the flow of genes from the wild to the cultivated subspecies because of their effective mimicry of the cultivated type. This opportunity for continuing gene flow from wild progenitors is a unique phenomenon among major cereals. Research has indicated the potential value of genes in the *monodii* (Hanna et al., 1985) and *stenostachyum* (Bramel-Cox et al., 1986) gene pools.

2.2 Characteristics of Major Growing Environments

Pearl millet is primarily grown in marginal, arid, and semiarid tropical and subtropical environments where its adaptive features, many of which trace back to its wild ancestors, make it the most reliable cereal for farmers in very unreliable environments. Its two major areas of cultivation are both arid to semiarid: the Sahelian and northern Sudanian zones south of the Sahara desert, stretching from Senegal in the west to Eritrea in the Horn of Africa, and the northwestern states of India. Both of these areas are on the edges of the major circulation systems that bring rain to much of sub-Saharan Africa (intertropical convergence zone) or to south Asia (the southwest monsoon). As a result, rainy seasons in both major pearl millet growing areas are short (70–120 days), mean total rainfall is low to moderate (250–700 mm), and both inter- and intra-annual rainfall is highly variable (see Kowal and Kassam, 1978, for a description of the agroclimatology of West African millet growing areas). Pearl millet is the sole cereal grown in the drier part of this range, and shares the wetter part with sorghum and, to an increasing degree, short duration maize in more edaphically favorable sites. In wetter areas, where farmers have a choice among cereals, they tend to use pearl millet to exploit the shallower and/or poorer soils (including those with high Al^{3+} saturation) and more drought-prone sites, such as the upper reaches of the toposequence, and sow the more productive sorghum or maize in the deeper, more fertile soil areas (van Staveren and Stoop, 1985). Water balance simulations in both India (e.g., van Oosterom et al., 1996a) and West Africa (e.g., Eldin, 1993) indicate that periods of low soil moisture availability, which limit potential growth, are the norm rather than the exception in most pearl millet growing environments.

Despite the above emphasis on water, poor soil fertility is the major limitation to productivity in much of the area in which pearl millet is grown (Charreau, 1972; Fussell et al., 1987), resulting frequently in incomplete use of soil water (Payne et al., 1990). The soils of much of the drier part of the main growing areas for this crop are eolian deposits from adjacent deserts, characterized by a high proportion of fine sand, a low inherent fertility, and a low cation exchange capacity to retain added nutrients (Kowal and Kassam, 1978). In addition, in West Africa, many of the sandy soils are highly leached (in previous geologic times), have a low pH, and, in some cases, an attendant aluminum toxicity problem, and very low available phosphorous levels that directly affect crop growth (Chase et al., 1989; Payne et al., 1991). In the wetter areas, soils sown to pearl millet are likely to be either or both shallow and with low inherent fertility, because this crop is better able to exploit such less-productive microenvironments than are maize or sorghum (van Staveren and Stoop, 1985). Farmers in Africa traditionally relied on long fallow periods to restore soil fertility, a practice that rapidly growing populations is rendering ineffective (Charreau, 1972), with a consequent large-scale decline in crop yields. Research on soil fertility improvement/maintenance in the West African region has documented consistent and sometimes striking yield responses in pearl millet to various techniques for improving soil fertility (e.g., Charreau and Nicou, 1971; Pieri, 1992; Subbarao et al., 2000). However, few if any farmers have traditionally had access to either the knowledge or the resources to use suggested means of maintaining soil fertility. In India, mixed cropping with legumes and applications of limited amounts of animal manure, add small but significant amounts of fertility, but these still meet only a part of crop needs in the average case. Limited additions of chemical fertilizer are increasing in India, particularly in more favorable environments, but nutrient deficiency remains the norm in the crop. Unfortunately, conventional genetic "solutions" to both water and nutrient limitations to productivity, while theoretically possible (at least to a limited degree), are generally considered to have too small a potential by most plant breeders, to warrant significant investment. Whether or not the availability of new molecular breeding techniques will occasion renewed interest in these problems (Hash et al., 2002) remains to be seen.

3 THE PHYSIOLOGICAL BASIS OF ADAPTATION

Because of the nature of the environments in which the crop is primarily grown, pearl millet has retained many of the adaptive mechanisms of its desert grass progenitors, rather than evolving into a high-seed-yield crop, as have species such as *Hordeum vulgare*. There is a clear trade-off between adaptation to marginal environments and yield potential, evident in various adaptive characteristics in pearl millet such as developmental asynchrony, short grain-filling period, small seed size, etc. Significant improvement in grain yield potential has been made within the limitations of pearl millet's primary adaptive characteristics, mainly by improving harvest index; modern Indian F_1 hybrids are capable of producing more than 3–4 t ha^{-1} of grain in 80 days under optimal conditions. However, only where pearl millet's short life cycle provides a clear advantage (e.g., where irrigation water is limited) is it competitive with inherently higher yielding tropical cereals in favorable agricultural environments.

Most of the known adaptive mechanisms of the crop are developmental or morphological; not as much is known about its biochemical or metabolic traits (Winkel and Do, 1992). Hopefully, the exploitation of synteny between pearl millet and the other cereals (Devos and Gale, 2000; Devos et al., 2000) will allow a broader range of adaptive mechanisms to be identified, evaluated, and exploited in the crop. This chapter will concentrate on the developmental mechanisms of adaptation, which the crop very effectively exploits.

3.1 Adaptation to Growing Season Length

Pearl millet is a primarily a quantitative or facultative short-day plant (Belliard and Pernes, 1984), but qualitative or absolute short-day types also exist, in which floral initiation occurs under natural conditions only when the plant experiences its threshold daylength. Traditional cultivars rely on sensitivity to daylength to adapt their growth cycle to the mean growing season length of their area of origin (Kouressy et al., 1998), which can vary from as little as 60 days in desert margin areas in northwestern India, to as long as 180 days in the northern Guinea zone of West Africa. Adaptation to season length is critical as potential evaporation rates in semiarid areas are high (6–9 mm day^{-1}) and soil moisture storage often limited, meaning that flowering too late involves a serious risk of severe drought stress during grain filling. However, flowering too early also has various risks, including poor seed set if flowering occurs during the heaviest part of the rainy season (as a result of a combination of failure of anther dehiscence under highly humid conditions and physical damage to receptive stigmas and/or grounding of the pollen cloud by falling rain drops), and insect, disease, and bird damage if grain filling occurs before the end of the rains (Kouressy et al., 1998). Local landrace cultivars have developed a finely tuned sensitivity to daylength, which assures that flowering occurs near the expected end of the rainy season, to utilize the entire growing period, but to avoid the problems noted above. The primary factor affecting season length is latitude. For example, the rainy season length in West and Central Africa varies from as little as 70 days at 15° north—the approximate limit of cultivation in the Sahelian zone—to more that 150 days at only 10° north—the approximate southern limit of wide-spread pearl millet cultivation (Kowal and Kassam, 1978). This difference in season length over only 5° degrees of latitude requires a very sensitive response to daylength to differentiate locally adapted cultivars over relatively small distances. A study of the flowering of four groups of farmers' landrace cultivars (total of 74 cultivars) collected in Mali between 10°33'N and 14°44'N (Kouressy et al., 1998) provides an excellent example of the sensitivity of the daylength response in local cultivars. This range in latitude represents a variation in rainy season length of approximately 75–140 days, based on long-term water balance estimations (Kouressy et al., 1998). For July and August, when plants are competent to undergo floral initiation, the difference in daylength (including civil twilight) between 10°N and 15°N ranges between 20 min (13 h 28 min vs. 13 h 48 min) on July 1, and 8 min (13 h 2 min vs. 13 h 10 min) on August 31 (Smithsonian Institution, 1984). The means for time to flowering of the four landrace groups (measured over two sowing dates in an intermediate latitude location) ranged from 63 to 150 days, in response to no more than 20-min difference in daylength during the likely period of floral initiation.

In addition, there is a much greater variation in the onset of the rains than in their cessation in the region (Sivakumar, 1988). Farmers traditionally sow with the first rains, whenever they occur, meaning that the time from sowing to the optimal flowering date, which is related to the cessation of the rains, also varies from year to year. Thus adaptation also requires a sufficiently sensitive response to maintain the optimal flowering date for the local environment, regardless of time of sowing. A number of sowing date experiments conducted in West Africa with local photoperiod-sensitive sorghum (e.g., Curtis, 1968; Andrews, 1973) and pearl millet landraces (e.g., Kouressy et al., 1998; Vaksmann and Traore, 1996, quoted by Kouressy et al., 1998) indicate that flowering date is commonly delayed by as little as 1 day for each week delay in sowing date, over as much as a 2-month range in sowing dates. Thus the same response to photoperiod that adjusts the duration of local cultivars to different season lengths in different latitudes provides an equally effective adjustment to variation in season length due to variation in sowing date.

3.2 Adaptation to Variable Moisture Environments

As much or more than adaptation to overall season length, pearl millet needs adaptation to periods of inadequate soil moisture at almost any time during the growing season. The crop appears to depend upon a largely opportunistic strategy to reproduce, which is consistent with its evolution in a highly unpredictable environment. This strategy combines: (1) short critical growth stages with a high photosynthetic rate and a high growth rate, to maximize growth during periods of favorable soil moisture with (2) a surprising degree of developmental plasticity for a cereal, to allow it to rapidly compensate for periods of lost development/growth due to drought stress, and with (3) a reasonable ability to avoid/tolerate high canopy temperatures resulting from reduced transpiration.

3.2.1 Length of Critical Growth Phases

The growth cycle of the pearl millet is conveniently divided into three growth stages (Maiti and Bidinger, 1981). The vegetative stage (GS 1) varies from as little as 25% to as much as 65% of the total crop cycle, depending on the time to floral induction, whereas both the floral morphogenesis (GS 2) and grain filling stages (GS 3) are of fixed duration and are relatively short. GS 1 is technically composed of a juvenile phase from germination to the point at which the plant is capable of undergoing floral initiation if environmental conditions are inductive, and a latency phase between the end of the juvenile phase and actual floral initiation (Belliard and Pernes, 1984). Thus the length of GS 1 depends on the environmental daylength and inherent genetic daylength response. Not a great deal is known about genetic variation in the length of the juvenile period; in very short duration/insensitive lines, it can be as little as 15 days from emergence under an inductive photoperiod (Belliard and Pernes, 1984; authors, unpublished). Even in what are normally considered long-duration genotypes, the true juvenile period is not likely to be more that 25 days at normal environmental temperatures (25°C mean), as these will flower in as little as 50 days under very short (e.g., 10 h) daylengths. Genotypes with a very short GS 1 (less than 20 days) are vulnerable to stress during this period, as they have a very short time, even under optimal conditions, to develop a sufficient leaf area to support adequate growth during the reproductive stages. Therefore cultivars with a growth duration less than the growing season length, especially where this is short (less than 75 days), are likely to be more vulnerable to intermittent stress than are "full season" cultivars, despite their advantage in escaping end of season stress (Eldin, 1993; van Oosterom et al., 1996a). On the other hand, a long vegetative stage (more than 35 days) allows for growth lost to periods of intermittent stress to be made up before floral initiation, with little effect on final grain yield (Mahalakshmi and Bidinger, 1985a, 1986). During GS 1—i.e., BC_4F_1 early in the season—there is less likely to be sufficient stored soil moisture to maintain growth during breaks in the rains, so this is the stage when interruptions to growth are likely (see van Oosterom et al., 1996a). A long GS 1 may also have a similar benefit under low soil nitrogen status, allowing the plant sufficient time to take up enough nitrogen to productively complete its life cycle (Coaldrake and Pearson, 1985a).

The floral morphogenesis stage (GS 2), between floral initiation and actual flowering, during which apical primordia develop reproductive structures, and existing vegetative primordia complete development, is generally considered a genetic constant in degree-day terms (°C days) over a base temperature of 10°C (Ong and Monteith, 1985). Published figures for GS 2 length vary among authors, with estimates for Indian cultivars in the order of 350°C days (e.g., Craufurd and Bidinger, 1988b; Huda, 1987; Ong, 1983b), which is equivalent to approximately 23 calendar days at a mean temperature of 25°C. (Too few cultivars have been sufficiently rigorously assessed in terms of degree days to draw conclusions on the degree of genetic variation in GS 2 length.) The short duration of GS 2 allows the plant to complete this critical growth stage on a relatively small amount of available moisture. For example, Dancette (1983) estimated water use rates of approximately 7.5 mm day^{-1} (6.5 mm day^{-1} pan evaporation × a pan coefficient of 1.15) during GS 2 in Bambey, Senegal; thus a 25-day GS 2 could be completed, with no stress, on only 188 mm of available water. As GS 2 would generally occur near the middle of the rainy season, when there should be some stored soil moisture available, as little as 100 mm of current rainfall—no more than 30 mm week^{-1}—would be sufficient for the crop to complete GS 2 without moisture stress, despite high rates of evaporation.

The length of GS 3, from flowering until physiological maturity of the grain, is also considered a constant, with most estimates at about 400°C days (Craufurd and Bidinger, 1988b; Huda, 1987; Ong, 1983b) or about 27 calendar days at a mean temperature of 25°C. With pan coefficients below 1.0 during grain filling (Dancette, 1983), grain filling can be completed, without moisture stress, on no more available water than required for GS 2. Stress during grain filling is likely if the rains terminate earlier than the normal date and the rains during the main part of the season have been insufficient to recharge the soil profile. However, the combination of a short GS 3 with an appropriate time of flowering for the local conditions does provide a reasonable level of adaptation to the effects of stress during grain filling.

3.2.2 Drought Escape

Because of the short length of its GS 2 and GS 3 growth stages, drought escape is a major factor determining relative cultivar performance in individual stress environments (e.g., Bidinger et al., 1987a), and is often a major cause of G×E interaction in multienvironment trials (van Oosterom et al., 1996b). For example, in the case of rains ending early, a 1-week difference in time to flowering between two cultivars is equivalent to nearly 30% of the grain filling period, a significant fraction that would escape stress in the early flowering cultivar, but that would be affected by stress in the later-flowering one. The assessment of true field drought tolerance or susceptibility in pearl millet (as distinct from drought escape) requires that the effects of differential drought escape among cultivars be considered, as these often account for a greater fraction of the observed genotype performance than do differences in tolerance itself (Bidinger et al., 1987b).

The general effects of the timing of single periods of stress, both before and after flowering, are reasonably well understood in pearl millet, and provide quantification of the powerful effects of drought escape. For example, early genotypes that flowered 20 days before the onset of a terminal (unrelieved, end-of-season) drought stress had four-fold lower yield reduction (12% vs. 51%) than later-flowering genotypes that flowered only 10 days before the onset of the same stress (Mahalakshmi et al., 1987). However, despite the very strong effects of drought escape in the crop, the ability to use drought escape in crop improvement is limited, for a variety of reasons. First, the effectiveness of escape depends upon the predictability of the occurrence of stress, which is characteristically low in arid and semiarid environments, as escape is a function of the specific type and timing of the stress in question. For example, early genotypes, which effectively escaped a terminal stress in the example above, suffered four-fold greater reduction in grain yield than did later flowering genotypes (21% vs. 5%) in a midseason stress, in which the early genotype began flowering 5 days before the stress was relieved and the later genotype flowered 5 days after the termination of the same stress (Mahalakshmi et al., 1987). Second, appropriate season length is a compromise of various factors, including the occurrence of drought stress. Once the best time of flowering is determined, then there is little more than can be performed, other than offering farmers a range of flowering times around the optimum, to allow them to select a specific cultivar for their own particular variation of the general season length requirement (e.g., available soil water storage), and/or to spread risk among several cultivars of slightly different season lengths.

3.2.3 Developmental Asynchrony and Plasticity

Pearl millet effectively utilizes its developmental asynchrony, almost certainly inherited from its wild progenitors, to adjust to periods of moisture stress prior to flowering that affect normal development and growth. Primary tillers in pearl millet appear at approximately 45–50 degree-day intervals (Craufurd and Bidinger, 1988b; Ong, 1883a, 1984), or approximately every 3 days at a mean temperature of 25°C. Secondary tillers are produced from primaries at similar rates, resulting in a potentially very large number of tillers, all at different stages of development (e.g., leaf number), in a very short period (Lambert, 1983; Ramond, 1968). Floral initiation among these tillers is not synchronous, although the interval between tiller appearance and floral initiation (GS 1) does decline somewhat in higher-order tillers, compared to lower-order ones (Craufurd and Bidinger, 1989). In contrast, the length of the period between floral initiation and flowering (GS 2) is similar for all tillers, which results in tillers that are at different stages of apical development at any given time in GS 2 (Craufurd and Bidinger, 1988b).

There is considerable circumstantial evidence that the sensitivity to drought stress prior to flowering (in terms of subsequent yield loss) increases in more advanced stages of apical development (Mahalakshmi et al., 1987). Shoots with a vegetative apex appear to have the ability to arrest development during a period of severe stress and resume development upon the relief of the stress with little effect on overall productivity, if there is sufficient moisture to complete subsequent development (Mahalakshmi and Bidinger, 1985a). Similarly, shoots whose apices are in early stages of panicle development appear to be less affected by stress than those in later stages (Mahalakshmi and Bidinger, 1985b). This probably occurs because floral primordia are in developmental, rather than growth stages (Craufurd and Bidinger, 1988b), and are less affected by the reduction in assimilate supplies as a consequence of stress. Therefore later-developing tillers are likely to be less affected by stress prior to flowering than are main shoots and early tillers. In addition, later tillers will have a longer period of recovery, if the stress terminates before flowering, to expand new leaves and produce sufficient assimilate to support full panicle growth.

In addition to direct effects of stress on shoot growth—reduced inter-node and leaf area expansion, failure of spikelets and florets to complete development, etc.—preflowering stress appears to alter the normal hierarchy among shoots. In the absence of stress, the main shoot itself (in longer duration, African cultivars) or the main shoot plus the primary tillers (in shorter duration, Indian cultivars) dominate the later developing tillers, resulting in the latter failing to continue growth and to produce a panicle (Craufurd and Bidinger, 1988a; Ong, 1984). It has been suggested that this dominance is a consequence of the competition for assimilates/nutrients from stem growth in the main shoot/first tillers, which begins before stem growth in the later tillers, because of the normal asynchrony of tiller development and growth (Craufurd and Bidinger, 1988a). In the presence of stress during the GS 2, main shoot growth is often considerably reduced, and the suppression of the later tillers is reduced once stress is relieved and growth resumes (Mahalakshmi and Bidinger, 1986). This response provides the opportunity for replacing grain yield lost to the stress on the more advanced main shoot (and first tillers) that have been seriously affected by stress, with an increased yield from less-advanced, later-flowering tillers, which have been less affected by the stress (Mahalakshmi and Bidinger, 1986). This primarily involves an increase in the number of tillers producing a panicle, rather than an increase in grain numbers per tiller panicle, suggesting that the reduction in dominance of the main shoot, rather than an improved light environment for the tillers, was the primary cause of the compensation. Surgical removal of the main shoot, in contrast, resulted in both an increase in the number of later tillers producing panicles and an increase in the numbers of grains per tiller panicle, suggesting that both effects were operating in this extreme case (Mahalakshmi and Bidinger, 1986). As a consequence, grain yield in pearl millet is often little affected by a midseason stress, provided that there is sufficient time and moisture, after the end of the stress period, for later-developing tillers (including those which arrested development in a vegetative stage) to reach maturity (Mahalakshmi and Bidinger, 1985a, 1985b).

However, the asynchrony of tiller development in pearl millet has different implications in terminal (end of season) drought stress as the later-flowering tillers are more severely affected by the stress than is the main shoot. To the degree that these represent significant portion of the overall crop yield under normal conditions, the yield loss under terminal stress is likely to be greater than probably would have occurred if flowering were more synchronous among shoots. In general, pearl millet's developmental mechanisms of adaptation are mainly effective for varying moisture conditions during the main part of the season, rather than at the final stages of crop growth. This is probably a reasonable trade-off, as natural selection, operating on genetic variation for daylength response, has likely adjusted flowering to maximize escape from terminal drought stress to the degree possible.

3.3 Drought Avoidance, Tolerance, and Water Use Efficiency

3.3.1 Water Uptake

Pearl millet appears to have the ability to effectively access available water in the soil profile before flowering, even at the cost of above ground growth, provided that the crop is continuing to assimilate sufficient carbon to continue to invest in root growth. The situation is less clear at the end of the season, when grain filling, rather than vegetative growth, is the major sink, stomatal control of water use is less effective (see below), and root length densities below $1.0–1.5$ cm cm^{-3} may not be sufficient to extract all of the available water from the profile (Payne et al., 1990; Winkel and Do, 1992). It is reasonable that the crop would have evolved means to efficiently use available soil water in an arid/semiarid climate as drought avoidance, rather than drought tolerance, is the obvious strategy when it is possible. The greater problem is the situation when there is little or no available soil water, and drought avoidance is no longer applicable.

Pearl millet is capable of producing a very extensive root system, under favorable conditions, because of relatively rapid root extension rates driven by the favorable temperatures in tropical soils (Gregory, 1986). Mean root penetration rates of 3.5 cm day^{-1} (Chopart, 1983) to 4.5 cm day^{-1} (Azam Ali et al., 1984) have been measured in sandy soils under field conditions, with maximum rates approaching 7 cm day^{-1} (Azam Ali et al., 1984). Maximum depth of root penetration appears to depend, partly at least, upon season length, with maximum rooting depths of 140 cm in short-duration (75 day) Indian cultivars (Azam Ali et al., 1984; Gregory and Squire, 1979), 180 cm in medium-duration (90 day) West African cultivars (Chopart, 1983), to more than 3 m in long-duration (120 day) cultivars (Begg, 1965). Lateral spread of roots under widely spaced hill planting conditions can reach as much as 2 m in all directions (Chopart, 1983), indicating that under the conditions of that experiment, an individual hill of pearl millet could have explored as much as 6 m^3 of soil for available water.

Root length densities in millet have been reported to range from 0.4–0.8 cm cm^{-3} at the soil surface to 0.2–0.4 cm cm^{-3} in the main part of the profile (between 40 and 80–120 cm, depending upon the cultivar/season length), but declining to 0.1 cm cm^{-3} below this depth (Azam Ali et al., 1984; Chopart, 1983; Gregory and Squire, 1979). Active root growth continues to at least flowering in short-duration Indian cultivars (Gregory and Squire, 1979) and through grain filling in longer-duration West African cultivars, although at reduced rates (Chopart, 1983, Do et al., 1989, quoted by Winkel and Do, 1992). In addition, there is evidence that there is considerable plasticity in the ratio of root growth to shoot growth in response to increasing aridity of the above-ground environment, to support continued water uptake from the soil (Squire et al. 1987). These authors cite increases in rooting depth and root length density, and particularly in the ratio of root length to shoot area, which corresponded to increases in the maximum daily saturation deficit in three experimental crops of the same cultivar. However, virtually nothing is known of other aspects of water uptake in pearl millet apart from the growth and morphology of the root system—water uptake rates, hydraulic conductivity, etc. (Winkel and Do, 1992).

3.3.2 Control of Water Loss

There are a number of studies that suggest that pearl millet attempts to maximize carbon gain when water is available, rather than conserving it for likely future periods of drought. On the one hand, this might be considered a reckless strategy, but it can be argued that it is appropriate for a variable, and unpredictable, moisture environment, in which the objective is to able to take immediate and full advantage of (often short) periods when moisture is available. This strategy would make particular sense for a crop with short critical developmental periods (GS 2, and especially GS 3), which, with a bit of luck, could be largely completed during periods of favorable soil moisture.

Stomatal conductance in well-watered pearl millet appears to respond to variation in potential evaporation in such a way as to keep canopy transpiration at as high a level as possible, consistent with maintaining leaf water potentials at favorable levels (Squire, 1979; Henson and Mahalakshmi, 1985). Midday conductance (i.e., under high irradiance levels) in well-watered plants under moderate vapor pressure deficits responds to short-term changes in evaporative demand, driven by changes in saturation deficit (Black and Squire, 1979; Squire et al., 1983), as well as to changes in leaf area (Black and Squire, 1979; Squire et al., 1986). When soil water is not freely available, conductance does not reach the same levels as in well-watered plants, and the responsiveness of conductance to saturation deficits (which are inevitably greater in such conditions) is much less evident (Do et al., 1996). Conductance under such conditions is more closely related to irradiance (Cantini et al., 1998; Henson et al., 1982a; Squire, 1979), than to saturation deficit, but leaf water potentials decline at midday (Henson et al., 1982a). Under these conditions, conductance is higher in the younger, upper leaves than in older leaves (Henson et al., 1982a; Squire, 1979). Finally, when soil water becomes seriously limiting to transpiration, stomatal opening in response to irradiance is only partial, and closure progressively occurs earlier in the day (Azam Ali, 1983; Do et al., 1996; Henson et al., 1982a), associated with loss of leaf turgor (Henson et al., 1982a). Senescence of older, lower leaves in the canopy and later-developing tillers begins in these conditions, reducing the total crop transpiration, and helping to maintain the water status of the younger, more photosynthetically efficient leaves in the upper canopy (Do et al., 1996; Wallace et al., 1993). This suite of responses to declining water availability supports the crop's strategy of maximizing the continued carbon assimilation of the crop for as long as possible, perhaps partially to maintain root growth (as well as shoot growth), and hence water supply, in sandy soils with low volumetric water contents.

One rather interesting feature of the stomatal behavior of pearl millet is a decrease in the sensitivity of the stomata to water stress after flowering, in leaves at similar water potentials (Henson et al., 1984). Differences in conductance pre- and postflowering were not related to differences in turgor, but were related to lower levels of abscisic acid in the leaves of postflowering plants (Henson et al., 1983; 1984), possibly because of enhanced ABA export associated with enhanced carbon export from the leaves to the developing grain in postflowering plants (Henson and Mahalakshmi, 1985). This reduction in sensitivity of the stomata to stress is consistent with the perception of pearl millet as pursuing a strategy to maximize assimilation, particularly in its most critical growth stage—grain filling.

3.3.3 Water Use Efficiency

Estimation of water use efficiency (WUE) for pearl millet under typical growing conditions is complicated by two factors: the weak relationship between dry matter and evapotranspiration (ET) in sparse canopies with low percentage ground cover (Payne, 2000), and by the effects of high vapor pressure deficits (VPD) on WUE (Squire et al., 1987). In moderate VPD environments and for crops with full ground cover (LAI >3–4), estimates of WUE (total aboveground dry mass/seasonal ET) for pearl millet range between 300 and 400 kg ha^{-1} cm^{-1} (Chaudhuri and Kanemasu, 1985; Singh et al., 1983), which is comparable to other C4 cereals. However, for traditional low-density, low-input millet crops in areas characterized by high VPD, published estimates of WUE range from as low as 50–150 kg ha^{-1} cm^{-1} in unfertilized crops to 150–250 kg ha^{-1} cm^{-1} with added fertility (Cisse and Vachaud, 1988; Klaij and Vachaud, 1992; Payne, 1997). In the absence of a full crop cover in traditional low-density millet crops, system ET is poorly related to crop growth; direct soil evaporation in such crops effectively replaces what in a more complete canopy would be transpired water, absorbed from soil surface layers by shallow roots. Soil evaporation in traditional millet crops accounts for as much as 30% of crop ET or 35–45% of seasonal rainfall (Wallace, 1991). The addition of fertility has minimal effects on seasonal ET but major effects on crop dry mass (Payne, 1997; Sivakumar and Salaam, 1999). As a result, WUE estimates vary with crop dry matter productivity, and ET is of little value in predicting crop yield (Payne, 2000). Therefore there are major opportunities to enhance field water use efficiency in such sparse canopy crops by increasing crop growth by improving nutrient supply and increasing plant population (Payne, 1997, 2000).

The other factor that reduces WUE of pearl millet in arid environments is a high ambient VPD, which can have a significant effect on WUE. Squire et al. (1987) estimated that for four crops of a single variety, pearl millet grown in different ambient VPDs, but otherwise under good management, WUE (total aboveground dry mass/seasonal transpiration) ranged from 640 kg ha^{-1} cm^{-1} at a mean daily maximum VPD of 1.4 kPa (glasshouse crop in the UK) to 210 kg ha^{-1} cm^{-1} at a mean daily maximum VPD of 4.0 kPa (dry season crop in Niger). While this range in VPD is very broad, Payne (1997) did report a more than twofold range in average daily VPD during the growing season across 4 years in Niger, and a threefold range in the estimated daily integral of the VPD. If the product of WUE and VPD is a relatively stable parameter (Squire et al., 1987), then differences in ambient VPD can be expected to cause at least a twofold variation in realized WUE between high and low VPD environments.

3.3.4 Drought and High Temperature Tolerance

Very little is known about the existence of mechanisms of tolerance to low water status and/or supraoptimal temperatures in pearl millet, despite its arid environment origins. The foregoing analysis of the basis of its adaptation to stress environments focused on avoidance and adjustment mechanisms, partly because of the lack of information on tolerance mechanisms. What is known is reviewed below; but this should be an area in which future research can exploit the synteny between pearl millet and crops such as rice and sorghum (Devos et al., 2000; Ventelon et al., 2001), whose tolerance mechanisms have been studied in much greater detail.

Osmotic adjustment has been shown to occur in pearl millet, but observed changes in osmotic potential under field conditions in several breeding lines were too small to significantly lower the water potential at which turgor became negligible (Do et al., 1996; Henson, 1982; Henson et al., 1982b). A growth cabinet comparison of pearl millet landraces from different rainfall zones did indicate a greater capacity to osmotically adjust in landraces originating from more arid locations (Blum and Sullivan, 1986), with the expected effects on the relationship between relative water content and stomatal conductance (Blum and Sullivan, unpublished). Therefore it is possible that a significant ability to osmotically adjust may exist in arid zone millet germ plasm, if not in modern varieties.

Pearl millet is known to have high optimum (30–35°C) and high maximum ($> 40°C$) temperatures for various physiological processes, including germination (Garcia-Huidobro et al., 1982), leaf extension (Ong, 1983a), stem elongation (Squire, 1989), and photosynthesis (McPherson and Slatyer, 1973). This would be expected from its arid zone origins, and would provide adaptation to the maximum temperatures common in the areas in which it is grown. Genetic differences in optimum/maximum temperatures for germination and vegetative growth have been reported (Mohammed et al., 1988a, 1988b), and in survival of seedling heat stress (Peacock et al., 1993), suggesting useful variation in physiological mechanisms of adaptation or tolerance to high temperature (Howarth et al., 1997b).

3.4 Adaptation to Low-Fertility Environments

Although low availability of plant nutrients is one of the major limitations to productivity in most of the areas in which pearl millet is grown (see above), not a great deal is known about the specific ways in which the crop adapts to, or reproduces in, soils with very limited nitrogen and phosphorous concentrations. However, there are a variety of adaptive mechanisms to low-fertility environments (Gourley et al., 1997) that could be important for pearl millet. Pearl millet has been shown to form associations with vesicular-arbuscular mycorrhiza and nonsymbiotic nitrogen fixing bacteria in the rhizosphere, which may enhance nutrient uptake/availability in extremely poor soils, and has been shown to be capable of growth at relatively low levels of tissue nitrogen; but apart from these, most of what follows on possible physiological mechanisms of adaptation is only speculative.

In West Africa, where the problem is the most critical, farmers traditionally attempt to manage the problem of low soil fertility largely by agronomic means, primarily long fallow periods to regenerate soil fertility (Charreau and Nicou, 1971). Farmers also understand the benefits of sowing with the first rains, to capture the initial flush of mineralized soil nitrogen before it is leached down the profile by subsequent rains (Blondel, 1971), as this often represents much of the nitrogen available to an unfertilized crop. They also utilize very low plant populations, which allow cultivars with a long vegetative period sufficient time to scavenge large soil volumes for nutrients and water. In India, where farmers are better able to augment soil fertility through rotation with legumes and application of animal manure, sowing with the first rains is also the norm, but cultivar vegetative periods are considerably shorter, and target plant populations are higher (AICPMIP, 1988). This could be interpreted as an alternative strategy, which substitutes plant numbers for time as a means of maximizing crop nutrient uptake, especially where deficiencies of immobile elements such as phosphorous are often less chronic than in West Africa.

3.4.1 Microbial Associations

Pearl millet, in common with a number of tropical cereals and grasses, supports populations of various rhizosphere bacteria (*Azospirillum*, *Azobacter*, etc.) that are capable of fixing nitrogen (Boddey and Dobereiner, 1982). However, there is no consensus as to the importance of such associations in enhancing nitrogen availability to the crop. Inoculation experiments with selected strains of several N-fixing bacteria have produced some positive results in some cases, particularly with sterilized soils, but not all cases (Lee et al., 1994; Smith et al., 1984; Wani et al., 1985). Direct comparisons of nitrogenase activity between inoculated pearl millet and several legumes in nonsterile soils suggest that actual amounts of N fixed in the millet case are likely to be of little agronomic significance even when soils are inoculated with selected strains (Bouton, 1988, Lee et al., 1994).

Similarly, millet roots are commonly colonized by vesicular-arbuscular mycorrhiza or VAM (Krishna et al., 1985) that are thought to enhance P uptake, particularly from soils with low levels of available P (Bolan, 1991). Possible mechanisms of VAM action include an exploration of a larger soil volume, solubilization of unavailable soil P, and enhanced uptake of P from the soil solution (Bolan, 1991). Reported work on pearl millet indicates considerable variability among genotypes in both amounts of natural VAM colonization and in P uptake and growth in response to inoculation (Krishna et al., 1985). Thus there is a potential role of VAM in adaptation to low P soils, but there are no data to quantify the importance of VAM colonization in low-fertility environments.

3.4.2 Nutrient Scavenging

The root system of pearl millet has been described in the previous section, in relation to water uptake. The rooting pattern of longer-duration African genotypes suggests that they should have the ability to effectively scavenge nutrients present in very low concentrations in the soil, both because of the large soil volume their roots can explore, the long vegetative period they have to do this, and the potential of VAM colonization to enhance nutrient uptake. However, low available soil nutrient supplies affect root as well as canopy growth (Cisse and Vachaud, 1988; Payne et al., 1996) so nutrient scavenging ability may be limited where it is most needed. This is clearly the case for water use, where low soil P levels decrease rooting volume and soil water uptake from especially deeper soil layers, and allow greater water percolation through the root zone (Payne, 2000). It has also been suggested that the long duration of the vegetative and reproductive phases (GS 1 and GS 2) in traditional West African landraces provide a means of slowly accumulating nutrients in the vegetative mass of the crop, to support the rapid growth of the panicle and grain at the end of the growing season (Siband, 1983). N and P harvest indices in the crop are relatively high (Wani et al., 1990), which provide some support for this hypothesis.

3.4.3 Tiller Asynchrony/Plasticity

The asynchrony of tiller development, and consequent hierarchy among the main shoot and successive tillers (see above), is also likely to be an adaptive mechanism to low soil fertility conditions. With low levels of available nutrients, and the resulting slow rate of leaf area development and limited assimilate supply, the competitive advantage of the main shoot appears to be strengthened, probably because its belowground—as well as aboveground—sinks are established before those of its tillers. Under conditions of extreme N stress, few or no tillers develop (Coaldrake and Pearson, 1985a), reducing competition below as well as above ground, thus allowing the main shoot full use of the limited supply of both nutrients, greatly improving the chances of the main shoot being able to form a panicle and produce grain. It is tempting to argue that the very strong main shoot dominance characteristic of many

West African landraces is a consequence of evolution in very poor soil environments. This may have been exacerbated by the West African farmers' practice of sowing of a number of seeds in a single hill, adding a competitive challenge to obtaining sufficient nutrients from the soil.

3.4.4 Ability to Reproduce at Low Nitrogen Levels

Relative growth rate has been linearly related to relative nitrogen accumulation rate in pearl millet under conditions of a constant N supply (Coaldrake and Pearson, 1985a). These authors estimated that the total plant N concentration required for maximum growth rate prior to floral initiation was 1.6% and thereafter 1.3%; but suggested that growth would continue until the N concentration dropped below 0.2%. At below-optimal N concentration levels, development as well as growth are affected; for example, thermal time requirements increase for leaf and tiller appearance, leaf expansion, floral initiation, and panicle branch and spikelet initiation (Coaldrake, 1985; Coaldrake and Pearson, 1985a; 1985b). Thus the crop is able to maintain development processes at reduced rates, and growth processes at rates proportional to N accumulation rates, over a seemingly very wide range of plant N concentrations—down to as little as 15% of that required for optimal growth. Provided that the associated delay in development processes, and consequently in flowering and maturity, does not extend the growing season beyond the limit on water availability, the crop seems able to produce at least some seed, if not a significant grain yield, over a very wide range of soil nitrogen fertility levels.

4   THE PHYSIOLOGICAL BASIS OF YIELD DETERMINATION

4.1   Biomass Productivity

Total biomass productivity in pearl millet, as in any crop, is a product of growth rate and growth duration. Under favorable environmental conditions, the crop has very high potential growth rates (Begg, 1965) as it is a C4 cereal, with relatively erect canopy, a potentially high leaf area index (LAI) because of its tillering habit (Begg, 1965; Craufurd and Bidinger, 1989), a high radiation use efficiency (RUE) (Squire et al., 1986), plus a high temperature optimum for assimilation (McPherson and Slatyer, 1973). These characteristics make it well able to use the high levels of incoming solar radiation characteristic of arid and semiarid tropical environments. However, in the areas in which it is most widely grown, potential growth rates are seldom achieved, because of sparse plant canopies and frequent nutrient and water stress limitations (Payne, 2000). Therefore actual growth rates are much more likely to be determined by environmental or management factors than by the plant's inherent assimilation capacity. As a consequence, these factors, combined with crop duration, are more likely to be the major determinants of total biomass productivity than are genetic differences in radiation interception or radiation use efficiency.

4.1.1   Seedling Growth

Early establishment of a leaf area sufficient to intercept a majority of the incoming radiation is essential for producing a large crop biomass. The individual component processes of establishment of an early crop canopy (germination, leaf emergence, and leaf extension) have relatively high rates in pearl millet. For example, germination requires a thermal time of less than $20°C$ days over a relatively wide range of temperatures in pearl millet, from a minimum of about $12°C$ to a maximum of $48°C$ (Garcia-Huidobro et al., 1982). Seedling emergence is equally quick, requiring as little as $30°C$ days over the estimated base temperature of $10°C$ (Ong, 1983a), or approximately 3 calendar days at optimal soil temperatures. Individual leaves are produced at $30–40°C$ day intervals (Craufurd and Bidinger, 1988b; Ong, 1983a, 1984; van Oosterom et al., 2001b), and leaf extension rates range from 2.5 to 4.5 mm $°C$ day^{-1} (van Oosterom et al., 2001b), with maximum rates in the range of 7–10 mm h^{-1} at optimum temperatures (Ong, 1983c). However, the base capital to which these processes are applied—the size of the seed, and its consequent limited reserves of both carbon and minerals—is very small, resulting in small first leaves (van Oosterom et al., 2001a) and a slow early canopy development, relative to maize or sorghum (Siband, 1979). Pearl millet's small seed size (0.6–1.3 mg per seed, compared to 2.0–3.5 mg in sorghum, and 25–35 mg in maize) is an essential component of its strategy of adaptation to stress environments, inherited from its wild progenitors (de Wet et al., 1992). Small seed size allows seed filling to be completed in less than 25 calendar days (compared to approximately 35 days for sorghum and 45 days for maize), and the production of large seed numbers per inflorescence (as many as 3000–4000 in large panicle types). However, this strategy has a significant cost in terms of early canopy development—LAI and fractional radiation interception in the crop are very small until about 20 days after emergence when individual main shoot leaf size begins to increase and tiller leaf area begins to develop (Craufurd and Bidinger, 1989; Squire et al., 1987; van Oosterom et al., 2001b). There is genetic variation in early leaf area development (Mohammed et al., 1988b; Soman and Bidinger, unpublished), possibly based on genetic differences in base temperature for this process (Mohammed et al., 1988b), but this has not been intentionally exploited to the authors' knowledge. In favorable environments, increasing plant populations provides an effective adjustment mechanism to compensate for slow individual plant leaf area development (Carberry et al., 1985; Craufurd and Bidinger, 1989), but this strategy is less applicable in low-fertility, moisture-deficit environments.

4.1.2 Canopy Development

Pearl millet's major means of developing a full plant canopy is through its tillering ability. The first tiller generally appears in the axil of the third leaf, at approximately 200°C days after emergence, with subsequent primary tillers appearing at 45–50°C day intervals (Craufurd and Bidinger, 1988b; Ong, 1984; van Oosterom et al., 2001b). Secondary tillers are produced from primaries at similar rates, resulting in a total tillering capacity that can reach 50 per plant, given maximum water, nutrients, and space (Craufurd and Bidinger, 1989; Ramond, 1968). However, under normal crop conditions, total tiller numbers range from 6 to 10, with those actually producing a panicle (nearly always primary tillers) ranging from 3 to 5 under favorable conditions (Craufurd and Bidinger, 1988b; Ong, 1983a; van Oosterom et al., 2001b), and 0 to 2 under unfavorable conditions. Leaf appearance rates (in °C days) on tillers are similar to those on main shoots (Craufurd and Bidinger, 1988b; Ong, 1984; van Oosterom et al., 2001b), but leaf numbers per tiller decline with tiller position (van Oosterom et al., 2001a; Ong, 1984). However, the largest leaf occurs at a progressively lower position in the tillers than in the main shoot, and its area can nearly equal that of the largest leaf on the main shoot (van Oosterom et al., 2001a), with the result that tiller leaf areas approach that of the main shoot, despite a lower leaf number. Canopy development can be simulated as a crop of independent tillers, each producing a potential leaf according to its position in the hierarchy, and an actual leaf area based on the results of competition for light among tillers (van Oosterom et al., 2001b).

4.1.3 Radiation Interception and Use

Actual canopy development depends on plant population; crop duration (in thermal time); water and nutrient availability; and environmental hazards such as hazards such as foliar diseases, defoliation by insects, etc.; and thus varies widely. Under very favorable conditions, LAIs can reach 6 or more, at which fractional radiation interception approaches 95% (Begg, 1965), even in short-duration varieties (e.g., Squire et al., 1984). However, more commonly, even in well-managed crops, LAIs range from less than 2, especially in hill-sown crops in West Africa (e.g., Begue et al., 1991; Wallace et al., 1993), to 3–4 in higher density row-sown crops (e.g., Craufurd and Bidinger, 1988b; Singh et al., 1983; van Oosterom et al., 2001b). Crops seriously affected by water stress or by low nutrient availability may not even reach the low end of this range.

Therefore the time course of fractional radiation interception, and thus mean seasonal interception, will vary widely among individual crops. Early season fractional interception is always low in pearl millet, because of low initial LAIs (Squire et al., 1987). The rate of increase in fractional interception will primarily depend on plant population and on rate of leaf area development, which is largely dependent on environment—temperature, water, and nutrients. Duration of interception is largely dependent on the length of the preflowering period, and therefore on both environment—daylength and temperature—and on the specific photoperiod response and/or maturity of a given cultivar. Squire et al. (1984) modeled mean seasonal radiation interception from three factors: (1) the maximum interception achieved, (2) the time from sowing to the time when half of the maximum interception is achieved, and (3) the total season length. These factors are alternatively largely under genetic control (crop duration), environmental control (rate of increase in fractional interception), or both (maximum fractional interception, which is approximately equal to the product of rate of increase in fractional interception and the length of the preflowering period). However, assuming that for a given environment, variety alternatives are of the appropriate duration for the rainy season, then environment will be the main determinant of fractional radiation interception for a given crop. Squire et al. (1986) illustrated this in a comparison of four crops of the same cultivar grown in different available moisture environments. Preflowering fractional interception among crops ranged from 22% to 34% and mean postflowering interception ranged from 26% to 80%, because of significant effects of stress on both the rate and duration of the increase in fractional interception, and the onset, rate, and duration of leaf senescence.

Reported maximum radiation use efficiency (RUE) values for pearl millet vary from 2.5 g MJ^{-1} intercepted radiation (Squire et al., 1986) to more than 4 g MJ^{-1} (Ram Niwas et al., 1999). Common seasonal mean values for field experiments range from 1.0 to 2.0 g MJ^{-1} (e.g., Bishnoi and Niwas, 1992; McIntyre et al., 1993; Muchow, 1989; Ong and Monteith, 1985), although higher values have been reported in low plant populations in which there is presumably little mutual shading of leaves (Begue et al., 1991). Radiation use efficiency appears to be less affected by environmental conditions, including increasing saturation deficit, increasing temperature, and modest nutrient and water stress, than is radiation interception (McIntyre et al., 1993; Squire et al., 1986).

4.2 Biomass Partitioning

Harvest index (HI) in pearl millet tends to inversely vary with season length total, and hence also to inversely vary with biomass production. Harvest index can exceed 0.40 in short duration cultivars and drop to as low as 0.15 in traditional, photoperiod-sensitive West African cultivars. This is because daylength-mediated increases in crop duration almost entirely occur in the vegetative period, prior to floral initiation (Carberry and Campbell, 1985; Craufurd and Bidinger, 1988b). Increases in the duration of vegetative growth stages result in the formation of additional vegetative sinks, mainly additional stem internodes, and additional secondary and tertiary tillers (Carberry and Campbell, 1985; Craufurd and Bidinger, 1988b). Therefore increases in biomass productivity are largely in the form of additional vegetative biomass rather than reproductive biomass, with limited effect on grain yield potential (Craufurd and Bidinger 1988a,b), and consequently with a negative effect on harvest index. In addition, in most quantitative daylength-sensitive geno-types, an increase in the vegetative period enhances main shoot advantage over tillers, probably because of a combination of the increased sink size of the main shoot, and the fact that stem internode growth begins earlier in the main shoot than in the tillers (Craufurd and Bidinger, 1988b). As a result, increases in main stem biomass (and in grain number in the main stem panicle) are largely offset by decreases in productive tiller number, total tiller biomass, and tiller grain number (Carberry and Campbell, 1985; Craufurd and Bidinger, 1988b). This effect likely explains the characteristic phenotype of many traditional photoperiod-sensitive West African cultivars, which is a single productive main shoot, with a low harvest index, and little or no contribution to grain yield from the tillers that often remain vegetative or senescent.

4.3 Grain Yield

Grain yield in pearl millet is primarily a function of grain number per unit area (Bidinger et al., 2001; Craufurd and Bidinger, 1989), as in other cereals. Grain numbers produced are related to crop growth, and ultimately to radiation interception, during GS 2—the period between floral initiation and flowering (Craufurd and Bidinger, 1989; Ong and Squire, 1984). Potential crop growth during this period is a function of the length of the period, in thermal time (Ong, 1983b), and the amount of radiation intercepted per unit thermal time (Squire et al., 1986). Therefore there are opportunities for increasing grain numbers (and potentially grain yield) by increasing the leaf area index and the fractional radiation interception at floral initiation, and by increasing the duration of GS 2. For example, fractional radiation interception during GS 2 in short-duration Indian genotypes can be increased by moderate increases in the length of the period prior to floral initiation (Alagarswamy and Bidinger, 1985; Craufurd and Bidinger, 1988b), or by increases in plant population (Craufurd and Bidinger, 1989; Carberry et al., 1985). Both of these options have been shown to be effective in increasing grain numbers per unit area in short-duration genotypes (Craufurd and Bidinger, 1989). Temperature dur-ing GS 2 also affects the numbers of calendar days required to complete the growth stage, and hence in radiation intercepted during the stage. For example, at a mean temperature of 22°C, 41 calendar days were required to complete GS 2 during which the crop intercepted a total of 205 MJ of incoming radiation, compared to a mean temperature of 28°C, where GS 2 lasted 31 calendar days and the crop intercepted 133 MJ of incoming radiation (Ong and Monteith, 1985). Less is known about the genetic opportunities to extend the length of the period between floral initiation and flowering itself. There is limited genetic variation for this (Bidinger, unpublished) and divergent selection for differences in the length of this period was effective in increasing grain numbers in an exploratory experiment (Bidinger, unpublished). However, increases in grain yields resulting from increasing grain numbers by various manipulations have not been proportional to the increase in grain numbers, as individual grain mass is commonly reduced in response to an increase in grain numbers (Alagarswamy and Bidinger, 1985; Bidinger et al., 2001; Craufurd and Bidinger, 1989).

However, in practice, grain and biomass yields of pearl millet are much more likely to be limited by low plant population densities (McIntire and Fussell, 1989; Payne, 1997), which limit radiation interception, and by nutrient and water stress, which limit radiation and water use efficiencies (Payne, 2000), than by a limited genetic yield potential. The crop finds its principle niche in marginal environments, and has many unique adaptive features giving it a competitive advantage over other cereals in such environments. Therefore the emphasis on genetic (and molecular) improvement of the crop should focus more on manipulating and strengthening these adaptive advantages, rather than on increasing yield potential for environments in which there is limited opportunity for such potential to be expressed.

4.4 Improvement in Yield and Adaptation

4.4.1 Breeding Opportunities

The reproductive characteristics of pearl millet present a wide range of opportunities for improvement of the crop. The crop is normally cross-pollinated because of its protogynous flowering habit, in which the stigmas in a panicle emerge and are receptive several days before anthers shed pollen (Mangat et al., 1999). This characteristic can be exploited to cross-pollinate panicles without the need for emasculation, by bagging panicles before stigma emergence (to prevent natural cross-pollination), and then exploiting the period between stigma and anther emergence to pollinate emerged stigmas with pollen from another source, before pollen shed on the anthers on the same panicle. Individual panicles contain from 1000 to as many as 3000 fertile florets (long panicle West African plant types), resulting in large seed numbers produced per pollination. Both the ease with which both large numbers of crosses or pollinations can be made and the resulting large F_1 population sizes per pollination make population breeding an very effective tool in the crop (Rai and Virk, 1999). Methods used include both intrapopulation and interpopulation improvement, backcross population breeding, and variety and topcross pollinator breeding from improved populations (see Witcombe, 1999).

Despite its natural outcrossing behavior, most pearl millet genotypes are self-compatible, and can be easily self-pollinated by simply bagging individual panicles to exclude foreign pollen. Seed set in self-pollinated panicles may be variable, and self-pollinated seed numbers are often less than cross-pollinated seed numbers. However, with the large floret number per panicle, realized self-pollinated seed numbers are usually more than sufficient for most breeding objectives. Self-pollination is usually performed to produce inbred parents, to exploit heterosis in synthetic varieties or, more commonly, F_1 hybrids, rather than to produce lines for direct use as varieties (Rai and Virk, 1999). The discovery of cytoplasmic-nuclear male sterility in the crop (Burton, 1958) provided the means to rapidly and effectively exploit heterosis in the breeding of both grain (Athwal, 1966) and forage (Burton, 1977) hybrid cultivars. Hybrid breeding methods in pearl millet are well documented (e.g., Talukdar et al., 1999) and hybrids now dominate millet production in India, except in the more marginal areas (Govila et al., 1997).

4.4.2 Grain Yield Improvement

Open-pollinated variety breeding has been practiced in pearl millet for 50 years, using mass, progeny, and recurrent selection, to produce both higher yielding random-mating populations or composites and more restricted open-pollinated varieties from them. Methods used and progress achieved has been reviewed by a number of authors (e.g., Andrews and Bramel-Cox, 1993; Lambert, 1983; Rai et al., 1997; Witcombe, 1999). Most population/variety breeding has been primarily based on empirical selection for grain yield, with secondary selection for maturity and plant type. Reported progress in improving grain yield in recurrent selection programs has been of the order of 2–4% per cycle (Singh et al., 1988; Rattunde and Witcombe, 1993), which is probably also a reasonable figure for other open-pollinated variety breeding methods.

The most effective means of increasing grain yield in pearl millet, for both optimal and suboptimal environments, has been the exploitation of heterosis (Talukdar et al., 1999). Reports of very high levels (>40%) of yield heterosis based on comparisons to inbred parents (e.g., Virk, 1988) tend to be misleading because pearl millet suffers from significant inbreeding depression (Burton, 1951). Many years of indirect comparisons of highly selected F_1 hybrids and improved open-pollinated varieties in the national yield trials of the All-India Coordinated Pearl Millet Improvement Program suggest a 10–

15% yield advantage with hybrids (AICPMIP, 1988). Direct comparisons of unselected topcross (inbred × variety) hybrids and their noninbred pollinators indicated a mean 20% yield advantage to the hybrids across a wide range of environments (Mahalakshmi et al., 1992) and 30% in more marginal environments (Yadav et al., 2000). Heterosis in these experiments was primarily expressed as an increase in growth rate, with effects on total biomass and grain yield being largely determined by seed parent effects on crop duration and partitioning (Bidinger et al., 1994; Yadav et al., 2000). Heterosis in pearl millet, as in other cereals, is a function of both general and specific combining abilities (Talukdar et al., 1999). Methods for assessing combining ability remain largely empirical, resulting in the need to make and evaluate many parental lines and experimental combinations between them (Talukdar et al., 1999). A better understanding of both general and specific combining abilities, at a molecular level, could significantly improve the effectiveness of hybrid parent breeding programs, and increase the likelihood of exploiting heterosis in breeding programs targeting marginal environments, while maintaining or improving the adaptive features necessary for such environments.

4.4.3 Improvement in Adaptation

Despite the overwhelming role of the environment in the determination of grain yield in the crop, there have been few, if any, attempts to directly breed for specific adaptive traits, apart from breeding for shorter time to maturity and disease resistance (see Hash et al., 1999). Screening techniques have been proposed for selection for improved adaptation to drought stress on different phases of the crop cycle and for heat stress in the vegetative stage (see Yadav and Weltzien, 1999). However, apart from the work on selection for terminal drought tolerance by Bidinger et al. (2000), there are virtually no data to support breeding for either component traits or whole plant adaptation to stress. There are likely several reasons for the lack of attempts to improve adaptation. First, adaptation is poorly understood, both physiologically and genetically, and there are few proven selection procedures or selection criteria for specifically improving adaptation. Second, naturally occurring stress environments make unrewarding selection environments because of the wide degree of spatial and temporal variation they exhibit, resulting in environment and genotype × environment variances being much greater than genotype variances. Third, pearl millet already possesses a broad range of adaptive mechanisms because of its evolution in stress environments, reducing the likelihood of significant gains from further empirical selection. It has been argued that the more effective approach, for traditional millet growing areas, may be to select for higher productivity in landrace materials from stress environments, which are likely to exhibit the greatest levels of adaptive mechanisms (Yadav et al., 2000). Selection for productivity is much better understood genetically and physiologically than is adaptation, and the management of selection environments for greater productivity is considerably easier. However, it has not been demonstrated that it is possible to combine significant levels of adaptation to stress and a high yield potential. Even if it were possible to do this, a conventional breeding program attempting to do this would have to carefully balance selection environments to retain all needed traits. Future molecular approaches may be more successful in introgressing traits for either adaptation to stress or for enhanced productivity into backgrounds that contain traits for one but not the other.

5 BIOTECHNOLOGICAL TOOLS FOR PEARL MILLET IMPROVEMENT

5.1 Molecular Maps and Molecular Markers

Molecular markers have been rapidly adopted by crop improvement researchers globally as an effective and appropriate tool for basic and applied studies addressing biological components in agricultural production systems (Jones et al., 1997; Mohan et al., 1997; Prioul et al., 1997) as they offer specific advantages in assessment of genetic diversity and in trait-specific crop improvement (Edwards, 1992; Kochert, n.d.; Paterson et al., 1991). Use of markers in applied breeding programs can range from facilitating appropriate choice of parents for crosses, to mapping/tagging gene blocks associated with economically important traits [often termed quantitative trait loci (QTLs)] (Doerge et al., 1997). Gene tagging and QTL mapping in turn permit marker-assisted selection (MAS) in backcross (Frisch et al., 1999a,b; Ribaut et al., 1997, 2002) and pedigree (Mohan et al., 1997) programs, and monitoring of response to selection in population improvement programs (marker-evaluated selection, MES). Plant numbers required to obtain the desired segregants can be readily predicted (Hash et al., 2000; Sedcole, 1977; Stam and Zeven, 1981), allowing well-structured plant breeding programs to be designed. Marker-assisted selection appears to be especially useful for crop traits that are otherwise difficult or impossible to deal with by conventional means (Moreau et al., 1998). Near-isogenic products of a marker-assisted backcrossing program in turn provide genetic tools for crop physiologists and crop protection scientists to use in improving understanding of the mechanisms of tolerance to various abiotic stresses such as extremes of temperature and water and/or nutrient availability (Jones et al., 1997; Prioul et al., 1997), in addition to mechanisms of resistance to biotic production constraints such as diseases, insect pests, nematodes, and parasitic weeds like *Striga*. Quantitative trait loci mapping of yield and quality components, and the components of other physiologically or biochemically complex pathways, can provide crop breeders with a better understanding of the basis for genetic correlations between economically important traits (linkage and/or pleiotropic relationships between gene blocks controlling associated traits, (e.g., flowering time and biomass, or inflorescence size and inflorescence number). This can facilitate more efficient incremental improvement of specific individual target traits like drought tolerance and P-acquisition ability. Further, specific genomic regions associated with QTLs of large effect for one target trait (e.g., a grain yield component) can be identified that have minimal effects on traits that are otherwise highly correlated, permitting an improvement in the first trait that need not be accompanied by counterbalancing reductions in others. Finally, these molecular marker tools can also be used in ways that allow more effective discovery and exploitation of the evolutionary relationships between organisms, through comparative genomics (Devos and Gale, 1997, 2000; Devos et al., 2000; Gale and Devos, 1998a,b).

5.2 Pearl Millet Markers, Linkage Maps, Mapping Populations, and Their Use

Over the past decade or so, the International Crops Research Institute for the Semi-Arid Tropics (ICRISAT) and its partners have made substantial investments in developing mapping populations (Hash and Witcombe, 1994) and DNA-based molecular marker systems including restriction fragment length polymorphism (RFLP) (Liu et al., 1994), simple tandem repeat (STS) (Devos et al., 1995), amplified fragment length polymorphism (AFLP), and simple sequence repeat (SSR) markers (Qi et al., 2000; Allouis et al., 2001), and a bacterial artificial chromosome (BAC) library (Allouis et al., 2001) for pearl millet. These genetic tools have been used to develop a DNA-marker-based linkage map for pearl millet (Liu et al., 1994), and to map QTLs conferring resistance to biotic stresses (Jones et al., 1995, 2002; Morgan et al., 1998) and tolerance to terminal drought stress (Yadav et al., 2002). They have also been used to identify QTLs for flowering time that appear to be largely responsible for G × E interactions for grain and stover yield under favorable growing conditions (Yadav et al., 2003), as well as for diversity assessment (Bhattacharjee et al., 2002; Liu et al., 1992), studies of recombination rates (Busso et al., 1995; Liu et al., 1996), the domestication syndrome (Poncet et al., 2000, 2002), and comparative genomics (Devos et al., 1998, 2000). Levels of DNA marker polymorphism in pearl millet are very high (except in the RFLP-derived STS markers), even between elite inbred parental lines of hybrids adapted to India. The current pearl millet DNA-marker-based genetic linkage map covers about 700 cM (Haldane function) distributed across the expected seven linkage groups for this diploid ($2n = 2x = 14$) species, but telomeric regions capping the chromosomes have not yet been mapped (KM Devos, pers. comm.), and these DNA-marker-based linkage groups have not been definitively associated with the chromosome map of this species (Minocha and Sidhu, 1981; Kaul and Sidhu, 1997), which was developed over the past 35 years using morphological markers (Anand Kumar and Andrews, 1993) and conventional cytogenetic methods (Jauhar and Hanna, 1998). Compared to most other grass genomes, that of pearl millet appears to have undergone a large number of structural rearrangements (Devos et al., 2000). It seems likely that these rearrangements could have been associated with the evolution and maintenance of adaptive gene complexes that permit this highly cross-pollinated crop and its wild progenitor to thrive in environments where they are routinely subject to severe abiotic stresses (e.g., seedling heat stress, sand blasting of seedlings, drought stress at all plant growth stages, and heat stress during flowering and grain filling). These structural rearrangements continue to be common in pearl millet, although marker relationships are nearly all colinear across the 10 pearl millet mapping populations skeleton mapped to date (Liu et al., 1994, 1996; Devos et al., 2000; Azhaguvel, 2001; Kolesnikova, 2001).

Several pearl millet mapping populations of moderate size (120–275 progenies) have been developed at ICRISAT-Patancheru as sets of F_4 progeny bulks and their F_3 testcrosses derived from individual skeleton-mapped F_2 plants (Hash and Witcombe, 1994; Hash and Bramel-Cox, 2000). These now involve some 10 pairs of genetically diverse inbred lines, of Asian, African, and American origin, selected for QTL mapping of disease resistances (Jones et al., 1995, 2002), abiotic stress tolerances (Howarth et al., 1997a,b; Yadav et al., 1999, 2000, 2002), grain and stover yield and quality components (Yadav et al., 2003), and morphological markers (Azhaguvel, 2001). Several of these populations have parents of contrasting Indian and West African origin (e.g., PT 732B × P 1449-2; H 77/833-2 × PRLT 2/89-33; 841B × 863B; and W 504 × P 310-17) that are expected to differ for many traits. Parental lines of all available skeleton-mapped pearl millet mapping populations at ICRISAT-Patancheru have recently been testcrossed to a set of four genetically, cytoplasmically, and phenotypically diverse elite male sterile lines (Hash et al., 2001) to allow ready assessment of their differences for many traits, in both inbred and hybrid form. These materials are currently being evaluated for differences in grain and stover yield components under favorable growing conditions, under a range of managed drought stress regimes, and under conditions of low phosphorus availability to explore the potential for mapping QTLs for these additional traits using the existing mapping populations.

5.3 Opportunities for Marker-Assisted Breeding

For the foreseeable future, use of markers in breeding pearl millet will probably be restricted to diversity assessment to aid in choice of parental materials,
fixation of oligogenic traits in early generations of pedigree selection programs, marker-assisted backcross maintenance breeding and enhancement of
elite hybrid parental lines, and postmortem evaluation of conventional breeding program successes and failures. Most effort to date in pearl millet has been
directed at marker-assisted backcross improvement of downy mildew resistance, terminal drought tolerance, and stover yield and quality of elite hybrid
parental lines used commercially in India.

Mapping of drought tolerance QTLs in pearl millet (Yadav et al., 1999,
2000, 2002) began as a secondary target trait in a project intended to identify
QTLs for seedling thermotolerance in pearl millet (Howarth et al., 1997a).
The first pearl millet mapping population with drought tolerance as a target
trait was based on the cross of thermotolerant, drought-sensitive elite inbred
pollinator line H 77/833-2 from Haryana Agricultural University and thermosensitive, drought-tolerant breeding line PRLT 2/89-33 from ICRISAT-
Patancheru (Hash and Witcombe, 1994). Studies of this population were
followed by development and evaluation of a second pearl millet mapping
population having terminal drought tolerance as its primary target trait. In
this case, the drought-sensitive parent was ICMB 841 (Singh et al., 1990) and
the drought-tolerant parent was 863B. Both ICMB 841 and 863B were bred at
ICRISAT-Patancheru and are elite maintainer lines of hybrid seed parents
that are extensively used in India. Both PRLT 2/89-33 and 863B are derived
from the Iniadi landrace of pearl millet (Andrews and Anand Kumar, 1996).
Mapping population development was as described by Hash and Witcombe
(1994), with RFLP skeleton mapping, trait phenotyping, and QTL mapping
as described by Yadav et al. (1999, 2000, 2002). The parental lines, skeleton
maps, and skeleton-mapped progenies from these two mapping populations
were then used as starting points in a series of marker-assisted backcrossing
(MABC) programs, initiated before or after completion of QTL mapping of
the target trait (terminal drought tolerance, and its components). These
MABC programs are described in detail below.

5.3.1 Conventional MABC Programs

Conventionally, MABC programs begin only after QTL mapping has
identified the map position and closely linked flanking markers for donor
parent gene blocks that substantially contribute to target trait phenotypic
variation in the mapping population. At that point, the breeder selects one or
more genotyped (and preferably phenotyped) progenies from the mapping
population that combine(s), as a minimum, heterozygosity for donor parent

markers in the vicinity of the target QTL with homozygosity for the recurrent parent marker genotype in most of the remainder of the mapped genome. There are then two broad avenues that can be pursued (along with many paths between these). The first makes extensive use of marker genotyping in nontarget regions of the genome to reduce the number of backcrosses required to recover a desirable segregant (Hospital et al., 1992, 1997; Frisch et al., 1999a,b; Ribaut et al., 2002). The other extreme is to marker genotype only at points immediately flanking (and inside) the target region, and use serial backcrossing to more rapidly recover the recurrent parent genotype in nontarget regions of the genome. Choice between these two extremes, and/or some intermediate path, will largely be determined by the type of molecular markers available and length of the vegetative phase of the crop life cycle. For species with a long juvenile phase in which microsatellite markers (SSRs) are available, extensive use of marker genotyping would make a lot of sense; however, for pearl millet, this has not been the case.

- *Advantages*: It is less likely that any conventional MABC program that is started will have to be abandoned, because the marker poly-morphism of the donor and recurrent parents is already characterized, and the markers identified appear to be linked to substantial differences in phenotypic performance (i.e., significant QTLs of large effect have purportedly been found).
- *Disadvantages*: There is substantial lag time before the MABC program starts. Further, this program is restricted to use as its starting point the best marker genotype segregant(s) present in the original mapping population (which is largely a function of genotyped mapping population size).
- *ICRISAT experience*: In pearl millet, we have a crop with a short life cycle, and short juvenile phase that can be further reduced by artificially reducing daylength to induce early flowering. Given that RFLP markers were the only codominant marker system initially available, this led ICRISAT to initiate a program of MABC for terminal drought tolerance improvement based on two mapping progenies from the cross H 77/833-2 × PRLT 2/89-33. Both selections were homozygous for two drought tolerance QTLs from linkage group 2 (LG2) and LG4 of PRLT 2/89-33, and at least heterozygous for a third drought tolerance QTL from LG6 of H 77/833-2 (Yadav et al., 2002). F_3 plants derived from these skeleton-mapped F_2 selections were backcrossed to elite pollinator line H 77/833-2, and the resulting BC_1F_1 progenies were backcrossed again, yielding BC_2F_1 progenies segregating 1:1:1:1 for the two QTLs from PRLT 2/89-33. Individual plants from these progenies were then genotyped at three RFLP markers flanking and centered over each of the three target

drought tolerance QTLs. Work on this approach halted when it became apparent that the procedure outlined below would provide acceptable finished products earlier and at lower cost.

5.3.2 Jump-Started Marker-Assisted Backcrossing

In this case, backcrossing begins during mapping population development itself, and perhaps even before marker polymorphism of the two parents has been fully characterized. The individual F_1 plant from which the mapping population will be derived (and itself the product of a cross between the trait donor and recurrent parent) is backcrossed to the parent weakest for the target trait. Alternatively, but less reliably, selfed progeny from the individual plant of the donor parent used in creating the mapping population can be used as the trait donor in the backcrossing program. This procedure uses probability theory (Sedcole, 1977) to ensure that every possible QTL for the target trait is carried forward as rapidly as possible through the backcrossing generations. This continues until such time as markers become available, when a minimum of two markers per chromosome or linkage group arm can be used to identify segregants in which individual donor chromosome arms have been transferred into the recurrent parent genetic background. Once QTL mapping has succeeded in identifying flanking markers for QTLs of large effect, these can be used to rapidly bring the MABC program to its logical conclusion—one or more derivatives of the recurrent parent, each carrying a small homozygous segment of the donor genome consisting of a QTL for the target trait (or one of its components) and two or three flanking markers.

- *Advantages*: The major advantage of this procedure is early and rapid recovery of the recurrent parent genotype in nontarget regions. This is made possible by the early onset of the backcrossing program—even before QTL mapping, skeleton mapping, or in extreme cases even determination of parental line marker-polymorphism, have been completed.
- *Disadvantages*: The downside of this procedure is that if the F_1 used as nonrecurrent parent does not have a marker and QTL genotype identical to that mapped, all of the efforts may go waste.
- *ICRISAT experience*: We have used this procedure to transfer the drought tolerance QTL identified on LG2 of PRLT 2/89-33 (Yadav et al., 2002) to H 77/833-2. The first BC_4F_3 finished products from this MABC program, homozygous for PRLT 2/89-33 marker alleles flanking the LG2 QTL, were evaluated for hybrid performance in the drought nursery at ICRISAT-Patancheru during 2002, and several showed substantial grain yield advantage over hybrids of their recur-

rent parent H 77/833-2 under a range of terminal drought stress conditions while having hybrid performance that was at least comparable to the commercial hybrid control (HHB 67 = 843A × H 77/833-2) under fully irrigated conditions (Hash et al., unpublished).

5.3.3 Contiguous Segmental Substitution Line Sets

A logical extension of the two procedures outlined above is the development of a contiguous segment substitution line ("contig line") set in which short overlapping genomic regions from across the entire nuclear genome of one inbred line are introgressed into the nuclear genome of another inbred line.

- *Advantages*: This procedure will also permit detection of QTLs associated with smaller portions of the phenotypic variability for the target trait than can be detected by phenotyping modest-sized mapping populations. Further, it results in a small set (say 25–35) near-isogenic homozygous lines that differ from each other by pairs of overlapping introgressed segments. For QTL mapping, it will be much less expensive, and probably even more effective, to phenotype this small set of near-isogenic substitution lines (or their hybrids on a set of diverse testers) than to similarly phenotype a much larger conventional mapping population. Finally, it will be possible to use the substitution line set to map QTLs for many traits that individually would not be worth the effort. An example of this is fertility restoration for the A_1 cytoplasmic–genetic male sterility system in pearl millet, which we have mapped to LG3 while developing a contig line set of ICMP 85410 substitutions in the background of elite maintainer line 843B (Hash et al., unpublished).
- *Disadvantages*: These substitution line sets are rather expensive (in terms of both human and operational resources) and time-consuming to produce. Therefore they are probably not worthwhile unless several of the derived lines are expected to prove economically useful as improved hybrid parental lines. This in turn will generally require multiple target traits and extremely diverse parents, at least one of which is extremely elite.
- *ICRISAT experience*: We have initiated development of a contiguous segment substitution line set based on the cross ICMB 841 × 863B (863B segments being introgressed into ICMB 841 background), and plan to use it for mapping drought tolerance QTLs of small effect along with additional QTLs for downy mildew resistance, combining ability for grain and stover yield components, and ruminant nutritional quality of stem internode, leaf sheath, and leaf blade fractions of pearl millet stover.

5.3.4 Recommendations for Marker-Assisted Breeding in Pearl Millet

In pearl millet, and any other crop having a relatively short vegetative growth phase, for most cost-effective MABC transfer of a small number of QTLs of large effect, we recommend advancing to BC_2F_2/BC_3F_1 pairs by selfing and backcrossing seven random plants in each of seven BC_2F_1 progenies (each derived from a single BC_1F_1 plant having a 50% probability of carrying any given marker or QTL from the donor parent). DNA restriction digests of the 49 advanced generation segregants (BC_2F_2/BC_3F_1 pairs), the donor and recurrent parent, and the Tift 23DB standard genotype will fit on two 30-well filters (for RFLP) or a single 64-track gel (for PCR-compatible markers) along with molecular weight markers on each end. This allows $>98\%$ probability of having advanced any target QTL, located anywhere in the donor parent genome, to BC_3F_1 in the recurrent parent genetic background before spending any resources on marker-genotyping of backcross progenies. Further, once the appropriate BC_3F_1 progeny has been identified for advancement, seven plants from it can be randomly advanced to BC_4F_1, and seven plants from each of these BC_4F_1 progenies randomly advanced to BC_4F_2/BC_5F_1 pairs (by selfing and backcrossing) before the next round of marker genotyping is necessary. This should be followed by two generations of selfing, and one more cycle of marker genotyping, to produce the desired homozygous substitution lines. If target QTLs have been identified by the time the BC_3F_1 selection must be performed, it is possible to get by with just 49 BC_4F_2/BC_5F_1 pairs (and 49 BC_5F_2 plants) per target QTL. If target QTLs have not yet been identified, then the amount of marker genotyping required in later generations will be much larger, and probably not economic except for high value traits of low heritability (Hospital et al., 1997; Moreau et al., 1998) despite the potential time savings, unless development of a full or partial contiguous segment substitution line set is intended. Ribaut et al. (2002) recommend that background marker genotyping for loci on nontarget linkage groups (to more rapidly recover recurrent parent genotype in these regions) be conducted in the BC_3F_1 generation to optimize economic response to marker-assisted selection. In the scenario outlined above, this background genotyping would take place on a substantially larger set of BC_3F_2/BC_4F_1 pairs derived from a very small number of selected BC_2F_2/BC_3F_1 pairs.

5.4 Future Opportunities for Pearl Millet

5.4.1 Opportunities to Exploit Synteny for Applied Pearl Millet Improvement

Syntenic relationships between the pearl millet genome and those of rice and foxtail millet have been described (Devos et al., 2000), allowing pearl millet workers to use the cereal genome circles (Devos and Gale, 1997, 2000; Gale and Devos, 1998a,b) to more readily access and apply to pearl millet improvement the vast amount of knowledge, including QTL and genomic sequence information that has been generated on better researched cereals such as rice, maize, and sorghum. As an example, pearl millet linkage group 2 (PMLG2), on which a major QTL for terminal drought tolerance has been detected (Yadav et al., 2002), appears to be largely comprised of sequences that have evolutionarily significant relationships with the short arms of rice linkage groups 2 (R2S) and 6 (R6S), and the long arms of rice linkage groups 10 (RL10) and 3 (R3L). Parts of rice linkage groups 3 and 10 are in turn related to the long arm of maize linkage group 1 (M1L), and located on M1L, R3L, and PMLG2 are several common RFLP loci including that detected by maize RFLP probe umc107. A short distance from *umc107* maize workers have mapped teosinte branched 1 (*tb1*), a gene thought to confer a shade-sensitive control of tillering in teosinte, the ancestor of maize. Further, they have explored the genetic sequence variation for this gene in many accessions of wild teosinte and cultivated maize. It is possible that *tb1* is a counterpart of the pearl millet drought tolerance QTL, which appears to confer drought-sensitive control of basal tillering (Yadav et al., 1999, 2000, 2002). By using information on conserved portions of the *tb1* sequence from teosinte and maize to design PCR primers, it should be possible to quickly identify its counterpart in pearl millet and assess whether its sequence variants are associated with differential response to terminal drought stress in mapping population parental line pairs PRLT 2/89-33 and H 77/833-2, and ICMB 841 and 863B, as well as in their backcross introgression lines in which the PMLG2 drought tolerance QTL from PRLT 2/89-33 has been transferred to H 77/833-2 background, and that from 863B has been transferred to ICMB 841 background. In the best-case scenario, this candidate-gene approach would permit us to quickly and inexpensively identify a perfect molecular marker for the terminal drought tolerance QTL on PMLG2, i.e., the responsible gene itself. Allele-specific single nucleotide polymorphism markers could then be developed to minimize the cost of marker-assisted selection for the favorable allele at this QTL.

5.4.2   Opportunities for Basic Genomics Research Using Pearl Millet

The short crop life cycle of pearl millet, combined with its breeder-friendly reproductive behavior (simple selfing and crossing, with large seed numbers per plant and availability of diverse cytoplasmic male sterility systems), its superior tolerance to heat stress, low pH, high levels of Al^{3+} saturation, and low soil nutrient availability, its relatively small (if incomplete) genetic map, and relatively close taxonomic relationships with maize and sorghum make pearl millet a potentially powerful tool for use in basic genomic studies to improve our understanding of abiotic stress tolerance in cereals. Pearl millet producers—as well as those of cereals grown in more favorable agroecologies—could greatly benefit if the world scientific community made greater use of this orphan crop as a tool for genetic research.

ACKNOWLEDGMENTS

Preparation of this manuscript has been supported by unrestricted core funding contributions from many donors to the International Crops Research Institute for the Semi-Arid Tropics (ICRISAT), and by a series of collaborative projects funded by the United Kingdom's Department for International Development (DFID) for the benefit of developing countries. The views expressed are not necessarily those of DFID or ICRISAT.

REFERENCES

AICPMIP (All India Coordinated Pearl Millet Improvement Project). Technology for Increasing Bajra Production in India. Cachan, France: Indian Council of Agricultural Research, 1988:34.

Alagarswamy G, Bidinger FR. The influence of extended vegetative period and d_2 dwarfing gene in increasing grain number per panicle and grain yield in pearl millet. Field Crops Res 1985; 11:265–279.

Allouis A, Qi X, Lindup S, Gale MD, Devos KM. Construction of a BAC library of pearl millet, *Pennisetum glaucum*. Theor Appl Genet 2001; 102:1200–1205.

Amoukou AI. Incompatibilite post-zygotique chez le mil *Pennisetum glaucum*. In: Hamon S, ed. Le Mil en Afrique, Diversity Genetique et Agro-physiologique: Potentialites et Contraintes pour l'Amelioration et la Culture. Paris: ORSTOM, 1993:107–117.

Anand Kumar K, Andrews DJ. Genetics of qualitative traits in pearl millet: A review. Crop Sci 1993; 33:1–20.

Andrews DJ. Effects of date of sowing on photosensitive Nigerian sorghums. Exp Agric 1973; 9:337–347.

Andrews DJ, Anand Kumar K. Use of the West African pearl millet landrace Iniadi in cultivar development. Plant Genet Res Newsl 1996; 105:15–22.

Andrews DJ, Bramel-Cox PJ. Breeding cultivars for sustainable crop production in low input dryland agriculture in the tropics. In: Buxton DR, Shibles R, Forsberg RA, Blad BL, Asay KH, Paulsen GM, Wilson RF, eds. International Crop Science I. Madison WI: CSSA, 1993:221–224.

Athwal DS. Current plant breeding research with special reference to *Pennisetum*. Indian J Genet 1966; 26A(symp. no.):73–85.

Azam Ali SN. Seasonal estimates of transpiration from a millet crop using a porometer. Agric Meteorol 1983; 30:13–24.

Azam Ali SN, Gegory PJ, Monteith JL. Effect of planting density on water use and productivity of pearl millet (*Pennisetum typhoides*) growing on stored water. Exp Agric 1984; 20:203–214.

Azhaguvel P. Linkage map construction and identification of QTLs for downy mildew (*Sclerospora graminicola*) resistance in pearl millet (*Pennisetum glaucum* (L.) R. Br.). Ph.D. thesis, Agricultural College and Research Institute, Tamil Nadu Agricultural University, Madurai, Tamil Nadu, India, 2001:168.

Begg JE. The growth and development of a crop of bulrush millet (*Pennisetum typhoides* S. & H.). J Agric Sci Camb 1965; 65:341–349.

Begue A, Desprat JF, Imberon J, Baret F. Radiation use efficiency of pearl millet in the Sahelian zone. Agric For Meteorol 1991; 56:93–110.

Belliard J, Pernes J. *Pennisetum typhoides*. Halevy AH, ed. Handbook of Flowering. Volume IV. Boca Raton, FL: Chemical Rubber Company, 1984:23–37.

Bhattacharjee R, Bramel PJ, Hash CT, Kolesnikova-Allen MA, Khairwal IS. Assessment of genetic diversity within and between pearl millet landraces. Theor Appl Genet 2002; 105:666–673.

Bidinger FR, Mahalakshmi V, Rao GDP. Assessment of drought resistance in pearl millet (*Pennisetum americanum* (L.) Leeke). I. Factors affecting yields under stress. Aust J Agric Res 1987a; 38:37–48.

Bidinger FR, Mahalakshmi V, Rao GDP. Assessment of drought resistance in pearl millet (*Pennisetum ameicanum* (L.) Leeke). II. Estimation of genotype response to stress. Aust J Agric Sci 1987b; 38:49–59.

Bidinger FR, Weltzien RE, Mahalakshmi V, Singh SD, Rao KP. Evaluation of landrace topcross hybrids of pearl millet for arid zone environments. Euphytica 1994; 76:215–226.

Bidinger FR, Chandra S, Mahalakshmi V. Genetic improvement of tolerance to terminal drought stress in pearl millet (*Pennisetum glaucum* (L.) R. Br.). In: Ribaut J-M, Poland D, eds. Molecular Approaches for the Genetic Improvement of Cereals for Stable Production in Water-limited Environments. Mexico, DF: CIMMYT, 2000:59–63.

Bidinger FR, Chandra S, Raju DS. Genetic variation in grain-filling ability in dwarf pearl millet [*Pennisetum glaucum* (L.) R. Br.] restorer lines. Theor Appl Genet 2001; 102:387–391.

Bishnoi OP, Niwas R. Interception of light energy and its efficiency for dry matter production in pearl miller under rainfed conditions. Haryana Agric Univ J Res 1992; 22:152–158.

Black CR, Squire GR. Effects of atmospheric saturation deficit on the stomatal conductance of pearl millet (*Pennisetum typhoides* S. and H.) and groundnut (*Arachis hypogea*). J Exp Bot 1979; 30:935–945.

Blondel D. Contribution a la connaissance de la dynamique de l'azote mineral on sol sableux (Dior) au Senegal. Agron Trop 1971; 26:1302–1334.

Blum A, Sullivan CY. The comparative drought resistance of sorghum and millet landraces from dry and humid regions: I. Growth and water relations. Ann Bot 1986; 57:835–846.

Boddey RM, Dobereiner J. Association of *Azospirillum* and other diazotrophs with tropical gramineae. Non-Symbiotic Nitrogen Fixation and Organic Matter in the

Tropics, Symposia Papers I, Transactions of the 12th International Congress of Soil Science. New Delhi, India: Indian Agricultural Research Institute, 1982:28–49.

Bolan NS. A critical review on the role of mycorrhizal fungi in the uptake of phosphorous by plants. Plant Soil 1991; 134:189–207.

Bonamigo LA. Pearl millet crop in Brazil: implementation and development in the *Cerrado* savannahs. In: Neto AL, de F, Amabile RF, Netto DAM, Yamashita T, Gocho H, eds. Proceedings International Pearl Millet Workshop, Planaltina, Brazil, June 9–10, 1999. Planaltina, Brazil: Embrapa Cerrados, 1999:31–66.

Bouton JH. Acetylene reduction activity of pearl millet inbred lines grown in soil. Plant Soil 1988; 110:143–144.

Bramel-Cox PJ, Andrews DJ, Frey KJ. Exotic germplasm for improving growth rate and grain yield in pearl millet. Crop Sci 1986; 26:687–690.

Brunken JN, de Wet JMJ, Harlan JR. The morphology and domestication of pearl millet. Econ Bot 1977; 31:163–174.

Burton GW. Quantitative inheritance in pearl millet (*Pennisetum glaucum*). Agron J 1951; 43:409–417.

Burton GW. Cytoplasmic male-sterility in pearl millet (*Pennisetum glaucum*) (L.) Br. Agron J 1958; 50:230.

Burton GW. Registration of Gahi 3 pearl millet. Crop Sci 1977; 345–346.

Busso CS, Liu CJ, Hash CT, Witcombe JR, Devos KM, de Wet JMJ, Gale MD. Analysis of recombination rate in female and male gametogenesis in pearl millet (*Pennisetum glaucum*) using RFLP markers. Theor Appl Genet 1995; 90:242–246.

Cantini C, Pierini F, Bacci L, Maracchi G. Methodes pour la measure de la transpiration du mil a chandelle: applications a une variete photoperiodique pour l'evaluation des effets de la fertilisation et de la date du semis sur l'efficience de la transpiration. In: Bacci L, Reyniers F-N, eds. Le Futur des Céréales Photopério-dique pour une Production Durable en Afrique Tropicale Semi-Aride. Florence, Italie: Ce.S.I.A.-Accademia dei Georgofili, 1998:45–57.

Carberry P, Campbell LC. The growth and development of pearl millet as affected by photoperiod. Field Crops Res 1985; 11:207–217.

Carberry P, Campbell LC, Bidinger FR. The growth and development of pearl millet as affected by plant population. Field Crops Res 1985; 11:193–205.

Charreau C. Problemes poses par l'utilisation agricole des sols tropicaux par des cultures annuelles. Agron Trop 1972; 27:905–929.

Charreau C, Nicou R. L'amelioration du profil cultural dans les sol sableux et sablo-argileux de la zone tropicale seche ouest-africane et ses incidences agronomique. Agron Trop 1971; 26:903–978, 1183–1247.

Chase RG, Wendt JW, Hossner LR. A study of crop growth variability on sandy Sahelian soils. Soil, Crop and Water Management Systems Rainfed Agriculture in the Sudano-Sahelian Zone. Patancheru, India: ICRISAT, 1989:229–240.

Chaudhuri UN, Kanemasu ET. Growth and water use of sorghum (*Sorghum bicolor* (L.) Moench) and pearl millet (*Pennisetum americanum* (L.) Leeke). Field Crops Res 1985; 10:113–124.

Chopart J-L. Etude du systeme racinaire du mil (*Pennisetum typhoides*) dans un sol sableux du Senegal. Agron Trop 1983; 38:37–51.

Cisse L, Vachaud G. Influence d'apports de matiere organique sur la culture de mil et d'arachide sur un sol sableux du Nord-Senegal: I. Bilan de consummation, production et development racinaire. Agronomie 1988; 8:315–326.

Coaldrake P. Leaf area accumulation of pearl millet as affected by nitrogen supply. Field Crops Res 1985; 11:185–192.

Coaldrake PD, Pearson CJ. Whole plant development and dry weight accumulation of pearl millet as affected by nitrogen supply. Field Crops Res 1985a; 11:171–184.

Coaldrake PD, Pearson CJ. Panicle differentiation and spikelet number related to the size of panicle in *Pennisetum americanum*. J Exp Bot 1985b; 36:833–840.

Craufurd PQ, Bidinger FR. Effect of the duration of the vegetative phase on crop growth, development and yield in two contrasting pearl millet hybrids. J Agric Sci Camb 1988a; 110:71–79.

Craufurd PQ, Bidinger FR. Effect of the duration of the vegetative phase on shoot growth, development and yield in pearl millet (*Pennisetum americanum* (L.) Leeke). J Exp Bot 1988b; 39:124–139.

Craufurd PQ, Bidinger FR. Potential and realized yield in pearl millet (*Pennisetum americanum*) as influenced by plant population density and life-cycle duration. Field Crops Res 1989; 22:211–225.

Curtis DL. The relation between the date of heading in Nigerian sorghums and the duration of the growing season. J Appl Ecol 1968; 4:215–216.

Dancette C. Besoins en eau du mil au Senegal, adaptation en zone semi-aride tropicale. Agron Trop 1983; 38:267–280.

Devos KM, Gale MD. Comparative genetics in the grasses. Plant Mol Biol 1997; 35:3–15.

Devos KM, Gale MD. Genomic relationships, the grass model in current research. Plant Cell 2000; 12:637–646.

Devos KM, Pittaway TS, Busso CS, Gale MD, Witcombe JR, Hash CT. Molecular tools for the pearl millet nuclear genome. Int Sorghum Millets Newsl 1995; 36:64–66.

Devos KM, Wang ZM, Beales J, Sasaki T, Gale MD. Comparative genetic maps of foxtail millet (*Setaria italica*) and rice (*Oryza sativa*). Theor Appl Genet 1998; 96:63–68.

Devos KM, Pittaway TS, Reynolds A, Gale MD. Comparative mapping reveals a complex relationship between the pearl millet genome and those of foxtail millet and rice. Theor Appl Genet 2000; 100:190–198.

de Wet JMJ. Pearl millet (*Pennisetum glaucum*) in Africa and India. Proceedings of the International Pearl Millet Workshop. Patancheru, India: ICRISAT, 1987:3–4.

de Wet JMJ, Bidinger FR, Peacock JM. Pearl millet (*Pennisetum glaucum*)—a cereal of the Sahel. In: Chapman GP, ed. Desertified Grasslands, Their Biology and Management. London: Academic Press, 1992:259–267.

Do F, Daouda OS, Marini P. Etude agrophysiologique des mecanismes de resistance du mil a la secheresse. Rev Res Amelior Prod Agr Milieu Aride 1989; 1:57–74.

Do F, Winkel T, Cournoc L, Louguet D. Impact of late season drought on water relations in a sparse crop of millet (*Pennisetum glaucum* (L.) R. Br.). Field Crops Res 1996; 48:103–113.

Doerge RW, Zeng Z-B, Weir BS. Statistical issues in the search for genes affecting quantitative traits in experimental populations. Stat Sci 1997; 12:195–219.

Edwards M. Use of molecular markers in the evaluation and introgression of genetic diversity for quantitative traits. Field Crops Res 1992; 29:241–260.

Eldin M. Analyse de l'effect des deficits hydriques sur la recolte du mil au Niger, consequences agronomiques. In: Hamon S, ed. Le Mil en Afrique, Diversity Genetique et Agro-physiologique: Potentialites et Contraintes pour l'Amelioriation et la Culture. Paris: ORSTOM, 1993:149–160.

Frisch M, Bohn M, Melchinger AE. Minimum sample size and optimal positioning of flanking markers in marker-assisted backcrossing for transfer of a target gene. Crop Sci 1999a; 39:967–975.

Frisch M, Bohn M, Melchinger AE. Comparison of selection strategies for marker-assisted backcrossing of a gene. Crop Sci 1999b; 39:1295–1301.

Fussell LK, Serafini PG, Bationo A, Klaij MC. Management practices to increase yield and yield stability of pearl millet in Africa. Proceedings of the International Pearl Millet Workshop. Patancheru, India: ICRISAT, 1987:255–268.

Gale MD, Devos KM. Comparative genetics in the grasses. Proc Natl Acad Sci 1998a; 95:1971–1974.

Gale MD, Devos KM. Plant comparative genetics after 10 years. Science 1998b; 282:656–659.

Garcia-Huidobro J, Monteith JL, Squire GR. Time, temperature and germination of pearl millet (*Pennisetum typhoides* S. and H.): I. Constant temperature. J Exp Bot 1982; 33:288–296.

Gourley LM, Watson CE, Schaffert RE, Payne WA. Genetic resistance to soil chemical toxicities and deficiencies. Proceedings of the International Conference on the Genetic Improvement of Sorghum and Pearl Millet. Lincoln, NE: INTSOR-MIL, 1997:461–480.

Govila OP, Rai KN, Chopra KR, Andrews DJ, Stegmeier WD. Breeding pearl millet hybrids for developing countries: Indian experience. Proceedings of the International Conference Genetic Improvement of Sorghum and Pearl Millet. Lincoln, NE: INTSORMIL, 1997:97–118.

Gregory PJ. Responses to temperature in a stand of pearl millet (*Pennisetum typhoides* S. and H.). VIII. Root development. J Exp Bot 1986; 34:744–756.

Gregory PJ, Squire GR. Irrigation effect on roots and shoots of pearl millet (*Pennisetum typhoides*). Exp Agric 1979; 15:241–252.

Hanna WW, Wells HD, Burton GW. Dominant gene for rust resistance in pearl millet. J Heredity 1985; 76:134.

Harinarayana G, Anand Kumar A, Andrews DJ. Pearl millet in global agriculture. In: Khairwal IS, Rai KN, Andrews DJ, Harinarayana G, eds. Pearl Millet Breeding. New Delhi: Oxford and IBH, 1999:479–506.

Hash CT, Bramel-Cox PJ. Marker applications in pearl millet. In: Haussmann BIG, Geiger HH, Hess DE, Hash CT, Bramel-Cox P, eds. Training Manual for a Seminar held at IITA, Ibadan, Nigeria, 16–17 August 1999. Patancheru, India: ICRISAT, 2000:112–127. http://www.icrisat.org/text/research/grep/homepage/mol/S2_5Hash.pdf.

Hash CT, Witcombe JR. Pearl millet mapping populations at ICRISAT. In: Witcombe JR, Duncan RR, eds. Use of Molecular Markers in Sorghum and Pearl Millet Breeding for Developing Countries. London: Overseas Development Administration, 1994:69–75.

Hash CT, Singh SD, Thakur RP, Talukdar BS. Breeding for disease resistance. In: Khairwal IS, Rai KN, Andrews DJ, Harinarayana G, eds. Pearl Millet Breeding. New Delhi: Oxford and IBH, 1999:337–380.

Hash CT, Yadav RS, Cavan GP, Howarth CJ, Liu H, Qi X, Sharma A, Kolesnikova-Allen MA, Bidinger FR, Witcombe JR. Marker-assisted backcrossing to improve terminal drought tolerance in pearl millet. In: Ribaut J-M, Poland D, eds. Molecular Approaches for the Genetic Improvement of Cereals for Stable Production in Water-limited Environments. Mexico, DF: CIMMYT, 2000:114–119.

Hash CT, Abdu Rahman MD, Bhasker Raj AG, Zerbini E. Molecular markers for improving nutritional quality of crop residues for ruminants. In: Spangenberg G, ed. Molecular Breeding of Forage Crops, Developments in Plant Breeding 10. Dordrecht, Netherlands: Kluwer Academic Publishers, 2001:203–217.

Hash CT, Schaffert RE, Peacock JM. Prospects for using conventional techniques and molecular biological tools to enhance performance of "orphan" crop plants on soils low in available phosphorus. Plant Soil 2002; 245:135–146.

Henson IE. Osmotic adjustment to water stress in pearl millet in a controlled environment. J Exp Bot 1982; 33, 78–87.

Henson IE, Mahaklakshmi V. Evidence for panicle control of stomatal behavior in water-stressed plants of pearl millet. Field Crops Res 1985; 11:281–290.

Henson IE, Alagarswamy G, Bidinger FR, Mahalakshmi V. Stomatal responses of pearl millet (*Pennisetum americanum* (L.) Leeke) to leaf water status and environmental factors in the field. Plant Cell Environ 1982a; 5:65–74.

Henson IE, Mahalakshmi V, Bidinger FR, Alagarswamy G. Osmotic adjustment to water stress in pearl millet (*Pennisetum americanum* (L.) Leeke) under field conditions. Plant Cell Environ 1982b; 5:147–154.

Henson IE, Alagarswamy G, Mahalakshmi V, Bidinger FR. Stomatal response to water stress and its relationship to bulk leaf water status and osmotic adjustment in pearl millet (*Pennisetum americanum* (L.) Leeke). J Exp Bot 1983; 34:442–450.

Henson IE, Mahalakshmi V, Alagarswamy G, Bidinger FR. The effect of flowering on stomatal response to water stress in pearl millet (*Pennisetum americanum* (L.) Leeke). J Exp Bot 1984; 35:219–226.

Hospital F, Chevalet C, Mulsant P. Using markers in gene introgression breeding programs. Genetics 1992; 132:1199–1210.

Hospital F, Moreau L, Lacoudre F, Charcosset A, Gallais A. More on the efficiency of marker-assisted selection. Theor Appl Genet 1997; 95:1181–1189.

Howarth C, Cavan G, Skøt K, Yadav R, Weltzien RE, Hash T. Mapping quantitative trait loci for seedling thermotolerance and other traits in pearl millet. Proceedings of the International Conference on Genetic Improvement of Sorghum and Pearl Millet. Lincoln, NE: INTSORMIL, 1997a:650–651.

Howarth CJ, Pollock CJ, Peacock JM. Development of laboratory based methods for assessing seedling thermotolerance in pearl millet. New Phytol 1997b; 137:129–139.

Huda AKS. Simulating the growth of sorghum and pearl millet in the semi-arid tropics. Field Crops Res 1987; 15:309–325.

ICRISAT (International Crops Research Institute for the Semi-Arid Tropics) and FAO (Food and Agricultural Organization of the United Nations). The World Sorghum and Millet Economies: Facts, Trends and Outlook. Patancheru, India: ICRISAT and Rome, Italy: FAO, 1996:68.

Jauhar PP, Hanna WW. Cytogenetics and genetics of pearl millet. Adv Agron 1998; 64:1–26.

Jones ES, Liu CJ, Gale MD, Hash CT, Witcombe JR. Mapping quantitative trait loci for downy mildew resistance in pearl millet. Theor Appl Genet 1995; 91:448–456.

Jones ES, Breese WA, Liu CJ, Singh SD, Shaw DS, Witcombe JR. Mapping quantitative trait loci for downy mildew resistance in pearl millet: Field and glasshouse screens detect the same QTL. Crop Sci 2002; 42:1316–1323.

Jones N, Ougham H, Thomas H. Markers and mapping: we are all geneticists now. New Phytol 1997; 137:165–177.

Kaul J, Sidhu JS. Establishment of chromosome map of pearl millet through the use of interchanges. J Cytol Genet 1997; 32:113–123.

Klaij MC, Vachaud G. Seasonal water balance of sandy soil in Niger cropped with pearl millet, based on profile soil moisture measurements. Agric Water Manag 1992; 21:313–330.

Kochert G. Introduction to RFLP mapping and plant breeding applications. New York: The Rockefeller Foundation International Program on Rice Biotechnology, no date.

Kolesnikova M. Mapping new quantitative trait loci (QTL) for downy mildew resistance in pearl millet. PhD thesis, Russian Academy of Science, Moscow, 2001:249.

Kowal JM, Kassam AH. Agricultural Ecology of Savanna—A Study of West Africa. Oxford: Clarendon Press, 1978:403.

Kouressy M, Niangado O, Vaksmann M, Reyniers F-N. In: Bacci L, Reyniers F-N, eds. Le Futur des Céréales Photopériodique pour une Production Durable en Afrique Tropicale Semi-Aride. Florence, Italie: Ce.S.I.A.-Accademia dei Georgofili, 1998:59–75.

Krishna KR, Dart PJ, Andrews DJ. Genotype dependent variation in mycorrhizal colonization and response to inoculation of pearl millet. Plant Soil 1985; 86:113–125.

Lambert C. Influence de la precocite sur le developpement du mil (*Pennisetum typhoides* Stapf et Hubbard) en conditions naturelles. I. Elaboration de la touffe. Agron Trop 1983; 38:7–15.

Lee K-K, Wani SP, Yoneyama T, Trimurtulu N, Harikrishnan R. Associative N_2-fixation in pearl millet and sorghum: levels and responses to inoculation. Soil Sci Plant Nutr 1994; 40:477–484.

Liu C, Witcombe J, Pittaway TS, Nash M, Hash CT, Gale M. Restriction fragment length polymorphism in pearl millet, *Pennisetum glaucum*. Complexes d'Espèces, Flux de Gènes et Ressources Génétiques des Plantes. Cachan, France: Lavoisier-Technique et Documentation, 1992:233–241.

Liu CJ, Witcombe JR, Pittaway TS, Nash M, Hash CT, Busso CS, Gale MD. An

RFLP-based genetic map of pearl millet (*Pennisetum glaucum*). Theor Appl Genet 1994; 89:481–487.

Liu CJ, Devos KM, Witcombe JR, Pittaway TS, Gale MD. The effect of genome and sex on recombination rates in *Pennisetum* species. Theor Appl Genet 1996; 93:902–908.

Mahalakshmi V, Bidinger FR. Flowering response of pearl millet to water stress during panicle development. Ann Appl Biol 1985a; 106:571–578.

Mahalakshmi V, Bidinger FR. Water stress and time of floral initiation in pearl millet. J Agric Sci Camb 1985b; 105:437–445.

Mahalakshmi V, Bidinger FR. Water deficit during panicle development in pearl millet: yield compensation by tillers. J Agric Sci Camb 1986; 106:113–119.

Mahalakshmi V, Bidinger FR, Raju DS. Effect of timing of water stress on pearl millet (*Pennisetum americanum*). Field Crops Res 1987; 15:327–339.

Mahalakshmi V, Bidinger FR, Rao KP, Raju DS. Performance and stability of pearl millet topcross hybrids and their variety pollinators. Crop Sci 1992; 32:928–932.

Maiti RK, Bidinger FR. Growth and Development of the Pearl Millet Plant. Research Bulletin No. 6. Patancheru, India: ICRISAT, 1981:14.

Mangat BK, Maiti RK, Khairwal IS. Pearl millet biology. In: Khairwal IS, Rai KN, Andrews DJ, Harinarayana G, eds. Pearl Millet Breeding. New Delhi: Oxford and IBH, 1999:1–28.

McIntire J, Fussell LK. On-farm experiments with millet in Niger: crop establishment, yield loss factors and economic analysis. Exp Agric 1989; 25:217–233.

McIntyre BD, Flower DJ, Riha SJ. Temperature and soil water status effects on radiation use and growth of pearl millet in a semi-arid environment. Agric For Meteorol 1993; 66:211–227.

McPherson HG, Slatyer RO. Mechanisms regulating photosynthesis in *Pennisetum* typhoides. Aust J Biol Sci 1973; 26:329–339.

Minocha JL, Sidhu JS. Establishment of Linkage Group and Chromosome Maps in *Pennisetum typhoides*. Final Research Report of PL-480 Project (GF-In-500, A7-CR-397). Ludhiana, India: Department of Genetics, Punjab Agricultural University, 1981:110.

Mohammed HA, Clark JA, Ong CK. Genotypic differences in temperature response of tropical crops: I. Germination characteristics of groundnut (*Arachis hypogea* L.) and pearl millet (*Pennisetum typhoides* S. & H.). J Exp Bot 1988a; 39:1121–1128.

Mohammed HA, Clark JA, Ong CK. Genotypic differences in temperature response of tropical crops. II. Light interception and dry matter production of pearl millet. J Exp Bot 1988b; 39:1137–1143.

Mohan M, Nair S, Bhagwat A, Krishna TG, Yano M. Genome mapping, molecular markers and marker-assisted selection in crop improvement. Mol Breed 1997; 3:87–103.

Moreau L, Charcosset A, Hospital F, Gallais A. Marker-assisted selection efficiency in populations of finite size. Genetics 1998; 148:1353–1365.

Morgan RN, Wilson JP, Hanna WW, Ozias-Akins P. Molecular markers for rust and pyricularia leaf spot disease resistance in pearl millet. Theor Appl Genet 1998; 96:413–420.

Muchow RC. Comparative productivity of maize, sorghum and pearl millet in a semi-arid tropical environment I. Yield potential. Field Crops Res 1989; 20:191–205.

Ong CK. Response to temperature in a stand of pearl millet (*Pennisetum typhoides* S. & H.) I. Vegetative development. J Exp Bot 1983a; 34:322–336.

Ong CK. Response to temperature in a stand of pearl millet (*Pennisetum typhoides* S. & H.) II. Reproductive development. J Exp Bot 1983b; 34:337–348.

Ong CK. Response to temperature in a stand of pearl millet (*Pennisetum typhoides* S. & H.) IV. Extension of individual leaves. J Exp Bot 1983c; 34:1731–1739.

Ong CK. Response to temperature in a stand of pearl millet (*Pennisetum typhoides* S. & H.) V. Development and fate of tillers. J Exp Bot 1984; 35:83–90.

Ong CK, Monteith JL. Response of pearl millet to light and temperature. Field Crops Res 1985; 11:141–160.

Ong CK, Squire GR. Response to temperature in a stand of pearl millet (*Pennisetum typhoides* S. & H.) VII. Final number of spikelets and grains. J Exp Bot 1984; 35:233–1240.

Paterson AH, Tanksley SD, Sorrells ME. DNA markers in plant improvement. Adv Agron 1991; 46:39–90.

Payne WA. Managing yield and soil water use of pearl millet in the Sahel. Agron J 1997; 89:481–890.

Payne WA. Optimizing water use in sparse stands of pearl millet. Agron J 2000; 92:808–814.

Payne WA, Wendt CW, Lascano RJ. Root zone water balance of three low input millet fields in Niger, West Africa. Agron J 1990; 82:813–818.

Payne WA, Lascano RJ, Hossner LR, Wendt CW, Onken AB. Pearl millet growth as affected by phosphorus and water. Agron J 1991; 83:942–948.

Payne WA, Bruck H, Sattlemacher B, Shetty SVR, Renard C. Root growth and soil water extraction of three pearl millet varieties during different phenological stages. Roots and Nitrogen in Cropping Systems of the Semi-Arid Tropics. Patancheru, India: ICRISAT, 1996:251–259.

Peacock JM, Soman P, Jayachandran R, Rani AU, Howarth CJ, Thomas A. Effects of high soil surface temperature on seedling survival in pearl millet. Exp Agric 1993; 29:215–225.

Pieri CJMG. Fertility of Soils. A Future for Farming in the West African Savannah. Berlin: Springer-Verlag, 1992:349.

Poncet V, Lamy F, Devos KM, Gale MD, Sarr A, Robert T. Genetic control of domestication traits in pearl millet (*Pennisetum glaucum* L. Poaceae). Theor Appl Genet 2000; 100:147–159.

Poncet V, Martel E, Allouis S, Devos KM, Lamy F, Sarr A, Robert T. Comparative analysis of QTLs affecting domestication traits between two domesticated × wild pearl millet (*Pennisetum glaucum* L. Poaceae) crosses. Theor Appl Genet 2002; 104:965–975.

Prioul J-L, Quarrie S, Causse M, de Vienne D. Dissecting complex physiological functions through the use of molecular quantitative genetics. J Exp Bot 1997; 48:1151–1163.

Qi X, Pittaway T, Allouis S, Lindup S, Liu H, Gale M, Devos K. Development of simple sequence repeats (SSRs) markers from small and large insert libraries in pearl millet. Plant and Animal Genome VIII, 2000. http://www.intl-pag.org/8/ abstracts/ pag8875.html.

Rai KN, Virk DS. Breeding methods. In: Khairwal IS, Rai KN, Andrews DJ, Harinarayana G, eds. Pearl Millet Breeding. New Delhi: Oxford and IBH, 1999:185–212.

Rai KN, Anand Kumar K, Andrews DJ, Gupta SC, Ouendeba B. Breeding pearl millet for grain yield and stability. Proceedings of the International Conference on Genetic Improvement of Sorghum and Pearl Millet. Lincoln, NE: INTSORMIL, 1997:71–83.

Ram Niwas, Sheoran K, Sastry CVS. Radiation efficiency and its efficiency in dry biomass production of pearl millet cultivars. Ann Agric Res 1999; 20:286–291.

Ramond C. Pour une meillure connaissance de la croissance et du développement des mils *Pennisetum*. Agron Trop 1968; 23:844–863.

Rattunde HFW, Witcombe JR. Recurrent selection for increased grain yield and resistance to downy mildew in pearl millet. Plant Breed 1993; 110:63–72.

Ribaut JM, Hu X, Hoisington D, Gonzalez de Leon D. Use of STSs and SSRs as rapid and reliable preselection tools in a marker-assisted selection-backcross scheme. Plant Mol Biol Rep 1997; 15:154–162.

Ribaut J-M, Jiang C, Hoisington D. Simulation experiments in efficiencies of gene introgression by backcrossing. Crop Sci 2002; 42:557–565.

Sedcole JR. Number of plants necessary to recover a trait. Crop Sci 1977; 17:667–668.

Siband P. Nutrition minerale des plantules de mil, mais et sorgho au cours de premier jours de vegetation. Agron Trop 1979; 34:242–249.

Siband P. Essai d'analyse du fonctionnement du mil (*Pennisetum typhoides*) en zone Sahelienne. Agron Trop 1983; 38:27–36.

Singh Piara, Kanemasu ET, Singh Phool. Yield and water relations in pearl millet genotypes under irrigated and dryland conditions. Agron J 1983; 75:886–890.

Singh P, Rai KN, Witcombe JR, Andrews DJ. Population breeding methods in pearl millet improvement *Pennisetum americanum*. Agron Trop 1988; 43:185–193.

Singh SD, Singh P, Rai KN, Andrews DJ. Registration of ICMA 841 and ICMB 841 pearl millet parental lines with A1 cytoplasmic-genic male-sterility system. Crop Sci 1990; 30:1378.

Sivakumar MVK. Predicting rainy season potential from the onset of the rains in southern Sahelian and Sudanian climatic zones of West Africa. Agric For Meteorol 1988; 42:295–305.

Sivakumar MVK, Salaam SA. Effect of year and fertilizer on water use efficiency of pearl millet (*Pennisetum glaucum*) in Niger. J Agric Sci Camb 1999; 132:139–148.

Smith RL, Schank SC, Milam JR, Baltenspenger AA. Response of *Sorghum* and *Pennisetum* species to the N_2-fixing bacteria *Azospirillum brasilense*. Appl Environ Microbiol 1984; 46:1331–1336.

Smithsonian Institution Meteorological Tables, 6th edition. Washington DC: Smithsonian Institution Press, 1984:540.

Squire GR. The response of the stomata of pearl millet (*Pennisetum typhoides* S. and H.) to atmospheric humidity. J Exp Bot 1979; 118:925–933.

Squire GR. Response to temperature in a stand of pearl millet (*Pennisetum typhoides* S. & H.) IX. Expansion processes. J Exp Bot 1989; 40:1389.

Squire GR, Black CR, Ong CK. Response to saturation deficit of leaf extension in a stand of pearl millet (*Pennisetum typhoides* S. & H.) II. Dependence on leaf water status and irradiance. J Exp Bot 1983; 34:859–865.

Squire GR, Marshall B, Terry AC, Monteith JL. Response to temperature in a stand of pearl millet (*Pennisetum typhoides* S. & H.) VI. Light interception and dry matter production. J Exp Bot 1984; 36:599–610.

Squire GR, Marshall B, Ong CK. Development and growth of pearl millet (*Penniseetum typhoides*) in response to water supply and demand. Exp Agric 1986; 22:289–299.

Squire GR, Ong CK, Monteith JL. Crop growth in semi-arid environments. Proceedings of the International Pearl Millet Workshop. Patancheru, India: ICRISAT, 1987:219–231.

Stam P, Zeven AC. The theoretical proportion of the donor genome in near-isogenic lines of self-fertilizers bred by backcrossing. Euphytica 1981; 30:227–238.

Stapf O, Hubbard CE. Pennisetum. Prain D, ed. Flora of Tropical Africa 1934; Vol 9. London: Crown Agents, 1934:954–1070.

Subbarao GV, Renard C, Payne WA, Bationo A. Long term effects of tillage, phosphorous fertilization and crop rotation on pearl millet-cowpea productivity in the West African Sahel. Exp Agric 2000; 36:243–264.

Talukdar BS, Khairwal IS, Singh R. Hybrid breeding. In: Khairwal IS, Rai KN, Andrews DJ, Harinarayana G, eds. Pearl Millet Breeding. New Delhi: Oxford and IBH, 1999:269–301.

Vaksmann M, Traore SB. Adequation entre risque climatique et choix varietal du mil. In: Reyniers F-N, Netoyo L, eds. Bilan Hydrique Agricole et Secheresse en Afrique Tropicale. Paris: John Libbey Eurotext, 1996:113–123.

van Oosterom EJ, Whitaker ML, Weltzien RE. Integrating genotype by environment interaction analysis, characterization of drought patterns, and farmer preferences to identify adaptive plant traits in pearl millet. In: Cooper M, Hammer GL, eds. Plant Adaptation and Crop Improvement. Wallingford, UK: CAB International, 1996a:383–402.

van Oosterom EJ, Mahalakshmi V, Bidinger FR, Rao KP. Effect of water availability and temperature on the genotype-by-environment interaction of pearl millet in semi-arid tropical environments. Eupytica 1996b; 89:175–183.

van Oosterom EJ, Carberry PS, O'Leary GJ. Simulating growth, development and yield of tillering pearl millet I. Leaf area profiles on main shoots and tillers. Field Crops Res 2001a; 72:51–66.

van Oosterom EJ, Carberry PS, Hargreaves JNG, O'Leary GJ. Simulating growth, development and yield of tillering pearl: II. Simulation of canopy development. Field Crops Res 2001b; 72:67–91.

van Staveren JP, Stoop WA. Adaptation to toposequence land types in West Africa of different sorghum genotypes in comparison with local cultivars of sorghum, millet and maize. Field Crops Res 1985; 11:13–35.

Ventelon M, Deu M, Garsmeur O, Doligez A, Ghesquière A, Lorieux M, Rami JF,

Grivet L. A direct comparison between the genetic maps of sorghum and rice. Theor Appl Genet 2001; 102:379–386.

Virk DS. Biometrical analysis in pearl millet—a review. Crop Improv 1988; 15:1–29.

Wallace JS. The measurement and modeling of evaporation from semi-arid land. In: Sivakumar MVK, Wallace JS, Renard C, Giroux C, eds. Soil Water balance in the Sudano-Sahelian Zone. Wallingford, UK: IAHS Press, 1991:131–148.

Wallace JS, Lloyd CR, Sivakumar MVK. Measurements of soil, plant and total evaporation from millet in Niger. Agric For Meteorol 1993; 63:149–169.

Wani SP, Chandrapalaiah S, Dart PJ. Response of pearl millet cultivars to inoculation with nitrogen-fixing bacteria. Exp Agric 1985; 21:175–182.

Wani SP, Zambre MA, Lee KK. Genotypic diversity in pearl millet (*Pennisetum glaucum*) for nitrogen, phosphorous and potassium use efficiencies. In: van Buesichem ML, ed. Plant Nutrition—Physiology and Applications. Dordrecht, Netherlands: Kluwer Academic Publishers, 1990:595–601.

Winkel T, Do F. Caracteres morphologiques et physiologiques de resistance du mil (*Pennisteum glaucum* (L.) R. Br.) a la secheresse. Agron Trop 1992; 46:339–351.

Witcombe JR. Population improvement. In: Khairwal IS, Rai KN, Andrews DJ, Harinarayana G, eds. Pearl Millet Breeding. New Delhi: Oxford and IBH, 1999:213–256.

Yadav OP, Weltzien RE. Breeding for adaptation to abiotic stress. In: Khairwal IS, Rai KN, Andrews DJ, Harinarayana G, eds. Pearl Millet Breeding. New Delhi: Oxford and IBH, 1999:317–336.

Yadav OP, Weltzien-Rattunde E, Bidinger FR, Mahalakshmi V. Heterosis in landrace-based topcross hybrids of pearl millet across arid environments. Euphytica 2000; 112:285–295.

Yadav RS, Hash CT, Bidinger FR, Howarth CJ. QTL analysis and marker-assisted breeding for traits associated with drought tolerance in pearl millet. In: Ito O, O'Toole J, Hardy B, eds. Genetic Improvement of Rice for Water-Limited Environments. Los Baños, Philippines: IRRI, 1999:211–223.

Yadav RS, Hash CT, Bidinger FR, Dhanoa MS, Howarth CJ. Identification and utilisation of quantitative trait loci (QTLs) to improve terminal drought tolerance in pearl millet (*Pennisetum glaucum* (L.) R. Br.). In: Ribaut J-M, Poland D, eds. Molecular Approaches for the Genetic Improvement of Cereals for Stable Production in Water-limited Environments. Mexico, DF: CIMMYT, 2000:108–113.

Yadav RS, Hash CT, Cavan GP, Bidinger FR, Howarth CJ. Quantitative trait loci associated with traits determining grain and stover yield in pearl millet under terminal drought stress conditions. Theor Appl Genet 2002; 104:67–83.

Yadav RS, Bidinger FR, Hash CT, Yadav YP, Yadav OP, Bhatnagar SK, Howarth CJ. Mapping and characterization of QTL × E interactions for traits determining grain and stover yield in pearl millet. Theor Appl Genet 2003; 106:512–520.

6

Comparative Ecophysiology of Cowpea, Common Bean, and Peanut

Anthony E. Hall
University of California, Riverside
Riverside, California, U.S.A.

1 INTRODUCTION

The comparative ecophysiology of yield and environmental adaptation of grain legumes is discussed. Emphasis is placed on opportunities for enhancing yield potential and resistance to abiotic stresses through breeding-improved cultivars and developing management methods that complement the new cultivars. This work concentrates on biotic stresses that interact with resistance to abiotic stresses and factors influencing yield potential. Symbiotic associations involving rhizobia and mycorrhizal fungi are particularly relevant to the function of grain legumes in cropping systems and are examined.

Few comparisons have been made about the ecophysiology of different grain legume species. Such comparisons can be valuable in that what has been learned about one species may be relevant to another. Also, in the future, genetic engineering should make it possible to transfer genes and, possibly, traits between species. Cowpea (*Vigna unguiculata* L. Walp.) is emphasized and compared with common bean (*Phaseolus vulgaris* L.), peanut (*Arachis hypogaea* L.), and some other grain legumes to illustrate similarities and contrasts, and concepts that are not covered well in the cowpea literature.

Comprehensive reviews are available on breeding with cowpea (Hall et al., 1997a; Ehlers and Hall, 1997), common bean (Kelly et al., 1999; Kelly, 2000), and peanut (Knauft and Wynne, 1995); they contain some information on the ecophysiological traits conferring adaptation. Cowpea, common bean, and peanut are annual legumes which are adapted to warm conditions and are sensitive to chilling. Consequently, they are cultivated either in the tropics or during the warm season in subtropical or temperate zones. Cowpea, common bean, and peanut are mainly grown to produce dry grain. Cowpea is also grown to produce fresh southern peas, edible pods, edible leaves, hay, or forage, and as a green-manure crop. Common bean is likewise grown to produce edible pods, while peanut is often grown as a dual-purpose grain/hay crop. Cowpea and common bean are grown either as sole crops or as intercrops with other species, such as cereals. I will mainly focus on the sole-crop system because I feel this will be the dominant cropping system for grain legumes in the future. Future directions that are relevant to the breeding of all grain legumes are described at the end.

Prior to considering the phenological, morphological, and physiological traits that confer adaptation, I will provide a practical recommendation concerning a major principle that is particularly relevant to cowpea and common bean. Improved cultivars should produce the type of product that is desired by farmers, markets, and consumers. For grain legumes, consumer acceptance of the grain can be strongly influenced by the size and shape of the grain and the nature of the seed coat including the following factors: its color, how easily it takes up water during cooking, its cracking tendency, and whether it leaks pigments when cooked (Hall et al. 1997a). Selection of parents and progeny must place substantial emphasis on grain quality traits. For cowpea, this can be accomplished by making a large initial segregating generation (i.e., the F_2), and then practicing rigorous selection for grain quality traits within this and all subsequent generations. This aspect is of paramount importance because there have been too many cases where cowpea cultivars or advanced breeding lines have been developed which have outstanding adaptation and yields but which have not been adopted by farmers because of flaws in the quality of the grain (Kelly et al., 1999 discuss this issue for common bean).

2   GRAIN YIELD WITH OPTIMAL SOIL WATER AND OTHER CONDITIONS

Currently, the yield potential of grain legumes is not as high as that of cereals. This may be attributed to the inherent differences that constrain carbohydrate production in grain legumes, in comparison with cereals, such as the metabolic costs of fixing atmospheric nitrogen and producing grain with high protein or oil contents. Moreover, less research effort has been devoted to most grain legume species than to the major cereal species. In addition, evaluation of segregating lines and yield testing tend to be less efficient with grain legumes because of the need for larger plots and more hand labor and time during harvest than with small grain cereals. Currently, most cowpea and common bean cultivars cannot be combine-harvested as a standing crop due primarily to the fragility of the grain (Hall and Frate, 1996). Empirical approaches based on yield testing are an essential part of most breeding programs with annual grain crops. Selection for ecophysiological traits should be regarded as complementing and not replacing empirical approaches. If lack of research and insufficient yield testing and information on ecophysiological traits have been major constraints, substantial opportunities may still exist for enhancing the yield potential of grain legumes through the development of improved cultivars and management methods.

In optimal conditions, grain yield of grain legumes per unit ground area (Y) can be modeled by using the following equation (Hall, 2001):

$$Y = \sum_{i=d}^{i=1} \text{PFD}_i \times \text{GC}_i \times Q_i \times \text{CP}_i \tag{1}$$

where Σ is the summation of daily values of PFD $\times$ GC $\times$ Q $\times$ CP over the period of days (d) when photosynthesis significantly contributes carbohydrate to the developing grains, which often ranges from an early flowering stage to the date when most grains are physiologically mature. According to the model, grain yield depends on the flux density of photosynthetically active photons (PFD; expressed in mol[photon] m^{-2} day^{-1}), the proportion of ground covered by the crop (GC; in m^2 m^{-2}) as this determines the interception of PFD, the efficiency of conversion of intercepted photons into carbohydrate through photosynthesis (Q in g[CH$_2$O] mol[photon]$^{-1}$), and the proportion of carbohydrate that is partitioned to grain (CP in g[CH$_2$O] g [CH$_2$O]$^{-1}$) on each day. The CP is conceptually related to harvest index (HI), which is often calculated as the ratio of grain yield to total shoot biomass at harvest. The CP will have a variable metabolic energy component for cases where grain carbohydrate substantially varies in oil or protein content. For grains with high protein content, a negative interaction may occur between CP and Q because of the breakdown of photosynthetic enzymes in leaves providing the amino acids, which are translocated to grains and synthesized into proteins (Sinclair and deWit, 1975).

Achieving high grain yields requires that cultivars have a reproductive period (approximated by d) of optimal length, as determined by the length of the growing season or cropping system considerations. Cultivar and management methods should be designed to insure that ground cover is nearly complete (GC approaches 1.0) during the reproductive period. Cultivars should have maximal canopy efficiency for photosynthetic conversion of intercepted photons into carbohydrates (Q) and optimal values for partitioning of carbohydrate into grain (CP). In ideal environments, most of the carbohydrates for developing grains come from current photosynthesis, but some do come from carbohydrate that was previously stored in stems or roots. Methods for optimizing d and CP while maximizing Q and GC are discussed below.

2.1 Phenology and Reproductive Duration

Typically, the reproductive period of grain legumes consists of one or more discrete flushes encompassing overlapping periods of development by individual pods. Erect indeterminate cowpeas are able to produce two distinct flushes of pods (Gwathmey et al., 1992a). In California, the crop is harvested after producing either only the first flush of pods or after accumulating both flushes of pods (Hall and Frate, 1996). The duration of the reproductive period is influenced by temperature, as shown by studies where cowpeas were subjected to different night temperatures while growing in the same field (Nielsen and Hall, 1985a). The rate of development of individual cowpea pods was strongly influenced by night temperature. The period from anthesis to maturity of individual pods decreased linearly from 21 days with minimal night temperature of 15.5°C, typical of cool subtropical zones, to 14 days with minimal night temperature of 26.6°C, which is typical of hot tropical environments (Nielsen and Hall, 1985b). Genotypic variation during the pod development period is observed in cowpea, which is positively associated with individual seed weight (Wien and Ackah 1978). For very small seeds (50 mg per seed), the pod development period was 17 days, whereas with moderately large seeds (200 mg per seed), the pod development period was 21 days. The duration of the flush of pod production also depends on the rate at which reproductive nodes are produced and the total number of reproductive nodes that are produced. The rate of production of nodes is the same as the rate of production of leaves. Leaf appearance rate (RLA, in day^{-1}) in cowpea has exhibited a linear dependence on mean air temperature above a base temperature (T_b), with no effect of photoperiod (Craufurd et al., 1997).

$$RLA = (T - T_b)/P \qquad (2)$$

where T represents the mean air temperature; P is the phyllochron, which is constant for a genotype but varied among genotypes with a mean value of about 42°C day; while T_b varied between 9 and 12°C for different genotypes

(Craufurd et al., 1997). According to Eq. (2), a genotype with a T_b of 10°C would produce a leaf (and a node) about every 4 days when subjected to a cool mean temperature of 20°C, and about every 2 days when subjected to very hot conditions with a mean temperature of 30°C. Craufurd et al. (1997) reported that peanut and soybean (*Glycine max* L. Merr.) can have greater phyllo-chrons of about 56°C day, which means that they would produce leaves and nodes at a slower rate than cowpea.

Because the rate of production of nodes increases with increasing temperature, day-neutral cultivars exhibit fewer days to first flowering because they begin flowering at a specific nodal position on the main stem. A heat-unit model was developed to describe this effect (Ismail and Hall, 1998). Rate of development ($1/D$, the inverse of the period from sowing to flowering as expressed in days) increased linearly with average daily air temperature above a base temperature (T_b) up to a threshold level of 25–30°C, depending on the genotype (and day length), where the rate of development reached a plateau and remained constant.

$$1/D = (T - T_b)/\mathrm{HU} \tag{3}$$

where T is the mean air temperature and HU is the heat unit requirement, which can vary substantially among genotypes. The base temperature, T_b, exhibits less genotypic variation than HU. For a set of early flowering, erect, indeterminate breeding lines developed for subtropical conditions in California, HU was 734°C day and T_b was 8.5°C. With a cool mean air temperature of 20°C, these lines would begin flowering 64 days after sowing. In contrast, with a very hot mean air temperature of 30°C, the response would depend on whether the genotype is heat-sensitive and exhibits suppression of floral development under hot long-day conditions. In these hot conditions, heat-tolerant lines would begin flowering 34 days after sowing under either long or short days. Under short days, heat-sensitive lines would also exhibit rapid development in these conditions and begin flowering 34 days after sowing. In contrast, under hot long-day conditions, the rate of reproductive development of heat-sensitive lines would reach a plateau and not increase as much, such that they would begin flowering at about 44 days after sowing.

The period from first flowering to physiological maturity also decreases with hotter temperatures. Consider an early flowering, erect, indeterminate cowpea cultivar, e.g., "California Blackeye 5" (CB5), growing in a subtropical zone (Shafter, CA) and a tropical zone (Bambey, Senegal). During the summer growing season, night temperatures are much higher in Bambey (with average nightly minimum of 23°C) than in Shafter (average nightly minimum of 15°C; see Ismail and Hall, 1998 and Hall, 2001 for additional temperature data from these and other locations in the world.) This cultivar begins flowering about 50 days after sowing in Shafter and can reach physiological maturity of the first flush of pods about 100 days after sowing, resulting in a reproductive period of about 50 days. In contrast, the same cultivar begins flowering about 34 days after sowing in Bambey and reaches physiological maturity about 65 days after sowing, resulting in a reproductive period of about 31 days. Under optimal soil conditions, with careful management of pests, no significant diseases, and high levels of solar radiation, this cultivar achieved very high first-flush grain yields of about 5 ton/ha at Shafter and about 2.5 ton/ha at Bambey. Thus in sunny, cloud-free conditions in California, this cultivar has the potential ability to produce about 100 kg grain ha^{-1} per day of reproductive period. Much of the difference in yield between California and Senegal can be explained by the large difference in the length of reproductive period (50 vs. 31 days). However, the productivity per day of CB5 was slightly less in Senegal at only 81 kg grain ha^{-1} per day, but then CB5 is not well adapted to this environment. The relationship between grain yield and length of the reproductive period has been evaluated in other studies in California. Well-managed cowpeas were sown on different days and years, and showed a strong positive linear correlation between grain yield and the duration of the reproductive period as it varied due to temperature differences with a regression slope of about 84 kg grain ha^{-1} per day of reproductive period (Turk et al., 1980).

In optimal soil conditions, it is apparent that grain yield can be positively correlated with the length of the period from first flowering to physiological maturity. Cultivars of grain legumes can be bred with different durations of the reproductive period. This can be done by selecting for different numbers of days to first flowering and different plant habits. Genetic variation is present in cowpea for days to first flowering, and genotypes have been classified for this trait as it varies due to temperature and photoperiod (Patel and Hall, 1990; Ehlers and Hall, 1996). In all genotypes, temperature influences the rate of node production. In contrast, photoperiod determines the first node at which floral buds are initiated in genotypes that are sensitive to photoperiod in this way. These cowpea genotypes have a short-day response to photoperiod, and floral buds are initiated when the photoperiod becomes less than a critical day length of about 12.5 hr (Lush et al., 1980; Lush and Evans, 1980). Interaction between photoperiod and temperature can influence the rate of floral bud development in certain genotypes (Ehlers and Hall, 1996). In these cases, floral bud development is slowed down or completely suppressed in hot long-day conditions, but not in cooler long-day or short-day conditions. In general, cowpea plants that have earlier first flowering and are more erect have a shorter first-flush reproductive period than plants that have later first flowering and are more prostrate, which have a greater tendency to produce more branches and more reproductive nodes. For subtropical target production regions in the continental United States, breeders of cowpea for dry grain or fresh southern pea production have emphasized the development of erect cultivars that do not have a strong photoperiod requirement for first flowering. Erect cultivars are needed because they exhibit synchronous production of mature pods, which facilitates mechanical harvesting. Cultivars with a strong photoperiod requirement are not effective because they would not begin flowering until late September, because they must first undergo some days when the photoperiod is shorter than a critical value of about 12.5 hr. Generally, conditions become too cool after this date in the continental United States, such that pod filling by this warm-season species either ceases or is very slow. Exceptions to this are the low-elevation valleys in the southern United States, such as the Coachella Valley of California, which can remain warm through November.

Two approaches of developing cowpea cultivars for subtropical zones with greater yields due to an extended reproductive period are examined. The first approach involves managing the crop to produce two flushes of pods (Gwathmey et al., 1992a). Record grain yields for cowpea have been achieved on large plots at Shafter, CA, through this technique. As was described earlier, cowpea in California has the potential to produce about 5 ton/ha on the first flush of pods. These plants will begin producing a second flush of flowers about 15 days after the first flush has ended, and then will take about 30 days to potentially produce another 2 ton/ha (Ismail and Hall, 1998), giving a total grain yield of 7 ton/ha. In the rain-free environment of the summer in the San Joaquin Valley of California, these pods can accumulate without suffering much damage after which both flushes are harvested. There are two problems with this approach. The first is that in some soil conditions, plants die after producing the first flush of pods (this phenomenon has been observed in many fields in California and also in Bambey, Senegal; Hall et al., 1997b). The soil organism that causes premature death in cowpea is probably *Fusarium solani* f. sp. *phaseoli* and it may also be responsible for the premature death observed in soybean and common bean. A delayed-leaf-senescence (DLS) trait, which was discovered in cowpea, can overcome this problem and appears to be simply inherited and is easy to incorporate; however, it confers a modest yield penalty on the first-flush grain yield of about 400 kg/ha (Ismail et al., 2000). A possible explanation is that the delayed-leaf-senescence trait confers resistance to premature death by causing more carbohydrate to be translocated to stem bases and, presumably, roots during the first flush of podding, thereby reducing the carbohydrate available for the first-flush yield (Gwathmey et al., 1992b). In some way, roots with higher levels of carbohydrates appear to have greater resistance to fungal disease. Another problem with this approach is that it makes inefficient use of the growing season in that there is a period between the two flushes of about 15 days when the plants are not producing flowers. Eliminating this inactive period by improved management has the potential to enhance grain yields. Unfortunately, no management methods that can eliminate or reduce the length of the inactive period have, so far, been identified. The inactive period when no flowers are produced appears to be internally programmed in that it still occurs when young or older pods are continually removed during the duration of the first flush (Kwapata and Hall, 1990a; Gwathmey et al., 1992b). An alternative approach for achieving high grain yields would be to develop erect day-neutral cultivars that produce one pod flush of extended duration, which has the potential to produce a large number of pods. A hypothetical approach for doing this would involve selecting plants with a longer juvenile period that begin flowering later on a higher main-stem node. These plants would produce a larger vegetative base at first flowering, which would provide more reproductive nodes on branches, thereby supporting a single, longer and greater first flush of reproduction than current cultivars.

Cowpea breeders in tropical zones have several options for breeding cultivars with extended reproductive durations in that they can use either day-neutral or photoperiod-sensitive cultivars, and either erect or prostrate cultivars—depending on the patterns of the rainfall season and photoperiod, and degree of mechanization. Usually, prostrate cultivars are very sensitive to photoperiod, whereas erect cultivars can be either day-neutral or have varying degrees of sensitivity to photoperiod (Ehlers and Hall, 1996). With day-neutral, erect cultivars, a longer reproductive period may be achieved with a genotype that begins flowering early on a low node on the main stem by incorporating a gene causing "skipping." For most cowpea genotypes, as soon as floral buds are initiated on the main stem, subsequent nodes on the main stem are also reproductive. However, under hot short-day conditions, the main stem of certain genotypes produces a few vegetative nodes, then two to four reproductive nodes, then one or two vegetative nodes, and then more reproductive nodes—this has been called "skipping" (Ehlers and Hall, 1996). The vegetative nodes produce branches, and therefore "skipping" increases the number of reproductive nodes that occur on branches and thereby lengthens the reproductive period.

Photoperiod sensitivity may also be useful for lengthening the reproductive period in some environments. For common bean, it has been established that plasticity in days to first flowering may be adaptive under rainfed conditions in the semiarid highlands of Mexico, and that some photoperiod-sensitive cultivars have this plasticity (Acosta-Gallegos and White, 1995). In this region, common beans are sown at the onset of the summer rains, which can vary from early June to late July. The crop must be harvested before temperatures become too cool, or water becomes scarce owing to the cessation of summer rains, which often occurs in late September or early October. An adapted cultivar would have a shorter period from sowing to first flowering when sown later in this environment, and photoperiod-sensitive cultivars of common bean exhibit this type of phenological plasticity. Similar but more complex circumstances occur in the Savanna zone of West Africa, which is a major cowpea production zone (Wien and Summerfield, 1980). The sowing date is determined by the onset of the rains, which can be highly variable, whereas the end of physiological activity is determined by water limitations due to the cessation of the summer rains and is less variable. A cowpea cultivar that is well adapted to rainfed production in the Savanna zone would be plastic and begin flowering at the same date in the summer, irrespective of variation in the date of sowing. Achieving this plasticity also requires a degree of sensitivity to photoperiod in which long days either prevent the initiation of floral buds, or through an interaction with elevated night temperatures that slow down the development of floral buds during the early part of the cropping season (Ehlers and Hall, 1996). The complexity arises in that different local land races of cowpea grown by farmers in the Savanna zones of West Africa were adapted to different latitudes and photoperiods, and had different sensitivities to photoperiod (Wien and Summerfield, 1980). This enabled the different land races to begin flowering at the end of the rains in their specific regions, although they experienced different photoperiods just prior to flowering in these different locations.

For both common bean (Kornegay et al., 1993) and cowpea (Ehlers and Hall, 1996; Hall et al., 1997a), photoperiod sensitivity mainly depends on a few major genes, such that the trait can be manipulated by breeding. In common bean, there is substantial genotypic variation for the photoperiod response of flowering (White and Lang, 1989), and as for cowpea (Ehlers and Hall, 1996), this response can be influenced by temperature, which can make inheritance more complex (White et al., 1996). Most studies of environmental effects on days to first flowering in common bean did not discuss whether any suppression of floral bud development occurred as it does in cowpea under hot long-day conditions. However, the qualitative studies of Shonnard and Gepts (1994) indicate that some genotypes of common bean may exhibit high-temperature-induced suppression of floral bud development during long days. Therefore the increased photoperiod sensitivity of days to first flower observed in some common bean genotypes under hotter conditions (White et al., 1996) may be attributed to either floral bud suppression, which is induced under hot long-day conditions, as also occurs in some cowpea genotypes (Ehlers and Hall, 1996), or an interactive effect of photoperiod and temperature on floral bud initiation.

For peanut, studying the length of the period from first flowering to maturity is constrained by difficulties in defining and determining maturity for an indeterminate, nonsenescent plant whose fruits mature below the ground. Consequently, inheritance of phenological traits is poorly understood in peanut, and these traits are difficult to manipulate during breeding (Knauft and Wynne, 1995). Peanut has been considered as a photoperiod-insensitive species, but in some cases, the extent of flowering may be slightly less under long hot days than short hot days (Knauft and Wynne, 1995).

For stable production environments, an important strategy in breeding grain legumes is to develop a cultivar that has a duration from sowing to grain maturity which fits into the available growing season, and also a date of first flowering that divides the growing season into vegetative and reproductive stages that have optimal durations.

2.2 Canopy Photosynthetic Efficiency (Q)

In grain legumes and other crop species, it has proven difficult to increase the photosynthetic efficiency of the canopy in converting PFD to carbohydrate (Q) by breeding. One approach for increasing Q under optimal soil conditions is to breed plants with higher rates of photosynthetic assimilation of carbon dioxide per unit leaf area because their leaves have either more open stomata and/or greater mesophyll capacity for fixing carbon dioxide. Genotypic differences in the potential grain yield of cowpea have been positively associated with leaf stable carbon isotope discrimination (Δ), indicating that more productive genotypes had higher internal carbon dioxide concentration in their leaves (Hall et al., 1997b; Condon and Hall, 1997). A genotype with higher internal carbon dioxide concentration would have higher stomatal conductance in relation to its mesophyll capacity for fixing carbon dioxide. This positive association between grain yield and Δ has been observed in other crop species with C_3 photosynthetic metabolism (reviewed by Hall et al. 1994a; Condon and Hall, 1997). The higher Δ in more productive genotypes of cowpea, cotton (*Gossypium barbadense* L.), and wheat (*Triticum* sp.) was probably due to their having more open stomata, which could have resulted in greater rates of photosynthesis due to diffusion effects (Condon and Hall, 1997), or beneficial effects on the plant resulting from greater evaporative cooling (Lu et al., 1998), but which could also be associated with either no change or a reduction in the mesophyll capacity for photosynthesis. Selection for higher Δ would only result in an increase in photosynthetic capacity if it is accompanied by an even greater increase in stomatal conductance. An alternative explanation for the higher grain yields in cowpea, cotton, and wheat is that Δ has also been positively correlated with HI (Menéndez and Hall, 1996). The mechanism for this association is unknown but genotypes with greater HI may also have more open stomata because of feedback effects associated with their greater reproductive sink activity. The evolutionary basis of the positive genotypic association observed between biomass production and stomatal conductance with several crop species has been discussed by Condon and Hall (1997). They speculated that the evolution of these crop species favored conservative stomatal function—that is, a tendency for stomata to be at least partially closed under many circumstances. This could have arisen if plant performance during very dry years, where conservative stomatal performance would be adaptive, had disproportionate influences on seed production and long-term evolutionary success over many years owing to soil "seed banks" becoming considerably less effective after 1 year. Interestingly, F_1 hybrids of cowpea usually have strong early vegetative vigor, as well as Δ values that are higher than midparent means under well-watered field conditions (Ismail and Hall, 1993, but also see Ismail et al., 1994 for a discussion of pot rooting volume effects on Δ). The extent in which grain yield can be increased by selecting for higher Δ and/or greater stomatal conductance will be constrained when these parameters approach optimal levels.

For peanut, in contrast with cowpea and wheat, biomass production was shown to be negatively correlated with Δ because of a positive genotypic correlation between biomass production and mesophyll capacity for photosynthesis (Wright et al., 1993). For peanut, a strong positive association was observed between Δ and the ratio of leaf area/leaf dry weight, which, for selection purposes, is much easier to measure than either Δ or photosynthetic capacity (Nageswara Rao and Wright, 1994). Consequently, in peanut, there is an opportunity to select for increases in photosynthetic capacity of leaves by selecting for reductions in leaf area/leaf dry weight and this may increase Q. In several crop species, however, there has been a trade-off in that genotypes with greater photosynthetic capacity had smaller leaves (Bhagsari and Brown, 1986), but this may not be the case with the peanut genotypes used by Wright et al. (1993). However, in peanut, there is a strong negative genotypic correlation between biomass production and HI, such that genotypes with the highest biomass production usually do not produce the greatest grain yields (Wright et al., 1993).

In common bean, it is not clear whether a general relation exists between genotypic differences in potential biomass production and gas exchange traits. In some cases, genotypic differences in Δ positively correlated with stomatal conductance and biomass production (Ehleringer, 1990; White et al., 1990). However, for other sets of common beans, genotypic differences in photosynthetic rate, leaf thickness, and stomatal conductance that were positively correlated with relative growth rates measured during the first month after sowing were observed (Sexton et al., 1997). Small-seeded Mesoamerican lines tended to have greater photosynthetic rate, leaf thickness, stomatal conductance, and relative growth rate than large-seeded Andean lines. The authors concluded that the previously reported negative association of seed size with relative growth rate in common bean appears to be a function of higher photosynthetic capacity, and consequently, faster assimilation rate and faster relative growth rates for small-seeded lines (Sexton et al., 1997). However, where greater growth rates and biomass production in common bean are associated with higher Δ (White et al., 1990), this would mean that the genotypes producing more biomass would have a greater ratio of stomatal conductance to photosynthetic capacity, and are therefore unlikely to have a greater photosynthetic capacity.

Leaf canopy architecture can influence Q through the effects on the distribution of light within the canopy. Leaflets of cowpea and other grain legumes tend to be diaheliotropic under optimal soil conditions (Shackel and Hall, 1979). The leaflets track the sun, such that they are perpendicular to the direct beam of solar radiation, which results in substantial absorption of sunlight in the upper canopy and nonuniform distribution of light within the canopy. Cultivars with more uniform light distribution in the canopy may have higher Q values. Comparison between cowpea genotypes with broad leaflets and genotypes with smaller, much narrower leaflets showed no differences in light distribution within canopies or shoot dry matter production (Wien, 1982). Under drought stress, cowpea leaflets became paraheliotropic and were oriented parallel to the direct beams of solar radiation (Shackel and Hall, 1979). Canopies with paraheliotropic leaflets can have more uniform light distribution than canopies with diaheliotropic leaflets. A small number of cowpea accessions have paraheliotropic leaflets under optimal soil conditions, and this trait might be used to enhance Q and thus yield potential, providing it is not associated with other conservative stress-response traits such as partial stomatal closure. But I anticipate that the effect on Q of genotypic differences in leaf movement and orientation would not be large in cowpea. Seventy-five common bean cultivars were screened for variation in leaflet movements (Kao et al., 1994). Mesoamerican types were slightly more paraheliotropic, which could enhance Q, compared with Andean types. However, there was strong genotype $\times$ environment interaction and some inconsistency in the responses of lateral and terminal leaflets that could constrain selection for this trait in breeding.

Another approach for increasing the uniformity of light distribution would be to select for leaves that reflect more PFD due to their having either less chlorophyll content or reflective trichomes, but I anticipate that this may also only have a small effect on Q. There appears to be no advantage from selecting for high chlorophyll content per unit leaf area in that a cowpea mutant with yellowish leaves and 36% less chlorophyll content (Kirchhoff et al., 1989b) had similar quantum efficiencies (Habash et al., 1994), rates of leaf photosynthesis per unit of intercepted PFD (Kirchhoff et al., 1989c), and grain yield (Kirchhoff et al., 1989a) as its parent, which had dark green leaves. Possibly, leaves with light green appearance may have optimal levels of chlorophyll.

Breeding for some canopy traits can result in Q decreasing below maximum values. Under optimal conditions, the maximum Q value of cowpea may be about 600 mg dry matter/mol of intercepted photosynthetically active photons (Kwapata et al., 1990). This could result in crop growth rates of 300 kg dry matter ha^{-1} day^{-1} for a canopy that completely covers the ground and sunny conditions with PFD of 50 mol photon m^{-2} day^{-1}. Surprisingly, a cowpea genotype with many pods displayed above the canopy had a Q value that was only 46% of the value of a genotype whose pods were retained within the canopy but had a similar genetic background (Kwapata et al., 1990). The authors hypothesized that this effect was a result of absorption of PFD by pods, which had much lower rates of photosynthesis than leaves. Green cowpea pods can exhibit even a net loss of carbon dioxide when exposed to full sunlight (Littleton et al., 1981). Removal of young pods would be expected to reduce pod interference with PFD interception by leaves and was shown to substantially increase the Q rate of a genotype that displayed its pods above the canopy, and have little effect on the Q rate of a genotype that retained its pods within the canopy (Kwapata et al., 1990). Cowpea cultivars having canopies with pods displayed above the leaves have advantages with respect to providing fewer oviposition sites for pod borer (*Maruca testulalis* Geyer). Pod borer is a major pest of cowpea in tropical zones such as the Savannas of West Africa (Hall et al., 1997a). Under rainy conditions, pods displayed above the canopy are also less likely to be damaged by various wet and dry pod rot organisms because they dry out more rapidly after the rain ends than pods retained inside the canopy. Moreover, pods displayed above the canopy are easier to harvest by hand than pods retained within the canopy. Consequently, there is some merit in having cultivars with pods displayed above the canopy for rainfed production in wetter areas of Africa, where the crop is manually harvested. However, such cultivars have likely reduced ability to produce biomass compared with canopies in which the pods are covered by leaves. In California, cowpea is grown under irrigation in environments where it rarely rains, there are few problems with pod rots, pod borer is not present, and the crop is mechanically harvested. Consequently, cultivars for use in California should and do have canopies where pods do not project above the leaves.

2.3 Partitioning of Carbohydrate to Grain (CP) and Harvest Index (HI)

Grain legume breeders have not been as successful as cereal breeders in achieving increases in grain yield by enhancing the partitioning of carbohydrate to grain (CP) and thereby increasing HI. Earlier analysis indicated that, under optimal conditions, cultivars of cowpea can achieve Q values of about 600 mg dry matter/mol of intercepted photons, which with PFD of about

50 mol photon m^{-2} day^{-1} would result in biomass production of 300 kg ha^{-1} day^{-1}. Under optimal conditions, rates of grain production of about 100 kg grain ha^{-1} day^{-1} of the reproductive period have been achieved in sunny conditions with PFD of about 50 mol photon m^{-2} day^{-1}. This suggests that CP values were about 0.3, but this is an underestimation, in that biomass production would have decreased in the last half of the reproductive period due to leaf senescence.

Earlier studies of partitioning simply measured the ratio of grain yield to total shoot biomass at harvest (HI), which is somewhat different from, but conceptually similar to, CP. For cowpea, Kwapata and Hall (1990b) suggested that selecting bush types with greater HI could produce cultivars with potential for greater productivity under high plant densities, and that this selection could be conducted in early generations with widely spaced plants. This hypothesis concerning HI and yield potential was tested by using pairs of cowpea breeding lines with similar genetic backgrounds and either have, or do not have, heat tolerance during reproductive development. The heat-tolerance trait was shown to be associated with shorter internodes, dwarfing, and greater HI (Ismail and Hall, 1998, 1999). One of the semidwarf lines with heat tolerance, California Blackeye No. 27 (CB27), was released as a cultivar in California (Ehlers et al., 2000). Three semidwarf and three standard-height cowpea lines with similar indeterminate plant habit and phenology were evaluated at row spacings of 51, 76, and 102 cm in four productive field environments that varied in soil conditions and the extent of early vegetative vigor (Ismail and Hall, 2000). Semidwarf lines produced relatively greater first-flush grain yield than standard lines at narrower row spacings (15%, 11%, and 4% greater yield than standard lines at 51-, 76-, and 102-cm row spacing, respectively). Semidwarf lines produced greater grain yield than standard lines at narrow row spacing in soil conditions that promoted moderate to vigorous early plant growth. The smaller grain yield of the standard lines was caused by their impaired reproduction on branches compared with the semidwarf lines when competition for light was strong. Genotypic mean grain yield averaged over the two narrower row spacings (51 and 76 cm) and four environments was positively correlated with HI ($r = 0.97, n = 6$) with the slope of regression predicting an increase of 587 kg/ha in grain yield per 0.10 change in HI. Among the six genotypes, the range in first-flush grain yield was 2992–3597 kg/ha, and the range in HI was 0.41–0.51. Genotypic mean grain yield was also positively correlated with the average number of pods per peduncle on the first five reproductive nodes on the main stem ($r = 0.86, n = 6$). An even stronger positive correlation was observed between HI and the number of pods per peduncle ($r = 0.94, n = 6$). These correlations suggest that grain yield was responding to increases in HI that resulted from increases in pod setting ability. In contrast, the genotypic mean grain yield was negatively correlated with vegetative shoot biomass ($r = -0.90$, $n = 6$). The slope of regression for the combined data from 51- and 76-cm row spacing predicted that an increase of 578 kg/ha in grain yield would be accompanied by a 1000 kg/ha decrease in vegetative shoot biomass, and therefore a substantial decrease in total shoot biomass and possibly GC or Q. At the narrowest row spacing (51 cm), however, the average total shoot biomass production of the semidwarf lines over the four environments was very similar (7534 kg/ha) to that of the standard lines (7528 kg/ha). This indicates that, with dense plant spacing, GC and Q may have been similar for standard and semidwarf lines. These studies are consistent with the hypothesis of Kwapata and Hall (1990b), stated above, that selecting for high HI could produce cowpea cultivars with the potential for high grain yields under high plant densities. However, selecting for higher HI only would be effective until an optimal HI had been achieved.

Erect or semierect semidwarf grain legume cultivars with high HI may only be suitable for intensive single-flush production under high plant densities in environments with low levels of abiotic and biotic stresses, such as drought and weed competition. Dwarf determinate accessions with very high HI are also available in cowpea, but they have not been used in breeding for subtropical zones because the plants are very small and have less adaptive plasticity than the indeterminate semidwarf types. For tropical zones, neither dwarf nor semidwarf cowpeas may be effective because the higher night temperatures accelerate plant development and cause more extreme dwarfing than occurs in subtropical zones.

Empirical breeding in over 50 years for grain yield in peanut in the United States has resulted in plants with greater HI because of an earlier transition to reproductive growth, shorter main stem lengths, smaller vegetative mass, and greater CP (Wells et al., 1991). This is similar to what has occurred in breeding cowpea for intensive production in California (Ismail and Hall, 2000). Peanut cultivars with extremely high CP and HI may, however, be less tolerant to some of the stresses that damage leaves in midseason (Knauft and Wynne, 1995). Optimal levels of CP and HI may have been achieved in breeding peanut and cowpea for some intensive production environments, such that further increases in CP and HI may not result in additional increases in grain yield, except for possible effects of increases in atmospheric [CO_2] (refer to the discussion in Section 6 on future directions for plant breeding).

Achieving high grain yields probably requires that GC be close to 1.0 (100% ground cover) during the reproductive period. In many cases, this may require using narrower rows and higher plant densities than were used in the past, especially when growing erect semidwarf cultivars. For common bean, however, early canopy closure may have disadvantages in some regions, in that it can enhance the extent of infection by *Sclerotinia sclerotiorum* Lib. De Bary and thus, the extent of the white mold disease (Kelly, 2000).

An ideotype approach to breeding common beans, which was pioneered by M.W. Adams (reviewed by Kelly, 2000), bears similarities with the approach described above for cowpea and provides guidelines concerning canopy architecture. This approach initially emphasized erect determinate types but subsequently focused on erect indeterminate types when it was discovered that the grain yields of indeterminate Mesoamerican-type common beans were greater and more stable than those of determinate common beans. Small leaflets that achieve a more vertical orientation (paraheliotropic) were considered useful in that this trait would result in more uniform distribution of light in the canopy. In addition to enhancing Q, this trait might also reduce the extent of the white mold disease. Stem and root storage of starch and its complete remobilization to pods during grain filling was considered to be a desirable trait. This would enhance grain yield by enhancing CP and HI in single-flush production systems, but may enhance susceptibility to *F. solani* f. sp. *phaseoli*, which may cause premature death in some cases. A strong stem was considered desirable to reduce lodging, which is needed in both common beans and cowpeas. Some consideration was given to the number of main stem nodes and basal branches that would be optimal. However, Kelly (2000) concluded that because there is a wide diversity in grain sizes required by different market classes of common beans and production environments, it is likely that different ideotypes would be needed for these different conditions. As an example, Kelly (2000) pointed out that decumbent (less erect) common bean cultivars are needed in the semiarid highlands of Mexico. Also, spreading cowpea cultivars have some advantages in West Africa in that they exhibited more stable yields of grain and greater yields of hay than erect cultivars in environments where the rainy season is not too short. However, spreading cultivars are not as early to first flowering and do not have as short a period to maturity of the first flush as the earliest erect cowpea cultivars.

3 ADAPTATION TO TEMPERATURE EXTREMES

3.1 Chilling

In subtropical zones, warm-season grain legumes are often sown early in the spring to ensure a long growing season and permit harvest prior to cool weather or rain in the fall. With early sowing, chilling soil temperatures that reduce maximal emergence of seedlings can occur. For cowpea, the threshold temperature for some chilling damage to occur during germination (El-Kholy et al., 1997) and emergence (Ismail et al., 1997) is about 18°C. Germination studies with peanut (Mohamed et al., 1988) indicate a similar threshold temperature for germination and emergence as cowpea, but some genotypic differences were reported. Common bean is more chilling-tolerant than either cowpea or peanut, with a threshold temperature for the initiation of chilling damage during germination and emergence of about 15°C (Scully and Waines, 1987).

Genotypic differences in chilling tolerance during emergence were observed in cowpea (Ismail et al., 1997). A hypothesis was developed to explain the chilling tolerance as a result of the presence in the seed of a specific dehydrin protein with a positive nuclear effect and an independent additive effect associated with the extent of electrolyte leakage from the seed as a negative maternal effect. Near-isogenic lines were bred with and without the specific dehydrin and by using reciprocal crosses to distinguish between nuclear and cytoplasmically inherited effects associated with differences in electrolyte leakage. Studies with these lines (Ismail et al., 1999b) supported the hypothesis that the presence of the dehydrin is associated with an increment of chilling tolerance during seedling emergence that is independent of electrolyte leakage effects. The maternal electrolyte leakage effect was not shown to be cytoplasmically inherited, and may involve a nuclear trait that influences the seed coat. The dehydrin protein, which has been associated with chilling tolerance, was purified (Ismail et al., 1999a). The cDNA corresponding to the dehydrin was isolated and its sequence determined (Ismail et al., 1999b). A set of recombinant inbred cowpea lines that had been used to develop a genetic-linkage map of cowpea (Menéndez et al., 1997) was screened with gene-specific oligonucleotides derived from the cDNA and an immunoblot assay for the presence of the dehydrin (Ismail et al., 1999b). The dehydrin structural gene and dehydrin presence/absence trait mapped to the same position on the genetic-linkage map (Ismail et al., 1999b; Ouédraogo et al., 2002).

Only 3 out of 61 U.S. cultivars contain the dehydrin protein, indicating that there is substantial opportunity to enhance chilling tolerance during emergence in U.S. cultivars (Ismail and Hall, 2002). The dehydrin can be readily incorporated into cowpea by using conventional breeding by selecting based on an immunoblot assay of a chip taken from a cotyledon with the seed still being capable of germination (Ismail et al., 1999b). For other warm-season grain legumes, chilling tolerance during seedling emergence might be enhanced by incorporating a similar seed dehydrin through conventional breeding, or if a gene conferring its expression is not present in the germplasm by transgenic approaches using the cowpea gene. The specific dehydrin protein only confers an increment of chilling tolerance during seedling emergence. Obtaining more complete chilling tolerance requires incorporating genes for other aspects of chilling tolerance, such as those that confer slow electrolyte leakage under chilling conditions as an indicator of greater membrane thermostability. Cowpea accessions are available with slow electrolyte leakage from seed under chilling conditions (Ismail and Hall, 2002). Breeding for this trait can be efficient, in that instruments that can simultaneously determine electrolyte leakage from 100 individual seeds (Ismail and Hall, 2002) are available.

When warm-season species, including cowpea, common bean, and peanut, are subjected to chilling nights followed by high levels of solar radiation, photoinhibitory damage can occur to photosystem II, and other components of the photosynthetic system can be damaged by photooxidation (Hall, 2001). I am not aware of any attempts to overcome this problem by breeding with grain legumes.

3.2 Heat

High temperatures can reduce yields of grain legumes in some production environments. In breeding to enhance heat tolerance, it is important to know the extent to which grain yield is being reduced by heat in the particular target production zone and the types of plant processes and temperatures that are responsible for the yield reductions. Reviews are available that provide information on these points (Hall, 1992, 1993 and www.plantstress.com). For many grain legumes, reproductive development is particularly sensitive to heat stress.

Heat resistance is defined as the ability of a cultivar to yield more than other cultivars when subjected to heat. For cowpea, heat damage can occur in a developmental sequence influencing a series of reproductive processes that, in turn, influence different components of grain yield. A yield component model can be used to describe these effects of heat, which is effective when the size of the reproductive sink is strongly limiting yield, such that there are few negative correlations among the yield components. It should be noted that yield component models of this type may be of little value in more optimal environments, where the size of the reproductive sink does not limit the yield because of the very effective yield-component compensation that can occur in grain legumes.

$$Y = (\text{\# flowers/m}^2) \times (\text{\# pods/flower}) \times (\text{\# seeds/pod})$$

$$\times (\text{g dry weight/seed}) \tag{4}$$

where Y is the grain yield in g m^{-2}. I will begin by discussing the effects of heat on pod set (# pods/flower or pods/peduncle) because this may represent the most damaging effect of heat stress on cowpea. We had observed that grain yield and the number of pods/ground area were negatively correlated with a degree day estimate of day-time heat stress, °C above 35°C during each day summed for the 30 days after the first appearance of floral buds (Turk et al., 1980). We then determined the extent that day or night, or root, or shoot temperatures were causing the effect by using growth chambers (Warrag and Hall 1984a,b). We discovered that the pod set can be reduced to zero by high night shoot temperature (30°C). This effect was surprising in that hotter day shoot temperatures (33°C) did not damage pod set. Additional growth chamber studies (Warrag and Hall, 1983) demonstrated that pod set can be substantially reduced by the combination of moderately hot nights (27°C) and very hot days (36°C). Reciprocal artificial pollination between plants grown under optimal and high night temperatures indicated that the low pod set was caused by male sterility in that the pistils did not appear to be damaged by high night temperature. The detrimental effects of high night temperature on pod set were also shown to occur in field conditions (Nielsen and Hall, 1985b). In these experiments, a unique approach was used in which field plots of cowpea plants were subjected to different increments of higher night temperatures during early stages of flowering by using enclosure systems placed over the plots only during the nighttime (Nielsen and Hall, 1985a). In these studies, pod set and grain yield exhibited similar linear decreases for minimum daily nighttime temperature increasing from 15°C, with 50% reductions occurring at 27°C. These results suggested that heat stress may be causing substantial reductions in grain yield in the warmer target production zones where cowpea are grown (Nielsen and Hall, 1985a).

Possible mechanisms for the sensitivity of pod set and pollen development to high night temperatures have been proposed. Limitations in carbohydrate supplies were suggested as being responsible for reproductive failure, but this does not appear to account for the low pod set induced by high night temperature in cowpea. Damage to pod set caused by high night temperature is greater for plants in long days when they would be expected to have greater supplies of photosynthate compared with short days (Mutters et al., 1989b). Also, elevated atmospheric carbon dioxide concentration resulted in higher overall carbohydrate levels in plants but did not increase their heat tolerance with respect to either flower production or pod set (Ahmed et al., 1993a). In field conditions, high temperatures have greater effects on the reproductive development and grain yield of cowpea than they do on shoot biomass production (Ismail and Hall, 1998), indicating that effects of heat on photosynthesis do not constitute the major factor behind the reductions in grain yield. It appears that high night temperature may have a direct damaging effect on reproductive development. Mutters and Hall (1992) demonstrated that there is a distinct period during the 24-hr cycle when pollen development in cowpeas is sensitive to high night temperatures. Plants subjected to high night temperature during the last 6 hr of a 12-hr night exhibited substantial decreases in pollen viability and pod set, whereas plant subjected to high temperature during the first 6 hr of the night exhibited no damage. Mutters and Hall (1992) hypothesized that these results could be explained if a heat-sensitive process in pollen development is under circadian control and only occurs in the late night period. The damaging effect of high night temperature on pod set was greater in long days than in short days, and red and far-red light treatments indicated that it is a phytochrome-mediated response (Mutters et al., 1989b). The phytochrome effects also involve circadian rhythms.

Processes influencing pod set were shown to be most sensitive to high night temperature 9–7 days before anthesis by experiments in which plants were transferred between growth chambers with either optimal or high night temperatures (Warrag and Hall, 1984b; Ahmed et al., 1992). This sensitive period takes place after meiosis and coincides with the release of pollen microspores from tetrads (Ahmed et al., 1992; Mutters and Hall, 1992). Damage due to high night temperature was associated with the premature degeneration of the tapetal layer that provides nutrients to developing pollen. The transfer of proline from the tapetal layer to pollen was inhibited by high temperatures (Mutters et al., 1989a). Healthy pollen contains large quantities of proline. Just prior to anthesis, proline constitutes 55% of the amino acids in anthers and pollen. In very hot field conditions, proline content of pollen in heat-sensitive cowpea genotypes was only half of the content in more optimal field conditions (Mutters et al., 1989a). Proline has been hypothesized to play a role in protecting pollen from heat-induced damage during germination (Hong-Qui and Croes, 1983). Heat-induced damage to pod set of cowpea has been associated with high levels of infertile pollen, but for some genotypes, anthers also may not dehisce (Mutters and Hall, 1992; Ahmed et al., 1992).

Another component of the heat-tolerance model for yield [Eq. (4)] can be strongly affected by heat. Early floral bud development can be damaged, such that sensitive plants produce few or no flowers. Two weeks or more of consecutive or interrupted hot nights during the first month after germination can cause complete suppression of floral bud development (Ahmed and Hall, 1993). In both growth chamber and field conditions, the damage can be much more pronounced under long days than short days (Dow El-Madina and Hall, 1986; Patel and Hall, 1990). However, responses to red and far-red light indicated that the effect was only partially consistent with the system being mediated by phytochrome (Mutters et al., 1989b). The damaging effect of high night temperature and long days on floral bud development also depends on light quality (Ahmed et al., 1993b). When growth chambers were used with relatively large amounts of fluorescent light and little incandescent light, such that the red/far-red ratio was high (the photon flux density at 660 ± 5

nm/photon flux density at 730 ± 5 nm was 1.9), floral buds were not suppressed in long-day, high night temperature conditions (but pod set was very low). In contrast, when growth chambers were used with a lighting system that had a lower red/far-red ratio of 1.2, which is similar to sunlight, floral buds were suppressed in long days with high night temperature and no flowers were produced, i.e., the same as occurs in field conditions. There are two implications of the results of these studies. First, the use of growth chambers with lighting systems that mainly depend on fluorescent lights can result in serious artifacts when studying plant reproductive responses to heat stress. Second, in field conditions, intense shading of floral buds can reduce the red/far-red ratio far below 1.2 and may intensify the floral bud suppression effect. In densely sown fields of cowpea, individual plants that are suffering from competition are tall and spindly and can exhibit floral bud suppression, although night temperatures are not too hot.

Another component of the heat-tolerance model for yield [Eq. (4)] can be strongly affected by heat and other stresses. Pods of different cowpea genotypes produce 9–20 ovules with many cultivars having 15, but they rarely produce this many seeds per pod. Under optimal conditions, two-thirds of the ovules may produce seed, whereas with high day or high night temperatures (Warrag and Hall, 1983) and other stresses, such as drought (Turk et al., 1980), fewer seed are produced per pod. For most cowpea cultivars and stresses, it is the ovules at the blossom end of the pod, which are furthest from the carbohydrate supply, that suffer embryo abortion and do not produce seed, resulting in the production of "pinched" pods. The acceleration of reproductive development caused by high night temperature (Nielsen and Hall, 1985b) would enhance the demand by embryos for carbohydrate and where the demand is not met by the supply, it could lead to increased embryo abortion. Drought could have a similar effect but by reducing photosynthesis and the supply of carbohydrate to developing embryos.

Temperature can influence seed quality. Cowpea seeds produced under high day temperatures can have asymmetrically twisted cotyledons (Warrag and Hall, 1984a). Germination of the seed is not influenced and this effect of heat stress may not be a major problem. In contrast, heat-induced brown discoloration of cowpea seed coats can occur with some cultivars and be a major problem in that consumers reject the grain. Higher night temperatures resulted in a progressively larger numbers of seed of accession TVu 4552 with larger areas of brown discoloration of their seed coats (Nielsen and Hall, 1985b). The important breeding line TVx 3236 has also exhibited brown discoloration of seed coats as it grows during the main cropping season in Senegal. The accelerated pod development associated with higher night temperatures is associated with the production of smaller seed (Nielsen and Hall, 1985b) influencing yield as indicated in Eq. (4).

The traditional method for breeding to enhance adaptation to hot conditions is to grow advanced lines in a hot target production environment and select lines that have greater grain yields than current cultivars (by definition, such lines would have greater heat resistance). This approach is not very effective with crops such as cowpea and common bean, where yield evaluation requires considerable land and labor. Also, the presence of other stresses can confound the evaluation of heat stress effects. For example, insect pests such as lygus bugs (*Lygus hesperus* Knight) and flower thrips (*Megalurothrips sjostedti* Trybom) can cause damage to developing flower buds of cowpea, an effect which appears similar to that caused by high night temperatures. Irrespective of these problems, some slow progress may have been made in enhancing heat resistance of cowpea in West Africa by breeding programs that made selections based on grain yield in hot target production environments (Ehlers and Hall, 1998).

Efficient approaches to breeding for heat resistance involving early generation selection for specific traits that confer heat tolerance have been developed. The first step in this approach is to discover accessions with heat-tolerance traits. We have screened cowpea accessions in field environments with very high night and day temperatures for heat tolerance during reproductive development. Our current most effective field nursery for screening for heat tolerance is located in the Coachella Valley of California, with sowing in mid-June. Plants are provided with optimal irrigation, fertilizer, and pest management practices. For the 3-week period beginning 1 week prior to the start of flowering, minimum daily (night) air temperatures are 23–27°C, and maximum daily air temperatures are 42–50°C. This is one of the hottest crop production environments on earth. The plants also experience long days (14.5 hr) and sunny skies. We screened hundreds of cowpea accessions and only three, accounting for less than 1%, showed the ability to abundantly produce flowers and set pods in this very hot environment. Two of the heat-tolerant accessions, TVu4552 and Prima, came from hot environments in West Africa (Warrag and Hall, 1983), while one, MN13, was bred for the very cool conditions of Minnesota (Davis et al., 1986) and was presumably selected in nurseries where plants were not exposed to heat. Some chilling-tolerant snap beans also have heat tolerance during reproductive development (Dickson, 1993). However, we suspect that, for cowpea line MN13, the association is between heat tolerance and extreme earliness (Ehlers and Hall, 1996), and not between heat tolerance and chilling tolerance. Screening for heat tolerance was conducted by the International Institute of Tropical Agriculture (IITA) at Kano in northern Nigeria, using screenhouses to reduce attacks by insect pests. With sowing set in March, the plants experience minimum daily (night) air temperatures between 24 and 27°C, maximum daily air temperatures between 38 and 42°C, and short-day conditions during the early flowering stage. In this hot short-day environment, lines TVu4552, IT88D-641-1, IT88D-867-11, and IT97K-472-12 produced many pods, while many other cowpea lines showed infertile pollen and little or no pod set (B.B. Singh, personal communication, March, 2001).

Most cowpea breeding programs do not have field nurseries or screen-houses with consistently high night and day temperatures, but have otherwise optimal growing conditions. Consequently, we have developed a glasshouse environment for screening cowpea to detect differences in reproductive-stage heat tolerance. The plants are subjected to minimum and maximum 24-hr air temperatures of 27 and 36°C, respectively, and sunny long-day conditions. This glasshouse environment is very effective for screening cowpea for reproductive-stage heat tolerance, but only a few hundred plants can be grown in it, compared with the thousands of plants that can be screened in field nurseries. One advantage of using a hot glasshouse compared with most field conditions is that air temperatures are relatively stable overtime, and genotypes that begin flowering at various dates can be reliably screened. In contrast, the Coachella Valley field nursery is mainly effective for screening genotypes that begin flowering at about the same time because temperatures vary substantially from day to day and exhibit seasonal changes.

In breeding for heat tolerance, we grow plants from the first segregating generation in a heat-screening nursery, and select single plants that are able to produce many flowers. We have shown that heat tolerance at early flowering is consistent with it being conferred by a single recessive gene with high heritability (Hall, 1993). Selection in the first segregating generation fixes the ability to produce flowers in most, but not all, selected plants. We also select for seed quality traits in the first segregating generation and all subsequent generations, and we feel it is effective. Genetic studies demonstrated that heat-induced seed coat browning, which is an undesirable trait, is consistent with the effect of a single dominant gene that is not linked to the gene conferring heat tolerance at early flowering (Patel and Hall, 1988). Consequently, both the absence of seed coat browning and the ability to produce flowers can be usually fixed via selection during the first segregating generation. In addition, we also selected single plants with high numbers of pods per peduncle (3–4) in the first segregating generation, but selection for this trait is not very effective at this stage with single plants. Heat tolerance during pod set was shown to be consistent with the effect of a single dominant gene, but with strong environmental effects and low, narrow-sense and realized heritabilities of 0.26 (Marfo and Hall, 1992). We also suspect that, with some crosses, at least two major genes and some minor genes are involved in conferring the ability to set pods under hot conditions. Consequently, we conduct family selection for high numbers of pods per peduncle on advanced lines in several subsequent generations. In all generations, we practice some selection for well-filled pods as an indicator of less embryo abortion, but this selection may not be very effective. Embryo abortion is a complex factor that is influenced by many stresses, plant pod load, and age. Negative correlations among yield components can occur for this trait. For example, in some environments, there is a tendency for plants that set very many pods to have poorly filled pods, so it is necessary to select for both high pod set and pods that are not too pinched. Two cowpea accessions, TN88-63 and B89-600, exhibited no heat-induced reductions in the number of seeds per pod, even when they had a substantial pod load (Ehlers and Hall, 1998). These genotypes may provide an opportunity for breeding to enhance heat tolerance during embryo development. It should be noted that negative correlations among yield components is not as big a problem for cowpea under high night temperatures as it would be under more optimal conditions. Heat stress tends to reduce reproductive development of cowpea more than biomass production (Ismail and Hall, 1998), so that the plants are strongly limited by the size of the reproductive sink and thus exhibit few negative correlations among yield components in very hot conditions. Advanced lines that have been selected in the heat-screening and other nurseries are then evaluated for yield and other agronomic traits in multi-location trials conducted in experiment stations and commercial fields in the target production zone.

We have demonstrated that these methods can be effective in incorporating heat tolerance for subtropical environments with hot long-day conditions, such that heat resistance is enhanced. We did this by breeding six pairs of lines that either have, or do not have, a set of heat-tolerance genes in similar genetic backgrounds. These pairs of lines were evaluated in eight subtropical field environments with average night temperatures ranging from being cool to being very hot, but with other conditions being similar and near optimal (Ismail and Hall, 1998). The heat-sensitive lines, which included a commercial cultivar, CB5, exhibited a 13.5% decrease in grain yield per °C increase in average minimum night temperature above 16.5°C for the 3-week period starting 1 week prior to first flowering. The heat-tolerant lines had 50% greater grain yield and numbers of pods per peduncle than the heat-sensitive lines with average minimum night temperatures of 21°C, but similar grain yields as the heat-sensitive lines under cool night temperatures (Ismail and Hall, 1998; also see www.plantstress.com). Minimum night temperatures exceeding 21°C occur in several commercial production zones (Nielsen and Hall, 1985a). One of the heat-tolerant lines, CB27, was released as a cultivar in California (Ehlers et al., 2000). It should be noted that heat tolerance, by itself, will not justify the release of a new cultivar; the cultivar must have greater grain yield than current cultivars when grown in the target production environment (i.e., greater heat resistance is needed). In addition to greater grain yields than current cultivars when conditions are hot at flowering,

CB27 has also greater grain yields in some fields because of its resistance to a broader range of biotypes of root-knot nematodes and Fusarium wilt than previous California cultivars.

When breeding to incorporate heat tolerance or any other trait, it is important to evaluate the potential negative effects of the trait. In hot environments, the reproductive-stage heat tolerance genes caused cowpea to be more compact and dwarfed because their internodes are shorter. At a minimum night temperature of 18°C, the heat-sensitive cowpea lines had 50% longer main stems, and at 22°C, they had 50% more vegetative biomass than the heat-tolerant lines (Ismail and Hall, 1998; photographs of these plants are presented in www.plantstress.com). The performance of heat-tolerant semi-dwarf cowpea lines was compared with that of standard-height cowpea lines under different row spacing (Ismail and Hall, 2000). The heat-tolerant semi-dwarf lines were less effective than the standard-height lines at the wide row spacing of 102 cm used by some farmers, more effective with the widely used 76-cm row spacing, and even more effective with a narrow row spacing of 51 cm. Natural selection likely would not favor this type of heat tolerance in that the compact plant habit is not very competitive, such that careful weed management is needed.

Our approach for breeding heat-resistant cowpeas is effective in hot subtropical zones, but it is still not clear whether the approach and the genes will be effective in hot tropical zones. The six pairs of heat-tolerant and heat-sensitive cowpea lines used in the studies in California were evaluated in three experiments in northern Ghana and three experiments in the Peanut Basin of Senegal (Hall et al., 2002). The average minimum daily (night) air temperatures for the 3-week period beginning 1 week prior to the start of flowering ranged from 21 to 26°C in these six experiments. However, in all of the experiments, there were no differences in grain yield between the heat-tolerant and the heat-sensitive lines, and the average grain yields in the different experiments were small, ranging from 562 to 1866 kg/ha with an overall average of 1185 kg/ha. The low yields are partially attributed to the susceptibility of the California lines to both wet and dry pod rots under humid tropical conditions. Controlled environment studies provided some clues, which can be used to try to explain these results. In some cases in tropical zones, the day length may not be long enough to trigger heat × photoperiod induced damage to floral bud development (the critical effective day length, defined as the period from sunrise to sunset plus any twilight effects, is estimated to be about 12.5 hr). With effective day length of less than 12.5 hr, heat may have little damaging effect on floral bud development and the number of flowers produced (Dow El-Madina and Hall, 1986; Patel and Hall, 1990), and only have a partial damaging effect on pod set (Mutters et al., 1989b). Consequently, heat-tolerance genes may be expected to have less beneficial effect on grain yield in hot tropical environments with short days than in hot subtropical environments with long days. In developing the heat-tolerant lines used in the experiments in California and West Africa, accessions Prima and Tvu4552 were used as sources of heat tolerance because they are extremely tolerant in long days. These accessions were also shown to exhibit heat tolerance for pod set under short days. In controlled environment chambers under hot 12-hr day conditions, Prima was shown to have higher grain yield per plant than IT84S-2246 because it has better ability to maintain peduncle and flower production and greater pod set (Craufurd et al., 1998). In hot short-day conditions in screenhouses in West Africa, TVu4552 exhibited much higher pod set and grain production than many other cowpea accessions and lines (B.B. Singh, personal communication, March 2001). Advanced breeding lines were developed by using Prima and TVu4552 as sources of heat tolerance, and these breeding lines were evaluated in hot glasshouses under both short-day and long-day conditions (Ehlers and Hall, 1998). The heat-tolerant lines had high grain yields under both short-day and long-day conditions, whereas heat-sensitive lines that gave zero grain yield under hot long-day conditions exhibited variable yields under hot short-day conditions ranging from being 20% to 100% of those of the heat-tolerant lines (Ehlers and Hall, 1998; Hall et al., 2002). In the hot short-day glasshouse, grain yields were positively correlated with the number of pods per peduncle and pods produced per day, and excessive suppression of floral buds was not detected (Ehlers and Hall, 1998). This suggests that enhanced pod set was a major factor in the heat tolerance exhibited by some genotypes under hot short-day conditions. My overall conclusions are that the genes that confer heat tolerance during pod set in the lines bred in California could enhance pod set in hot tropical conditions, but that they need to be combined with additional genes that confer local adaptation so that the pods become filled, and the effect on grain yield would not be as large as has been achieved in hot subtropical zones.

Common bean exhibited heat-induced damage to reproduction in hot subtropical (long-day) field conditions that bear some similarities to what has been observed in cowpea. Heat-sensitive common bean genotypes exhibited substantial flower bud abortion and low pod filling (Shonnard and Gepts, 1994), presumably due to ineffective fertilization (Gross and Kigel, 1994). Sensitivity to heat during early floral bud development may only occur in long days for common bean and may be consistent with the action of a single dominant gene (White et al., 1996) as it is in cowpea (Hall, 1993). However, in the study of White et al. (1996), it was not reported whether the genotypic differences in days to first flowering resulted from differences in time of floral bud initiation or differences in extent of floral bud suppression. In common bean, elevated night temperature was more detrimental to reproductive development than high day temperature (Konsens et al., 1991) as it is in cowpea (Warrag and Hall 1984a,b). High night temperatures caused heat-induced reductions in number of seeds per pod and number of pods per plant, while even higher day temperatures had little effect, providing night temperatures were optimal (Konsens et al., 1991). This is similar to what was observed in cowpea, except that in cowpea, the number of seeds per pod can be reduced by both high night and high day temperature (Warrag and Hall, 1983; Ehlers and Hall, 1998). A major difference between the responses to heat is that, for most cowpea genotypes, it is the ovules at the blossom end of the pod that fail to produce seed, and the effect probably involves embryo abortion. In contrast, for common bean, it is the ovules at the peduncle end of the pod that fail to produce seed (Dickson and Petzoldt, 1989), and the effect probably involves impaired pollen tube growth and fertilization (Gross and Kigel, 1994).

Detailed studies have been conducted in which common beans were subjected to high night temperature (27°C), moderate day temperature (32°C), and long photoperiods (16 hr) for short durations (1 or 5 days) during different stages of reproductive development (Gross and Kigel, 1994). The most heat-sensitive stage was for pollen development occurring just after meiosis, as it was for cowpea (Warrag and Hall, 1984b; Ahmed et al., 1992). Pollen developed under heat stress were not viable, and anthers did not dehisce. As for cowpea, heat at this stage did not affect the gynoecium function. Similar detrimental effects on pollen and anther development of common bean were likewise observed under hot short-day (12 hr) conditions (Porch and Jahn, 2001). Gross and Kigel (1994) observed reductions in ovule fertilization and seed set that were greater at positions further away from the stigma and probably reflected heat stress constraints to pollen tube growth. In contrast with the results for cowpea, Gross and Kigel (1994) detected sensitivity of the gynoecium to heat occurring at anthesis, which contributed to reductions in the number of seeds per pod.

The ability of plants to acclimate to heat may be an important aspect of genotypic differences in resistance to heat if the production environment only experiences occasional periods of hot weather. Studies of high-temperature acclimation in common bean emphasized slow electrolyte leakage as a measure of membrane thermostability and heat tolerance (Li et al., 1991). For 74 common bean lines, genotypic differences were detected in the ability to acclimate to heat (24 hr at 37°C day and night air temperature) and exhibit longer periods (in min) before extremely high temperatures (50°C) caused 50% ion leakage from tissues (Li et al., 1991). Because neither the acclimation thermal regime nor the heat-stress temperature would be experienced by common bean in any reasonable target production environment, it is important to determine whether these differences in heat tolerance have any relevance to heat resistance (cultivar ability to produce dry grain or fresh pods in hot environments). For a small subset of the cultivars investigated (13), Li et al. (1991) reported a positive correlation between the number of pods produced 2 weeks after the heat stress (1 week at 37/35°C day/night air temperature), as a percentage of controls, and a measure of heat acclimation potential based on electrolyte leakage from leaf disks ($r = 0.78, n = 13$). This indicates a possible association between heat tolerance and heat acclimation potential, but it does not provide information on the heat resistance of these lines under field conditions. Others have noted (e.g., Dickson, 1993) that some common bean genotypes reported to have high heat acclimation potential with respect to electrolyte leakage can be highly sensitive to heat, in terms of reproductive development, and produce low pod yields when subjected to heat just prior to bloom. Prior to recommending the use of heat acclimation potential as a selection criterion for use in common bean breeding programs, it would appear necessary to conduct genetic experiments to determine the heritability of this trait and its association with grain or pod yield of plants subjected to heat in field environments. I am not aware of any reports of genetic studies on common bean dealing with heat acclimation potential as determined by electrolyte leakage. For cowpea, an association between reproductive-stage heat tolerance and slow electrolyte leakage from leaves has been reported (Ismail and Hall, 1999). In an extremely hot field environment, negative correlations were observed between grain yield and electrolyte leakage ($r = -0.79, n = 9$), and pod set and electrolyte leakage ($r = -0.89, n = 9$) among nine cowpea breeding lines. Genetic selection studies with cowpea, however, indicated that the realized heritability of leaf electrolyte leakage was low and associations with pod set and grain yield under hot conditions were only moderate (Thiaw, 2003).

Heat-acclimation effects may be related to the fact that seedlings subjected to high temperatures synthesize a novel set of proteins that have been called heat-shock proteins, and the plants become more tolerant, in terms of plant survival, to more extreme temperatures (Vierling, 1991). These proteins are thought to enable cells to survive the harmful effects of heat by two general types of mechanisms: as molecular chaperones, and by targeting proteins for degradation. As an example of chaperone activity, it has been shown that a specific small heat-shock protein cooperates with other heat-shock proteins to reactivate a heat-denatured protein (Lee and Vierling, 2000). Heat shock proteins do not appear to be the only mechanism whereby plants differ in heat tolerance. Heat shock protein profiles were examined in six common bean cultivars that differ in heat acclimation potential with respect to differences in heat killing time based on electrolyte leakage (Li and Udomprasert, 1993). No relationship was observed between the patterns of heat-shock proteins and heat acclimation potential. Also, genotypes of cowpea have been bred that have substantial differences in heat tolerance during reproductive development, but they produced the same set of low molecular weight heat shock proteins in their leaves when subjected to moderately high temperatures (A.S. El-Kholy, unpublished studies, 1996).

Progress has been made in breeding snap bean types of common beans with heat resistance through incorporating heat tolerance during reproductive development by selecting plants for high pod set in a glasshouse with high night temperatures (Dickson, 1993). With day/night temperatures of 35/ 27°C during the early floral bud and bloom period, narrow-sense heritabilities for pod production were only 0.11–0.30, but selection in these conditions over two generations produced lines with uniformly high pod production (Dickson, 1993). Some heat-resistant dry bean types of common bean cultivars have been developed (e.g., Beaver et al., 1999). For common bean, the detrimental effects of some diseases may be greater in some hot environments, so increased genetic resistance to these diseases may also be needed.

Peanut exhibits geocarpy in which flowers are produced above ground, and after fertilization pegs are produced that penetrate the soil, such that embryo development and pod formation occur in the soil. High day and night air temperatures occurring just prior to anthesis were shown to reduce pollen production and viability, and peg production (Vara Prasad et al., 1999a,b). As for cowpea (Mutters and Hall, 1992), peanut exhibited circadian variation in sensitivity to heat (Vara Prasad et al., 2000b). High air temperatures during the morning reduced fruit set, whereas high air temperatures during the afternoon had no effect on fruit set. Peanut flowers typically open early in the morning, self-pollination occurs just before opening, and fertilization is completed within 5–6 hr. The authors concluded that fruit number in peanut would be reduced when mean day air temperatures exceed 28°C (Vara Prasad et al., 2000b). High soil temperature reduced flower production, the proportion of pegs forming pods and individual seed weight, whereas high air temperature mainly reduced the proportion of flowers that set pegs (Vara Prasad et al., 2000a). Effects on pod yield of high soil and high air temperatures were additive.

In summary, pollen development and fertilization processes have been shown to be particularly sensitive to high temperatures in cowpea, common bean, and peanut, and to be mainly responsible for reductions in grain yield caused by high air temperatures. High temperatures occurring a few days prior to anthesis in either the late night (cowpea and common bean) or morning (peanut) appear to be most damaging. Heat-resistant cultivars of cowpea and common bean have been developed by selecting for high pod and seed set with plants in greenhouses subjected to high night air temperature (about 27°C daily minimum) plus high day air temperature (about 36°C daily maximum) and long days.

4 ADAPTATION TO WATER-LIMITED ENVIRONMENTS

Adaptation to water-limited environments can include both drought resistance—defined as where a cultivar has higher average grain yield than another cultivar when grown in the same water-limited environments—and drought escape. Cultivars with improved adaptation to water-limited environments can be bred by selecting lines with high average grain yield in tests conducted in several locations over several years in the target production environment. As was discussed before, this approach requires considerable effort and time with cowpea and common bean. Also, rainfall tends to be highly variable in semiarid zones, such that grain yields are highly variable and substantial genotype × environment interaction can occur. Selecting lines that have both a low coefficient of variation (calculated over all test environments using mean values for the genotype in individual tests) and a high average grain yield can take some of this variability into consideration, but many tests can be required to produce reliable results.

Conducting trials under both rainfed (stressed $= Y_s$) and irrigated (nonstressed $= Y_p$) conditions may speed up the process of empirical selection for grain yield. Based on trials of this type with common bean, Schneider et al. (1997) proposed that initial selection should be based on high geometric mean ($Y_s \times Y_p)^{0.5}$ grain yield, followed by selection for high Y_s. It should be noted that geometric mean yield selects genotypes with high yield in both stress an nonstress environments, whereas arithmetic mean yield ($Y_s + Y_p)/2$ can favor genotypes with high nonstress yields, and either ($Y_p - Y_s$) or ($1 - Y_s/Y_p$) can favor genotypes with high yields under stress (Fernandez, 1993). Kelly et al. (1999) reported that some progress in enhancing drought resistance was achieved with the procedure of Schneider et al. (1997). In addition, selection for phenological, morphological, physiological, and biochemical traits that confer adaptation to drought could complement a breeding program that is mainly based on empirical selection for grain yield and make it more efficient. Progress in defining traits that confer drought adaptation in cowpea, peanut, and common bean is discussed.

4.1 Drought Escape

Optimal days to first flowering and cycle length have proved to be effective indicators for cowpea adaptation to rainfed environments with terminal drought. Since 1968, droughts have occurred in the Sahelian zone of Africa that shortened the rainy season to less than 2 months during some years (Dancette and Hall, 1979; Khalfaoui, 1991). Traditional land races of cowpea are prostrate and begin flowering 50–70 days after sowing, and are ready for harvest of dry grain 80–100 days after sowing. In many of the dry years since 1968, these land races often suffered terminal droughts that began while they were flowering (Khalfaoui, 1991) and produced little grain. Erect lines were bred at the University of California, Riverside, that begin flowering about 35 days after sowing and are ready for harvest in 55–65 days. In dry years in the Sahelian zone of Senegal and Sudan with about 200 mm of rain, these "60-day" cowpea lines have produced grain yields of 500–1000 kg/ha, while local land race only produced about 150 kg/ha (Hall and Patel, 1985). The "60-day" cowpea lines also have reasonably high yield potential in that they produced about 2400 kg/ha of grain when grown at the wetter boundary of the Sahelian zone with rainfall of 452 mm (Hall and Patel, 1985). Note that, in environments with terminal drought, the cycle length of these "60-day" lines can be as short as 55 days, whereas in well-watered environments the cycle length can be about 65 days. Optimal row spacing for early erect cowpeas is about 50 cm, with 33 cm between plants in the row. Traditionally, the spreading land races were sown at very wide spacings of about 100 × 100 cm. Reasons for the wide spacing included shortage of seed attributed to lack of methods for preventing weevil attacks during storage, and shortage of labor for hand sowing during the critical period after the first major rains when considerable sowing must be carried out in the Sahel. These problems have become less pronounced in Senegal because effective methods are now available for storing cowpea seed, and the crop is sown by horse-drawn sowing machines. After many years of testing on experiment stations and farmers fields, one of the 60-day cowpea lines (1-12-3 in Hall and Patel, 1985) was released as the cultivar Ein El Gazal for use in the Sahelian zone of the Sudan (Elawad and Hall, 2002).

In Senegal, a 60-day semierect cowpea cultivar has been developed, Melakh, by selecting for early flowering and high average grain yield in the Sahelian zone (Cisse et al., 1997). This cultivar also has resistance to cowpea aphid-borne mosaic potyvirus and partial resistance to bacterial blight (*Xanthomonas campestris* pv. *vignicola* (Burkholder) Dye), cowpea aphid (*Aphis craccivora* Koch), and flower thrip. Extremely early cultivars, such as Ein El Gazal and Melakh, can be severely damaged by drought at flowering (Thiaw et al., 1993). A semierect cultivar was bred in Senegal by empirical yield testing procedures, Mouride, that begins flowering a few days later and has greater resistance to midseason drought than Melakh (Cisse et al. 1995). This cultivar has resistance to cowpea aphid-borne mosaic virus, bacterial blight, cowpea storage weevil (*Callosobruchus maculatus* (F.)), and the parasitic weed *Striga gesnerioides* (Willd.) Vatke, and tolerance to heat. Because of variation in the timing of droughts and occurrence of biotic stresses, it has been recommended that farmers in the Sahelian zone of Senegal grow both Melakh and Mouride to increase the probability that at least one good crop is obtained.

Spreading cultivars of cowpea that are initially erect and then become prostrate can have substantial resistance to midseason droughts and other stresses. The earliest of the spreading cultivars have a cycle length of about 75 days and can be effective for producing both grain and hay in all but the drier parts of the Sahelian zone. Varietal intercrops consisting of alternating rows of a spreading 75-day cultivar and an erect 60-day cultivar had higher and more stable yields of both grain and hay than the most productive sole crops, under the conditions of drought and infertile soil that are prevalent in the Sahelian zone (Thiaw et al., 1993).

The sensitivity of erect cowpeas to drought during pod development can be partially solved by incorporating a gene that confers delayed-leaf-senescence (DLS). Genotypes with DLS were able to recover after a midseason drought and produce a second flush of pods, whereas genotypes that did not have DLS senesced after producing the first flush of pods (Gwathmey and Hall, 1992). The DLS trait enables cowpea cultivars to more consistently produce two distinct flushes of pods. In fields where cowpeas have been grown for several years, most plants can die after producing the first flush of pods even with alternate year rotations. This premature death of cowpea is probably caused by the soil fungus *F. solani* f. sp. *phaseoli*. The DLS trait is associated with the maintenance of carbohydrate levels in roots and thus root health which, in some way, confers resistance to this pathogen. The DLS trait was effective in preventing the premature senescence of cowpea in several locations in California and in Bambey, Senegal (Hall et al., 1997b). At Bambey, Senegal, an early flowering cowpea line with the DLS trait began flowering in about 34 days and produced a first flush of pods and about 2000 kg/ha by 65 days, and then began flowering again at about 75 days and produced an additional 1000 kg/ha by about 100 days after sowing. Cultivars of this type have not yet been fully developed, but they could be well adapted to the wetter parts of the Sahelian zone and the dry parts of the Savanna zone where rainfall is 400–600 mm and there is a high probability of midseason droughts. In this environment, it would be necessary to harvest pods by hand throughout the growing season, which is the method currently used by farmers in much of sub-Saharan Africa. Genetic studies demonstrated that it is possible to breed cowpea lines that have early flowering, DLS, and heat tolerance at reproductive development, and that the DLS trait is highly heritable and appears to be conferred by a major gene (Ismail et al., 2000). Selection for DLS can be effective when carried out on a family basis with F_4 or later generations in field nurseries where the soil pathogen is present. When choosing families, it is important to select ones that have both DLS and a substantial first-flush pod load because plants with few pods exhibit a different type of DLS with no value. An example of the type of DLS that has no agronomic value is the male sterile plants with no pods that can be readily detected in breeding nurseries because of their green appearance at the end of the season when all of the other plants have produced many pods and have died.

Local cultivars of peanut grown in the Sahelian and Savanna zones of West Africa, prior to the beginning of severe droughts in 1968, were prostrate types with a 120-day cycle from sowing to harvest (Schilling and Misari, 1992). Since this time, erect earlier flowering cultivars have been bred for these zones that, in Senegal, are sown at close spacing using horse-drawn sowing machines. But the shortest cycle cultivars that have been developed still take 85–90 days from sowing to harvest (Subbarao et al., 1995), and are only well adapted to the wetter part of the Sahelian zone and the Savanna zone (Kahl-faoui, 1991). Genotypes with synchronous pod development and rapid partitioning of carbohydrate to pods may have improved adaptation to terminal droughts (Knauft and Wynne, 1995). Contrasting peanut genotypes have been compared with respect to their adaptation to hot, dry Sahelian conditions (Greenberg et al., 1992). Thirty-six genotypes were grown in five environments, where water supplies and temperature were varied by varying the sowing date and the irrigation treatments. Adaptation to hot dry Sahelian conditions was associated with below-average crop growth rates but much greater harvest index. Achieving a high harvest index also requires high levels of remobilization of carbon and nitrogen compounds to developing grain. Optimal rates and levels of remobilization are required, in that removal of nitrogen compounds from leaves leads to a reduction of their photosynthetic capacity (Sinclair and deWit, 1975). In addition, removal of carbohydrate from roots could result in decreases in nitrogen fixation (Subbarao et al., 1995) and increased sensitivity to soil organisms that cause premature senescence (Gwathmey et al., 1992b).

4.2 Drought Resistance

In some water-limited environments, there is no rain during the cropping season but considerable moisture is present in the soil during sowing. An example would be where grain legumes are grown after rice or after floodwaters have receded. Selecting for early appearance of mature pods can enhance the drought resistance of cowpea grown on stored soil moisture (Hall and Grantz, 1981). However, an optimal level of earliness must be sought because there can be an interaction between earliness and depth of rooting which can influence seasonal water extraction. For example, in pigeon pea (*Cajanus cajan* L. Millsp.), short-duration genotypes may develop shallow root systems of only 50-cm depth compared with long-duration genotypes that can develop 200-cm deep root systems (Subbarao et al., 1995). For cowpeas with similar cycle length, genotypic differences are present in the extent of rooting but it is not easy to screen for this trait. Typically, the evaluation of rooting characteristics has only been performed with a few cultivars when choosing parents for crosses or with a few promising advanced lines. However, a method has been developed that can evaluate rate of root growth of many cultivars or stable lines under field conditions (Robertson et al., 1985).

With this method, a herbicide is placed in a narrow horizontal band deep in the soil in-between rows of plants. Plants are scored daily for plant symptom responses as an indicator of the time taken for roots to grow and reach the herbicide. Using this technique, genotypic differences among cowpea accessions in rate of root development were detected (Hall and Patel, 1985) that were positively correlated with extent of soil water extraction (Robertson et al., 1985). A diverse set of peanut cultivars has been screened under field conditions by using this technique (Khalfaoui and Havard, 1993). Genotypic differences in days to appearance of herbicide symptoms were detected. Genotypes that exhibited symptoms earlier, and presumably had faster rates of root development, also had a longer cycle from sowing to maturity. It has been argued that increased root length density in deeper soil layers may be adaptive under water-limited conditions (Subbarao et al., 1995). However, studies comparing cowpea with pearl millet suggested that the micro-scale uniformity of root distribution may have a greater impact on the maintenance of plant water status than the overall density of the roots (Petrie and Hall, 1992a,b,c; Petrie et al., 1992).

Common beans have root systems that are less effective in extracting water stored deep in the soil than those of many cowpea accessions in fields with high bulk density soil at Riverside, CA. In many soils, more force is required to pull cowpea plants out of the ground than common beans. In general, an inadequate root system may be a weak link in the adaptation of many common bean cultivars. A unique study, in which root and shoot systems of contrasting common bean genotypes were separately evaluated using grafted plants, demonstrated that root characteristics can be much more important than shoot characteristics in conferring adaptation to drought (White and Castillo, 1989). Irrespective of their importance in drought adaptation, relatively little progress has been made in breeding for root traits because of the difficulty in screening large numbers of plants for these traits. For common beans, breeders can take advantage of an association which can be present between plant habit and rooting pattern (Kelly et al., 1999). The authors said that early flowering erect determinate cultivars tend to develop shallow root systems, erect indeterminate cultivars tend to have a more prominent tap root that can better exploit deep soil layers and can be effective with terminal droughts, and more prostrate indeterminate cultivars tend to have a more sprawling fibrous root system which can be effective under intermittent droughts.

Cowpea has exhibited substantial resistance to droughts occurring during the early vegetative stage. For example, cowpeas were subjected to a vegetative-stage drought that would have killed most other crop species, yet with rewatering, they recovered and produced very high grain yields (Turk et al., 1980). Seeds were sown into a soil that only had a small amount of water in the upper profile to permit germination and emergence, and little available water in the rest of the soil profile. These plants did not receive rain or irrigation for 43 days under hot high evaporative demand conditions. After this period of drought, the plants were severely stunted, with leaf area being reduced by 74% and shoot dry matter being reduced by 63% compared with well-watered plants. These severely stunted plants were then irrigated optimally. Surprisingly, they produced grain yields of 3978/ha similar to those of control plants (3916 kg/ha) that had been optimally irrigated about every week since sowing. The cycle length from sowing to harvest was the same for both treatments (107 days). This study also demonstrated that grain yield of cowpea can be strongly influenced by photosynthesis that occurs after the initiation of flowering because the plants subjected to vegetative stage drought produced considerably less biomass prior to flowering but the same high grain yields as well-watered plants.

A rapid method in screening cowpeas for resistance to drought during the early vegetative state has been developed (Singh et al., 1999). This method consists of sowing seed at close spacing (10 × 5 cm) in a 12-cm deep layer of soil and sand, and watering daily until emergence of the first trifoliate after which watering is stopped. Thereafter, the number of permanently wilted (dead) plants are counted on a daily basis and when all plants of susceptible control genotypes are dead, watering is resumed and the percentage of surviving plants is determined. Using this method, two different types of drought resistance were detected (Mai-Kodomi et al., 1999a). Type 1 resistant lines stopped growing and moisture was conserved in all plant tissues and plants stayed alive for at least 2 weeks. In Type 2 resistance, which appeared to be more effective in enhancing plant survival, trifoliates of resistant lines continued to grow slowly while unifoliates exhibited early senescence. Genetic studies indicated that the two types of drought resistance were conferred by different single dominant alleles that were closely linked or at the same locus (Mai-Kodomi et al., 1999b). Type 1 resistance appeared to be dominant over Type 2 resistance, and lines were not detected with both types of drought resistance. Because a shallow layer of soil is used, this screening method may detect shoot traits that enable plants to survive severe vegetative-stage droughts.

Drought during the vegetative stage can make cowpea (and common bean) more susceptible to ashy stem blight (*Macrophomina phaseolina* Tassi Goid) and lesser corn stalk borer (*Elasmopalpu lignosellus*). Host plant resistance to this disease and pest are not yet available in cowpea. Vegetative-stage drought can also strongly reduce biological nitrogen fixation (Elowad and Hall, 1987; Sinclair and Serraj, 1995).

The mechanisms whereby cowpea resists vegetative-stage drought may be related to the fact that leaf water potential does not decrease very much even under extreme drought. The lowest leaf water potential recorded for cowpea is −18 bar (−1.8 MPa) (e.g., Turk and Hall, 1980; Hall and Schulze, 1980a), whereas peanut has developed leaf water potentials under drought as low as −82 bar (−8.2 MPa) (Turner et al., 2000). Several mechanisms may partially explain the extreme dehydration avoidance of cowpea. Cowpea has stomata that are very sensitive to soil drought, partially closing even before differences in leaf water potential are detected (Bates and Hall, 1981). In addition, the stomata of cowpea are very sensitive to humidity, partially closing in dry air in a manner that optimizes stomatal function with respect to the control of transpiration in relation to photosynthetic uptake of carbon dioxide (Hall and Schulze, 1980b), and acts to maintain high leaf water potential. When subjected to soil drought, leaflets of cowpea become paraheliotropic and orientated parallel to the sun's rays, which cause them to become cooler and thus transpire less (Shackel and Hall, 1979). In peanut, genetic differences have been observed in drought-induced leaflet folding responses but the differences are difficult to quantify (Knauft and Wynne, 1995). Predawn leaf water potential remained high in cowpea at about −8 bar (−0.8 MPa), with soil drought, that were killing pearl millet (*Pennisetum glaucum* L.R. Br.) and inducing it to have a predawn leaf water potential of −30 bar (−3.0 MPa) (Petrie and Hall, 1992a). The only explanation developed for this difference between species was that cowpea had a more effective root system because it was more uniform on a micro-scale level than pearl millet, which had a clumped root system with higher root length density (Petrie and Hall, 1992b,c; Petrie et al., 1992).

Differences are observed among grain legumes in the extent of drought-induced osmotic adjustment. Cowpea exhibits very little drought-induced osmotic adjustment (Shackel and Hall, 1983) compared with peanut (Turner et al., 2000). Values for cowpea were 0–4 bar compared with 3–16 bar for peanut. A diverse set of 100 cowpea accessions was screened under extreme soil drought and found to exhibit only small differences in leaf solute potential ranging from −11 to −13 bar, which were not significant (Shackel and Hall, 1983). Among the grain legumes, substantial osmotic adjustment has only been detected in chickpea (*Cicer arietinum* L.), peanut, and pigeon pea (Turner et al., 2000). It is not known whether selecting for enhanced osmotic adjustment can enhance the drought resistance of grain legumes, although a positive correlation was reported for a set of eight chickpea breeding lines between grain yield under water-limited field conditions and osmotic adjustment values for leaves measured in controlled environments (Morgan et al., 1991). A proposed mechanism whereby osmotic adjustment may enhance adaptation to drought is that osmotic adjustment in root cells could enhance their turgor pressure and thus may increase the growth rates of the roots (Hall, 2001).

4.3 Transpiration Efficiency

Selection for genotypic differences in transpiration efficiency (the ratio of net photosynthesis to transpiration) has been considered as an approach for enhancing crop resistance to drought, by using the following model for grain yield under water-limited conditions (Hall, 2001).

$$Y = \sum_{i=d}^{i=1} ET_i \times T_i / ET_i \times W_i \times CP_i \tag{5}$$

where Σ is the summation of daily values of $ET \times T/ET \times W \times CP$ over the period of days (d) when photosynthesis significantly contributes carbohydrate to developing grains, which often starts from an early flowering stage to the date when most grains are physiologically mature. According to the model, grain yield (Y in g m^{-2}) depends on the amount of evapotranspiration (ET in g[H$_2$O] m^{-2}), the relative amount transpired (T/ET), the water-use efficiency (W in g[CH$_2$O]g[H$_2$O]$^{-1}$, which is conceptually similar to transpiration efficiency, the ratio of photosynthesis to transpiration), and the proportion of carbohydrate partitioned to grain (CP, which is conceptually similar to HI). This model would provide a useful guide for selection if all of the components in the model were independent or associated positively. As will be seen, this is not the case because negative correlations have been observed among W, CP, and T, as it is determined by the extent of root growth, leaf area, and stomatal opening.

Direct measurement of transpiration efficiency takes considerable effort in that it may have to be measured many times during the season, and measurement of W is constrained by difficulties in measuring both T and belowground biomass. A theory was developed which predicted that stable carbon isotope discrimination (Δ) can provide an indirect estimate of time-integrated internal carbon dioxide concentration (Farquhar et al., 1982). Consequently, the theory also predicts a negative association between Δ and time-integrated intrinsic transpiration efficiency (the ratio of net photosynthesis to leaf conductance to water vapor, P_n/g_w). This theory was shown to be effective, in that strong negative correlations have been observed between Δ and W in many cases (Hall et al., 1994a). This stimulated considerable research on the possible use of Δ as a selection criterion in plant breeding programs (Hall et al., 1994a). The association between crop performance and Δ has been reviewed for cowpea, common bean, and peanut (Condon and Hall, 1997).

In general, grain yield of cowpea cultivars has been positively associated with Δ (that is, negatively correlated with W) under well-irrigated and water-limited rainfed and stored soil–moisture conditions. When a set of diverse cowpea accessions was grown in different climatic zones, accessions adapted to a specific zone tended to have high $\varDelta$ when grown in this zone, but not when grown in other zones (Hall et al. 1994b). Genetic selection studies confirmed this positive association between grain yield and $\varDelta$ (Hall et al. 1993, 1997b, Condon and Hall 1997). For well-watered conditions, these results may be partially explained by genotypic variation in stomatal conductance with higher stomatal conductance, resulting in lower transpiration efficiency (and thus lower W) and higher $\varDelta$ but greater net photosynthesis, biomass production, and grain yield. For water-limited conditions, the positive correlations between grain yield and $\varDelta$ are more difficult to explain. A possible explanation is that the genotypes with higher $\varDelta$ had greater root growth, which enabled them to access more soil water and thus have greater ET and T. This would provide enhanced plant water status and higher stomatal conductance accounting for the higher $\varDelta$ and the greater shoot biomass production and grain yield, although transpiration efficiency is low (and therefore W is low). An additional possibility is that the greater grain yields are caused by higher harvest index (and thus greater CP), and that the greater reproductive sink for carbohydrates is, in some way, causing the stomata to be more open through a regulation mechanism, thus causing $\varDelta$ to be higher. This regulation mechanism could be present in both well-watered and droughted conditions, and a positive genetic correlation was observed between $\varDelta$ and HI (Menéndez and Hall 1996). Therefore, selection in cowpea for high $\varDelta$ may select plants with less conservative stomatal performance, i.e., plants whose stomata are more open for greater durations and this may indirectly reflect the fact that they have more effective root systems and/or greater partitioning of carbohydrate to grain.

For common bean under water-limited rainfed conditions on a deep fertile soil, White et al. (1990) observed positive associations between $\varDelta$ and both shoot biomass production and grain yield for 9 out of 10 common bean cultivars that were tested. This result is consistent with the positive genotypic association reported for stomatal conductance and $\varDelta$ in another set of common bean cultivars (Ehleringer, 1990). However, when White et al. (1990) studied the same 10 common bean cultivars in another water-limited rainfed environment with a shallow root zone due to an acid subsoil that restricted root penetration, there was no association among $\varDelta$ and shoot biomass production or grain yield. However, $\varDelta$ was positively associated with root length density, which substantially varied among the common bean cultivars. Consequently, a possible explanation for the genotypic differences in $\varDelta$ and grain yield is that, at the site where deep rooting was possible, genotypes with greater root length densities could have had greater access to water and therefore suffered smaller water deficits, which could have resulted in greater stomatal conductance (and therefore greater $\varDelta$), greater biomass, and greater grain yield. At the site where root depth was restricted, $\varDelta$ and root length density were again positively correlated but the highest grain yields were associated with intermediate values of Δ. In this case, it is possible that genotypes with very low root length density suffered from water deficits and had very low stomatal conductance (and therefore very low Δ), and grew slowly such that they had low yields. In contrast, genotypes with very high root length density initially had high stomatal conductance (and thus high Δ), rapid growth, and rapid water use, but depleted moisture in the shallow soil profile and became severely stressed during reproductive development such that CP was reduced, and they also produced small grain yields (White, 1993). These hypotheses have not been rigorously tested. However, it would appear that genotypic differences in root growth are an important component of drought resistance in common bean and can strongly influence the relations among Δ, edaphic conditions, and grain yield.

In peanut, the genotypic association between Δ and biomass production appears to be different from either cowpea or common bean. For peanut, the relationship between Δ and biomass production almost invariably has been found to be negative (Wright et al., 1993). There has been a positive association between transpiration efficiency (and thus W) and crop growth rate under both well-watered and water-limited conditions. Much of the genotypic variation for Δ arises from variation in photosynthetic capacity and the amount of rubisco per unit leaf area and not in stomatal conductance (Nageswara Rao et al., 1995). In addition, there does not appear to have been a trade-off between genotypic variation in photosynthetic capacity and leaf area growth rate. Specific leaf area (SLA = leaf area per unit dry weight) is easier to measure than Δ, and a strong positive association was observed between SLA and Δ (Nageswara Rao and Wright, 1994). Unfortunately, attempts to breed improved peanut cultivars by indirectly selecting for increased transpiration efficiency have been impeded by positive associations between harvest index (and thus CP) and either Δ or SLA (Wright et al., 1993). These associations have been found in studies with diverse peanut cultivars and also in populations from crosses involving parents with high and low Δ values. As a consequence of the association, there was no increase in pod yield with selection for lower Δ, despite large increases in biomass production. If the linkage between high transpiration efficiency and low harvest index proves to be unbreakable (this association was also observed in several other species including cowpea), the high biomass production associated with high transpiration efficiency should still be advantageous in production systems where fodder yield for animal production and pod yield for human consumption are both important.

In general, adaptation to water-limited environments requires that many traits be optimized: days to first flowering, cycle length, rooting rate and depth, transpiration efficiency, and harvest index. Where levels of several traits are not optimal, cultivars with improved adaptation may be developed by simultaneous selection for these traits so that they approach optimal levels. Days to first flowering, cycle length and harvest index can be effectively selected, but instead of selecting for root traits or transpiration efficiency, it may be more practically effective to select advanced lines based on their grain yield in the target production environment.

5 ADAPTATION TO INFERTILE SOILS

In many countries, grain legumes are often grown in infertile soils with major deficiencies of nitrogen and phosphate. Consequently, the biological fixation of atmospheric nitrogen (BNF) by legumes in symbiosis with rhizobia and their enhanced uptake of soil phosphate (P) when associated with endomycorrhizal fungi are critical aspects of their adaptation. The enhancement of soil fertility also benefits subsequent crops in the rotation. The extent to which BNF and P acquisition by grain legumes can be enhanced by breeding is examined.

5.1 Biological Nitrogen Fixation

The circumstances where inoculation of seed with selected strains of rhizobia can enhance BNF are considered. Typically, inoculant use is recommended when the legume crop is to be planted on a soil where this species has not been grown before and levels of effective native rhizobia are too low. Cowpea and peanut can exhibit effective symbiosis with a broad range of rhizobia and inoculation is not often necessary. In contrast, common bean can benefit from inoculation with rhizobia in some circumstances. Care is needed in choosing inoculants, however, in that some commercial inoculants have low quality (Date, 2000). Where effective inoculation technologies have been developed, the use of rhizobial inoculants in common bean production and research can be worthwhile. Cases where the use of inoculants has been effective with common bean are reviewed by Graham and Vance (2000).

Under well-watered conditions, cowpea has the potential for substantial BNF. In a 95-day growing season, cowpea has produced 3 ton/ha of grain and 4.7 ton/ha of hay while relying mainly on BNF for nitrogen (Elowad and Hall 1987), indicating the potential to fix about 200 kg/ha of atmospheric nitrogen. In California, no grain yield responses of cowpea to inoculation have been reported and the crop often is grown without applying nitrogen fertilizer or manure (Hall and Frate, 1996). In contrast, common bean often has exhibited low levels of BNF, while peanut is intermediate for this trait

(Hardason et al., 1993). In the United States, inoculants are rarely applied to common beans and substantial nitrogen fertilizer is usually applied. In many areas of the world, nitrogen fertilizer is applied to peanuts (Knauft and Wynne, 1995).

Substantial genotypic variation for BNF is present in common bean (Hardason et al., 1993). Bliss (1993) has bred genetic lines of common bean that have sufficient BNF to support grain yields of 1000–2000 kg/ha in infertile soils. Five of these lines (Bliss et al., 1989) and one common bean cultivar with high BNF (Henson et al., 1993) have been released. BNF appears to be a quantitatively inherited trait in common bean, and family rather than single-plant selection has been practiced (Bliss, 1993). With advanced lines, BNF can be assessed indirectly based on plant performance in field conditions on low N soils (Bliss, 1993). Selection for grain yield or grain nitrogen may be effective but in some conditions, it may be useful to select for total shoot dry matter or total shoot nitrogen as well (Elizondo Barron et al., 1999). For parents used in crosses and selected advanced lines, it is useful to determine the proportion of nitrogen in the plants because of BNF. This can be estimated based on the proportion of nitrogen in xylem sap that is ureides and thus was formed by BNF for some legumes, including common bean (Herridge and Rose, 2000) and cowpea (Elowad et al., 1987; Elowad and Hall, 1987), but not peanut.

BNF is extremely sensitive to drought with greater reductions occurring than the levels observed for photosynthesis and carbohydrate accumulation rate (Serraj et al., 1999).Sinclair and Serraj (1995) determined the ratio of the sensitivities to drought of BNF and transpiration (stomatal conductance) of nine grain legume species. Some species that transport BNF products as ureides (including cowpea) had BNF that was more drought-sensitive than transpiration. In contrast for species that transport BNF products as amides (including peanut), BNF had less sensitivity to drought than transpiration. For common bean, the study was somewhat inconclusive in that it transports BNF products as ureides and tends to have low BNF, but in this study, its BNF was less sensitive to drought than transpiration. Subsequent studies confirmed that the BNF of common bean can be less sensitive to drought than its transpiration, although it produces some ureides (Serraj and Sinclair, 1998). Serraj et al. (1999) provided a hypothesis to explain the sensitivity of BNF to drought in species such as soybean and cowpea. They argue that feedback by high concentrations of BNF products, such as ureides, may be responsible for drought-induced reductions in BNF. They also proposed that genetic variation existing in soybean may provide an opportunity to overcome this problem through plant breeding.

Generally, plants engaged in symbiotic nitrogen fixation have a higher requirement for phosphate than those that rely on inorganic nitrogen (Graham and Vance, 2000). Consequently, genotypes with enhanced P acquisition in P-deficient soil may also have enhanced BNF.

5.2 Phosphorus-Deficient Soils and Mycorrhizal Roots

In many—if not most—field conditions, the roots of cowpea and other grain legumes are mycorrhizal beginning about 3 weeks after germination because of associations with specific fungi. Vesicles and arbuscules resulting from this association are present in root cortical cells, and the hyphae of the fungi extend into the soil. In infertile soils, mycorrhizal cowpea plants can take up more P, zinc, and copper than nonmycorrhizal cowpea plants and produce more biomass and grain in either well-watered or water-limited conditions (de Faria, 1984; Kwapata and Hall, 1985). With optimal irrigation, mycorrhizal infection was inhibited by 80–91% when cowpea plants were grown under high levels of soil P, but with moderate levels of soil drought, normal levels of infection occurred at high soil P, despite high P concentrations in the plants (de Faria, 1984; Kwapata and Hall, 1985). Mycorrhizal infection did not influence biomass production or grain yield of plants grown with both high soil P and moderate soil drought. The mycorrhizal cowpea root system may be very effective in accessing soil P. In California field conditions, cowpea has shown no response to P fertilization in very low P soils where cotton (*Gossypium hirsutum* L.) and maize (*Zea mays* L.) showed substantial yield responses to P fertilization (Hall and Frate, 1996). Common bean is similar to cowpea in that mycorrhizal infection has resulted in improved acquisition of P and enhanced shoot growth in low soil P conditions, but little effect in high P soils (Lynch et al., 1991).

The initial and major response of common bean (Lynch et al., 1991) and cowpea (de Faria, 1984) to P deficiency is that plants produce fewer leaves and less leaf area. The net photosynthesis per unit leaf area does decrease, but at a much lower rate than the total leaf area. Low soil P availability is considered the greatest soil constraint to common bean production; consequently, substantial effort has been devoted to breeding common bean cultivars with improved productivity in low P soils (Lynch et al., 1998). Emphasis has been placed on enhancing the efficiency whereby common bean acquires soil P. Genetic variation in the ability to acquire soil P is present and was independent of mycorrhizal and soil-type effects (Yan et al., 1995a,b). Several mechanisms influencing soil P acquisition have been studied. Three Andean common bean genotypes exhibited greater extraction of P from calcium phosphates in artificial medium than three Mesoamerican genotypes (Yan et al., 1996). In low soil phosphorus conditions, one of these Andean genotypes acidified the rhizosphere more than the other five genotypes. It should be noted that this mechanism may be more important in high pH soils or where plants have been fertilized with rock phosphate. Many low P soils have low pH and in this situation other types of root exudates may enhance P acquisition (Lynch et al., 1998). Root architecture is important in P acquisition because of the relative immobility of P in the soil. The proportion of the root system that explores the region where soil P is highest, which is often the upper part of the soil profile, may be important. Genotypic differences have been detected in this trait under low soil P conditions (Lynch et al., 1998). A trade-off is involved, however, in that a root system that mainly explores the upper soil profile would not be adaptive in water-limited conditions where substantial water is available deep in the soil. A common bean genotype that is efficient in acquiring soil P also had a greater tendency to produce adventitious roots under low soil phosphorus conditions than a less efficient genotype (Lynch et al., 1998). The efficient genotype also produced more root hairs under low soil phosphorus than an inefficient genotype (Lynch et al., 1998). An additional complication is that root traits should complement mycorrhizal hyphae traits with respect to their abilities to access soil P in different parts of the rhizosphere (Koide, 2000).

Direct selection for root traits is usually difficult. Indirect selection for root traits using molecular markers could enhance the efficiency of breeding for improvements in phosphorus acquisition and some molecular markers for root traits have been detected (Lynch et al., 1998). The growth pouch technique of Omwega et al. (1988) was used to screen cowpea accessions for differences in root hairs. Cowpea genotypes have been detected (Melakh and IT82E-18) that produce more and longer root hairs under low soil P conditions than other cowpea genotypes using the growth pouch technique (J.D. Ehlers and W.C. Matthews, unpublished, 2001). The growth pouch technique for detecting differences in root hair production is very efficient in that thousands of plants can be screened in a few weeks and the screen is nondestructive in that selected plants can be removed from the pouches and grown to produce seed.

Some common bean breeding programs have selected for traits thought to confer enhanced performance in low P soils, but, according to Lynch et al. (1998), these programs have yet to produce genotypes with superior performance in low P soils than check cultivars. Papers describing genetic differences among peanut that may be relevant to performance in low P soils have been reviewed by Knauft and Wynne (1995).

Breeding for infertile soils needs to be integrated with other components of a comprehensive breeding program. An efficient way to partially accomplish this objective is to conduct at least some performance trials in fields in the target production environment with infertile soils. These trials can be expensive because of the substantial soil variability that inflates the error term, requiring many replications of both plots and experiments if reliable predictions are to be obtained from the results of such trials.

6 FUTURE DIRECTIONS FOR PLANT BREEDING

As more information is gained, it is likely that improved ideotypes will be developed for grain legume cultivars which will exhibit superior performance in specific target production zones. I am using a broad definition of an ideotype as being a plan of the phenotype of a cultivar that will perform optimally in a specific set of climatic, soil, biotic, and sociocultural conditions (Hall, 2001). Incorporating all of the traits defined by these improved ideotypes will be difficult even when the inheritance of many of the individual traits is known. Direct selection can be particularly difficult for traits that are difficult to screen, such as many of the root traits. Moreover, direct selection is not very effective for incorporating multiple traits. Indirect marker-assisted selection, with emphasis on molecular markers, has the potential to substantially increase the efficiency of breeding to incorporate multiple traits and traits which are difficult to screen. Prior to developing marker-assisted selection procedure, it is useful to have a high-density genetic-linkage map. Considerable progress has been made in this area with common bean (Gepts, 1999), and some progress has been made with cowpea (Ouédraogo et al., 2002). Common bean breeders are already using DNA marker-assisted selection (Kelly and Miklas, 1999).

Grain legume breeders should consider possible climate changes when developing breeding strategies (Hall and Ziska, 2000). Breeding for coordinated increases in both the reproductive sink and the photosynthetic source may be effective in increasing grain yield at the present time. However, increases are occurring in atmospheric $[CO_2]$ that could continue to make photosynthetic sources more effective per unit leaf area. Consequently, maintaining a balance between photosynthetic sources and reproductive sinks may require selecting plants for future environments that have even greater reproductive sinks. Breeding to enhance reproductive sinks will be particularly important for target production zones where stresses, such as high temperature, can cause greater damage to reproductive development than to the photosynthetic source. Breeding for heat tolerance during reproductive development has been effective (Hall, 1992, 1993). In cowpea, genes for heat tolerance during reproductive development enhance sink strength and harvest index (Ismail and Hall, 1998, 2000). Studies in elevated and current atmospheric $[CO_2]$s suggest that heat-tolerance genes may enhance the grain yield response of cowpea to elevated $[CO_2]$ under both optimal and high temperatures (Ahmed et al., 1993a; Hall and Allen, 1993). Selection for greater reproductive sink strength could indirectly enhance photosynthetic capacity and activity by minimizing the feedback effects that downregulate the photosynthetic system. Making full use of elevated $[CO_2]$ may also require selection to enhance those components of the photosynthetic system that limit photosynthesis per unit leaf area in these conditions.

Some radical approaches to grain legume breeding should be considered. Reproductive processes, especially production of pollen and pollination, are extremely sensitive to several stresses, such as chilling, high temperatures, and drought. The effects can be more severe than required to maintain a balance between photosynthetic source and sink, such that the seeds produced are adequately plump and viable. A potential solution to this problem is to develop cultivars that do not require pollination and other sexual processes while growing in farmers' fields. Cultivars with an appropriate type of apomixis would be able to produce viable seed from maternal tissue without requiring either meiosis of the embryo mother cell or pollen production and pollination of the embryo or endosperm. The advantages of cultivars with this type of apomixis are as follows.

1. Apomixis could confer resistance to stresses that damage pollination and other aspects of sexual production including chilling, heat, drought, and various insect pests (Hall et al., 1997a). Apomixis may also enhance grain yield in those grain legume species that may not have a completely effective pollination process, such as common bean (Ibarra-Perez et al., 1999).
2. Apomixis would fix hybridity in that F_1 hybrids would have true-breeding seeds. This would make possible the use of hybrid vigor, which would substantially increase yields of those grain legumes, such as cowpea (Hall et al., 1997a), that exhibit substantial heterosis for this trait. Farmers could reuse seed produced by apomictic hybrid cultivars in that the seed would retain its hybrid vigor. This would provide a significant advantage for the many poor farmers who grow grain legumes.
3. Seed production may be more effective in apomictic cultivars than with current grain legume cultivars, which are mainly self-pollinated but can exhibit a degree of cross-pollination that results in the production of genetically variable seed. Out-crossing can cause major problems in seed production of common bean (Ibarra-Perez et al., 1997) and cowpea.
4. Breeding programs would be accelerated by using apomictic lines because it would confer the ability to immediately fix superior heterozygous genotypes.

Sex would still be needed to permit the continual breeding of improved cultivars. This could be achieved by using facultative (switchable) apomixis systems, where the default state is apomictic. The sexual state would be switched on by the breeder by using specific chemical sprays or special environments. Several plant species have the genes needed for developing facultative apomictic breeding systems. Through genetic engineering, it may be possible to create and transfer the "cassette" of genes needed for facultative apomixis into grain legume cultivars (Jefferson, 1993). Grain legume cultivars with facultative apomixis could lead to a beneficial revolution in plant breeding, crop production, and the whole agricultural industry.

REFERENCES

Acosta-Gallegos JA, White JW. Phenological plasticity as an adaptation by common bean to rainfed environments. Crop Sci 1995; 35:199–204.

Ahmed FE, Hall AE. Heat injury during early floral bud development in cowpea. Crop Sci 1993; 33:764–767.

Ahmed FE, Hall AE, DeMason DA. Heat injury during floral development in cowpea (*Vigna unguiculata*, Fabaceae). Am J Bot 1992; 79:784–791.

Ahmed FE, Hall AE, Madore MA. Interactive effects of high temperature and elevated carbon dioxide concentration on cowpea (*Vigna unguiculata* (L.) Walp.). Plant Cell Environ 1993a; 16:835–842.

Ahmed FE, Mutters RG, Hall AE. Interactive effects of high temperature and high quality on floral bud development in cowpea. Aust J Plant Physiol 1993b; 20:661–667.

Bates LM, Hall AE. Stomatal closure with soil water depletion not associated with changes in bulk leaf water status. Oecologia (Berlin) 1981; 50:62–65.

Beaver JS, Miklas PN, Echavez-Badel R. Registration of "Rosada Nativa" pink bean. Crop Sci 1999; 39:1257.

Bhagsari AS, Brown RH. Leaf photosynthesis and its correlation with leaf area. Crop Sci 1986; 26:127–132.

Bliss FA. Breeding common bean for improved biological nitrogen fixation. Plant Soil 1993; 152:71–79.

Bliss FA, Pereira PAA, Araujo RS, Henson RA, Kmiecik KA, McFerson JR, Teixera MG, da Silva CC. Registration of five high nitrogen fixing common bean germplasm lines. Crop Sci 1989; 29:240–241.

Cisse N, Ndiaye M, Thiaw T, Hall AE. Registration of "Mouride" cowpea. Crop Sci 1995; 35:1215–1216.

Cisse N, Ndiaye M, Thiaw T, Hall AE. Registration of "Melakh" cowpea. Crop Sci 1997; 37:1978.

Condon AG, Hall AE. Adaptation to diverse environments: variation in water-use efficiency within crop species. In: Jackson LE, ed. Ecology in Agriculture. San Diego: Academic Press, 1997:79–116.

Craufurd PQ, Bojang M, Wheeler TR, Summerfield RJ. Heat tolerance in cowpea: effect of timing and duration of heat stress. Ann Appl Biol 1998; 133:257–267.

Craufurd PQ, Subedi M, Summerfield RJ. Leaf appearance in cowpea: effects of temperature and photoperiod. Crop Sci 1997; 37:167–171.

Dancette C, Hall AE. Agroclimatology applied to water management in the Sudanian and Sahelian zones of Africa. In: Hall AE, Cannell GH, Lawton HW, eds. Agriculture in Semi-Arid Environments, Ecological Studies. Vol. 34. Berlin: Springer-Verlag, 1979:98–118.

Date RA. Inoculated legumes in cropping systems of the tropics. Field Crops Res 2000; 65:123–136.

Davis DW, Marsh DB, Alvarez MN. MN 13 and MN 150 cowpea breeding lines. HortScience 1986; 21:1080–1081.

de Faria RM. Mycorrhizal influences on cowpea (*Vigna unuiculata* (L.) Walp.) under different levels of soil phosphorus and drought. Ph.D. dissertation, University of California, Riverside, 1984:141.

Dickson MH. Breeding for heat tolerance in green beans and broccoli. In: Kuo CG, ed. Adaptation of Food Crops to Temperature and Water Stress. Shanhua: Asian Vegetable Research and Development Center, Taiwan, Publ. No. 93-410, 1993:296–302.

Dickson MH, Petzoldt R. Heat tolerance and pod set in green beans. J Am Hort Sci 1989; 114:833–836.

Dow El-Madina IM, Hall AE. Flowering of contrasting cowpea (*Vigna unguiculata* (L.) Walp.) genotypes under different temperatures and photoperiods. Field Crops Res 1986; 14:87–104.

Ehleringer JR. Correlations between carbon isotope discrimination and leaf conductance to water vapor in common beans. Plant Physiol 1990; 93:1422–1425.

Ehlers JD, Hall AE. Genotypic classification of cowpea based on responses to heat and photoperiod. Crop Sci 1996; 36:673–679.

Ehlers JD, Hall AE. Cowpea (*Vigna unguiculata* (L.) Walp.). Field Crops Res 1997; 53:187–204.

Ehlers JD, Hall AE. Heat tolerance of contrasting cowpea lines in short and long days. Field Crops Res 1998; 55:11–21.

Ehlers JD, Hall AE, Patel PN, Roberts PA, Matthews WC. Registration of "California Blackeye 27" cowpea. Crop Sci 2000; 40:854–855.

Elawad HOA, Hall AE. Registration of "Ein El Gazal" cowpea. Crop Sci 2002; 4-2:1745–1746.

Elizondo Barron J, Pasini RJ, Davis DW, Stuthman DD, Graham PH. Response to selection for seed yield and nitrogen (N₂) fixation in common bean (*Phaseolus vulgaris* L.). Field Crops Res 1999; 62:119–128.

Elowad HOA, Hall AE. Influences of early and late nitrogen fertilization on yield and nitrogen fixation of cowpea under well-watered and dry field conditions. Field Crops Res 1987; 15:229–244.

Elowad HOA, Hall AE, Jarrell WM. Comparisons of ureide and acetylene reduction methods for estimating biological nitrogen fixation by glasshouse-grown cowpea. Field Crops Res 1987; 15:215–227.

El-Kholy AS, Hall AE, Mohsen AA. Heat and chilling tolerance during germination and heat tolerance during flowering are not associated in cowpea. Crop Sci 1997; 37:456–463.

Farquhar GD, O'Leary MH, Berry JA. On the relationship between carbon isotope discrimination and the intercellular carbon dioxide concentration in leaves. Aust J Plant Physiol 1982; 9:121–137.

Fernandez GCJ. Effective selection criteria for assessing plant stress tolerance. In: Kuo CG, ed. Adaptation of Food Crops to Temperature and Water Stress. Shanhua: Asian Vegetable Research and Development Center, Taiwan, Publ. No. 93-410, 1993:257–270.

Gepts P. Development of an integrated linkage map. In: Singh SP, ed. Common Bean Improvement in the Twenty-First Century. Dordrecht: Kluwer Acad, 1999:53–91.

Graham PG, Vance CP. Nitrogen fixation in perspective: an overview of research and extension needs. Field Crops Res 2000; 65:93–106.

Greenberg DC, Williams JH, Ndunguru BJ. Differences in yield determining processes of groundnut (*Arachis hypogaea* L.) genotypes in varied drought environments. Ann Appl Biol 1992; 120:557–566.

Gross Y, Kigel J. Differential sensitivity to high temperature of stages in the reproductive development of common bean (*Phaseolus vulgaris* L.). Field Crops Res 1994; 36:201–212.

Gwathmey CO, Hall AE. Adaptation to midseason drought of cowpea genotypes with contrasting senescence traits. Crop Sci 1992; 32:773–778.

Gwathmey CO, Hall AE, Madore MA. Adaptive attributes of cowpea genotypes with delayed monocarpic leaf senescence. Crop Sci 1992a; 32:765–772.

Gwathmey CO, Hall AE, Madore MA. Pod removal effects on cowpea genotypes contrasting in monocarpic senescence traits. Crop Sci 1992b; 32:1003–1009.

Habash DZ, Genty B, Baker NR. The consequence of chlorophyll deficiency for photosynthetic light use efficiency in a single nuclear gene mutation of cowpea. Photosynth Res 1994; 42:17–25.

Hall AE. Breeding for heat tolerance. Plant Breed Rev 1992; 10:129–168.

Hall AE. Physiology and breeding for heat tolerance in cowpea, and comparison with other crops. In: Kuo CG, ed. Adaptation of Food Crops to Temperature and Water Stress. Shanhua: Asian Vegetable Research and Development Center, Taiwan, Publ. No. 93-410, 1993:271–284.

Hall AE. Crop Responses to Environment. Boca Raton, FL: CRC Press, 2001:232.

Hall AE, Allen LH Jr. Designing cultivars for the climatic conditions of the next century. In: Buxton DR, Shibles R, Forsberg RA, Blad BL, Asay KH, Paulsen GM, Wilson RF, eds. International Crop Science I. Madison: Crop Sciences Society of America, 1993:291–297.

Hall AE, Frate CA. Blackeye bean production in California. Univ Calif Div Agric Sci Publ 1996; 21518:23.

Hall AE, Grantz DA. Drought resistance of cowpea improved by selecting for early appearance of mature pods. Crop Sci 1981; 21:461–464.

Hall AE, Patel PN. Breeding for resistance to drought and heat. In: Singh SR, Rachie

KO, eds. Cowpea Research, Production, and Utilization. New York: John Wiley and Sons, 1985:137–151.

Hall AE, Schulze E-D. Drought effects on transpiration and leaf water status of cowpea in controlled environments. Aust J Plant Physiol 1980a; 7:141–147.

Hall AE, Schulze E-D. Stomatal response to environment and a possible interrelation between stomatal effects on transpiration and CO_2 assimilation. Plant Cell Environ 1980b; 3:467–474.

Hall AE, Ziska LH. Crop breeding strategies for the 21st century. In: Reddy KR, Hodges HF, eds. Climate Change and Global Crop Productivity. Oxon: CABI, 2000:407–423.

Hall AE, Ismail AM, Ehlers JD, Marfo KO, Cisse N, Thiaw S, Close TJ. Breeding cowpea for tolerance to temperature extremes and adaptation to drought. Proceedings of World Cowpea research Conference III, 4–7 September 2000, IITA, Ibadan, Nigeria, 2002:14–21.

Hall AE, Ismail AM, Menendez CM. Implications for plant breeding of genotypic and drought-induced differences in water use efficiency, carbon isotope discrimination and gas exchange. In: Ehleringer JR, Hall AE, Farquhar GD, eds. Stable Isotopes and Plant Carbon–Water Relations. San Diego: Academic Press, 1993:349–369.

Hall AE, Richards RA, Condon AG, Wright GC, Farquhar GD. Carbon isotope discrimination and plant breeding. Plant Breed Rev 1994a; 12:81–113.

Hall AE, Singh BB, Ehlers JD. Cowpea breeding. Plant Breed Rev 1997a; 15:215–274.

Hall AE, Thiaw S, Ismail AM, Ehlers JD. Water-use efficiency and drought adaptation of cowpea. In: Singh BB, Mohan Raj DR, Dashiell KE, Jackai LEN, eds. Advances in Cowpea Research. Nigeria, Ibadan: IITA, 1997b:87–98.

Hall AE, Thiaw S, Krieg DR. Consistency of genotypic ranking for carbon isotope discrimination by cowpea grown in tropical and subtropical zones. Field Crops Res 1994b; 36:125–131.

Hardason G, Bliss FA, Cigales-Rivero MR, Henson RA, Kipe Nolt JA, Longeri L, Manrique A, Pena-Cabriales JJ, Pereira PAA, Sanabria CA, Tsai SM. Genotypic variation in biological nitrogen fixation by common bean. Plant Soil 1993; 152:59–70.

Henson RA, Pereira PAA, Carneiro JES, Bliss FA. Registration of "Ouro Negro," a high dinitrogen fixing, high-yielding common bean. Crop Sci 1993; 33:644.

Herridge D, Rose I. Breeding for enhanced nitrogen fixation in crop legumes. Field Crops Res 2000; 65:229–248.

Hong-Qui Z, Croes AF. Protection of pollen germination from adverse temperatures: a possible role for proline. Plant Cell Environ 1983; 6:471–476.

Ibarra-Perez FJ, Barnhart D, Ehdaie B, Knio KM, Waines JG. Effects of insect tripping on seed yield of common bean. Crop Sci 1999; 39:428–433.

Ibarra-Perez FJ, Ehdaie B, Waines JG. Estimation of outcrossing rate in common bean. Crop Sci 1997; 37:60–65.

Ismail AM, Hall AE. Inheritance of carbon isotope discrimination and water use efficiency in cowpea. Crop Sci 1993; 33:498–503.

Ismail AM, Hall AE. Positive and potential negative effects of heat-tolerance genes in cowpea lines. Crop Sci 1998; 38:381–390.

Ismail AM, Hall AE. Reproductive-stage heat tolerance, leaf membrane thermo-stability and plant morphology in cowpea. Crop Sci 1999; 39:1762–1768.

Ismail AM, Hall AE. Semidwarf and standard-height cowpea response to row spacing in different environments. Crop Sci 2000; 40:1618–1623.

Ismail AM, Hall AE. Variation in traits associated with chilling tolerance during emergence in cowpea germplasm. Field Crops Res 2002; 77:99–113.

Ismail AM, Hall AE, Bray EA. Drought and pot size effects on transpiration efficiency and carbon isotope discrimination of cowpea accessions and hybrids. Aust J Plant Physiol 1994; 21:23–35.

Ismail AM, Hall AE, Close TJ. Chilling tolerance during emergence of cowpea associated with a dehydrin and slow electrolyte leakage. Crop Sci 1997; 37:1270–1277.

Ismail AM, Hall AE, Close TJ. Purification and partial characterization of a dehydrin involved in chilling tolerance during seedling emergence of cowpea. Plant Physiol 1999a; 120:237–244.

Ismail AM, Hall AE, Close TJ. Allelic variation of a dehydrin gene co segregates with chilling tolerance during seedling emergence. Proc Natl Acad Sci 1999b; 96:13566–13570.

Ismail AM, Hall AE, Ehlers JD. Delayed-leaf-senescence and heat-tolerance traits mainly are independently expressed in cowpea. Crop Sci 2000; 40:1049–1055.

Jefferson RA. Beyond model systems—new strategies, methods and mechanisms for agricultural research. Ann N Y Acad Sci 1993; 700:53–73.

Kao W-Y, Comstock JP, Ehleringer JR. Variation in leaf movements among common bean cultivars. Crop Sci 1994; 34:1273–1278.

Kelly JD. Remaking bean plant architecture for efficient production. Adv Agron 2000; 71:109–143.

Kelly JD, Miklas PN. Marker-assisted selection. In: Singh SP, ed. Common Bean Improvement in the Twenty-First Century. Dordrecht: Kluwer Acad. Publ., 1999: 93–123.

Kelly JD, Schneider KA, Kolkman JM. Breeding to improve yield. In: Singh SP, ed. Common Bean Improvement in the Twenty-First Century. Dordrecht: Kluwer Acad. Publ., 1999:185–222.

Khalfaoui J-LB. Determination of potential lengths of the crop growing period in semi-arid regions of Senegal. Agric For Meteorol 1991; 55:251–263.

Khalfaoui J-LB, Havard M. Screening peanut cultivars in the field for root growth: a test by herbicide injection in the soil. Field Crops Res 1993; 32:173–179.

Kirchhoff WR, Hall AE, Isom WH. Phenotypic expression of a chlorophyll mutant in cowpea (*Vigna unguiculata* (L.) Walp.): environmental influences and effects on productivity. Field Crops Res 1989a; 21:19–28.

Kirchhoff WR, Hall AE, Roose ML. Inheritance of a mutation influencing chloro-phyll content and composition in cowpea. Crop Sci 1989b; 29:105–108.

Kirchhoff WR, Hall AE, Thomson WW. Gas exchange, carbon isotope discrimina-tion, and chloroplast ultrastructure of a chlorophyll-deficient mutant of cowpea. Crop Sci 1989c; 29:109–115.

Knauft DA, Wynne JC. Peanut breeding and genetics. Adv Agron 1995; 55:393–445.

Koide RT. Functional complementarity in the arbuscular mycorrhizal symbiosis. New Phytol 2000; 147:233–235.

Konsens I, Ofir M, Kigel J. The effect of temperature on the production and abscission of flowers and pods in snap bean (*Phaseolous vulgaris* L.). Annals Bot 1991; 67:391–399.

Kornegay J, White JW, Dominguez JR, Tejada G, Cajiao C. Inheritance of photoperiod response in Andean and Mesoamerican common bean. Crop Sci 1993; 33:977–984.

Kwapata MB, Hall AE. Effects of moisture regime and phosphorus on mycorrhizal infection, nutrient uptake, and growth of cowpeas (*Vigna unguiculata* (L.) Walp.). Field Crops Res 1985; 12:241–250.

Kwapata MB, Hall AE. Response of contrasting vegetable-cowpea cultivars to plant density and harvesting of young green pods: I. Pod production. Field Crops Res 1990a; 24:1–10.

Kwapata MB, Hall AE. Determinants of cowpea (*Vigna unguiculata*) seed yield at extremely high plant density. Field Crops Res 1990b; 24:23–32.

Kwapata MB, Hall AE, Madore MA. Response of contrasting vegetable cowpea cultivars to plant density and harvesting of young green pods: II. Dry matter production and photosynthesis. Field Crops Res 1990; 24:11–21.

Lee GJ, Vierling E. A small heat shock protein cooperates with heat shock protein 70 systems to reactivate a heat-denatured protein. Plant Physiol 2000; 122:189–197.

Li PH, Udomprasert N. Improving crop performance of *Phaseolus vulgaris* in high-temperature environments by heat acclimation potential. In: Kuo CG, ed. Adaptation of Food Crops to Temperature and Water Stress. Shanhua: Asian Vegetable Research and Development Center, Taiwan, Publ. No. 93-410, 1993:303–315.

Li PH, Davis DW, Shen Z-Y. High-temperature-acclimation potential of the common bean: can it be used as a selection criterion for improving crop performance in high temperature environments? Field Crops Res 1991; 27:241–256.

Littleton EJ, Dennett MD, Elston J, Monteith JL. The growth and development of cowpeas (*Vigna unguiculata*) under tropical field conditions. 3. Photosynthesis of leaves and pods. J Agric Sci Camb 1981; 97:539–550.

Lu Z, Percy RG, Qualset CO, Zeiger E. Stomatal conductance predicts yields in irrigated pima cotton and bread wheat grown at high temperatures. J Exp Bot 1998; 49:453–460.

Lush WM, Evans LT. Photoperiodic regulation of flowering in cowpeas (*Vigna unguiculata* (L.) Walp.). Ann Bot 1980; 46:719–725.

Lush WM, Evans LT, Wien HC. Environmental adaptation of wild and domesticated cowpeas (*Vigna unguiculata* (L.) Walp.). Field Crops Res 1980; 3:173–187.

Lynch J, Läuchli A, Epstein E. Vegetative growth of the common bean in response to phosphorus nutrition. Crop Sci 1991; 31:380–387.

Lynch JP, Yan X, Beebe SE. Breeding beans for low phosphorus soils. In: Schaffert RE, ed. Proceedings of a Workshop to Develop a Strategy for Collaborative Research and Dissemination of Technology in Sustainable Crop Production in Acid Savannas and other Problem Soils of the World, Purdue University, May 4–6, 1998:87–101.

Mai-Kodomi Y, Singh BB, Myers O, Yopp JH, Gibson PJ, Terao T. Two mechanisms of drought tolerance in cowpea. Indian J Genet 1999a; 59:309–316.

Mai-Kodomi Y, Singh BB, Terao T, Myers O, Yopp JH, Gibson PJ. Inheritance of drought tolerance in cowpea. Indian J Genet 1999b; 59:317–323.

Marfo KO, Hall AE. Inheritance of heat tolerance during pod set in cowpea. Crop Sci 1992; 32:912–918.

Menéndez CM, Hall AE. Heritability of carbon isotope discrimination and correlations with harvest index in cowpea. Crop Sci 1996; 36:233–238.

Menéndez CM, Hall AE, Gepts P. A genetic linkage map of cowpea (Vigna unguiculata) developed from a cross between two inbred, domesticated lines. Theor Appl Genet 1997; 95:1210–1217.

Mohamed HA, Clark JA, Ong CK. Genotypic differences in the temperature responses of tropical crops: I. Germination characteristics of groundnut (Arachis hypogaea L.) and pearl millet (Pennisetum typhoides S. & H.). J Exp Bot 1988; 39:1121–1128.

Morgan JM, Rodriguez-Maribona B, Knights EJ. Adaptation to water deficit in chickpea breeding lines by osmoregulation: relationship to grain yields in the field. Field Crops Res 1991; 27:61–70.

Mutters RG, Hall AE. Reproductive responses of cowpea to high temperatures during different night periods. Crop Sci 1992; 32:202–206.

Mutters RG, Ferreira LGR, Hall AE. Proline content of the anthers and pollen of heat-tolerant and heat-sensitive cowpea subjected to different temperatures. Crop Sci 1989a; 29:1497–1500.

Mutters RG, Hall AE, Patel PN. Photoperiod and light quality effects on cowpea floral development at high temperatures. Crop Sci 1989b; 29:1501–1505.

Nageswara Rao RC, Wright GC. Stability of the relationship between specific leaf area and carbon isotope discrimination across environments in peanut. Crop Sci 1994; 34:98–103.

Nageswara Rao RC, Udaykumar M, Farquhar GD, Talwar HS, Prasad TG. Variation in carbon isotope discrimination and its relationship to specific leaf area and ribulose 1,5-bisphosphate carboyxylase content in groundnut genotypes. Aust J Plant Physiol 1995; 22:545–551.

Nielsen CL, Hall AE. Responses of cowpea (Vigna unguiculata [L.] Walp.) in the field to high night temperature during flowering: I. Thermal regimes of production regions and field experimental system. Field Crops Res 1985a; 10:167–179.

Nielsen CL, Hall AE. Responses of cowpea (Vigna unguiculata [L.] Walp.) in the field to high night temperatures during flowering: II. Plant responses. Field Crop Res 1985b; 10:181–196.

Omwega CO, Thomason IJ, Roberts PA. A non-destructive technique for screening bean germplasm for resistance to Meloidogyne incognita. Plant Dis 1988; 72:970–972.

Ouédraogo JT, Gowda BS, Jean M, Close TJ, Ehlers JD, Hall AE, Gillaspie AG, Roberts PA, Ismail AM, Bruening G, Gepts P, Timko MP, Belize FJ. An improved genetic linkage map for cowpea (Vigna unguiculata L.) combining AFLP, RFLP, RAPD, biochemical markers, and biological resistance traits. Genome 2002; 45:175–188.

Patel PN, Hall AE. Inheritance of heat-induced brown discoloration in seed coats of cowpea. Crop Sci 1988; 28:929–932.

Patel PN, Hall AE. Genotypic variation and classification of cowpea for reproductive responses to high temperature under long photoperiods. Crop Sci 1990; 30:614–621.

Petrie CL, Hall AE. Water relations in cowpea and pearl millet under soil water deficits: I. Contrasting leaf water relations. Aust J Plant Physiol 1992a; 19:577–589.

Petrie CL, Hall AE. Water relations in cowpea and pearl millet under soil water deficits: II. Water use and root distribution. Aust J Plant Physiol 1992b; 19:591–600.

Petrie CL, Hall AE. Water relations in cowpea and pearl millet under soil water deficits: III. Extent of predawn equilibrium in leaf water potential. Aust J Plant Physiol 1992c; 19:601–609.

Petrie CL, Kabala ZJ, Hall AE, Simunek J. Water transport in an unsaturated medium to roots with different geometries. Soil Sci Soc Am J 1992; 56:1686–1694.

Porch TG, Jahn M. Effects of high-temperature stress on microsporogenesis in heat sensitive and heat-tolerant genotypes of *Phaseolus vulgaris*. Plant Cell Environ 2001; 24:723–731.

Robertson BM, Hall AE, Foster KW. A field technique for screening for genotypic differences in root growth. Crop Sci 1985; 25:1084–1090.

Schilling R, Misari SM. Assessment of groundnut research achievements in the savannah regions of West Africa. In: Nigam SN, ed. Groundnut—A Global Perspective: Proceedings of an International Workshop. Patancheru: ICRISAT, 1992: 97–112.

Schneider KA, Rosales-Serna R, Ibarra-Perez F, Cazares-Enriquez B, Acosta-Gallegos JA, Ramirez-Vallejo P, Wassimi N, Kelly JD. Improving common bean performance under drought stress. Crop Sci 1997; 37:43–50.

Scully B, Waines JG. Germination and emergence response of common and tepary beans to controlled temperature. Agron J 1987; 79:287–291.

Serraj R, Sinclair TR. N$_2$ fixation response to drought in common bean (*Phaseolus vulgaris* L.). Ann Bot 1998; 82:229–234.

Serraj R, Sinclair TR, Purcell LC. Symbiotic N$_2$ fixation response to drought. J Exp Bot 1999; 50:143–155.

Sexton PJ, Peterson CM, Boote KJ, White JW. Early-season growth in relation to region of domestication, seed size, and leaf traits in common bean. Field Crops Res 1997; 52:69–78.

Shackel KA, Hall AE. Reversible leaflet movements in relation to drought adaptation of cowpeas, *Vigna unguiculata* (L.) Walp. Aust J Plant Physiol 1979; 6:265–276.

Shackel KA, Hall AE. Comparison of water relations and osmotic adjustment in sorghum and cowpea under field conditions. Aust J Plant Physiol 1983; 10:423–435.

Shonnard GC, Gepts P. Genetics of heat tolerance during reproductive development in common bean. Crop Sci 1994; 34:1168–1175.

Sinclair TR, deWit CT. Photosynthate and nitrogen requirements for seed production by various crops. Science 1975; 189:565–567.

Sinclair TR, Serraj R. Legume nitrogen fixation and drought. Nature 1995; 378:344.

Singh BB, Mai-Kodomi Y, Terao T. A simple screening method for drought tolerance in cowpea. Indian J Genet 1999; 59:211–220.

Subbarao GV, Johansen C, Slinkhard AE, Nageswara Rao RC, Saxena NP, Chauhan

YS. Strategies for improving drought resistance in grain legumes. Crit Rev Plant Sci 1995; 14:469–523.

Thiaw S. Association between slow leaf-electrolyte-leakage under heat stress and heat tolerance during reproductive development in cowpea. Ph.D. dissertation, University of California, Riverside, 2003:100.

Thiaw S, Hall AE, Parker DR. Varietal intercropping and the yields and stability of cowpea production in semiarid Senegal. Field Crops Res 1993; 33, 217–233.

Turk KJ, Hall AE. Drought adaptation of cowpea: II. Influence of drought on plant water status and relations with seed yield. Agron J 1980; 72:421–427.

Turk KJ, Hall AE, Asbell CW. Drought adaptation of cowpea: I. Influence of drought on seed yield. Agron J 1980; 72:413–420.

Turner NC, Wright GC, Siddique KHM. Adaptation of grain legumes (pulses) to water-limited environments. Adv Agron 2000; 71:193–231.

Vara Prasad PV, Craufurd PQ, Summerfield RJ. Sensitivity of peanut to timing of heat stress during reproductive development. Crop Sci 1999a; 39:1352–1357.

Vara Prasad PV, Craufurd PQ, Summerfield RJ. Fruit number in relation to pollen production and viability in groundnut exposed to short episodes of heat stress. Annals Bot 1999b; 84:381–386.

Vara Prasad PV, Craufurd PQ, Summerfield RJ. Effect of high air and soil temperature on dry matter production, pod yield and yield components of groundnut. Plant Soil 2000a; 222:231–239.

Vara Prasad PV, Craufurd PQ, Summerfield RJ, Wheeler TR. Effects of short episodes of heat stress on flower production and fruit set of groundnut (*Arachis hypogaea* L.). J Exp Bot 2000b; 51:777–784.

Vierling E. The roles of heat shock proteins in plants. Annu Rev Plant Physiol Plant Mol Biol 1991; 42:579–620.

Warrag MOA, Hall AE. Reproductive responses of cowpea to heat stress: genotypic differences in tolerance to heat at flowering. Crop Sci 1983; 23:1088–1092.

Warrag MOA, Hall AE. Reproductive responses of cowpea (*Vigna unguiculata* (L.) Walp.) to heat stress: I. Responses to soil and day air temperatures. Field Crops Res 1984a; 8:3–16.

Warrag MOA, Hall AE. Reproductive responses of cowpea (*Vigna unuiculata* (L.) Walp.) to heat stress: II. Responses to night air temperature. Field Crops Res 1984b; 8:17–33.

Wells R, Bi T, Anderson WF, Wynne JC. Peanut yield as a result of fifty years of breeding. Agron J 1991; 83:957–961.

White JW. Implications of carbon isotope discrimination studies for breeding common bean under water deficits. In: Ehleringer JR, Hall AE, Farquhar GD, eds. Stable Isotopes and Plant Carbon–Water Relations. San Diego: Academic Press, 1993:387–398.

White JW, Castillo JA. Relative effect of root and shoot genotypes on yield of common bean under drought stress. Crop Sci 1989; 29:360–362.

White JW, Lang DR. Photoperiod response of flowering in diverse genotypes of common bean (*Phaseolus vulgaris*). Field Crops Res 1989; 22:113–128.

White JW, Castillo JA, Ehleringer JR. Associations between productivity, root growth and carbon isotope discrimination in *Phaseolus vulgaris* under water deficit. Aust J Plant Physiol 1990; 17:189–198.

White JW, Kornegay J, Cajiao C. Inheritance of temperature sensitivity of the photoperiod response in common bean (*Phaseolus vulgaris* L.). Euphytica 1996; 91:5–8.

Wien HC. Dry matter production, leaf area development, and light interception of cowpea lines with broad and narrow leaflet shape. Crop Sci 1982; 22:733–737.

Wien HC, Ackah EE. Pod development period in cowpeas: varietal differences as related to seed characters and environmental effects. Crop Sci 1978; 18:791–794.

Wien HC, Summerfield RJ. Adaptation of cowpeas to West Africa: effects of photoperiod and temperature responses in cultivars of diverse origin. In: Summerfield RJ, Bunting AH, eds. Advances in Legume Science. London: Her Majesty's Stationary Office, 1980:405–417.

Wright GC, Hubick KT, Farquhar GD, Nageswara Rao RC. Genetic and environmental variation in transpiration efficiency and its correlation with carbon isotope discrimination and specific leaf area in peanut. In: Ehleringer JR, Hall AE, Farquhar GD, eds. Stable Isotopes and Plant Carbon–Water Relations. San Diego: Academic Press, 1993:247–267.

Yan X, Beebe SE, Lynch JP. Genetic variation for phosphorus efficiency of common bean in contrasting soil types: II. Yield response. Crop Sci 1995; 35:1094–1099.

Yan X, Lynch JP, Beeebe SE. Genetic variation for phosphorus efficiency of common bean in contrasting soil types: I. Vegetative response. Crop Sci 1995; 35:1086–1093.

Yan X, Lynch JP, Beebe SE. Utilization of phosphorus substrates by contrasting common bean genotypes. Crop Sci 1996; 36:936–941.

7

The Physiological Basis of Soybean Yield Potential and Environmental Adaptation

Tara T. VanToai
USDA-ARS Soil Drainage Research Unit
Columbus, Ohio, U.S.A.

James E. Specht
University of Nebraska
Lincoln, Nebraska, U.S.A.

The soybean [*Glycine max* (L.) Merr.] was introduced into North America in 1765 (Hymowitz and Harlan, 1983). For many years thereafter it was used as a forage crop. However, its value as an oilseed crop grew after the turn of the century and, by 1924, nearly 181,300 ha (25% of the total U.S. crop) were harvested for seed. After the 1940s, nearly 100% of the crop was harvested for seed. By the 1960s, its value as a source of protein began exceeding its value as a source of oil. Nearly 30 million hectares of U.S. soybeans were harvested for seed in 2001.

Improvement of soybean seed yield, via both genetic and agronomic means, was initiated by personnel at U.S. agricultural experiment stations in the early 1930s and has continued unabated since then. The progress to date has been remarkable. Specht et al. (1999) reported that from 1924 to 1997 soybean yields on the average U.S. farm had increased at either a *linear* rate of

22.6 kg ha^{-1} yr^{-1} or at an *exponential* rate of 1.48% yr^{-1} (i.e., the R^2 values of 0.92 were nearly identical). Still, during just the last 25 years of that period (1972 to 1997), the linear improvement in yield has been 31.4 kg ha^{-1} yr^{-1}, a 40% faster rate than that of the prior 50 years.

Specht et al. (1999) attributed the continuing improvement in on-farm soybean yields over time to a combination of the three following factors: (1) genetic improvement in (a) yield potential, (b) adaptation to abiotic stresses, and (c) resistance/tolerance to specific biotic stresses; (2) agronomic improvement in management and production practices; and (3) a continuing rise in atmospheric CO_2 concentration. In this chapter, we review some of the physiological factors that may have contributed to genetic improvement in soybean yield potential and to improved adaptation to abiotic stress.

1 GENETIC BASIS OF YIELD IMPROVEMENT

Genetic improvement in yield is ordinarily estimated by evaluating the yield performance of a historical set of cultivar releases in a common environment, and then regressing cultivar yield on year of cultivar release. In the northern United States, Specht and Williams (1984) detected a 12.5 kg ha^{-1} yr^{-1} rate of *genetic* yield gain between 1943 and 1976, and noted that this was about 50% of the 23.7 kg ha^{-1} yr^{-1} *realized* yield gain during the same period. Similar rates of annual genetic yield gain were detected by Boerma (1979) from 1914 to 1973 in the southern United States (13.7 ha^{-1} yr^{-1}), by Ustun et al. (2001) from 1954 to 1987 in the mid-southern United States (14.4 ha^{-1} yr^{-1}), and by Voldeng et al. (1997) from 1934 to 1992 in Ontario, Canada (11 kg ha^{-1} yr^{-1}). However, genetic yield improvement in this Canadian Province was effectively nil prior to 1976, but was 30 kg ha^{-1} yr^{-1} thereafter. Karmaker and Bhatnagar (1996) reported a 22 kg ha^{-1} yr^{-1} estimate of annual genetic gain in India from 1969 to 1993. In the U.S. proprietary cultivar sector, Specht et al. (1999) calculated genetic gains of 25 to 30 kg ha^{-1} yr^{-1} since 1980. Wilcox (2001) recently evaluated 60 years of annual yield performance in the northern U.S. public breeding sector. He reported annual genetic gains in yield of about 20 to 30 kg ha^{-1} yr^{-1} in the various maturity groups over the entire 60 years. However, genetic gains were greater, 36 to 46 kg ha^{-1} yr^{-1}, when the analysis was limited to the past 20 years. Clearly, the foregoing estimated rates of current *genetic* yield improvement in the United States are of sufficient magnitude to account for at least 50%, if not more, of the estimated 31.4 kg ha^{-1} yr^{-1} rate of *on-farm* yield improvement that is currently being realized by "average" U.S. soybean producers.

It is important to recognize that the productivity of the evaluation environment has a significant impact on the estimated rate of genetic improvement (Specht et al., 1999; see also Slafer et al., 1993). The absolute yield difference between modern and obsolete cultivars is usually small in environments of low productivity, but widens significantly in more productive environments. In effect, as the major controlling *abiotic* factors in the production environment are incrementally optimized, the yield of the new cultivars rises more steeply than the yield of the old cultivars. Indeed, Mederski and Jeffers (1973) noted that the yield responsiveness to greater soil water availability was lower for older compared to newer cultivars. Frederick et al. (1991) confirmed this finding by showing that in high-yield water-abundant environments, modern cultivars had much greater seed and straw yields, whereas in the low-yield water-scarce environments, modern and ancestral cultivars did not differ in either trait. A yield response differential between modern vs. old cultivar may also arise from alteration of a *biotic* component of the production environment. For example, when two pre-1976 soybean cultivars, two post-1976 ones, and two very recent releases were grown at plant populations ranging from 33 to 100 plant m^{-2}, the yield of modern cultivars increased with greater plant density, whereas yield of the older cultivars did not (Specht et al., 1999).

While some may consider this old vs. new cultivar yield response differential to be merely genotype×environment ($G \times E$) interaction, others will recognize it for what it really is—an observable indication of *genetic improvement in yield potential*. This terminology, when used in a *relative* sense (Specht et al., 2001), denotes magnitudinal differences in genotypic *yield response per unit of optimization of some given abiotic factor* (e.g., change in yield per unit change in seasonal water, nitrogen, or other factor). When used in an *absolute* sense (Evans, 1993b), it is more narrowly defined as the genotypic *yield per se* differences observed when a historical or other reference set of genotypes are all grown in a common but *fully optimized* (nonstressed) production environment (Ludlow and Muchow, 1990; Evans and Fischer, 1999).

So what is the ultimate yield limit for soybean? Based on theoretical and empirical comparisons with maize, Specht et al. (1999) hypothesized that 8000 kg ha^{-1} was the likely *maximum* seed yield a soybean crop could achieve when grown in a completely optimized environment. Sinclair (1999) has suggested a seed yield limit of about 7000 kg ha^{-1}. Yield contest winners in the irrigated soybean production category (harvested area of at least 2 ha) have not yet achieved yields greater than 6660 kg ha^{-1} (83% of 8000 kg ha^{-1}). Insufficiency in the genetic yield potential of currently available cultivars might be a reason, or perhaps soybean yield contest winners have not yet been able to fully optimize their (relatively small scale) production environments. Of greater interest is the fact that the soybean yield achieved by the average U.S. soybean producer was, as recently as 1997, only about 2500 kg ha^{-1} (31% of 8000 kg ha^{-1}). Today, the average U.S. soybean producer adopts new cultivar releases almost as rapidly as do the yield contest winners. This led Specht et al. (1999)

to suggest that much of the genetic yield potential of currently available cultivars was not being realized on the average farm due to constraints (mostly abiotic ones) in the production environment. Some of those constraints are probably mitigable by better crop management (i.e., agronomic technologies). However, can abiotic production constraints be significantly alleviated by genetic technology? The answer depends on, as Evans and Fischer (1999) put it, whether on-farm yield improvement is ultimately driven only by genetic improvement in yield potential or by a combination of that kind of improvement *plus* improvement in genotypic adaptation to abiotic stress.

In most breeding programs, the best breeding lines chosen for release as new cultivars are almost invariably those that rank the highest in *overall* yield performance (i.e., yield averaged over all environments). As already noted, yield differences between breeding lines tend to be smaller in stressed (low-yield) environments, but larger in optimized (high-yield) environments, particularly so when water is the predominant yield-influencing factor (Frederick et al. 1991; Specht et al., 1986, 2001). When this $G \times E$ interaction is of the non-*crossover* type, genotypic yield differences between low- and high-yielding environments vary only in magnitude, not sign. Thus, selection based on overall genotypic yield rank would be expected to improve yield in the both kinds of environments, albeit less so in the stressed environments (Rosielle and Hamblin, 1981). On the other hand, if this $G \times E$ is of the *crossover* type, then inversions in genotypic yield rank between a stressed and optimized environment would result in an overall genotypic yield rank that obviously would not be representative of genotypic rank in either environment. In view of this, Evans and Fischer (1999) felt that reliance on genotypic yield in the optimized environment as the sole selection criterion would likely reduce yield stability (i.e., lead to new cultivars yielding less than older cultivars in stressed environments). However, Rossielle and Hamblin (1981) noted that if genetic variance in the stressed environment is lower than that in the optimized environment, then selection based on overall yield rank would still improve yield in both environments, unless there is a highly negative genetic correlation of yield between the two environments. Specht et al. (2001) empirically documented greater genetic variances in soybean yield in water-optimized environments and also showed that the genotypic yield correlation between water-optimized and water-stressed environments was highly positive.

Has long-term selection for yield per se had an effect on soybean yield stability? Wilcox et al. (1979) reported stability regression coefficients ranging from 0.88 to 1.17 for old and new cultivars adapted to the northern United States, but noted that only one coefficient (the highest one) was significantly different from unity. Voldeng et al. (1997), working with old and new cultivars adapted to southern Canada, reported an even wider range in coefficient values (0.39 to 2.23), but none of these values were statistically different from unity. Ustun et al. (2001), using soybean cultivars adapted to the mid-southern United States, showed that ancestral cultivars had stability regression coefficients of less than unity (*less* responsive to environmental optimization), whereas modern cultivars had coefficients of at least unity (maximum stability) or greater (*more* responsive to environmental optimization). Given these findings, it would seem that long-term selection based on overall yield performance has substantially improved soybean yield response to optimized environments, while *not* apparently resulting in new cultivars that yield *less* than older cultivars in stressed environments. Consequently, the average U.S. producer risks little by choosing to grow a new soybean cultivar in the absence of a priori knowledge of what the coming growing season will bring in terms of the controlling abiotic factors. Still, direct improvement in soybean adaptation to abiotic stress may be needed in some regions or specific production systems in which abiotic stress is routinely severe, and physiological traits that might help achieve this goal will be discussed in more detail in subsequent sections of this chapter.

If selection for overall mean yield performance in soybean has been successful in the development of new cultivars with greater yield potential without significant concomitant reduction in absolute stress tolerance (vis-à-vis older cultivars), then it is worthwhile to discuss how this genetic improvement in yield can be continued. Most of the genotypic variability in soybean yield has been shown to be due to additive gene effects and additive×additive epistasis (Burton, 1987). Genetic improvement in soybean yield is thus best achieved by fixing, at each and ultimately every gene locus, the most favorable allele available in the elite germ plasm pool. Breeders continually recombine their own elite germ plasm pools with those of other breeders to create the novel genotypes that will accomplish this fixation goal. Some researchers have argued that these "working" elite germ plasm pools do not contain all existing favorable alleles. They point to the genetic bottlenecks that likely resulted from just one (or a few) crop domestication events and from breeder use of just a few ancestor genomes to fuel the initial cultivar development efforts (Tanksley and McCouch, 1997). It is thus possible that many of the favorable alleles present in "banked" exotic germ plasm are not present in working elite germ plasm. U.S. soybean breeders are well aware of the narrow genetic base of modern U.S. cultivars (Gizlice et al., 1994, 1996). However, it is empirically difficult to introduce genetic diversity, even that of Chinese germ plasm (Cui et al., 2000), into the U.S. elite germ plasm pool, without dampening of the rate of genetic improvement that breeders are currently achieving in their cultivar development programs. Increasing the genetic diversity of U.S. elite germ plasm is frequently the stated goal of elite×exotic matings, but the implicit goal is actually *introgressing new favorable exotic alleles for yield into elite germ plasm*, and this has been difficult to do in practice. Tanksley and Nelson (1996) proposed *advanced backcross (BC)– quantitative trail locus (QTL) analysis* to achieve that goal. It involves mating an exotic donor parent (DP) with an elite recurrent parent (RP), then deriving a large number of BC1, BC2, or BC3 inbred lines (BILs) in which the retained DP genomic fractions have been reduced to 25%, 12.5%, or 6.25%, respectively (Wehrhahn and Allard, 1965). Any BIL possessing an exotic DP allele with a significant additive effect on yield should be readily identifiable in RP/ BIL yield comparison tests. As DP genomic fragments will have been (or can later be) "tagged" with flanking molecular markers, the favorable exotic alleles can be rapidly introgressed into other elite materials. This technique has been used successfully in tomato (Tanksley et al., 1996), and its utility in soybean is currently being evaluated (Specht, unpublished 2001–2002 data). The use of RP/BIL germ plasm renders more powerful the genetic analysis of differentials in genotypic yield response to environments varying for some given abiotic factor (e.g., water) because the yield response is then interpretable in terms of the presence of marker-delineated DP genomic fragments in the BIL.

2 PHYSIOLOGICAL BASIS OF YIELD IMPROVEMENT

The physiology of soybean seed yield is not well understood. Frederick and Hesketh (1994) noted 10 years ago that it remained to be seen whether selection based on any one physiological trait would ever offer greater annual genetic gain than what is currently being achieved via selection for yield per se. Still, a retrospective identification of physiological traits that have been significantly altered over the course of many years of breeding effort can offer insights as to how future empirical plant breeding might be mechanistically complemented.

An examination of the physiological basis of genetic improvement in yield is best begun from a simple yet basic theoretical framework. Sinclair (1993) noted that the potential for achieving further future improvements in crop yield (Y) was best understood in light of its relationship of seed yield with crop biomass (B) accumulation and final harvest index (HI) as in: $Y = B \times \mathrm{HI}$. Biomass accumulation was in turn best understood as an integrative function of daily crop interception (I) of solar radiation, crop radiation-use efficiency (RUE), and crop development time in days (DT) as in: $B = I \times \mathrm{RUE} \times \mathrm{DT}$.

Much of the genetic improvement in cereal crop seed yield over the past half-century has been attributed to a cumulative increase in crop HI (Evans, 1993a,b; Sinclair, 1998; Evans and Fischer, 1999). However, HI in many of those crops is now near its theoretical asymptotic limit of about 0.5 to 0.6, leading Sinclair (1993) to suggest that future genetic improvement of crop HI was probably rather limited. In soybean, the contribution of HI to past or future genetic improvement is not clear. Gay et al. (1980) reported that in a new vs. old *indeterminate* cultivar comparison, the new cultivar had a longer grain-filling period, whereas in a new vs. old *determinate* cultivar comparison, the new cultivar partitioned more photosynthate to seed. Specht et al. (1986) documented the asymptotic nature of the response of apparent HI to soil water and although HI in all 16 tested cultivars tended to plateau at seed yields of about 2500 kg ha^{-1}, the cultivars with the highest seed yields also had the highest HI values. Frederick et al. (1991) noted that old and new cultivars did not differ in their HI values, nor in their HI responses to water, simply because their seed and haulm (i.e., mature stems and pod walls) yields exhibited parallel responses to soil water variability. Sinclair (1993) observed that direct selection of greater HI in soybean can be problematic because shorter-season cultivars tend to have the highest HI, yet typically produce less seed yield than cultivars utilizing more of the growing season. Genetic alteration of soybean stem growth habit from an indeterminate to determinate form (or even to a semideterminate form) also does not seem to have much of an impact on apparent HI (Frederick and Hesketh, 1994). If HI in modern cultivars is in fact near its asymptotic genetic limit, then, as Specht et al. (1986) noted, significant future increases in seed yield would have to come from increases in *total biomass* (B). This would have to occur via genetic enhancement in components I, RUE, and/or DT of the above equation.

To improve I, crop interception of the incoming solar radiation would need to be increased. The rapidity with which crop leaf area completely covers the ground surface during early crop growth is certainly a suitable breeding target despite the fact that this trait is also greatly influenced by managerial choices of row spacing, plant population, and planting date. Greater early-season plant vigor might lead to a faster rate of leaf appearance, thereby reducing the seasonal time needed for the crop to attain a leaf area index (LAI) of 3.3, which Sinclair (1993) noted was necessary to intercept 90% of the incident solar radiation. Genetic variation for vigorous early growth and rapid canopy closure has been shown to exist (Mian et al., 1998).

There are two primary sources of *genetic* variation in RUE (Sinclair, 1993). One is the photosynthetic pathway for initial CO_2 carboxylation (i.e., C3 vs. C4 species), and the other is the intrinsic energy content of the seed biomass (i.e., protein and/or oil accumulating species vs. mostly carbohydrate accumulating species). Soybean, a C3 accumulator of protein and oil, has a maximum RUE about 1.1 g biomass MJ^{-1} of intercepted energy, whereas maize, a C4 accumulator of mostly carbohydrate, has a maximum RUE of about 1.7 g MJ^{-1} (Sinclair and Muchow, 1999). Obviously, a biotechnological transformation of soybean into a C4 species would likely improve its

RUE, but this is not possible yet (see Ku et al., 1999). Improving soybean RUE by selecting for less protein and oil in its seed would, however, destroy the very characteristics for which soybean is valued in the marketplace. While RUE is functionally related to the crop's (radiation-saturated) leaf CO_2 exchange rate (CER), that relationship is curvilinear such that the RUE response flattens out at higher CER values (Sinclair and Muchow, 1999). Thus, while crop biomass accumulation is indeed responsive to genetic (or other) increases in maximum photosynthetic rates at low CER levels, that responsiveness declines to near zero at the higher CER levels found in crops grown in productive environments. Sinclair (1993) noted that avoiding stress-induced *decreases* in CER and, thence, RUE may be a sounder approach for realizing more of the yield potential of a genotype.

Soybean breeders routinely develop and release crop cultivars that fully exploit the (average) available growing season in the production latitudes (usually termed maturity zones) for which those new cultivars are targeted. Therefore, the crop development time (DT) component in the above RUE equation has likely been maximized to the extent it can be. On the other hand, the continued rise in atmospheric CO_2 is expected to gradually lengthen growing seasons in more northern latitudinal zones, thereby offering an opportunity for genetic improvement in DT. Still, breeders routinely examine breeding lines for latitudinal adaptation and so will need no new selection tools to exploit a slowly lengthening growing season.

Seed yield is obviously the product of the number of seeds produced by plants in a given unit of area multiplied by the mean seed weight. Does a yield component analysis (e.g., seed yield = plants/area×nodes/plant×pods/ node×seed/pod×weight/seed, all components as means) offer any insights about genetic improvement in yield? Specht and Williams (1984) showed that soybean 100-seed weight had increased by 0.1 g yr^{-1} over the course of the 75 years (1902–1977) of 240 cultivar releases in the United States and Canada. When vegetable cultivar types were dropped from their analysis, the increase was smaller, 0.04 g yr^{-1}, but still statistically significant. Other studies involving fewer cultivars or shorter release periods have not, however, uncovered hard evidence that *genetic* yield improvement has altered either mean seed weight or seed number (Specht et al., 1999). Egli (1994) noted that total seed growth rate (TSGR) per unit land area (g m^{-2} day^{-1}) was the product of the number of seed produced per unit land area and the mean individual seed growth rate (ISGR) (g seed^{-1} day^{-1}). Seed yield was thus a product of TSGR and the seed-fill duration in days. While the duration of soybean seed fill is usually strongly related to final seed yield and a new cultivar was shown to have a longer seed-filling period than an old cultivar (Gay et al., 1980), its measurement is laborious, making its genetic manipulation difficult. Egli (1994) found no evidence of a relationship of ISGR with yield and observed that any relationship of TSGR with yield must then be based on variation in total seed produced per unit area, a variable that Shibles et al. (1975) had previously noted was often associated with yield. However, targeting specific yield components, such as seed number or seed size, for selective change in order to indirectly improve yield has not been very successful (Egli, 1994) primarily because yield components have an allometric (nonindependent) relationship with each other. Indeed, cultivars with *equivalent* yields in a given production environment frequently display disparate seed number/size combinations. According to Board (1985), soybean pod number is determined primarily during the early reproductive (R1–R4) stages (Fehr and Caviness, 1977), whereas seed size is determined during the effective filling period (EFP) between the R5 and the R7 reproductive stages. Yield components, because of this sequential development, are more useful in the analysis of the ontogenetic sensitivity of plant yield response to seasonally timed abiotic stresses (cf., Shaw and Laing, 1966; Sinoit and Kramer, 1977; Korte et al., 1983a,b; Kadhem et al., 1985a,b; Desclaux et al., 2000). Let us now examine some published research data on the physiological basis of genetic improvement in soybean accumulation of carbon and nitrogen dry matter.

2.1 Dry Matter Accumulation

According to Frederick and Hesketh (1994), soybean genetic improvement was associated with assimilate supply during the seed-filling period (SFP), but not with changes in harvest index. A similar association was also reported between assimilate supply during the SFP and the yield improvement in modern soybean cultivars (Specht et al.,1999). In a 2-year study of two old and two new soybean cultivars, Kumudini et al. (2001a) showed similar patterns of dry matter accumulation until the beginning of the SFP. After the onset of the SFP, the newer, higher-yielding cultivars accumulated more dry matter at a greater rate than the old cultivars, which they attributed to a greater leaf area index (LAI) during SFP in the newer cultivars. They reported that the genetic improvement in yield was 78% due to increased total dry matter and 22% due to a higher HI. Shiraiwa and Hashikawa (1995) also showed that two modern Japanese cultivars had more than double the dry matter increase during the SFP, when compared to two older cultivars. Contrary to the above findings, in a 4-year study of 14 short-season soybean cultivars representing six decades of breeding and selection (1934 to 1992) in Ontario, Canada, Morrison et al. (1999) reported a yield improvement of 0.5% yr^{-1} that was significantly correlated with increases in harvest index, leaf photosynthetic rate, and stomatal conductance and with a decrease in leaf area. The modern cultivars tended to have a higher chlorophyll content and thicker leaves with lower specific leaf area (SLA). The correlation between leaf chlorophyll content and SLA with year of release, however, was not significant. Buttery and Buzzell (1982), using 21 cultivars bred and released over a 30-year period, reported a significant correlation between yearly yield improvement and lower leaf area, but showed no relationship between leaf photosynthetic rate and yield. Larson et al. (1981) detected no correlation between apparent photosynthesis and yield in an evaluation of cultivars released between 1927 and 1973. The discrepancy between yield and leaf photosynthesis relationships among the different studies could be attributed to the different plant densities (Morrison et al., 1999). In general terms, modern cultivars tend to have a higher photosynthetic rate, net assimilate ratio, and canopy apparent photosynthesis during the SFP when compared to older cultivars (Dornhoff and Shibles, 1970; Buttery et al., 1981; Ashley and Boerma, 1989).

Dry matter partitioning to the seed can affect seed yield by limiting the rate and the duration of seed dry matter accumulation. The lower seed yield of older cultivars is consistent with a hypothesis that older cultivars are source-limited for assimilates during SFP. According to Specht et al. (1999), older cultivars had shorter leaf area duration than newer cultivars, which explains the lower assimilate supply of the older cultivars during the latter part of the SFP. Other evidences of a limitation in assimilate supply during the SFP were obtained from partial depodding experiments. When pods of alternate nodes were manually removed, older cultivars maintained leaf area for a longer duration and increased the 100-seed weight more significantly than newer cultivars.

2.2 Nitrogen Accumulation and Seed Protein

One direct impact of the genetic improvement in HI that has occurred in crops over the past century is obvious—a shuffling of more photosynthetic carbon from the nonseed fraction to the seed fraction. However, as Sinclair (1998) pointed out, in most crops, the seed fraction has a substantially greater nitrogen concentration than the nonseed fraction. Thus, modern cultivars with high HI must accumulate substantially more nitrogen than low HI older cultivars do, just to sustain the production of (the N-richer) seed dry matter. This is true even if the high-yielding new cultivars and the low-yielding old cultivars have an identical HI.

In general, soybean seed nitrogen does correlate with plant leaf nitrogen content (Specht et al., 2001). A number of studies have provided evidence of genotypic variation in nitrogen accumulation and variation in its partitioning to seeds. Nitrogen supply to the leaf canopy is often a limiting factor in soybean photosynthesis. Because the great demand for nitrogen by soybean seeds causes a rapid decline in nitrogen content of vegetative parts, nitrogen partitioning during seed filling may affect dry matter production and seed yield as well as nitrogen accumulation. Specht et al. (1999) found the patterns of N accumulation to be similar to those for dry matter accumulation. Shiraiwa and Hashikawa (1995) reported that newer Japanese cultivars accumulated more N during the SFP than did older cultivars. Kumudini et al. (2001b) also showed that newer cultivars continued N accumulation during the SFP. Nitrogen accumulation and partitioning in plants can also affect dry matter production. Partial depodding experiments showed increased root N content in older cultivars, but did not affect the root N content of more recent releases (Specht et al., 1999). Apparently, recently released cultivars are more effective than older cultivars in transporting N from the roots to the seeds. Voldeng et al. (1997) reported that during 58 years of Canadian cultivar releases, seed protein content declined by 4 g kg^{-1}, whereas oil rose by 4 g kg^{-1}, directional changes that Wilcox et al. (1979) observed in an earlier study. On average, a genetically mediated 1% increase in seed oil content will result in a 2% decrease in protein content (and vice versa), reflecting the highly negative genetic correlation that exists between these two seed constituents. Canadian soybean breeders have been able to increase seed yield while still ensuring that the combined percentage of seed protein and oil content did not change appreciably (Voldeng et al., 1997). Although the six decades of genetic improvement in seed yield was accompanied by a decline in seed protein that totaled to 24 g protein kg^{-1} seed, that decline was more than offset by a 36 kg increase in seed protein yield on a per hectare basis. The extra seed N per hectare required by the new cultivars clearly resulted from greater N fixation rather than from greater soil N (Specht et al., 1999). This would support the concept that new cultivars are better able to supply assimilates during the SFP than old cultivars, thus allowing more assimilate supply to roots and nodules for N fixation.

3 PHYSIOLOGICAL BASIS OF ENVIRONMENTAL TOLERANCE

According to Boyer (1982), the yield gap between average on-farm crop yield and the crop yield potential in optimized conditions was attributable to the following causes: (1) diseases and insects, which are sporadically devastating at the local level, generally do not depress average U.S. yields by more than 4.1% and 2.6%, respectively; (2) weed competition, which depresses soybean yield by 4%; and (3) inappropriate soil conditions and/or unfavorable weather, which were estimated to decrease yield potential by as much as 69%. The reduction in yield attributable to the various physiochemical factors (i.e., abiotic stresses) has long been the subject of much study from physiologists and breeders (Nilsen and Orcutt, 1996; Ludlow and Muchow, 1990; Raper and Kramer, 1987; Araus et al., 2002), and it is the topic we turn to next.

3.1 Drought Stress

In the yield potential section of this chapter, a simple equation was used to describe the dependence of biomass accumulation on solar radiation: $B = I \times$ RUE $\times$ DT. Unfortunately, for terrestrial plants, the acquisition of the atmospheric carbon dioxide needed for that biomass accumulation is inextricably coupled with transpiratory loss of H_2O, making the relationship between biomass accumulation and transpiration an obligatory, yet highly linear one (Sinclair, 1994). Biomass (B) accumulation can be treated as an integrative function of daily crop transpiration (T); crop water-use efficiency (WUE, defined as the $\Delta B/\Delta T$ coefficient relating B to T); and days of crop development time (DT): $B = T \times$ WUE $\times$ DT. In the literature, this equation is often condensed to $B =$ WUE $\times$ Tr, where Tr is the seasonal total transpiration (i.e., Tr $= T \times$ DT). Tanner and Sinclair (1983) noted that evaporative demand was invariably driven by the leaf-to-air vapor pressure deficit, which was highly dependent on site-specific meteorological conditions. They partitioned WUE into biological (crop) and physical (meteorological) components: WUE $= k/$vpd, where k was defined as a species-specific coefficient of crop seasonal water use and vpd was defined as a seasonal integration of the average daily atmospheric vapor pressure deficit experienced by the crop during transpiration. The mechanistic factors the authors coalesced into their k coefficient were described in detail in their classic book chapter (Tanner and Sinclair, 1983), summarized in a shorter journal article published a year later (Sinclair et al., 1984), and updated in another book chapter 10 years later (Sinclair, 1994). One of the key factors included in the constant k was the term $1 - C_{int}/C_{atm}$, where C_{int} denotes the mole fraction of the internal (leaf) CO_2 and C_{atm} denotes the atmospheric CO_2. Tanner and Sinclair (1983) noted that the maximal value for this term was about 0.3 for many C3 species and about 0.7 for many C4 species. More will be said about this term later.

Water availability is obviously the paramount abiotic factor affecting crop productivity (Boyer, 1982). Indeed, in nearly all crop production areas, insufficiency in the amount of water available for crop transpiration is the abiotic stress factor that is the most recurrent, if not annually prevalent, reducer of crop yield (Waggoner, 1994). Just what is the impact of water stress year in and year out? Specht et al. (1999) noted that in the 25-year period from 1972 to 1997, the yield difference between irrigated (I) and rainfed (R) soybean production in Nebraska averaged about 800 kg ha^{-1}. However, because the $I - R$ yield difference never went to zero in high rainfall years, some of the 800 kg ha^{-1} average difference was probably due to confounding differences in I vs. R husbandry (i.e., irrigated land is more intensively managed). To minimize some of this confounding, Specht et al. (1999) regressed the annual yield difference (I minus R) on either the annual I yield (result: $R^2 = 0$) or the annual R yield (result: $R^2 = 0.36$; $b = -0.39$). This analysis suggested that 36% of the year-to-year variation in the magnitude of $I-R$ yield difference was attributable to the annual variation in R yield. Within the same region, soybean I and R systems are presumably exposed to the same weather factors in any given year, except that in the former, rainfall deficits are mitigated by irrigation. One could logically conclude that in Nebraska, at least, about one-third (i.e., 36%) of the variation in the size of annual $I-R$ yield difference (800 kg ha^{-1}) is attributable to a fluctuating insufficiency of water for crop transpiration in rainfed soybean agriculture.

Passioura (1977, 1994) succinctly observed that seed yield in water-limited production environments was best understood as the product of three largely independent entities: $Y = Tr \times WUE \times HI$ (all as defined above). If water-limited yield can indeed be improved by genetic manipulation of some specified physiological trait, then the alteration of that trait must achieve that yield-enhancing effect via a demonstrable enhancement of Tr, WUE (i.e., its biological k component), and/or HI. Any plant trait having no obvious connection to any of those three entities is unlikely to have much influence on water-limited yield (Passioura, 1983). It is important to distinguish between intensified rainfed agriculture vs. subsistence agriculture when discussing crop drought tolerance. Producers in the former category will reject cultivars selected to perform better than average in drought years, i.e., smaller reduction in yield per unit of lessened transpiration, *if those same cultivars lack yield potential and thus perform poorly in favorable years*. In other words, the yield premium accrued by drought-tolerant cultivars in droughty environments must not be substantially less than the yield penalty these cultivars accrue in high rainfall years (vis-à-vis cultivars offering high yield potentials). This is especially true in areas where a low rainfall year is no more predictable than a high rainfall year.

The physiological nature of soybean response to water deficits has been the subject of many published studies (Lawn, 1982a,b,c; Muchow 1985a,b; Sinclair and Ludlow, 1986; Neyshabouri and Hatfield, 1986; Cox and Jolliff, 1987; Hoogenboom et al., 1987; Frederick et al., 1990; Djekoun and Planchon, 1991; Morgan 1992). The general theme that has emerged from these studies is that the soybean exploits available soil water more rapidly, to deeper depths, and to lower soil water potentials than do other grain legumes primarily because soybean stomatal closure occurs at lower (more negative) water potentials. While stomatal closure at higher water potentials in those other crops conserves water, it simultaneously lessens carbon dioxide uptake.

Given the tight linkage between transpiration and biomass accumulation (see above equation) and the near-identical WUE values that grain legumes exhibit in a common environment, water *not* transpired by crop is effectively equivalent to biomass *not* produced. Because of its opportunistic behavior with respect to exploiting soil water for transpiration purposes, a soybean crop is generally more productive than the other grain legumes, even in those environments in which water stress, although harsh at times, occurs only in intermittent intervals. An opportunistic behavior does entail more risk because a soybean crop could, in the event of a protracted period of water stress, exhaust all available soil water before completing its reproductive development. Moreover, although soybean stomata tend to remain open at leaf water potentials that ordinarily cause stomatal closure in other grain legumes, the soybean is actually less tolerant of dehydration than other grain legumes (Sinclair and Ludlow, 1986). However, this liability is apparently offset by the soybean's greater ability to maintain its turgor pressure by an active (as opposed to passive) adjustment of osmolytes (solutes) in the soybean plant cells (Morgan, 1992).

Osmoregulation (i.e., the lowering of osmotic potential) is an important means of maintaining cell turgor allowing important turgor-driven processes such as growth by cell expansion and stomatal opening/closing to continue despite falling water potentials. Most plants are known to accumulate solutes during drought stress (Guo and Ooserhuis, 1995, 1997; Gorham et al., 1984; Paul and Cockburn, 1989; Sacher and Staples, 1985; Nguyen and Lamant, 1988). See Hare et al. (1998) for a recent review of this topic. In soybean, pinitol, a methylated cyclitol, accumulates several time more than other solutes. Streeter et al. (2001) reported that pinitol concentrations were similar among U.S. soybean cultivars and breeding lines, but varied substantially among Chinese plant introductions (PIs). A survey of the geographical distribution of the PIs showed that pinitol accumulation and rainfall distribution were negatively correlated. More importantly, correlation was not detected with any other solutes including proline, a well-known osmoprotectant. Similar negative relationships between pinitol accumulation and rainfall distribution were also reported in populations of maritime pine (*Pinus-spinaster*) adapted to dry and humid areas (Nguyen and Lamant, 1988). The differential accumulation of pinitol was only detected in well-watered plants. Under drought stress, both groups of plants accumulated similar levels of pinitol. Therefore, the normal accumulation of pinitol, not its induction by drought stress, is the key to its role in stress tolerance. It is hypothesized that the normal accumulation of this osmoprotectant is an adaptive mechanism for survival in dry environments, where the plants must tolerate some water deficit on a regular, possibly daily basis. The actual mechanism for the action of pinitol as an osmoprotectant is not well established. Paul and Cockburn

(1989) reported that pinitol accumulated principally in chloroplasts. This is consistent with the positive relationship between cyclitol accumulation and CO_2 assimilation under drought (Sheveleva et al., 1997). The role of proline as a major osmoprotectant in the cytoplasm is well established in many species (Ford, 1984; Paul and Cockburn, 1989; Nanjo et al., 1999), but its role in the tolerance of soybean to drought and other stresses has not been documented. Hong et al. (2000) reported that genetic manipulation to increase proline accumulation increases the tolerance of the *Arabidopsis* plants to osmotic stress, while de Ronde et al. (2000) reported that antisense-mediated down-regulation of proline accumulation in soybean resulted in plants unable to survive a 6-day drought stress treatment. Although soybean does not accumulate glycinebetaine (a nontoxic cytoplasmic osmolyte) except in trace amounts, it has been shown that exogenous application of this osmolyte to soybean leaves reduced transpiration and increased phytomass (Agboma et al., 1997).

Ludlow and Muchow (1990) published a list of 16 "traits" that they rationalized as having demonstrable or putative benefits with respect to seed yield performance in water-limited production environments. They then evaluated those traits with respect to their possible enhancing impact on the Tr, WUE, and HI components of the above equation. The reader can refer to Ludlow and Muchow (1990) for a summary of their mostly equivocal postulations of the effect of each of the 16 traits on HI, as only those traits impacting Tr and WUE will be reviewed here. The reader needing additional detail can consult Ludlow and Muchow (1990) or Turner et al. (2001) for specifics relative to grain legumes, or to additional references we cite below.

Of the 16 traits, only two, greater leaf transpiration efficiency (TE) and possibly increased leaf reflectivity, were expected to possibly improve WUE ($\Delta B / \Delta T$). Transpiration efficiency (TE) is defined as the mass or moles of C or CO_2 fixed per unit of water lost from the *leaf*. Farquhar et al. (1982) showed that C_{int} is related to the degree with which ^{13}C (a naturally occurring stable isotope of carbon) is discriminated against (vis-à-vis ^{12}C) during photosynthesis in C3 species. They offered convincing arguments that carbon-13 isotope discrimination during biomass accumulation is inversely proportional to TE such that *less discrimination* of ^{13}C vis-à-vis ^{12}C during biomass accumulation is indicative of a *greater TE*. Specht et al. (2001) used this technique to assess the TE of over 200 genotypes in fully watered and extremely water-stressed conditions. They found that a genotype grown in the water-stressed condition exhibited a significantly greater TE than it did in the fully watered condition, indicating that water stress itself induced a greater TE. In fact, on a genotypic basis, the TE for a well-watered genotype was highly correlated with the TE of the same water-stressed genotype ($r = 0.71$). Moreover, there was a moderately positive correlation of the yield of a genotype with the TE of that same genotype, when the latter was expressed in either the fully watered or drought-stressed environment. However, a QTL analysis of genotypic differences in TE revealed that most of the genetic variation in TE was arising from the segregation of genes governing maturity and stem growth habit, and that the presumed QTLs for TE were simply pleiotropic extensions of the QTLs for those traits. This finding, plus the weak correlation ($r = 0.26$) of genotypic TE with an empirically determined genotypic yield response to water (i.e., *beta*), would be unlikely to convince breeders of its worth of directly selecting for TE, particularly given the costly expense of conducting ^{13}C measurements (Specht et al., 2001).

Sinclair (1994) stated that, in crops that have undergone intense and long-term selection for yield, WUE is probably already near its maximum. Although genotypic variability in WUE has been found, the observed ranges of that variability have not generally exceeded the values of 0.3 (C3 species) and 0.7 (C4 species) that Sinclair (1994) noted were probably near-maximal ones for the $1-C_{int}/C_{atm}$ term. Enhanced carboxylation activity would certainly decrease C_{int} (and thereby increase coefficient k in WUE). An enhancement in daily ΔB without appreciable change in daily ΔT would obviously improve WUE, plus contribute positively to RUE as well. However, significant increases in carboxylation activity in soybean are not likely short of a (possible future) transformation of soybean into a C4 species via the approaches that are currently being attempted in rice (Ku et al., 1999).

Greater leaf reflectivity to incoming solar radiation should lower leaf temperature and thus the vpd component of WUE, so long as no offsetting concomitant decrease in biomass accumulation occurs because of the non-absorption of the solar radiation that was reflected away. If daily ΔB in the WUE numerator is not appreciably changed by greater leaf reflectivity, then a fractionally smaller daily ΔT in the WUE denominator would indeed improve WUE. This might be of particular value in a terminal stress environment, assuming that the unused water (resulting from a lessening of daily ΔT per unit of daily ΔB) can be retained and be made available for crop use later in crop development. The value of such conserved water in an intermittent stress environment would depend on the duration of each period of stress. Soybean leaf reflectivity can be genetically manipulated to achieve a denser population of trichomes on the leaf surface and this does seem to improve WUE (Clawson et al., 1986). However, the denser leaf pubescence must be gray, not tawny, in color. In addition, isogenic analysis of normal vs. dense pubescence in multiple-cultivar backgrounds has revealed that densely pubescent lines do not always yield better than their normal counterparts in the field (Zhang et al., 1992).

The reflectivity of a leaf to incoming radiation, and thereby its temperature, is influenced by leaf orientation, which in turn is governed by reversibly rapid diurnal leaf movements. These leaf movements have been categorized as either *diaheliotropic* or *paraheliotropic*, in that the plane of the leaf lamina is moved to make it more *perpendicular* or more *parallel* to the sun's incident rays (Berg and Heuchelin, 1990; Donahue and Berg, 1990). Leaf movement is driven by variation in the turgor potential of the pulvinus (or pulvinule) organ at the base of the leaf blade. Genetic differences in soybean leaf movements have been reported (Wofford and Allen, 1982; Kao and Tsai, 1998). Paraheliotropism provides a rapid reversible mechanism for reducing the amount of energy available for transpiration, but it also obviously reduces the amount of energy available for photosynthesis, so the trait is unlikely to improve WUE. Note that paraheliotropic leaf movement and stomatal closure operate in parallel (both reduce transpiration, but at the expense of photosynthesis). Indeed, both mechanisms seem to commence at a leaf water potential of about -0.4 MPa (Berg and Heuchelin, 1990).

Given the previous discussion on how leaf orientation or reflectivity impacts leaf temperature, it is worthwhile digressing for a moment to discuss canopy temperature. The temperature of well-irrigated fully transpiring plant canopy is usually lower than the air temperature, but the inverse is usually true for a water-stressed minimally transpiring canopy. The temperature differential between the canopy and air ($Td = Tc - Ta$) can thus serve as an indicator of plant water stress and that differential can be estimated with a portable, handheld infrared thermometer. However, attempts to utilize Td to divergently select for "hot" and "cool" segregants in soybean populations were not very successful due to low heritabilities (0% to 20%) and the imprecision inherent in Td measurement (McKinney et al., 1989a). Moreover, the so-called hot genotypes (exhibiting less transpiration) were neither more drought-tolerant nor more yield-stable than the cool genotypes (McKinney et al., 1989b).

If improvement in WUE is not readily achievable, then the equation $Y = Tr \times WUE \times HI$ suggests that the next best means of improving water-limited yield is rendering available to the plant (for Tr purposes) any stored soil water that is not currently being exploited. Of the 16 traits listed in Ludlow and Muchow (1990), modifications in eight were postulated to result in the transpiration of more water. Two of the eight—calibrating crop phenology or adjustment of photoperiod sensitivity to better match predictable meteorological patterns in the seasonal water supply—have probably already been optimized by soybean breeders in existing growing regions. The same is probably true for a third, increasing developmental plasticity (e.g., indeterminate flowering) to mitigate the uncertainty in exact timing of meteorological events. Three other traits expected to maintain transpiration were maintenance of leaf area (notably a minimum LAI of 3.3) during water stress, tolerance to higher temperatures with respect to stomatal closure, and increased capacity for osmotic adjustment (to be discussed in more detail below). It should be pointed out that the goal in modifying these three traits is sustaining transpiration in the face of drought stress and thereby sustaining biomass accumulation (via the inextricable linkage of the two processes). The main problem with any trait that sustains (or possibly even enhances) transpiration is that it must invariably be coupled with another trait that will make more soil water available to accommodate that transpiration. The remaining two (of the eight) traits—early-season vegetative vigor and a root system of the depth and the density needed to optimal exploitation of soil water—have the potential for doing just that.

Early-season plant vigor was mentioned previously in this chapter as a means of ramping up leaf area to capture more of the solar radiation that early in the season can be intercepted by leaves instead of falling needlessly on the soil surface. Early-season shoot and leaf vigor, if accompanied by a more vigorous surface root development, could divert into transpiration, a greater fraction of the available surface soil water whose fate is ordinarily evaporation from the soil surface.

Most researchers believe that a rooting system capable of extracting water from deeper soil layers (or inadequately explored soil volumes more near the surface) is critical for achieving higher yields in water-limited environments. Implicit in this rationalization, however, are several assumptions: (1) that this water is actually present and available in the inadequately unexplored or deeper soil volumes; (2) that this water is routinely replenished before the next cropping cycle; and (3) that the cost in photosynthetic assimilate diverted from the shoot to the root system to tap those soil volumes is not an appreciable one (Ludlow and Muchow 1990; Sinclair, 1994). The downward progress of a root system as a function of time and depth is termed the extraction front velocity (EFV). Soybean roots have EFV values of about 30–35 mm day^{-1}, but the downward progression of roots tends to cease at about the beginning of the R5 seed-filling stage (Hoogenboom et al., 1987). Genetic variation for root system phenotype relative to drought stress has been reported (Goldman et al., 1989; Hudak and Patterson, 1996), and research work is continuing in this area. Passioura (1983) questioned the value of deep roots on the basis of whether the *transpiratory water cost* of producing the dry matter biomass to create those deeper roots could be fully offset by the *amount of water actually recovered* by those roots and added to the transpiratory stream. The solution to this dilemma is not clear although Passioura (1994) believed that drought-mediated regulation of both leaf and root growth, whether in the *relative* (to the shoot) or *absolute* sense, was probably mediated by hormones such as abscisic acid (cf., Bunce, 1990) and, therefore, needs more serious study. On the other hand, while acknowledging that there was no evidence of a causal connection between leaf water potential and stomatal closure, Bunce (1999) presented data that was not supportive of the thesis that an abscisic acid (ABA) signal emanating from drying roots was responsible for stomatal closure. Instead, his data strongly suggested control of stomatal conductance by ABA produced by local leaf water potential rather than root signals.

The remaining six traits in the list of 16 offered by Ludlow and Muchow (1990) were not postulated to enhance Tr or WUE. These included: (1) greater mobilization of preanthesis dry matter; (2) lower root hydraulic conductivity; (3) lower lethal water status; (4) reduced stomatal conductance; (5) leaf movements (discussed earlier); and (6) lower epidermal conductance (cf., Sinclair and Ludlow, 1986; Paje et al., 1988).

For many years, it was assumed that water crossed biological membranes via simple diffusion although membrane permeability in some tissues was much too high to be due to simple diffusion. Based on research conducted during the 1990s, it now appears that certain membrane-intrinsic proteins (MIPs), more commonly known as aquaporins or water channels, regulate and facilitate the permeation of water across biological membranes and may thus be intimately involved in plant response to water stress. These proteins are certainly drawing the attention of many researchers, given the number of recent published reviews on the possible role of aquaporins in root water uptake and hydraulics (Stuedle, 2000; Baiges et al., 2002; Sperry et al., 2002; Tyerman et al., 2002). Perhaps the most readable recent review in this subject area is that by Javot and Maurel (2002). In the apoplastic region of the plant, water moves passively in response to water potential gradients that establish hydrostatic tension in the shoot-to-root xylem pathway. In the symplastic region of the plant, cell membranes allow the establishment of osmotic gradients along the cell-to-cell pathway (i.e., via active pumping of solutes into root cells than in turn creates a positive hydrostatic root pressure). In plants with a high rate of transpiration, hydrostatic tension predominates to move water along the apoplastic pathway from root to shoot. When stomata are closed, osmotic forces predominate, with water movement occurring via the cell-to-cell pathway. Plant roots have the remarkable capacity to rapidly alter their hydraulic conductivity, mediated apparently by changes in cell membrane permeability. It is becoming clear that aquaporin-rich membranes are involved in the intense flow of water across roots and probably act as a "gating" mechanism for facilitating and controlling root water uptake in the symplast. A comprehensive knowledge of nature and function of root cell membrane aquaporins is still lacking; thus, it is too soon to know if plant aquaporins will be consequential in terms of genetically improving soybean tolerance to abiotic stress.

Although soybean photosynthesis is sensitive to water stress, N_2 fixation activity in soybean actually declines in advance of virtually all other physiological processes when the soil dries (Sinclair et al., 1987). The sensitivity of N_2 fixation to water stress appears to be related to a water stress-induced accumulation of ureides in leaves, which feedback inhibits the N_2 fixation process in the nodules. This has led to the postulation that ureide degradation in leaves may be the key to improving soybean tolerance to dry soils (Vadez and Sinclair, 2000, 2001). Genetic variation in N_2 fixation capacity under water stress exists (Patterson and Hudak, 1996), but the recent focus has been on using leaf ureide amount under stress conditions to identify genotypes with greater N_2 fixation tolerance to water deficits (Sinclair et al., 2000; Vadez and Sinclair, 2002).

3.2 Flooding

After drought, flooding due to excess water is the second most damaging constraint on crop growth and affects about 16% of the production areas worldwide (Boyer, 1982). Soil can become flooded when it is poorly drained or when rainfall or irrigation is excessive. Other terms, such as soil saturation, waterlogging, anoxia, and hypoxia are also commonly used to describe flooding conditions. Flooding causes premature senescence which results in leaf chlorosis, necrosis, defoliation, reduced nitrogen fixation, cessation of growth and reduced yield. The severity of the flooding stress is affected by many factors including flooding duration, crop variety, growth stage, soil type, fertility levels, pathogens, and flooding conditions (Sullivan et al., 2001). In general, flooding can be divided into either waterlogging, where only the roots are flooded, or complete submergence, where the entire plants are under water. While plants develop adaptive mechanisms to allow them to survive long-term waterlogging (Bacanamwo and Purcell, 1990), most plants die within one or two days of submergence (Sullivan et al., 2001). Soybean plants adapt to prolonged root-flooding conditions by producing adventitious roots close to the soil surface (Bacanamwo and Purcell, 1990). Other anatomical traits associated with adaptation to flooding involve the development of stem hypertrophy and formation of lysigenous aerenchyma tissues (Bacanamwo and Purcell, 1990). The internal gas-filled channels are used   to transport O_2 from the shoots to submerged roots to sustain aerobic respiration and growth (Armstrong and Webb, 1985; Drew et al., 1985). Internal gas-space network is also used for the upward escape of potentially toxic substances, including ethanol and acetaldehyde, products of anaerobic fermentation (Chircova and Gutman, 1972), as well as CO_2.

   The lack of oxygen has been proposed as the main problem associated with flooding (Armstrong,1978; Jackson and Drew, 1984; Kozlowski, 1984). Plants are able to survive short periods of oxygen deprivation by switching from aerobic to anaerobic metabolism. Anaerobiosis induces rapid dissociation of polysomes (Lin and Key, 1967) and the biosynthesis of four anaerobic proteins in soybean (Russell et al., 1990), which are much fewer than the 20+ anaerobic proteins detected in maize roots (Sachs et al., 1980). Anaerobic proteins in maize are enzymes of glycolysis, glucose metabolism, or fermentation (Peschke and Sachs 1994) and may play an important role in the tolerance of this species to anaerobiosis (Sachs et al., 1996). The simpler molecular response of soybean to anaerobic stress was interpreted as indicative of the less tolerant of this species to anaerobiosis as compared to maize (Russell et al., 1990).

Tolerance to anoxia and hypoxia has been used synonymously with tolerance to flooding stress (Russell et al., 1990). However, tolerance to field flooding appears to be much more complex than tolerance to artificially induced hypoxia and anoxia. While field flooding of soybean for as short as three days often results in injurious symptoms listed above, soybeans can thrive in stagnant water in the greenhouse. Soybeans grown in hydroponic medium continuously bubbled with nitrogen gas, where the dissolved oxygen level was not detectable, showed no symptoms of stress (Boru et al., 2002). Soybean, therefore, is much more tolerant to excessive water and the lack of oxygen than previously reported. The reasons underlying the dramatic differences between responses to flooding in the greenhouse and flooding in the field are not known. However, growth reduction and yield loss in flooded fields could arise from root rot diseases, nitrogen deficiency, nutrient imbalance, and/or the accumulation of toxic levels of CO_2 in the root zone.

According to Ponnamperuma (1972), the concentration of CO_2 in flooded soils can reach a level as high as 50% of the total gases and could be toxic to plants. In central Ohio, the soil CO_2 concentrations of nonflooded soybean fields were around 1%, but increased to 30% to 35% (v/v) after two weeks of flooding (Boru et al., 2002). In experiments where soybean was grown in root zones with 15% and 30% CO_2 balanced with nitrogen gas— conditions similar to those in flooded fields—the plants were stunted, chlorotic, and necrotic. No detrimental effects were observed in similarly treated rice plants. The results suggested that the high susceptibility of soybean to flooding as compared to rice might be due to its greater sensitivity to CO_2.

Flooding tolerance is usually defined as minimal or no yield loss. According to VanToai et al. (1994), waterlogging for 4 weeks during the early flowering stage reduced the average grain yield of 84 U.S. soybean cultivars by 25%. Yield reduction, however, varied from 9% in the most flood-tolerant cultivar to 75% in the most flood-sensitive cultivar. Flooding tolerance can also be defined as high yield under flooding stress. According to this definition, the most flood-tolerant variety in this study produced 3.7 Mg ha^{-1}, while the least produced 1.27 Mg ha^{-1}. When the cultivars were ranked for flooding tolerance based on both definitions, seven of the 10 most flood-tolerant cultivars were the same; seven of the 10 least flood-tolerant cultivars were also the same. Thus, the two definitions of flooding tolerance, either high yield under flooding or minimal yield difference between nonflooded and flooded conditions, appear to be compatible. Flooding tolerance is independent from nonflooded yield, indicating that genetic variability for flooding tolerance exists and could be improved through plant breeding and selection.

VanToai et al. (2001) identified a single deoxyribonucleic acid (DNA) marker, which was associated with improved plant growth (from 11% to 18%) and grain yields (from 47% to 180%) of soybean in waterlogged environments. The identified marker was uniquely associated with waterlogging tolerance and was not associated with maturity, normal plant height, or grain yield. Field testing of near-isogenic lines with and without the marker failed to show any yield increase associated with the marker under soil waterlogging conditions. It was not clear whether the near-isogenic lines that were developed based on one single marker, not flanking markers, possess the "gene" for flooding tolerance. Nevertheless, flooding tolerance, similar to drought tolerance, is a complex trait. Its improvement would require combined efforts of many disciplines including crop management, soil fertility, plant pathology, plant physiology, plant breeding, molecular biology, and genetic transformation (Zhang et al., 2000).

3.3 Oxidative Stress

In addition to flooding, soybean plants are also sensitive to the return to normoxic (normal atmospheric) conditions following flooding stress, caused by reactive oxygen species (VanToai and Bolles, 1991). Exogenous application of ascorbate, a reactive oxygen scavenger, to the flooding medium reduced the postanoxic injuries in soybean seedlings that were returned to normoxic conditions after 2 to 3 hr of submergence in deoxygenated water. Oxidative stress is also associated with other stresses including salt, cold, and heat (Zhu, 2000). According to Tsugane et al. (1999), a mutant with enhanced capacity to detoxify reactive oxygen was also more tolerant to salt. Moreover, plants genetically engineered to overproduce reactive oxygen scavenging osmolytes grow and survive better in salt, cold, and hot environments (Hayashi et al., 1997).

3.4 Salt Stress

Salinity reduces the growth of many plant species including soybean (Shannon, 1997). While the roots are in direct contact with the surrounding saline medium, root growth of soybean and other nonhalophytic plant species are impaired to a lesser degree than shoot growth (Munns and Termaat, 1986).

Salt stress results in an accumulation of Na^+ and Cl^- to toxic concentrations in soybean tissues (Lauchli and Wieneke, 1979) as well as a decrease in root hydraulic conductivity (Joly, 1989). The lower water flux through the plant reduces leaf water potential which together with the high osmotic potential and specific ion toxicity result in reduced soybean emergence and seedling growth (Wang and Shannon, 1999) as well as in lowered LAI, radiation absorption, and radiation-use efficiency at whole plant levels (Wang et al., 2001).

At the cellular levels, salt stress reduces the cell size of soybean root as well as the rate of cell division resulting in shorter roots (Azaizeh et al., 1992). Among the different root cell types, meristemic cells are the most sensitive to salt stress (Huang and Van Steveninck, 1990). Injuries of soybean meristemic cells observed at NaCl concentrations of 150 mM or higher include separation of cell membrane from cell wall, disorganization of mitochondria, and cytoplasm, followed by nuclear deformation and apoptosislike DNA degradation (Liu et al., 2000).

Salt tolerance is a complex trait involving tolerance to ionic and osmotic stresses as well as to oxidative stresses as described above (Zhu, 2000; Zhu, 2001). The regulation of salt stress tolerance appears to be mediated by cytosolic Ca which is induced by salt stress (Lauchli, 1990; Knight et al., 1997). External Ca has been known to enhance plant tolerance to salt stress. The mechanism by which Ca relieves salt toxicity is likely achieved through the signaling pathways that regulate the influx, efflux, and compartmentalization of Na, K, Ca, and other ions. Tolerant plants can either restrict the influx of Na or Cl or enhance the influx of K (Warne and Hickok, 1987) and/ or Ca. Much progress on the mechanism of salt tolerance has been accomplished through studies of *Arabidopsis* (Zhu, 2000). Osmotolerant mutants were identified that could germinate under saline conditions as well as under conditions of high KCl, K_2SO_4, LiCl, and mannitol. However, the identification of mutants that are sensitive to Na, but not to Mg, Ca, Cl, and NO_3, or SO_4 or manitol confirms that salinity tolerance is independent from osmotic tolerance.

Much less information is available on the tolerance of soybean to salt stress. Salt-tolerant soybean biotypes appear to be able to exclude the influx, to reduce translocation of Cl and Na to the shoots, and to maintain high root hydraulic conductivity when exposed to toxic NaCl concentrations (An et al., 2001; Abel, 1969).

3.5 Cold Stress

As a temperate species, soybean plants do not withstand subfreezing temperatures. The ability to tolerate cold stress can be induced in many plant species by a period of exposure to low nonlethal temperatures known as cold acclimatization (Guy, 1990; Guy et al., 1985; Levitt, 1980). Cold acclimation occurs through complex biochemical and physiological changes that include reduction or cessation of growth, reduction of tissue water content, increase in ABA content, alterations in membrane lipid composition, and accumulation of compatible osmolytes such as proline, betaine, and soluble sugars (Joshi, 1999). The expression of the cold regulated (COR) genes associated with acclimation has been well documented (Bohnert et al., 1995; Somerville, 1995; Thomashow, 1998). Some of the COR genes are also inducible by drought and salinity as well as by ABA treatment, suggesting that certain portions of the environmental stress-signaling pathways are common among the stresses and are triggered by stress-inducible ABA. Treatment with 7.5×10^{-5} molar ABA significantly increase both the degree of cold hardiness and the rate of hardening of winter wheat (*Triticum aestivum* L. cv. Norstar), winter rye (*Secale cereale* L. cv. Cougar), and bromegrass (*Bromo inermis* Leyiss) cultured cells more than that induced by low temperature alone. However, according to Chen and Gusta (1983), soybean and other temperate-season species do not cold harden upon exposure to cold or treatment with ABA. Unlike the heat shock response, soybean plants exposed to cold temperatures continue to synthesize housekeeping proteins, and only quantitative differences in protein synthesis were detected in soybean cultivars differing in cold sensitivity (Cabane et al., 1992). Kim et al. (2001) isolated a novel cold-inducible DNA-binding protein from soybean whole constitutive expression-induced COR gene expression and enhanced cold tolerance of nonacclimated transgenic plants. However, mutants that are constitutively freezing-tolerant accumulate a high level of proline, but no other COR gene products. According to the results, distinct signaling pathways activate different aspects of cold acclimation and the activation of one pathway may be sufficient for cold tolerance without the activation of other pathways.

3.6 Heat Stress

As a temperate crop species, soybean plants experience heat stress when the leaf temperatures exceed 32–33°C. Symptoms of heat injuries include cessation of cytoplasmic streaming, protein denaturation, changes in membrane lipid composition, reduced membrane stability, and reduced photosynthesis efficiency (Srinivasan et al., 1996). Leakage of electrolytes has been used frequently as a screening test for tolerance to heat stress based on the effects of heat on membrane integrity (Srinivasan et al., 1996; Bouslama and Scha-paugh, 1984). The tolerance index calculated from the electrolyte leakage results provides consistent and reproducible genotypic differences among the 20 soybean cultivars tested. However, no correlation was detected between tolerance index and dry land yield or dry land/irrigated yield stability ratio (Bouslama and Schapaugh, 1984).

Tolerance of soybean plants to injuries and death by extreme heat increases dramatically when plants are exposed to a short treatment of moderately elevated temperature (Vierling, 1991; Burke, 1998). The acquired thermotolerance is correlated with the accumulation of the low molecular weight heat-shock proteins (17 to 30 kDa) that are the most abundant proteins induced by heat (Lin et al., 1984, 1985; Kimpel and Key, 1985; Nagao et al., 1985). The heat-shock proteins act as molecular chaperones in normal cellular metabolism, but become indispensable in stabilizing membrane structures and functions under heat stress.

Among the many cellular structures and metabolic processes that are perturbed by heat, photosynthesis is the most heat-sensitive process in soybean (Alfonso et al., 2000). Heat stress induces inactivation of the oxygen-evolving machinery of photosystem II (PSII) which is the most heat-susceptible component of the chloroplast membranes (Berry and Bjorkman, 1980). Chlorophyll accumulation in seedling cotyledons was used to characterize acquired thermotolerance of soybean (Burke, 1998).

An atrazine-resistant (STR7) soybean mutant was isolated, which was also intrinsically more tolerant to heat. The heat tolerance of the STR7 mutant was correlated with an increased saturated C16:0 content and reduced levels of C16:1 fatty acids. Differential scanning calorimetry revealed that the temperature of the onset of phase transition (Tc), which disrupts membrane organization and function, was $5°C$ higher for STR7 thylakoid membranes than for wild-type membranes. Additionally, genetic manipulation to reduce fatty acid unsaturation in higher plants has resulted in thermotolerant phenotypes. These results together suggest that plant tolerance to higher temperature might be improved by reducing the content of chloroplast-specific dienoic and monoic fatty acids.

4 CONCLUSION

Today's soybean cultivars are more efficient at producing and allocating carbon resources to seeds than were their predecessors. Genetic improvements through soybean breeding and selection for high yield under a wide range of environmental conditions have resulted in modern cultivars that withstand high plant density as well as provide high yield in rainfed conditions. To sustain and enhance crop yield improvement in the future, innovations in the form of genetic and agronomic technologies must be continually interjected into the on-farm agricultural enterprise. Those technologies, originating from both basic and applied research, will keep us moving ever closer to the ceiling seed yield of 7000–8000 kg ha^{-1} (Sinclair,

1999; Specht et al., 1999). In any event, continued development of genomic tools and technologies will become critically important in the future to mechanistically complement conventional empirical methods, given that genetic gains in yield will become increasingly more difficult to achieve after each incremental step brings us closer to the biological soybean yield limit.

REFERENCES

Abel GH. Inheritance of the capacity for chloride inclusion and exclusion by soybeans. Crop Sci 1969; 9:697–698.

Agboma PC, Sinclair TR, Jokinen K, Peltonensainio P, Pehu E. An evaluation of the effect of exogenous glycinebetaine on the growth and yield of soybean—Timing of application, watering regimes and cultivars. Field Crops Res 1997; 54:51–64.

Alfonso M, Yruela I, Almarcegui S, Torrado E, Perez MA, Picorel R. Unusual tolerance to high temperatures in a new herbicide-resistant D1 mutant from *Glycine max* (L.) Merr. cell cultures deficient in fatty acid desaturation. Planta 2000; 212:573–582.

An P, Inanaga S, Kafkafi U, Lux A, Sugimoto Y. Different effect of humidity on growth and salt tolerance of two soybean cultivars. Biol Plant 2001; 44:405–410.

Araus JL, Slafer GA, Reynolds MP, Royo C. Plant breeding and drought in C3 cereals: What should we breed for? Ann Bot 2002; 89:925–940.

Armstrong W. Root aeration in the wetland condition. In: Hook DD, Crawford RMM, eds. Plant Life in Anaerobic Environments. Ann Arbor, MI: Ann Arbor Science, 1978:269–297.

Armstrong W, Webb T. A critical oxygen pressure for root extension in rice. J Exp Bot 1985; 36:1573–1582.

Ashley DA, Boerma HR. Canopy photosynthesis and its association with seed yield in advanced generations of a soybean cross. Crop Sci 1989; 29:1042–1045.

Azaizeh H, Gunse B, Steudle E. Effects of NaCl and CaCl2 on water transport across root cells of maize (*Zea mays* L.) seedlings. Plant Physiol 1992; 99:886–894.

Bacanamwo M, Purcell LC. Soybean root morphological and anatomical traits associated with acclimation to flooding. Crop Sci 1999; 39:143–149.

Baiges I, Schaffner AR, Affenzeller MJ, Mas A. Plant aquaporins. Physiol Plant 2002; 115:15–182.

Berg VS, Heuchelin S. Leaf orientation in soybean seedlings: I. Effect of water potential and photosynthetic photon flux density on paraheliotropism. Crop Sci 1990; 30:631–638.

Berry J, Bjorkman O. Photosynthetic response and adaptation to temperature in higher plants. Annu Rev Plant Physiol 1980; 31:491–543.

Board JE. Yield components associated with soybean yield reduction at nonoptimal planting dates. Agron J 1985; 77:135–140.

Boerma HR. Comparison of past and recently developed soybean cultivars in maturity groups VI, VII and VIII. Crop Sci 1979; 19:611–613.

Bohnert HJ, Nelson DE, Jensen RG. Adaptation to environmental stresses. Plant Cell 1995; 7:1099–1111.

Boru G, VanToai T, Alves J, Hua D, Knee M. Susceptibility of soybean to flooding is caused by elevated CO_2 levels in the root zone. Ann Bot 2002. In Press.

Boru G, VanToai T, Alves JD. Flooding injuries in soybean are caused by elevated carbon dioxide levels in the root zone. Fifth National Symposium on Stand Establishment, 1997:205–209.

Bouslama M, Schapaugh WT. Stress tolerance in soybeans: I. Evaluation of three screening techniques for heat and drought tolerance. Crop Sci 1984; 24:933–937.

Boyer JS. Plant productivity and environment. Science 1982; 218:443–448.

Bunce JA. Abscisic acid mimics effects of dehydration on area expansion and photosynthetic partitioning in young soybean leaves. Plant Cell Environ 1990; 13:295–298.

Bunce LA. Leaf and root control of stomatal closure during drying in soybean. Physiol Plant 1999; 106:190–195.

Burke JJ. Characterization of acquired thermotolerance in soybean seedlings. Plant Physiol Biochem 1998; 36:601–607.

Burton JW. Quantitative genetics: Results relevant to soybean breeding. In: Wilcox JR, ed. Soybeans: Improvement, Production, and Uses. Madison, WI: American Society of Agronomy, 1987:211–247.

Buttery BR, Buzzell RI. Some differences between soybean cultivars observed by growth analysis. Can J Plant Sci 1982; 52:13–20.

Buttery BR, Buzzell RI, Findlay WI. Relationships among photosynthetic rate, bean yield and other characters in field-grown cultivars of soybean. Can J Plant Sci 1981; 61:191–198.

Cabane M, Vincens P, Boudet AM. Protein synthesis at low temperatures in 2 soybean cultivars differing by their cold sensitivity. Physiol Plant 1992; 85:573–580.

Chen THH, Gusta LV. Abscisic acid-induced freezing resistance in cultured plant cells. Plant Physiol 1983; 73:71–75.

Chircova TV, Gutman TC. The physiological role of branch lenticels of willow and poplar under conditions of root anaerobiosis. Fiziol Rast 1972; 19:352–359.

Clawson KL, Specht JE, Blad BL. Growth analysis of soybean isolines differing in pubescence density. Agron J 1986; 78:164–172.

Cox WJ, Joliff GD. Crop–water relations of sunflower and soybean under irrigated and dryland conditions. Crop Sci 1987; 27:553–558.

Cui Z, Carter TE Jr, Burton JW. Genetic diversity patterns in Chinese soybean cultivars based on coefficient of parentage. Crop Sci 2000; 40:1780–1793.

de Ronde JA, Spreeth MH, Cress WA. Effect of antisense L-D1-pyrroline-5-carboxylate reductase transgenic soybean plants subjected to osmotic and drought stress. Plant Growth Regul 2000; 32:13–26.

de Wit CT. Photosynthesis: Its relationship to overpopulation. In: San Pietro A, ed. Harvesting the Sun. New York: Academic Press, 1967:315–320.

Desclaux D, Huynh TT, Roumet P. Identification of soybean plant characteristics that indicate the timing of drought stress. Crop Sci 2000; 40:716–722.

Djekoun A, Planchon C. Tolerance to low leaf water potential in soybean genotypes. Euphytica 1991; 55:247–253.

Donahue R, Berg VS. Leaf orientation of soybean seedlings: II. Receptor sites and light stimuli. Crop Sci 1990; 30:638–643.

Dornhoff GM, Shibles RM. Varietal differences in net photosynthesis of soybean leaves. Crop Sci 1970; 10:42–45.

Drew MC, Saglio PH, Pradet A. Larger adenylate charge and ATP/ADP ratios in aerenchymatous roots of *Zea mays* in aerobic media as a consequence of improved internal oxygen transport. Planta 1985; 165:51–58.

Egli DG. Seed growth and development. In: Boote KJ, Bennet JM, Sinclair TR, Paulsen GM, eds. Physiology and Determination of Crop Yield. Madison, WI: American Society of Agronomy, 1994:127–148.

Evans LT. Crop Evolution, Adaptation and Yield. Cambridge: Cambridge Univ. Press, 1993a.

Evans LT. Processes, genes, and yield potential. In: Buxton DR, Shibles RM, Forsberg RA, Bland BL, Asay KH, Paulsen GM, Wilson RF, eds. International Crop Science I. Madison, WI: Crop Science Society of America, 1993b:687–696.

Evans LT, Fischer RA. Yield potential: Its definition, measurement, and significance. Crop Sci 1999; 39:1544–1551.

Farquhar GD, O'Leary MH, Berry JA. On the relationship between carbon-isotope discrimination and intercellular carbon dioxide concentration. Aust J Plant Physiol 1982; 11:121–137.

Fehr WR, Caviness CE. Stages of soybean development. Iowa Coop Ext Serv Spec Rep 1977; 80.

Ford CW. Accumulation of low molecular weight solutes in water-stressed tropical legumes. Phytochemistry 1984; 23:1007–1015.

Frederick JR, Hesketh JD. Genetic improvement in soybean: Physiological attributes. In: Slafer GA, ed. Genetic Improvement of Field Crops. New York: Marcel Dekker, Inc., 1994:237–286.

Frederick JR, Woolley JT, Hesketh JD, Peters DB. Water deficit development in old and new soybean cultivars. Agron J 1990; 82:76–81.

Frederick JR, Woolley JT, Hesketh JD, Peters DB. Seed yield and agronomic traits of old and modern soybean cultivars under irrigation and soil water-deficit. Field Crops Res 1991; 27:71–82.

Gay S, Egli DB, Reicosky DC. Physiological aspects of yield improvement in soybeans. Agron, 1980, 72–387.

Gizlice Z, Carter TE Jr, Burton JW. Genetic base for North American public soybean cultivars released between 1947 and 1988. Crop Sci 1994; 34:1143–1151.

Gizlice Z, Carter TE Jr, Gerig TM, Burton JW. Genetic diversity patters in North American public soybean cultivars based on coefficient of parentage. Crop Sci 1996; 36:753–765.

Goldman IL, Carter TE Jr, Patterson RP. Differential genotypic response to drought stress and subsoil aluminum in soybean. Crop Sci 1989; 29:330–334.

Gorham J, McDonnell E, Wyn Jones RG. Pinitol and other solutes in salt-stressed Sesbania aculeata. Z Pflanzenphysiol 1984; 114:173–178.

Guo C, Ooserhuis DM. Pinitol occurrence in soybean plants as affected by temperature and plant growth regulators. J Exp Bot 1995; 46:253–299.

Guo C, Ooserhuis DM. Effect of water-deficit stress and genotypes on pinitol occurrence in soybean plants. Environ Exp Bot 1997; 37:147–152.

Guy CL. Cold acclimation and freezing stress tolerance: Role of protein metabolism. Annu Rev Plant Physiol Plant Mol Biol 1990; 41:187–223.

Guy CL, Niemi KJ, Brambl R. Altered gene expression during cold acclimation of spinach. Proc Natl Acad Sci 1985; 82:3673–3677.

Hare PD, Cress WA, Van Staden J. Dissecting the roles of osmolyte accumulation during stress. Plant Cell Environ 1998; 21:535–553.

Hayashi H, Mustardy L, Deshnium P, Ida M, Murata N. Transformation of *Arabidopsis thaliana* with the codA gene for choline oxidase: Accumulation of glycinebetaine and enhanced tolerance to salt and cold stress. Plant J 1997; 12:133–142.

Hong Z, Lakkineni K, Zhang Z, Verma DPS. Removal of feedback inhibition of l-pyrroline-5-carboxylate synthetase results in increased proline accumulation and protection of plants from osmotic stress. Plant Physiol 2000; 122:1129–1136.

Hoogenboom G, Peterson CM, Huck MG. Root growth of soybean as affected by drought stress. Agron J 1987; 79:607–614.

Huang CX, Van Steveninck RFM. Salinity induced structural changes in meristemic cells of barley roots. New Phytol 1990; 115:17–22.

Hudak CM, Patterson RP. Root distribution and soil moisture depletion pattern of a drought-resistant—soybean plant introduction. Agron J 1996; 88:478–485.

Hymowitz T, Harlan JR. Introduction of Soybeans to North America by Samuel Bowen in 1765. Econ Bot 1983; 37:371–379.

Jackson MB, Drew MC. Effects of flooding on growth and metabolism of herbaceous plants. In: Kozlowski TT, ed. Flooding and Plant Growth. Orlando, FL: Academic Press, Inc., 1984:47–128.

Javot H, Maurel C. The role of aquaporins in root water uptake. Ann Bot 2002; 90:301–313.

Joly RJ. Effects of sodium chloride on the hydraulic conductivity of soybean root systems. Plant Physiol 1989; 91:1262–1265.

Joshi AK. Genetic factors affecting abiotic stress tolerance in crop plants. In: Pessarakli M, ed. Handbook of Plant and Crop Stress. 2d ed. New York: Marcel Dekker, 1999.

Kadhem FA, Specht JE, Williams JH. Soybean irrigation serially timed during stages R1 to R6: I. Agronomic responses. Agron J 1985; 77:291–298.

Kadhem FA, Specht JE, Williams JH. Soybean irrigation serially timed during stages R1 to R6: II. Yield component responses. Agron J 1985; 77:299–304.

Kao W-Y, Tsai T-T. Tropic leaf movements, photosynthetic gas exchange, leaf d13 and chlorophyll *a* fluorescence of three soybean species in response to water availability. Plant Cell Environ 1998; 21:1055–1062.

Karmaker PG, Bhatnagar PS. Genetic improvement of soybean varieties released in India from 1969 to 1993. Euphytica 1996; 90:95–103.

Kim JC, Lee SH, Cheong YH, Yoo CM, Lee SI, Chun HJ, Yun DJ, Hong JC, Lee SY, Lim CO, Cho MJ. A novel cold-inducible zinc finger protein from soybean, SCOF-1, enhances cold tolerance in transgenic plants. Plant J 2001; 25:247–259.

Kimpel JA, Key JL. Presence of heat shock mRNAs in field grown soybeans. Plant Physiol 1985; 79:672–678.

Knight H, Trewavas AJ, Knight MR. Calcium signaling in *Arabidopsis thaliana* responding to drought and salinity. Plant J 1997; 12:1067–1078.

Korte LL, Specht JE, Williams JH, Sorenson RC. Irrigation of soybean genotypes during reproductive ontogeny: II. Yield component responses. Crop Sci 1983; 23:528–533.

Korte LL, Williams JH, Specht JE, Sorenson RC. Irrigation of soybean genotypes during reproductive ontogeny: I. Agronomic responses. Crop Sci 1983; 23:521–527.

Kozlowski TT. Extent, causes, and impacts of flooding. In: Kozlowski TT, ed. Flooding and Plant Growth. Orlando, FL: Academic Press, Inc., 1984:1–8.

Ku MSB, Agarie S, Nomura M, Fukayama H, Tsuchida H, Ono K, Toki S, Miyao M, Matsuoka M. High-level expression of maize phosphoenolpyruvate carboxylase in transgenic rice plants. Nat Biotechnol 1999; 17:76–80.

Kumudini S, Hume DJ, Chu G. Genetic improvement in short-season soybeans: I. Dry matter accumulation, partitioning, and leaf area duration. Crop Sci 2001; 41:391–398.

Kumudini S, Hume DJ, Chu G. Genetic improvement in short-season soybeans: I. Dry matter accumulation, partitioning, and leaf area duration. Crop Sci 2001; 41:391–398.

Larson EM, Hesketh JD, Wooley JT, Peters DB. Seasonal variation in apparent photosynthesis among plant stands of different cultivars. Photosynth Res 1981; 2:3–20.

Lauchli A. Calcium, salinity and the plasma membrane. In: Leonard RT, Hepler PK, eds. Calcium in Plant Growth and Development. Vol. 4. Rockville, MD: American Society of Plant Physiologists, 1990:26–35.

Lawn RJ. Response of four grain legumes to water stress in south-eastern Queensland: I. Physiological response mechanisms. Aust J Agric Resour 1982; 33:481–496.

Lawn RJ. Response of four grain legumes to water stress in south-eastern Queensland: II. Plant growth and soil water extraction pattern. Aust J Agric Resour 1982; 33:497–509.

Lawn RJ. Response of four grain legumes to water stress in south-eastern Queensland: III. Dry matter production, yield, and water use efficiency. Aust J Agric Resour 1982; 33:511–521.

Lauchli A, Wieneke J. Studies on growth and distribution of Na^+, K^+, and Cl^- in soybean varieties differing in salt tolerance. Z Pflanzenernahr Bodenkd 1979; 143:3–13.

Lemke-Keyes CA, Sachs MM. Genetic variation for seedling tolerance to anaerobic stress in maize germplasm. Maydica 1989; 34:329–337.

Levitt J. Chilling, freezing, and high temperature stress. In: Kozlowski TT, ed. Responses of Plants to Environmental Stress. Vol. 1. New York: Academic Press, 1980:3–55.

Lin CY. Dissociation and reassembly of polyribosomes in relation to protein synthesis in the soybean root. J Mol Biol 1967; 26:237–247.

Lin CY, Roberts JK, Key JL. Acquisition of thermotolerance in soybean seedlings. Plant Physiol 1984; 74:152–160.

Lin CY, Chen YM, Key JL. Solute leakage in soybean seedlings under various heat shock regimes. Plant Cell Physiol 1985; 26:1493–1498.

Liu T, van Staden J, Cress WA. Salinity induced nuclear and DNA degradation in meristematic cells of soybean (*Glycine max* (L.)) roots. Plant Growth Regul 2000; 30:49–54.

Ludlow MM, Muchow RC. A critical evaluation of traits for improving crop yields in water-limited environments. Adv Agron 1990; 43:107–153.

McKinney NV, Schapaugh WT, Kanemasu ET. Selection for canopy temperature differential in six populations of soybean. Crop Sci 1989; 29:255–259.

McKinney NV, Schapaugh WT, Kanemasu ET. Canopy temperature, seed yield, and vapor pressure deficit relationships in soybean. Crop Sci 1989; 29:1038–1041.

Mederski HJ, Jeffers DL. Yield response of soybean varieties grown at two soil moisture stress levels. Agron J 1973; 65:410–412.

Mian MAR, Ashley DA, Vencill WK, Boerma HR. QTLs conditioning early growth in a soybean population segregating for growth habit. Theor Appl Genet 1998; 97:1210–1216.

Morgan JM. Adaptation to water deficits in three grain legume species—Mechanisms of turgor maintenance. Field Crops Res 1992; 29:91–106.

Morrison MJ, Voldeng HD, Cober ER. Physiological changes from 58 years of genetic improvement of short-season soybean cultivars in Canada. Agron J 1999; 91:685–689.

Muchow RC. Canopy development in grain legumes grown under different soil water regimes in a semi-arid tropical environment. Field Crops Res 1985; 11:99–109.

Muchow RC. Stomatal behaviour in grain legumes grown under different soil water regimes in a semi-arid tropical environment. Field Crops Res 1985; 11:291–307.

Munns R. Whole-plant responses to salinity. Aust J Plant Physiol 1986; 13:143–160.

Nagao RT, Czarnecka E, Gurley WB, Schoffl F, Key JL. Genes for low-molecular-weight heat shock proteins of soybean: Sequence analysis of a multigene family. Mol Cell Biol 1985; 5:3417–3428.

Nanjo T, Kobayashi M, Yoshiba Y, Sanada Y, Wada K, Tsukaya H, Kkubari Y, Yamaguchi-Shinozaki K, Shinozaki K. Biological functions of proline in morphogenesis and osmotolerance revealed in antisense transgenic *Arabidopsis thaliana*. Plant J 1999; 18:185–193.

Neyshabouri MR, Hatfield JL. Soil water deficit effects on semi-determinate and indeterminate soybean growth and yield. Field Crops Res 1986; 15:73–84.

Nguyen A, Lamant A. Pinitol and myo-inositol accumulation in water-stressed seedlings of maritime pine. Phytochemistry 1988; 27:3423–3427.

Nilsen ET, Orcutt DM. Physiology of plants under stress. Abiotic Factors. New York: John Wiley & Sons, 1996.

Paje MCM, Ludlow MM, Lawn RJ. Variation among soybean (*Glycine max* (L.) Merr.) accessions in epidermal conductance of leaves. Aust J Agric Resour 1988; 39:373–636.

Passioura JB. Grain yield, harvest index, and water use of wheat. J Aust Inst Agric Sci 1977; 43:117–121.

Passioura JB. Roots and drought resistance. Agric Water Manag 1983; 7:265–280.

Passioura JB. The yield of crops in relation to drought. In: Boote KJ, Bennet JM, Sinclair TR, Paulsen GM, eds. Physiology and Determination of Crop Yield. Madison, WI: American Society of Agronomy, 1994:343–359.

Patterson RP, Hudak CM. Drought-avoidant soybean germplasm maintains nitrogen–fixation capacity under water stress. Plant Soil 1996; 186:39–43.

Paul MJ, Cockburn W. Pinitol, a compatible solute in *Mesembryanthemum crystallinum* L. J Exp Bot 1989; 40:1093–1098.

Peschke VM, Sachs MM. Characterization and expression of transcripts induced by oxygen deprivation in maize (*Zea mays* L.). Plant Physiol 1994; 104:378–394.

Ponnamperuma FN. The chemistry of submerged soils. Adv Agron 1972; 24:29–95.

Raper CD Jr, Kramer PJ. Stress physiology. In: Wilcox JR, ed. Soybeans: Improvement, Production, and Uses. Madison, WI: American Society of Agronomy, 1987:211–247.

Rosielle AA, Hamblin J. Theoretical aspects of selection for yield in stress and nonstress environments. Crop Sci 1981; 21:943–945.

Russell DA, Wong DM, Sachs MM. The anaerobic response of soybean. Physiol Plant 1990; 92:401–407.

Sacher RF, Staples RC. Inositol and sugars in adaptation of tomato to salt. Plant Physiol 1985; 77:206–210.

Sachs MM, Freeling M, Okimoto R. The anaerobic proteins in maize. Cell 1980; 20:761–767.

Sachs MM, Subbaiah CC, Sabb IN. Anaerobic gene expression and flooding tolerance in maize. J Exp Bot 1996; 47:1–15.

Shannon MC. Adaptation of plants to salinity. Adv Agron 1997; 60:75–120.

Shaw RH, Laing DR. Moisture stress and plant response. In: Peirre WH, ed. Plant Environment and Efficient Water Use. Madison, WI: American Society of Agronomy, 1966:73–94.

Sheveleva E, Chmara W, Bohnert HJ, Jensen RG. Increased salt and drought tolerance by D-ononitol production in transgenic *Nicotiana tabacum* L. Plant Physiol 1997; 115:1211–1219.

Shibles RM, Anderson IC, Gibson AH. Soybean. In: Evans LT, ed. Crop Physiology. London: Cambridge Univ. Press, 1975:151–190.

Shiraiwa T, Hashikawa U. Accumulation and partitioning of nitrogen during seed filling old and modern soybean cultivars in relation to seed production. Jpn J Crop Sci 1995; 64:754–759.

Sinclair TR. Crop yield potential and fairy tales. In: Buxton DR, Shibles RM, Forsberg RA, Bland BL, Asay KH, Paulsen GM, Wilson RF, eds. International Crop Science I. Madison, WI: Crop Science Society of America, 1993:707–711.

Sinclair TR. Limits to crop yield? In: Boote KJ, Bennet JM, Sinclair TR, Paulsen GM, eds. Physiology and Determination of Crop Yield. Madison, WI: American Society of Agronomy, 1994:509–532.

Sinclair TR. Historical changes in harvest index and crop nitrogen accumulation. Crop Sci 1998; 38:638–643.

Sinclair, TR. Limits of crop yield. In: Plants and Population: Is There Time? Proceedings of the National Academy of Science Colloquium, Dec, 5–6, 1998. http://www.lsc.psu.edu/NAS/Panelists/Sinclair%20Comments.html, 1999.

Sinclair TR, Ludlow MM. Influence of soil water supply on the plant water balance of four tropical grain legumes. Aust J Plant Physiol 1986; 13:329–341.

Sinclair TR, Muchow RC. Radiation use efficiency. Adv Agron 1999; 65:215–265.

Sinclair TR, Muchow RC, Bennett JM, Hammond LC. Relative sensitivity of nitrogen and biomass accumulation to drought in field-grown soybean. Agron J 1987; 79:986–991.

Sinclair TR, Purcell LC, Vadez V, Serraj R, King CP, Nelson R. Identification of soybean genotypes with N_2 fixation tolerance to water stress. Crop Sci 2000; 40:1803–1809.

Sinclair TR, Tanner CB, Bennett JM. Water-use efficiency in crop production. Bioscience 1984; 34:36–40.

Sinoit N, Kramer PJ. Effect of water stress during different stages of growth of soybean. Agron J 1977; 69:274–278.

Slafer GA, Satorre EH, Andrade FH. Increases in grand yield in bread wheat from breeding and associated physiological changes. In: Slafer GA, ed. Genetic Improvement of Field Crops. New York: Marcel Dekker, Inc., 1993:1–68.

Somerville C. Direct tests of the role of membrane lipid composition in low-temperature-induced photoinhibition and chilling sensitivity in plant and cyanobacteria. Proc Natl Acad Sci 1995; 84:739–743.

Specht JE, Williams JH. Contribution of genetic technology to soybean productivity—Retrospect and prospect. In: Fehr WR, ed. Genetic Contribution to Yield Gains of Five Major Crop Plants. CSSA Spec Publ 1984; Vol. 7:. Madison, WI: Crop Science Society of America, 1984:49–74.

Specht JE, Chase K, Macrander M, Graef GL, Chung J, Markwell JP, Germann M, Orf JH, Lark GK. Soybean response to water: A QTL analysis of drought tolerance. Crop Sci 2001; 41:493–509.

Specht JE, Hume DJ, Kumudini SV. Soybean yield potential—A genetic and physiological perspective. Crop Sci 1999; 39:1560–1570.

Specht JE, Williams JH, Weidenbenner CJ. Differential responses of soybean genotypes subjected to a seasonal soil water gradient. Crop Sci 1986; 26:922–934.

Sperry JS, Hacke UG, Oren R, Comstock JP. Water deficits and hydraulic limits to leaf water supply. Plant Cell Environ 2002; 25:251:263251:263.

Srinivasan A, Takeda H, Senboku T. Heat tolerance in food legumes as evaluated by cell membrane thermostability and chlorophyll fluorescence techniques. Euphytica 1996; 88:35–45.

Streeter JG, Lohnes DG, Fioritto RJ. Patterns of pinitol accumulation in soybean plants and relationships to drought tolerance. Plant Cell Environ 2001; 24:429–438.

Stuedle E. Water uptake by roots: Effects of water deficit. J Exp Bot 2000; 51:1531–1542.

Sullivan M, VanToai T, Fausey N, Beuerlein J, Parkinson R, Soboyejo A. Evaluating on-farm flooding impacts on soybean. Crop Sci 2001; 41:1–8.

Tanksley SD, McCouch SR. Seedbanks and molecular maps: Unlocking genetic potential from the wild. Science 1997; 277:1063–1066.

Tanksley SD, Nelson JC. Advanced backcross QTL analysis: A method for the simultaneous discovery and transfer of valuable QTLs from unadapted germ plasm into elite breeding lines. Theor Appl Genet 1996; 92:191–213.

Tanksley SD, Grandillo S, Fulton TM, Zamir D, Esed Y, Petiard V, Lopez J, Beck-

Bunn T. Advanced backcross QTL analysis in a cross between and elite processing line of tomato and its wild relative *L. pimpinellifolium*. Theor Appl Genet 1996; 92:213–224.

Tanner CB, Sinclair TR. Efficient water use in crop production: Research or re-search. In: Taylor H, et al., ed. Limitations to Efficient Water Use in Crop Production. Madison, WI: American Society of Agronomy, 1983:1–28.

Thomashow MF. Role of cold-responsive genes in plant freezing tolerance. Plant Physiol 1998; 118:1–7.

Tsugane K, Kobayashi K, Niwa Y, Ohba Y, Wada K, Kobayashi H. A recessive *Arabidopsis* mutant that grows photoautotrophically under salt stress shows enhanced active oxygen detoxification. Plant Cell 1999; 1:1195–1206.

Turner NC, Wright CG, Siddique HHM. Adaptation of grain legumes (pulses) to water-limited environments. Adv Agron 2001; 71:193–231.

Tyerman SD, Niemietz CM, Bramley H. Plant aquaporins: Multifunctional water and solute channels with expanding roles. Plant Cell Environ 2002; 25:173–194.

Ustun A, Allen FL, English BC. Genetic progress in soybean of the U.S. Midsouth. Crop Sci 2001; 41:993–998.

Vadez V, Sinclair TR. Ureide degradation pathways in intact soybean leaves. J Exp Bot 2000; 51:1459–1465.

Vadez V, Sinclair TR. Leaf ureide degradation and N-2 fixation tolerance to water deficit in soybean. J Exp Bot 2001; 52:153–159.

Vadez V, Sinclair TR. Sensitivity of N_2 fixation traits in soybean cultivar Jackson to manganese. Crop Sci 2002; 42:791–796.

VanToai TT, Bolles CS. Postanoxic injury in soybean (*Glycine max*) seedlings. Plant Physiol 1991; 97:588–592.

VanToai TT, Beuerlein JE, Schmitthenner AF, St. Martin SK. Genetic variability for flooding tolerance in soybeans. Crop Sci 1994; 34:1112–1115.

VanToai TT, St. Martin SK, Chase K, Boru G, Schnipke V, Schmitthenner AF, Lark KG. Identification of a QTL associated with tolerance of soybean to soil water-logging. Crop Sci 2001; 41:1247–1252.

Vierling E. The roles of heat-shock proteins in plants. Annu Rev Plant Physiol Plant Mol Biol 1991; 42:579–620.

Voldeng HD, Cober ER, Hume DJ, Gillard C, Morrison MJ. Fifty-eight years of genetic improvement of short-season soybean cultivars in Canada. Crop Sci 1997; 37:428–431.

Waggoner PE. How Much Land Can Ten Billion People Spare for Nature? Task Force Report No. 121. Ames, IA: Council for Agricultural Science and Technology, Feb. 1994.

Wang D, Shannon MC. Emergence and seedling growth of soybean cultivars and maturity groups under salinity. Plant Soil 1999; 214:117–124.

Wang D, Shannon MC, Grieve CM. Salinity reduces radiation absorption and use efficiency in soybean. Field Crops Res 2001; 69:267–277.

Warne TR, Hickok LG. Single gene mutants tolerant to NaCl in the fern *Ceratopteris*: Characterization and genetic analysis. Plant Sci 1987; 52:49–55.

Wehrhahn C, Allard RW. The detection and measurement of the effects of individual genes involved in the inheritance of a quantitative character in wheat. Genetics 1965; 51:109–119.

Wilcox JR. Sixty years of improvement in publicly developed elite soybean lines. Crop Sci 2001; 41:1711–1716.

Wilcox JR, Schapaugh WT Jr, Cooper RL, Fehr WR, Niehaus MH. Genetic improvement of soybeans in the Midwest. Crop Sci 1979; 19:803–805.

Wofford T, Allen F. Variation in leaflet orientation among soybean cultivars. Crop Sci 1982; 22:999–1004.

Zhang J, Specht JE, Graef GL, Johnson BL. Pubescence density effects on soybean seed yield and other agronomic characteristics. Crop Sci 1992; 32:641–648.

Zhang J, VanToai T, Huynh L, Preiszner J. Development of flooding-tolerant *Arabidopsis thaliana* by autoregulated cytokinin production. Mol Breed 2000; 6:135–144.

Zhu JK. Genetic analysis of plant salt tolerance using *Arabidopsis*. Plant Physiol 2000; 124:941–948.

Zhu JK. Plant salt tolerance. Trends Plant Sci 2001; 6:66–71.

8

Physiological Basis of Yield and Environmental Adaptation in Cotton

Derrick M. Oosterhuis and James McD. Stewart
University of Arkansas
Fayetteville, Arkansas, U.S.A.

1 INTRODUCTION

In the early part of the biotechnology era, great expectations were placed on the potential to modify major biochemical pathways and physiological parameters of crops, e.g., creation of nitrogen-fixing corn. Thus far, that expectation has not materialized. As our knowledge of plant metabolic and molecular systems responsible for physiological function have grown, so too has the awareness that those systems are highly integrated and not easily modified by simple approaches. In cotton, as in most crops, genetic yield parameters are multigenic and rarely susceptible to modification through genetic engineering with one or two genes. To provide the reader a realistic background to the application of biotechnology to cotton yield improvement, the physiological aspects of yield development and adaptation to environmental stresses are emphasized in this chapter. Where appropriate, potential applications of biotechnology and/or molecular techniques to enhance yield components are discussed.

The cotton crop is grown primarily for its fiber, although oil is extracted from the seeds and the seeds are also used for cattle feed. Productivity, as defined by dry matter production, is influenced by numerous factors such as plant structure, translocation, and partitioning of assimilate among plant parts, management, and biotic and abiotic stresses. Photosynthesis and carbohydrate production are the fundamental building blocks of most crops (Constable and Oosterhuis, 2003). This is evident in cotton where the primary product for harvest and profit is cellulose, i.e., the fiber, which is 99% carbohydrate. Understanding the physiological and molecular bases for development of yield in the cotton crop, and crop adaptations to a dynamic environment, are critical in continued efforts to genetically improve fiber and seed yield for efficient and profitable production. There are many physiological variables associated with yield. However, the relationship between a given physiological variable and yield may vary across environments as well as between different genotypes (Heitholt, 1999). This review will attempt to distinguish between genetic and environmental variations in the relationship between a physiological variable and yield.

Several reviews of cotton physiology are available (McArthur et al., 1975; Hearn, 1976; Mauney and Stewart, 1986; Cothren, 1999) with some concentrating on aspects of the unique yield physiology of cotton ranging from biochemical to agronomic (Benedict, 1984; Hearn and Constable, 1984; Mauney and Stewart, 1986; Wells, 1989; Hearn and Fitt, 1992; Munro, 1995; Hake et al., 1996; Heitholt, 1999). Of these, only the last concentrated exclusively on yield physiology. The following discussion describes the physiological basis of yield in cotton with emphasis on dry matter production, allocation to fruiting forms, and effect of environment in the form of plant adaptations to environmental perturbations. The possible influence of breeding and biotechnology will also be indicated. The majority of cultivated cotton today is *Gossypium hirsutum* L. or upland cotton, which together with *G. barbadense* L. represents more than 95% of the world's production (Niles and Feaster, 1984); therefore, this chapter will deal mainly with the development and physiology of upland cotton. Furthermore, the biological basis of fiber quality, an integral component of cotton profitability, will not be addressed in this review.

2 GROWTH AND DEVELOPMENT OF THE COTTON PLANT

The cotton plant has one of the most complicated growth habits of all major row crops (Oosterhuis and Jernstedt, 1999). This is because cotton is a perennial cultivated in row-crop agriculture as an annual and has an indeterminate growth habit with a complicated sympodial fruiting pattern (Mauney, 1986). Furthermore, the crop is extraordinarily sensitive to changes in the environment to which it responds primarily by changes in vegetative growth or fruit abscission. Cotton differs from most annual crops in that it exhibits many of the woody perennial and xerophytic characteristics found in its ancestors (Hearn, 1976). However, domestication of cotton from its wild ancestry continues today as breeders select developmental traits and performance characteristics more typical of annual plant-growth habit than perennial growth habit. Recognition that cotton is in a transitional state of domestication provides insight into potential physiological systems and pathways wherein genetic and molecular modifications might lead to crop improvement. For example, perennial plants typically invest in some type of storage reserve to initiate regrowth during the next growth cycle. In cotton this takes the form of starch stored in the stems and roots (Pace et al., 1999) that at the end of the cropping season remains in the field. If the metabolic pathways and genetic regulatory systems controlling deposition and remobilization of starch were to be modified such that the carbohydrate were allocated to the fruiting structures during the first growth cycle, then, theoretically, yield could be increased without placing additional demands for photosynthate on the plant. Support for this hypothesis can be found in reports by Haigler et al. (2000a,b). They stated that genetically engineered cotton plants with up-regulated sucrose phosphate synthase stored less starch in the lower stem. The modified plants also had fiber-quality parameters consistent with improved availability and utilization of photosynthate compared to nontransgenic control plants. The increased value of some of the quality parameters would contribute directly to yield, e.g., micronaire, fiber length, and uniformity.

The growth and development of the cotton plant is a genetically regulated sequence of events, with each seed containing all the necessary information to produce a plant with the potential of regenerating seed. However, external factors, both biotic and abiotic, have major influences on extent and type of plant growth. The development of the cotton plant has been described by Oosterhuis (1990). Plant development proceeds through a number of stages, which for practical reasons may be divided into five main phases of growth: (1) germination and emergence, (2) seedling establishment, (3) leaf and canopy development, (4) flowering and boll development, and (5) maturation (Fig. 1). The overall growth pattern, i.e., dry matter production of the cotton plant, follows a typical sigmoidal curve with a relatively slow rate during emergence and early root growth, followed by an exponential increase in growth rate during canopy and flowering, and finally by a declining rate during boll maturation (Bassett et al., 1970; Constable and Gleeson, 1977). The transitions between successive developmental phases are subtle and not morphologically distinct. Each stage is assumed to have a unique subset of genes that are being expressed to drive the stage-specific

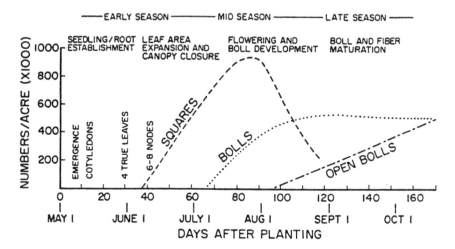

Figure 1 Seasonal development of cotton in the mid-South showing the production pattern of squares, bolls, and open bolls. (From Oosterhuis, 1990.)

physiological processes. Awareness of the phase-dependent differences in genetic and physiological requirements for optimum growth allows many yield problems to be avoided through proper crop management and also provides insight into targets for molecular modification. The sensitivity and, consequently, the productivity, of each phase to environmental fluctuations vary based on the interaction of developmentally expressed genes and those induced by the varying environment.

Cotton has a typical tap root system, and root growth follows a sigmoidal pattern that continues to increase until about the time maximum plant height is achieved soon after flowering (Taylor and Klepper, 1974). At this time fruit begins to form and as the number of bolls increases, carbohydrates are increasingly directed toward the fruit as it becomes the dominant sink for available photoassimilate. As older tertiary roots senesce total root length begins to decline, even though new roots continue to be formed (Hons and McMichael, 1986). For high-yielding cotton under certain soil conditions this decreased root activity can lead to nutrient shortages, e.g., potassium (Oosterhuis, 1995).

The indeterminate nature of the cotton plant means that it is capable of producing a large number of potential fruiting positions. For example, an average plant population of 100,000 plants/ha in the mid-South will have about 22 main-stem nodes with 10 effective fruiting branches each with an average of three fruiting positions on the lower sympodia and fewer on upper sympodia for a total of approximately 20 potential bolls per plant. However, natural abscission of floral buds and bolls in response to limited resources and other environmental stresses reduces this number by about 70% (Guinn, 1982b). Only a small proportion of floral buds produced by the cotton plant reaches maturity as mature cotton bolls, and this proportion decreases as the season progresses and the plant gets older (Munro, 1995). Presumably, this is because the developing bolls have priority in the demand for nutrients and water and are stronger nutritional sinks than the younger and smaller squares and bolls. Fruit abscission is an important natural process by which the plant adjusts its fruit load to match the supply of inorganic and organic nutrients to the prevailing environment.

3 REPRODUCTIVE GROWTH AND YIELD OF COTTON

3.1 Maximum Lint Yield

Baker and Hesketh (1969) used models to predict the theoretical maximum lint yield of cotton, assuming all net photosynthate was translocated to the fruit, to be 5.9 bales/acre (3172 kg/ha) for a typical growing season in Mississippi. They also predicted that this would increase to 7.6 bales/acre (4086 kg/ha) for a season of cloudless days, and to 7.75 bales/acre (4166 kg/ha) if all flowers matured. In contrast, cotton lint yields in the U.S. Cotton Belt have been considerably lower; for example, the average yield for Arkansas for the past 5 years has been 833 kg lint/ha (National Cotton Council, 2001) indicating that numerous parameters restrict the cotton crop from achieving its potential. Maximum yields from Arkansas producers in exceptional years are in the 3 bales/acre range, which represents only about half the theoretical maximum achievable yield in the mid-South.

3.2 Reproductive Growth

A delicate balance exists between vegetative and reproductive growth in cotton. Under certain environmental or management conditions, assimilate can be preferentially partitioned into nonreproductive growth instead of harvestable reproductive yield. These circumstances usually involve excessive amounts of water and nitrogen, and are circumvented by the use of growth retardants, primarily inhibitors of gibberellin synthesis (Carlson, 1987; Oosterhuis et al., 2000). Initiation of floral buds in modern cotton cultivars is not influenced by photoperiod and begins as early as 10 to 14 days after emergence when the primordium of the first sympodium begins to form (Mauney, 1966). Position of the first fruiting node on the main stem is strongly influenced by day and night temperature as well as by the intensity and quality of the light (Mauney, 1986) and usually occurs at about node 5 to 7. The vertical fruiting interval (VFI) is between 2.2 and 4.0 days and the horizontal flowering interval (HFI) is between 5.8 and 8.5 days (McClelland and Neely, 1931; Kerby and Buxton, 1978), although on average the VFI and HFI are considered to be 3 and 6, respectively (Tharp, 1960). A variety of phenomena are involved in the initiation and development of reproductive growth (Wells and Meredith, 1984b). Various environmental parameters affect flowering, including low night temperature ($< 32\,°C$) and long photoperiods (14 h vs. 8 h), both of which promote flowering at a lower main-stem node (Mauney, 1966). Water deficit generally decreases rate of flowering (Guinn and Mauney, 1984a), although a preflowering stress may increase the subsequent rate of flowering (Stockton et al., 1961; Singh, 1975). Development of the crop reproductive sink depends on the calendar date of the first flower, the time interval between successive flowers, and the rate of boll growth (Hearn and Room, 1979).

3.3 Flowering and Fruiting Pattern

Cotton has a distinctive flowering and fruiting pattern. Flowers are produced on sympodial branches, and once flowering commences it proceeds spirally up the main stem as well as outward along the sympodial branches away from the main stem. This unique pattern is due to an indeterminate growth habit and the three-eights phyllotaxy of the cotton plant (McClelland and Neely, 1931). The number of fruiting sites on a plant is related to the number of main-stem nodes (Hearn, 1969a). As mentioned earlier, cotton has a high capacity for producing floral buds, although in a cropping environment a high-yielding cultivar normally produces only about 150 to 200 flowers/m^2 (Guinn et al., 1981; Guinn, 1985; Heitholt, 1993). Of these flowers, only about 30% are retained and mature due to the sensitivity of the cotton crop to the environment (Guinn, 1982). Therefore, factors that reduce flower production or increase abscission will have a direct and immediate adverse impact on yield potential.

On the surface, efforts to reduce the rate of abscission would appear to be a promising area of research for the application of biotechnology and molecular genetics to improve cotton yield. However, a limited amount of shedding may be necessary for good yields, but in some cases a temporary stress during boll set may cause shedding of fruit that subsequent favorable environmental conditions could support. On the other hand, good environmental conditions during the first half of the cropping season can result in more fruit load than the crop can adequately support under a stressful late-season environment when the plants do not have the option of shedding to adjust the demand/supply ratio. Because of the finely regulated balance between fruit retention and available nutrition and carbohydrate, genetic or biotechnological modifications targeting only the shedding of fruiting forms (formation of abscission zone) would probably be of little benefit to crop yield, and in some instances may be detrimental.

3.4 Distribution of Fruit

The distribution of bolls on the plant varies due to abscission from physiological and environmental causes (Guinn, 1982; Benedict, 1984). Abscission of squares and bolls is a natural and necessary process as the plant is incapable of nurturing to maturity all the squares initiated. Boll retention declines later in the season because of the greater sink demand of existing bolls and because the crop assimilatory capacity decreases (Verhalen et al., 1975). Fruiting site position within the canopy influences boll retention and boll development (Jenkins et al., 1990). Plant diagrams are used to "map" the positions of bolls on the plant and are useful management tools to follow development and to assess the success of production inputs. A large percentage (> 70%) of the total yield is derived from the central portion of the canopy, approximately between main-stem nodes 9 and 14 (Jenkins et al., 1990), depending on the length of the growing season, plant density, nutrition, insect control, and water availability. These nodes coincide with the distribution of maximum leaf area within the canopy (Oosterhuis and Wullschleger, 1988). Above these nodes, fewer bolls are produced, and they tend to take longer to mature, are smaller, and are inferior in seed and fiber quality. The decline in vigor of the fruiting structures at higher nodes is evident in bud size prior to flowering (Holt et al., 1994).

3.5 Cotton Yield Components

The production of seedcotton is the main reason farmers grow cotton. The components of yield and their relative contribution to yield involve complex interactions among stage of development, genotype, and environment. Agronomically, yield of seedcotton consists of two main components: number of bolls per unit area and average boll weight. However, lint, the main economic product, can be defined by two basic components, the number of seeds produced per hectare and the weight of fiber produced on the seed (Lewis et al., 2000). The number of seeds per hectare is determined by the number of plants per hectare, the number of bolls per plant, and the number of seeds per boll. Although the number of seeds per boll is primarily determined by genotype, the other two parameters are strongly influenced by management practices and the environment. The weight of fibers per seed is a function of the number of fibers per seed and the average weight per fiber (Lewis, 2001). The number of fibers per seed can be approximated by dividing the total weight of fiber per seed by the mean fiber length × micronaire. Micronaire provides an estimation of the mean linear density of the fibers.

The weight of fiber per seed, particularly the number of fibers per seed, is heavily influenced by genotype because the mean weight per fiber, developmentally, is a function of both primary and secondary wall growth (Lewis, 2001). Furthermore, lint weight per seed and seed size are highly related, i.e., larger seed have more fiber weight per seed. However, the positive correlation between high lint weight per seed and seed size is not necessarily related to high lint yield per unit area of land. Thus, lint per seed needs to be increased with attention to seed size. Historically this has been approached through selection for increases in lint percent (lint weight/seed weight as a percentage). Larger seed usually result in lower lint percentage (Bourland, personal communication).

A relatively small increase in the weight of fiber per seed could have a highly significant impact on yield. For example, in the south-central and southeastern U.S. Cotton Belt, the long-term average number of seeds per hectare is approximately 17 million (Lewis, 2001). Thus, an increase in the weight of the fiber per seed of only 5 mg would result in a yield increase of about 84 kg/ha. Some older studies have suggested a genetic correlation between smaller seed size and higher yield (Bourland, personal communication). Lewis (2001) suggested that under stressful conditions, there is a greater detrimental influence on seed number than on lint weight per seed, presumably due to the greater metabolic cost of the seed components (e.g., triglycerides) compared to the fiber (predominantly carbohydrates), thereby making the modern cultivars more vulnerable to stress than the older, larger seeded cultivars. Furthermore, the smaller seeds may be weaker sinks for partitioned assimilate and thus have fewer reserves to resist a stress during boll development.

Cotton yields in the mid-South increased steadily during the eighties, but leveled off in the nineties, and even decreased in recent years (Lewis, 2001). This has been accompanied by extreme year-to-year variability, with record yields one year and disastrous results the next. Oosterhuis (1995) suggested that these yield responses were due to unusually hot, dry periods during boll development coupled with the poor ability of the smaller seeded modern cultivars to resist environmental stress (Lewis, 2001).

3.6 Genetic Manipulation of Yield Components

Selection for lint yield over the years has generally resulted in an increase in the percentage lint in seedcotton (Scholl and Miller, 1976). Furthermore, cultivars with larger lint yields generally have smaller-sized seed (Bridge et al., 1971). However, the largest increase in yield has been associated with bolls produced per unit area of land rather than with boll weight (Bridge et al., 1971; Ramey, 1972; Scholl and Miller, 1976; Wells and Meredith, 1984b). A correlation coefficient of 0.85 for the relationship between lint yield and bolls per unit area was reported by Bridge et al. (1971). However, a negative relationship between yield and fiber quality has been reported (Scholl and Miller, 1976; Meredith, 1984).

Efforts to directly impact the development of cotton fibers via biotechnology have been primarily to improve quality parameters or impart new characteristics, rather than yield of fiber (John, 1999). The number of fibers per boll can be increased by increasing the effective fiber producing seed surface area per boll and per unit of land. Effective seed surface area can be increased by one or more of four ways. Assuming other yield parameters are held constant, the first way is to assure that lint fibers initiate uniformly and more or less synchronously on the surface of all the seeds of the boll. This would narrow the gap between total surface area and effective surface area. At our present level of knowledge, we do not know how to modify this parameter other than by selection through empirical observation, but breeders have not looked at this trait directly. Realistically, this can be measured only indirectly through the quality parameter of fiber uniformity ratio. However, because the uniformity ratio is the average end result of many interacting genetic and environmental factors during fiber development, harvesting, and processing, it is useless as a measure of uniformity of fiber initiation.

A second approach to increase the fiber-bearing surface area of a boll would be to increase the average seed size, and a third approach would be to increase the number of seeds per boll. Either of these two approaches can be accomplished with the genetic diversity currently available in cotton germplasm. The fourth method encompasses many other factors not directly related to fiber initiation and development, but ultimately in the cropping situation is the most important: that is to increase the number of bolls per unit area of land. As indicated earlier, the trend in yield improvement for several years has been to increase bolls per area by selection for small, rapidly developing bolls with smaller seeds than in the past. With the present high-yielding cultivars any managerial, genetic, or biotechnological innovation that increases the number of bolls per area of land will have a beneficial impact on yield. It is this fact that drives all the currently recommended management strategies discussed in various parts of this review.

Assuming that the number of bolls can be held constant, then any increase in seed surface area per boll has the potential to increase yield. At present it is difficult to discern from past research whether increasing seed size or seed number would be more effective in increasing effective fiber-producing surface area. Intuitively, more seed per boll would seem a more effective approach than larger seed because seed volume increases as a cube function whereas surface area increases as a square. Larger seed would seem to require more nutritional support to develop an embryo than smaller seed, hence would partition relatively less of the available carbohydrate to the fiber. With either larger seeds or more seeds per boll the boll size would increase and would probably require a longer period to mature.

4 YIELD DEVELOPMENT AND DRY MATTER PRODUCTION

Yield formation is a function of the production of dry matter or carbohydrates by photosynthesis in the leaves and the partitioning of the resultant assimilate to the fruit over time. Therefore, yield can be described as a function of crop growth rate, leaf area duration, and harvest index (Heitholt, 1999). Other factors that are involved in yield formation are the amount of photosynthetically active radiation captured per unit area, radiation-use efficiency, and leaf physiological activity. All of these components have been studied as a means of enhancing cotton yield, but only the harvest index has changed significantly in modern cultivars compared to older cultivars (Wells and Meredith, 1984a). Modern cultivars make an earlier transition from vegetative to reproductive growth and have greater reproductive to vegetative ratios (Wells and Meredith, 1984b). Partitioning to the roots, although very important in early season plant establishment, will not be dealt with in this discussion because in most cotton-production systems in the United States, with adequate seedbed preparation and judicious use of fertilizer and irrigation, the roots should not be a limitation to boll growth. Roots constitute about 13% of the total dry matter in the cotton plant (Hearn, 1969b).

4.1 Crop Growth Rate

The maximum dry matter production, or crop growth rate (CGR), by a crop utilizing all incident radiation has been calculated to be 77 $g/m^2/day$ (Loomis and Williams, 1963). For C3 plants, the maximum short-term CGRs are about 36 $g/m^2/day$ (Jones, 1983). For cotton, the theoretical maximum CGR under ideal environmental conditions was calculated to be 42 $g/m^2/day$ (Baker and Hesketh, 1969). However, actual growth rates reported for irrigated cotton during flowering are about 17 $g/m^2/day$ in California (Kerby et al., 1987, 1990), 19 $g/m^2/day$ in Uganda (Hearn, 1972), and 25–30 $g/m^2/day$ in Arizona (Mauney, 1986). These values for dry matter production would be reduced under conditions of environmental stress, particularly water deficit, i.e., by as much as 60% during the boll fill period due to large reductions in leaf area index (LAI) (Krieg, personal communication). Dry matter accumulation by cotton follows a sigmoid curve with the largest percentage increase occurring in the period from early boll formation to maturity (Olson and Bledsoe, 1942; Bassett et al., 1970; Mullins and Burmester, 1991), and

60% of total dry matter is produced between 10 and 16 weeks after sowing (Oosterhuis et al., 1983). These authors reported the total dry matter production at maturity was 8.5 t/ha.

The growth rate of a floral bud is exponential, its weight increasing from about 10 to 130 mg in the 4 weeks prior to flowering (Hearn and Constable, 1984). The dry weight of the developing boll subsequently increases rapidly following a sigmoid pattern with a maximum growth rate of accumulation of 0.28 g/day about 20 days after flowering (Mutsaers, 1976), reaching full size in 20–25 days with a final weight of 3–8 g/boll at maturity (van Iersel et al., 1994). Maturation of the boll from anthesis to the time of carpel dehiscence usually takes about 50 days, but this varies with genotype and environmental conditions. In the U.S. mid-South, about 660 bolls of a typical cultivar are required to produce a kilogram of lint and about 145,000 bolls per 218 kg bale of lint (Oosterhuis, 1990).

4.2 Leaf Area and Yield

Leaf area has long been recognized as an important determinant of yield. The development of leaf area is relatively slow during the first six weeks after emergence, and then more rapid during early fruiting and canopy closure when maximum LAI is achieved. During the early vegetative period much of the available photoassimilate is partitioned into root development, whereas with the onset of flowering the dominant sink changes to the reproductive parts, i.e., developing bolls. For maximizing yield, it is imperative that development of leaf area not be delayed so that maximum utilization of incoming solar radiation for photosynthesis is attained as soon as possible. Maximum LAI is usually achieved 70–80 days after sowing, a time that coincides with peak flower production (Munro, 1995). Hearn and Constable (1984) maintained that peak LAI occurred 3–5 weeks after flowering. Ashley et al. (1965) estimated than an LAI of 5.0 was necessary to sustain growth of early maturing fruit for maximum yield in an irrigated system, a value supported by Wells and Meredith (1984a,b) who found that an LAI of 3.9–6.3 was necessary to not limit carbon assimilation rates. Kerby et al. (1987) reported that an LAI greater than 3.0 was needed to sustain maximum dry matter gain during early growth of cotton, and an LAI near 3.9 was necessary for maximum dry matter production during the remainder of the season. However, others (Hearn, 1972; Constable and Gleeson, 1977) claim that maximum CGR for cotton is achieved at an LAI of about 3.0. Hearn and Constable (1984) reported that LAI varies from 0.5 for a severely water-stressed crop to more than 6.0 for a well-fertilized and irrigated crop grown in a warm area. The LAI of modern cultivars is higher than for obsolete cultivars (Heitholt et al., 1998; Wells and Meredith, 1984c).

Ashley et al. (1965) reported that after the initiation of reproductive growth, the amount of leaf area remained positively correlated with the weight of young fruit (squares and small bolls) but negatively correlated with larger (< 1.25 cm) bolls, but Johnson and Addicott (1967) found a constant ratio of leaf area to seed cotton. At the level of individual fruit, the weight of a boll is positively correlated with the area of the subtending sympodial leaf (Fig. 2) such that the larger the sympodial leaf area the heavier the subtended boll, i.e., a boll weighing 5 g required a sympodial leaf of 100 cm^2 (Oosterhuis and Wullschleger, unpublished). Although leaf area is a major variable affecting the ability of the plant to gain carbon, other factors such as a greater leaf nitrogen content could also be important (Sinclair and Horie, 1989).

As discussed earlier, rapid elaboration of leaf area is important during the early crop development to capture as much light as possible as quickly as possible. However, as the canopy develops, especially under adequate water and fertility, the LAI reaches the point that additional leaf area does not capture additional light and is not beneficial to yield. LAI can be excessive when it shades lower bolls and their supporting leaves such that the average photosynthetic rate (per unit area of leaf) of the canopy declines. Management strategies are designed to obtain optimum LAI as quickly as possible, and then maintain it. Only indirectly, primarily through adequate water,

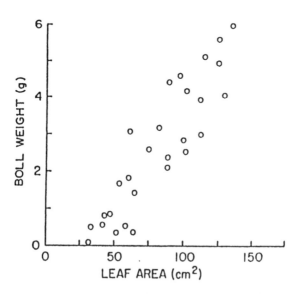

Figure 2   Relationship between boll weight and area of the subtending sympodial leaf. (From Oosterhuis and Wullschleger, unpublished data.)

fertility, and pest control, is attention given to leaf photosynthetic rate. However, this is a parameter that can be manipulated genetically.

One possibility for genetically increasing photosynthetic rate would be to increase leaf thickness, especially the length of the palisade layer, so that the number of chloroplasts per unit area available for light interception is optimum. Some of the wild *Gossypium* species of Baja, California, possess a double palisade layer that, according to Fryxell (1986), allows the plants to retain photosynthetic capacity while minimizing water loss. Plant selections from populations derived from a hybrid of one of these species with cotton have palisade cells longer than normal cultivars and a higher chlorophyll content per unit of leaf area (J. M. Stewart, unpublished data). The photosynthetic rate of these plants has not been measured.

4.3 Partitioning and Harvest Index

As in most crops, the partitioning of resources between vegetative and reproductive structures constitutes a major determinant of yield in cotton. Selection for increased harvest index has been one of the major trends in modern cultivars that has contributed to increased yields (Wells and Meredith, 1984b). During the early part of the season the cotton plant partitions its resources amongst the vegetative components, favoring the developing root system early and later the leafy canopy. The appearance of fruiting forms constitutes an emerging sink for available carbohydrate and nutrients that gradually becomes dominant as the boll load increases. Agronomically, harvest index is the major determinant of economic yield and, along with boll number per unit area and average boll weight, determines the lint yield. The harvest index in cotton has improved from as low as 0.08 in undomesticated cotton plants growing in the wild (Fryxell, 1979) to values of 0.47 and 0.53 in modern cultivars and as high as 0.66 in certain germplasm lines (Kerby et al., 1990). Differential partitioning also occurs within the cotton fruit between fiber and seed, i.e., deposition of cellulose or oil and protein, respectively, depending on the availability of assimilate and environmental stress. Lewis et al. (2000) suggested that modern cultivars have smaller seeds with less tolerance to stress, which may be related to a change in the partitioning of the available carbon between cellulose and triglycerides in the fiber and seed, respectively. Brown et al. (2002) subsequently reported that a stress during boll development, a period of fierce competition for available carbohydrate and other resources, results in a change in dry matter partitioning between the seed and the fiber. The partitioning of dry matter to reproductive structures compared to leaves and stems is influenced by environmental parameters, such as drought and temperature, as well as by production practices such as plant density and the availability of water and nitrogen.

5 CARBON DIOXIDE ASSIMILATION CAPACITY

Photosynthesis is the cornerstone of agricultural productivity, with the leaf playing a critical role due to its function in interception of radiant energy and as the major site for photosynthesis. Therefore, as has been mentioned, leaf area duration and the efficiency of the leaf area in CO_2 fixation are central for dry matter production. Crop photosynthesis has usually been studied by measuring either single leaf photosynthesis or canopy photosynthesis. Although Peng and Krieg (1991) and others have reported reasonable correlation between single leaf and canopy photosynthetic rates per unit leaf area but not per unit soil area, productivity is more closely related to canopy photosynthesis than to single leaf photosynthesis (Zelitch, 1982). Measurements of leaf photosynthesis estimate the maximum potential of a genotype because the uppermost, fully expanded leaves are more physiologically active and occupy an optimum position on the plant (Elmore, 1980; Wullschleger and Oosterhuis, 1990d; Constable and Rawson, 1980a), whereas canopy photosynthesis measurements more realistically estimate the CO_2 uptake of the whole stand (Peters et al., 1974; Wells, 1988) by describing the photosynthetic activity per unit ground area and integrating leaf morphology and canopy architecture. Furthermore, the nature of the canopy is dynamic with continually changing average leaf ages (Constable and Rawson, 1980a; Wullschleger and Oosterhuis, 1992) and light interception profiles. Therefore, seasonal estimates of canopy photosynthesis can better record changes in photosynthate production due to these factors as well as due to genotypic and environmental factors (Christy and Porter, 1982; Wullschleger and Oosterhuis, 1990b; Peng and Krieg, 1991).

5.1 Canopy Photosynthetic Rate

The potential rate of net photosynthesis of individual cotton leaves is about 30 $\mu mol\ CO_2/m^2/s$ for a recently expanded leaf well supplied with water, nutrients, and light (Hearn and Constable, 1984), although field measurements of single leaf photosynthesis are often less, i.e., 20 $\mu mol\ CO_2/m^2/s$ (Wullschleger and Oosterhuis, 1990b). Canopy photosynthesis in cotton typically peaks soon after the start of flowering and boll development (Wells et al., 1986; Peng and Krieg, 1991; Bourland et al., 1992) at about 1.0 mg $CO_2/m^2/s$ (Wells et al., 1986) or 15–30 $\mu mol/m^2/s$ (Wullschleger and Oosterhuis, 1990a; Peng and Krieg, 1991). Thereafter, canopy photosynthesis of cotton declines as the assimilate needs for the developing boll load are peaking (Constable and Rawson, 1980a; Wells et al., 1986; Wullschleger and Oosterhuis, 1990a; Pettigrew et al., 2000). Wells (1995) reported that 60% of the bolls opened after canopy photosynthesis was at 25% of its maximum or less. This decline in canopy photosynthesis has been related to the indeterminate nature of cotton that results in increased shading of lower leaves (Mauney, 1986) and increasing average leaf age in the canopy (Constable and Rawson, 1980a; Wullschleger and Oosterhuis, 1992). Furthermore, Peng and Krieg (1991) found that reductions in canopy photosynthetic rates were due to decreases in photosynthesis in both young and old leaves. However, Wells et al. (1986) showed that canopy photosynthesis was closely related to the ability to intercept light and less closely to the photosynthetic efficiency of individual leaves. These patterns are consistent with results from classical growth analysis (Hearn, 1969b; Constable and Gleeson, 1977) where net assimilation rate decreased with canopy age. This lack of synchrony between the declining assimilatory capacity of the crop and the requirements of the developing boll load emphasizes the importance of maintaining a healthy, vigorous crop after flowering with adequate water and nutrient availability and good pest control (mainly insects) to allow maximum photoassimilate production.

5.2 Leaf Photosynthetic Contributions and Efficiency

The indeterminate growth habit of cotton causes new leaves to be continuously initiated in the upper canopy while leaves lower on the plant are successively shaded (Mauney, 1986). In contrast to many grain crops, where the fruit are located near relatively young leaves, the growth habit of cotton dictates that developing fruits are more closely associated with older leaves. The ability of these leaves to produce sufficient photosynthate in concert with fruit requirements may be a significant determinant of yield, particularly within dense canopies characteristic of field conditions. Muramoto et al. (1967) reported that the photosynthetic mechanism of leaves deteriorates due to both aging and shading about 30 days after unfurling. Others found that individual leaf photosynthesis peaked about 18–20 days after unfolding and then declined steadily until senescence and death (Fig. 3) (Constable and Rawson, 1980a; Krieg, 1988; Oosterhuis and Wullschleger, 1988; Wullschleger and Oosterhuis, 1990b). For sympodial leaves subtending a fruiting position, this pattern is not well synchronized with the requirements of the developing boll necessitating imports of carbon from distal sources.

Numerous studies (Kearney, 1929; Morris, 1965; Ashley, 1972; Benedict et al., 1973; Peoples and Matthews, 1981; Constable and Rawson, 1980b; Landivar et al., 1983; Wullschleger and Oosterhuis, 1990b) have shown that the supply of photosynthate for successful yield development in cotton requires the contribution of assimilate from several sources. Results of these experiments indicate that the major suppliers of carbon for fruit development are the leaf subtending the boll (Ashley, 1972; Benedict and Kohel, 1975), the leaf subtending the adjacent fruiting position, and the main-stem leaf sub-

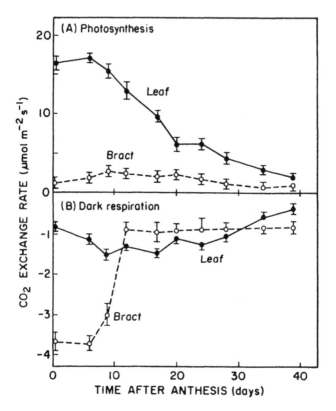

Figure 3 Photosynthetic (A) and respiratory (B) CO_2 activity of leaves and bracts during development of reproductive structures at main-stem node 8. At the time of anthesis the subtending leaf was 18 days old. Error bars represent the standard errors (±) of the mean for 12 replicates. (From Wullschleger and Oosterhuis, 1990c.)

tending the sympodium (Horrocks et al., 1978; Constable and Rawson, 1982). The leaf subtending the fruit is considered to be the major supplier of carbon to the developing boll (Ashley, 1972, Wullschleger and Oosterhuis, 1990d). However, this sympodial leaf is small compared to the main-stem leaf, and therefore the absolute amount of ^{14}C exported is also relatively small (Constable and Rawson, 1982). Main-stem leaves play an important role in the development of the sympodial branch (Oosterhuis and Urwiler, 1988), and subsequently as an additional source of carbon to the developing boll (Constable and Rawson, 1982; Wullschleger and Oosterhuis, 1990d). However, it is becoming increasingly evident that these leaves alone cannot meet the seasonal carbon demands necessary to sustain observed increases in fruit dry weight (Constable and Rawson, 1980b; Wullschleger and Oosterhuis, 1990b,d). This appears particularly true for fruiting positions in the lower canopy where a significant proportion of the carbon requirements for fruit development must be imported (Landivar et al., 1983; Wullschleger and Oosterhuis, 1990d). Other sources of carbon for the developing fruit include the green capsule wall, the bracts subtending the fruiting structure, and storage in the stem and root.

The pattern of ^{14}C distribution is strongly determined by phyllotaxis and associated vascular connections from the main stem to each sympodial branch (Brown, 1968), such that fruit in vertical alignment with a leaf labeled with ^{14}C received 2 to 4 times more label than fruit on the opposite side of the main stem (Constable and Rawson, 1982). Furthermore, a logarithmic relationship exists between the amount of carbohydrate transported (^{14}C-labeled) and the distance a sink is from a carbohydrate source (Constable and Rawson, 1982).

The pattern of leaf nitrogen content was similar to that of photosynthetic activity (Zhu and Oosterhuis, 1992), presumably due to the associated changes in ribulose bisphosphate carboxylase. Aging of cotton leaves is accompanied by a progressive decline in chloroplast ultrastructure, notably a decrease in the number of grana and thylakoids per grana and an increase in the number and size of plastoglobuli (Bondada and Oosterhuis, 1998). There is also a concomitant increase in epicuticular wax content and composition (Bondada et al., 1997) that could possibly influence leaf gas exchange.

5.3 Nonlaminar Sources of CO$_2$ Fixation

The significance of photosynthesis by reproductive structures in cotton and their contribution to yield development have been debated for many years. Bracts, the most evident feature of a square, reach their maximum size at about anthesis (Zhao and Oosterhuis, 1999a). These bracts completely enclose the developing floral bud until the corolla emerges on the day prior to flowering, and they continue to enclose the developing boll for about a week after anthesis (Oosterhuis and Jernstedt, 1999). These modified leaves function in protection of the young fruiting structures and as a source of photoassimilates. Kearney (1929) demonstrated that bract removal at anthesis caused substantial reductions in fruit weight, weight of fiber per seed, and fiber length. Subsequently, Morris (1965), Brown (1968), Elmore (1973), Constable and Rawson (1980b), and Zhao and Oosterhuis (1999a) documented that bracts were indeed capable of CO$_2$ fixation, but that their contribution to yield was relatively minor compared to leaves. Although bracts have a relatively low photosynthetic activity (i.e., 13%) compared to that of their subtending leaf, the photoassimilates they produce may be exported entirely to the boll because of their close proximity to the ovary base (Wullschleger et al., 1990c).

The bracts also play an important role in development of the square by supplying about 50% of the required carbohydrate (Constable and Rawson, 1980b). Zhao and Oosterhuis (1999a) showed that a 20-day-old square obtained 56% of its ^{14}C-assimilate from the bracts, 27% from the subtending leaf, and 17% from the main-stem leaf. These authors demonstrated that removal of bracts from 5-day-old squares resulted in a 20% decrease in final boll weight. The relative contribution of the bracts may increase during water deficit (Wullschleger et al., 1990) because the fruiting structure apparently can maintain a high water potential (Ψ_w) even when the leaves have a much lower Ψ_w (Trolinder et al., 1993) due to the isolation of the fruit from the apoplast of the rest of the plant (van Iersel and Oosterhuis, 1996) and acquisition of water from the phloem (van Iersel et al., 1994b). Bract contribution may also be increased during other adverse environmental conditions including shading and nitrogen stress (Wullschleger et al., 1990). Hence, bracts are intimately involved in the nourishment of the developing boll, particularly during square development and when stresses prevail during boll growth.

Additional sources of carbohydrate for boll development include the green capsule wall, respired CO_2, and stem storage. Photosynthesis from the capsule wall is of limited importance because stomatal conductance decreases considerably about 2 weeks after fertilization (Wullschleger and Oosterhuis, 1990c) due to occlusion of the stomates with wax and closing of the substomatal cavities by expanding parenchyma cells (Bondada et al., 1994; Bondada and Oosterhuis, 2000). Brown (1968) proposed that reassimilation of respired CO_2 by the capsule wall may be significant, but Wullschleger and Oosterhuis (1990c), using $^{14}CO_2$, demonstrated that the capsule wall was only a minor source of CO_2 fixation from the ambient air. However, Wullschleger and Oosterhuis (1991a) showed that although cotton fruiting forms were sites of significant respiratory CO_2 loss, they can recycle internal CO_2, and, thereby, function in retention of otherwise lost assimilate for reproductive development. Storage in stems may provide an additional source of carbohydrate for yield development. Wells (1995) reported considerable carbohydrate storage in the taproot, lateral roots, and lower main stem, with these organs containing 2 to 5 times more starch (w/w) than leaves 100 days after planting. Furthermore, these stem reserves have been shown to increase with plant age (Woon, 1978). The redistribution of stored carbohydrates in the stems and branches to reproductive structures could benefit cotton production during the later stages of growth (Pace et al., 1999).

Based on simulation studies with the GOSSYM/COMAX model, Landivar et al. (1983) suggested that higher photosynthetic rates could result in a 50% yield increase, provided nitrogen and water were adequate.

Increased leaf longevity and the maintenance of photosynthetic activity by leaves during reproductive development were also suggested as potential means to increase yield. Constable and Rawson (1980b) indicated that a larger photosynthetic leaf surface might also increase yield, but cautioned that selection for this trait could be counterproductive if leaf shading occurred, particularly for those leaves at the bottom of the canopy.

As more knowledge is gained concerning the molecular genetic regulation of chloroplast replication and number per cell, it may be possible that this parameter will be amenable to modification through biotechnology. Alternatively, specific biochemical components of the chloroplast may be manipulated to increase the efficiency of capture and utilization of light energy.

5.4 Respiration

Dark respiration for cotton leaves is comparatively small, peaking at about $-1.5\ \mu mol/m^2/s$ (compared to a photosynthetic rate of 16.5 $\mu mol\ CO_2/m^2/s$) for new leaves and then declining to about $-0.4\ \mu mol\ CO_2/m^2/s$ for a 40-day-old leaf (Fig. 3) (Hearn and Constable, 1984; Wullschleger and Oosterhuis, 1990c). Bract respiration on the other hand ranged from $-0.8\ \mu mol\ CO_2/m^2/s$ at anthesis (when the subtending leaf was 18 days old with a photosynthetic rate of 16 $\mu mol\ CO_2/m^2$) to -1.5 and $-0.4\ \mu mol\ CO_2/m^2/s$ at 14 and 40 days after anthesis, respectively. Respiratory activity of cotton fruit is substantial at about $-18.7\ \mu mol\ CO_2/m^2/s$ 6 to 12 days after anthesis, but declining until 34 days and then stabilizing at $-2.6\ \mu mol\ CO_2/m^2$. Diurnal patterns of dark respiration for the cotton fruit are age dependent and closely correlated with dry weight, capsule wall area, and stomatal conductance (Wullschleger and Oosterhuis, 1990c). Stomata on the capsule wall are functional for about 2 weeks then lose this capacity due to occlusion by epicuticular wax and closing of the substomatal cavity with parenchyma (Bondada et al., 1994). Respiratory losses within the cotton plant do not appear to be substantial in the context of the carbon budget for the overall productivity of carbon partitioning within the canopy.

5.5 Inefficiency of Carbon Partitioning in the Cotton Canopy

Although ^{14}C-labeled photosynthates readily translocate throughout the cotton canopy (Brown, 1968; Ashley, 1972), several studies have raised questions concerning the extent to which these assimilates contribute to fruit development (Ashley, 1972; Kerby and Buxton, 1981). These authors suggested that ^{14}C-assimilates produced in the upper portion of the canopy were more likely to go to vegetative growth and only minor amounts moved to the fruit. Wullschlger and Oosterhuis (1990c) found that upper canopy leaves were highly photosynthetically active, while $^{14}CO_2$ assimilation of leaves in the lower one-third of the canopy was reduced more than threefold. Ashley (1972) concluded that at a large portion of the canopy leaf area was inefficient in supplying assimilate for fruit development. These observations, coupled with those of Constable and Rawson (1980a) and Wullschlger and Oosterhuis (1990b), suggest that cotton leaves are limited in their ability to support fruit growth, particularly during periods of peak reproductive development. This is related to the perennial nature of cotton as well as its unique flowering pattern and sensitivity to fruit abscission. A more compact plant with shorter branches and altered leaf morphology for improved light penetration should improve the efficiency of carbon partitioning within the plant to developing fruit. Obviously, redistribution of assimilates must be a fundamental process within the cotton plant. Constable and Rawson (1982) showed that ^{14}C label applied to presquaring plants was subsequently found in bolls on those plants, confirming redistribution.

5.6 Leaf Morphology and Photosynthesis

Leaf shapes of commercially released cotton cultivars range from normal-lobed leaves to okra leaf and even supra okra leaf (Jones, 1982). Okra-leaf types have been reported to exhibit earlier maturity (Heitholt and Meredith, 1998), less boll rot due to more air movement (Andries et al., 1969), greater flower production (Wells and Meredith, 1986; Heitholt, 1993), some insect resistance (Wilson, 1990), improved insecticide penetration (James and Jones, 1985), higher leaf carbon exchange rates attributed to greater specific leaf weight and leaf chlorophyll concentration (Pettigrew et al., 1993), higher canopy photosynthesis per unit leaf area (Kerby et al., 1980; Peng and Krieg, 1991), and shorter sympodial plastochrons (Kerby and Buxton, 1978). Okra-leaf cottons have reduced leaf area compared to normal-leaf cotton (Rao and Weaver, 1976), although Baker and Myhre (1969) reported that okra leaves are as efficient photosynthetically as a canopy of normal leaves if both are at the same plant density and intercepting the same amount of light. Because chloroplast biogenesis occurs early in leaf development and of the developmental pattern of lobed leaves, they possibly possess more chloroplasts per cell than normal-shaped cotton leaves; hence, their specific photosynthetic rate per area of leaf may be greater than normal. However, Wells et al. (1986) showed that canopies of near-isogenic lines with okra or superokra leaf exhibited reduced canopy photosynthesis compared to normal-leaf lines due to reduced canopies and less light interception in the okra and superokra lines. Also, Threadgill et al. (1972) reported that okra-leaved cotton obstructed downward movement of spray droplets more than normal-leaf cotton.

Despite apparent attributes of okra-leaf cotton, yield response has been variable and generally disappointing. Karmi and Weaver (1979) found greater yield in an okra-leaf isoline compared to its normal counterpart, whereas others (Rabey and Oosterhuis, 1974; Heitholt and Meredith, 1998) found no yield advantage from okra-leaf cotton. In Australia, okra-leaf cotton has been used widely and successfully (Thomson, 1995), and recent okra-leaf introductions have performed well in some U.S. regions (Anonymous, 2001). An okra-leaf cultivar developed in Australia has become the dominant cultivar grown in south Texas and accounted for 44% of the acreage in 2001. Landivar et al. (1983) suggested from simulation studies that okra leaf may be superior for yield under optimum conditions, and normal leaf under adverse conditions. Perhaps a better understanding of the attributes of okra-leaf cotton, and improved selections of the okra-leaf trait in proven genetic backgrounds, can offer new potential for yield increases in commercial cotton production.

Low light intensity has been found to reduce yield and increase shedding of squares and bolls (Eaton and Ergle, 1954; Guinn, 1974; Zhao and Oosterhuis, 1999b). Pettigrew (1994) showed that raising the light level by physically manipulating the canopy increased boll weight. Breeding for canopies with increased light penetration to lower leaves for better utilization of the intercepted light may lead to increased lint yield.

Because cotton leaves are heliotropic and hence follow the movements of the sun, especially in the morning, to maximize photosynthesis (Lang, 1973; Fukai and Loomis, 1976), this may provide an additional means of improving net photosynthetic contributions. However, Constable (1986) reported that although cotton leaves faced the sun in the early morning and late afternoon, the light levels were low at these times, and therefore daily photosynthesis was increased only by 9% over leaves that stayed at a constant leaf display.

5.7 Genetic Aspects

The genetic association between photosynthesis and yield has been addressed in various studies (Ford et al., 1983; Nelson, 1988). Results of these investigations indicated that although specific selection programs can increase individual leaf photosynthesis, seldom are these improvements reflected in seed yield or harvest index (Ford et al., 1983). Day and Chalabi (1988) reconciled these observations by reporting that less than 25% to 40% of genetic improvements in leaf photosynthesis are ultimately expressed in canopy photosynthesis due to leaf shading and generally low irradiance for many leaves within the canopy. Apparently, success in manipulating the physiological functioning of single leaves will not necessarily ensure benefits at the whole canopy level. However, modification of the physiological functioning of leaves in conjunction with canopy modification could possibly improve overall crop performance.

Firstly, manipulation of canopy morphological characteristics could have a positive influence on yield. During early seedling and canopy development a rapid increase in leaf area index is desirable, not only to maximize light interception as quickly as possible. but also to rapidly shade weeds that might compete with the crop. As discussed previously, during mid to late crop development self-shading causes the onset of senescence and decline in photosynthetic ability in the older leaves. The ideal morphological development of the cotton plant in the canopy would be to have broad, lobeless juvenile leaves that develop rapidly at the first six main-stem nodes. In subsequent development of additional main-stem nodes the main-stem leaves, as well as the sympodial leaves, would become increasingly more lobed (approaching okra to superokra-type leaves by midseason). With this phenotype, light would better penetrate the canopy, delay senescence in the older leaves, and maintain a useful photosynthetic capacity in the older leaves. Although this phenotype would be difficult to develop, it is within the realm of possibility through selection from among diploid *Gossypium* species already expressing this type of developmental pattern to some extent, e.g., *G. gossypioides*. Also, at the molecular level gene sets are differentially expressed between juvenile and reproductive plants. Thus, as our knowledge of the regulation of plant development increases, regulator genes and gene promoters will probably be identified that can be modified and/or utilized to proscribe the desired phenotype.

In conjunction with canopy architecture, leaves genetically modified to delay senescence might continue to contribute to yield. One possible approach to delaying leaf senescence through genetic engineering would be to incorporate genes for the enzymes involved in cytokinin production. Such genes would be regulated by a promoter with both leaf-specific and ethylene-inducible control sequences, so that normal development of the plant would not be impacted by abnormal phytohormone production during regularly programmed development.

6 DEVELOPMENT AND RETENTION OF SQUARES AND BOLLS

Cotton square and boll shedding has received much attention and generated much controversy during the past 50 years. The focus on boll retention is partly because boll number is strongly correlated with yield (Grimes et al., 1969; Guinn et al., 1981; Wells and Meredith, 1984b). There is the concern that lost squares and bolls represent lost yield, such that if shedding were decreased, then productivity would be increased. On the other hand, evidence indicates that boll shedding may be an important natural process by which the plant adjusts its fruit load to match the supply of inorganic and organic nutrients. A limited amount of shedding is probably normal and necessary for good yields. During a cropping season a cotton plant commonly sheds about 70% of its squares and young bolls under typical growing conditions. However, excessive shedding of squares or bolls can result in significantly altered plant morphology by effectively increasing vegetative growth (Holman and Oosterhuis, 1999). Fruit removal changes the major "sink" for assimilates from the squares and bolls to vegetative components, such as the terminal and leaves, resulting in excessive vegetative growth (Sadras, 1995).

6.1 Growth and Retention of Floral Buds

Floral buds (i.e., pinhead squares) are first visible in the apical part of the plant about 4 to 5 weeks after planting. This is followed in approximately 25 days by flower opening and anthesis (Tharp, 1965) and subsequent development of the fruit (boll) to maturity in about 6 weeks. Squares are composed of three large green bracts, the epicalyx, enclosing the developing flower bud (Oosterhuis and Jernstedt, 1999). The sensitivity of cotton fruiting forms to shedding varies during the season as well as with the size and age of the square or boll (Hake et al., 1989). Rates of shedding are typically low at the beginning of the season and increase to almost 100% at the time of cutout (Guinn, 1982b). Furthermore, squares are more prone to shed as their size increases, flowers are not shed, and bolls are particularly sensitive for about 4 to 6 days following anthesis (Guinn, 1982b; Zhao and Oosterhuis, 1999b). Large squares, flowers, and medium-sized bolls are very resistant to environmental shed, possibly due to high indole acetic acid (IAA) content (Rodgers, 1977). Large bolls are stronger sinks and rarely shed after 15 days, possibly due to a thicker peduncle (Guinn, 1982b) and tougher, well-established vascular connections that feed the growing boll (De Coene, 1950; van Iersel and Oosterhuis, 1994b). McMichael and Guinn (1980) reported that the sensitivity of cotton floral buds to water deficit was greatest during the first week after visible square appearance. Under typical environmental stress, the cotton plant will shed only small- to medium-size squares and small bolls.

6.2 Physiology of Fruit Retention and Abscission

Square sheds have been attributed to both biotic and environmental factors. However, considerable debate has surrounded the relative importance of these two factors. One school of thought holds that small squares shed strictly because of biotic stress, namely, insects, and not due to physiological causes, while the other school maintains that environmental causes, the so-called physiological shedding, are more important. Mauney and Henneberry (1978) reported that small squares that have shed are often found to have insect punctures. Most likely, a combination of both factors contribute to shedding, with insects causing all ages of squares or small bolls to shed, and environmental stresses resulting in the shed of only specific size fruit (Hake et al., 1989).

Many theories have been put forward to explain fruit abscission in cotton, most of them focusing either on the nutritional (Mason, 1922) or hormonal (Rodgers, 1977; Guinn and Brummett, 1987) theories or a combination of both (Guinn, 1982). If the production of carbohydrates cannot supply the demand, then abscission of young bolls occurs (Wadleigh, 1944; Conner et al., 1972; Hearn, 1972). Actually, nutritional stresses alter the hormone balance that in turn causes fruit abscission. Inadequate supply of organic nutrients increases ethylene production and, therefore, conditions that decrease photosynthesis tend to increase shedding (Guinn, 1982b). Conversely, increasing the CO_2 concentration from 350 to 1000 ppm significantly decreased boll shedding (Guinn, 1974). However, Heitholt and Schmidt (1994) found no correlation between carbohydrate or nitrogen concentrations in the ovaries and boll retention. In general, the concentrations of IAA and abscisic acid (ABA) vary considerably in the peduncles of young bolls, with IAA increasing after flowering in retained bolls and ABA increasing sharply in shed bolls (Rodgers, 1977; Guinn and Brummett, 1987). Changes in ethylene evolution and ABA content parallel boll shedding rates (Guinn, 1982b). Gibberellic acid (GA) is probably also involved in boll retention, e.g., Rodgers (1977) found more GA in retained bolls than in bolls that were shedding. A combination of hormonal activity seems to mediate boll retention or shedding. However, the control of shedding does not reside in any single hormone, nutrient, or environmental factor, but is rather regulated by a complex interaction of hormones, nutrients and environment (Guinn, 1982b).

The main environmental triggers of shedding of squares and young bolls includes low light/cloudy overcast weather, high temperatures, water stress, nutrient deficiencies, and insect damage. The perennial nature and indeterminate growth habit of the cotton plant means that both vegetative and reproductive types of growth occur during boll formation, resulting in competition for available assimilate. This, coupled with the sensitivity of square and boll development to adverse environmental conditions, often results in fruit shedding. The quantity and quality of photosynthetically active radiation reaching the cotton fruiting structure has been proposed as an important factor regulating square and boll retention (Constable, 1991; Kasperbauer, 1994; Sadras, 1996). Low light, such as occurs during overcast cloudy weather and with excessive vegetative growth, decreases the rate of photosynthesis such that less carbohydrate is available for boll growth. The associated hormonal imbalance leads to increased boll abscission, usually after a delay of 3 or 4 days (Guinn, 1982b). In dense, high plant populations, the flowers are buried deep in the canopy and excessive abscission will occur. Extremely high temperatures will also cause square abscission and delay fruiting. If temperatures reach such extremes that leaves can no longer be sufficiently cooled by transpiration, i.e., below about 30°C, to allow photosynthesis to operate efficiently (Reddy et al., 1991), then small bolls may abscise. This abscission of small bolls due to excessive temperatures is regulated largely by photosynthate supply and demand, and is also related to night respiratory loss of carbohydrates. As the boll load builds on the plant, abscission of small bolls due to high temperature can be expected to increase. Pollen sterility, due to high night temperatures, is another cause of small boll shed, which occurs 17 to 19 days following night temperatures greater than 29°C (Stewart, 1986). Water deficit can cause fruit abscission (McMichael et al., 1980) due to a reduction in photosynthesis from prematurely aged leaves, smaller leaf size, and stomatal closure (which increases leaf temperature), and also decreased nutrient uptake from the soil. The net effect of severe moisture deficiency on cotton bolls is decreased boll set during the stress, a long delay before boll set fully recovers, and premature cutout. Shedding can also result from excess water. In fine-textured soils, the oxygen level decreases when overwatered, especially if the temperature is warm and organic matter content is high. Cotton plants close their stomates when soil oxygen is low, thereby reducing photosynthesis and evaporative cooling, both of which increase fruit shedding. Lack of oxygen causes an increase in ethylene that in turn promotes abscission.

Cotton growers have long recognized the role nitrogen plays in productivity and have learned to manage the level of nitrogen fertilization to provide adequate nitrogen for boll filling while depleting plant nitrogen prior to defoliation. Nitrogen-deficient cotton plants stop developing new nodes and squares, enter premature cutout, and increase shedding of young bolls. Excess nitrogen is also associated with fruit shed when conditions favor rank growth due to excessive shading and increased insect attractiveness of plants.

6.3 Flower Production and Yield

The production of flowers per unit land area is closely related to yield (Bruce and Romkens, 1965; Micinski et al., 1992; Heitholt, 1993, 1995), a relationship that is obvious under cultural practices that limit plant size (Heitholt, 1999). However, boll retention is an integral aspect of high yields, and researchers have shown that both boll retention as well as total flower numbers per unit area are important determinants of yield (Guinn and Mauney, 1984b). Cook and El-Zik (1993) reported that yield increases from irrigation were associated with increased numbers of flowers and a longer flowering period. These studies indicate the importance of flower production when selecting for high yields. An exception is okra-leaf cotton, which produces more flowers than normal-leaf cottons (Wells and Meredith, 1986) but does not generally produce more yield (Heitholt, 1993). In the Mississippi river delta the pattern of flower production begins with the first flower at about 65 days after planting, peak flowering at about 80 days, and a decline in flowering as the boll load matures (Oosterhuis, 1990). Peak flowering may last a few weeks (Heitholt, 1995) with a high-yielding cotton crop producing about 7.5 flowers/m^2/day with a seasonal total of 150 flowers/m^2 (Heitholt, 1999). Obviously the length of the season for flower production will impact the importance of boll retention, such that geographical locations with short seasons will have a shorter period for flowering and a lower total flower number per unit area. Furthermore, these crops will also be less able to compensate for low flower numbers from poor management or environmental stress. Rain or sprinkler irrigation on open flowers early in the morning dramatically reduces set of those flowers because water causes pollen grains to rupture before they can germinate (Stewart, 1986). It is generally thought that fruit lost because of ruptured pollen is compensated by increased fruit set on days before and after the irrigation. However, recent evidence from Burke (2001) showed that rain in open flowers can result in an overall decrease in boll set and yield reduction.

6.4 Importance of Fruiting Position

The distribution of bolls on the plant varies due to abscission, different genotypic responses in abscission related to cultural practices, and environment. A large percentage of the yield comes from the central portion of the canopy, approximately between main-stem nodes 7 and 14 (Jenkins et al., 1990), which coincides with the distribution of leaf area within the canopy (Oosterhuis and Wullschleger, 1988). Fewer bolls are produced above these nodes, and they tend to be smaller and take longer to mature. Furthermore, there is a hierarchy of size along the branches with fruit being progressively smaller, and of lower quality, away from the main stem (Jenkins et al., 1990).

6.5 Yield Compensation

The importance of adequate early-season fruit set to cotton yield is well documented; however, cotton has the ability to compensate for fruit loss and, in some cases, to even exhibit a yield increase (Brook et al., 1992; Sadras, 1995). To achieve compensation, the cotton crop must set a higher percentage of existing flowers, produce more flowers on distal sympodial fruiting positions, produce more flowers on monopodial branches, or produce more main-stem nodes for additional fruiting sites and/or increased boll size (Hearn and Room, 1979; Heitholt, 1999). The success and form of compensation depends largely on the length of the season and the availability of resources. In a short season, typical of the Mississippi river delta, compensation must come from existing floral buds as the limited season length does not permit the development of additional floral buds. Holman and Oosterhuis (1999) showed that insect damage during squaring resulted in increased leaf and canopy photosynthetic rates as well as reallocation of carbon within the plant to the terminal causing more vegetative growth and taller plants. However, even though early square loss in the U.S. mid-South results in changes in carbon exchange and allocation, poor late-season growing conditions often limits yield compensation. Early-season square loss is usually associated with increased vegetative growth including increased biomass partitioning to the leaves (Kennedy et al., 1986) and to the roots under stressful environments (Constable and Rawson, 1982; Sadras, 1996b). On the other hand, late-season fruit loss will usually reduce yield (Jones et al., 1996). Sadras (1995) suggested the hypothesis that successful yield compensation to fruit loss was more likely when yield potential was limited but not extremely low, and less likely when yield potential was high. This is based on the premise that crops growing in low yield potential environments will likely have inadequate LAI, radiation capture, fruiting sites, and poor root development. Hearn and Room (1979) proposed a method for taking crop development, tolerance, and compensation into account in cotton pest management using a yield development threshold defined as the minimum course of crop yield development that is consistent with both commercial aims and biological feasibility. Compensation is usually not taken into account in crop models and yield projections, and it is suggested that a compensatory function should be included to take into account the dynamic fruiting nature of the cotton plant.

Again, looking at the cotton plant as an unfinished work in domestication, it is possible to visualize a plant growth pattern and structure somewhat modified from that currently existing. Ideally, we would like the plant to perform like an annual in the allocation of photosynthate to harvestable yield, but retain sufficient buffering capacity from its perennial ancestry to compensate for unexpected fruit losses should they occur. Primarily this means allocation of most photosynthate to fruiting structures during the cropping season. If loss of fruit occurs, e.g., from hail, insects, drought, etc., the plant should rapidly restore fruiting structures when the stress causing the loss is past. Among a number of developmental options, conversion of second axillary buds at main-stem and sympodial nodes (Mauney, 1986) from vegetative to reproductive buds might partially compensate for boll loss on sympodial branches at that node. Under vigorous flowering conditions, cotton already has the ability to produce a second flower and boll at each sympodial node. In a best-case scenario, two fruiting structures from the two buds at each node might increase the overall boll load of each plant if developed in conjunction with other modifications that increased net photosynthesis of the crop.

7 CROP MONITORING AND MANAGEMENT

Recent advances in crop monitoring techniques have allowed crop managers to closely follow the fruiting development of the crop and thereby detect any stress effects during the season on fruit set and retention. One of the more widely used crop monitoring programs is COTMAN (COTton MANagement), which utilizes frequent recordings of height and main-stem nodes as well as the number of fruiting branches and square retention (Danforth and O'Leary, 1999) to follow crop growth. Numerous end-of-season management decisions are made based on the decline in nodes above the uppermost white flower (NAWF), which is caused by the stronger demand of the developing fruit load compared to the development of new vegetative nodes on the main stem. Bourland et al. (1992) showed that the decline in canopy photosynthesis soon after flowering closely followed the decline in the NAWF. Close attention to the relationship between reproductive growth (fruiting branches) and vegetative growth (nodes above the reproductive growth) allows detection of stress on the crop and points to various management decisions about plant growth regulators, termination of insecticide application, and crop termination (Bourland et al., 1997; Oosterhuis et al., 1999).

8 BIOTECHNOLOGY AND YIELD

The objective of crop management is to provide the level of fertility and protection from biotic and abiotic stress necessary for the crop to attain its genetic yield potential. Breeding and biotechnology have two goals. One is to provide genetic tools to reduce the level of management input required to protect the crop, while the second is to increase the inherent genetic yield potential of the crop. To date the application of biotechnology to cotton has addressed the first goal because of the ability to transfer the cost of chemical crop protection to genetic crop protection. Biotechnology products emphasizing genetic suppression of major insect pests (Bt), and the ease and convenience of weed control via herbicide tolerance (Roundup, Buctril, etc.) are well established in popular U.S. cotton cultivars. No genetically modified cultivars are registered that address the second goal.

Because genetic enhancement of yield potential is an "output" trait, such an engineered trait would probably have highly variable performance from year to year. Producers are unlikely to pay a premium for engineered seed of uncertain performance beyond improvements obtained by classical breeding. Accordingly, biotech companies for the most part are not willing to invest in develop costs that might take many years to recoup. Although publicly funded scientists can and do develop traits that affect cotton output traits, as well as input traits, in the current regulatory environment they generally cannot afford the cost of meeting the regulatory requirements for testing and releasing transgenic plants into the environment. Accordingly, many of the traits that have been examined by the public sector that have the potential to aid the producer probably will not be incorporated into commercial cultivars.

Throughout this review, the underlying fact and conclusion is that yield is strongly related to the ability of the plant to fix CO_2 and provide a plentiful supply of carbohydrate and other essential nutrition to developing fruit. On a whole-plant basis, and especially on a crop canopy basis, any modification that can increase the effective rate of photosynthesis, the efficiency of carbohydrate utilization, or the partitioning ratio, has the potential to increase yield. A few specific modifications of plant structure and growth patterns that might improve allocation of carbohydrate are mentioned based on current performance of the cotton plant. However, to approach modification of developmental events and partitioning patterns in a logical and effective manner through biotechnology, we must understand the role and function of regulatory genes (those genes that dictate and control the expression of structural genes involved in metabolic pathways). It is through manipulation of the regulatory genes that we will begin to control the complex traits that determine plant structure, function and yield. In the meantime perhaps the historical experience accumulated on the commercial cultivars being grown today will lead to less stringent and costly regulatory requirements to release cotton genetically modified with traits that enhance genetic yield potential.

REFERENCES

Andries JA, Jones JE, Sloane LW, Marshall JG. Effects of okra leaf shape on boll rot, yield and other important characteristics of upland cotton, *Gossypium hirsutum* L. Crop Sci 1969; 9:705–710.

Anonymous. Cotton Varieties Planted, 2001 Crop. Memphis, TN: USDA, AMS, Cotton Division, Aug 2001.

Ashley DA. ^{14}C-labeled photosynthate translocation and utilization in cotton plants. Crop Sci 1972; 12:69–74.

Ashley DA, Doss BD, Bennett OL. Relations of cotton leaf area index to plant growth and fruiting. Agron J 1965; 57:61–64.

Baker DN, Hesketh JD. Respiration and the carbon balance in cotton (*Gossypium hirsutum* L.). In: Brown JM, ed. Proceedings Beltwide Cotton Production Research Conference. Memphis, TN: National Cotton Council of America, 1969:60–64.

Baker DN, Myhre DL. Effects of leaf shape and boundary layer thickness on photosynthesis in cotton (*Gossypium hirsutum*). Physiol Plant 1969; 22:1043–1049.

Bassett DM, Anderson WD, Werkhoven CHE. Dry matter production and nutrient uptake in irrigated cotton (*Gossypium hirsutum* L.). Agron J 1970; 62:299–303.

Benedict CR. Physiology. In: Kohel RJ, Lewis CF, eds. Cotton. Agronomy Series No. 24. Madison, WI: ASA, CSSA, SSSA, 1984:151–200.

Benedict CR, Kohel RJ. Export of ^{14}C-assimilate in cotton leaves. Crop Sci 1975; 15:367–372.

Benedict CR, Smith RH, Kohel RJ. Incorporation of ^{14}C-photosynthate into developing cotton bolls, *Gossypium hirsutum* L. Crop Sci 1973; 13:89–91.

Bondada BR, Oosterhuis DM. Decline in photosynthesis as related to alterations in chloroplast ultrastructure during cotton ontogeny. Photosynthetica 1998; 35:467–471.

Bondada BR, Oosterhuis DM. Comparative epidermal ultrastructure of cotton leaf, bract, and capsule wall. Ann Bot 2000; 86:1143–1152.

Bondada BR, Oosterhuis DM, Wullschleger SD, Kim KS, Harris WM. Anatomical considerations related to photosynthesis in cotton (*Gossypium hirsutum* L.) leaves, bract and capsule wall. J Exp Bot 1994; 45:111–118.

Bondada BR, Oosterhuis DM, Norman RJ. Cotton leaf age, epicuticular wax, and nitrogen-15 absorption. Crop Sci 1997; 37:807–811.

Bourland FM, Oosterhuis DM, Tugwell NP. Concept for monitoring the growth and development of cotton plants using main-stem node counts. J Agric Prod 1992; 5: 532–538.

Bourland FM, Oosterhuis DM, Tugwell NP, Cochran MJ, Danforth DM. Interpretation of Plant Growth Curves Generated by COTMAN. University of Arkansas, Agricultural Experiment Station, Fayetteville, Arkansas, Special report, 1997; 181:8.

Bridge RR, Meredith WR Jr, Chism JF. Comparative performance of obsolete varieties and current varieties of upland cotton. Crop Sci 1971; 11:29–32.

Brook KD, Hearn AB, Kelly CF. Response of cotton to damage by insects in Australia: compensation for early season fruit damage. J Econ Entomol 1992; 85:1378–1386.

Brown KJ. Translocation of carbohydrate in cotton: movement to the fruiting bodies. Ann Bot 1968; 32:703–713.

Brown RS, Oosterhuis DM, Coker DL, Fowler L. Partitioning at the whole plant, boll and seed levels in relation to genotype and environment for predicting yield and stress. Proceedings Beltwide Cotton Conferences. Memphis, TN: National Cotton Council, 2002. CD-Rom.

Bruce RR, Romkens MJM. Fruiting and growth characteristics of cotton in relation to soil moisture tension. Agron J 1965; 57:135–140.

Burke J, Oosterhuis DM, Coker DL, Fowler L. Evaluation of sprinkler induced flower losses and yield reductions. Proceedings Beltwide Cotton Conferences. Memphis, TN: National Cotton Council, 2002. CD-Rom.

Carlson DR. The impact of Pix™ a plant growth regulator, on the physiology of cotton. Agronomy Abstracts. Madison, WI: American Society of Agronomy, 1987: 89.

Christy AL, Porter CA. Canopy photosynthesis and yield in soybean. In: Govindjee, ed. Photosynthesis: Development, Carbon Metabolism and Plant Productivity 1982; Vol. II:. New York: Academic Press, 1982:499–511.

Conner JW, Krieg DR, Gipson JR. Accumulation of simple sugars in developing cotton bolls as influenced by night temperatures. Crop Sci 1972; 12:752–754.

Constable GA. Growth and light receipt by mainstem leaves in relation to plant density in the field. Agric For Meteorol 1986; 37:279–292.

Constable GA. Mapping the production and survival of fruit on field-grown cotton. Agron J 1991; 83:374–378.

Constable GA, Gleeson AC. Growth and distribution of dry matter in cotton. Aust J Agric Res 1977; 28:249–256.

Constable GA, Oosterhuis DM. Temporal dynamics of cotton leaves and canopies. In: Stewart JM, Oosterhuis DM, Heitholt J, eds. Cotton Physiology Book II, 2004. In press.

Constable GA, Rawson HM. Distribution of ^{14}C label from cotton leaves: consequences of changed water and nitrogen status. Aust J Plant Physiol 1982; 9:735–747.

Constable GA, Rawson HM. Effect of leaf position, expansion and age on photosynthesis, transpiration and water use efficiency in cotton. Aust J Plant Physiol 1980a; 7:89–100.

Constable GA, Rawson HM. Photosynthesis, respiration and transpiration of cotton fruit. Photosynthetica 1980b; 14:557–653.

Cook CG, El-Zik KM. Fruiting and lint yield of cotton cultivars under irrigated and non-irrigated conditions. Field Crop Res 1993; 33:411–421.

Cothren JT. Cotton Physiology. In: Smith CW, Cothren JT, eds. Cotton: Origin, History, Technology, and Production. John Wiley and Sons, Inc., New York, 1999:207–268.

Danforth DM, O'Leary PF. Cothman™ Expert System 5.0. University of Arkansas. Cotton Incorporated, Gary, NC.

Day W, Chalabi ZC. Use of models to investigate the link between the modification of photosynthetic characteristics and improved yields. Plant Physiol Biochem 1988; 26:511–517.

De Coene R. Anatomie, histologie et histogénèse de la capsule du cotonnier *Gossypium hirsutum* L. La Cellule 1950; 53:135–150.

Eaton FM. Physiology of the cotton plant. Annu Rev Plant Physiol 1955; 6:299–328.

Eaton FM, Ergle DR. Effects of shade and partial defoliation on carbohydrate levels and growth, fruiting and fiber properties of cotton plants. Plant Physiol 1954; 29:39–49.

Elmore CD. Contributions of the capsule wall and bracts to developing cotton fruit. Crop Sci 1973; 13:751–752.

Elmore CD. The paradox of no correlation between leaf photosynthetic rates and crop yields. In: Hesketh JD, Jones JW, eds. Predicting Photosynthesis for Ecosystem models. Vol. LI. Boca Raton, FL: CRC Press, 1980:155–167.

Elmore CD, Hesketh JD, Muramoto H. A survey of rates of growth, leaf aging and leaf photosynthetic rates among and within species. J Ariz Acad Sci 1967; 4:215–219.

Ford DM, Shibles R, Green DE. Growth and yield of soybean lines selected for divergence leaf photosynthetic ability. Crop Sci 1983; 23:517–520.

Fryxell PA. The Natural History of the Cotton Tribe. Malvaceae (Tribe Gossypieae). College Station, TX: Texas A&M University Press, 1979.

Fryxell PA. Ecological adaptations of *Gossypium* species. In: Mauney JR, Stewart JM, eds. Cotton Physiology. Memphis, TN: The Cotton Foundation, National Cotton Council, 1986:1–7.

Fakai A, Loomis RS. Leaf display and light environments in row-planted cotton communities. Agric Meteorol 1976; 17:343–379.

Grimes DW, Dickens WL, Anderson WD. Functions for cotton (*Gossypium hirsutum* L.) production from irrigation and nitrogen fertilization variables. L1. Yield components and quality characteristics. Agron J 1969; 61:773–776.

Guinn G. Abscission of cotton floral buds and bolls as influenced by factors affecting photosynthesis and respiration. Crop Sci 1974; 14:291–293.

Guinn G. Abscisic acid and abscission of young cotton bolls in relation to water availability and boll load. Crop Sci 1982a; 22:580–583.

Guinn G. Causes of square and boll shedding in cotton. U.S. Department of Agriculture, Tech Bull, 1982b.

Guinn G. Fruiting of cotton. III. Nutritional stress and cutout. Crop Sci 1985; 25:981–985.

Guinn G, Brummett DL. Concentrations of abscisic acid and indole acetic acid in cotton fruits and their abscission zones in relation to fruit retention. Plant Physiol 1987; 83:199–202.

Guinn G, Mauney JR. Fruiting of cotton. I. Effects of moisture status on flowering. Agron J 1984a; 76:90–94.

Guinn G, Mauney JR. Fruiting of cotton. II. Effects of plant moisture status and active boll load on boll retention. Agron J 1984b; 76:94–98.

Guinn G, Mauney JR, Fry KE. Irrigation scheduling and plant population effects on growth, bloom rates, boll abscission, and yield of cotton. Agron J 1981; 73:529–534.

Haigler CH, Hequet EF, Kreig DR, Strauss RE, Wyatt BG, Cai W, Jaradat T, Keating K, Srinivas NS, Wu C, Holaday AS, Jividen GJ. Transgenic cotton with improved fiber micronaire, strength, and length and increased fiber weight. Proceedings Beltwide Cotton Conferences. Memphis, TN: National Cotton Council, 2000a:483.

Haigler CH, Martin LK, Tummala J, Cai W, Keating K, Anconetan R, Holaday AS, Jividen GJ, Gannaway JG. Mechanisms by which fiber quality and fiber and seed weight can be improved in transgenic cotton growing under cool night temperatures. Proceedings Beltwide Cotton Conferences. Memphis, TN: National Cotton Council, 2000b:483–484.

Hake K, Guinn G, Oosterhuis DM. Environmental triggers of square and boll shed. Newsl Cotton Physiol Educ Prog 1989; 1(2):1–4.

Hake SJ, Kerby TA, Hake K. Cotton Production Manual. Publication No. 3352. Oakland, CA: University of California, 1996:417.

Hearn AB. Growth and performance of cotton in a desert environment. I. Morphological development of the crop. J Agric Sci 1969a; 73:65–74.

Hearn AB. The growth and performance of cotton in a desert environment. II. Dry matter production. J Agric Sci Camb 1969b; 73:75–86.

Hearn AB. The growth and performance of cotton in a desert environment. III. Crop performance. J Agric Sci Camb 1969c; 73:75–86.

Hearn AB. Growth and performance of rain-grown cotton in a tropical upland environment. II. The relationship between yield and growth. J Agric Sci Camb 1972; 79:137–145.

Hearn AB. Crop physiology. In: Arnold MH, ed. Agricultural Research for Development. London, UK: Cambridge University Press, 1976:77–122.

Hearn AB, Constable GA. Cotton. In: Goldsworthy PR, Fisher NM, eds. The Physiology of Tropical Field Crops. Chichester, UK: John Wiley & Sons, 1984:495–527.

Hearn AB, Fitt GP. Cotton cropping systems. In: Pearson CJ, ed. Ecosystems of the WorldField Crop Ecosystems. Vol. 18. Amsterdam, Netherlands: Elsevier, 1992: 85–142.

Hearn AB, Room PM. Analysis of crop development for cotton pest management. Prot Ecol 1979; 1:265–277.

Heitholt JJ. Cotton boll retention and its relationship to yield. Crop Sci 1993; 33: 486–490.

Heitholt JJ. Cotton flowering and boll retention in different planting configurations and leaf shapes. Agron J 1995; 87:994–998.

Heitholt JJ. Cotton: factors associated with assimilation capacity, flower production, boll set, and yield. In: Smith DL, Hamel C, eds. Crop Yield, Physiology and Processes. Berlin: Springer Verlag, 1999:235–269.

Heitholt JJ, Meredith WR. Yield, flowering and leaf area index of okra-leaf and normal-leaf cotton isolines. Crop Sci 1998; 38:643–648.

Heitholt JJ, Schmidt JH. Receptacle and ovary assimilate concentrations and subsequent boll retention in cotton. Crop Sci 1994; 34:125–131.

Heitholt JJ, Meredith WR, Rayburn ST. Leaf area index response of four obsolete and four modern cotton cultivars to two nitrogen levels. J Plant Nutr 1998; 21: 2319–2328.

Holman EM, Oosterhuis DM. Cotton photosynthesis and carbon partitioning in response to floral bud loss due to insects. Crop Sci 1999; 39:1347–1351.

Holt SJ, Stewart JM, McNew RW. Floral bud development in greenhouse grown cotton. Crop Sci 1994; 34:519–527.

Hons FM, McMichael BL. Planting pattern effects on yield and root growth of cotton. Field Crops Res 1986; 13:147–159.

Horrocks RD, Kerby TA, Buxton DR. Carbon source for developing bolls in normal and superokra leaf cotton. New Phytol 1978; 80:335–340.

James D, Jones JE. Effects of leaf and bract isolines on spray penetration and insecticidal efficacy. In: Nelson TC, Brown JM, eds. Proceedings Beltwide Cotton Production Research Conferences. Memphis, TN: National Cotton Council, 1985: 395–396.

Jenkins JN, McCarty JC Jr, Parrott WL. Effectiveness of fruiting sites in cotton: boll size and boll set percentage. Crop Sci 1990; 30:857–860.

John ME. Genetic engineering strategies for cotton fiber modification. In: Basra AS, ed. Cotton Fibers: Developmental Biology, Quality Improvement and Textile Processing. New York, NY: Food Products Press, 1999:271–292.

Johnson RE, Addicott FT. Boll retention in relation to boll and leaf development in cotton (*Gossypium hirsutum* L.). Crop Sci 1967; 7:571–574.

Jones JE. The present state of the art and science of cotton breeding for leaf morphological types. Proceedings Beltwide Cotton Production Research Conference. Memphis, TN: National Cotton Council, 1982:93–99.

Jones HG. Plant and microclimate. A quantitative approach to environmental plant physiology. Cambridge, UK: Cambridge University Press, 1983.

Jones MA, Wells R, Guthrie DS. Cotton response to seasonal patterns of flower removal: I. Yield and fiber quality. Crop Sci 1996; 36:633–638.

Karmi E, Weaver JB. Growth analysis of American upland cotton, *Gossypium hirsutum* L., with different leaf shapes and colors. Crop Sci 1979; 12:317–320.

Kasperbauer MJ. Cotton plant size and fiber development responses to FR/R reflected from the soil surface. Physiol Plant 1994; 91:317–321.

Kearney TH. Development of the cotton boll as affected by removal of the involucre. J Agric Res 1929; 38:381–393.

Kennedy CW, Smith WC, Jones JE. Effect of early season square removal on three leaf types of cotton. Crop Sci 1986; 26:139–145.

Kerby TA, Buxton DR. Effect of leaf shape and plant population on rate of fruiting position appearance in cotton. Agron J 1978; 70:535–538.

Kerby TA, Buxton DR. Competition between adjacent fruiting forms in cotton. Agron J 1981; 73:867–871.

Kerby TA, Buxton DR, Matsuda K. Carbon source–sink relationships within narrow-row cotton canopies. Crop Sci 1980; 20:208–213.

Kerby TA, Cassman KG, Keeley M. Genotypes and plant densities for narrow row cotton systems. II. Leaf area and dry matter partitioning. Crop Sci 1990; 30:649–653.

Kerby TA, Keely M, Johnson S. Growth and Development of Acala Cotton. Madison, WI: University of California Experimental Station, Division of Natural Resources, Bull 1921 1987:13.

Krieg DR. Whole plant responses to water deficits: carbon assimilation and utilization. In: Taylor HM, Jordan WR, Sinclair TR, eds. Limitations to Efficient Water Use in Crop Production. Madison, WI: ASA, CSSA and SSSA, 1983:319–330.

Krieg DR. Leaf age-gas exchange characteristics. Proceedings Beltwide Cotton Production Research Conferences. Memphis, TN: National Cotton Council, 1988:55–57.

Landivar JA, Baker DN, Jenkins JN. Application of GOSSYM to genetic feasibility studies. I. Analyses of fruit abscission and yield in okra-leaf cotton. Crop Sci 1983; 23:497–504.

Lang ARG. Leaf orientation of a cotton plant. Agric Meteorol 1973; 1:37–51.

Lewis H. A review of yield and quality trends and components in American Upland cotton. Proceedings Beltwide Cotton Conferences. Memphis, TN: National Cotton Council, 2001:1447–1453.

Lewis H, May L, Bourland F. Cotton yield components and yield stability. Proceedings Beltwide Cotton Conferences. Memphis, TN: National Cotton Council, 2000:532–536.

Loomis RS, Williams WA. Maximum crop productivity: an estimate. Crop Sci 1963; 3:67–72.

Mason TG. Growth and abscission in Sea Island cotton. Ann Bot 1922; 36:457–483.

Mauney JR. Floral initiation of upland cotton (*Gossypium hirsutum* L.) in response to temperature. J Exp Bot 1966; 17:452–459.

Mauney JR. Carbohydrate production and distribution in cotton canopies. In: Mauney JR, Stewart JM, eds. Cotton Physiology. Memphis, TN: The Cotton Foundation, National Cotton Council, 1986:183–191.

Mauney JR. Vegetative growth and development of fruiting sites. In: Mauney JR, Stewart JM, eds. Cotton Physiology. Memphis, TN: The Cotton Foundation, National Cotton Council, 1986:11–28.

Mauney JR, Henneberry TJ. Plant bug damage and shed of immature cotton squares in Arizona. In: Brown JM, ed. Proceedings Beltwide Cotton Production Research Conferences. Memphis, TN: National Cotton Council, 1978:41–42.

Mauney JR, Stewart JM, eds. Cotton Physiology. Memphis, TN: The Cotton Foundation, National Cotton Council, 1986:786.

Meredith WR Jr. Quantitative genetics. In: Kohel RJ, Lewis CF, eds. Cotton. Agronomy Series No. 24. Madison, WS: American Society of Agronomy, 1984: 131–150.

McArthur JA, Hesketh JD, Baker DN. Cotton. In: Evans LT, ed. Crop Physiology. New York: Cambridge University Press, 1975:297–325.

McClelland CK, Neely JW. The order, rate, and regularity of blooming in the cotton plant. J Agric Res 1931; 42:751–763.

McMichael BL, Guinn G. The effect of moisture deficit on square shedding. In: Brown JM, ed. Proceedings Beltwide Cotton Production Research Conferences. Memphis, TN: National Cotton Council of America, 1980:38.

McMichael BL, Guinn G, Fry KE. The effects of moisture deficits on square shedding. In: Brown JM, ed. Proceedings Beltwide Cotton Production Research Conferences. Memphis, TN: National Cotton Council, 1980:38.

Micinski S, Colyer PD, Nguyen KT, Koonce KL. Cotton white flower counts and yield with and without early-season pest control. J Prod Agric 1992; 5:126–130.

Morris DA. Photosynthesis by the boll wall and bractioles of the cotton plant. Emp Cotton Grow Rev 1965; 42:49–51.

Mullet JE. Chloroplast development and gene expression. Annu Rev Plant Physiol Plant Mol Biol 1988; 39:475–502.

Mullins GL, Burmester CH. Dry matter, nitrogen, phosphorus, and potassium accumulation by four cotton varieties. Agron J 1991; 82:729–736.

Munro JM. Cotton and its production. In: Matthews GA, Tunstall J, eds. Insect Pests of Cotton. Wallingford, UK: CAB International, 1995:3–26.

Muramoto H, Heskweth JD, Elmore CD. Leaf growth, leaf ageing, and leaf photosynthetic rates of cotton plants. Proceedings Beltwide Cotton Production Research Conferences. Memphis, TN: National Cotton Council, 1967:161–165.

Mutsaers HJW. Growth and assimilate conversion of cotton bolls. 2. Influence of temperature on boll maturation period and assimilate conversion. Ann Bot 1976; 40:317–324.

Cotton Econ Rev 2001; 32(9):1–4. National Cotton Council of America.

Nelson CJ. Genetic association between photosynthetic characteristics and yield: review of the evidence. Plant Physiol Biochem 1988; 26:543–554.

Niles GA, Feaster CV. Breeding. In: Kohel RJ, Lewis CF, eds. Cotton. Agronomy Series No. 2. Madison, WI: American Society of Agronomy, 1984:202–231.

Olson LC, Bledsoe RP. The Chemical Composition of the Cotton Plant and the Uptake of Nutrients at Different Growth Stages. Bulletin 222. Atlanta, GA: Georgia Agricultural Experiment Station, Tifton, GA, 1942.

Oosterhuis DM. Growth and development of the cotton plant. In: Miley WN, Oosterhuis DM, eds. Nitrogen Nutrition in Cotton: Practical Issues. Proceedings Southern Branch Workshop for Practicing Agronomists. Madison, WI: American Society of Agronomy, 1990:1–24.

Oosterhuis DM. A postmortem of the disappointing yields in the 1993 Arkansas cotton crop. Proceedings 1994 Arkansas Cotton Research Meeting and Summaries of Research, Special Report 166. University of Arkansas Agricultural Experiment Station, 1995; 166:22–26.

Oosterhuis DM. Foliar fertilization of potassium. In: Oosterhuis DM, Berkowitz GA, eds. Frontiers in Potassium Nutrition. Atlanta, GA: Potash and Phosphate Institute and Crop Science Society of America, 1999:87–100.

Oosterhuis DM, Jernstedt J. Morphology and anatomy of the cotton plant. Chap. 2.1. In: Smith W, Cothren JS, eds. Cotton: Origin, History, Technology and Production. John Wiley & Sons, Inc., New York, 1999:175–206.

Oosterhuis DM, Chipamaunga J, Bate GC. Nitrogen uptake of field grown cotton. I. Distribution in plant components in relation to fertilization and yield. Exp Agric 1983; 19:91–102.

Oosterhuis DM, Kosmidou K, Cothren JT. Managing cotton growth and development with growth regulators. Proceedings Second World Cotton Research Conference, Athens, Greece, Sept 6–12, 1998:46–68. Published P. Petmdis Thersaloniki, Greece.

Oosterhuis DM, Tugwell NP, Teague TG, Danforth DM. A new method of assessing plant stress using the ratio of the change in square shedding to number of main-stem nodes. Proceedings 1999 Cotton Research Meeting and Summaries of Research in Progress. Special Report 193. Memphis, TN: Arkansas Agricultural Experiment Station, Fayetteville, Arkansas, 1999:136–141.

Oosterhuis DM, Urwiler MJ. Cotton main-stem leaves in relation to vegetative development and yield. Agron J 1988; 80:65–67.

Oosterhuis DM, Wullschleger SD. Carbon partitioning and photosynthetic efficiency during boll development. Proceedings Beltwide Cotton Production Research Conferences. Memphis, TN: National Cotton Council, 1988:57–60.

Pace PF, Cralle HT, Cothren JT, Senseman SA. Photosynthate and dry matter partitioning in short- and long-season cotton cultivars. Crop Sci 1999; 39:1065 1069.

Patterson DT, Bunce JA, Alberte RS, van Volkenburgh E. Photosynthesis in relation to leaf characteristics of cotton from controlled and field environments. Plant Phyiol 1977; 59:384–387.

Peng S, Krieg DR. Single leaf and canopy photosynthesis response to plant age in cotton. Agron J 1991; 83:704 708.

Peoples TR, Matthews MA. Influence of boll removal on assimilate partitioning in cotton. Crop Sci 1981; 21:283–286.

Peters DB, Clough DF, Graves RA, Stahl GR. Measurements of dark respiration, evaporation, and photosynthesis in field plots. Agron J 1974; 66:460 462.

Pettigrew WT. Source-to-sink manipulation effects on cotton lint yield and yield components. Agron J 1994; 86:731–735.

Pettigrew WT, Heitholt JJ, Vaughn KC. Gas exchange differences and comparative anatomy among cotton leaf-type isolines. Crop Sci 1993; 33:1295–1299.

Pettigrew WT, McCarty JC, Vaughn KC. Leaf-senescence-like characteristics contribute to cotton's premature photosynthesis decline. Photosynth Res 2000; 65:187–195.

Rabey GG, Oosterhuis DM. High Density, Variety and Population Trial. Annual Report, Cotton Research Institute, Gatooma. Rhodesia (Zimbabwe): Government Printers, 1974.

Ramey HH Jr. Yield response of six cultivars of upland cotton, *Gossypium hirsutum L.*, in two cultural regimes. Crop Sci 1972; 12:353–354.

Rao MJ, Weaver JB. Effect of leaf shape on response of cotton to plant population, nitrogen rate and irrigation. Agron J 1976; 68:599–601.

Reddy VR, Reddy KR, Baker DN. Temperature effects on growth and development of cotton during the fruiting period. Agron J 1991; 83:211–217.

Rodgers JP. Plant growth substances in relation to fruit development and fruit abscission in cotton. Ph.D. dissertation, University of Rhodesia, Salisbury, Rhodesia (Zimbabwe), 1977.

Sadras VO. Compensatory growth in cotton after loss of reproductive forms. Field Crops Res 1995; 40:1–18.

Sadras VO. Population-level compensation after loss of vegetative buds: interactions among damaged and undamaged cotton neighbours. Oecologia 1996; 106:417–423.

Scholl RL, Miller PA. Genetic association between yield and fiber strength in upland cotton. Crop Sci 1976; 16:780–783.

Sinclair TR, Horie T. Leaf nitrogen, photosynthesis and crop radiation-use efficiency: a review. Crop Sci 1989; 29:90–98.

Singh SP. Studies on the effects of soil moisture stress on yield of cotton. Ind J Plant Physiol 1975; 18:49 55.

Stewart JM. Integrated events in the flower and fruit. In: Mauney JR, Stewart JM, eds. Cotton Physiology. Memphis, TN: The Cotton Foundation, National Cotton Council, 1986:261 300.

Stockton JR, Doneen LD, Walhood VT. Boll shedding and growth of the cotton plant in relation to irrigation frequency. Agron J 1961; 53:272 275.

Taylor HM, Klepper B. Water relations of cotton. 1. Root growth and water use as related to top growth and soil water content. Agron J 1974; 66:584–588.

Tharp WH. The Cotton Plant. How It Grows and Why Its Growth Varies. (Agricultural Handbook No. 178). Washington, DC: USDA, 1960.

Thomson NJ. Commercial utilization of the okra leaf mutant of cotton. The Australian experience. In: Constable CA, Forrester NW, eds. Challenging the Future. Proceedings World Cotton Research Conference 1. Brisbane, Queensland, Feb 24–28, 1994. Australia: CSIRO, 1995:393–401.

Threadgill ED, Miles GE, Douglas AG, Anderson KL. Penetration of topical sprays in cotton canopies. Proceedings 25th Annual Meeting Southern Weed Science Society, Champaign, IL, 1972:486–488.

Trolinder NL, McMichael BL, Upchurch DR. Water relations of cotton flower petals and fruit. Plant Cell Environ 1993; 16:755–760.

Van Iersel M, Oosterhuis DM. Drought effects on the water relations of cotton fruits, bracts and leaves during ontogeny. Environ Exp Bot 1996; 36:51–59.

Van Iersel MW, Harris WM, Oosterhuis DM. Phloem in developing cotton fruits: 6(5)carboxyfluorescein as a tracer for functional phloem. J Exp Bot 1994a; 46:321–328.

Van Iersel MW, Oosterhuis DM, Harris WM. Apoplastic water flow to cotton leaves and fruits during development. J Exp Bot 1994b; 45:163–169.

Verhalen LM, Managhani R, Morrison WC, McNew RW. Effect of blooming date on boll retention and fiber properties in cotton. Crop Sci 1975; 15:47–52.

Wadleigh CH. Growth status of the cotton plant as influenced by the supply of nitrogen. (Bulletin No. 466). Arkansas Agricultural Experiment Station, Fayetteville, Arkansas, 1944:138.

Weir BL, Kerby TA, Hake KD, Roberts BA, Zelinski LJ. Cotton fertility. In: Johnson Hake S, Kerby TA, Hake KD, eds. Cotton Production Manual. Publication No. 3352. Memphis, TN: University of California, Oakland, California, 1996:210–227.

Wells R. Response of leaf ontogeny and photosynthesis to reproductive growth in cotton. Plant Physiol 1988; 87:274–279.

Wells R. Measurements of lint production in cotton and factors affecting yield. Mod Methods Plant Anal 1989; 10:278–294.

Wells R. Seasonal patterns of carbohydrate storage in cotton. In: Richter DA, Armour J, eds. Proceedings Beltwide Cotton Conferences. Memphis, TN: National Cotton Council, 1995:1137.

Wells R, Meredith WR. Comparative growth of obsolete and modern cultivars. I. Vegetative dry matter partitioning. Crop Sci 1984a; 24:858–862.

Wells R, Meredith WR. Comparative growth of obsolete and modern cultivars. II. Reproductive dry matter partitioning. Crop Sci 1984b; 24:863–868.

Wells R, Meredith WR. Comparative growth of obsolete and modern cultivars. III. Relationship of yield to observed growth characteristics. Crop Sci 1984c; 24:868–872.

Wells R, Meredith WR. Normal vs okra leaf yield interactions in cotton. 11. Analysis of vegetative and reproductive growth. Crop Sci 1986; 26:223–228.

Wells R, Meredith WR, Williford JR. Canopy photosynthesis and its relationship to plant productivity in near-isogenic cotton lines differing in leaf morphology. Plant Physiol 1986; 82:635–640.

Wilson FD, George BW, Dean P. Cotton leaf shape affects insects. Agric Res 1982; 30:11, 16.

Wilson FD. Relative resistance of cotton lines to pink bollworm. Crop Sci 1990; 30, 500–504.

Wilson LJ. Resistance of okra-leaf cotton genotypes to two-spotted spider mites (Acari: Tetranychidae). J Econ Entomol 1994; 87:1726–1735.

Woon CK. Seasonal changes in stem carbohydrate and petiole nitrate in cotton. Ph.D. dissertation, University of Arizona. Diss Abstr Int 1978; 38:5133–5134.

Wullschleger SD, Oosterhuis DM. Water use efficiency as a function of leaf age and position within the cotton canopy. Plant Soil 1989; 120:79–85.

Wullschleger SD, Oosterhuis DM. Canopy development and photosynthesis of cotton as influenced by nitrogen fertilization. J Plant Nutr 1990a; 13:1141–1154.

Wullschleger SD, Oosterhuis DM. Photosynthesis of individual field-grown leaves during ontogeny. Photosynth Res 1990b; 23:163–170.

Wullschleger SD, Oosterhuis DM. Photosynthetic and respiratory activity of fruiting forms within the cotton canopy. Plant Physiol 1990c; 94:463–469.

Wullschleger SD, Oosterhuis DM. Photosynthetic carbon production and use by cotton leaves and bolls. Crop Sci 1990d; 30:1259–1264.

Wullschleger SD, Oosterhuis DM. Evidence for light-dependent recycling of respired carbon dioxide by the cotton fruit. Plant Physiol 1991a; 97:574–579.

Wullschleger SD, Oosterhuis DM. Photosynthesis, transpiration, and water-use-efficiency of cotton leaves and fruit. Photosynthetica 1991b; 25:505–515.

Wullschleger SD, Oosterhuis DM. Canopy leaf area development and age-class dynamics in cotton. Crop Sci 1992; 32:451–456.

Wullschleger SD, Oosterhuis DM, Rutherford S. Importance of bracts in the carbon economy of cotton. Ark Farm Res 1990; 39(3):44.

Zelitch I. The close relationship between photosynthesis and crop yield. BioScience 1982; 32:796–802.

Zhao D, Oosterhuis DM. Carbon contribution of leaves and bracts to developing cotton floral buds. Photosynthetica 1999a; 36:279–290.

Zhao D, Oosterhuis DM. Cotton responses to shade at different growth stages. 1. Lint yield and fiber quality. Exp Agric 1999b; 36:27–39.

Zhu B, Oosterhuis DM. Nitrogen distribution within a sympodial branch of cotton. Plant Nutr 1992; 15:1–14.

9

Genome Mapping and Genomic Strategies for Crop Improvement

Prasanta K. Subudhi
Louisiana State University
Baton Rouge, Louisiana, U.S.A.

Henry T. Nguyen
University of Missouri–Columbia
Columbia, Missouri, U.S.A.

1 INTRODUCTION

Plant-breeding programs made significant strides in improving the productivity of major crop plants in the last century. This was largely made possible by the application of conventional genetic principles. While genetic improvement of qualitative traits was successful as a result of the simple inheritance pattern, progress in manipulating quantitative traits, such as stress tolerance, yield, and yield components, has lagged behind. The major constraint is the lack of knowledge about the mechanisms and complex biological pathways that are involved in the plants' response in target environments. During the past two decades, developments in molecular biology provided new opportunities in a variety of ways that would have far-reaching impact on crop-breeding programs. Some key areas of crop improvement include: (1) assessment of genetic diversity, (2) new gene discovery, (3) genetic dissection of

403

complex traits, and lastly, the most practical being (4) marker-assisted selection (MAS) of useful traits. Designing new crop cultivars through introgression of useful genes using molecular markers has been routine in many
private- and public-sector breeding programs. Molecular linkage maps constructed virtually in all major crop plants have been instrumental in mapping
and marker-assisted selection of agronomically important genes. Our understanding about inheritance and expression of complex traits is progressing
rapidly through positional cloning of quantitative trait loci (QTL). With the
acquisition of overwhelming amount of genomic sequences as a result of
whole genome and expressed sequence tags (EST) sequencing initiatives in
many crop plants, there has been a dramatic shift in approaches followed by
plant biologists to interpret the intricacy of plant biology. Defining the
functions of all genomic or EST sequences, their networking, and regulation
in the plant system remains as a major challenge for postgenomic researchers.
We are now better equipped to apply genome mapping, reverse genetics tools,
and expression profiling in a wide array of organisms to understand complex
biological processes (Osterlund and Paterson, 2002). Technological breakthrough in genomics promises to provide a comprehensive picture of the
various plant growth and development responses to ensure real gain in
improving and stabilizing crop productivity. High-throughput genomics
tools, coupled with innovative bioinformatics techniques, offer new avenues
to exploit model systems to make rapid advances.

This chapter provides an overview of modern genomic tools and
strategies from the crop improvement perspective. Application of the high-
throughput molecular tools to understand the basic plant biology will also be
discussed.

2 OVERVIEW OF MOLECULAR MARKER TECHNOLOGY

Before the advent of DNA markers, morphological attributes and protein loci
(isozymes) were typically used in linkage studies in many crop species.
Although they are comparable to DNA markers, their usefulness has been
limited because of the limited number of loci and limited level of polymorphism associated with them. The era of DNA markers began two decades
ago when Botstein et al. (1980) highlighted the potential of DNA markers and
unleashed the possibility of innumerable restriction fragment length polymorphic (RFLP) markers due to variation of restriction sites in natural populations. Soon after the discovery of RFLP, numerous marker types were
recognized and widely reviewed for their competence in crop improvement
(Tanksley et al., 1989; Paterson et al., 1991; Burrow and Blake 1998; Joshi et
al., 1999). Some of these marker systems are random amplified polymorphic
DNA (RAPD) (Williams et al., 1990), amplified fragment length polymorphism (AFLP) (Zabeau and Vos. 1993; Vos et al., 1995), microsatellites or

simple sequence repeat (SSR) (Litt and Luty, 1989), and sequence tagged sites (STS) (Olson et al., 1989). These markers are based on polymerase chain reaction (PCR) amplification of genomic DNA and detection of size differences in the electrophoretically separated PCR products. Several variations of RAPD technique are available. The technique of DNA amplification fingerprinting (DAF) (Caetano-Anolles et al., 1991) and arbitrary primed PCR (AP-PCR) (Welsh and McClelland, 1990) are essentially equivalent in principle to the RAPD technique with minor modifications. Sequence characterized amplified region (SCAR) markers (Martin et al., 1991) were developed by end sequencing of RAPD fragments and use of longer primers. Restriction digestion of PCR products resulted in cleaved amplified polymorphic sequences (CAPS) (Koniecyzn and Asubel, 1993). STS markers are similar to SCAR markers in principle and application, except that they are generated from the sequence information of the RFLP markers that are linked to specific traits. As molecular markers have been reviewed extensively, our focus will be on comparing the different marker systems for efficiency, cost effectiveness, and applications to crop improvement.

Among the DNA markers described above, the RFLP technique has been widely used for DNA analysis in crop plants. This method is robust, reproducible, and transferable between laboratories. The codominant nature of RFLP makes it very reliable and informative for linkage analysis and breeding. Numerous DNA probes have been generated and mapped in many plant species, and are easily available to researchers. RFLP is also a valuable tool for meaningful comparisons of mapping information across populations or species level (Paterson, 1996). Despite its usefulness, factors such as lengthy protocol, requirement for large quantity of DNA, technical complexity, and use of radioisotopes have limited its usefulness in plant-breeding programs, where thousands of segregating individuals are assayed in a limited time frame for selection of desirable genotypes. These limitations necessitated the development of the PCR-based marker technology, in which the demand for genomic DNA is several folds less, and a less demanding protocol and automation make it very efficient for generating a detailed map of small populations in short time. Because the cost of DNA assay and time efficiency are deciding factors for integrating molecular markers in plant-breeding programs, Ragot and Hoisington (1993) compared the RAPD and RFLP technology, and concluded that RAPD analysis is generally more cost- and time-efficient for studies involving small sample sizes, while RFLPs have the advantage for larger sample sizes.

Unlike RFLP, RAPD and AFLP are usually dominant markers, thus preventing the accurate detection of heterozygotes. It is possible to identify heterozygotes using RAPD and AFLP tightly linked in repulsion, but it may not be advisable to design such mapping experiments as the genotyping cost escalates significantly due to the requirement of a large population. New software is being developed to identify homo- and heterozygotes in AFLP fingerprints (Pot and Pouwels, 2002). The level of polymorphism detected by AFLP is lower than those with other mapping techniques, such as RFLP and SSR. However, because of its robustness, reliability, and efficiency, it is now gaining popularity among researchers for many types of investigation, including mapping, fingerprinting, and phylogenetics (Ridout and Donini, 1999). It combines the robustness of RFLP with the technical simplicity of the PCR-based techniques. In lieu of radioactive or fluorescent labeling, the amplified loci can be visualized by silver staining (Cho et al., 1996).

The developmental cost of microsatellites is very high, but the simpler protocols, abundance, and increased level of polymorphism make it very popular among plant breeders (Gupta et al., 1996). Because SSR markers are amenable to automation from DNA isolation to gel loading to imaging and data generation, it certainly has great potential for application in molecular-marker-assisted breeding programs. Considering the advantages of micro-satellite markers, their usage in germplasm improvement has witnessed considerable expansion in comparison with other classes of markers in many crop species. It is obvious that, in most plant-breeding programs, where a large number of individuals are assayed with a limited amount of leaf tissue (e.g., seedling stage), PCR-based genotyping is the only suitable alternative.

Partial sequencing of random cDNA clones undertaken in many important crop species results in numerous EST markers that are suitable for both RFLP and PCR-based assays for genome analysis. Because these are expressed genetic elements, their placement onto the existing molecular linkage maps is of critical importance in identifying candidate genes for useful traits. Currently, in addition to RFLP analysis, a most promising new method for mapping appears to be a combination of AFLP and multiplexed microsatellite analysis, whereas STS and SSR may be the choice for marker-assisted selection. These high-throughput marker systems, coupled with appropriate genetic stocks (e.g., near-isogenic lines) or strategy (e.g., bulked segregant analysis) (Michelmore et al., 1991), now allow for a specific region of the genome to be saturated with DNA markers with considerably less effort. AFLP, as a high-throughput marker, has proven its advantages in many aspects of genome analysis from gene mapping to positional cloning (Buschges et al., 1997).

2.1 Development of Novel DNA Markers

There has been a consistent effort to develop high-throughput molecular marker technologies. Toward this end, the most recent additions are single nucleotide polymorphism (SNP) (Brookes, 1999) and the markers based on miniature inverted repeat transposable element (MITE) (Casa et al., 2000).

Recent advances in sequencing technology have accelerated the discovery of SNP and insertions/deletions. SNP is projected as a powerful tool for genome investigations. They are simple, ubiquitous, and abundant, and have a great potential for automation (Roskey et al., 1996), making those excellent landmarks for navigating the genome. SNP information can be exploited in a similar manner as other molecular markers. They can be used for linkage analysis, fingerprinting, and marker-assisted breeding. Each of these applications involves the analysis of a large number of samples and will ultimately require rapid, inexpensive, and automated methods for analyzing DNA sequence variants. Available public SNP databases (http://www.ncbi.nlm. nih.gov/SNP/) provide easy access to the rapidly growing mass of SNP data for effective utilization. SNP discovery is becoming relatively easy because of the presence of abundant nucleotide diversity (Rafalski, 2002). Moreover, as the amount of linkage disequilibrium increases, whole-genome scanning for SNPs becomes a powerful tool in detecting association with useful traits. Bhattramakki et al. (2000a) discovered a high rate of polymorphism in SNP (1/80 bp) and indels (1/240 bp) in maize, and could map 164 out of 311 loci for which SNP information was obtained. In another study, they identified 655 indels by resequencing 502 maize loci across 8 maize inbreds, out of which single-nucleotide indels accounted for more than half (54.8%), followed by two- and three-nucleotide indels (Bhattramakki et al., 2002). In barley, Kanazin et al. (2002) scored 112 SNPs in 38 loci by using five barley genotypes. With the development of rapid, simple, low-cost, and high-throughput SNP genotyping methodologies, SNP will be an attractive marker system for genome mapping and marker-assisted selection (Ye et al., 2001).

A new type of molecular marker, based on the presence and the absence of the MITE family, Heartbreaker (*Hbr*), in the maize genome (Casa et al., 2000) has proven useful in a number of applications including genotyping and assessment of genetic relationships (Casa et al., 2002). Hbr markers are highly polymorphic, highly reproducible, and evenly distributed in the maize genome. Unlike the universality of AFLP primers, MITE-specific primers are to be designed based on the sequence information of the MITE family that can be obtained after database searches. This technology is highly cost-effective and time-efficient as it needs only a few sets of primers and a large number of markers can be simultaneously generated by multiplexing the products of several MITE families, amplified with distinct fluorescent-tagged primers. It is not only the high-throughput characteristics but also the preference for genic regions (Shirasu et al., 2000; Zhang et al., 2000) makes it an efficient marker system to perform chromosome walks and pursue map-based cloning in crop species with larger genomes.

The primary challenge faced by plant breeders today is how to develop superior cultivars by transferring a repertoire of useful genes in a cost-effective, timely, and precise manner. Although the greatest benefit of using molecular markers in breeding is the considerable savings in time required to attain a certain genetic gain for the agronomic traits that are difficult to evaluate (Burr et al., 1983; Tanksley et al., 1989), integration of molecular marker technology in plant-breeding programs has not been widespread because of the high cost of DNA-based assays (De Verna and Alpert, 1990). The DNA-marker technology developed during the last two decades undoubtedly revolutionalized the genetic analysis of both simple and complex traits, and opened up new avenues for crop improvement with added efficiency and precision. However, further technical progress to simplify and automate the DNA analysis with reduced cost and minimal infrastructural demands will be required to empower the plant-breeding programs.

3 COMPARATIVE MAPPING WIDENS THE EXPLOITABLE CROP GENE POOL

Extensive genome colinearity established at genome and genic levels allows the leveraging of information and resources from plants such as rice and *Arabidopsis* that are extensively studied to other species (Schmidt, 2000, 2002). Genome colinearity could be exploited to facilitate fine mapping and map-based cloning experiments in crop plants with larger genomes (Druka et al., 2000). The microsynteny studies provide invaluable information on genome organization. Resemblances of diverse species in both gene order and genic content benefits crop improvement programs by increasing the opportunity to widen the crop gene pool to unprecedented levels. Sequence information obtained from *Arabidopsis* (Arabidopsis Genome Initiative, 2000) and rice (Barry 2001; Goff et al., 2002; Yu et al., 2002) is an important resource for genetic investigations in important crop species. Use of a physical map of the model species can be a powerful tool to generate physical maps of selected genomic regions of other species (Draye et al., 2001; Chen et al., 2002). By using sequence information from species with small genomes, it should be possible to accelerate the positional cloning of orthologous genes (Grotewald and Peterson, 1994). Even in the case of hard-to-manipulate traits such as abiotic stress tolerance, progress is anticipated through the exploitation of comparative genomics information generated from model organisms using high throughput structural and functional genomics tools (Cushman and Bohnert, 2000).

Despite the large differences in DNA content among grass species, high levels of macrosynteny and moderately high level of microsynteny (Tarchini et al., 2000; Keller and Feuillet, 2000) have been maintained among grass genomes (Gale and Devos, 1998). This suggests that the rice genome sequence will be central for understanding the biology of crop species (McCouch, 1998;

Gale et al., 2001), because it can facilitate the analysis of structure and function of genomes in the other grasses (Dubcovsky et al., 2001; Freeling, 2001). To facilitate the exchange of information between rice and other grass relatives, a comparative grass genome mapping database, Gramene (http://www.gramene.org) has recently been created (Ware et al., 2002). This database stores information about maps, sequences, genes, genetic markers, mutants, QTL, controlled vocabularies, and relevant publications in grass species.

4 CROP GENOME MAPS AND THEIR INTEGRATION TO PHYSICAL MAPS

The benefits of both genetic and physical molecular maps have already been realized and well demonstrated. The immediate utilities accruing from genome maps include DNA fingerprinting, gene mapping, marker-assisted selection, and positional cloning of useful genes. Genome-wide physical mapping in model crop species with bacteria-based, large-insert clones [e.g., bacterial artificial chromosome (BAC), P1 artificial chromosomes (PACs)] also promises to revolutionize genomic investigations in species with large and complex genomes. Our discussion hereafter is confined to the current status of the genetic and physical maps in some important crop species.

4.1 Rice (*Oryza sativa* L.)

Rice genome research is in the forefront of crop genomics. Molecular resources developed for rice genomics exert a tremendous impact to the genetic improvement of other field crops as they provide an important tool for comparative grass genomics research. These resources include dense genetic maps, integrated physical–genetic map, comprehensive transcript map, yeast artificial chromosome maps, bacterial artificial chromosome (BAC) libraries, and databases for ESTs and BAC end sequences. The most saturated molecular genetic map of rice carrying 2275 DNA markers was constructed in an F2 population of "Nipponbare," a japonica rice, crossed with an indica rice (Kasalath) covering 1521cM genetic distance (Harushima et al., 1998) and provides the basic framework for the physical mapping and sequencing project. Currently, this map has a total of 3267 Markers: (http://rgp.dna.affrc.go.jp/publicdata/geneticmap2000/index.html). Moreover, another dense rice molecular map comprising 726 RFLP markers, developed in a backcross mapping population derived from a cross between *O. sativa* and *Oryza longistaminata* is also available (Causse et al., 1994). Other molecular resources include EST development (Kurata et al., 1994; Yamamoto and Sasaki, 1997), a yeast artificial chromosome (YAC) library (Umehara et al., 1995), a PAC library (http://rgp.dna.affrc.go.jp/), and several BAC libraries (Wang et al., 1995; Zhang et al., 1996; http://www.genome.clemson.edu/where/nippon_ bac/index.html). YAC-based physical maps of rice were reported earlier (Kurata et al. 1997; Saji et al., 2001; http://rgp.dna.affrc.go.jp/publicdata/ physicalmap99/yacall.html). The most recent comprehensive YAC-based transcript map consists of 6591 ESTs with 80.8% coverage of the rice genome (Wu et al., 2002).

In the United States, a physical map of the rice genome has been developed by fingerprinting and aligning of overlapping contigs of BACs of whole rice genome and extensively integrated with the genetic map (Chen et al., 2002). Also, Tao et al. (2001) developed a genome-wide BAC-based map of the rice genome consisting of 298 BAC contigs with a coverage of 419 Mbp of rice genome using a high-throughput DNA sequence electrophoresis-based restriction fingerprinting method in conjunction with several large-insert BAC libraries. These resources should be valuable for the ongoing rice genome-sequencing project and map-based cloning and functional analysis of agriculturally important genes in rice and its relatives.

Nearly 500 microsatellite primers were earlier placed on the rice molecular map (Temnykh et al., 2001) (://www.gramene.org/microsat/ RM_primers.html). To increase the density and utility of the rice SSR map, McCouch et al. (2002) recently developed an additional 2240 SSRs, using the sequence information from the rice genome program of Monsanto Corp. (http://www.rice-research.org/rice_ssr.html).

4.2 Maize (*Zea mays*)

The most comprehensive maize genome map developed using an intermated B73×Mo17 (IBM) population carries 1736-loci (1156 cDNAs, 545 random genomic clones, 16 simple sequence repeats, 14 isozymes, and 5 unknown clones) (Davis et al., 1999). In addition, large numbers of SSR loci have been positioned on this map (1855 primer pairs, 1797 distinct loci) (http://www. agron.missouri.edu/ssr.html). The IBM genetic map is accessible at the website: http://cafnr.missouri.edu/mmp/ibmmaps.htm. With National Science Foundation (NSF) support, the construction of an integrated genetic and physical map is progressing (Cone et al., 2002). Three quality BAC libraries have been constructed and characterized (Yim et al., 2002). Using high-density filter hybridization in conjunction with RFLP landmarks, BAC fragments have been identified to demarcate bin positions on the physical map for which BAC contigs are assembled through BAC fingerprinting (Cone et al., 2002). Sequencing gene-rich regions of the maize genome has been started (http://maize.danforthcenter.org/).

4.3 Wheat (*Triticum aestivum* L. em Thell)

Bread wheat is an ideal species for cytogenetic analysis because the polyploid nature of its genome enables it to tolerate structurally and numerically manipulated chromosomes. This feature has been instrumental in developing a high-resolution, physical map of the wheat genome and its integration into molecular linkage maps (Werner et al., 1992). Using deletion lines involving all 21 wheat chromosomes (Endo and Gill, 1996), physical maps have been constructed for each of the seven homoeologous chromosome groups of wheat (Werner et al., 1992; Kota et al.1993; Gill et al., 1993, 1996a,b; Hohmann et al., 1994; Delaney et al., 1995a,b; Mickelson-Young et al., 1995; Weng et al., 2000). Ma et al. (2001) physically located 18 RFLP markers on homoeologous group 1 and 3 chromosomes of wheat by using wheat ditelosomic lines and in situ hybridization. Most of these markers were physically placed onto chromosome arms in the same order as in the genetic maps.

Chromosome walking and positional cloning have been a formidable task to carry out in wheat because of its overwhelming genome size and complexity. Stein et al. (2000) demonstrated that subgenome chromosome walking in wheat could result in saturation of the genomic regions to support positional cloning. A physical contig constructed by them in the diploid wheat (*Triticum monococcum*) genome containing the *Lr10* leaf rust disease resistance locus can be used for isolation of orthologous regions in bread wheat. Earlier attempts (Röder et al., 1995; Bryan et al., 1997; Stephenson et al., 1998) resulted in the development of only 100 SSRs. Subsequently, another 230 microsatellites have been added by Röder et al. (1998).

4.4 Sorghum (*Sorghum bicolor* L. Moench)

Abundant DNA polymorphism between cultivated *S. bicolor* and wild *S. propinquum* expedited the placement of more than 2500 RFLP loci on a high-denity linkage map (Bowers et al., 2000), which was later used as a framework by Draye et al. (2001) to assemble a sorghum physical map by employing both hybridization and BAC fingerprinting. Using heterologous probes, many orthologous loci from maize, rice, wheat, sugarcane, and *Arabidopsis* were mapped on the physical map facilitating the alignment of the genetic maps of those species to some degree. These aligned maps would make a detailed comparison of genome structure, function, and evolution possible between sorghum and its relatives (Draye et al., 2001).

Menz et al. (2002) developed a high-density sorghum genetic map with 2929 loci (2454 AFLP, 136 SSR, and 203 cDNA and genomic clones from different grass species) using a recombinant inbred line (RIL) population derived from the BTx623×IS3620C cross. This map has been integrated into a sorghum physical map (Klein et al., 2000), by using a high-throughput PCR-based method that involves fingerprinting and AFLP-based contig assembly and mapping. Development of SSRs in sorghum was very slow (Brown et al., 1996; Taramino et al., 1997). Bhattramakki et al. (2000b) reported development of 313 SSR primers with success of amplification and mapping in 266 and 165 cases, respectively.

4.5 Soybean (*Glycine max* L. Merr.)

Significant progress has been made in soybean genomics to target important genes in this model legume species. Specifically, the development of the integrated genetic linkage map (Cregan et al., 1999) and physical framework of the soybean genome (Marek and Shoemaker, 1997; Marek et al., 2001) provides a deeper insight into its genome structure and organization. The integrated genetic linkage map of soybean with more than 1400 markers, including 606 SSRs, is an important tool to accelerate soybean breeding. Currently, the soybean genetic map contains a total of 1845 markers, including 1010 SSRs, 718 RFLPs, 73 RAPDs, 23 classical traits, and 10 others (Cregan, personal communication, 2003). Rapidly growing soybean EST database (Shoemaker et al., 2002) and mapping of these ESTs onto the soybean genetic map (Matthews et al., 2001) promise to be a valuable public resource for efficient gene discovery, study of evolution, and comparative analysis between genera to identify candidate genes for important biological functions. Construction of a high-resolution physical map of soybean is in progress, and because of its economic importance, sequencing of the soybean genome is likely to be initiated in the near future. Meanwhile, generation of localized physical maps should facilitate map-based cloning of economically important genes in soybean (Lewers et al., 2002).

5 STRICTURAL GENOMICS LAYS THE FOUNDATION FOR FUTURE PLANT GENOMIC RESEARCH

Genome mapping and sequencing are two important tools of structural genomics and have been the major focus of plant genomics for the past two decades. The most important achievement of the past years is that the plant genome has been fully decoded in both rice and *Arabidopsis* and partially in many crop species through several EST sequencing projects. Besides enhancing our understanding of the genome organization, progress in structural genomics is also laying the foundation for comparing allelic diversity beyond the crossability barrier. Genomics is now moving with focus on gene function. Given the progress in structural genomics tools and resources, functional genomics is now well positioned to add meaning to the sequence data by understanding the function and interaction of each and every gene and protein (Heiter and Boguski, 1997). To realize the huge and complex magnitude of this ambitious goal, a global approach is being explored for the simultaneous examination of all components rather than traditional gene-by-gene approach.

A useful introduction to goals, new methods, and vocabulary in genomics research is provided in an update in *Plant Physiology* (Bouchez and Höfte, 1998). Well-documented reviews in *Science* (vol. 285, July 1999 issue) highlight the potential of genomics tools to improve grain quality traits (Mazur et al., 1999), nutritional content of major crops to improve human health (DellaPenna, 1999), and to understand the genetic mechanisms that control growth and development and responses to environment (Summerville and Summerville, 1999). An excellent summary of the promise of genomics for plant biology and agriculture can be found in the *Proceedings of the National Academy of Sciences USA* (vol. 95, 1998) as the report of a colloquium entitled "Protecting our food supply: The value of plant genome initiatives." The first ever large-scale plant genome initiative of NSF started in 1998 with an award of $85 million for 23 projects (http://www.nsf.gov/bio/pubs/awards/genome99.htm). The scope of these projects was updated by Walbot (1999). In subsequent years, several projects have been funded with various objectives in diverse crop species:

http://www.nsf.gov/bio/dbi/dbi_pgr.htm
http://www.nsf.gov/bio/pubs/awards/genome99.htm
http://www.nsf.gov/bio/pubs/awards/genome00.htm
http://www.nsf.gov/bio/pubs/awards/genome01.htm
http://www.nsf.gov/bio/pubs/awards/genome02.htm

These projects, along with many public domain genomics efforts (http://ars-genome.cornell.edu), will provide a huge reservoir of new genomics information and new tools for rapid gene tagging, map-based cloning and gene expression analysis projects.

5.1 Genome Sequencing Reveals the Blueprint of Plant Life

The complete sequence of a plant genome is considered as the "biologist's periodic table" (Lander, 1996), because it encompasses all genes that are required for existence and functioning of a living organism. Large-scale sequencing projects, numerous genome initiatives in several plant species, have dramatically changed experimental plant biology by providing additional avenues for gene discovery. Whole-genome sequencing offers insights into gene structure, genome organization, and evolution (Rounsley et al.,

1998). As complete genome sequencing is unlikely to be realized in the near future for many important crop species, sequencing of ESTs remains an important goal for genome exploration and genome function studies (Rudd, 2003). Gene-rich chromosomal regions in species with large genomes, such as wheat, are ideal targets for partial genome sequencing (Keller and Feuillet, 2000).

An EST is an unedited single-pass DNA sequence of random cDNA clones. Large-scale EST projects have been initiated in several crop species and EST sequences for a large number of genes have been generated in a very cost-efficient manner. EST information is available in the dbEST section of Gen Bank (Baguski et al., 1993) (http://www.ncbi.nlm.nih.gov/dbEST/index. html). Plant molecular informatics center at the University of Minnesota maintains a database of plant ESTs from several species based on similarity search operations (http://www.cbc.med.umn.edu/ResearchProjects/seq. proc.html). The Institute of Genomic Resources (TIGR) also maintains a web database with all known transcripts and ESTs of both rice and Arabidopsis (Rice Gene Index: http://www.tigr.org/tdb/ogi/; Arabidopsis EST assemblies: http://www.tigr.org/tdb/at/at.html). An EST database represents a useful resource to identify genes involved in plant metabolism (Ohlrogge and Benning, 2000) and to pursue a candidate gene approach to map resistance genes or QTL in an efficient manner (Wang et al., 2001b).

The significant achievements of *Arbidopsis* genomics led to the initiation of genome research in other plant species. The *Arabidopsis* EST project, initiated in the early 1990s (Hofte et al., 1993; Newman et al., 1994), proceeded to complete sequencing of its whole genome in 2000, through an international collaboration collectively termed as Arabidopsis Genome Initiative (AGI) (Arabidopsis Genome Initiative, 2000; http://www.arabidopsis.org/). Public sequence databases, GenBank, European Molecular Biology Laboratory (EMBL), and DNA database of Japan (DDBJ), which can be conveniently accessed via the services available at the National Center of Biotechnology Information (NCBI) (http://www.ncbi/nlm.nih.gov/), contain all annotated genes.

Besides *Arabidopsis*, whole genome of rice (*O. sativa*) has been sequenced (http://www.ncbi.nlm.nih.gov/cgi-bin/Entrez/map_search?chr = rice.inf). Both public and commercially funded efforts are directed at the rice genome sequencing. There are two publicly funded rice genome-sequencing projects. The first one, the International Rice Genome Sequencing Project (IRSGP), is a collaboration of publicly funded investigators from 10 countries whose goal is to sequence the *O. sativa* ssp. *japonica* genome completely and accurately (http://rgp.dna.affrc.go.jp/cgi-bin/statusdb/seqcollab.pl). The fully annotated sequence of japonica rice is expected to be completed by

2004. The research groups for sequencing share BAC and PAC libraries, constructed from a single plant of *O. sativa* ssp. *japonica* variety "Nippon-bare," by adopting a clone-by-clone strategy. Sequence-ready contigs are generated by end-sequencing; fingerprinting and marker-aided PCR screening and all annotated sequences are immediately released for public use. To date, sequencing of three rice chromosomes 1, 4, and 10 has been completed (Feng et al., 2002; Sasaki et al., 2002; the rice chromosome 10 sequencing consortium, 2003). Analysis of sequence of chromosomes 1 and 4 revealed 6756 and 4658 protein coding genes, respectively, whereas 3471 genes were identified from the sequence information of chromosome 10. The second public sector effort is the whole-genome shotgun sequencing project funded by the Chinese Academy of Sciences, Beijing, China. This project has reported a draft sequence assembly of *O. sativa* L. ssp. *indica* variety 93–11 using whole-genome shotgun sequencing (Yu et al., 2002). The number of estimated genes is within the range of 46,022 and 55,615 in the genome size of 466 Mbp with a functional coverage of 92%.

The first commercial effort by Monsanto Corp. (Barry, 2001) revealed 399 Mbp of rice genomic sequence and 7000 SSRs that are generated using japonica cultivar "Nipponbare." The second commercial effort, a collaboration between Myriad Genetics and Syngenta, recently completed the sequencing of the rice genome of the same rice cultivar "Nipponbare" (http://www.syngenta.com/en/media/article.asp?article_id = 126) (Goff et al., 2002). Unlike the Beijing Genome Initiative, most of their sequence contigs were anchored onto the genetic and physical maps. Both have publicly stated their willingness to share the rice genomics information with the academic scientific community through collaboration agreements. Because synteny and gene homology between rice and other cereal genomes are so extensive (Bevan, 2003), the information derived from rice will have great potential to provide clues for genetic improvement of important cereal crops.

Medicago truncatula, an annual relative of alfalfa with a diploid genome with $2n = 16$, is also targeted for whole-genome sequencing (http://www.ncbi.nlm.nih.gov/cgi-bin/Entrez/map00?taxid = 3880, http://www.genome.ou.edu/medicago.html). Its genome size is ~4 times larger than that of *Arabidopsis thaliana*. Approximately 500 Mbp of its genome is targeted as a whole-genome shotgun project at the University of Oklahoma. The initial goal of the *M. truncatula* genome project is to generate approximately 1-fold whole-genome shotgun sequence data from a double-stranded, pUC-based genomic library. To date, approximately 150,000 ESTs have been generated in this project. An integrated physical and genetic map of *M. truncatula*, constructed by Kulikova et al. (2001), is targeted for saturation with more than 3000 nonredundant EST sequences.

6 FUNCTIONAL GENOMICS BRIDGES THE GAP BETWEEN GENOME AND PHENOME

In the postgenomic era, major emphasis has been on the utilization of the genome sequence data to elucidate the function of each and every annotated gene by quantitative determination of the spatial and temporal expression patterns of specific mRNAs, proteins, and important metabolites. Functional genomics emphasizes on analysis of function using molecular tools such as microarrays, insertional mutagenesis, proteomics, and metabolomics. Expression analyses are carried out at three levels, such as mRNA, proteins, and metabolites, which ultimately lead to the expression of a phenotype. Transcriptomics deals with the measurement of mRNA levels, while proteomics and metabolomics aim at determining the protein profile and important metabolites, respectively. Oliver et al. (2002) discussed the applications and limitations of different high-throughput techniques for mRNA, protein, and metabolite analyses that can monitor changes to discover new metabolic pathways in plants.

6.1   Variation in Transcript Profiles Provides Initial Evidence for Phenotypic Differences

Transcript profiling plays an important role in associating rapidly accumulating DNA sequences to phenotypes. Classical tools, such as Northern blotting, RNA dot blotting, and reverse transcription-polymerase chain reaction (RT-PCR), allow the monitoring of single or few genes at one time. Several transcription profiling techniques are currently available for monitoring the gene expression on a global scale (Kuhn, 2001; Donson et al., 2002). Some of these techniques are differential display (Liang and Pardee, 1992), differential screening (Kiyosue et al., 1994), cDNA-AFLP (Bachem et al., 1996), representational difference analysis (RDA) (Hubank and Schatz, 1994), serial analysis of gene expression (SAGE) (Velculescu et al., 1995), and DNA microarray (Schena et al., 1995). Our discussion is confined to both commonly used and emerging technologies, such as differential display, cDNA-AFLP, SAGE, and microarray.

Differential display, although simple and less expensive, requires a limited quantity of RNA and suffers from several drawbacks. Poor reproducibility, difficulty in cloning of a desired fragment, underrepresentation of rare transcripts, and high rate of false positives are the major concerns. Several variations of differential display addressing these concerns have been reported (Matz and Lukyanov, 1998). Both reproducibility and sensitivity could be improved with the application of AFLP to cDNA fragments (Bachem et al., 1996, 1998). This technique, known as cDNA-AFLP, involves systematic scanning of the whole transcriptome by selective amplification under stringent PCR conditions, because of the ligation of anchors to restricted cDNA fragments. This is now widely used to discover differentially expressed genes associated with various biological responses (Durrant et al., 2000; Jones et al., 2000; van der Biezen et al., 2000; Ditt et al., 2001; Dubos and Plomion, 2003; Trindade et al., 2003). By systematically sequencing the cDNA-AFLP fragments from diverse tissues and treatments, Donson et al. (2002) generated a reference database of the *Arabidopsis* transcriptome for discovery of the differentially expressed genes. This database can be interrogated for matching with a specific cDNA-AFLP pattern without sequencing any fragments. cDNA-AFLP holds a great promise for development of transcript-derived markers for genetic analyses, map-based cloning, and MAS (Brugmans et al., 2002). Further improvement of the cDNA-AFLP method now allows the discovery of rare transcripts and relative quantitation of the expression level of all the genes in a transcription profiles using gel imaging softwares (Breyne and Zabeau, 2001). The fact that this fragment-based technology can be adapted to investigate genome-wide expression in any organism without any sequence resources and to analyze both discovered and undiscovered genes, makes it a suitable alternative for the systematic analyses of transcriptional programs. With this technique, however, it is extremely time-consuming to generate a global profile of a particular biological process. Moreover, it does not allow for comparing or pooling the datasets from separate experiments such as microarray or SAGE technology. A number of improved and automated methods sharing similar principles as cDNA-AFLP for transcript analysis are available (Prashar and Weisman, 1996; Shimkets et al., 1999; Sutcliffe et al., 2000). Bruce et al. (2000) used Gene Calling, an automated cDNA-AFLP method to study the transcription factors involved in the activation of flavanoid biosynthesis. These methods are both time-consuming and resource-intensive, especially when used to confirm the sequence identity of the target genes. Therefore high-throughput methods need to be developed to accelerate the functional genomics research.

SAGE offers the capability to accurately quantify the expression levels of thousands of genes at a time. It is a powerful technology, but its use is limited to those organisms for which a large sequence database is available and is not very well suited for comparing a large number of samples (Bouchez and Höfte, 1998). The major advantages of SAGE are analysis of unknown gene expression and the comparison and compilation of separately obtained data sets. Another unique feature of SAGE is that transcript profiles can be given a digital format for easy data processing and construction of gene expression databases. SAGE has been used less frequently for plant functional genomics studies (Matsumura et al., 1999, 2003; Thomas et al., 2002). SAGE, in combination with microarray analysis, may be a powerful tool for the discovery and isolation of differentially expressed genes.

The most powerful tool currently available for functional genomics studies is DNA microarray technology, which allows genome-wide monitoring of transcriptional nature (Deyholos and Galbraith, 2001; Panda et al., 2003). It is a technology based on miniaturization and hybridization using fluorescent-labeled probes. Using two different fluorescent labels, hybridization signals of probes prepared from both test and reference materials can be detected and compared simultaneously in a single hybridization experiment. The potential of this technology is now increasingly felt because of exponential growth of DNA sequence data and availability of cDNA clones from many diverse organisms. Assigning of function to each and every gene annotated from the genome sequence and the study of interaction among them now can be tremendously facilitated using this technology. Microarray was first used by Schena et al. (1995), who compared the expression pattern of 48 EST clones between roots and leaves in *Arabidopsis*. Two DNA chip formats currently used are the cDNA array (Schena et al., 1995) and the in situ synthesized oligonucleotide array (Lockhart et al., 1996). Detailed technical aspects of generating and using DNA microarrays have been extensively reviewed (Lemieux et al., 1998; Duggan et al., 1999; Eisen and Brown, 1999; Lipshutz et al., 1999). We will focus primarily on the applications and limitations of this technology.

The oligonucleotide arrays, popularly known as gene chips, are made by printing of oligonucleotides directly onto microscope slides using photolithography technique. Gene chips are particularly suitable for genotyping, detection of gene mutations (SNPs), mapping and discovery of novel polymorphisms, and positional cloning of useful genes (Wang et al., 1998). The major disadvantage in constructing oligo-arrays is that the accurate sequence information of many plant species is not available. Thus the use of oligonucleotide arrays in plant species has been limited. In the case of rice, whole-genome oligoarrays may be expected in the near future. On the other hand, cDNA microarrays are widely preferred by plant scientists for transcription profiling because cDNA and EST clones are readily available or can be easily generated.

Both microarrays have multifarious applications, but the major application in plant research has been in gene expression analysis. Any kind of alteration in gene expression pattern that is naturally occurring or induced between samples can be analyzed. DNA microarrays, coupled with unique genetic materials, such as mutants, NILs, deletion lines, substitution lines, and transgenic lines, can identify gene(s) responsible for a specific plant phenotype (Kehoe et al., 1999). It provides a tool to better understand the expression of candidate genes that function in metabolic pathways (Devaux et al., 2001). Aharoni and Vorst (2002) provided an up-to-date account of DNA microarray studies in plants since the first report of Schena et al. (1995). While most of these studies used cDNA microarrays, few reports are available using *Arabidopsis* oligoarrays (Cho et al., 1999; Harmer et al., 2000; Spigelman et al., 2000; Zhu and Wang, 2000). Some of cDNA microarray studies include characterization of changes in mRNA accumulation during seed development (Girke et al., 2000), and changes in gene expression in a mutant (Helliwell et al., 2001). It also provided deeper insights into plant response to pathogens (Dong, 2001), oxidative stress (Desikan et al., 2001), and drought and salinity (Kawasaki et al., 2001; Ozturk et al., 2002).

Microarray technology has witnessed fast progress in recent years and will be continuously employed in expression analysis of an increasing number of plant species as diverse as *Arabidopsis*, rice, maize, potato, soybean, tomato, ice plant, and loblolly pine, as evidenced from the interest in funding by the NSF (Walbot, 1999). Many genomics programs have been initiated to carry out genome sequencing, development of EST databases, and microarray technology. This genome-scale hybridization technology has improved throughput, sensitivity, and versatility for identifying differentially expressed genes, but the high cost of equipment, software, and reagents limited its application by plant biologists. The large-scale adoption of this technology, however, will depend on commercial availability of chips and future technological innovations to make it affordable. Besides aiming at high-throughput analysis by improved miniaturization, progress in automation, development of an efficient detection system and software will be required to reduce the cost. Another challenge faced by DNA microarray researchers is the difficulty in performing a meaningful comparison and exchange of data generated in various laboratories following different experimental conditions (Brazma et al., 2000). Some progress has been made in this direction by the formation of a collaborative working group called the Microarray Gene Expression Database collaborative group (http://www.mged.org) to periodically discuss the issues relating to efficient usage and exchange of microarray data. In view of the magnitude of data generated by microarray studies, renewed focus on development of technology for meaningful data analysis and integration with other related functional genomics tools, such as proteomics and metabolomics, will be critical to understand the complex plant biological machinery (Aharoni and Vorst, 2002). The value of this technology will be recognized when the transcription profiles generated by individual researchers are deposited in a public database, so that the unknown gene can be associated with function by matching it with an expression profile in the database (Summerville and Summerville, 1999; Brazma et al., 2000; Yazaki et al., 2002).

Plant breeding will be immensely benefited by obtaining a global perspective on plant growth and developmental responses through gene expression studies. Identification of genes involved in response to environmental changes and pathogen attack will provide important clues to expedite the development of plants for stress environments (Schaffer et al., 2000). A possible application for crop improvement is the selection of desirable genotypes, based on a unique expression pattern associated to complex agronomic traits, such as yield and stress tolerance. In addition to studying the effect of transgenes or specific mutation on global gene expression, comparison of expression profiles in closely related species will help to elucidate the genetic basis of many agriculturally important traits.

6.2   Proteins are the Closest Link to Trait Expression

Understanding the phenotypic variability in living organisms is a real challenge to all biologists today. Modern genomics procedures are gradually comprehending the diversity at DNA and RNA levels, but the physiological or biochemical consequences of these differences and the coordination between the genes still remain a mystery. Proteomics refers to large-scale analyses of the protein component of an organism (Pandey and Mann, 2000), and it represents an essential complement to bridge the gap between the genotype and phenotype in post-genomic era. Analysis of proteins was possible with the availability of two-dimensional (2-D) gel electrophoresis in the 1970s. Recent advances in mass spectrometry (MS) and commercialization of MS equipment, together with the increasing amount of EST and genomic sequence data, have been the driving force for the development of this new field.

Plant proteomics is still in its infancy and with the recent release of the whole-genome sequence of the model crops, *Arabidopsis* and rice, tremendous opportunities are anticipated in this field. Thiellement et al. (1999) published the first review in plant proteomics. Many of these studies involved the use of 2-D gel patterns to identify possible markers for different genotypes and phenotypes and phylogenetic relationships. An attempt was made to map the proteome of different tissues from rice and *Arabidopsis* (Tsugita et al., 1996). Several recent studies focused on specific subcellular proteomes, such as the plasma membrane (Rouquie et al., 1997; Santoni et al., 1999), roots (Chang et al., 2000), mitochondria (Vener et al., 2001; Heazlewood et al., 2003), chloroplasts (Ferro et al., 2000; Yamaguchi and Subramanian, 2000; Yamaguchi et al., 2000; Peltier et al., 2000, 2001; van Wijk, 2000), maize leaves (Porubleva et al., 2001), and on symbiosis between legume roots and nitrogen-fixing bacteria (Panter et al., 2000; Natera et al., 2000). Proteome analysis in *Arabiopsis* was performed by Gallardo et al. (2001) to understand the complex developmental process of germination and revealed some new proteins involved in this process. De Vienne et al. (1999) presented protein quantity loci (PQL) methodology, citing an example of the phosphoglycerate mutase variation in maize, and illustrated the candidate gene/protein approach for traits responsive to drought stress. Comparative proteomics was applied to study the phylogenetic relationship in different species and genera of Brassicaceae (Marqués et al., 2001). Few reports discussed changes in protein expression in relation to environmental stresses such as anoxia and drought (Chang et al., 2000; Salekdeh et al., 2002).

The functions of more than half of the plant's proteins are still unknown. The proteomics approach can be employed in numerous ways from straightforward identification of proteins to characterization of post-translational modifications and protein–protein interactions (van Wijk, 2001). Proteomics can help map translated genes and loci controlling their expression, and identify proteins accounting for the variation of complex phenotypic traits (Zivy and deVienne, 2000). Proteomics, in combination with microarray technology, can assess whether the mechanisms of gene regulation are at the level of transcription, or translation and protein accumulation.

Proteomics is now moving toward a high-throughput platform to investigate protein expression patterns on a global scale (MacBeath and Schreiber, 2000; Figeys and Pinto, 2001). This is further supported by the development of high-resolution, two-dimensional (2-D) polyacrylamide gel electrophoresis, highly sensitive biological mass spectrometry, and the rapidly growing protein and DNA databases (Gevaert and Vandekerckhove, 2000). Emerging technologies for plant proteomics have been reviewed (Rossignol, 2001; Kersten et al., 2002). Despite poor understanding of the relationship between mRNA, protein, and biological functions, development of high-performance, reliable, and efficient systems for protein purification, protein digestion, and other related activities should accelerate the proteomics research in the future (Figeys and Pinto, 2001; Roberts, 2002).

6.3 Metabolites are the Ultimate Level of Genome Expression

The term "metabolome" refers to the entire complement of all small-molecular-weight metabolites inside a cell or tissue of interest. Metabolites represent the ultimate expression of genomic information, which has remained unexplored to date. The combination of biochemical profiling with genetic analysis created the new field of metabolomics, or metabolic genomics. One major component of metabolomics is to obtain a snapshot of metabolite composition and their concentration in a given tissue at a given time. It is based on the premise that differences in gene function can be assessed by comparing the metabolic profile of samples that are temporally, spatially, or genetically separated. Recent developments in functional genomics have compelled the need for global profiling of gene expression at the level of metabolites (Stitt and Fernie, 2003). Enhanced sensitivity of analytical capabilities due to technological advances in chromatographic procedures and mass spectral-based analytical procedures makes unbiased and accurate determination of metabolites possible. Fiehn (2002) discussed the current approaches of metabolic profiling and the potential applications with special emphasis on data mining and mathematic modeling of metabolism. Glassbrook and Ryals (2001) also reviewed some of the most widely used analytical chemistry technique, and recent advances promise to bring the biochemical profiling of small molecules into the main stream as a tool for functional genomics studies. Major limitations, however, include the lack of high-throughput procedures for handling of large numbers of samples and the absence of a single analytical technique capable of profiling all low molecular weight metabolites.

Metabolic profiling is primarily used in the biomedical field and is gradually evolving as a new functional genomics tool. A few studies in plant systems are available (Schnable et al., 1994; Fiehn et al., 2000; Roessner et al., 2001). Roessner et al. (2001) compared the metabolic profiles of wild-type potato tubers to those of transgenic tubers that were altered in the expression of a single sugar-metabolizing gene that led to understanding the regulation of plant metabolism. Similarly, comparing the composition of the cuticular waxes associated with each mutant plant and wild type, Schnable et al. (1994) could assess the metabolic function associated with each of the mutant genes. Traditional methods, such as gas chromatography (GC)/mass spectrometry (MS), are being used to profile many metabolites including lignins and simple phenylpropanoids. More sophisticated techniques have been developed involving HPLC/MS to profile many of the more difficult classes of metabolites including isoflavonoids, related phenolic conjugates (Sumner et al., 1996), and saponins (Huhman et al., 2002).

Although the progress in metabolomics is slow, its potential as a new functional genomics tool is recognized. The use of metabolomic data for crop improvement is being explored. Metabolite profiling can provide a thorough insight into the biochemical functions of plant genes and their regulatory networks involved in plant metabolism (Fiehn et al., 2000; Trethewey, 2001; Fiehn, 2002). Manipulation of biochemical pathways using metabolic engineering will result in plant cultivars improved for yield, stress tolerance, or useful pharmaceutical production (Giddings et al., 2000).

6.4 Insertional Mutagenesis Unambiguously Proves the Functionality of DNA Sequences

A large-scale accumulation of sequence data from EST and genome-sequencing projects in several important species warrants the establishment of the functionality of those discovered gene sequences. Reverse genetics studies, in which the function of a gene is determined by mutating it by transposons or

T-DNA of *Agrobacterium tumefaciens* and studying the resulting phenotype, play a major role in functional genomics. Unfortunately, no effective tool to replace a targeted gene in flowering plants is currently available (Britt and May, 2003). Recent developments in large-scale insertional mutagenesis open up new possibilities for functional genomics investigations in *Arabidopsis*, rice, and maize. Martienssen (1998) outlined two approaches to assign gene function using transposons. The first approach uses high copy transposons to generate a pool of heavily mutagenized individuals followed by PCR identification of plants carrying the insertion in specific genes by PCR. The second approach involves the generation of a large library of plants, each carrying a unique insertion followed by amplification and sequencing of each insertion site. These methods have been demonstrated in *Arabidopsis* and maize. Ramachandran and Sundaresan (2001) recently reviewed the use of most widely used transposons and the successful exploitation of these elements in heterologous plant species as insertional mutagens and as a reverse genetics tool for functional genomics.

Vast resources of gene knockouts are currently available that can be subjected to different types of reverse genetic screens to ascertain the functions of the sequenced genes. Most *Arabidopsis* genes were already be represented in T-DNA and transposon insertion lines generated in different laboratories (Krisan et al., 1999, Speulman et al., 1999; Parinov and Sundaresan, 2000; Tissier et al., 1999). The seeds of these transgenic lines are available at the *Arabidopsis* Biological Resource Center (ABRC) (http://aims.cps.msu.edu/aims), which is organizing their distribution to interested researchers in cooperation with Nottingham *Arabidopsis* Stock Center (NASC; http:/nasc.nott.ac.uk). Despite large collections of insertion lines, targeting each and every gene in the genome through insertional mutagenesis may remain an elusive goal. Krisan et al. (1999) estimated that 600,000 insertion lines would need to be generated and screened to identify an insertion in a 1-kbp gene with 99% probability. In rice, a large population of T-DNA tagged lines generated by Jeon et al. (2000) will be useful for discovering new genes. Use of maize Ac transposable element (Enoki et al., 1999) and a rice retrotransposon Tos17 (Yamazaki et al., 2001) have been proposed for functional analysis of rice genome. In maize, Cowperthwaite et al. (2002) identified insertions in a number of putative maize genes by sequencing transposed *Ac* elements (*tac* sites) from two insertion libraries and comparing the sequences to existing databases.

The use of genome-wide mutant libraries for testing the functionality of a gene is a welcome development. There have been attempts to develop such mutant stocks in maize and alfalfa and other important crop plants to be used as functional genomic tools. The Ac/Ds and MuDR/Mu maize transposons have been used recently to generate mutant stocks (Walbot 2000; http://

www.zmdb.iastate.edu/zmdb/RescueMuPopulations.html). The maize targeted mutagenesis project funded by the NSF plant genome program
generated ~ 50,000 plants and is providing seeds for the phenotypic evaluation of "hits" (maize-targeted mutagenesis database URL, http://
mtm.cshl.org). Although a powerful tool, insertional mutagenesis, by itself,
is not sufficient to assign function to a gene sequence and must, therefore, be
integrated into more genome-scale approaches to realize its full potential
(Bouche and Bouchez, 2001).

7 BIOINFORMATICS ACCELERATES THE UTILIZATION
AND INTEGRATION OF GENOME INFORMATION

The major goal of bioinformatics is acquisition, organization, display, and
retrieval of a large assembly of complex biological data. The activities related
to bioinformatics are typically genomics-based. Tinker (2002) reviewed these
classical roles of bioinformatics in furthering genomics science, but emphasized the role of bioinformaticists in converting the genomic data into usable
format for crop improvement. During the past two decades, there has been an
exponential increase in molecular data, such as sequences, maps, markers,
ESTs, gene expression profiles, and proteomics data. Tools have been developed and continuously refined to efficiently access and analyze the data.
Because proper management and exploitation of data is fundamental to both
generators and users of genomic data, the role of a bioinformaticist in developing tools and resources is critical for future advances in plant genomics.
Different types of data management systems and relevant issues for their
development have been recently reviewed (Reiser et al., 2002). A review on
computational methods and softwares for sequence analysis is also available
(Rhee, 2000). Extensive listings of these resources are available elsewhere
(Cartinhour, 1997; Baxevanis, 2001), and descriptions of these resources are
presented in the January 2001 annual database issue of *Nucleic Acids Research* (http://nar.oupjournals.org/content/vol29/issue1/). The development
of new bioinformatic tools to facilitate the exchange of genomic data and
ideas among genomics researchers will be essential to sustain the growth of
this exciting field of plant genomics in the future.

8 MARKER-ASSISTED SELECTION IMPROVES PRECISION
AND EFFICIENCY IN CROP BREEDING

Technological progress in plant breeding has remained relatively stagnant,
whereas the DNA technology evolved rapidly during the last two decades.
Molecular marker technology-empowered genome analysis offers a new level

of resolution to crop improvement efforts. It has been well accepted as a valuable adjunct to classical breeding tools. The potential of this new technology has been reviewed widely (Burr et al., 1983; Beckmann and Soller, 1986; Tanksley et al., 1989; Paterson et al., 1991; Young 1992; Paterson 1996, 1998). DNA marker technology is impacting plant breeding significantly in: (1) variety identification for intellectual property rights, (2) assessment of genetic diversity, (3) accelerated back-crossing, and (4) mapping of and marker-assisted selection for agronomically important traits.

Molecular marker-based tracking of useful genes in segregating population improved the efficiency of selection with considerable saving in time and resources, and sometimes eliminated the cumbersome selection processes. To exploit the entire diversity in DNA sequence existing in a crop gene pool, high-resolution genetic maps are being constructed for nearly every important crop species at an unprecedented speed. Molecular mapping and QTL analysis have been performed for many simple and complex traits associated with many aspects of crop productivity. Some of these traits include plant height, maturity (Beavis et al., 1991; Lin et al., 1995), oil, protein and starch content (Goldman et al., 1993; Diers et al., 1992), aroma, kernel elongation (Lorieux et al., 1996; Ahn et al., 1993), seed hardness (Keim et al., 1990), seed size and seed weight (Fatokun et al., 1992; Doganlar et al., 2000), soluble solids (Osborn et al., 1987; Paterson et al., 1988), insect resistance (Bonierbale et al., 1994; Byrne et al., 1998), tuber shape (Van Eck et al., 1994), tolerance to low phosphorous stress (Reiter et al., 1991), water use efficiency (Martin et al., 1989), aluminum tolerance (Miftahudin et al., 2002; Nguyen et al., 2001), and yield and yield components (Xiao et al., 1998; Austin and Lee, 1998). Evidence from literature indicates DNA mapping has been successful in disease tolerance studies where, in many cases, reliable phenotyping procedures exist. The quantitative nature of resistance was observed in some cases (Wang et al., 1994), which, however, poses a real challenge to plant breeders. Drought tolerance, although complex and difficult to evaluate in natural environments, has been subjected to QTL analysis in rice, sorghum, maize, and pearl millet (Ribaut et al., 1996; Subudhi et al., 2000; Zhang et al., 2001; Yadav et al. 2002). Several component traits of drought tolerance, viz. osmotic adjustment (Lilley et al., 1996; Morgan and Tan 1996; Teulat et al., 1998), plant water status, water-soluble carbohydrate (Teulat et al., 2001), abscisic acid accumulation (Lebreton et al., 1995; Quarrie et al., 1997; Tuberosa et al., 1998), stomatal behavior (Lebreton et al., 1995; Price et al., 1997), root and root morphology (Champoux et al., 1995; Yadav et al., 1997; Zheng et al., 2000), and anthesis silking interval (Ribaut et al., 1996) have been mapped.

The application of the marker-assisted selection in crop improvement has not been striking for complex quantitative traits. It has been largely limited to simply inherited traits (Hittalmani et al. 1995; Huang et al., 1997;

Mudge, 1999; Mudge et al., 1997; Davierwala et al., 2001; Wang et al., 2001a). Marker-assisted selection procedure that could be applied for several useful traits has been outlined in important field crops (Zheng et al., 1995; http://maswheat.ucdavis.edu/). Current progress in mapping consistent QTL for yield and stress tolerance characteristics can potentially lead to successful MAS for complex traits in the future.

8.1   Strategies for Improving Complex Traits

Quantitative traits are often the most valuable traits for crop improvement but have, so far, proven refractory to contemporary methodologies. However, the fact that quantitative traits can be genetically dissected into Mendelian units with greater resolution (Alpert and Tanksley, 1996), making them amenable for map-based cloning (Frary et al., 2000), represents a significant development. Genomics is providing tools and techniques for investigating and making genetic gains in complex quantitative traits. In addition to MAS, molecular markers can be used to obtain information about: (1) the number, effect, and chromosomal location of genes affecting traits; (2) effect of multiple copies of individual genes (gene dosage); (3) interaction between /among genes controlling a trait (epistasis); (4) whether individual genes affect more than one trait (pleiotropy); and (5) stability of gene function in different environments ($G \times E$ interaction) (Paterson et al., 1991; Tanksley, 1993; Kearsey, 2002). While success has been achieved in some areas, progress in others will require the development of appropriate genetic materials coupled with suitable field evaluation strategies.

Several strategies can be employed for improved understanding and manipulation of complex traits. The genes involved in quantitative trait expression may be structural genes or regulatory genes in particular biochemical pathways (Byrne and McMullen, 1996). The major role of regulatory loci in the metabolic pathway for quantitative trait expression has been demonstrated in maize by using a QTL mapping technique (Byrne et al., 1996). Availability of high-density EST maps may accelerate the identification of these candidate genes for successful MAS. Use of exotic libraries consisting of marker-defined genomic regions from wild species and introgressed into the background of an elite line, proposed by Zamir (2001), can help plant breeders improve the agricultural performance of modern crop varieties. These libraries can also act as tools for the discovery and characterization of genes that underlie traits of agricultural value.

Plant breeders frequently ignore the germplasm collections of wild species and primitive landraces because they do not show obviously valuable phenotypes; they are difficult to screen, or are often not adapted to a specific environment. The domestication and selection pressure by early humans and modern plant breeders narrowed the genetic variation restricting the future genetic gain in every crop species. The genomic tools can now be employed to reveal the genetic potential of wild and cultivated germplasm resources for exploitation with higher efficiency (Tanksley and McCouch, 1997). While high-yielding lines contain a large number of positive alleles for high yield, inferior parents also often have some superior alleles to improve yield performance (deVincente and Tanksley 1993; Xiao et al., 1996, 1998). Screening of exotic germplasm can now be performed using the advanced backcross QTL method that allows a subset of alleles from the wild or exotic plant to be examined in the genetic background of an elite cultivar (Bernacchi et al., 1998a, 1998b). Ribaut and Bertan (1999) proposed a modified MAS procedure called single large-scale marker-assisted selection (SLS-MAS) to improve polygenic traits in maize, where selection pressure for the target gene was combined with maintenance of genetic diversity in the rest of the genome. Zhu et al. (1999) focused on determining optimum combinations of QTL alleles and accumulation of favorable alleles, rather than pyramiding alleles detected in a reference mapping population.

8.2 High-Throughput Technologies for Marker-Assisted Selection

Molecular marker technology needs another revolution for its successful and routine implementation in plant-breeding programs. Success of such endeavor will depend on the development of high-throughput, rapid, and cost-effective assays capable of processing thousands of segregating individuals with a minimal tissue requirement. Automated high-throughput plant DNA purification has been developed to cater to the needs of plant breeders for marker-assisted breeding, seed quality testing, SNP discovery and scoring, and analysis of transgenic plants (Shenoi et al., 2002; Gauch, 2002). The usefulness of AFLP for crop improvement can now be well appreciated because of the development of a highly automated AFLP data production system (Van Eijk, 2002) and a specialized software, AFLP-Quanter (Pro) (Pot and Pouwels, 2002) that is capable of sorting out the homo- and heterozygotes in AFLP fingerprints.

Abundance and adaptability to large-scale screening make SNPs the most favored marker system for crop breeders. One key bottleneck in high-throughput genotyping of SNPs is DNA amplification. Multiplex-PCR with amplification of several target loci and several genotypes in parallel is being explored (Fan et al., 2000, Hirschhorn et al., 2000; Lindblad-Toh et al. 2000; Cyranoski, 2001; Buetow et al., 2001). New technologies that eliminate the need for electrophoresis are also under development. One promising technique, known as "molecular beacons" (Tyagi and Kramer, 1996), discriminates SNPs that fluoresce only when there is a match with the exact DNA sequence. SNP genotyping technology is moving rapidly and, hopefully, this high-throughput marker will be routinely used in the future for many crop species (Rafalski, 2002).

8.3 Technological Advances in Genomics Should Expedite Cloning of Useful Genes

Application of genetic engineering technology to rational plant improvement is currently limited by a shortage of cloned genes for important traits. It is the resistance genes for which cloning efforts have been more successful. More than 20 resistance genes have been cloned by either map-based cloning or transposon tagging (Michelmore, 2000). Cloning of genes that can make noticeable improvement in crop yield and stress adaptation has, so far, remained elusive. However, in the postgenome era, a better understanding of quantitative trait variation is foreseen because QTL cloning can now be expedited as a result of the growing interest in insertional mutagenesis, development of novel DNA polymorphisms, complete decoding of rice and *Arabidopsis* genomes along with sustained public sector funding for genome research in many important plant species.

Currently, map-based cloning is the most preferred strategy for isolation of agriculturally important genes. Despite several successes in plants (Martin et al., 1993; Song et al. 1995; Buschges et al., 1997; Cai et al., 1997; Cao et al., 1997; Ori et al., 1997), it is far from a routine procedure in plant biology. A few groups have succeeded in cloning quantitative trait loci. Frary et al. (2000) cloned *fw2.2*, a fruit weight QTL in tomato. Fridman et al. (2000) delimited a QTL for tomato sugar content (Brix9-2-5) to a 484-bp region within an invertase gene. Two QTL (*Hd1*, *Hd6*) with effect on photoperiod sensitivity were cloned in rice using advanced backcross progeny (Yano et al., 2000; Takahashi et al., 2001). El-Din El-Assal et al. (2001) cloned a QTL that partly accounts for the flowering response to the photoperiod between two *Arabidopsis* as a result of a single amino-acid substitution. In rice, there is also progress toward cloning of *pms1* locus for photosensitive genetic male sterility (Liu et al., 2001).

New techniques and new strategies are continuously developed and tested for gene discovery. A modified cDNA selection procedure (Childs et al., 2001) should be useful for genome-wide gene discovery. Because of the availability of extensive literature documenting similarity of genome information among various crop species and their relatives, the usefulness of comparative genomics approach is increasingly appreciated. Genome analysis of organisms with a complex and large genome is becoming more efficient with the use of model systems where genomic information is more advanced.

Sequence data obtained from sequencing projects can reveal a large number of putative genes for validation of their functional identity and discovery of orthologous genes in related species. Sequence data coupled with fine-mapping strategy can help associate QTL regions to variation at single nucleotide level (Lyman et al., 1999). Cloning of useful genes can be further accelerated as cross-species linked maps and integrated high-density genetic and physical contig maps become available. On the other hand, the candidate gene approach using the information from biochemical pathways and linkage disequilibrium strategy may be ideal and cost effective to elucidate the molecular basis of quantitative trait variation in many agronomic traits (Morgante and Salamini, 2003). Therefore skillful integration of both functional and structural genomic information relating to allelic diversity, genome organization, and evolution, by innovative bioinformatic tools will be essential (Young, 2000).

9 CONCLUSIONS AND FUTURE PROSPECTS

The need for future gains in agricultural productivity is crucial to sustain the burgeoning world population, which is projected to reach 10 billion by the middle of the current century. It seems appropriate to take stock of the recent developments in the field of genomics and to assess their potential for crop improvement. Progress in genomics and its related disciplines benefits both basic and applied plant biology. There is slow but gradual assimilation of genomics data into crop breeding efforts. A fuller appreciation of the mutual advantage from collaboration between genomics researchers and plant breeders can play a vital role in transforming the methods of crop development (Stuber et al., 1999; Koncz, 2003). New genomics tools and techniques should help design experiments on a genomic scale, providing massive amount of data for enhanced understanding of plant growth and adaptation in both favorable and adverse environments.

Knowledge gained from plant genome analysis has enormous potential to revolutionize breeding programs in a number of ways. The most immediate application of this genomics technology is molecular breeding of agronomic traits (Thomas, 2003). In addition, positional cloning of useful genes facilitated by abundance of genomic information, such as sequences, ESTs, maps and markers, and expression data, can be rewarding for both development of transgenic crops and developing an understanding of gene function and gene interaction. Genetic enhancement is theoretically possible in every imaginable trait. While marker-assisted selection has been successfully applied to many simply inherited traits, progress in improving complex traits through MAS has been marginal. Most agriculturally important traits are inherently complex (Tanksley, 1993), with epistasis, pleiotropy, and genotype×environment interaction confounding the effect of the target genes. Instead of using MAS indiscriminately for every available trait, a closer scrutiny of several factors—such as project objectives, time period, inheritance pattern, speed, and amount of genetic gain and, most importantly, the cost involved—should be carried out to identify possible phenotypes for MAS application. Reliable identification of QTL for a complex trait in a reasonably large mapping population using a well-distributed linkage map should be given top priority in the initial phases. Further validation of QTL expression and accurate estimation of QTL effect in different genetic backgrounds and different environments using specialized genetic stocks such as NILs may later be necessary to realize a noticeable genetic gain (Young, 1999). Although time-consuming, integration of these procedures with germplasm improvement schemes should be advantageous. Finally, with the implementation of cost-effective simple molecular assays capable of large-scale genotyping in a reasonable amount of time, the impact of genomics revolution will be visible in most plant-breeding programs (Koebner and Summers, 2003). As the technological progress continues, it is reasonable to expect that diagnostic expression profiles generated through the use of DNA chips would find its place in marker-assisted breeding in the near future. Continuing leverage of knowledge from genomics and bioinformatics will be crucial for the development of useful selection tools for germplasm development. The rapidly growing databases of gene sequences will streamline the discovery of new genes and their allelic diversity in germplasm stored in gene banks for successful utilization in crop improvement programs.

Discovery of all of the genes, their function, and their regulation is the ultimate goal of genome biologists. Systematic genome analyses in many important food crops (e.g., wheat, barley, rye, oat, and maize, to name a few) have been challenging because of their large and complex genomes. However, a major step toward this goal, the availability of complete genome sequence in two model plant species *Arabidopsis* and rice, should facilitate the transfer of genomic information to a majority of crop plants. The role of comparative genomics in widening the exploitable crop gene pool is growing and offers unique opportunities to design new and improved crops. There have been significant advances in many areas of genomics and related sciences, which promise to unravel the products and pathways for a phenotype. In particular, the developments in structural genomics have been instrumental in driving other branches of genomics science. While we continue to be successful in acquiring the genome sequence data in various organisms, better means to utilize and translate these data for crop improvement are being explored. It is now opening up new experimental opportunities to close the gap between sequence information and phenotype through functional genomics analyses (Wilson et al., 2003).

New powerful functional genomics tools are gradually replacing the traditional, time-consuming gene-by-gene analysis, allowing researchers to rapidly learn the role of numerous genes and their interactions simultaneously. Correlation of specific pattern of transcript profile with a phenotype is being explored to provide further insights into quantitative trait variation (Yazaki et al., 2002; Schadt et al., 2003). However, the full potential of the global gene expression technology will be realized as the technology advances to efficiently collect, display, and analyze massive data obtained from quantitative gene expression studies (Eisen and Brown, 1999). Concurrently, the proteomics approach is increasingly adapted by genomics researchers as a result of increasing power and sensitivity with a series of innovations in two-dimensional electrophoresis and mass spectrometry. Well-annotated proteomics databases are now emerging in a number of areas to provide a platform for systematic research. Thus construction of integrated databases with information on such things as transcription profiles, protein profiles, metabolic profiles, gene sequence, map location, integrated linkage and physical maps, cross-species comparative maps, mutant phenotypes, and allelic variation should be anticipated in the immediate future (Bouchez and Höfte, 1998; Stein, 2003).

Given the complexity associated with plant growth and developmental responses, it is unlikely that we will be able to obtain answers in the near future to all intriguing questions that are central to crop improvement. While it may be possible to ultimately determine the function of every single gene from the sequence repository of model organisms, quantification of the component of the quantitative trait variation resulting from the complex interaction of genes and their downstream products will remain a demanding task for future genomics researchers. Despite these overwhelming challenges, the current pace of scientific progress and technological developments leading to comprehension of the intricacy of plant biology is reassuring.

ACKNOWLEDGMENTS

The authors wish to thank Professor Manjit S. Kang of Louisiana State University Agricultural Center and Dr. Babu Valliyodan of the University of Missouri for reviewing this manuscript and making useful comments.

REFERENCES

Aharoni A, Vorst O. DNA microarrays for functional plant genomics. Plant Mol Biol 2002; 48:99–118.

Ahn SN, Anderson JA, Sorrels ME, Tanksley SD. Homoeologous relationships of rice, wheat, and maize chromosomes. Mol Gen Genet 1993; 241:483–490.

Alpert KB, Tanksley SD. High-resolution mapping and isolation of a yeast artificial chromosome contig containing fw2.2: a major fruit weight quantitative trait locus in tomato. Proc Natl Acad Sci USA 1996; 93:15503–15507.

Arabidopsis Genome Initiative. Analysis of the genome sequence of the flowering plant *Arabidopsis thaliana*. Nature 2000; 408:796–815.

Austin DF, Lee M. Detection of quantitative trait loci for grain yield and yield components in maize across generations in stress and nonstress environments. Crop Sci 1998; 38:1296–1308.

Bachem CW, van der Hoeven RS, de Bruijn SM, Vreugdenhil D, Zabeau M, Visser RG. Visualization of differential gene expression using a novel method of RNA fingerprinting based on AFLP: analysis of gene expression during potato tuber development. Plant J 1996; 9:745–753.

Bachem CW, Oomen RJFJ, Visser RG. Transcript imaging with cDNA-AFLP: a step-by-step protocol. Plant Mol Biol Rep 1998; 16:157–173.

Baguski MS, Lowe TMJ, Tolstoshev CM. dbEST—database for "Expresses sequence tags". Nat Genet 1993; 4:332–333.

Barry GF. The use of the Monsanto draft rice genome sequence in research. Plant Physiol 2001; 125:1164–1165.

Baxevanis AD. The molecular biology database collection: an updated compilation of biological database resources. Nucleic Acid Res 2001; 29:1–10.

Beavis WD, Grant D, Albertson M, Fincher R. Quantitative trait loci for plant height in four maize populations and their associations with qualitative genetic loci. Theor Appl Genet 1991; 83:141–145.

Beckmann JS, Soller M. Restriction fragment length polymorphisms and genetic improvement of agricultural species. Euphytica 1986; 35:111–124.

Bernacchi D, Beck-Bunn T, Emmatty D, Eshed Y, Inai S, Lopez J, Petiard V, Sayama H, Uhlig J, Zamir D, Tanksley SD. Advanced backcross QTL analysis of tomato: II. Evaluation of near-isogenic lines carrying single-donor introgressions for desirable wild QTL-alleles derived from *Lycopersicon hirsutum* and *L. pimpinellifolium*. Theor Appl Genet 1998a; 97:170–180.

Bernacchi D, Beck-Bunn T, Eshed Y, Lopez J, Petiard V, Uhlig J, Zamir D, Tanksley SD. Advanced backcross QTL analysis in tomato: I. Identification of QTLs for traits of agronomic importance from *Lycopersicon hirsutum*. Theor Appl Genet 1998b; 97:381–397.

Bevan M. Surprises inside a green grass genome. Science 2003; 300:1514–1515.

Bhattramakki D, Ching A, Dolan M, Register J, Tingey S, Rafalski A. Single nucleotide polymorphisms (SNPs) in corn: early lessons. Maize Genet Coop Newslett No. 74, 2000, 54.

Bhattramakki D, Dong J, Chhabra AK, Hart GE. An integrated SSR and RFLP linkage map of *Sorghum bicolor* (L.) Moench. Genome 2000; 43:988–1002.

Bhattramakki D, Dolan M, Hanafey M, Wineland R, Vaske D, Register JC III, Tingey S, Rafalski A. Insertion–deletion polymorphisms in 3′ regions of maize genes occur frequently and can be used as highly informative genetic markers. Plant Mol Biol 2002; 48:539–547.

Bonierbale MW, Plaisted RL, Pineda O, Tanksley SD. QTL analysis of trichome-mediated insect resistance in potato. Theor Appl Genet 1994; 87: 973–987.

Botstein D, White RL, Skolnick M, Davis RW. Construction of a genetic linkage map in man using restriction fragment length polymorphisms. Am J Hum Genet 1980; 32:314–331.

Bouche N, Bouchez D. Arabidopsis gene knockout: phenotypes wanted. Curr Opin Plant Biol 2001; 4:111–117.

Bouchez D, Höfte H. Functional genomics in plants. Plant Physiol 1998; 118:725–732.

Bowers JE, Schertz KF, Abbey C, Anderson S, Chang C, Chittenden LM, Draye X, Hoppe AH, Jessup R, Lennington J. A high-density 2399-locus genetic map of sorghum. Plant and Animal Genome VIII Conference, San Diego, http://www.intl-pag.org/pag/8/abstracts/pag8712.html.

Brazma A, Robinson A, Cameron G, Ashburner M. One-stop shop for microarray data. Nature 2000; 403:699–700.

Breyne P, Zabeau M. Genome-wide expression analysis of plant cell cycle modulated genes. Curr Opin Plant Biol 2001; 4:136–142.

Britt AB, May GD. Re-engineering plant gene targeting. Trends Plant Sci 2003; 8(2):90–95.

Brookes AJ. The essence of SNPs. Gene 1999; 234:177–186.

Brown SM, Hopkins MS, Mitchell SE, Senior ML, Wang TY, Duncan RR, Gonjalez-Candelas F, Kresovich S. Multiple methods for the identification of polymorphic simple sequence repeats (SSRs) in sorghum (*Sorghum bicolor* L. Moench). Theor Appl Genet 1996; 93:190–198.

Bruce W, Folkerts O, Garnaat C, Crasta O, Roth B, Bowen B. Expression profiling of the maize flavonoid pathway genes controlled by estradiol-inducible transcription factors CRC and P. Plant Cell 2000; 12:65–79.

Brugmans B, de Carmen AF, Bachem CWB, van Os H, van Eck HJ, Visser RGF. A novel method for the construction of genome wide transcription maps. Plant J 2002; 31:211–222.

Bryan GJ, Collins AJ, Stephenson P, Orry A, Smith JB, Gale MD. Isolation and characterization of microsatellites from hexaploid bread wheat. Theor Appl Genet 1997; 94:557–563.

Buetow KH, Edmonson M, MacDonald R, Clifford R, Yip P, Kelley J, Little DP, Strausberg R, Koester H, Cantor CR, Braun A. High throughput development and characterization of a genome wide collection of gene-based single nucleotide polymorphism markers by chip-based matrix-assisted laser desorption/ionization time-of-flight mass spectrometry. Proc Natl Acad Sci USA 2001; 98:581–584.

Burr B, Evola SV, Burr FA. The application of restriction fragment length polymorphism to plant breeding. In: Setlow JK, Hollaender A, eds. Genetic Engineering, Principles and Methods. Vol. 5. New York: Plenum Press, 1983:45–59.

Burrow MD, Blake TK. Molecular tools for the study of complex traits. In: Paterson AH, ed. Molecular Dissection of Complex Traits. Boca Raton: CRC Press, 1998:13–29.

Buschges R, Hollricher K, Panstruga R, Simons G, Wolter M, Frijters A, van Daelen R, van der Lee T, Diergaarde P, Groenendijk J, Topsch S, Vos P, Salamini F, Schulze-Lefert P. The barley *Mlo* gene: a novel control element of plant pathogen resistance. Cell 1997; 88:695–705.

Byrne PF, McMullen M. Defining genes for agricultural traits: QTL analysis and the candidate gene approach. Probe 1996; 7:24–27.

Byrne PF, McMullen MD, Snook ME, Musket TA, Theuri JM, Widstrom NW, Wiseman BR, Coe EH. Quantitative trait loci and metabolic pathways: genetic control of the concentration of maysin, a corn earworm resistance factor, in maize silks. Proc Natl Acad Sci USA 1996; 93:8820–8825.

Byrne PF, McMullen M, Wiseman BR, Snook ME, Musket TA, Theuri JM, Widstrom NW, Coe EH. Maize silk maysin concentration and corn earworm antibiosis: QTLs and genetic mechanisms. Crop Sci 1998; 38:461–471.

Caetano-Anolles G, Bassam BJ, Gresshoff PM. DNA amplification fingerprinting using very short arbitrary oligonucleotide primers. Biotechnology (NY) 1991; 9: 553–557.

Cai D, Kleine M, Kifle S, Harloff HJ, Sandal NN, Marcker KA, Klein-Lankhorst RM, Salentijn EM, Lange W, Stiekema WJ, Wyss U, Grundler FM, Jung C. Positional cloning of a gene for nematode resistance in sugar beet. Science 1997; 275: 832–834.

Cao H, Glazebrook J, Clarke JD, Volko S, Dong X. The Arabidopsis NPR1 gene that controls systemic acquired resistance encodes a novel protein containing ankyrin repeats. Cell 1997; 88:57–63.

Cartinhour SW. Public informatics resources for rice and other grasses. Plant Mol Biol 1997; 35:241–251.

Casa AM, Brouwer C, Nagel A, Wang L, Zhang Q, Kresovich S, Wessler S. The MITE family Heartbreaker (*Hbr*): molecular markers in maize. Proc Natl Acad Sci USA 2000; 97:10083–10089.

Casa AM, Mitchell SE, Smith OS, Register JC, Wessler SR, Kresovich S. Evaluation of *Hbr* (MITE) markers for assessment of genetic relationships among maize (*Zea mays* L.) inbred lines. Theor Appl Genet 2002; 104:104–110.

Causse MA, Fulton TM, Cho YG, Ahn SN, Chunwongse J, Wu K, Xiao J, Yu Z, Ronald PC, Harrington SE, et al. Saturated molecular map of the rice genome based on an interspecific backcross population. Genetics 1994; 138:1251–1274.

Champoux MC, Wang G, Sarkarung S, Mackill DJ, O'Toole JC, Huang N, McCouch SR. Locating genes associated with root morphology and drought avoidance in rice via linkage to molecular markers. Theor Appl Genet 1995; 90:969–981.

Chang WW, Huang L, Shen M, Webster C, Burlingame AL, Roberts JK. Patterns of protein synthesis and tolerance of anoxia in root tips of maize seedlings acclimated to a low-oxygen environments and identification of proteins by mass spectrometry. Plant Physiol 2000; 122:295–318.

Chen M, Presting G, Barbazuk WB, Goicoechea JL, Blackmon B, Fang G, Kim H, Frisch D, Yu Y, Sun S, Higingbottom S, Phimphilai J, Phimphilai D, Thurmond S, Gaudette B, Li P, Liu J, Hatfield J, Main D, Farrar K, Henderson C, Barnett L, Costa R, Williams B, Walser S, Atkins M, Hall C, Budiman MA, Tomkins JP, Luo M, Bancroft I, Salse J, Regad F, Mohapatra T, Singh NK, Tyagi AK, Soderlund C, Dean RA, Wing RA. An integrated physical and genetic map of the rice genome. Plant Cell 2002; 14:537–545.

Childs KL, Klein RR, Klein PE, Morishige DT, Mullet JE. Mapping genes on an integrated sorghum genetic and physical map using cDNA selection technology. Plant J 2001; 27:243–255.

Cho RJ, Mindrinos M, Richards DR, Sapolsky RJ, Anderson M, Drenkard E, Dewdney J, Reuber TL, Stammers M, Federspiel N, Theologis A, Yang W-H, Hubbel E, Au M, Chung EY, Lashkari D, Lemieux B, Dean C, Lipshutz RJ, Ausubel FM, Davis RW, Oefner PJ. Genome wide mapping with biallelic markers in *Arabidopsis thaliana*. Nat Genet 1999; 23:203–207.

Cho YG, Blair MW, Panaud O, McCouch SR. Cloning and mapping of variety-specific rice genomic NA sequences: amplified fragment length polymorphisms (AFLP) from silver stained polyacrylamide gels. Genome 1996; 39:373–378.

Cone KC, McMullen MD, Bi IV, Davis GL, Yim Y-S, Gardiner JM, Polacco ML, Sanchez-Villeda H, Fang Z, Schroeder SG, Havermann SA, Bowers JE, Paterson AH, Soderlund CA, Engler FW, Wing RA, Coe EH Jr. Genetic, physical, and informatics resources for maize. On the road to an integrated map. Plant Physiol 2002; 130:1598–1605.

Cowperthwaite M, Park W, Xu Z, Yan X, Maurais SC, Dooner HK. Use of the transposon Ac as a gene-searching engine in the maize genome. Plant Cell 2002; 14:713–726.

Cregan PB, Jarvik T, Bush AL, Shoemaker RC, Lark KG, Kahler AL, VanToai TT, Lohnes DG, Chung J, Specht JE. An integrated genetic linkage map of the soybean. Crop Sci 1999; 39:1464–1490.

Cushman JC, Bohnert HJ. Genomic approaches to plant stress tolerance. Curr Opin Plant Biol 2000; 3:117–124.

Cyranoski D. Japan speeds up mission to unravel genetic diseases. Nature 2001; 410(6832):1013.

Davierwala AP, Reddy APK, Lagu MD, Ranjekar PK, Gupta VS. Marker assisted selection of bacterial blight resistance genes in rice. Biochem Genet 2001; 39:261–278.

Davis GL, McMullen MD, Baysdorfer C, Musket T, Grant D, Staebell M, Xu G, Polacco M, Koster L, Melia-Hancock S, Houchins K, Chao S, Coe EH Jr. A maize map standard with sequenced core markers, grass genome reference points and 932 expressed sequence tagged sites (ESTs) in a 1736-locus map. Genetics 1999; 152:1137–1172.

DeVerna JW, Alpert KB. RFLP technology Proc Hort Biotechnol Symp Univ of California, Davis. New York: Wiley-Liss, 1990:247–259.

de Vienne D, Leonardi A, Damerval C, Zivy M. Genetics of proteome variation for QTL characterization: application to drought stress responses in maize. J Exp Bot 1999; 50:303–309.

Delaney D, Nasuda S, Endo TR, Gill BS, Hulbert SH. Cytologically based physical maps of group-3 chromosomes of wheat. Theor Appl Genet 1995b; 91:780–782.

Delaney D, Nasuda S, Endo TR, Gill BS, Hulbert SH. Cytologically based physical maps of group-2 chromosomes of wheat. Theor Appl Genet 1995a; 91:568–573.

DellaPenna D. Nutritional genomics: manipulating plant micronutrients to improve human health. Science 1999; 285:375–379.

Desikan R, Mackerness SA, Hancock JT, Neil SJ. Regulation of the *Arabidopsis* transcriptome by oxidative stress. Plant Physiol 2001; 127:159–172.

Devaux F, Marc P, Jacq C. Transcriptomes, transcription activators and microarrays. FEBS Lett 2001; 498:140–144.

deVicente MC, Tanksley SD. QTL analysis of transgressive segregation in an interspecific tomato cross. Genetics 1993; 134:585–596.

Deyholos MK, Galbraith DW. High density microarrays for gene expression analysis. Cytometry 2001; 43:229–238.

Diers BW, Keim P, Fehr WR, Shoemaker RC. RFLP analysis of soybean seed protein and oil content. Theor Appl Genet 1992; 83:608–612.

Ditt RF, Nester EW, Comai L. Plant gene expression response to *Agrobacterium tumefaciens*. Proc Natl Acad Sci USA 2001; 98:10954–10959.

Doganlar S, Frary A, Tanksley SD. The genetic basis of seed-weight variation: tomato as a model system. Theor Appl Genet 2000; 100:1267–1273.

Dong X. Genetic dissection of systematic acquired resistance. Curr Opin Plant Biol 2001; 4:309–314.

Donson J, Fang Y, Espiritu-Santo G, Xing W, Salazar A, Miyamoto S, Armendarez V, Volkmuth W. Comprehensive gene expression analysis by transcript profiling. Plant Mol Biol 2002; 48:75–97.

Draye X, Lin YR, Qian X, Bowers JE, Burow GB, Morrell PL, Peterson DG, Presting GG, Ren S, Wing RA, Paterson AH. Toward integration of comparative genetic, physical, diversity, and cytomolecular maps for grasses and grains, using the sorghum genome as a foundation. Plant Physiol 2001; 125:1325–1341.

Druka A, Kudrna D, Han F, Kilian A, Stefenson B, Frisch D, Tomkins J, Wing R, Kleinhofs A. Physical mapping of the barley stem rust resistance gene *rpg4*. Mol Gen Genet 2000; 264:283–290.

Dubcovsky J, Ramakrishna W, San Miguel PJ, Busso CS, Yan L, Shiloff BA, Bennetzen JL. Comparative sequence analysis of collinear barley and rice bacterial artificial chromosomes. Plant Physiol 2001; 125:1342–1353.

Duggan DJ, Bittner M, Chen Y, Meltzer P, Trent JM. Expression profiling using cDNA microarrays. Nat Genet 1999; 21(suppl):10–14.

Dubos C, Plomion C. Identification of water-deficit responsive genes in maritime pine (*Pinus pinaster* Ait.) roots. Plant Mol Biol 2003; 51:249–262.

Durrant W, Rowland O, Piedras P, Hammond-Kosack K, Jones J. cDNA-AFLP reveals a striking overlap in race specific resistance and wound response gene expression profiles. Plant Cell 2000; 12:963–977.

Eisen MB, Brown PO. DNA arrays for analysis of gene expression. Methods Enzymol 1999; 303:179–205.

El-Din El-Assal S, Alonso-Blanco C, Peeters AJ, Raz V, Koornneef M. A QTL for flowering time in Arabidopsis reveals a novel allele of CRY2. Nat Genet 2001; 29:435–440.

Endo TR, Gill BS. The deletion stocks of common wheat. J Hered 1996; 87:295–307.

Enoki H, Izawa T, Kawahara M, Komatsu M, Koh S, Kyozuka J, Shimamoto K. Ac as a tool for the functional genomics of rice. Plant J 1999; 19:605–613.

Fan JB, Chen X, Halushka MK, Berno A, Huang X, Ryder T, Lipshutz RJ, Lockhart DJ, Chakravarti A. Parallel genotyping of human SNPs using generic high-density oligonucleotide tag arrays. Genome Res 2000; 10:853–860.

Fatokun CA, Menancio-Hautea DI, Danesh D, Young ND. Evidence for orthologous seed weight genes in cowpea and mung bean based on RFLP mapping. Genetics 1992; 132:841–846.

Feng M, et al. Sequence and analysis of rice chromosome 4. Nature 2002; 420:316–320.

Ferro M, Seigneurin-Berny B, Rolland N, Chapel A, Salvi D, Garin J, Joyard J. Organic solvent extraction as a versatile procedure to identify hydrophobic chloroplast membrane proteins. Electrophoresis 2000; 21:3517–3526.

Fiehn O, Kopka J, Dormann P, Altmann T, Trethewey RN, Willmitzer L. Metabolic profiling for plant functional genomics. Nat Biotechnol 2000; 18:1157–1161.

Fiehn O. Metabolomics—the link between genotypes and phenotypes. Plant Mol Biol 2002; 48:155–171.

Figeys D, Pinto D. Proteomics on a chip; promising developments. Electrophoresis 2001; 22:208–216.

Frary A, Nesbitt TC, Grandillo S, Knaap E, Cong B, Liu J, Meller J, Elber R, Alpert KB, Tanksley SD. fw2.2: a quantitative trait locus key to the evolution of tomato fruit size. Science 2000; 289(5476):85–88.

Freeling M. Grasses as a single genetic system: reassessment 2001. Plant Physiol 2001; 125:1191–1197.

Fridman E, Pleban T, Zamir D. A recombination hotspot delimits a wild-species quantitative trait locus for tomato sugar content to 484 bp within an invertase gene. Proc Natl Acad Sci USA 2000; 97:4718–4723.

Gale M, Moore G, Devos K. Rice-the pivotal genome in cereal comparative genetics. Novartis Found Symp 2001; 236:46–53. Discussion 53-8.

Gale MD, Devos KM. Comparative genetics in the grasses. Proc Natl Acad Sci USA 1998; 95:1971–1974.

Gallardo K, Job C, Groot SPC, Puype M, Demol H, Vandekerckhove J, Job D. Proteomic analysis of *Arabidopsis* seed germination and priming. Plant Physiol 2001; 126:835–848.

Gauch S. Rapid purification of nucleic acids from plant and animal tissues streamlines molecular analysis. Plant, Animal and Microbe Genomes X. Conference, January 12–16, 2002, Town & Country Convention Center, San Diego, CA, http://www.intl-pag.org/pag/10/abstracts/PAGX_W324.html.

Gevaert K, Vandekerckhove J. Protein identification methods in proteomics. Electrophoresis 2000; 21:1145–1154.

Giddings G, Allison G, Brooks D, Carter A. Transgenic plants as factories for biopharmaceuticals. Nat Biotechnol 2000; 18:1151–1155.

Gill KS, Gill BS, Endo TR. A chromosome region specific mapping strategy reveals gene-rich telomeric ends in wheat. Chromosoma 1993; 102:374–381.

Gill KS, Gill BS, Endo TR, Boyko EV. Identification and high-density mapping of gene-rich regions in chromosome group 5 of wheat. Genetics 1996a; 143:1001–1012.

Gill KS, Gill BS, Endo TR, Taylor T. Identification and high-density mapping of gene-rich regions in chromosome group 1 of wheat. Genetics 1996b; 144:1883–1891.

Girke T, Todd J, Ruuska S, White J, Benning C, Ohlrogge J. Microarray analysis of developing *Arabidopsis* seeds. Plant Physiol 2000; 124:1570–1581.

Glassbrook N, Ryals J. A systematic approach to biochemical profiling. Curr Opin Plant Biol 2001; 4:186–190.

Goff SA. A draft sequence of the genome (*Oryza sativa* L. ssp. *japonica*). Science 2002; 296:92–100.

Goldman IL, Rocheford TR, Dudley JW. Quantitative trait loci influencing protein and starch concentration in the Illinois long term selection maize strains. Theor Appl Genet 1993; 87:217–224.

Grotewald E, Peterson T. Isolation and characterization of a maize gene encoding chalcone flavonone isomerase. Mol Gen Genet 1994; 242:1–8.

Gupta PK, Balyan HS, Sharma PC, Ramesh B. Microsatellites in plants: a new class of molecular markers. Curr Sci 1996; 70:45–54.

Harmer SL, Hogenesch JB, Straume M, Chang H-S, Han B, Zhu T, Wang X, Kreps JA, Kay SA. Orchestrated transcription of key pathways in *Arabidopsis* by the circadian clock. Science 2000; 290:2110–2113.

Harushima Y, Yano M, Shomura A, Sato M, Shimano T, Kuboki Y, Yamamoto T, Lin SY, Antonio BA, Parco A, Kajiya H, Huang N, Yamamoto K, Nagamura Y, Kurata N, Khush GS, Sasaki T. A high-density rice genetic linkage map with 2275 markers using a single F2 population. Genetics 1998; 148:479–494.

Heazlewood JL, Howell KA, Whelan J, Miller AH. Toward an analysis of the rice mitochondrial proteome. Plant Physiol 2003; 132:230–242.

Heiter P, Boguski M. Functional genomics: it's all you read it. Science 1997; 278:601–602.

Helliwell CA, Chin-Atkins AN, Wilson IW, Chapple R, Dennis ES, Choudhury A. The Arabidopsis *amp 1* gene encodes a putative glutamate carboxypeptidase. Plant Cell 2001; 13:2115–2125.

Hirschhorn JN, Sklar P, Lindblad-Toh K, Lim YM, Ruiz-Gutierrez M, Bolk S, Langhorst B, Schaffner S, Winchester E, Lander ES. SBE-TAGS: an array-based method for efficient single-nucleotide polymorphism genotyping. Proc Natl Acad Sci USA 2000; 97:12164–12169.

Hittalmani S, Foolad MR, Mew T, Rodriguez RL, Huang N. Development of a PCR-based marker to identify rice blast resistance gene, Pi-2 (t), in a segregating population. Theor Appl Genet 1995; 91:9–14.

Hofte H, Desprez T, Amselem J, Chiapello H, Rouze P, Caboche M, Moisan A, Jourjon MF, Charpenteau JL, Berthomieu P. An inventory of 1152 expressed sequence tags obtained by partial sequencing of cDNAs from *Arabidopsis thaliana*. Plant J 1993; 4:1051–1061.

Hohmann U, Endo T, Gill KS, Gill BS. Comparison of genetic and physical maps of group 7 chromosomes from *Triticum aestivum* L. Mol Gen Genet 1994; 245:644–653.

Huang N, Angeles ER, Domingo J, Magpantay G, Singh S, Zhang G, Kumaravadivel N, Bennett J, Khush GS. Pyramiding of bacterial blight resistance genes in rice:

marker-assisted selection using RFLP and PCR. Theor Appl Genet 1997; 95:313–320.

Hubank M, Schatz DG. Identifying the differences in mRNA expression by representational difference analysis of cDNA. Nucleic Acids Res 1994; 22:5640–5648.

Huhman DV, Sumner LW. Metabolic profiling of saponins in *Medicago sativa* and *Medicago truncatula* using HPLC coupled to an electrospray ion-trap mass spectrometer. Phytochemistry 2002; 59:347–360.

Jeon JS, Lee S, Jung KH, Jun SH, Jeong DH, Lee J, Kim C, Jang S, Yang K, Nam J, An K, Han MJ, Sung RJ, Choi HS, Yu JH, Choi JH, Cho SY, Cha SS, Kim SI, An G. T-DNA insertional mutagenesis for functional genomics in rice. Plant J 2000; 22:561–570.

Jones CS, Davies HV, Taylor MA. Profiling of changes in gene expression during raspberry (*Rubus idaeus*) fruit ripening by application of RNA fingerprinting techniques. Planta 2000; 211:708–714.

Joshi SP, Ranjekar PK, Gupta VS. Molecular markers in plant genome analysis. Curr Sci 1999; 77:230–240.

Kanazin V, Talbert H, See D, DeCamp P, Nevo E, Blake T. Discovery and assay of single-nucleotide polymorphisms in barley (*Hordeum vulgare*). Plant Mol Biol 2002; 48:529–537.

Kawasaki S, Bochert C, Deyholos M, Wang H, Brazille S, Kawai K, Galbraith D, Bohnert H. Gene expression profiles during the initial phase of salt stress in rice. Plant Cell 2001; 13:889–906.

Kearsey MJ. QTL analysis: problems and (possible) solutions. In: Kang MS, ed. Quantitative Genetics, Genomics, and Plant Breeding. New York, NY 10016, USA: CABI Publishing, 2002:45–58.

Kehoe DM, Villand P, Summerville S. DNA microarrays for studies of higher plants and other photosynthetic organisms. Trends Plant Sci 1999; 4:38–41.

Keim P, Diers BW, Shoemaker RC. Genetic analysis of soybean hard seededness with molecular markers. Theor Appl Genet 1990; 79:465–469.

Keller B, Feuillet C. Colinearity and gene density in grass genomes. Trends Plant Sci 2000; 5:246–251.

Kersten B, Bürkle L, Kuhn EJ, Giavalisco P, Konthur Z, Lueking A, Walter G, Eickhoff H, Schneider U. Large-scale plant proteomics. Plant Mol Biol 2002; 48: 133–141.

Kiyosue T, Yamaguchi-Shinozaki K, Shinozaki K. Cloning of cDNAs for genes that are early-responsive to dehydration stress (ERDs) in *Arabidopsis thaliana* L.: identification of three ERDs as HSP cognate genes. Plant Mol Biol 1994; 25:791–798.

Klein PE, Klein RR, Cartinhour SW, Ulanch PE, Dong J, Obert JA, Morishige DT, Schlueter SD, Childs KL, Ale M, Mullet JE. A high throughput AFLP-based method for constructing integrated genetic and physical maps: progress toward a sorghum genome map. Genome Res 2000; 10:789–807.

Koebner RMD, Summers RW. 21st century wheat breeding: plot selection or plate detection? Trends Biotechnol 2003; 21(2):59–63.

Koncz C. From genome projects to molecular breeding [Editorial overview]. Curr Opin Biotechnol 2003; 14(2):133–135.

Koniecyzn A, Asubel FM. A procedure for mapping *Arabidopsis* mutations using co-dominant ecotype-specific PCR-based markers. Plant J 1993; 4:403–410.

Kota RS, Gill KS, Gill BS. A cytogenetically based physical map of chromosome 1B in common wheat. Genome 1993; 36:548–554.

Krisan PJ, Young JC, Sussman MR. T-DNA as an insertional mutagen in *Arabidopsis*. Plant Cell 1999; 11:2283–2290.

Kuhn E. From library screening to microarray technology: strategies to determine gene expression profiles and to identify differentially regulated genes in plants. Ann Bot 2001; 87:139–155.

Kulikova O, Gualtieri G, Geurts R, Kim DJ, Cook D, Huguet T, de Jong JH, Fransz PF, Bisseling T. Integration of the FISH pachytene and genetic maps of *Medicago truncatula*. Plant J 2001; 27:49–58.

Kurata N, Nagamura Y, Yamamoto K, Harushima Y, Sue N, Yu J, Antonio BA, Shomura A, Shimizu T, Lin SY, Inoue T, Fukuda A, Shimano T, Kuboki Y, Toyama T, Miyamoto Y, Kirihara T, Hayasaka K, Miyayo A, Monna L, Zhong HS, Tamura Y, Wang ZX, Momma T, Umehara Y, Yano M, Sasaki T, Minobe Y. A 300-kilobase interval genetic map of rice including 883 expressed sequences. Nat Genet 1994; 8:365–372.

Kurata N, Umehara Y, Tanoue H, Sasaki T. Physical mapping of the rice genome with YAC clones. Plant Mol Biol 1997; 35:101–113.

Lander ES. The new genomics: global views of biology. Science 1996; 274:536–539.

Lebreton C, Lazic-Jancic V, Steed A, Pekic S, Quarrie SA. Identification of QTL for drought responses in maize and their use in testing causal relationships between traits. J Exp Bot 1995; 46:853–865.

Lemieux B, Aharoni A, Schena M. Overview of DNA chip technology. Mol Breed 1998; 4:277–289.

Lewers K, Heinz R, Beard H, Marek L, Matthews B. A physical map of a gene-dense region in soybean linkage group A2 near the black seed coat and Rhg4 loci. Theor Appl Genet 2002; 104:254–260.

Liang P, Pardee AB. Differential display of eukaryotic messenger RNA by means of the polymerase chain reaction. Science 1992; 257:967–971.

Lilley JM, Ludlow MM, McCouch SR, O'Toole JC. Locating QTL for osmotic adjustment and dehydration tolerance in rice. J Exp Bot 1996; 47:1427–1436.

Lin YR, Schertz KF, Paterson AH. Comparative analysis of QTLs affecting plant height and maturity across the Poaceae, in reference to an interspecific sorghum population. Genetics 1995; 141:391–411.

Lindblad-Toh K, Tanenbaum DM, Daly MJ, Winchester E, Lui WO, Villapakkam A, Stanton SE, Larsson C, Hudson TJ, Johnson BE, Lander ES, Meyerson M. Loss-of-heterozygosity analysis of small-cell lung carcinomas using single-nucleotide polymorphism arrays. Nat Biotechnol 2000; 18:1001–1005.

Lipshutz RJ, Fodor SP, Gingeras TR, Lockhart DJ. High-density synthetic oligonucleotide arrays. Nat Genet 1999; 21(suppl):20–24.

Litt M, Luty JA. A hypervariable microsatellite revealed by in vitro amplification of a dinucleotide repeat within the cardiac muscle actin gene. Am J Hum Genet 1989; 44:397–401.

Liu N, Shan Y, Wang FP, Xu CG, Peng KM, Li XH, Zhang Q. Identification of an 85-kb DNA fragment containing pms1, a locus for photoperiod-sensitive genic male sterility in rice. Mol Genet Genomics 2001; 266:271–275.

Lyman RF, Lai C, Mackay T. Linkage disequilibrium mapping of molecular polymorphisms at the scabrous locus associated with naturally occurring variation in bristle number in *Drosophila melanogaster*. Genet Res 1999; 74:303–311.

Lockhart DJ, Dong H, Byrne MC, Follettie MT, Gallo MV, Chee MS, Mittmann M, Wang C, Kobayashi M, Horton H, Brown EL. Expression monitoring by hybridization to high-density oligonucleotide arrays. Nat Biotechnol 1996; 14:1675–1680.

Lorieux M, Petrov M, Huang N, Guiderdoni E, Ghesquire A. Aroma in rice: genetic analysis of a quantitative trait. Theor Appl Genet 1996; 93:1145–1151.

Ma XF, Ross K, Gustafson JP. Physical mapping of restriction fragment length polymorphism (RFLP) markers in homoeologous groups 1 and 3 chromosomes of wheat by in situ hybridization. Genome 2001; 44:401–412.

MacBeath G, Schreiber SL. Printing protein as microarrays for high throughput function determination. Science 2000; 289:1760–1763.

Marek LF, Shoemaker RC. BAC contig development by fingerprint analysis in soybean. Genome 1997; 40:420–427.

Marek LF, Mudge J, Darnielle L, Grant D, Hanson N, Paz M, Huihuang Y, Denny R, Larson K, Foster-Hartnett D, Cooper A, Danesh D, Larsen D, Schmidt T, Staggs R, Crow JA, Retzel E, Young ND, Shoemaker RC. Soybean genomic survey: BAC-end sequences near RFLP and SSR markers. Genome 2001; 44:572–581.

Marqués K, Sarazin B, Chané-Favre L, Zivy M, Thiellement H. Comparative proteomics to establish genetic relationships in the Brassicaceae family. Proteomics 2001; 1:1457–1462.

Martienssen RA. Functional genomics: probing plant gene function and expression with transposons. Proc Natl Acad Sci USA 1998; 95:2021–2026.

Martin B, Nienhuis J, King G, Schafer A. Restriction fragment length polymorphisms associated with water use efficiency in tomato. Science 1989; 243:1725–1728.

Martin GB, Williams JGK, Tanksley SD. Rapid identification of markers linked to a *Pseudomonas* resistance gene in tomato by using random primers and near-isogenic lines. Proc Natl Acad Sci USA 1991; 88:2336–2340.

Martin GB, Brommonschenkel SH, Chunwongse J, Frary A, Ganal MW, Spivey R, Wu T, Earle ED, Tanksley SD. Map-based cloning of a protein kinase gene conferring disease resistance in tomato. Science 1993; 262:1432–1436.

Matsumura H, Nirasawa S, Terauchi R. Transcript profiling in rice (*Oryza sativa* L.) seedlings using serial analysis of gene expression (SAGE). Plant J 1999; 20:719–726.

Matsumura H, Nirasaw S, Kiba A, Urasaki N, Saitoh H, Ito M, Kawai-Yamada M, Uchimiya H, Terauchi R. Overexpression of Bax inhibitor suppresses the fungal elicitor-induced cell death in rice (*Oryza sativa* L) cells. Plant J 2003; 33:425–434.

Matthews BF, Devine TE, Weisemann JM, Beard HS, Lewers KS, MacDonald MH, Park YB, Maiti R, Lin JJ, Kuo J, Pedroni MJ, Cregan PB, Saunders JA. Incorporation of sequenced cDNA and genomic markers into the soybean genetic map. Crop Sci 2001; 41:516–521.

Matz MV, Lukyanov SA. Different strategies of differential display: areas of application. Nucleic Acids Res 1998; 26:5537–5543.

Mazur B, Krebbers E, Tingey S. Gene discovery and product development for grain quality traits. Science 1999; 285:372–375.

McCouch SR, Teytelman L, Xu Y, Lobos KB, Clare K, Walton M, Fu B, Maghirang R, Li Z, Xing Y, Zhang Q, Kono I, Yano M, Fjellstrom R, DeClerck G, Schneider D, Cartinhour S, Ware D, Stein L. Development and mapping of 2240 new SSR markers for rice (*Oryza sativa* L.). DNA Res 2002; 9:199–207.

McCouch S. Toward a plant genomics initiative: thoughts on the value of cross-species and cross-genera comparisons in the grasses. Proc Natl Acad Sci USA 1998; 95:1983–1985.

Menz MA, Klein RR, Mullet JE, Obert JA, Unruh NC, Klein PE. A high-density genetic map of *Sorghum bicolor* (L.) Moench based on 2926 AFLP, RFLP and SSR markers. Plant Mol Biol 2002; 48:483–499.

Michelmore RW, Paran I, Kesseli RV. Identification of markers linked to disease-resistance genes by bulked segregant analysis: a rapid method to detect markers in specific genomic regions by using segregating populations. Proc Natl Acad Sci USA 1991; 88:9828–9832.

Michelmore RW. Genomics approaches to plant disease resistance. Curr Opin Plant Biol 2000; 3:125–131.

Mickelson-Young L, Endo T, Gill BS. A Cytological ladder map of the wheat homoelogous group 4 chromosomes. Theor Appl Genet 1995; 90:1007–1011.

Miftahudin, Scoles GJ, Gustafson JP. AFLP markers tightly linked to the aluminum-tolerance gene *Alt3* in rye (*Secale cereale* L.). Theor Appl Genet 2002; 104:626–631.

Morgan JM, Tan MK. Chromosomal location of a wheat osmoregulation gene using RFLP analysis. Aust J Plant Physiol 1996; 23:803–806.

Morgante M, Salamini F. From plant genomics to breeding practice. Curr Opin Biotechnol 2003; 14:214–219.

Mudge J, Cregan PP, Kenworthy JP, Kenworthy WJ, Orf JH, Young ND. Two microsatellite markers that flank the major cyst nematode resistance locus. Crop Sci 1997; 37:1611–1615.

Mudge J. Marker-assisted selection for soybean cyst nematode resistance and accompanying agronomic traits. Ph.D. thesis, University of Minnesota, 1999.

Natera SH, Guerreiro N, Djordjevic MA. Proteome analysis of differentially displayed proteins as a tool for the investigation of symbiosis. Mol Plant–Microb Interact 2000; 13:995–1009.

Newman T, de Bruijn FJ, Green P, Keegstra K, Kende H, McIntosh L, Ohlrogge J, Raikhel N, Somerville S, Thomashow M. Genes galore: a summary of methods for accessing results from large-scale partial sequencing of anonymous *Arabidopsis* cDNA clones. Plant Physiol 1994; 106:1241–1255.

Nguyen VT, Burow MD, Nguyen HT, Le BT, Le TD, Paterson AH. Molecular mapping of genes conferring aluminum tolerance in rice (*Oryza sativa* L.). Theor Appl Genet 2001; 102:1002–1010.

Ohlrogge J, Benning C. Unravelling plant metabolism by EST analysis. Curr Opin Plant Biol 2000; 3:224–228.

Oliver DJ, Nikolau B, Wurtele ES. Functional genomics: high throughput mRNA, protein, and metabolite analyses. Metab Eng 2002; 4:98–106.

Olson M, Hood L, Cantor C, Botstein D. A common language for physical mapping of the human genome. Science 1989; 245(4925):1434–1435.

Ori N, Eshed Y, Paran I, Presting G, Aviv D, Tanksley SD, Zamir D, Fluhr R. The I2C family from the wilt disease resistance locus I2 belongs to the nucleotide binding, leucine-rich repeat superfamily of plant resistance genes. Plant Cell 1997; 9:521–532.

Osborn TC, Alexander DC, Fobes JF. Identification of restriction fragment length polymorphisms linked to genes controlling soluble solids content in tomato fruit. Theor Appl Genet 1987; 73:350–356.

Osterlund MT, Paterson AH. Applied plant genomics: the secret is integration. Curr Opin Plant Biol 2002; 5:141–145.

Ozturk ZN, Talame V, Deyholos M, Michalowski CB, Galbraith DW, Gozukirmizi N, Tuberosa R, Bohnert HJ. Monitoring large-scale changes in transcript abundance in drought- and salt-stressed barley. Plant Mol Biol 2002; 48:551–573.

Panda S, Sato TK, Hampton GM, Hogenesch JB. An array of insights: application of DNA chip technology in the study of cell biology. Trends Cell Biol 2003; 13(3):151–156.

Pandey A, Mann M. Proteomics to study genes and genomes. Nature 2000; 405(6788):837–846.

Panter S, Thomson R, de Bruxelles G, Laver D, Trevaskis B, Udvardi M. Identification with proteomics of novel proteins associated with the peribacteroid membrane of soybean root nodules. Mol Plant–Microb Interact 2000; 13:325–333.

Parinov S, Sundaresan V. Functional genomics in *Arabidopsis*: large-scale insertional mutagenesis complements the genome sequencing project. Curr Opin Biotechnol 2000; 11:157–161.

Paterson AH, Lander ES, Hewitt JD, Peterson S, Lincoln SE, Tanksley SD. Resolution of quantitative traits into Mendelian factors using a complete linkage map of restriction fragment length polymorphisms. Nature 1988; 335:721–726.

Paterson AH, Tanksley SD, Sorrells ME. DNA markers in plant improvement. Adv Agron 1991; 46:39–90.

Paterson AH. DNA marker assisted crop improvement. In: Paterson AH, ed. Genome Mapping in Plants. Austin, TX: RG Landes Co., 1996:69–81.

Paterson AH. QTL mapping in DNA marker-assisted Plant and animal improvement. In: Paterson AH, ed. Molecular Dissection of Complex Traits. Boca Raton, FL: CRC Press, 1998: 131–143.

Peltier J, Friso G, Kalume DE, Roepstorff P, Nilsson F, Adamska I, van Wijk KJ. Proeomics of the chloroplasts: systematic identification and targeting analysis of lumenal and peripheral thylakoid proteins. Plant Cell 2000; 12:319–342.

Peltier J, Ytterberg J, Liberles D, Roepstorff P, van Wijk KJ. Identification of a 350 kDa ClpP protease complex with 10 different Clp isoforms in chloroplasts of *Arabidopsis thaliana*. J Biol Chem 2001; 276:16318–16327.

Porubleva L, Vander Velden K, Kothari S, Oliver DJ, Chitnis PR. The proteome of maize leaves: use of gene sequences and EST data for identification of proteins with peptide mass fingerprints. Electrophoresis 2001; 22:1724–1738.

Pot J, Pouwels D. Hands-on-experience AFLP-QUANTAR(pro) software. Plant, Animal and Microbe Genomes X. Conference, January 12–16, 2002, Town & Country Convention Center, San Diego, CA, http://www.intl-pag.org/pag/10/abstracts/PAGX_W320.html.

Prashar Y, Weisman SM. Analysis of differential gene expression by display of 3' end restriction fragments of cDNAs. Proc Natl Acad Sci. USA 1996; 93:659–663.

Price A, Young EM, Tomos AD. Quantitative trait loci associated with stomatal conductance, leaf rolling, and heading date mapped in upland rice (*Oryza sativa*). New Phytol 1997; 137:83–91.

Quarrie SA, Laurie DA, Zhu J. QTL analysis to study the association between leaf size and abscisic acid accumulation in droughted rice leaves and comparisons across cereals. Plant Mol Biol 1997; 35:155–165.

Rafalski A. Applications of single nucleotide polymorphisms in crop genetics. Curr Opin Plant Biol 2002; 5:94–100.

Ragot M, Hoisington DA. Molecular markers for plant breeding: comparisons of RFLP and RAPD genotyping costs. Theor Appl Genet 1993; 86:975–984.

Ramachandran S, Sundaresan V. Transposons as tools for functional genomics. Plant Physiol Biochem 2001; 39:243–252.

Reiser L, Mueller LA, Rhee SY. Surviving in a sea of data: a survey of plant genome data resources and issues in building data management systems. Plant Mol Biol 2002; 48:59–74.

Reiter RS, Coors JG, Sussman MR, Gabelman WH. Genetic analysis of tolerance to low-phosphorous stress in maize using restriction fragment length polymorphisms. Theor Appl Genet 1991; 82:561–568.

Rhee SY. Bioinformatic resources, challenges, and opportunities using *Arabidopsis* as a model organism in a post-genomic era. Plant Physiol 2000; 124:1460–1464.

Ribaut JM, Bertan J. Single large-scale marker-assisted selection (SLS-MAS). Mol Breed 1999; 5:531–541.

Ribaut JM, Hoisington DA, Deutsch JA, Jiang C, González de León D. Identification of quantitative trait loci under drought conditions in tropical maize: 1. Flowering parameters and the anthesis-silking interval. Theor Appl Genet 1996; 92:905–914.

Ridout CJ, Donini P. Use of AFLP in cereals research. Trends Plant Sci 1999; 4:76–79.

Roberts JKM. Proteomics and a future generation of plant molecular biologists. Plant Mol Biol 2002; 48:143–154.

Röder MS, Korzun V, Wendehake K, Plaschke J, Tixier MH, Leroy P, Ganal MW. A microsatellite map of wheat. Genetics 1998; 149:2007–2023.

Röder MS, Plaschke J, König SU, Börner A, Sorrells ME, Tanksley SD, Ganal MW. Abundance, variability and chromosomal location of microsatellites in wheat. Mol Gen Genet 1995; 246:327–333.

Roessner U, Luedermann A, Brust D, Fiehn O, Linke T, Willmitzer L, Fernie A. Metabolic profiling allows comprehensive phenotyping of genetically or environmentally modified plant systems. Plant Cell 2001; 13:11–29.

Roskey MT, Juhasz P, Smirnov IP, Takach EJ, Martin SA, Haff LA. DNA sequencing by delayed extraction-matrix-assisted laser desorption/ionization time of flight mass spectrometry. Proc Natl Acad Sci USA 1996; 93:4724–4729.

Rossignol M. Analysis of plant proteome. Curr Opin Biotechnol 2001; 12:131–134.

Rounsley S, Lin X, Ketchum KA. Large scale sequencing of plant genome. Curr Opin Plant Biol 1998; 1:136–141.

Rouquie D, Peltier JB, Mansion M, Tournaire C, Dumas P, Rossignol M. Construction of a directory of tobacco plasma membrane proteins by combined two-dimensional gel electrophoresis and protein sequencing. Electrophoresis 1997; 18: 654–660.

Rudd S. Expressed sequence tags: alternative or complement to whole genome sequences? Trends Plant Sci, July 01, 2003 (10. 1016/S1360–1385(03)00131-6).

Saji S, Umehara Y, Antonio BA, Yamane H, Tanoue H, Baba T, Aoki H, Ishige N, Wu J, Koike K, Matsumoto T, Sasaki T. A physical map with yeast artificial chromosome (YAC) clones covering 63% of the 12 rice chromosomes. Genome 2001; 44:32–37.

Salekdeh GH, Siopongo J, Wade LJ, Ghareyazie B, Bennet J. Proteomic analysis of rice leaves during drought stress and recovery. Proteomics 2002; 2:1131–1145.

Santoni V, Rabilloud T, Doumas P, Rouquie D, Mansion M, Kieffer S, Garrin J, Rossignol M. Towards the recovery of hydrophobic proteins on two-dimensional electrophoresis gels. Electrophoresis 1999; 20:705–711.

Sasaki, et al. The genome sequence and structure of rice chromosome 1. Nature 2002; 420:312–316.

Schadt EE, Monks SA, Drake TA, Lusis AJ, Che N, Colinayo V, Ruff TG, Milligan SB, Lamb JR, Cavet G, Linsley PS, Mao M, Stoughton RB, Friend SH. Genetics of gene expression surveyed in maize, mouse and man. Nature 2003; 422:297–302.

Schaffer R, Landgraf J, Perez-Amador M, Wisman E. Monitoring genomewide expression in plants. Curr Opin Biotechnol 2000; 11:162–167.

Schena M, Shalon D, Davis RW, Brown PO. Quantitative monitoring of gene expression patterns with a complementary DNA microarray. Science 1995; 270: 467–470.

Schmidt R. Synteny: recent advances and future prospects. Curr Opin Plant Biol 2000; 3:97–102.

Schmidt R. Plant genome evolution: lessons from comparative genomics at the DNA level. Plant Mol Biol 2002; 48:21–37.

Schnable PS, Stinard PS, Wen TJ, Heinen S, Weber D, Zhang L, Hansen JD, Nikolau BJ. The genetics of cuticular wax biosynthesis. Maydica 1994; 39:279–287.

Shenoi H, Burnham J, Koller S, Bitner R. Automated purification of plant DNA. Plant, Animal and Microbe Genomes X. Conference, January 12–16, 2002, Town & Country Convention Center, San Diego, CA, poster 124.

Shimkets RA, Lowe DG, Tai JT, Sehl P, Jin H, Yang R, Predki PF, Rothberg BE, Murtha MT, Roth ME, Shenoy SG, Windemuth A, Simpson JW, Simons JF, Daley MP, Gold SA, McKenna MP, Hillan K, Went GT, Rothberg JM. Gene expression analysis by transcript profiling coupled to a gene database query. Nat Biotechnol 1999; 17:798–803.

Shirasu K, Schulman AH, Lahaye T, Schulze-Lefert P. A contiguous 66-kb barley

DNA sequence provides evidence for reversible genome expansion. Genome Res 2000; 10:908–915.

Shoemaker R, Keim P, Vodkin L, Retzel E, Clifton SW, Waterston R, Smoller D, Coryell V, Khanna A, Erpelding J, Gai X, Brendel V, Raph-Schmidt C, Shoop EG, Vielweber CJ, Schmatz M, Pape D, Bowers Y, Theising B, Martin J, Dante M, Wylie T, Granger C. A compilation of soybean ESTs: generation and analysis. Genome 2002; 45:329–338.

Song WY, Wang GL, Chen LL, Kim HS, Pi LY, Holsten T, Gardner J, Wang B, Zhai WX, Zhu LH. A receptor kinase-like protein encoded by the rice disease resistance gene, Xa21. Science 1995; 270(5243):1804–1806.

Speulman E, Metz PL, van Arkel G, Lintel Hekkert B, Stiekema WJ, Pereira A. A two-component enhancer–inhibitor transposon mutagenesis system for functional analysis of the Arabidopsis genome. Plant Cell 1999; 11:1853–1866.

Spigelman JI, Mindrinos MN, Fankhauser C, Richards D, Lutes J, Chory J, Oefner PJ. Cloning of the *Arabidopsis* RSF1 gene by using a mapping strategy based on high-density DNA arrays and denaturing high performance liquid chromatography. Plant Cell 2000; 12:2485–2498.

Stein LD. Integrating biological databases. Nat Rev Genet 2003; 4:337–345.

Stein N, Feuillet C, Wicker T, Schlagenhauf E, Keller B. Subgenome chromosome walking in wheat: a 450-kb physical contig in *Triticum monoccum* L. spans the Lr10 resistance locus in hexaploid wheat (*Triticum aestivum* L.). Proc Natl Acad Sci USA 2000; 97:13436–13441.

Stephenson P, Bryan G, Kirby J, Collins A, Devos K, Busso C, Gale MD. Fifty new microsatellite loci for the wheat genetic map. Theor Appl Genet 1998; 97:946–949.

Stitt M, Fernie AR. From measurements of metabolites to metabolomics: an "on the fly" perspective illustrated by recent studies of carbon–nitrogen interactions. Curr Opin Biotechnol 2003; 14:136–144.

Stuber CW, Polacco M, Senior ML. Synergy of empirical breeding, marker-assisted selection, and genomics to increase crop yield potential. Crop Sci 1999; 39:1571–1583.

Subudhi PK, Rosenow DT, Nguyen HT. Quantitative trait loci for the stay-green trait in sorghum (*Sorghum bicolor* L. Moench): consistency across genetic backgrounds and environments. Theor Appl Genet 2000; 101:733–741.

Summerville C, Summerville S. Plant functional genomics. Science 1999; 285:380–383.

Sumner LW, Pavia NL, Dixon RA, Geno PW. HPLC-continuous-flow liquid secondary ion mass spectrometry of flavonoid glycosides in leguminous plant extracts. J Mass Spectrom 1996; 31:472–485.

Sutcliffe JG, Foye PE, Erlander MG, Hilbush BS, Bodzin LJ, Durham JT, Hasel KW. TOGA: an automated parsing technology for analyzing expression of nearly all genes. Proc Natl Acad Sci USA 2000; 97:1976–1981.

Takahashi Y, Shomura A, Sasaki T, Yano M. *Hd6*, a rice quantitative trait locus involved in photoperiod sensitivity, encodes the α subunit of protein kinase CK2. Proc Natl Acad Sci USA 2001; 98:7922–7927.

Tanksley SD, McCouch SR. Seed banks and molecular maps: unlocking genetic potential from the wild. Science 1997; 277(5329):1063–1066.

Tanksley SD, Young ND, Paterson AH, Bonierbale MW. RFLP mapping in plant breeding: new tools for an old science. Biotechnology 1989; 7:257–264.

Tanksley SD. Mapping polygenes. Annu Rev Genet 1993; 27:205–233;

Tao Q, Chang YL, Wang J, Chen H, Islam-Faridi MN, Scheuring C, Wang B, Stelly DM, Zhang HB. Bacterial artificial chromosome-based physical map of the rice genome constructed by restriction fingerprint analysis. Genetics 2001; 158:1711–1724.

Taramino G, Tarchini R, Ferrario S, Lee M, Pe ME. Characterization and mapping of simple sequence repeats (SSRs) in *Sorghum bicolor*. Theor Appl Genet 1997; 95:66–72.

Tarchini R, Biddle P, Wineland R, Tingey S, Rafalski A. The complete sequence of 340 kb of DNA around the rice Adh1–adh2 region reveals interrupted co linearity with maize chromosome 4. Plant Cell 2000; 12:381–391.

Temnykh S, DeClerck G, Lukashova A, Lipovich L, Cartinhour S, McCouch SR. Computational and experimental analysis of microsatellites in rice (*Oryza sativa* L.): frequency, length variation, transposon associations, and genetic marker potential. Genome Res 2001; 11:1441–1452.

Teulat B, Borries C, This D. New QTLs identified for plant water status, water-soluble carbohydrate and osmotic adjustment in a barley population grown in a growth chamber under two water regimes. Theor Appl Genet 2001; 103:161–170.

Teulat B, This D, Khairallah M, Borries C, Rogot C, Sourdille P, Leroy P, Monneveux P, Charrier A. Several QTLs involved in osmotic adjustment trait variation in barley (*Hordeum vulgare* L.). Theor Appl Genet 1998; 96:688–698.

The Rice Chromosome 10 Sequencing Consortium. In-depth view of structure, activity and evolution of rice chromosome 10. Science 2003; 300:1566–1569.

Thiellement H, Bahrman N, Damerval C, Plomion C, Rossignol M, Santoni V, de Vienne D, Zivy M. Proteomics for genetic and physiological studies in plants. Electrophoresis 1999; 20:2013–2026.

Thomas WTB. Prospects for molecular breeding of barley. Ann Appl Biol 2003; 142:1–12.

Thomas SW, Glaring MA, Rasmussen SW, Kinane JT, Oliver RP. Transcript profiling in the barley mildew pathogen *Blumeria graminis* by serial analysis of gene expression (SAGE). Mol Plant–Microb Interact 2002; 15:847–856.

Tinker NA. Why quantitative geneticists should care about bioinformatics. In: Kang MS, ed. Quantitative Genetics, Genomics, and Plant Breeding. New York, NY 10016, USA: CABI Publishing, 2002:33–44.

Tissier AF, Marillonnet S, Klimyuk V, Patel K, Torres MA, Murphy G, Jones JD. Multiple independent defective suppressor–mutator transposon insertions in *Arabidopsis*, a tool for functional genomics. Plant Cell 1999; 11:1841–1852.

Trethewey RN. Gene discovery via metabolic profiling. Curr Opin Biotechnol 2001; 12(2):135–138.

Trindade LM, Horvath B, Bachem C, Jacobsen E, Visser RG. Isolation and functional characterization of a stolon specific promoter from potato (*Solanum tuberosum* L.). Gene 2003; 303:77–87.

Tsugita A, Kamo M, Kawakami T, Ohki Y. Two dimensional electrophoresis of plant proteins and standardization of gel patterns. Electrophoresis 1996; 17:855–865.

Tuberosa R, Sanguineti MC, Landi P, Salvi S, Casarini E, Conti S. RFLP mapping of quantitative trait loci controlling abscisic acid concentration in leaves of drought-stressed maize (*Zea mays* L.). Theor Appl Genet 1998; 97:744–755.

Tyagi S, Kramer FR. Molecular beacons: probes that fluoresce upon hybridization. Nat Biotechnol 1996; 14:303–308.

Umehara Y, Inagaki A, Tanoue H, Yasukochi Y, Nagamura Y, Saji S, Otsuki Y, Fujimura T, Kurata N, Minobe Y. Construction and characterization of a rice YAC library for physical mapping. Mol Breed 1995; 1:79–89.

van der Biezen EA, Juwana H, Parker JE, Jones JD. cDNA-AFLP display for the isolation of *Peronospora parasitica* genes expressed during infection in *Arabidopsis thaliana*. Mol Plant–Microb Interact 2000; 13:895–898.

van Eck HJ, Jacob JME, Stam P, Ton J, Stiekema WJ, Jocobsen E. Multiple alleles for tuber shape in diploid potato detected by qualitative and quantitative genetic analysis using RFLPs. Genetics 1994; 137:303–309.

van Eijk M. Highly automated AFLP and SNP detection (W319). Plant, Animal and Microbe Genomes X. Conference, January 12–16, 2002, Town & Country Convention Center, San Diego, CA, http://www.intipag.org/pag/10/abstracts/PAGX_W319.html.

van Wijk KJ. Proteomics of the chloroplasts: experimentation and prediction. Trends Plants Sci 2000; 5:420–425.

van Wijk KJ. Challenges and prospects of plant proteomics. Plant Physiol 2001; 126:501–508.

Velculescu VE, Zhang L, Vogelstein B, Kinzler KW. Serial analysis of gene expression. Science 1995; 270:484–487.

Vener AV, Harms A, Sussman MR, Vierstra RD. Mass spectrometric resolution of reversible protein phosphorylation in photosynthetic membranes of *Arabidopsis thaliana*. J Biol Chem 2001; 276:6959–6966.

Vos P, Hogers R, Bleeker M, Reijans M, van de Lee T, Hornes M, Frijters A, Pot J, Peleman J, Kuiper M, Zabeau M. AFLP: a new technique for DNA fingerprinting. Nucleic Acids Res 1995; 23:4407–4414.

Walbot V. Genes, genomes, genomics. What can plant biologists expect from the 1998 National Science Foundation plant genome research program? Plant Physiol 1999; 119:1151–1155.

Walbot V. Saturation mutagenesis using maize transposons. Curr Opin Plant Biol 2000; 3:103–107.

Wang DG, Fan JB, Siao CJ, Berno A, Young P, Sapolsky R, Ghandour G, Perkins N, Winchester E, Spencer J, Kruglyak L, Stein L, Hsie L, Topaloglou T, Hubbell E, Robinson E, Mittmann M, Morris MS, Shen N, Kilburn D, Rioux J, Nusbaum C, Rozen S, Hudson TJ, Lander ES. Large-scale identification, mapping, and genotyping of single-nucleotide polymorphisms in the human genome. Science 1998; 280(5366):1077–1082.

Wang GL, Mackill DJ, Bonman JM, McCouch SR, Champoux MC, Nelson RJ. RFLP mapping of genes conferring complete and partial resistance to blast in a durably resistant rice cultivar. Genetics 1994; 136:1421–1434.

Wang GL, Holsten TE, Song WY, Wang HP, Ronald PC. Construction of a rice bacterial artificial chromosome library and identification of clones linked to the Xa-21 disease resistance locus. Plant J 1995; 7:525–533.

Wang XY, Chen PD, Zhang SZ. Pyramiding and marker-assisted selection for powdery mildew resistance genes in common wheat. Yi Chuan Xue Bao 2001a; 28:640–646.

Wang Z, Taramino G, Yang D, Liu G, Tingey SV, Miao GH, Wang GL. Rice ESTs with disease-resistance gene- or defense-response gene-like sequences mapped to regions containing major resistance genes or QTLs. Mol Genet Genomics 2001b; 265:302–310.

Ware D, Jaiswal P, Ni J, Pan X, Chang K, Clark K, Teytelman L, Schmidt S, Zhao W, Cartinhour S, McCouch SR, Stein L. Gramene: a resource for comparative grass genomics. Nucleic Acids Res 2002; 30:103–105.

Welsh J, McClelland M. Fingerprinting genomes using PCR with arbitrary primers. Nucleic Acids Res 1990; 18:7213–7218.

Weng Y, Tuleen NA, Hart GE. Extended physical maps and a consensus physical map of the homoeologous group-6 chromosomes of wheat (*Triticum aestivum* L. em Thell.). Theor Appl Genet 2000; 100:519–527.

Werner JE, Endo TR, Gill BS. Toward a cytogenetically based physical map of the wheat genome. Proc Natl Acad Sci USA 1992; 89:11307–11311.

Williams JGK, Kubelik AR, Livak KJ, Rafalski JA, Tingey S. DNA polymorphisms amplified by arbitrary primers are useful as genetic markers. Nucleic Acids Res 1990; 18:6531–6535.

Wilson ID, Barker GI, Edwards KJ. Genotype to phenotype: a technological challenge. Ann Appl Biol 2003; 142:33–39.

Wu J, Maehara T, Shimokawa T, Yamamoto S, Harada C, Takazaki Y, Ono N, Mukai Y, Koike K, Yazaki J, Fujii F, Shomura A, Ando T, Kono I, Waki K, Yamamoto K, Yano M, Matsumoto T, Sasaki T. A comprehensive rice transcript map containing 6591 expressed sequence tag sites. Plant Cell 2002; 14:525–535.

Xiao J, Grandillo S, Ahn SN, McCouch SR, Tanksley SD, Li JM, Yuan LP. Genes from wild rice improve yield. Nature (Lond) 1996; 384(6606):223–224.

Xiao J, Li JM, Grandillo S, Ahn SN, Yuan LP, Tanksley SD, McCouch SR. Identification of trait-improving quantitative trait loci alleles from a wild rice relative. *Oryza rufipogon*. Genetics 1998; 150:899–909.

Yadav RS, Courtois B, Huang N, McLaren G. Mapping genes controlling root morphology and root distribution in a doubled-haploid population of rice. Theor Appl Genet 1997; 94:619–632.

Yadav RS, Hash CT, Bidinger FR, Cavan GP, Howarth CJ. Quantitative trait loci associated with traits determining grain and stover yield in pearl millet under terminal drought-stress conditions. Theor Appl Genet 2002; 104:67–83.

Yamaguchi K, Subramanian AR. The plastid ribosomal proteins. Identification of all the proteins in the 50s subunit of an organelle ribosome (chloroplast). J Biol Chem 2000; 275:28466–28482.

Yamaguchi K, von Knoblauch K, Subramanian AR. The plastid ribosomal proteins. Identification of all the proteins in the 30s subunit of an organelle ribosome (chloroplast). J Biol Chem 2000; 275:28455–28465.

Yamamoto K. Sasaki T. Large-scale EST sequencing in rice. Plant Mol Biol 1997; 35:135-144.

Yamazaki M, Tsugawa H, Miyayo A, Yano M, Wu J, Yamamoto S, Matsumoto T, Sasaki T, Hirochika H. The rice retrotransposon Tos17 prefers low-copy number sequences as integration targets. Mol Genet Genomics 2001; 265:336-344.

Yano M, Katayose Y, Ashikari M, Yamanouchi U, Monna L, Fuse T, Baba T, Yamamoto K, Umehara Y, Nagamura Y, Sasaki T. *Hd1*, a major photoperiod sensitivity quantitative trait locus in rice, is closely related to the Arabidopsis flowering time gene CONSTANS. Plant Cell 2000; 12:2473-2484.

Yazaki J, Kishimoto N, Ishikawa M, Kikuchi S. Rice expression database: the gateway to rice functional genomics. Trends Plant Sci 2002; 7:563-564.

Ye S, Dhillon S, Ke X, Collins AR, Day INM. An efficient procedure for genotyping single nucleotide polymorphisms. Nucleic Acids Res 2001; 29:e88.

Yim Y-.S, Davis GL, Duru NA, Musket TA, Linton EW, Messing JW, McMullen MD, Soderlund CA, Polacco ML, Gardiner JM, Coe EH Jr. Characterization of three maize bacterial artificial chromosome libraries toward anchoring of the physical map to the genetic map using high-density bacterial artificial chromosome filter hybridization. Plant Physiol 2002; 130:1686-1696.

Young ND. Restriction fragment length polymorphisms (RFLPs) and crop improvement. Exp Agric 1992; 28:385-397.

Young ND. A cautiously optimistic vision for marker-assisted breeding. Mol Breed 1999; 5:505-510.

Young ND. The genetic architecture of resistance. Curr Opin Plant Biol 2000; 3:285-290.

Yu J. A draft sequence of the rice genome (*Oryza sativa* L. ssp. *indica*). Science 2002; 296:79-92.

Zabeau M, Vos P. Selective restriction fragment amplification: a general method for DNA fingerprinting. European patent application number: 92402629.7. Publication no. 534 858 A1, 1993

Zamir D. Improving plant breeding with exotic genetic libraries. Nat Rev Genet 2001; 2(12):983-989.

Zhang HB, Choi S, Woo SS, Li Z, Wing RA. Construction and characterization of two-rice bacterial artificial chromosome library from the parents of a permanent recombinant inbred mapping population. Mol Breed 1996; 2:11-24.

Zhang J, Zheng HG, Aarti A, Pantuwan G, Nguyen TT, Tripathy JN, Sarial AK, Robin S, Babu RC, Nguyen BD, Sarkarung S, Blum A, Nguyen HT. Locating genomic regions associated with components of drought resistance in rice: comparative mapping within and across species. Theor Appl Genet 2001; 103:19-29.

Zhang Q, Arbuckle J, Wessler SR. Recent, extensive, and preferential insertion of members of the miniature inverted-repeat transposable element family heartbreaker into genic regions of maize. Proc Natl Acad Sci USA 2000; 97:1160-1165.

Zheng HG, Babu RC, Pathan MS, Ali L, Huang N, Courtois B, Nguyen HT. Quantitative trait loci for root-penetration ability and root thickness in rice: Comparison of genetic backgrounds. Genome 2000; 43:53-61.

Zheng K, Huang N, Bennett J, Khush GS. PCR-based marker-assisted selection in rice breeding. IRRI discussion paper series No. 12. International Rice Research Institute, P.O. Box 933, Manila, Philippines, 1995.

Zhu H, Briceno G, Dovel R, Hayes PM, Liu BH, Liu CT, Ullrich SE. Molecular breeding for grain yield in barley: an evaluation of QTL effects in a spring barley cross. Theor Appl Genet 1999; 98:772–779.

Zhu T, Wang X. Large scale profiling of the *Arabidopsis* transcriptome. Plant Physiol 2000; 124:1472–1476.

Zivy M, deVienne D. Proteomics: a link between genomics, genetics and physiology. Plant Mol Biol 2000; 44:575–580.

10

Marker-Assisted Utilization of Exotic Germ Plasm

Ilan Paran
The Volcani Institute
Bet Dagan, Israel

1 RATIONAL FOR MARKER-ASSISTED UTILIZATION OF EXOTIC GERM PLASM

Modern cultivars of crop plants usually represent a small fraction of the variation that exists in their gene pool. For example, it has been estimated that less than 5% of the genetic variation in tomato exists in the cultivated species, while the rest of the variation remains distributed among its wild relatives (Miller and Tanksley, 1990). Similarly, very narrow genetic base exists in other self-pollinated crops such as soybean and rice where breeding has been mainly confined within the primary gene pool (Singh and Hymowitz, 1999; Wang et al., 1992).

Until recently, plant breeders related to exotic germ plasm mostly as a rich source for genes that resist biotic and abiotic stresses. Numerous disease-resistance genes were identified and introgressed to elite cultivars. Results for these gene transfers were reviewed for several crops such as wheat (Friebe et al., 1996), rice (Brar and Khush, 1997), and soybean (Singh and Hymowitz, 1999). However, wild accessions were rarely considered as a source for valuable genes for quantitatively inherited traits related to yield because of their inferior performance and the possible linkage of favorable genes with unfavorable ones.

The potential of the utilization of wild species as a source of novel phenotypes was revealed by the existence of transgressive variation often observed in the offspring of wide crosses (reviewed by Rieseberg et al., 1999). Quantitative trait locus (QTL) mapping for many traits identified transgressive alleles at QTLs in species such as tomato (DeVicente and Tanksley, 1993) and sunflower (Kim and Rieseberg, 1999). At these loci, allelic effects opposite to those predicted by the parental values were detected, indicating that alleles with favorable effects on yield-related traits exist in the wild parents. Tanksley and McCouch (1997) suggested that a change from the old paradigm of selecting for a phenotype to the new paradigm of identifying favorable genes is needed for better exploitation of exotic germ plasm. This will be achieved by the use of wide crosses combined with molecular mapping to identify the favorable alleles and to further introgress them into elite genetic backgrounds.

This chapter will review the recent literature reporting on using molecular markers for the identification and introgression of genes from exotic germ plasm that control quality and yield-related traits in several major crop species. Results of mapping and introgression of disease-resistance genes will not be included.

2 TYPES OF POPULATIONS

2.1 Substitution Lines

Substitution lines (SL) are generated by replacing a chromosome segment from an exotic (donor) parent with its homolog in a recipient parent. The most extensive use of SL has been in the bread wheat, which, as a hexaploid, can tolerate deficiencies or addition of chromosomes and is an ideal target for cytological manipulations. Substitution lines are usually produced by backcrossing monosomic lines with addition lines containing the alien chromosomes. Other techniques for the generations of SL were summarized by Jiang et al. (1994). Initially, intervarietal SL were produced using monosomic series of Chinese Spring (Kuspira and Unrau, 1957). Since then, several other SL were produced (Fedak, 1999). In addition to substitution of chromosomes, other methods for alien chromosome (or arm) transfer have been used in wheat, including addition lines (Feldman, 1975; Friebe et al., 2000) and translocation lines (Crasta et al., 2000; Friebe et al., 1996).

A major constrain to the integration of alien chromosome segments into the wheat genome is the extent to which recombination occurs between the donor and recipient genomes. When the donor parent contains at least one genome in common with wheat, recombination will occur between the homologous genomes. For example, recombinant SL were developed from a cross of Bethlehem, a modern bread wheat cultivar with the tetraploid wild emmer wheat (*Triticum dicoccoides*) that shares the A and B genomes with the bread wheat (Millet et al., 1988). However, if the genome of the two parents are not homologous, homologous pairing and recombination will be prevented by the *Ph1* locus on chromosome 5B of wheat. Introduction of recombination in this case is possible by creating translocation lines and using methods such as *ph* mutants and ionizing radiation (Fedak, 1999; Jauhar and Chibbar, 1999).

Whole-genome SL were also produced in diploid species such as *Brassica napus* (Howell et al., 1996), *B. oleracea* (Ramsay et al., 1996), and tomato (Eshed and Zamir, 1995; Eshed et al., 1996; Monforte and Tanksley, 2000a,b). These lines were produced by backcrossing and marker-assisted selection so that each line was selected to contain a single marker-defined chromosome segment introgressed from a donor parent in an otherwise uniform genetic background of the recurrent parent. Recently, a complete set of oat–maize chromosome addition lines was reported and used for mapping maize sequences (Okagaki et al., 2001).

Law (1996) originally suggested the use of SL as a tool for the identification of QTLs. The main advantages of SL for QTL mapping are as follows: (1) SL provide a permanent genetic resource that can be replicated in different environments, thus enabling to reduce the environmental variance associated with the QTL. (2) Population size for QTL detection can be reduced because each SL is fixed and the genome of the donor parent is represented in a small number of SL. The increase of statistical power of QTL detection was demonstrated by Eshed and Zamir (1995) that identified more QTLs per trait in tomato than in other studies despite the use of much smaller population (six plants in each of 50 SL). (3) Efficient high-resolution mapping of QTLs by subsequent crossing of the SL is possible because segregation is confined to a single QTL. (4) SL allow the detection of additive QTL that are important for breeding purposes because only one chromosome segment is introgressed from the donor parent in each line. Detection of dominant QTLs as well as epistasis is possible by crossing the SL with the recurrent parent or with parents that differ in their genetic background (Eshed and Zamir, 1995, 1996). (5) Following QTL detection, the SL can immediately be utilized for breeding if the recurrent parent is a common variety.

The use of SL for QTL mapping has several limitations: First, the development of SL requires a large investment in time and resources; second, the position of the QTL within the substituted segment cannot be determined; and third, the existence of multiple vs. single QTL in the substituted segment can not be distinguished from each other. However, these latter two limitations can be overcome by additional crossing of the SL and decreasing the length of the introgression. For example, by creating SL with variable length of the introgressed region of chromosome 2 of tomato, Eshed and Zamir (1995) identified three distinct closely linked QTL for fruit mass.

2.2 Advanced Backcross

Tanksley and Nelson (1996) suggested the use of advanced backcross (AB) QTL analysis as a method for concomitant QTL identification and variety development and as a way to identify and introgress beneficial alleles at QTLs of exotic germ plasm. In advanced backcross populations, QTL analysis is performed in BC_2 or BC_3 segregating generations. Some of the advantages of this method are similar to those of SL. These include: (1) Reduction or elimination of negative genes from the donor parent that can interfere with yield measurements by using phenotypic selection to create the population. (2) Reduction of epistasis between donor alleles. (3) Production of near-isogenic lines (NILs) for the QTL is possible in 1-2 years from the initial QTL detection. The limitations of AB-QTL analysis are: (1) Advanced backcross generations are not balanced (e.g., in BC_2 the frequency of the homozygous and heterozygous classes at each marker are 0.75 and 0.25, respectively). The unbalanced allele frequencies and the reduction of genetic variance in advanced backcross generations implies that the statistical power of AB-QTL detection is reduced compared to balanced selfed populations (Kaeppler, 1997). Furthermore, map construction in unbalanced populations has less statistical power than in balanced ones. Therefore marker order may need to rely on information from other maps (Moncada et al., 2001). (2) The phenotypic selection employed during backcrossing may result in a loss of portions of the donor parent genome.

Advanced backcross quantitative trait locus analysis was employed mainly in tomato and rice. In tomato, a series of studies involving crosses with four wild species has been performed (Bernacchi et al., 1998a; Fulton et al., 1997, 2000; Tanksley et al., 1996). In rice, AB-QTL analysis for yield related traits was performed with one wild species crossed with two different recurrent parents (Xiao et al., 1988; Moncada et al., 2001). In addition, an AB analysis for heading date was performed in a cross of two subspecies of rice, *Oryza indica* and *O. japonica* (Yamamoto et al., 2000). Advanced backcross quantitative trait locus analysis for fruit related traits is currently performed in pepper in a cross between a common cultivar of *Capsicum annuum* and a wild accession of *C. frutescens* (Rao and Paran, unpublished).

2.3 Backcross Inbred Lines

Backcross inbred lines (BIL) are produced by repeated backcrossing followed by several generations of inbreeding. Wehrhahm and Allard (1965) originally described the use of BIL as a method to introgress quantitative traits and to estimate their gene number. Theoretical calculations have shown that a set of BIL will contain lines genotypically identical to the recurrent parent except for one or few introgressed regions from the donor parent. The BIL method has been used as a breeding method to improve quantitative traits in several crop species (Bliss, 1981; Hartman and St Clair, 1998; Thurling, 1982). As a method for QTL identification, it has the advantage of consisting homozygous lines but has the limitation of AB analysis in reduction of power in QTL detection because of unbalanced allele frequency. Unlike AB populations, phenotypic selection is usually not performed during backcrossing. Therefore a more complete genome coverage is expected in BIL.

Recently, the BIL method was used for QTL validation in tomato for fruit quality parameters (Azanza et al., 1995) and in soybean for seed protein concentration (Sebolt et al., 2000). Backcross inbred lines were also developed for whole-genome QTL identification in a rice cross of *O. indica* × *O. japonica* for seed dormancy, heading date and tolerance to phosphorus-deficiency (Lin et al., 1998; Wissuwa et al., 1998); and in *B. napus* for yield traits and oil content (Butruille et al., 1999). Marker-characterized BIL population was also developed in the cultivated tomato by its crossing with the nightshade species *Solanum lycopersicoides* (Chetelat and Meglic, 2000).

3 EXPERIMENTAL RESULTS

3.1 Tomato

Quantitative trait locus mapping and introgression studies were performed in crosses of the cultivated tomato species *Lycopersicon esculentum* with six wild species: *L. chmielewskii* (Azanza et al., 1995), *L. pennellii* (Eshed and Zamir, 1995; Eshed et al., 1996), *L. hirsutum* (Bernacchi et al., 1998a,b; Monforte and Tanksley, 2000a,b), *L. parviflorum* (Fulton et al., 2000), *L. peruvianum* (Fulton et al., 1997), and *L. pimpinellifolium* (Bernacchi et al., 1998b; Tanksley et al., 1996). These studies have identified hundreds of QTLs for many agronomic traits, in particular those related to fruit yield and quality. The success of the series of AB-QTL analyses in crosses with four wild species can be evaluated based on a number of parameters: (1) The percentage of QTLs and traits for which favorable alleles were identified in the donor parent. The data that was summarized by Fulton et al. (2000) revealed that for 14 traits that were studied in all four crosses, 17–42% of the QTL per cross had favorable alleles in the wild parent. The percentages ranged for different traits with a minimum of 4% for red yield to 93% for soluble solids content. The high percentage of positive QTL alleles present in the wild parents for the latter trait reflects their phenotypic superiority for this trait compared to the cultivated tomato. For 60–70% of all traits per cross, at least one QTL was detected with a favorable allele that was originated in the wild parent. (2) The conservation of QTLs across species. For 16 traits measured in all four AB-QTL studies, 39% of the QTLs detected in *L. parviflorum* were potentially orthologous to QTLs detected in at least one other species (potential orthology was declared if two QTLs were mapped to the same 20 cM region on the same chromosome). Similarly, 30% and 19% of the QTL detected in *L. hirsutum* and in *L. peruvianum* were potentially orthologous to QTLs in *L. pimpinellifolium* and in *L. hirsutum* or *L. pimpinellifolium*, respectively. The relatively low level of favorable QTL conservation among the species indicates that it is should be possible to pyramid multiple exotic favorable alleles for a given trait in a common elite cultivar. (3) Validation of QTL effects in subsequent generations. Bernacchi et al. (1998b) constructed near isogenic lines for 25 previously identified QTLs (QTL-NILs) in two AB-QTL studies. These QTL-NILs differed in a single chromosomal region that was identified as containing the QTL. For most of the QTL-NILS (88% and 68% at significance levels of 0.1 and 0.05, respectively), the QTL effect was repeated as predicted by the AB-QTL analysis, indicating that most wild species QTL alleles are additive. However, independent NILs constructed for the same region often differed in their phenotype. In addition, the NILs often showed a QTL effect for traits for which no QTLs were detected in the AB analysis, indicating the importance of factors such as epistasis and variable amount of linkage drag in QTL detection in the NILs compared to the initial AB study. These results demonstrated that although the AB-QTL method is robust with respect to QTL detection, multiple QTL-NILs should be constructed for the targeted regions to maximize the chance to develop superior lines containing the QTLs.

The importance of epistasis in controlling the expression of wild species QTLs was demonstrated by Eshed and Zamir (1996) who crossed 10 *L. pennellii* introgression lines in a half-diallele design and compared the phenotypic values of 45 double heterozygotes for four yield-associated traits to the values of the single heterozygote ILs. Out of the 180 tested interactions, 28% had significant ($P < 0.05$) less than additive deviation from complete additivity of the corresponding single ILs; that is, the phenotypic effect of the double heterzygotes were lower than the sum of the effects of the single heterozygotes. This type of interaction implies that when pyramiding QTLs in a single genotype, one should select for a favorable combination of QTLs for which the less than additive epistasis is minimized.

Monforte et al. (2001) compared NILs for chromosome 4 that contain introgressions from three green-fruited species: *L. pennellii*, *L. peruvianum*, and *L. hirsutum* that affect various agronomic traits. Although the different introgressions affected the traits in a similar manner, differences in gene action and epistatic interactions were observed, indicating the possibility that the three species carry different alleles at the QTL. For example, while *L. hirsutum* chromosome 4 NILs negatively interacted with *L. hirsutum* chromosome 1 introgression for soluble solids concentration (SSC) and fruit color, no interaction was detected between *L. peruvianum* chromosome 4 NIL and the same *L. hirsutum* chromosome 1 introgression. Unlike the less than additive interactions observed by Eshed and Zamir (1996), combining QTL for SSC from different wild species, i.e., *L. hirsutum*, chromosome 4 introgression and *L. chmielewskii* chromosome 1 introgression had a combined increase in SSC that was greater than the sum of the effects of the individual introgressions. Furthermore, some undesirable effects in the individual introgressions were compensated in the double-introgression line. These results demonstrate that combining appropriate QTLs from different exotic sources can improve the phenotypes of the cultivated tomato.

Numerous QTLs that increase soluble solids concentration were detected in tomato wild species. A detailed molecular characterization of one such QTL, *Brix9-2-5*, derived from *L. pennellii*, was performed by Fridman et al. (2000). This QTL was transferred from the wild species to multiple genetic backgrounds and its effect was maintained in all backgrounds and environments. High-resolution mapping using NILs that differ for the QTL delimited it to a recombination hot spot of 484 bp within an intron of an apoplastic invertase gene.

Several genes that have positive effect on soluble solids content were also detected in the green-fruited wild species *L. hirsutum*. Schaffer et al. (2000) constructed a line containing an introgression from this species that had an increased starch content in the immature fruit and a concomitant increased soluble solids concentration. Measurements of enzyme activities from the sucrose to starch metabolic pathway in the high starch line indicated that the activity of ADPglucose pyrophosphorylase was significantly increased in this line compared to a standard tomato line. It was subsequently shown that the *L. hirsutum* introgression carries the ADPglucose pyrophosphorylase gene *LS1* in chromosome 1. This work demonstrated that allelic variants of sugar metabolism enzymes with increased activity exist in wild tomato germ plasm, which can be utilized to increase soluble solids concentration in the cultivated tomato.

While the main sugar components in the fruit of the cultivated tomato are glucose and fructose that accumulate in approximately equal concentrations, the fruit of *L. hirsutum* is characterized by a relatively low level of these sugars and a relatively high level of sucrose (Schaffer et al., 1999). In addition, the ratio of fructose to glucose is increased to above 1.5:1 in the fruit of the wild species. The trait of high fructose-to-glucose ratio was transferred to the cultivated tomato (Schaffer et al., 1999) and was determined to be controlled by a major gene, *Fgr*, that was mapped to chromosome 4 (Levin et al., 2000).

Recently, a major fruit size QTL, *fw2.2*, contributing to the large increase of fruit size during domestication of tomato was isolated by positional cloning using a cross of the cultivated tomato and a wild *L. pennellii* accession (Frary et al., 2000). *fw2.2* is a negative regulator of cell division and is transcribed at low level at pre-anthesis in all floral organs of both the cultivated and wild plants. Transcript comparison between NILs that differ for *fw2.2* alleles indicated that its expression in the small-fruited NIL is higher than in the large-fruited NIL and that it is predominantly expressed in the placental cortex of the fruit (Cong and Tanksley, 2002). As a consequence of changing fruit size, *fw2.2* affects other components of yield such as inflorescence number and fruit number, but it does not affect total yield as the reduction in fruit size is compensated by an increase of fruit number (Nesbitt and Tanksley, 2001).

3.2 Rice

Many QTL studies were performed in rice mainly using F_2 and recombinant inbred lines (RILs) populations using crosses between different subspecies of *Oryza sativa*, e.g., *japonica* and *indica* (summarized by McCouch and Doerge, 1995; Yano and Sasaki, 1997). Quantitative trait locus studies involving exotic germ plasm were recently reported in two AB-QTL studies in which the same wild species accession (*O. rufipogon*) was crossed with different recurrent parents and the offspring were evaluated in different cultivation environments. In the first study (Xiao et al., 1988), *O. rufipogon* was crossed with an inbred (V20A) which is a parent of a widely used F1 hybrid in China and the population was grown in favorable rice growing conditions. Although the wild species was inferior in its phenotype with respect to the 12 yield associated traits, it contained numerous favorable QTL alleles (51% of the total QTLs detected in the study) for all traits studied. About one half of these QTLs had no deleterious effects on other traits. Of particular interest were two QTLs that increased grain yield in 17% and 18% without negative effects on maturity and plant height. A similar percentage (56%) of favorable wild QTLs were detected in the second study in which *O. rufipogon* was crossed to the cultivar Caiapo and the population was grown in an upland cultivation environment (Moncada et al., 2001).

While *O. rufipogon* has the AA genome similar to the cultivated rice (*O. sativa*) and is easily crossed with it, other wild species having other genomes are more distantly related to the cultivated rice. Interspecific hybrids from such crosses can be produced by embryo rescue, which allows subsequent production of introgression lines (Brar and Khush, 1997). Jena et al. (1992) produced 52 BC_2F_8 introgression lines using *O. officinalis* (CC genome) as the donor parent. Whole-genome restriction fragment length polymorphism (RFLP) analysis of these lines revealed mostly small introgressions (average size of 3 cM) in 11 out of the 12 rice chromosomes covering 14% of the donor genome. Similar limited introgressions were detected by RFLP analysis in crosses involving other distantly related wild species having the genomes FF and GG (Brar et al., 1996). Restriction fragment length polymorphism analysis of an introgression line from a cross of *O. sativa* and *O. australiensis* (EE genome) allowed tagging of genes for resistance to brown planthopper and to earliness and indicated that a limited amount of crossing over occurs between chromosomes of the two species (Ishii et al., 1994).

3.3 Wheat

Addition and substitution lines of bread wheat with several wild species were produced, mostly aiming to introgress disease-resistance genes (Fedak, 1999). Many of these lines were characterized by molecular markers such as random amplified polymorphic DNA (RAPD), restriction fragment length polymorphism (RFLP), simple sequence repeat (SSR), and fluorescence in situ hybridization (FISH) technology (summarized by Fedak, 1999; Jauhar and Chibbar, 1999). Numerous QTL studies were performed in wheat; however, most studies involved intervarietal F_2 or recombinant inbred-line populations (Gupta et al., 1999). One example of introgression and marker utilization of quality gene originated in exotic germ plasm is the high grain protein content gene identified in a wild accession of *Triticum turgidum* var. *dicoccoides*. This gene was introgressed into durum wheat and mapped to chromosome 6B by using recombinant inbred chromosome lines (Joppa et al., 1997). This gene showed an increase of protein content in the background of several durum cultivars and a positive effect on pasta quality (Kovacs et al., 1998). Furthermore, the high grain protein gene was introgressed to bread wheat showing a positive effect on grain protein content (Humphreys et al., 1988). Additional PCR-based marker saturation of the region containing the high grain protein gene should enable further transfer of this gene to durum and bread wheat cultivars (Khan et al., 2000).

3.4 Soybean

Glycine soja is considered as the wild progenitor of the cultivated soybean *G. max* and it is rich in protein concentration compared to the cultivated species. Diers et al. (1992) identified two QTLs for protein and oil concentration in *G.*

soja for which the wild species alleles were associated with increased and decreased protein and oil concentration, respectively. These QTLs were subsequently transferred to the background of *G. max* and their effect on protein and oil concentration and on other agronomic traits was reexamined (Sebolt et al., 2000). The effect of increased protein concentration of the QTL in chromosome 1 was confirmed in three out of four genetic backgrounds; however, the QTL was also associated with negative effects on oil concentration and yield. The effect of the second QTL in chromosome E was not detected in these experiments.

Recently, a QTL for increased yield originated from *G. soja* was identified and mapped in linkage group B2 (Concibido et al., 2002). Heterozygous BC_2F_1 lines containing the wild allele at the QTL had 9.3% yield increase (an average of 10 locations) compared to homozygous BC_2F_1 lines containing the cultivated parent allele. The QTL was subsequently transferred to six additional commercial soybean lines; however, only in two backgrounds the wild QTL allele showed 5% and 9% yield increase over haplotypes containing the cultivated parent allele. Despite the limited adaptability of this QTL across genetic backgrounds, this study demonstrated the potential of using exotic germ plasm for yield improvement in soybean.

3.5 Maize

Teosinte (*Zea mays* ssp. *parviglumis*), is considered as the wild progenitor of the cultivated maize. Quantitative trait locus mapping in maize×teosinte crosses revealed the presence of five chromosomal regions that control the major morphological differences between the two species (Doebley and Stec, 1993). The effect of two QTLs in chromosomes 1L and 3L controlling plant and inflorescence architecture was studied by marker-assisted transfer of maize QTL alleles into the teosinte background and teosinte QTL alleles into the maize background (Doebley et al., 1995). By coincidence of map position, similarity of phenotypes and complementation tests, it was determined that the QTL in chromosome arm 1L is the maize mutant *teosinte branched1* (*tb1*) that affects plant architecture. *tb1* acts by repressing organ growth, causing shorter branches in maize compared to teosinte. The study of the two QTLs in the background of both maize and teosinte allowed to demonstrate the importance of genetic background as a factor that influence the expression of the QTLs. While the phenotypic effect of *tb1* was stronger in the teosinte background compared to the maize background, the phenotypic effect of the QTL on chromosome arm 3L was equal in both backgrounds but reduced compared to a F_2 population of maize × teosinte. In addition, the maize alleles at both QTLs exhibited greater dominance in the maize background than in the teosinte background. Subsequent cloning of *tb1* (Doebley et al.,

1997) and sequence polymorphism analysis in its coding and regulatory regions in a sample of maize and teosinte lines revealed that while much of the diversity (39%) found in the coding region in teosinte remains in maize, only very little (3%) of the diversity within the nontranscribed region found in teosinte exits in maize (Wang et al., 1999). This implies that selection during maize domestication occurred mainly in the noncoding region of *tb1*.

Most QTL studies in maize were performed in crosses of parents representing different heterotic groups of the cultivated maize (Stuber, 1995). Only very limited information is available on marker utilization of exotic germ plasm. To determine the potential of exotic germ plasm as a reservoir of favorable alleles, Ragot et al. (1995) have analyzed grain yield related QTLs in five F_2 populations from crosses of temperate maize with Latin American accessions. Out of 70 QTL identified over all populations and traits, 12 (17%) had favorable alleles of exotic origin. Such favorable QTL alleles were detected for 10 out of 11 traits tested. Although this study used small populations, low mapping resolution, and presented no data on introgression of QTLs, it did demonstrate that exotic maize germ plasm contains favorable QTL alleles for grain yield.

Advanced backcross quantitative trait locus analysis in several maize test crosses involving exotic germ plasm is currently being evaluated as a method for hybrid maize improvement (Harjes et al., 2001; Painter et al., 1997). Preliminary results from one such cross with *Zea diploperennis* indicated that the frequency of exotic favorable QTL alleles is trait dependent (Harjes et al., 2001). While a relatively low frequency (10%) of exotic favorable alleles were detected for yield, 80–90% of the favorable QTL alleles for protein and oil content originated from the exotic parent.

Quantitative trait loci that affect the transition from vegetative to generative growth were identified in a cross of an early open-pollinated population Gaspe Flint collected from Canada with an elite maize inbred line N28 (Vladutu et al., 1999). Gaspe Flint QTL alleles at two linked regions in chromosome 8L that cause reduction in node number, height, and days to anthesis were introgressed from the donor parent to the background of N28 by 20 backcrosses and selection for earliness to create the line E20. Selection for homozygous recombinants with variable length of the introgressed 8L segment allowed to separate the effect of the two linked QTLs from each other. The two QTLs reacted additively with respect to node number, while significant deviation from additivity was observed for height and days to anthesis.

4 CONCLUDING REMARKS

The success of identification and introgression of genes with favorable effect on quantitatively inherited traits related to quality and yield in exotic germ plasm of several crop species should encourage the search for such genes for additional traits such as tolerance to abiotic stress and in additional exotic germ plasm collections. While numerous mapping studies have been reported for crosses of cultivated parents with wild species and efficient methods of identification and introgression of exotic QTLs were developed, much less is known on the efficiency of marker-assisted selection for these QTLs. Confounding effects, in particular interaction of QTLs with other genes, remain a major obstacle to predict the performance of the QTL after its transfer to various backgrounds. For each introgressed QTL, an experimental design that examines different genetic backgrounds and different combination of introgressed QTLs will be needed to determine the optimal combination of introgressed QTL and cultivated background to maximize the expression of the trait of interest.

REFERENCES

Azanza F, Kim D, Tanksley SD, Juvik JA. Genes from *Lycopersicon chmielewskii* affecting tomato quality during fruit ripening. Theor Appl Genet 1995; 91:495–504.

Bernacchi D, Beck-Bunn T, Eshed Y, Lopez J, Petiard V, Uhlig J, Zamir D, Tanksley SD. Advanced backcross QTL analysis in tomato: I. Identification of QTLs for traits of agronomic importance from *Lycopersicon hirsutum*. Theor Appl Genet 1998a; 97:381–397.

Bernacchi D, Beck-Bunn T, Emmatty D, Eshed Y, Inai S, Lopez J, Petiard V, Sayama H, Uhlig J, Zamir D, Tanksley SD. Advanced backcross QTL analysis of tomato: II. Evaluation of near-isogenic lines carrying single-donor introgressions for desirable wild QTL-alleles derived from *Lycopersicon hirsutum* and *L pimpinellifolium*. Theor Appl Genet 1998b; 97:170–180.

Bliss FA. Utilization of vegetable germplasm. Hortic Sci 1981; 16:3–6.

Brar DS, Dalmacio R, Elloran R, Aggarwal R, Angeles R, Khush GS. Gene transfer and molecular characterization of introgression from wild *Oryza* species into rice. Rice Genetics III. Manila: International Rice Research Institute, 1996:477–486.

Brar DS, Khush GS. Alien introgression in rice. Plant Mol Biol 1997; 35:35–47.

Butruille DV, Guries RP, Osborn TC. Linkage analysis of molecular markers and quantitative trait loci in populations of inbred backcross lines of *Brassica napus* L. Genetics 1999; 153:949–964.

Chetelat RT, Meglic V. Molecular mapping of chromosome segments introgressed from *Solanum lycopersicoides* into cultivated tomato (*Lycopersicon esculentum*). Theor Appl Genet 2000; 100:232–241.

Concibido VC, La Vallee B, Mclaird P, Pineda N, Meyer J, Hummel L, Yang J, Wu K, Delannay X. Introgression of a quantitative trait locus for yield from *Glycine soja* into commercial soybean cultivars. Theor Appl Genet 2002; 106:575–582.

Cong B, Tanksley SD. Expression analysis of *fw2.2* gene regulating fruit weight during tomato fruit development. Plant, Animal and Microbe Genome X Conference, San Diego, 2002:266.

Crasta OR, Francky MG, Bucholtz DB, Sharma HC, Zhang J, Wang RC, Ohm HW, Anderson JM. Identification and characterization of wheat–wheatgrass translocation lines and localization of barley yellow dwarf virus resistance. Genome 2000; 43:698–706.

DeVicente MC, Tanksley SD. QTL analysis of transgressive segregation in an interspecific tomato cross. Genetics 1993; 134:585–596.

Diers BW, Keim P, Fehr WR, Shoemaker RC. RFLP analysis of soybean seed protein and oil content. Theor Appl Genet 1992; 83:608–612.

Doebley J, Stec A. Inheritance of the morphological differences between maize and teosinte: comparison of results for two F_2 populations. Genetics 1993; 134:559–570.

Doebley J, Stec A, Gustus C. *Teosinte branched1* and the origin of maize: evidence for epistasis and evolution of dominance. Genetics 1995; 141:333–346.

Doebley J, Stec A, Hubbard L. The evolution of apical dominance in maize. Nature 1997; 386:485–488.

Eshed Y, Zamir D. Introgression line population of *Lycopersicon pennellii* in the cultivated tomato enables the identification and fine mapping of yield associated QTL. Genetics 1995; 141:1147–1162.

Eshed Y, Zamir D. Less than additive epistatic interactions of quantitative trait loci in tomato. Genetics 1996; 143:1807–1817.

Eshed Y, Gera G, Zamir D. A genome-wide search for wild-species alleles that increase horticultural yield of processing tomatoes. Theor Appl Genet 1996; 93: 877–886.

Fedak G. Molecular aids for integration of alien chromatin through wide crosses. Genome 1999; 42:584–591.

Feldman M. Alien addition lines of common wheat containing *Triticum longissimum* chromosomes. Proceeding 12th International Botanical Congress 1975; 2:506.

Frary A, Nesbitt TC, Frary A, Grandillo S, van der Knapp E, Cong B, Liu J, Meller J, Elber R, Alpert KB, Tanksley SD. *fw2.2*: A quantitative trait locus key to the evolution of tomato fruit size. Science 2000; 289:85–88.

Fridman E, Pleban T, Zamir D. A recombination hotspot delimits a wild-species quantitative trait locus for tomato sugar content to 484 bp within an invertase gene. Proc Natl Acad Sci USA 2000;97), 4718–4723.

Friebe B, Jiang J, Raupp WJ, McIntosh RA, Gill BS. Characterization of wheat-alien translocations conferring resistance to diseases and pests: current status. Euphytica 1996; 91:59–87.

Friebe B, Qi LL, Nasuda S, Zhang P, Tuleen NA, Gill BS. Development of a complete set of *Triticum aestivum–Aegilops* speltoides chromosome addition lines. Theor Appl Genet 2000; 101:51–58.

Fulton TM, Beck-Bunn T, Emmatty D, Eshed Y, Lopez J, Petiard V, Uhlig J, Zamir D, Tanksley SD. QTL analysis of an advanced backcross of *Lycopersicon peruvianum* to the cultivated tomato and comparisons with QTLs found in other wild species. Theor Appl Genet 1997; 95:881–894.

Fulton TM, Grandillo S, Beck-Bunn T, Fridman E, Frampton A, Lopez J, Petiard

V, Uhlig J, Zamir D, Tanksley SD. Advanced backcross QTL analysis of a *Lycopersicon esculentum* × *Lycopersicon parviflorum* cross. Theor Appl Genet 2000; 100:1025–1042.

Gupta PK, Varshney RK, Sharma PC, Ramesh B. Molecular markers and their application in wheat breeding. Plant Breed 1999; 118:369–390.

Harjes CE, Smith ME, Ruff TG, Tadmor Y, Tanksley SD, McCouch SR. Advanced backcross QTL analysis and introgression of perennial teosinte (*Zea diploperennis*) alleles to maize. Plant and Animal Genome VII Conference, San Diego, 2001: 260.

Hartman JB, St Clair DA. Variation for insect resistance and horticultural traits in tomato inbred backcross populations derived from *Lycopersicon pennellii*. Crop Sci 1998; 38:1501–1508.

Howell PM, Marshall DF, Lydiate DJ. Towards developing intervarietal substitution lines in *Brassica napus* using marker assisted selection. Genome 1996; 39:348–358.

Humphreys DG, Procunier JD, Mauthe W, Howes NK, Brown PD, McKenzie RIH. Marker assisted selection for high protein concentration in wheat. In: Fowler DB, ed. Wheat Protein Symposium, Saskatoon, 1998:255–258.

Ishii T, Brar DS, Multani DS, Khush GS. Molecular tagging of genes for brown planthopper resistance and earliness introgressed from *Oryza australiensis* into cultivated rice, *O sativa*. Genome 1994; 37:217–221.

Jauhar PP, Chibbar RN. Chromosome-mediated and direct gene transfers in wheat. Genome 1999; 42:570–583.

Jena KK, Khush GS, Kochert G. RFLP analysis of rice (*Oryza sativa* L) introgression lines. Theor Appl Genet 1992; 84:608–616.

Jiang J, Friebe B, Gill BS. Recent advances in alien gene transfer in wheat. Euphytica 1994; 73:199–212.

Joppa LR, Du C, Hart GE, Hareland GA. Mapping gene(s) for grain protein in tetraploid wheat (*Triticum turgidum* L) using a population of recombinant inbred chromosome lines. Crop Sci 1997; 37:1586–1589.

Kaeppler SM. Power analysis for quantitative trait locus mapping in populations derived by multiple backcrosses. Theor Appl Genet 1997; 95:618–621.

Khan IA, Procunier JD, Humphreys DG, Tranquilli G, Schlatter AR, Marcucci-Poltri S, Frohberg R, Dubcovsky J. Development of PCR-based markers for a high grain protein content gene from *Triticum turgidum* ssp. *dicoccoides* transferred to bread wheat. Crop Sci 2000; 40:518–524.

Kim SC, Rieseberg LH. Genetic architecture of species differences in annual sunflowers: implications for adaptive trait introgression. Genetics 1999; 153:965–977.

Kovacs MIP, Howes NK, Clarke JM, Leisle D. Quality characteristics of durum wheat lines deriving high protein from *Triticum dicoccoides* (6b) substitution. J Cereal Sci 1998; 27:47–51.

Kuspira J, Unrau J. Genetic analysis of certain characters in common wheat using whole chromosome substitution lines. Can J Plant Sci 1957; 37:300–326.

Law CN. The location of genetic factors affecting a quantitative character in wheat. Genetics 1966; 53:487–498.

Levin I, Gilboa N, Yeselson E, Shen S, Schaffer AA. *Fgr*, a major locus that modulates the fructose to glucose ratio in mature tomato fruit. Theor Appl Genet 2000; 100:256–262.

Lin SY, Sasaki T, Yano M. Mapping quantitative trait loci controlling seed dormancy and heading date in rice, *Oryza sativa* L, using backcross inbred lines. Theor Appl Genet 1998; 96:997–1003.

McCouch SR, Doerge RW. QTL mapping in rice. Trends Genet 1995; 11:482–487.

Millet E, Rong JK, Feldman M. Production of wild emmer recombinant substitution lines in a modern bread wheat cultivar and their use in wheat mapping. Proceedings of 9th International Wheat Genetics Symposium, Saskatoon, Canada, 1988:127–130.

Miller JC, Tanksley SD. RFLP analysis of phylogenetic relationships and genetic variation in the genus *Lycopersicon*. Theor Appl Genet 1990; 80:437–448.

Moncada P, Martinez CP, Borrero J, Chatel M, Gauch H Jr, Guimarae E, Tohme J, McCouch SR. Quantitative trait loci for yield and yield components in an *Oryza sativa* × *O rufipogon* BC2F2 population evaluated in an upland environment. Theor Appl Genet 2001; 102:41–52.

Monforte AJ, Tanksley SD. Development of a set of near isogenic and backcross recombinant inbred lines containing most of the *Lycopersicon hirsutum* genome in a *L esculentum* genetic background: a tool for gene mapping and gene discovery. Genome 2000a; 43:803–813.

Monforte AJ, Tanksley SD. Fine mapping of a quantitative trait locus (QTL) from *Lycopersicon hirsutum* chromosome 1 affecting fruit characteristics and agronomic traits: breaking linkage among QTLs affecting different traits and dissection of heterosis for yield. Theor Appl Genet 2000b; 100:471–479.

Monforte AJ, Friedman E, Zamir D, Tanksley SD. Comparison of a set of allelic QTL-NILs for chromosome 4 of tomato: deductions about natural variation and implication for germplasm utilization. Theor Appl Genet 2001; 102:572–590.

Nesbitt TC, Tanksley SD. *fw2.2* directly affects the size of developing tomato fruit with secondary effects on fruit number and photosynthate distribution. Plant Physiol 2001; 127:575–583.

Okagaki RJ, Kynast RG, Livingston SM, Russell CD, Rines HW, Phillips RL. Mapping maize sequences to chromosomes using oat–maize chromosome addition materials. Plant Physiol 2001; 125:1228–1235.

Painter S, McCouch SR, Smith ME, Tanksley SD. Discovery of desirable QTL alleles from teosinte. Proceedings of the Fifty-Second Northeastern Corn Improvement Conference, Ottawa, 1997:37–38.

Ragot M, Sisco PH, Hoisington DA, Stuber CW. Molecular-marker-mediated characterization of favorable exotic alleles at quantitative trait loci in maize. Crop Sci 1995; 35:1306–1315.

Rieseberg LH, Archer MA, Wayne RK. Transgressive segregation, adaptation and speciation. Heredity 1999; 83:363–372.

Ramsay LD, Jennings DE, Bohuon EJR, Arthur AE, Lydiate DJ, Kearsey MJ, Marshall DF. The construction of a substitution library of recombinant backcross lines in *Brassica oleracea* for the precision mapping of quantitative trait loci. Genome 1996; 39:558–567.

Schaffer AA, Petreikov M, Miron D, Fogelman M, Spiegelman M, Bnei-Moshe Z,

Shen S, Granot D, Hadas R, Dai N, Levin I, Bar M, Friedman M, Pilowsky M, Gilboa N, Chen L. Modification of carbohydrate content in developing tomato fruit. HortScience 1999; 34:12–15.

Schaffer AA, Levin I, Oguz I, Petreikov M, Cincarevsky F, Yeselson Y, Shen S, Gilboa N, Bar M. ADPglucose pyrophosphorylase activity and starch accumulation in immature tomato fruit: the effect of a *Lycopersicon hirsutum*-derived introgression encoding for the large subunit. Plant Sci 2000; 152:135–144.

Sebolt AM, Shoemaker RC, Diers BW. Analysis of a quantitative trait locus allele from wild soybean that increases seed protein concentration in soybean. Crop Sci 2000; 40:1438–1444.

Singh RJ, Hymowitz T. Soybean genetic resources and crop improvement. Genome 1999; 42:605–616.

Stuber CW. Mapping and manipulating quantitative traits in maize. Trends Genet 1995; 11:477–481.

Tanksley SD, Nelson JC. Advanced backcross QTL analysis: a method for the simultaneous discovery and transfer of valuable QTLs from unadapted germplasm into elite breeding lines. Theor Appl Genet 1996; 92:191–203.

Tanksley SD, Grandillo S, Fulton TM, Zamir D, Eshed Y, Petiard V, Lopez J, Beck-Bunn T. Advanced backcross QTL analysis in a cross between an elite processing line of tomato and its wild relative *L pimpinellifolium*. Theor Appl Genet 1996; 92:213–224.

Tanksley SD, McCouch SR. Seed banks and molecular maps: unlocking genetic potential from the wild. Science 1997; 277:1063–1066.

Thurling N. The utilization of backcrossing in improving the seed yield of spring rape (*Brassica napus*). Z Pflanzenzucht 1982; 88:43–53.

Vladutu C, McLaughlin J, Phillips RL. Fine mapping and characterization of linked quantitative trait loci involved in the transition of the maize apical meristem from vegetative to generative structures. Genetics 1999; 153:993–1007.

Wang ZW, Second G, Tanksley SD. Polymorphism and phylogenetic relationships among species in the genus *Oryza* as determined by analysis of nuclear RFLPs. Theor Appl Genet 1992; 83:565–581.

Wang RL, Stec A, Hey J, Lukens L, Doebley J. The limits of selection during maize domestication. Nature 1999; 398:236–239.

Wehrhahm C, Allard RW. The detection and measurement of the effects of individual genes in the inheritance of a quantitative character in wheat. Genetics 1965; 51:109–119.

Wissuwa M, Yano M, Ae N. Mapping QTLs for phosphorus-deficiency tolerance in rice (*Oryza sativa* L). Theor Appl Genet 1998; 97:777–783.

Xiao J, Li J, Grandillo S, Ahn SN, Yuan L, Tanksley SD, McCouch SR. Identification of trait-improving quantitative trait loci alleles from a wild rice relative, *Oryza rufipogon*. Genetics 1988; 150:899–909.

Yamamoto T, Lin H, Sasali T, Yano M. Identification of heading date quantitative trait locus *Hd6* and characterization of its epistatic interaction with *Hd2* in rice using advanced backcross progeny. Genetics 2000; 154:885–891.

Yano M, Sasaki T. Genetic and molecular dissection of quantitative traits in rice. Plant Mol Biol 1997; 35:145–153.

11

Heterosis of Yield: Molecular and Biochemical Perspectives

Charles W. Stuber
North Carolina State University
Raleigh, North Carolina, U.S.A.

1 INTRODUCTION

Heterosis (or hybrid vigor) is defined in terms of the superiority of F_1 performance over some measure of the performance of one or both of its parents. The application of this phenomenon is a multibillion-dollar enterprise manifested in hundreds of millions of hectares of field and vegetable crops throughout the world (usually in terms of yield of grain, fruit, or some aspect of biomass). The use of heterosis represents the single greatest applied achievement of the discipline of genetics, and it is one of the primary reasons for the success of the commercial maize industry as well as the success of plant breeding endeavors in many other crop and horticultural plants. However, the causal factors at the physiological, biochemical, and molecular levels are today almost as obscure as they were at the time of the conference on heterosis held in 1952 (Stuber, 1994, 1999).

Although the effects of the heterosis phenomenon have been quantified in a wide variety of plant studies, the underlying genetic basis also has not been clearly elucidated despite many attempts to do so. The genetic theories for heterosis (which have changed little since 1952) include: (1) dominance, including linked-dominant favorable factors; (2) true overdominance, which is nearly impossible to distinguish from pseudo-overdominance (i.e., closely linked loci at which alleles having dominant or partially dominant advantageous effects are in repulsion phase linkage); and (3) certain types of epistasis. Although dominance seems to have emerged as the most widely accepted hypothesis (Crow, 2000), small contributions to hybrid superiority from overdominance and some types of epistasis cannot be ruled out. Elucidation of these theories continues to provide a challenge to researchers. Biometrical approaches can only evaluate average (or net) genetic effects on heterosis; however, studies that have used biochemical, molecular, or physiological approaches appear to provide some limited insights into a better understanding of this phenomenon. Although the biochemical and/or physiological responses reported may only represent manifestations of heterosis at a level other than at the overall mature plant stage (as measured in the field), these responses should be closer to the gene level and may ultimately help to more clearly elucidate the genetic basis of heterosis. A better understanding of the biochemical and molecular mechanisms underlying this phenomenon should enhance the ability of plant breeders to generate new genotypes which may be used directly as hybrids or form the basis for future selection programs.

2   GENETIC DIVERSITY AND HETEROSIS

For most plant species, the successful utilization of heterosis requires the development and testing of inbred lines for use as parents of superior hybrids. This is a very costly process and requires many years of traditional breeding activity. Much of this effort is devoted to field evaluations of newly generated lines in various single-cross combinations to identify those lines with superior combining ability. In an effort to make these breeding processes more efficient, a great many investigations have been conducted in the search for reliable methods for: (1) identifying source populations from which to generate new lines that will predictably produce superior hybrids and (2) predicting hybrid performance without generating and testing hundreds or thousands of single-cross combinations.

These desires to increase plant breeding efficiency have stimulated numerous studies by plant breeders and geneticists that have focused on the development of valid measures of genetic diversity (or genetic distance) among parental lines or cultivars that could be used to reliably predict performance of F_1 hybrids generated from specific line or cultivar crosses. Early studies in maize, beginning in about 1970, attempted to use marker diversity (initially isozymes, and later RFLPs [restriction fragment length polymorphisms]) of inbred lines as a measure of genetic diversity to predict hybrid performance from crosses among untested lines. The number of markers used in these studies varied from fewer than 11 isozymes to 230 RFLPs. These investigations showed that genetic distances based on marker data agreed well with pedigree data for assigning lines to heterotic groups. However, in those studies that included field evaluations, it was concluded that isozyme and RFLP genotypic data were of limited usefulness for predicting the heterotic performance in crosses of unrelated maize inbred lines (Stuber, 1992, 1994, 1998, 1999, 2001; Tsaftaris, 1995). Similar results have been reported in earlier studies in other crops such as rapeseed, rice, barley, wheat, peanut, dry bean, and tomato (see review by Stuber, 1994).

Varying results have been reported in a number of more recent studies that utilized DNA-based molecular markers including SSRs (simple sequence repeats), RAPDs (random amplifications of polymorphic DNA), AFLPs (amplified fragment-length polymorphism), and RFLPs to measure genetic distances which were then evaluated for their possible associations with heterosis. For example, studies in rice involving both intra- and inter-subspecific crosses (indica lines and japonica lines) showed little correlation between marker (RFLP, SSR, and RAPD) heterozygosity and heterosis in inter-subspecific crosses. However, the results showed significant positive correlations in intra-subspecific crosses (Xiao et al., 1996; Zhao et al., 1999). Studies made within indica and within japonica sets of rice lines by Zhang et al. (1996) showed low to intermediate correlations between hybrid performance and heterozygosity of RFLP and SSR markers. Their study included a broad spectrum of the cultivated rice germ plasm pool, and they concluded that the relationship between marker heterozygosity and heterosis is variable and depends on the genetic materials used and the diversity of the germ plasm. In a half-diallel study of indica rice, Liu and Wu (1998) found that SSR marker heterozygosity of each parental pair was significantly associated with general combining ability, but not with specific combining ability. They concluded that neither genetic diversity nor heterozygosity was a good indicator for predicting heterosis. They also reported that four favorable alleles and six favorable heterogenic patterns significantly contributed to heterosis of grain yield in the hybrids studied, whereas six unfavorable alleles and six unfavorable heterogenic patterns significantly reduced heterosis. In another diallel study of eight elite rice lines, differential gene expression analyses based on RFLP and SSR data showed little consistency, and both negative and positive correlations of heterosis with marker heterozygosity were found (Xiong et al., 1998).

A study of hybrids generated from 20 wheat lines showed that genetic distance based on RAPD markers was not significantly correlated with hybrid yield performance (Liu et al., 1999). They did report, however, that it was possible to differentiate wheat lines with different performances using RAPD markers, and that the classification of parents from these markers was of predictive value for developing superior hybrids. In another wheat study involving 722 hybrids, genetic distance based on RFLPs was not correlated with heterosis in any of the years tested (Barbosa-Neto et al., 1996).

In a study of 123 single-cross sunflower hybrids, correlations of genetic distances (estimated from AFLP fingerprints) with hybrid seed yield were estimated. Although the correlations were significant, it was concluded that the genetic distance was a poor predictor of hybrid seed yield (Cheres et al., 2000). The correlation of genetic distance (based on RFLP markers) with heterosis for yield in a study of 120 F_2 cotton hybrids was only 0.08 (Meredith and Brown, 1998), indicating that genetic distance was not a good predictor of heterosis in this crop. A study of 24 soybean hybrids showed that there was no association between genetic distance estimated by RFLP markers and seed yield heterosis (Cerna et al., 1997). Although isozyme markers did show a relationship with heterosis for yield, the number of assayable isozyme loci in soybean is too small to be useful in a breeding program. A recent study in maize (Ajmone-Marsan et al., 1998) agreed with earlier results in maize and showed that correlations of genetic distance (based on molecular markers) with single-cross performance for grain yield were positive but too small to be of predictive value.

3   MOLECULAR, BIOCHEMICAL, AND PHYSIOLOGICAL CONSIDERATIONS

3.1   Genome Expression at Enzyme and Protein Levels

Although rare, examples of heterosis associated with a single genetic locus have been reported. For example, Schwartz and Laughner (1969) have demonstrated single gene heterosis for alcohol dehydrogenase activity in maize. For maize glutamic dehydrogenase, Pryor (1974) found a single molecular form in F_1 hybrids. More recently, Hall and Wills (1987) reported an example of conditional overdominance at an alcohol dehydrogenase locus in yeast. They suggested that their finding could be considered equivalent to that of sickle cell hemoglobin in man and shows promise as a tool for investigating the physiological basis for overdominance.

These studies demonstrate a hybrid response at the individual enzyme level; however, the overall plant response involves the interactions of many enzymes associated with numerous genetic loci. For example, Hageman et al. (1967) made a thorough study of the role of nitrate reductase in the heterotic response of maize hybrids and found considerable variation in the activity of this enzyme among the inbreds tested. In spite of the wide range in activity, they were unable to causally associate nitrate reductase activity with heterosis. Results such as these indicate the difficulty in the dissection of the individual components of heterosis, which, undoubtedly, form a complex interactive system.

Srivastava (1981) strongly suggests that heterotic hybrids are endowed with a more balanced metabolism than the inbred parents. He also indicates that the metabolic advantage of the gene products at the cellular or organelle level could be viewed as being produced by nonallelic complementation rather than by intergenic complementation pertaining to only one gene locus. Demonstrations of multimeric *hybrid enzymes* or *isozyme spectrum* due to nonallelic or intergenomic interaction provide evidence for a buffering mechanism at the subcellular level in the hybrids.

As stated by de Vienne and Rodolphe (1985), protein complementation (or interallelic complementation) represents an extreme expression of molecular interactions since only the heteropolymeric form is active. However, the complementation is not always positive. The presence in an oligomer of both mutant and wild (normal) protomers can result in complete inactivation of the enzyme. Molecular interactions between subunits that are all active have been reported in the literature. For example, Scandalios et al. (1972) studied maize catalase, which is a tetrameric enzyme, and showed that different combinations of alleles from two loci resulted in isozymes which differed in biochemical properties, such as specific activity, temperature sensitivity, photosensitivity, inhibitor sensitivity, etc. In most instances, heterotetramers generated by either intragenic or intergenic hybridization exhibit clear interaction phenomena, sometimes including molecular heterosis. It is of interest to note, however, that electrophoretic mobility of isoenzymes generally shows additivity, that is, a heterodimer usually migrates exactly between the two homodimeric forms. Tsaftaris (1995) points out that molecular interactions among subunits of polymeric proteins occur not only for enzymatic proteins, but also (and more importantly) for transcription factor proteins and regulatory proteins, in general.

As stated earlier, the large repository of isozyme and molecular markers distributed throughout the genome may be useful for studying the correlation between parental diversity and F_1 performance; however, these markers provide no information about the genes and the molecular and physiological mechanisms involved in the phenotypic expression of complexly inherited characters, such as yield or yield heterosis (Tsaftaris and Polidoros, 2000). Only indirectly, through linkage to relevant quantitative trait loci (QTL), such as those associated with the expression of yield or yield heterosis, can these markers provide information on the inheritance of quantitative characters.

It has been speculated that the molecular foundations of phenotypic diversity might reside in the variability of gene expression, which would result in variation in different biochemical or physiological processes, and ultimately would be related to hybrid vigor. Variation in gene expression can be assessed through RNA amount polymorphisms (RAP) and through protein-amount polymorphisms (PAP). In a study of both highly heterotic and lowly heterotic maize hybrids, Tsaftaris and Polidoros (1993) found that the polymorphisms of individual RNA amounts were similar to those of protein-amount polymorphisms.

Leonardi et al. (1987, 1988, 1991) correlated protein-amount polymorphisms (PAP) with field performances of a number of single-cross maize hybrids. Significant correlations between PAP and hybrid values led these researchers to conclude that the loci controlling protein amounts would themselves be QTLs and, particularly those with multiple effects, directly affected the expression of hybrid vigor. In their discussion, they suggested that the greater the two parental genotypes differ in their profile of gene expression during development, the greater would be the possibilities of interactions at the molecular level. These interactions would be expressed as nonadditive, and heterosis could be considered as a measure of these interactions. In addition, it was suggested that protein-amount polymorphisms, as well as the frequency of nonadditivity, might allow for the definition of predictors of heterosis based on molecular indicators.

Conclusions from the data on PAP, RAP, and enzyme amounts indicate that quantitative variation controlling the expression of certain loci may be important in vigor manifestation, and they also underline the significance of regulatory mechanisms involved in the quantitative modulation of gene expression in manifesting vigor (Tsaftaris and Polidoros, 2000). It will be necessary then for the regulatory mechanisms responsible for this extended variation in the amounts of mRNAs or proteins to be analyzed, their respective modes of action to be defined, as well as their relationships to QTLs to be found through gene-mapping endeavors.

3.2 DNA Methylation

DNA methylation can be considered as a genome-wide regulatory mechanism that is involved in the expression of many genes important in the manifestation of heterosis (Tsaftaris and Kafka, 1998; Tsaftaris et al., 1999). Hepburn et al. (1987) studied DNA methylation in plants and reported several examples in the literature that provided evidence for a relationship between methylation and suppression of gene activity. Extrapolations of these observations (Hepburn, personal communication) might implicate the degree of DNA methylation as a molecular regulator of heterosis. This hypothesis provides for a gradual accumulation of methylation during selfing, which is then released and/or repatterned when the selfed lines are crossed to generate hybrids. Phillips presented a model for heterosis (that remains to be tested) invoking a relationship between gene expression and methylation in which methylation might turn on genes previously inactive, thus leading to creation of recessive and dominant genes, respectively (presented at the Workshop on Molecular Basis of Heterosis in Plants, March 19–20, 1991, University of Minnesota).

Tsaftaris et al. (1997) implicated cytosine methylation in genomic DNA as a mechanism for controlling gene regulation in maize. They reported that DNA methylation was found to be genotype, tissue, and developmental stage-specific. They also found that hybrids were less methylated than inbreds, which is consistent with the hypothesis proposed by Hepburn (see previous paragraph), and that heterotic hybrids are less methylated than related nonheterotic hybrids. In addition, their data showed that inbreds varied in their methylation status and improved lines were less methylated than older low yielding lines. These findings support the hypothesis made by researchers, such as Hepburn et al. (1987), that systematic selfing for the development of inbreds, with emphasis only on combining ability of the inbreds (that concurrently leads to severe inbreeding depression), is also leading to a gradual accumulation of more methylated sites, which then could be released and/or repatterned when the selfed lines are crossed to generate hybrids. Tsaftaris and Polidoros (2000) suggest that shifting the emphasis to line performance per se during selection could moderate line inbreeding depression. Thus, while approaching homozygosity, degenerative alleles (possibly methylated and/or even mutated genes) could be selected out during the development of inbred lines, thus leading to lower methylation levels of the resulting lines.

Tsaftaris et al. (1997) also reported that stressful environments (such as high-density planting) increased the level of methylation, which, in turn, suppressed genetic expression throughout the genome. F_1 hybrids, however, were found to be more resistant to such a density-induced increase in methylation than their parental inbreds. They suggested that an epigenetic DNA modification such as this (which may alter gene expression) could be heritable because such a change can occur in a plant cell that may become a gamete.

3.3 Metabolic Balance and Physiological Bottlenecks

Several researchers have approached the biochemical/physiological explanations of heterosis from a more complex view. Hageman et al. (1967) proposed the metabolic balance concept as a determinant of the hybrid response and suggested that traits such as plant growth and grain yield reflect the end results of a series of biochemical reactions, each of which is governed by one or more specific enzymes. According to Hageman and Lambert (1988) and Schrader (1985), the metabolic concept of heterosis is dependent upon the coordination of all reactions and systems for efficient growth under a given environment. The primary elements of this concept are the following: (1) inbred lines in a crop such as maize tend to have unbalanced metabolic systems with some enzymes controlled at high levels, some at medium levels, and some genetically limited to low or ineffective levels of activity, (2) highly homozygous lines of maize may have some important enzyme reactions that are severely limiting metabolism, and (3) the specific limitations probably differ among individual lines, which may be overcome in a heterozygote by the appropriate choice of complementary inbred parents, resulting in a better-balanced metabolic system in the hybrid.

The metabolic balance concept is analogous to the concept of limiting factors or physiological bottlenecks (as outlined by Mangelsdorf, 1952), in which the bottlenecks may reside at different loci (and hence in different metabolic pathways) in different inbreds. Schrader (1985) outlined a simplified model system for two inbreds (I and II) that differ for units of enzyme activity for two of three metabolic pathways—A, B, and C (see the three columns under "Relative enzyme units" in Table 1). With Schrader's model, the hybrid will exhibit intermediate enzyme activities and will have a more balanced metabolism than either of the inbred parents. A limitation or bottleneck in pathway A in inbred I and a limitation or bottleneck in pathway B in inbred II are overcome in the F_1 hybrid (Rhodes et al., 1992). I have added three columns under "Relative effect" to Schrader's model (Table 1) showing hypothetical results when the combined effects of these three enzyme pathways are either: (1) constrained by the pathway with the lowest "Relative enzyme units" under column headed "Lowest,"(2) functioning in an additive fashion, column headed "Additive," or (3) functioning in a multiplicative fashion, column headed "Multiplic."

In the Heterosis Conference held in 1952, Mangelsdorf (1952) suggested several postulates that are relevant to the metabolic concept of heterosis and genotype by environment interactions as follows:

1. At each moment throughout its life, the physiological processes of even the most vigorous organism are limited to their prevailing rates by bottlenecks or limiting factors.

Table 1 Hypothetical Model for Two Inbreds and a Heterotic F_1

Genotype	Relative enzyme units[a]			Relative effect		
	Pathway A	Pathway B	Pathway C	Lowest	Additive	Multiplic
Inbred I	1	3	2	1	6	6
Inbred II	3	1	2	1	6	6
F_1 hybrid	2	2	2	2	6	8

[a] After Schrader, 1985.

2. The physiological bottleneck at any given moment results from the interaction of a particular locus (referred to as the bottleneck locus) with the remainder of the genotype and the environment at that moment.

3. The physiological bottleneck may be ameliorated, or even removed, by correcting the particular feature of the environment contributing to the bottleneck.

4. The bottleneck may be ameliorated by substituting a more effective allele at the bottleneck locus, provided that such an allele is available.

5. The amelioration, or removal of a bottleneck, by improving the environment or by substituting a better allele at the bottleneck locus, will permit an increase in the rate of the essential physiological process. This increase may be either small or large, depending upon the point at which the next ensuing bottleneck becomes limiting. The substitution of a more efficient allele at a bottleneck locus in a certain genotype under a particular environment may result in a large gain. The substitution of the same allele in a different genotype or under another environment may result in little or no gain. Difficulties would be encountered in analyzing the inheritance of genes affecting yield and other quantitative characters, which are subjected to the influence of a varied and fluctuating array of genetic-environmental bottlenecks.

6. The differences between the weakest inbred and the most vigorous hybrid are merely those of degree. Each represents an integration of the many genetic-environmental bottleneck effects under which each is subjected. The weak inbred has been throttled down by one or more bottlenecks to a low level. The superior hybrid is able to grow much greater, even attaining what might be termed extreme vigor. However, both the weak inbred and the vigorous hybrid have, throughout their lives, been restricted to their respective levels by their genetic-environmental bottlenecks.

7. Success in developing higher-yielding genotypes depends largely upon the ability of the breeder to substitute more effective alleles at the bottleneck loci and to accomplish this without establishing new and equally serious bottlenecks at other loci.

As Rhodes et al. (1992) state, this concept postulated by Mangelsdorf (1952) predicts that for each inbred, there is likely to be at least one major bottleneck locus and potentially a series of less serious (minor) bottleneck loci, the latter of which will manifest themselves only when the bottleneck at the major locus is ameliorated. For a hybrid to exhibit heterosis, it is likely that the two inbred parents must have bottlenecks at different loci, either in different metabolic pathways or at different steps in the same metabolic pathway. Thus analyses of segregating populations generated from a vigorous hybrid would be expected to reveal at least two major independently segregating loci determining productivity, with perhaps a series of minor loci contributing relatively smaller effects. This concept is not inconsistent with the results reported by Stuber et al. (1992) in which isozyme and RFLP markers were used to identify genetic factors affecting heterosis in the maize cross, B73×Mo17.

If the goal of the breeder is to identify, map, and then manipulate these potential bottleneck loci in each inbred (Mangelsdorf, 1952), then DNA-based marker-facilitated technology should provide the tool to accomplish this objective. Obviously, map location does not necessarily define function. The more difficult challenge to the plant physiologist or molecular biologist is to define the precise imperfections in metabolism that these loci may determine, that is, to equate function to location. The use of markers derived from candidate genes with known functions for quantitative trait locus (QTL) mapping may be helpful in this endeavor. As Rhodes et al. (1992) state, "Only when both location and function have been elucidated will the question of dominance versus overdominance likely be resolved for each locus contributing to heterosis."

Although the concept of metabolic balance resembles the dominance hypothesis, the heterozygote does not necessarily have to exhibit a character (enzyme level) that is equal to (dominance), approaches (partial dominance), or which is greater than (overdominance) either of the parents in order to explain heterosis. Schrader (1985) reported that the heterozygote usually exhibits intermediate enzyme levels with respect to the inbred parents. Among maize inbreds, two- to threefold differences in the levels of enzymes, which catalyze important biochemical reactions, are not uncommon. Generally, the observed mean enzyme level of an F_1 maize hybrid closely approximates the mean of the two parents (Schrader, 1985). Therefore additivity usually characterizes the enzyme level response found.

If additivity is the rule rather than the exception for the character enzyme level, does this imply that genetic variability for this character is for the most part irrelevant to heterosis? Kacser and Burns (1981) attempt to resolve this apparent dilemma by considering the relationship between the character "enzyme level" and the character "metabolic flux." They describe flux as a function of environmental parameters, which are assumed to be constant, and a composite of genetically determined enzyme parameters that are proportional to enzyme concentration and are modifiable by mutation. It is so defined that if any one activity is reduced to zero, the flux will be reduced to zero, resulting in a metabolic block. With only a single enzyme contributing to the character metabolic flux, a 50% reduction in activity will reduce the flux by one-half, and the character will show additivity. However, as the number of enzymes of a pathway increases, a 50% reduction in any one enzyme will have a progressively smaller effect on flux. Therefore the character metabolic flux will show increasing levels of dominance depending upon the number of enzymes in the pathway. This concept reveals that alleles that affect the character enzyme level in a fundamentally additive fashion can effect the character metabolic flux in an additive, partially dominant, or dominant fashion depending on the number of enzymes in the pathway. Although the effects of genes governing traits such as productivity may be additive at the enzyme level, the consequence of nonlinearity between enzyme level and metabolic flux in complex metabolic pathways may therefore appear as a dominant effect (Kacser and Burns, 1981).

An obvious implication of these concepts for plant physiologists is that physiological selection criteria (such as metabolic flux measurements) may not necessarily provide a more sensitive indicator of heterosis than net productivity per se because the latter integrates flux over time. Plant breeders have been successful in evaluating and exploiting heterosis by comparing the relative productivity of heterozygotes with homozygotes. This success arises from the fact that productivity is an exponential and time-dependent function of the flux via the prevailing bottleneck in metabolism, wherever this may be located. The poor progress in defining the causal factors of heterosis at the biochemical level stems from the fact that bottlenecks can potentially reside at any one of a great many loci essential for growth, and that the immediate effects on physiological processes of allele substitution at these loci are very difficult to evaluate experimentally (Rhodes et al., 1992).

3.4 Complementary Effects of Parents

Sinha and Khanna (1975) reviewed the physiological, biochemical, and genetic concepts of heterosis that prevailed more than 25 years ago and suggested that processes such as germination, respiration, and photosynthesis can be split into components that show Mendelian dominance in F_1 hybrids. In heterotic hybrids, the parents usually bring together contrasting but complementary characters that could have multiplicative effects. This then leads to the types of gene interactions that result in complementation of associated physiological and biochemical processes. For example, during germination, complementation could occur with regard to water absorption, amylase activity, phytase activity, respiratory activity, and the number of leaf primordia. In photosynthesis, complementation could involve carboxylases and cyclic and noncyclic photophosphorylation. Sinha and Khanna (1975) further suggested that interaction of genes at the processes level, such as those involved in the components of photosynthesis, is complementary and provides only a limited advantage over the parents. However, the end products of complementary effects interact multiplicatively. It is this interaction that eventually makes the F_1 hybrids outstandingly superior to their parents.

In a QTL fine-mapping study using materials generated from the cross of maize inbreds B73 and Mo17, Graham et al. (1997) showed that a major QTL on chromosome 5, which appeared to act in an overdominant fashion, could be partitioned into at least two smaller QTLs. Dominant gene action was observed at each of these two QTLs, with the favorable dominant allele from B73 in one QTL and the favorable dominant allele from Mo17 in the other QTL. Thus what had appeared to be a single overdominant locus was partitioned into two loci linked in repulsion phase with dominant gene action at each. Further interpretations of the data indicated that the longer ear of Mo17 with more kernels per row positively complemented the shorter B73 ear.

An investigation of a sorghum cross by Gibson and Schertz (1977) provided an example of a hybrid that displayed the complementary effects of its parents. The hybrid has a high crop growth rate during grain filling (characteristic of its female parent) and a highly effective conversion of dry matter to grain (as did its male parent). In a study of cotton, Wells and Meredith (1986) attempted to better understand physiological alterations related to heterosis by monitoring growth and partitioning of dry matter in four upland cotton cultivars and their F_1 progenies. The heterotic growth pattern observed in this study, although related to larger leaf-area production, did not appear to be the result of greater proportional leaf-area partitioning. Instead, the final increase of leaf area in the hybrids (when compared with the parents) appeared to be the result of early growth responses. In a study of three cotton cultivars (different from the above study) and their F_1 progenies, Wells et al. (1988) examined the relationship between leaf area and canopy photosynthesis. Their data showed that hybrids produced bigger plants that intercepted more light than their respective parents, and hence have increased photosynthetic rates on a per-plant basis. They found that during early growth stages, apparent photosynthesis (based on CO_2 exchange rates) was associated significantly with leaf area per plant. However, later in growth, this association weakened as mutual shading of the leaves occurred. The conclusions were similar to those above in that much of the heterotic response appeared to be related to early seedling growth and early leaf-area development.

Leaf areas of four maize inbreds (CM7, CM49, CL3, and CG8) and four hybrids (CM49 × CL3, CM49 × CG8, CM7 × CM49, and CM7 × CG8) were analyzed by Pavlikova and Rood (1987) in terms of numbers of leaves and lengths and widths of individual leaf blades. The lengths and widths were further analyzed in terms of numbers, widths, and lengths of long abaxial epidermal cells. Their results showed incomplete dominance for increasing leaf number and overdominance for increasing individual leaf blade area. They concluded that the overdominance for increasing individual leaf area resulted from the complementation of incomplete dominance for increasing cell length and width and overdominance for cell number in length and width. They concluded further that both complementation of dominance effects and overdominance for increased mitotic activity were involved in heterosis for leaf area in the maize hybrids studied. Pavlikova and Rood (1987) also suggested that this type of leaf-area analysis may be useful to breeders attempting to increase the leaf area of maize because it reduces the complex, polygenic trait of leaf area into easily measured components that display less complex inheritance patterns.

3.5 Complementation at the Organelle Level

It has been noted that heterosis in plants often results in enhanced conservation of energy Srivastava (1981). If this phenomenon is found to prevail in mitochondria and chloroplasts, it could have importance for increasing crop yield through the manipulation of organelle genomes. The role of mitochondrial heterosis and complementation as an explanation for the basic mechanism of heterosis was first reported by McDaniel and Sarkissian (1966). They reported that mitochondria isolated from seedlings of maize hybrids and their inbred parents showed differential efficiencies of oxidative phosphorylation in the synthesis of ATP. Mitochondria from a nonheterotic hybrid showed the same phosphorylative efficiency as those of its inbred parents. This was interpreted as the classical expression of heterosis, at the organelle level, judged by the efficiency of enzyme catalysis; and the phenomenon was discussed extensively in a review paper by McDaniel (1986). Attempts to corroborate the mitochondrial complementation approach in other laboratories and in other plant species or to use this phenomenon in plant breeding have met with minimal success.

Chloroplast heterosis and complementation associated with several parameters have been observed in several economically important crops (Srivastava, 1981). The term "chloroplast complementation" is generally used to indicate that greater activity is observed from a 1:1 mixture of isolated parental chloroplasts when compared with the midparental value. Higher photosynthetic rates in isolated chloroplasts of hybrids of several crop species have been observed in the seedling stages. Srivastava (1981) further states that, "It is likely that heterotic hybrids are endowed with superior systems of chloroplasts and mitochondria, and such superiority is provided by genomic and intergenomic interactions."

3.6 Phytohormones and Heterosis

The phytohormone, gibberellin (GA), has been investigated in several crop plants and in trees with respect to its role in the regulation of hybrid vigor. In a study of four commercially important maize inbreds and their 12 F_1 hybrids, Rood et al. (1988b, 1990) found that the hybrids contained higher concentrations of endogenous GAs than their parental inbreds. Accelerated shoot growth of the inbreds generated by the addition of exogenous GA_3 indicated that a deficiency of endogenous GA limited the growth of the inbreds, which could be a factor in inbreeding depression. Conversely, they suggested that the increased endogenous concentration of GA in the hybrids, relative to the inbreds, could provide a phytohormonal basis for the heterosis of shoot growth in maize.

Rood et al. (1992) also studied the association of GA with shoot growth in four inbred lines and two F_1 hybrids of sorghum. The results were similar to what they had found in maize. At some stages, the sorghum inbreds appeared to be limited in growth due to a partial deficiency of endogenous GA. Positive responses to exogenous GA_3 application suggested that the GA deficiency was a limitation to their growth. Thus Rood et al. (1992) concluded that GAs are involved in the heterosis for shoot growth in sorghum. In another investigation, Rood (1995) found that the rate of GA metabolism was positively correlated with growth in sorghum, being faster in a rapid-growing hybrid and slower in slower-growing parental inbreds.

Rood (1995) concluded that in both maize and sorghum, and more broadly in various field crop plants and trees, high GA content and rapid GA metabolism are associated with, and probably at least partially responsible for, rapid shoot growth and, particularly, rapid height growth. He also concluded that hybrids are particularly fast-growing and the correlation between GA content, GA metabolism, and growth rate in inbreds vs. hybrids further suggests that heterosis for rapid growth rate is partially mediated by GA.

3.7 Effects of Environmental Stress on Heterosis

Environmental variability may affect the relationships of specific biochemical, physiological, or molecular components of heterosis. However, in a study of the influence of temperature on heterosis for several maize seedling growth traits, Rood et al. (1988a) found that the level of heterosis for these traits could not be explained simply by the ability of a hybrid to better tolerate cool temperatures. They concluded that hybrids derived from a group of four elite inbred lines displayed heterosis similarly under either favorable or cool temperature conditions.

In an effort to determine how QTLs might behave under several environmental stress variables, we conducted a major study in which QTLs affecting heterosis for grain yield and several other traits were mapped under eight combinations of stress- and nonstress-related variables (Stuber, 1999; LeDeaux and Stuber, 1997). The variables were low (drought) and optimum moisture, low (deficient) and optimum nitrogen, and low (about 36,000 plants ha^{-1}) and high (about 72,000 plants ha^{-1}) plant density. For the two traits, grain yield and ear height, most of the marker loci were found to be linked to QTLs that affect at least one of these two traits. However, only a few were found to show significant interaction with environmental stresses. Although the yield level of the least stressful combination of variables was 10 times greater that that of the most stressful combination, the QTL mapping pattern differed very little under the two regimes. Veldboom and Lee (1996) conducted a maize mapping study in two different environments and also found very few differences in the QTL mapping patterns. However, they did not impose the same severity of stress that we did in the studies conducted in North Carolina. Results from these studies indicate that lines and hybrids bred for superior performance in nonstress environments should also perform well in stress environments, at least for the stresses used in these investigations.

Additional conclusions from the study reported above (LeDeaux and Stuber, 1997) included the following: (1) most QTLs for grain yield (in the materials from the B73 × Mo17 derived populations) act dominantly or (pseudo)overdominantly, but most QTLs for ear height act additively or show partial dominance, and (2) in a separate analysis that focused on locus × locus interaction effects, several pairs of unlinked QTLs were found to significantly affect yield in both of the backcross populations evaluated, indicating that epistasis is a factor contributing to heterosis for grain yield in maize.

Blum et al. (1990) studied two grain sorghum hybrids and their parental lines in the greenhouse under a gradient of ambient temperatures and two water regimes (well irrigated and drought up to heading). Significant heterosis was found for biomass, grain yield per plant, and grain number per panicle. No heterosis occurred for harvest index, indicating that heterosis in grain yield was due to heterosis in biomass. Neither growth duration nor leaf area could explain heterosis in biomass. When carbon exchange rate (CER) data were subjected to a stability analysis, the two hybrids had greater CER than their respective parents, especially under conditions favoring high CER. When extreme stress conditions developed, the hybrid's performance depended on its genetic background more than on heterosis.

In another sorghum study, Blum (1989) measured gas exchange of detached turgid leaves of four hybrids and their parents as leaf temperatures rose steadily from 32°C to 43°C over a 4-hr period. Three of the hybrids showed heterosis for carbon exchange rate (CER) and over a greater temperature range than their respective parents. However, Blum does not suggest that CER is the sole explanation for grain yield heterosis in sorghum.

In additional studies, Blum et al. (1990) reported that sorghum hybrids produced more dry matter in proportion to their increase in leaf area. This is supported by the finding that sorghum hybrids had the same ratio of dry matter to leaf area as their parents. Heterosis must come from a greater or a more efficient source of assimilates in the hybrid in order to account for a larger plant with a greater leaf area. The heterotic yield component in this case was grain number per panicle and not kernel weight. The greater grain yield in the hybrids was achieved with about the same harvest index as the parents, that is, hybrids did not excel in relative dry matter partitioning to grain. Irrespective of the moisture regime, hybrids produced more grain in proportion to their greater biomass.

4 CONCLUSIONS

Because of the complexity of the biochemical, molecular, and genetic factors (and their respective interactions) influencing the phenomenon of heterosis, no simple conclusions are apparent. There is considerable documentation suggesting that much of the phenotypic variation associated with quantitative traits can be attributed to variation in levels of gene expression. Thus it would logically follow that the variation in the expression of heterosis must be associated with those genetic factors (perhaps, QTLs) that function in the regulation of gene expression. When variation of gene regulation is coupled with the exposure of the targeted organism to a wide array of environmental effects, the complexity of the understanding of the heterosis phenomenon is greatly magnified. Obviously, there are many opportunities for further research.

From a plant breeder's standpoint, it can be concluded that selection criteria based on molecular, biochemical, or physiological measurements will not necessarily provide a more sensitive indicator of heterosis than net productivity per se (whether it be for grain yield or some other quantitatively inherited trait) because net productivity integrates such things as metabolic flux over time. Plant breeders will continue to capitalize on the heterosis phenomenon, in spite of the relative lack of information regarding the biochemical, physiological, molecular, and genetic factors involved in the expression of the phenomenon. As new technology at all levels (molecular, cellular, tissue, and organismal) unfolds, exciting new methodologies should become available to make the exploitation of the heterosis concept more efficient and more precise as well as to greatly enhance the efficiency of plant breeders (Stuber et al., 1999).

REFERENCES

Ajmone-Marsan P, Castiglioni P, Fusari F, Kuiper M, Motto M. Genetic diversity and its relationship to hybrid performance in maize as revealed by RFLP and AFLP markers. Theor Appl Genet 1998; 96:219–227.

Barbosa-Neto JF, Sorrells ME, Cisar G. Prediction of heterosis in wheat using coefficient of parentage and RFLP-based estimates of genetic relationship. Genome 1996; 39:1142–1149.

Blum A. The temperature response of gas exchange in sorghum leaves and the effect of heterosis. J Exp Bot 1989; 40:453–460.

Blum A, Ramaiah S, Kanemasu ET, Paulsen GM. The physiology of heterosis in sorghum with respect to environmental stress. Ann Bot 1990; 65:149–158.

Cerna FJ, Cianzio SR, Rafalski A, Tingey S, Dyer D. Relationship between seed yield heterosis and molecular heterozygosity in soybean. Theor Appl Genet 1997; 95:460–467.

Cheres MT, Miller JF, Crane JM, Knapp SJ. Genetic distance as a predictor of heterosis and hybrid performance within and between heterotic groups in sunflower. Theor Appl Genet 2000; 100:889–894.

Crow JF. The rise and fall of overdominance. Plant Breed Rev 2000; 17:225–257.

de Vienne D, Rodolphe F. Biochemical and genetic properties of oligomeric structures: a general approach. J Theor Biol 1985; 116:527–568.

Gibson PT, Schertz KF. Growth analysis of a sorghum hybrid and its parents. Crop Sci 1977; 17:387–391.

Graham GI, Wolff DW, Stuber CW. Characterization of a yield quantitative trait locus on chromosome five of maize by fine mapping. Crop Sci 1997; 37:1601–1610.

Hageman RH, Lambert RJ. The use of physiological traits for corn improvement. In: Sprague GF, Dudley JW, eds. Corn and Corn Improvement. 3d ed. Madison, WI: Am Soc Agronomy, 1988:431–461.

Hageman RH, Leng ER, Dudley JW. A biochemical approach to corn breeding. Adv Agron 1967; 19:45–86.

Hall JG, Wills C. Conditional overdominance at an alcohol dehydrogenase locus in yeast. Genetics 1987; 117:421–427.

Hepburn AG, Belanger FC, Mattheis JR. DNA methylation in plants. Dev Genet 1987; 8:475–493.

Kacser H, Burns JA. The molecular basis of dominance. Genetics 1981; 97:639–666.

LeDeaux JR, Stuber CW. Mapping heterosis QTLs in maize grown under various stress conditions. Symposium on the Genetics and Exploitation of Heterosis in Crops [Abstr]. Mexico City: CIMMYT, 1997.

Leonardi A, Damerval C, de Vienne D. Inheritance of protein amounts: comparison of two-dimensional electrophoresis patterns of leaf sheaths of two maize lines (*Zea mays* L) and their hybrids. Genet Res 1987; 50:1–5.

Leonardi A, Damerval C, de Vienne D. Organ-specific variability and inheritance of maize proteins revealed by two-dimensional electrophoresis. Genet Res 1988; 52: 97–103.

Leonardi A, Damerval C, Herbert Y, Gallais A, de Vienne D. Association of protein amount polymorphism (PAP) among maize lines with performances of their hybrids. Theor Appl Genet 1991; 82:552–560.

Liu XC, Wu JL. SSR heterogenic patterns of parents for marking and predicting heterosis in rice breeding. Mol Breed 1998; 4:263–268.

Liu ZQ, Pei Y, Pu ZJ. Relationship between hybrid performance and genetic diversity based on RAPD markers in wheat, *Triticum aestivum* L. Plant Breed 1999; 118:119–123.

Mangelsdorf AJ. Gene Interaction in Heterosis. In: Gowen JW, ed. Heterosis. Ames, IA: Iowa State College Press, 1952:321–329.

McDaniel RG. Biochemical and physiological basis of heterosis. CRC Crit Rev Plant Sci 1986; 4:227–246.

McDaniel RG, Sarkissian IV. Heterosis: complementation by mitochondria. Science 1966; 152:1640–1642.

Meredith WR, Brown JS. Heterosis and combining ability of cottons originating from different regions of the United States. J Cotton Sci: Vol. 2. The Cotton Foundation, Washington, DC, 1998:77–84.

Pavlikova E, Rood SB. Cellular basis of heterosis for leaf area in maize. Can J Plant Sci 1987; 67:99–104.

Pryor AJ. Allelic glutamic dehydrogenase isozymes in maize—a single hybrid isozyme in heterozygotes. Heredity 1974; 32:397–401.

Rhodes D, Ju GC, Yang W-J, Samaras Y. Plant metabolism and heterosis. Plant Breed Rev 1992; 10:53–91.

Rood SB. Heterosis and the metabolism of gibberellin A_{20} in sorghum. Plant Growth Regul 1995; 16:271–278.

Rood SB, Buzzell RI, MacDonald MD. Influence of temperature on heterosis for maize seedling growth. Crop Sci 1988a; 28:283–286.

Rood SB, Buzzell RI, Major DJ, Pharis RP. Gibberellins and heterosis in maize: quantitative relationships. Crop Sci 1990; 30:281–286.

Rood SB, Buzzell RI, Mander LN, Pearce D, Pharis RP. Gibberellins: a phytohormonal basis for heterosis in maize. Science 1988b; 241:1216–1218.

Rood SB, Witbeck JET, Major DJ, Miller FR. Gibberellins and heterosis in sorghum. Crop Sci 1992; 32:713–718.

Scandalios JG, Liu EH, Campeau MA. The effects of intragenic and intergenic complementation on catalase structure and function in maize: a molecular approach to heterosis. Arch Biochem Biophys 1972; 153:695–705.

Schrader LE. Selection for metabolic balance in maize. In: Harper JE, Schrader LE, Howell RW, eds. Exploitation of Physiological and Genetic Variability to Enhance Crop Productivity. Baltimore, MD: Waverly Press, 1985:79–89.

Schwartz D, Laughner WJ. A molecular basis for heterosis. Science 1969; 166:626–627.

Sinha HK, Khanna R. Physiological, biochemical, and genetic basis of heterosis. Adv Agron 1975; 27:123–174.

Srivastava HK. Intergenomic interaction, heterosis, and improvement of crop yield. Adv Agron 1981; 34:117–195.

Stuber CW. Biochemical and molecular markers in plant breeding. Plant Breed Rev 1992; 9:37–61.

Stuber CW. Heterosis in plant breeding. Plant Breed Rev 1994; 12:227–251.

Stuber CW. Case history in crop improvement: Yield heterosis in maize. In: Paterson AH, ed. Molecular Analysis of Complex Traits. CRC Press, Inc, Boca Raton, FL, 1998:197–206.

Stuber CW. Biochemistry, molecular biology, and physiology of heterosis. In: Coors JG, Pandey S, eds. The Genetics and Exploitation of Heterosis in Crops. Madison, WI: Am Soc Agronomy, 1999:173–183.

Stuber CW. Breeding multigenic traits. In: Phillips R, Vasil I, eds. DNA-Based Markers in Plants. Kluwer Academic Publishers, Dordrecht, The Netherlands, 2001:115–137.

Stuber CW, Lincoln SE, Wolff DW, Helentjaris T, Lander ES. Identification of genetic factors contributing to heterosis in a hybrid from two elite maize inbred lines using molecular markers. Genetics 1992; 132:823–839.

Stuber CW, Polacco M, Senior ML. Synergy of empirical breeding, marker-assisted selection, and genomics to increase crop yield potential. Crop Sci 1999; 39:1571–1583.

Tsaftaris AS. Molecular aspects of heterosis in plants. Physiol Plant 1995; 94:362–370.

Tsaftaris AS, Kafka M. Mechanisms of heterosis in crop plants. J Crop Prod 1998; 1:95–111.

Tsaftaris AS, Kafka M, Polidoros AN. Epigenetic modifications of total genomic maize DNA: the role of growth conditions. In: Tsaftaris AS, ed. Genetics, Biotechnology and Breeding of Maize and Sorghum. Cambridge, UK: Royal Society of Chemistry, 1997:125–130.

Tsaftaris AS, Kafka M, Polidoros AN, Tani E. Epigenetic changes in maize DNA and heterosis. In: Coors JG, Pandey S, eds. The Genetics and Exploitation of Heterosis in Crops. Madison, WI: Am Soc Agronomy, 1999:195–203.

Tsaftaris AS, Polidoros AN. Studying the expression of genes in maize parental inbreds and their heterotic and non-heterotic hybrids. In: Bianci A, Lupotto E, Motto M, eds. Proc XVI Eucarpia Maize and Sorghum Conference, Bergamo, Italy, 1993, pp. 283–292.

Tsaftaris AS, Polidoros AN. DNA methylation and plant breeding. Plant Breed Rev 2000; 18:87–176.

Veldboom LR, Lee M. Genetic mapping of quantitative trait loci in maize in stress and nonstress environments: I. Grain yield and yield components. Crop Sci 1996; 36:1310–1319.

Wells R, Meredith WR Jr. Heterosis in upland cotton. I. Growth and leaf area partitioning. Crop Sci 1986; 26:1119–1123.

Wells R, Meredith WR Jr, Williford JR. Heterosis in upland cotton. II Relationship of leaf area to plant photosynthesis. Crop Sci 1988; 28:522–525.

Xiao J, Li J, Yuan L, McCouch SR. Genetic diversity and its relationship to hybrid performance and heterosis in rice as revealed by PCR-based markers. Theor Appl Genet 1996; 92:637–643.

Xiong LZ, Yang GP, Xu CG, Zhang Q, Saghai-Maroof MA. Relationships of differential gene expression in leaves with heterosis and heterozygosity in a rice diallel cross. Mol Breed 1998; 4:129–136.

Zhang Q, Zhou ZQ, Yang GP, Xu CG, Liu KD, Saghai-Maroof MA. Molecular marker heterozygosity and hybrid performance in indica and japonica rice. Theor Appl Genet 1996; 93:1218–1224.

Zhao MF, Li XH, Yang JB, Xu CG, Hu RY, Liu DJ, Zhang Q. Relationship between molecular marker heterozygosity and hybrid performance in intra- and inter-subspecific crosses of rice. Plant Breed 1999; 118:139–144.

12

Genetic Engineering for Enhancing Plant Productivity and Stress Tolerance

Tuan-hua David Ho

Washington University
St. Louis, Missouri, U.S.A.
and Institute of Botany Academia Sinica
Taipei, Taiwan, Republic of China

Ray Wu

Cornell University
Ithaca, New York, U.S.A.

1 INTRODUCTION

Yield enhancement of crops has been the most important goal in agriculture, and genetic manipulations in plants are pivotal in this effort. Until the early 1980s, traditional breeding had been virtually the only effort in altering plant genetic makeup so that desirable agricultural traits could be obtained. Although these breeding efforts have generated numerous new plant varieties that contribute enormously to the success of agriculture, recent advances in genetic engineering based on recombinant DNA and plant transformation technologies have allowed a much faster and more precise approach in manipulating plant genetics. Furthermore, the genetic engineering approach is not limited to introducing genes among plant species that can be cross-pollinated. Virtually any genes from diverse sources can be introduced to desirable plants. Both the Agrobacterium-based and biolistic plant transformation techniques have been well developed, and most crops, including wheat, rice, maize, soybean, and potatoes, are now routinely transformed by using these methods. These technical advancements, in conjunction with the vast amount of knowledge gained in plant biochemistry and physiology, have allowed researchers to alter plant functions by over- or under-expressing specific genes involved in key metabolic pathways, so that desirable metabolites are accumulated for the enhancement of plant productivity.

It has long been recognized that field-grown plants are constantly subjected to adverse environmental conditions, such as drought, flooding, extreme temperatures, excessive salts, heavy metals, high-intensity irradiation, and infection by pathogenic agents. Because of their immobility, plants have to make necessary metabolic and structural adjustments to cope with these stressful environmental conditions. It has been estimated that about two-thirds of the potential yield of major U.S. crops are routinely lost due to unfavorable environment (Boyer, 1982). Therefore, an effective way to improve plant productivity is to grow plant varieties more tolerant to environmental stresses. In addition to the breeding efforts in generating stress-tolerant varieties, many attempts in genetic engineering for tolerance to environmental stresses or resistance to pathogens have obtained promising preliminary results.

Despite the complexity involved in cellular metabolism and its regulation, the flux of individual pathways can be easily altered by manipulating the levels or activities of one or two key enzymes. This is well within the capability of current plant-transformation methods, in which up to a few genes can be easily introduced to the recipient plants. In addition to over-expression of transgenes driven by strong constitutive promoters, down-regulation of resident genes can be achieved by antisense suppression, co-suppression, or dsRNA interference (RNAi). For altering enzymatic activities, mutant genes encoding enzymes with properties different from those of wild type can be introduced. Furthermore, it has been demonstrated that introducing certain regulatory genes, such as those encoding key transcription factors, into plants by transformation, can alter the expression levels of multiple downstream genes, hence affecting multiple processes (Borevitz et al., 2000; Jaglo-Ottosen et al., 1998; Lloyd et al., 1992). Although gene replacement has not yet been demonstrated for higher plants, initial success in metabolic engineering for yield enhancement and stress resistance has already been achieved.

In this article, we review the recent progress in metabolic engineering for enhancing the production of major metabolites, such as starches and lipids, as well as secondary metabolites, such as terpenoids and phenylpropanoids. In addition to the increase in biomass production, overproduction of these metabolites enhances the nutritional value as well as the industrial use of plant products. Because stress responses have a profound impact on plant productivity, we also evaluate recent attempts to engineer tolerance to various stresses in plants. This article critically examines the current progress in this promising area and will hopefully stimulate more interest in expanding the current scope of research and development for improving plant productivity through genetic engineering.

2 GENETIC ENGINEERING FOR ENHANCING THE PRODUCTION OF MAJOR METABOLITES

Photosynthesis is the center of plant metabolism, and efficiency of photosynthesis is directly related to biomass production. Plants with the Calvin-cycle-based C_3 photosynthesis have been shown to suffer yield reduction when grown in high-light-intensity and high-temperature environments. Because of the oxygenase activity associated with ribulose 1,5-bisphosphate carboxylase (Rubisco), the photorespiration process takes place under these environmental conditions leading to the loss of fixed C and N. C_4 and crasulacean acid metabolism (CAM) photosynthetic pathways have evolved among plant species capable of concentrating scarce atmospheric CO_2 by the action of phosphoenol pyruvate carboxylase (PEPC). Phosphoenol pyruvate carboxylase has a much higher affinity toward CO_2 than Rubisco. However, certain important crops, especially those grown in subtropical and tropical areas, such as rice and cassava, are C_3 plants. Breeding efforts involving interspecies hybridizations have thus far failed to produce rice varieties capable of carrying out C_4 photosynthesis. Following the genetic engineering approach, Ku et al. (1999, 2001) have successfully transformed rice with maize genes encoding two key enzymes of C_4 photosynthesis: PEPC and pyruvate dikinase (PPDK). They have shown that photosynthesis efficiency among the transgenic rice plants increased by as much as 35% compared to the wild type. The increased photosynthetic capacity in these plants was mainly associated with an enhanced stomatal conductance and a higher internal CO_2 concentration (Ku et al., 2001). The results suggest that both PEPC and PPDK play a key role in organic acid metabolism in the guard cells to regulate stomatal opening. It is surprising that this relatively simple introduction of one or two key enzymes in C_4 photosynthesis has such a profound effect on photosynthesis in transgenic rice. Much work appears to be needed to investigate the mechanisms underlying this promising observation.

The demand for starches has been steadily increasing in the last few decades for specialized food and industrial use. A key rate-limiting enzyme in plant starch biosynthesis is ADPG pyrophosphorylase. This enzyme is known to be sensitive to activation by fructose 1,6-bisphosphate and also to inhibition by AMP and Pi. An *E. coli* mutant of ADPG pyrophosphorylase, less sensitive to this type of metabolic regulation, causes an increase in glycogen accumulation (Leung et al., 1986). Stark et al. (1992) generated transgenic potato overexpressing this mutant *E. coli* enzyme and found that tubers accumulated up to 60% more starch than those transformed with the wild-type enzyme. This work has clearly demonstrated that manipulation of enzyme kinetics is as effective, if not more so, as altering the level of enzyme protein in engineering a biochemical pathway.

Plant seeds are the major source of storage lipids, and vegetable oils from the half dozen major oil crops account for 85% of the oil and fat production in the world. In addition to human consumption, up to one-third of plant oils have been extensively used in industry. However, the majority of the more than 200 known fatty acids are not synthesized in appreciable amounts in the major oil crops. For the improvement of nutritional value and industrial usage, it is beneficial to alter the fatty acid composition in major oil crops utilizing the diverse genetic resources available in other oil producing plants. Thus research efforts on genetic engineering of plant lipids have been extensively carried out by both academic and industrial scientists (Broun et al., 1999; Thelen and Ohlrogge, 2002; Topfer et al., 1995; Voelker and Kinney, 2001). Three major efforts have been the focus of lipid metabolic engineering: (1) alteration of the length of fatty acids, (2) regulation of the degree of fatty acid desaturation, and (3) addition of special functional groups to fatty acids. Among the fatty acid biosynthesis enzymes, the condensing enzyme, 3-ketoacyl-ACP synthase (KAS) and thioesterase (TE), which is responsible for the release of fatty acid from acyl carrier protein, are the two logical targets in metabolic engineering for altering the length of fatty acids. At least three KASs exist in most plants: KASI for C_{4-14} fatty acids, KASII for C_{14-16} fatty acids, and KASIII for acetylCoA. Dehesh et al. (2001) overexpressed KASIII in transgenic tobacco, rapeseeds, and Arabidopsis, and observed an increase in the content of palmitate (16:0), indicating that this enzyme has a universal role in fatty acid biosynthesis. However, these transgenic plants also contained lower levels of lipids probably because of lack of sufficient malonyl CoA, a substrate for KASIII. Two major isoforms for TE encoded by the FatA and B genes are present in plants with FatB TE capable of releasing short-chain fatty acids. Voelker et al. (1996) cloned a gene encoding a TE specific for lauric acid (C_{12}) from bay plants. Overexpression of this TE leads to much higher content of lauric acid, which is very useful in cosmetics, detergents, and as synthetic creams in food. The degree of desaturation in plant fatty acids is also of much interest to the food industry because polyunsaturated fatty acids tend to oxidize easily generating an undesirable odor. Antisense inhibition of oleate (18:1) desaturase synthesis resulted in a significant decrease in polyunsaturated fatty acids in soybean oil with more than 80% of them being the monounsaturated oleic acid (Hitz et al., 1995).

Plants normally produce all the amino acids needed for their growth. Blocking the synthesis of these amino acids is lethal to plants but has no effect in humans because animals lack these biosynthetic pathways. Roundup (glyphosate), a commonly used herbicide, inhibits the key enzyme, EPSP synthase, in the shikimate pathway for the synthesis of three essential amino acids, phenylalanine, tyrosine, and tryptophan (Smart et al., 1985). Glyphosate functions as a competitor against a substrate, PEP, of EPSP synthase. By introducing a glyphosate-tolerant mutant EPSP synthase gene into plants, Shah et al. (1986) were able to produce the first herbicide-resistant transgenic plants. This is the foundation for the much-debated Roundup Ready soybean that has been grown by U.S. farmers since 1996.

3 GENETIC ENGINEERING FOR ALTERING SECONDARY METABOLISM

Plants produce a vast and diverse assortment of secondary metabolites, many of which are potentially useful in medicine, agriculture, and industry. Three major classes of secondary metabolites exist: terpenoids, phenylpropanoids, and alkaloids. While alkaloids are composed of thousands of diverse individual compounds, specific biochemical pathways and several key enzymes leading to the production of terpenoids and phenylpropanoids have been elucidated. One major subclass of the phenylpropanoids is composed of flavonoids, which have long been known to function as floral pigments for the attraction of insect pollinators, signaling molecules for plant–microbe interactions, and self-defense compounds. Attempts to overexpress a key enzyme, chalcone synthase, often led to the surprising result of "co-suppression,", i.e., where there was a decrease in flower pigmentation, rather than the more intense color originally intended. By more carefully manipulating one of the other enzymes in the pathway, researchers have now successfully generated a variety of pigmentations in selected flowers (for reviews, see Jergensen, 1995; Tanaka et al., 1998). Isoflavonoids in legumes have been discovered to be beneficial to human health. A key enzyme for isoflavonoid synthesis, 2-hydroxy isofavanone synthase, has recently been cloned (Steele et al., 1999). This has created the possibility of generating isoflavone nutraceuticals in a range of food crops other than legumes (Dixon and Steele, 1999). Monoterpenoids are the major components of essential oils of the mint family. In an attempt to enhance essential oil yield and composition in mint, Mahmoud and Croteau (2001) transformed peppermint plants with a sense construct of the cDNA of 1-deoxy-D-xylulose-5-phosphate reductoisomerase, an enzyme involved in the formation of the building block of terpenoids, isopentenyl diphosphate. They observed a substantial increase (about 50%) in the amount of total essential oil in these transgenic plants. These authors have also generated transgenic plants expressing an antisense construct of the cDNA of menthofuran synthase to successfully suppress the production of the less-desirable monoterpene, menthofuran. Thus this work has demonstrated that the level and composition of essential oils can be effectively manipulated through metabolic engineering.

The production of vitamins or their precursors in plants has been engineered so that plant products are more beneficial to human health. It has been estimated that as many as a million children in Third-World countries develop blindness each year due to lack of sufficient vitamin A in their rice-based diet. Potrykus and his associates (Ye et al., 2000) have transformed rice plants with three genes encoding three enzymes involved in the biosynthesis of β-carotene, the precursor of vitamin A. These transgenes, originally isolated from *Erwinia* bacteria and daffodil plants, were linked to promoters of endosperm-specific seed storage-protein genes. The endosperm of seeds produced by the transgenic rice plants is yellow because of the presence of β-carotene. Although the level of β-carotene is still lower than the recommended daily dose, improving the system to produce higher levels of β-carotene is an attainable goal. Another example to enhance nutritional value of plants is related to vitamin E. Most oilseed crops contain a high proportion of γ-tocopherol, which has only 10% of vitamin E activity of α-tocopherol. Shintani and DellaPenna (1998) transformed *Arabidopsis* with a gene encoding γ-tocopherol methyltransferase capable of converting γ-tocopherol to the more active α-tocopherol. Indeed, the transgenic plants yielded 10-fold higher vitamin E activity than wild-type plants.

4 GENETIC ENGINEERING FOR ENHANCING STRESS TOLERANCE

Plant productivity is often influenced by unfavorable conditions, in particular drought, salinity, and low temperatures. A common feature among these stresses is the dehydration of plant cells. Therefore most of the genetic engineering efforts in conferring tolerance to these stresses have been centered on producing compatible osmolytes to prevent further water loss. Attempts have been made to overproduce enzymes that are responsible for the synthesis of compatible osmolytes. These include the overproduction of proline (by pyrroline-5-caboxylase), polyols (by myo-inositol O-methyltransferase and mannitol-1-phosphate dehydrogenase), glycine betaine (by choline dehydrogenase and choline oxidase), trehalose (by trehalose-6-phosphate synthase and trehalose-6-P phosphatase), and polyamines (arginine decarboxylase and ornithine decarboxylase), which have generated promising stress-tolerant plants (for review see Bajaj et al., 1999).

It should be pointed out that although plant biochemists and physiologists have attempted for over 10 years to understand how these compatible osmolytes confer dehydration stress tolerance in transgenic plants, no firm conclusions have been made. It seems that the increased cellular level of a single osmolyte may have multiple effects on cellular metabolism. Increasing osmolarity in the cells seems to be only one of several effects, because in several cases, the level of a particular osmolyte in the transgenic plant is not high enough to account for osmotic adjustment as the main cause for the observed increase in stress tolerance. For example, when the fusion genes *OtsA* and *OtsB*, which encode for trehalose-6-P synthase and trehalose-6-P phosphatase (Seo et al., 2000), were introduced into rice by *Agrobacterium*-based transformation, the level of trehalose was increased after salt or drought stress from approximately 30 μg/g leaf FW to 300 μg/g FW cellular water (Garg et al., 2001). This increase is far from sufficient, based on an increase in cellular osmolarity, to account for the observed high level of salt and drought tolerance in the transgenic plants. Trehalose overproduction in rice also resulted in many other cellular changes. For example, under stress conditions, the cellular levels of glucose in shoots became higher than those in control plants. The Na^+ level in the shoots of transgenic plants increased much less than those in control plants after salt stress. On the other hand, the levels of Zn^{2+} and phosphate ion in both leaves and roots were higher relative to controls after salt stress. In addition, transgenic plants showed a higher photosynthetic efficiency (Garg and Wu, unpublished).

In addition, a group of proteins normally expressed during the desiccation stage of seed formation, termed late embryogenesis abundant (LEA) proteins, have been targets for genetic engineering. Based on their sequences, the LEA proteins are classified into several subgroups with LEAIII being the most interesting, because it contains one long stretches of amphipathic α-helical structures (Dure et al., 1989). Dure (1993) has proposed that these proteins may be able to sequester ions accumulated in dehydrated tissues, hence benefiting the stressed plant. Overexpressing a barley LEAIII protein, HVA1, in tobacco, rice, and wheat confers tolerance to both drought and salinity stresses (Xu et al., 1996; Sivamani et al., 2000; and ThD Ho, unpublished). In the case of transgenic wheat, the elevated stress tolerance level is apparently related to higher water use efficiency, i.e., more biomass produced per unit of water consumed by plants (Sivamani et al., 2000). Although it was thought that LEA proteins were unique to higher plants, proteins with similar sequences have now been identified from many organisms, ranging from prokaryotes to humans (see "Representative LEA family proteins" at website: < http://www.sanger.ac.uk/cgi-bin/Pfam/getallproteins.pl?name = LEA&-acc = PF02987&-verbose = true&type = seed&zoom_factor = 0.5&list = View + Graphic >). Indeed, a LEA-like protein has been shown to be expressed during anhydrobiosis (roughly equivalent to dehydration stress in plants) in the nematode, *Aphelenchus avenae* (Browne et al., 2002). Furthermore, over-expression of barley HVA1 in yeast also confers higher level of tolerance to NaCl (Zhang et al., 2000). Although the exact function of LEA proteins is still not clear, their unique structural features and potential biotechnology applications certainly warrant more research effort in the future.

Maintenance of ionic homeostasis is also a common response to dehydration stresses. This is usually accomplished by the balance between ion transports and sequestrations among different cellular compartments. To manipulate ionic distribution, Apse et al. (1999) overexpressed a vacuolar Na^+/H^+ antiport, AtNHX1, in transgenic *Arabidopsis*. These transgenic plants were able to undergo normal development even in the presence of 200 mM NaCl. This elevated salinity tolerance was correlated with higher-than-normal levels of AtNHX1 transcripts, protein, and vacuolar Na^+/H^+ antiport activity. Consequently, it was observed that transgenic plants contained more Na^+ ions and managed to sequester them in vacuoles.

A third common denominator among dehydration stresses is the generation of reactive oxygen species. It is likely that the electron transport systems in plant membranes are easily disturbed by any stressful condition causing the production of harmful reactive oxygen species. Thus to detoxify these reactive oxygen species would minimize the damages caused by the stresses and be beneficial to the stressed plants. Indeed, overexpression of oxidative-stress-related genes, such as superoxide dismutase and glutathione *S*-transferase, has been shown to confer higher level of tolerance not only to ozone stress but also to salt, drought, and cold stresses (for reviews see Bajaj et al., 1999). Payton et al. (2001) have shown that photosynthesis efficiency can be maintained during moderate chilling stress by increasing chloroplastic antioxidant enzyme activities through genetic engineering. In addition, over-production of the osmolyte mannitol has been shown to be able to reduce reactive oxygen species in chloroplasts, which may also contribute to the pro-tection against stress-induced damages (Shen et al., 1997). It has been recently suggested that targeting the detoxification pathways, including the efficient removal of reactive oxygen species, could be an efficient approach to obtain plants with multiple stress tolerance (Bartels, 2001).

As mentioned above, there are multiple components in typical plant responses to a single physical stress such as drought, salinity, or low temper-ature. Genetic engineering efforts involving multiple genes, although techni-cally feasible, could become laborious and time consuming. It is conceivable that certain intermediates in the stress-mediated signaling cascade could function as a "master" switch capable of regulating multiple downstream events. Analyses of the promoters of stress-induced genes have led to the iso-lation of transcription factors, such as CBF1 and CREB, which may serve as trans-acting factors for regulating multiple downstream genes. Jaglo-Ottosen et al. (1998) overexpressed CBF1 in *Arabidopsis* and observed that multiple cold response (COR) genes were induced even in the absence of cold stress. In addition, the transgenic plants displayed freezing-stress tolerance without prior cold acclimation. Similarly, Kasuga et al. (1999) transformed *Arabidopsis* with the *DREB1A* gene, and the resulting transgenic plants acquired higher levels of tolerance to drought, salt, and freezing stresses. More recently, Lee et al. (2003) transformed tomato with *Arabidopsis CBF1* gene and observed that the transgenic tomato plants were more tolerant to drought, cold, salinity, and oxidative stresses. These authors also detected a higher level of transcriptional activity and catalase activity in these transgenic plants, which was most likely responsible for the higher level of tolerance to oxidative stress.

5 EMERGING APPROACHES FOR PLANT BIOCHEMICAL ENGINEERING

To date, most genetic engineering efforts have employed strong constitutive promoters, such as CaMV 35S and rice actin 1 gene promoters, to drive the expression of transgenes that are integrated into the nuclear genome. While this is usually sufficient if the transgene product is needed throughout the plant such as the case for Roundup Ready soybeans, overexpression of a transgene at high level in all the tissues all the time may actually lead to suboptimal growth or yield penalty due to the diversion of cellular building blocks for the overproduction of a single gene product. This problem is further Compounded if the transgene product has a profound effect on other processes in plants. In the case of Kasuga et al. (1999) transforming *Arabidopsis* with 35S::DREB1A mentioned above, the resulting stress-tolerant plants suffered from severe growth retardation because of the use of the strong constitutive 35S promoter. In contrast, expression of the same DREB1A using the stress-inducible rd29A promoter produced transgenic plants with little or no growth defects. Similar results were observed by Lee et al. (2003) who generated transgenic tomato plants overexpressing CBF1 gene driven by the 35S promoter. The transgenic tomato plants they generated were tolerant to multiple stresses but severely stunted with poor fruit and seed sets. Surprisingly, exogenous applications of gibberellin restored the growth defects without affecting the elevated stress tolerance. Furthermore, these authors also used an ABA/stress-inducible promoter, ABRC1, developed by Shen et al. (1996) and Su et al. (1998), to drive the expression of the *CBF1* gene, and were able to obtain stress-tolerant tomato plants without apparent yield penalties. Therefore it is apparent that the use of an inducible promoter, rather than a strong constitutive promoter, is desirable for transgenic plants where the transgene product is needed only under specific conditions.

Many important plant metabolites, such as starches, fatty acids, some amino acids, carotenoids, and some plant hormones, are synthesized in the chloroplasts (or plastids in general). In addition, plastids provide high-capacity storage for certain metabolites. Proteins encoded by nuclear transgenes can be imported into chloroplasts if a chloroplast import targeting sequence is included. However, it would be more efficient to deliver the transgenes into the chloroplast genome directly. The concept of chloroplast genetic engineering first emerged in the 1980s, but only recently have genes conferring useful agricultural traits, including herbicide resistance, pathogen resistance, drought and salt tolerance, been introduced to the chloroplast genome. This was done via biolistic transformation (for review, see Daniell et al., 2002). Two additional potential benefits are associated with the chloroplast genome transformation: (1) multiple copies of transgenes can stably exist in the chloroplast population, and (2) reduced probability of the transgene escaping due to cross pollination because chloroplast genes are rarely transmitted through pollen. However, it should be pointed out that the mechanism of exporting proteins from chloroplasts is virtually unknown, thus the chloroplast transformation strategy is limited to engineering biochemical processes occurring in chloroplasts.

Finally, plant metabolic engineering is not just limited to modifications of biochemical pathways normally existing in plants. For the production of pharmaceuticals, antibodies have been produced and assembled in and secreted from plant tissues (Ma et al., 1995). Although plant-based antibody production is still being evaluated, it holds the promise of generating precious antibodies by relatively inexpensive agricultural practice. Edible vaccines are another promising approach in which genes encoding components of a pathogen are expressed in transgenic plants (for review see Walmsley and Arntzen, 2000). These antigens when ingested in sufficient quantities can trigger mucous immune responses. Thus plants that are usually consumed in large quantities without cooking, such as tomato and banana, are ideal for this purpose. It has been shown that several plant-derived antigenic proteins have delayed or prevented the onset of disease in animals and have proven to be safe and functional in human clinical trials. Research and testing of edible vaccines are still in their infancy. A great deal of work is still needed to expand the scope to the prevention of a large number of diseases. In addition to the production of pharmaceuticals, plants can also be engineered to produce industrial raw materials. Besides the specific fatty acids and their derivatives, briefly mentioned above, production of biodegradable plastics, such as polyhydroxyalkanoate (PHA) and polylactide (PLA), has also been accomplished by transforming plants with bacterial enzymes capable of catalyzing the formation of these plastics using plant sugars as precursors (Poirier et al., 1992; and for review see Gerngross, 1999). Again, this approach, if fully explored, would certainly add extra value to the currently existing agricultural commodities.

6 CONCLUSIONS AND PERSPECTIVES

It is clear that recombinant DNA technology and methods for efficient plant transformation have propelled the recent advancements in plant metabolic engineering for the improvement of productivity and stress tolerance. Most of the accomplishments to date have been quite promising, and some of the products have already been commercialized with a significant impact in current agricultural practice. With *Arabidopsis* and rice genome information readily available, genomics-based strategies for gene discovery coupled with high-throughput transformation processes would undoubtedly further enhance the development of many new plant-based products by applying metabolic engineered plant products. However, much basic information about plant biochemistry, plant physiology, and genetics is still needed for this effort. For example, the functions of more than 50% of the putative open reading frames in the relative simple yeast genome are still not clear. Therefore much more effort is needed to determine the function of most genes in plant genomes. Despite the intense research effort in plant biochemistry in the last several decades, information about biochemical pathways and their enzymes remains largely incomplete. A search of a popular biochemical pathway website: < http://www.genome.ad.jp/kegg/ > (KEGG: Kyoto Encyclopedia of Genes and Genomes) revealed that many key enzymes have not yet been studied in plants. In addition, the studies of the vast and diverse plant secondary metabolites have only increased in recent years. Therefore despite the promising recent accomplishments in plant metabolic engineering, future success relies on a two-pronged approach utilizing the recently available genomic information in combination with the traditional biochemical and physiological explorations. Finally, it should be recognized that proper field tests of transgenic plants are the ultimate way to determine the value of any transgenic plant, because only those plants that give higher yields under field conditions after repeated trials are useful.

ACKNOWLEDGMENTS

Work carried out in THDH's laboratory cited in this work was supported by grants from NSF, USDA and Academia Sinica, Taipei, Taiwan. Work carried out in RW's laboratory cited in this work was partially supported by a grant from the Rockefeller Foundation.

REFERENCES

Apse MP, Aharon GS, Snedden WA, Blumwald E. Salt tolerance conferred by over-expression of a vacuolar Na^+/H^+ antiport in Arabidopsis. Science 1999; 285:1256–1258.

Bajaj S, Targolli J, Liu LF, Ho T, Wu R. Transgenic approaches to increase dehydration-stress tolerance in plants. Mol Breed 1999; 493–503.

Bartels D. Targeting detoxification pathways: an efficient approach to obtain plants with multiple stress tolerance? Trends Plant Sci 2001; 6:284–286.

Borevitz JO, Xia Y, Blount J, Dixon RA, Lamb C. Activation tagging identifies a conserved MYB regulator of phenylpropanoid biosynthesis. Plant Cell 2000; 12:2383–2394.

Boyer J. Plant productivity and environment: potential for increasing crop plant productivity, genotypic selection. Science 1982; 218:443–448.

Broun P, Gettner S, Somerville C. Genetic engineering of plant lipids. Annu Rev Nutr 1999; 19:197–216.

Browne J, Tunnacliffe A, Burnell A. Anhydrobiosis: plant desiccation gene found in a nematode. Nature 2002; 416:38.

Daniell H, Khan MS, Allison L. Milestones in chloroplast genetic engineering: An environmentally friendly era in biotechnology. Trends Plant Sci 2002; 7:84–91.

Dehesh K, Tai H, Edwards P, Byrne J, Jaworski JG. Overexpression of 3-ketoacyl-acyl-carrier protein synthase in plants reduces the rate of lipid synthesis. Plant Physiol 2001; 125:1103–1114.

Dixon RA, Steele CL. Flavonoids and isoflavonoids—a gold mine for metabolic engineering. Trends Plant Sci 1999; 4:394–400.

Dure L III. A repeating 11-mer amino acid motif and plant desiccation. Plant J 1993; 3:363–369.

Dure L III, Crouch M, Harada J, Ho THD, Mundy J, Quatrano R, Thomas T, Sung ZR. Common amino acid sequence domains among LEA proteins of higher plants. Plant Mol Biol 1989; 12:475–486.

Garg A, Kim JK, Ranwala A, Wu R. Accumulation of trehalose in transgenic indica rice using bifunctional fusion enzymes of trehalose-6-P synthetase and trehalose-6-P phosphatase of *E. coli*. Rice Genet Newsl 2001; 18: 87–89.

Gerngross T. Can biotechnology move us toward a sustainable society? Nat Biotechnol 1999; 17:541–544.

Hitz W, Yadav N, Reiter R, Mauvais C, Kinney AJ. Reducing polyunsaturation in oils of trangenic canola and soybean. In: JC K, P M, eds. Plant Lipid Metabolism. London: Kluwer, 1995:506–508.

Jaglo-Ottosen KR, Gilmour SJ, Zarka DG, Schabenberger O, Thomashow MF. Arabidopsis CBF1 overexpression induces COR genes and enhances freezing tolerance. Science 1998; 280:104–106.

Jergensen R. Cosuppression, flower color patterns, and metastable gene expression states. Science 1995; 268:686.

Kasuga M, Liu Q, Miura S, Yamaguchi-Shinozaki K, Shinozaki K. Improving plant drought, salt, and freezing tolerance by gene transfer of a single stress-inducible transcription factor. Nat Biotechnol 1999; 17, 287–291.

Ku MS, Agarie S, Nomura M, Fukayama H, Tsuchida H, Ono K, Hirose S, Toki S, Miyao M, Matsuoka M. High-level expression of maize phosphoenolpyruvate carboxylase in transgenic rice plants. Nat Biotechnol 1999; 17:76–80.

Ku MS, Cho D, Li X, Jiao DM, Pinto M, Miyao M, Matsuoka M. Introduction of genes encoding C4 photosynthesis enzymes into rice plants: physiological consequences. Novartis Found Symp 2001; 236:100–111.

Lee J-T, Prasoad V, Yang P-T, Wu J-F, Ho T-HD, Charng Y-Y, Chan M-T. Expression of Arabidopsis CBF1 regulated by an ABA/stress inducible promoter in transgenic tomato confers stress tolerance without affecting yield. Plant Cell Environ 2003; 26:1181–1190.

Leung P, Lee Y-M, Greenberg E, Esch K, Boylan S, Preiss J. Cloning and expression of the *Escherichia coli glgC* gene from a mutant containing an ADPglucose pyrophosphorylase with altered allosteric properties. J Bacteriol 1986; 167:82–88.

Lloyd AM, Walbot V, Davis RW. Arabidopsis and Nicotiana anthocyanin production activated by maize regulators R and C1. Science 1992; 258:1773–1775.

Ma JK, Hiatt A, Hein M, Vine ND, Wang F, Stabila P, van Dolleweerd C, Mostov K, Lehner T. Generation and assembly of secretory antibodies in plants. Science 1995; 268:716–719.

Mahmoud SS, Croteau RB. Metabolic engineering of essential oil yield and composition in mint by altering expression of deoxyxylulose phosphate reductoisomerase and menthofuran synthase. Proc Natl Acad Sci USA 2001; 98:8915–8920.

Payton P, Webb R, Kornyeyev D, Allen R, Holaday A. Protecting cotton photosynthesis during moderate chilling at high light intensity by increasing chloroplastic antioxidant enzyme activity. J Exp Bot 2001; 52:2345–2354.

Poirier Y, Dennis D, Klomparins K, Somerville C. Polyhydroxybutyrate, a biodegradable thermoplastic, produced in transgenic plants. Science 1992; 256:520–622.

Seo HS, Koo YJ, Lim JY, Song JT, Kim CH, Kim JK, Lee JS, Choi YD. Characterization of a bifunctional enzyme fusion of trehalose-6-phosphate synthetase and trehalose-6-phosphate phosphatase of *Escherichia coli*. Appl Environ Microbiol 2000; 66:2484–2490.

Shah D, Horsch R, Klee H, Kishore G, Winter J, Tumer N, Hironaka C, Sanders P, Gasser C, Aykent S. Engineering herbicide tolerance in transgenic plants. Science 1986; 233:478–481.

Shen Q, Zhang P, Ho T-HD. Modular nature of abscisic acid (ABA) response complexes: composite promoter units that are necessary and sufficient for ABA induction of gene expression in barley. Plant Cell 1996; 8:1107–1119.

Shen B, Jensen RG, Bohnert HJ. Increased resistance to oxidative stress in transgenic plants by targeting mannitol biosynthesis to chloroplasts. Plant Physiol 1997; 113:1177–1183.

Shintani D, DellaPenna D. Elevating the vitamin E content of plants through metabolic engineering. Science 1998; 282:2098–2100.

Sivamani E, Baheildin A, Wrath JA. Niemi T, Dyer W, Ho THD, Qu R. Improved biomass productivity and water use efficiency under water deficit conditions in transgenic wheat constitutively expressing the barley *HVA1* gene. Plant Sci 2000; 155:1–9.

Smart CC, Johanning D, Muller G, Amrhein N. Selective overproduction of 5-enolpyruvylshikimic acid 3-phosphate synthase in a plant cell culture which tolerates high doses of the herbicide glyphosate. J Biol Chem 1985; 260:16338–16346.

Stark D, Timmerman K, GF B, Preiss J, Kishore G. Regulation of the amount of starch in plant tissues by ADP glucose pyrophosphorylase. Science 1992; 258:287–292.

Steele CL, Gijzen M, Qutob D, Dixon RA. Molecular characterization of the enzyme catalyzing the aryl migration reaction of isoflavonoid biosynthesis in soybean. Arch Biochem Biophys 1999; 367:146–150.

Su J, Shen Q, Ho T-HD, Wu R. Dehydration-stress-regulated transgene expression in stably transformed rice plants. Plant Physiol ,1998; 117:913–922.

Tanaka Y, Tsuda S, Kusumi T. Metabolic engineering to modify flower color. Plant Cell Physiol 1998; 39:1119–1126.

Thelen JJ, Ohlrogge JB. Metabolic engineering of fatty acid biosynthesis in plants. Metab Eng 2002; 4:12–21.

Topfer R, Martini N, Schell J. Modification of plant lipid synthesis. Science 1995; 268:681.

Voelker T. Plant acyl-ACP thioesterases: chain-length determining enzymes in plant fatty acid biosynthesis. Genet Eng 1996; 18:111–133.

Voelker T, Kinney AJ. Variations in the biosynthesis of seed-storage lipids. Annu Rev Plant Physiol Plant Mol Biol 2001; 52:335–361.

Walmsley AM, Arntzen CJ. Plants for delivery of edible vaccines. Curr Opin Biotechnol 2000; 11:126–129.

Xu D, Duan X, Wang B, Hong B, Ho T-HD, Wu R. Expression of a late embryogenesis abundant protein gene, HVA1, from barley confers tolerance to water deficit and salt stress in transgenic rice. Plant Physiol 1996; 110:249–257.

Ye X, Al-Babili S, Kloti A, Zhang J, Lucca P, Beyer P, Potrykus I. Engineering the provitamin A (beta-carotene) biosynthetic pathway into (carotenoid-free) rice endosperm. Science 2000; 287:303–305.

Zhang L, Ohta A, Takagi M, Imai R. Expression of plant group 2 and group 3 lea genes in *Saccharomyces cerevisiae* revealed functional divergence among LEA proteins. J Biochem (Tokyo) 2000; 127:611–616.

13

Genome Mapping and Marker-Assisted Selection for Improving Cotton (*Gossypium* spp.) Productivity and Quality in Arid Regions

Yehoshua Saranga

The Hebrew University of Jerusalem
Rehovot, Israel

Andrew H. Paterson

University of Georgia
Athens, Georgia, U.S.A.

1 GENOME MAPPING AND MARKER-ASSISTED SELECTION FOR CROP ADAPTATION TO ARID LANDS

About one-third of the world's arable land suffers from chronically inadequate supplies of water for agriculture, and in virtually all agricultural regions, crop yields are periodically reduced by drought (Kramer, 1980; Boyer, 1982). Global climatic trends may accentuate this problem (cf. Le Houerou, 1996). Efficient irrigation technologies help to reduce the gap between potential and actual yield; however, diminishing water supplies in many regions impel intrinsic genetic improvement of crop productivity under arid conditions (cf. Blum, 1988) as a sustainable and economically viable solution to this problem.

Water loss from a plant (transpiration) is an unavoidable consequence of photosynthesis (Cowan, 1986), whereby the energy of solar radiation is used for carbon fixation. While increased transpiration without a corresponding increase in photosynthesis reduces "water-use efficiency" [WUE, the ratio between dry matter (DM) production and water consumption at the whole-plant level, or between rates of CO_2 fixation and transpiration at the leaf level], it is also a benefit in dissipating excess heat (Cornish et al., 1991; Radin et al., 1994). Water stress and heat stress almost invariably co-occur under arid-region field conditions. The resulting need for a balance between tolerance of heat and drought complicates strategies for manipulating plant water use to improve productivity under arid conditions.

The development of drought-tolerant crops has been hindered by low heritability of complex traits such as yield, and by lack of knowledge of physiological parameters that reflect genetic potential for improved productivity under water deficit. A merger of physiology and genetics promises to improve basic understanding of plant response to arid conditions, offering new avenues for crop improvement. Using genetic mapping to dissect the inheritance of different complex traits in the same population is a powerful means to distinguish common heredity from casual associations between such traits (cf. Paterson et al., 1988). In principle, this can permit a direct test of the role of specific physiological traits in genetic improvement of plant productivity under abiotic stresses, such as those imposed by arid conditions.

Genetic mapping has been previously used to identify quantitative trait loci (QTLs) responsible for improved productivity under arid conditions (Agrama and Moussa, 1996; Tuinstra et al., 1996; Ribaut et al., 1997). Separately, QTLs have also been reported that confer physiological variations thought to be associated with stress tolerance, such as osmotic adjustment (OA) (Lilley et al., 1996; Morgan and Tan, 1996), WUE (measured indirectly as carbon isotope ratio [$^{13}C/^{12}C$, expressed with differential notation as $\delta^{13}C$]: Martin et al., 1989; Mansur et al., 1993; or ash content: Mian et al., 1996, 1998), abscisic acid levels (Quarrie et al., 1994; Tuberosa et al., 1998), stomatal conductance (Ulloa et al., 2000), and various measures of plant water status (Lebreton et al., 1995; Teulat et al., 1998a). However, we are aware of only two prior studies in which productivity and physiological differences were genetically mapped in the same populations. These studies found productivity to be unrelated to $\delta^{13}C$ (Mansur et al., 1993) or to relative water content (Teulat et al., 1998b).

This chapter will focus on a long-term research project ongoing in our laboratories since 1994, the overall goal of which is to establish a scientific framework for improving crop yield and quality under arid conditions. Cotton (*Gossypium* spp.) has been selected as the model crop for this. In the first phase we examined variation in WUE and found remarkable interaction between genotypes and environments (field vs. greenhouse); nevertheless, $\delta^{13}C$ consistently associated with WUE under both environments. In the second phase of the study we used QTL mapping to dissect the genetic control of productivity and selected physiological parameters under water-limited and well-watered environments, respectively. Interrelationships among QTLs related to productivity and plant physiological status suggested that predawn osmotic potential and productivity under arid conditions share a partly common genetic basis. We also showed that different cotton species have evolved different, complementary alleles that can be assembled into new genotypes with improved productivity under water deficit, and set the stage for molecular dissection of key chromosomal regions that confer adaptation to arid conditions. In the current phase, marker-assisted selection is being used to develop near-isogenic lines (NILs) for selected genomic regions.

2 COTTON—A MODEL FOR STUDY OF CROP ARID-LAND ADAPTATION

With a farm-gate value of about $20 billion yearly (worldwide), cotton (*Gossypium hirsutum* L. and *G. barbadense* L., abbreviated hereafter as *GH* and *GB*, respectively) is usually grown during the summer in arid and semiarid regions where water availability is often limited. Regardless of whether it is irrigated or not, cotton is often exposed to drought, which adversely affects both yield and lint (fiber) quality.

The cotton genus, *Gossypium*, originates from wild perennial plants adapted to semiarid, semitropical environments that experienced periodic drought and temperature extremes (Kohel, 1974), so adaptations to heat and drought stress are expected to exist. In exotic cotton genotypes, Quisenberry et al. (1982) demonstrated variation in DM accumulation, heat tolerance, root growth, and WUE. However, such exotic genotypes present limitations in breeding programs due to unfavorable traits such as photoperiodic response, low yield, and poor quality. An ongoing "conversion" program (McCarty and Jenkins, 1992) only partly ameliorates these problems. Some improvement in WUE can be realized simply by manipulating crop maturity to best exploit the natural climatic cycle (Jordan, 1982), but additional improvements of WUE must come from other traits.

Modern cotton cultivars are the result of intensive selection to produce large quantities of seed epidermal hairs ("fibers" or "lint") suitable for mechanical harvesting and processing. This selection has unintentionally narrowed genetic variability for drought tolerance and WUE (Rosenow et al., 1983). However, considerable variation persists within and between the two prevailing cultivated cotton species, *GH* and *GB*, in physiological traits such as WUE (Yakir et al., 1990; Saranga et al., 1998) and photosynthetic rate (Pettigrew and Meredith, 1994).

Cotton breeding, usually done under optimal soil and water conditions, has resulted in fairly good adaptation to the high temperatures characterizing arid regions, but poor adaptation to water limitation also prevailing in these regions. In field-grown cotton, stomatal conductance (g_s) is typically extremely high and stomata exert little influence over transpiration rate (Ackerson and Krieg, 1977; Radin et al., 1987). In hot environments stomatal behavior of upland cotton does not maximize WUE, unless the crop is water stressed (Hutmacher and Krieg, 1983; Radin, 1989). Advances in the yield of irrigated pima cotton over the past 50 years have been associated with heat avoidance via increased g_s (Cornish et al., 1991; Radin et al., 1994). This heat-avoidance mechanism is effective for irrigated production in very hot environments; however, it is detrimental to WUE that is essential to obtain high productivity under conditions of limited water availability.

Both *GH* and *GB* are tetraploid ($2n = 4x = 52$), composed of "A" and "D" subgenomes that appear to have diverged from a common ancestor about 4–11 million years ago and then rejoined in a common nucleus about 1–2 million years ago (Wendel, 1989). Virtually all genes in tetraploid cotton are represented by one or more copies in each subgenome, in similar (albeit not identical) chromosomal orders in the two subgenomes (Reinisch et al., 1994) and their diploid ancestors (Brubaker et al., 1999). The fact that cotton, like many of the world's major crops, is polyploid, adds to its value a fringe benefit that its intensive study may shed new light on the role of polyploidy in plant adaptation to the abiotic environmental stress.

3 VARIATION IN WATER-USE EFICIENCY UNDER FIELD AND GREENHOUSE CONDITIONS

Variation in WUE among cotton cultivars, *GH*, *GB*, and interspecific hybrids (ISHs; *GH* × *GB*), was examined under two irrigation regimes, well-watered and water-limited (50% water application of the well-watered plants) in two field trials (consistent with commercial cotton production) and two greenhouse trials. Field trials were conducted in the western Negev in Israel during summer and plants examined over their entire life cycle, whereas greenhouse trials were conducted during winter and plants grown in pots until bud development or onset of flowering.

Based on field trials, analysis of variance revealed a significant effect of cultivars on both DM-based WUE and seed-cotton-based WUE; whereas the effects of either Cultivar × Year or Cultivar × Irrigation interactions were not significant for either variable (Saranga et al., 1998). Two *GH* cultivars (Siv'on and Vered) and the ISH (195×08) exhibited significantly greater WUE (about

2.5 g DM·L^{-1} H$_2$O) than two GB cultivars (F-177 and S-7; about 2.1 g DM·L^{-1} H$_2$O). Another 10 cultivars (GH cvs. G-1, 9003, CO-320, DP-6166, GC-356, Maxxa and ST-907; GB cvs. FC-18 and 362; and ISH cv. 326) were examined within the framework of only one trial, and did not exhibit further variation in WUE. One GH cultivar (G-414), selected for its low WUE in preliminary observations, reached WUE similar to the GB cultivars. While Siv'on and Vered are derived from American Acala germplasm, which was adapted to Israeli conditions, G-414 is an Acala-type cultivar that originated from central Asia and has not undergone extensive selection in Israel.

In the greenhouse, significant Cultivar × Year interaction called for a separate analysis of cultivar effects on these variables in each trial (Saranga et al., in preparation). In both trials, however, the GH cultivars (Siv'on, and G-1) revealed lower values of WUE (2.85 g DM·L^{-1} H$_2$O, on average) than either the GB cultivars (F-177 and BD-13; 3.52 g DM·L^{-1} H$_2$O) or the ISH (cvs. 195×10 and 326; 3.63 g DM·L^{-1} H$_2$O). The ISHs exhibited higher DM and WUE than the GB cultivars in Trial 1, whereas no difference between these cultivars was found in Trial 2.

The different rankings for WUE obtained in field and greenhouse conditions demonstrated the existence of Genotype × Environment interactions. In comparisons between outdoor and greenhouse environments, genotypic ranking changed also in wheat (Richards and Condon, 1993) and sorghum (Henderson et al., 1998). Field and greenhouse conditions differ from one another in several ways that can alter genotypic ranking. Masle et al. (1993) reported that Ecotype × Environment interaction for δ^{13}C in *Arabidopsis thaliana* was affected by light intensity and atmospheric humidity. Rooting environment, as determined by pot size, influenced δ^{13}C and caused contrasting genotypic responses in cowpea (Ismail et al., 1994). Changes in genotypic ranking in greenhouse versus field environments could also be due to Genotype × Plant age interactions, as found in wheat (Condon et al., 1992), cowpea (Ismail and Hall, 1993), and sorghum (Henderson et al., 1998).

Possible causes of the Genotype × Environment interactions observed in our studies were examined in a subsequent experiment (Saranga et al., in preparation). Two cultivated species (GH cv. Siv'on and GB cv. F-177) and their F1 progeny (ISH) were grown during summer in pots in a phytotron under four combinations of temperature (28/34 and 16/22 C, night/day respectively) and solar radiation (100% and 30%) levels and examined for their WUE. Both Cultivar × Temperature and Cultivar × Radiation interactions affected WUE. Under the high temperature–high radiation conditions (similar to field conditions) the GH cultivar and the ISH exhibited greater WUE than the GB cultivar, whereas under the low temperature–low radiation conditions (similar to greenhouse winter conditions) the GB cultivar and the ISH where superior to the GH cultivar in terms of WUE. Under the other two environments (high temperature–low radiation, and low temperature–high radiation) no differences between cultivars were found.

The different rankings of GH versus GB for WUE, obtained under field and greenhouse environments and confirmed in the corresponding phytotron treatments, suggest that the respective species have evolved different environmental adaptations. In both environments, however, the ISHs exhibited WUE values either similar to or higher than the highest WUE among the two species, suggesting that combining genes from the two species offers the potential to improve WUE. This conclusion is further supported by genetic mapping (see later).

Consistency of genotypic ranking is essential for effective breeding. The assessment of WUE (either directly or indirectly as $\delta^{13}C$) in the greenhouse out of season could enable several generations of selection per year. This, however, would not be effective if Genotype × Environment interaction for WUE or $\delta^{13}C$ is high, resulting in different genotypic rankings. The occurrence of Genotype × Environment interactions, found in our studies, emphasizes the need to examine cotton germplasm under the relevant field conditions, when attempting to improve WUE.

4 ASSOCIATIONS BETWEEN WUE AND GAS EXCHANGE IN THE FIELD AND GREENHOUSE

The associations between WUE and gas exchange were also studied in the same field and greenhouse experiments. In the field, instantaneous net photosynthetic rate (Pn) positively correlated across different cultivars with seed-cotton-based WUE. Similar trends, although not always significant, were also observed for DM-based WUE in both years (Saranga et al., 1998). In contrast to this, our greenhouse-grown cotton revealed negative correlation in both trials between WUE and instantaneous, midday values of Pn (Saranga et al., in preparation). While the gas exchange data associated high WUE with low Pn, the long-term WUE data were positively correlated with DM. This discrepancy may indicate that instantaneous measurements under greenhouse conditions were not representative of the long-term integrative WUE.

In the greenhouse, g_s values were generally low ($0.1–0.4\ mol\cdot m^{-2}\cdot s^{-1}$) as compared to the field ($0.4–1.1\ mol\cdot m^{-2}\cdot s^{-1}$). Stomatal behavior of cotton differs between greenhouse and field environments in two important aspects, greenhouse plants exhibiting much lower g_s (as shown here) and higher sensitivity to water stress than field-grown plants (Radin, 1992). In addition, root restriction, which may have occurred in our greenhouse trial, was reported to reduce leaf carbon assimilation (Schaffer et al., 1997; Whilley et al., 1999) and increase WUE (Ismail et al., 1994). These differences may have underlain the different relationships between WUE and Pn observed under the different environments.

5 CARBON ISOTOPE RATIO AS AN INDICATOR FOR WUE

The existence of genetic variation in WUE has been known since the pioneering studies of Briggs and Shantz in 1914 (reviewed by Hall et al., 1994). However, exploitation of this variation for breeding was limited for many years due to the difficulties involved in accurate evaluation of WUE. The demonstration that $\delta^{13}C$ could provide an indirect measure of plant WUE (Farquhar et al., 1982; Farquhar and Richards, 1984) stimulated considerable research on the potential use of $\delta^{13}C$ as a selection criterion in plant breeding programs. This association occurs because of independent linear relationships of $\delta^{13}C$ and WUE to the ratio of intercellular to ambient partial pressures of CO_2 (C_i/C_a) in leaves (Farquhar and Richards, 1984). Experimental evidence of the correlation between C_i/C_a and both WUE and $\delta^{13}C$ has been provided for a number of crops (for reviews, see Farquhar et al., 1989; Ehleringer et al., 1993). It is worth noting, however, that although both $\delta^{13}C$ and WUE are related to C_i/C_a, the relationships between $\delta^{13}C$ and WUE are complex. Environmental factors, such as water availability, temperature, and vapor pressure deficit, can influence these relationships differently in different plants (Hall et al., 1994); for example, in peanut WUE was correlated with $\delta^{13}C$ only under well-watered conditions but not under extreme drought (Wright et al., 1993).

A major limitation of the carbon isotope approach is the fact that $\delta^{13}C$ is influenced by any change in C_i. For example, both low g_s and high Pn will decrease C_i/C_a and increase WUE and $\delta^{13}C$; however, DM production increases only with high Pn. Therefore, when the goal is improving crop plants, it is essential to understand the physiological basis for improved WUE, to ensure that it is combined with high productivity.

In spite of the differences between field and greenhouse conditions, under both environments WUE was positively related to $\delta^{13}C$ (Saranga et al., 1998) showing decreasing discrimination as WUE increases, as expected. In the field, $\delta^{13}C$ correlated with DM-based WUE (significant in Trial 1 and nearly significant in Trial 2) and seed-cotton-based WUE (significant in both trials). In the greenhouse, WUE was significantly correlated with $\delta^{13}C$ in Trial 1, whereas in Trial 2, *GH* cultivars diverted from the trend line of all other cultivars. Marked differences in $\delta^{13}C$ values were found between the field-grown ($-26‰$ to $-28‰$) and greenhouse-grown ($-23.5‰$ to $-27.5‰$) plants. The more negative $\delta^{13}C$ values in the field were consistent with the higher g_s values discussed earlier.

Breeding for improved WUE requires a reliable and easily obtainable selection criterion. DM-based WUE is based on overall plant productivity and, therefore, reflects the physiological WUE; however, seed-cotton-based WUE is more relevant to crop productivity. For the cultivars examined in our field study, the various estimates of WUE were highly correlated and could, therefore, be used interchangeably. However, estimates of both DM-based and yield-based WUE are labor intensive and, thus, not suitable for breeding purposes. Moreover, an evaluation of these parameters for single field-grown plants, as would be required for breeding purposes, is impossible due to the difficulty in estimating the water use of a single plant. Photosynthetic rate was inconsistently correlated with WUE, emphasizing the limitations of short-term gas-exchange measurements as a basis for estimating crop WUE.

The application of the carbon isotope approach for breeding "water-efficient" cotton varieties seems feasible, but attention must be paid to the separation between water-conserving and high-productivity WUE-type adaptations. Variations in $\delta^{13}C$ values reflect changes in discrimination associated primarily with ci/ca. A decrease in intercellular CO_2 concentration and the consequent improvement in WUE can be due to lower g_s or higher Pn. The $\delta^{13}C$ analysis alone cannot distinguish between these two options. It is, therefore, significant that in the cotton plants we used, WUE was associated with DM, and thus $\delta^{13}C$ seemed to provide a useful and consistent indication of DM-related WUE across different growth conditions and physiological characteristics.

6 OVERVIEW OF GENETIC MAP

Segregating populations were studied in two field trials each with two irrigation regimes, well-watered and water-limited (50% water application of the well-watered plants). The first experiment consisted of 900 interspecific F_2 cotton plants (self-fertilized progenies of a F_1 hybrid, *GH* cv. Siv'on × *GB* cv. F-177), of which 430 plants that produced sufficient seed for the subsequent experiment were completely phenotyped and genotyped. The second experiment consisted of 214 F_3 families (self fertilized progenies of the F_2, 107 from each treatment to eliminate any possible consequences of differential selection in the F_2) were selected to represent the entire population with an emphasis on families for which parents exhibited extreme values of $\delta^{13}C$. Three replicates of five plants per plot were grown from each F_3 family under each irrigation regime. Plants were phenotyped and mapped for productivity (DM; seed-cotton yield; harvest index; boll weight; boll number), related physiological traits (osmotic potential at dawn, OP; $\delta^{13}C$ in leaves during boll development; midday canopy temperature, chlorophyll *a*, chlorophyll *b*) and fiber properties (length, length uniformity, elongation, strength, fineness, and yellowness).

Plant performance under arid conditions showed a high degree of genetically determined variation (Paterson et al., 2003; Saranga et al., submitted). For most traits, phenotypic variation between the parental lines was small with the exception of $\delta^{13}C$, boll weight, fiber length, and fiber strength. However, variation among F3 families was severalfold larger than that found between the parental lines, suggesting that the parental lines possessed different alleles that can confer transgressive phenotypes to their progenies. This hypothesis was well supported by QTL mapping results (see the following), with only harvest index failing to show either overdominant QTLs, or favorable alleles from each parent at different QTLs, forming the basis for the observed phenotypic transgression.

A total of 161 QTLs were detected, 82 QTLs for 10 traits reflecting crop productivity and related physiological variables and 79 QTLs for 6 traits reflecting lint quality (Saranga et al., 2001). Among the QTLs detected, 102 (63%) showed no significant difference in their effects between well-watered and water-limited conditions. Of particular interest were 33 QTLs (20%) that influenced plant productivity (11 QTLs), physiological traits (5 QTLs), or fiber quality (17 QTLs) only in the water-limited treatment, but showed no differences between GH and GB alleles in the well-watered treatment. Favorable alleles at such loci may be assembled into genotypes that incorporate adaptations to water-limited conditions, but also retain high levels of productivity under well-watered conditions. Thirteen QTLs (8%; 7, 4, and 2 for productivity, physiology, and quality, respectively) that influenced plant performance only under well-watered conditions may be useful for basic research to identify specific metabolic lesions that render some genotypes especially sensitive to water deficit. Thirteen QTLs (8%; 1, 3, and 9 for productivity, physiology, and quality, respectively) influenced the relative values (ratio of phenotype under water-limited to well-watered conditions) indicating differences in stability of plant performance between the two environments.

7 ASSOCIATION BETWEEN PRODUCTIVITY AND PHYSIOLOGICAL TRAITS

To test the extent to which different traits were under related genetic control, we evaluated the correspondence of QTL likelihood intervals. The 82 QTLs discovered in this study were located in 45 nonoverlapping genomic regions (Saranga et al., submitted). While some correspondence is expected (and found) among interrelated traits such as seed-cotton yield (SC) and harvest index, or SC and yield components (boll weight and boll number), we were especially interested in relationships between physiological parameters and measures of plant productivity. Such associations were found in 12 genomic regions. In most cases, the same parent conferred the favorable allele for each of the associated traits.

There was a particularly strong relationship between QTLs for OP and QTLs reflecting plant productivity. An association between SC and OP occurred in three genomic regions (chromosomes [Chrs] 6 and 25, and linkage group [LG] A02) and between boll weight and OP in one genomic region (LG A01), with the favorable allele (GH in LG A01, Chr 06 and LG A02, and GB in Chr 25) associated with both high SC (or boll weight) and low OP. These results are further supported by significant correlations between OP and SC (r = −0.28, $P < 0.001$) and OP and DM (r = −0.17, $P < 0.05$) in the water-limited treatment of the replicated year 2 trial. The likelihood that 3 of 12 OP QTLs would be associated with 3 of 14 SC QTLs by chance, in a genome the size of cotton's, is about 0.5% (Larsen and Marx, 1985; Lin et al., 1995). An especially important type of correspondence between OP and SC QTLs is highlighted on Chrs 6 and 25. The OP and SC QTLs on Chrs 6 and 25 mapped to homoeologous (corresponding) locations on the two different subgenomes of tetraploid cotton. Such an event is highly unlikely to happen by chance and thus further supports the veracity of the association between OP and SC.

Osmotic potential determinations were based on leaves sampled at dawn during the boll development period, when irrigation in both treatments had permitted overnight recovery of plant water status. Therefore, differences in OP may have resulted from active osmotic adjustment (OA), rather than solute concentration due to water loss (Morgan, 1984). While disagreement and even confusion may characterize some discussions on what constitutes a significant and an effective mechanism of crop drought resistance, the importance of OA is receiving growing recognition (Zhang et al., 1999). Prior studies of OA have been largely based on phenotypic associations between OA and yield under drought stress (cf. Ludlow et al., 1990; Morgan, 1995; El Hafid et al., 1998; Kumar and Singh, 1998; Tangpremsri et al., 1995). Our results add a new dimension to previously reported relationships between these traits, in that we have shown that there appears to exist not only a phenotypic correlation but also a partly common genetic basis of OA and productivity.

Surprisingly, $\delta^{13}C$ showed little clear association with productivity. QTLs conferring high $\delta^{13}C$ overlapped with one QTL conferring high SC (LG D01) and two QTLs conferring high boll weight (Chr 25 and LG D02). In a genome the size of cotton's, the likelihood that these associations would occur by chance is high, about 83% and 81%, respectively (Larsen and Marx, 1985; Lin et al., 1995). $\delta^{13}C$ correlated with SC (r = 0.27, $P < 0.001$) only in the well-watered treatment of year 1; however, it did not correlate with boll weight. $\delta^{13}C$ reflects a complex physiological response (Farquhar and Lloyd, 1993), specifically an integrated season-long measure of quantitative changes in the relationships between g_s and photosynthetic capacity, often used as an indicator for WUE in plants (Condon and Hall, 1997; see also Ehleringer et al., 1993, and references therein). However, high WUE is not necessarily associated with productivity, since a plant's WUE can be modified by different mechanisms. For example, either increased carbon assimilation rates (at a given g_s) or reduced g_s (and transpiration) would enhance WUE, but only the former would increase productivity. Our data indicate that selection for high $\delta^{13}C$ alone may reduce water consumption but is not expected to either increase or decrease productivity. The two other studies in which both yield and $\delta^{13}C$ were mapped in soybean (Mansur et al., 1993; Specht et al., 2001) both support our finding. Phenotypic correlation between $\delta^{13}C$ and productivity are inconsistent; apparently, $\delta^{13}C$ can be negatively, positively, or not at all associated with productivity, depending upon the growing environment and plant factors responsible for the phenotypic variation in $\delta^{13}C$ (see review by Hall et al., 1994). In *GB* cotton, lint yield was positively correlated with g_s (Radin et al., 1994), which in turn correlated negatively with $\delta^{13}C$ (Lu et al., 1996), whereas in *GH*, yield and $\delta^{13}C$ were positively correlated (Gerik et al., 1996). We cannot preclude the possibility that under different environmental conditions or in a different plant population $\delta^{13}C$ can show association with productivity.

A second case of QTLs on homoeologous regions involved $\delta^{13}C$ and chlorophyll content. Surprisingly, QTL alleles associated with higher $\delta^{13}C$ under water-limited conditions also coincided with *lower* chlorophyll *b* (Chr 22 and LG D05) and chlorophyll *a* (LG D05). The degree of overlap between these QTLs would be expected to occur by chance in only 0.02% of cases. These findings are further supported by the results of a subsequent study in which F4 and F5 progenies of the plants from the current study were examined (Cohen, 2001). Among six interspecific lines and the two parents examined in two field trials under two irrigation regimes, the line having the highest $\delta^{13}C$ values also had the lowest chlorophyll concentrations. Since chlorophyll concentration often directly correlates with photosynthetic capacity (Araus et al., 1997), the combination of high chlorophyll concentration with low $\delta^{13}C$ may indicate high g_s. This finding corresponds with a previous study with *GB* showing that increases in lint yield obtained over the last few decades were associated with increases in g_s (Radin et al., 1994).

The polyploidy of cotton was reflected in two cases in which corresponding "homoeologous" loci on each of the two different "subgenomes" appeared to account for common sets of traits (Chrs 6 and 25 for SC and OP; Chr 22 and LGD05 for $\delta^{13}C$ and chlorophyll contents). The discovery that each of two homoeologous locations account for genetic variation in the same phenotypes suggests that subsequently to polyploid formation in cotton, new functionally significant mutations (alleles) have arisen at each of the two homoeologous loci.

8 LINT QUALITY

The results obtained in this study generally support previous studies of fiber quality (Jiang et al., 1998) and other traits (Wright et al., 1998, 1999; Jiang et al., 2000a,b) in suggesting that the D-subgenome, from an ancestor that does not produce spinnable fiber, plays an important role in the genetic determination of fiber quality in tetraploid cotton. Among the 79 marker-trait associations reported here, 45 (57%) are located on D-subgenome chromosomes (Paterson et al., 2003). Although this modest excess of D-subgenome QTLs (chi-square $= 1.76$, $P = 0.2$) falls short of statistical significance, it continues to point to the notion that the D-subgenome of cotton not only contributes, but may even contribute a higher level of phenotypically relevant variability to tetraploid cottons than the A-subgenome.

Only six pairs of fiber-quality QTLs appear to map to homoeologous locations (fiber fineness QTLs on Chrs 2 and 14, Chrs 9 and 23, and Chrs 6 and 25; fiber strength QTLs on LGs A02-D03 and LGs A03-D02, and fiber yellowness QTLs on Chrs 6 and 25), so few that such associations are readily explained by chance (using the methods described in Paterson et al., 1995). The paucity of homoeologous associations supports the previously suggested notion that the A-subgenome (for which diploid forms do produce spinnable fiber) may already have contained favorable alleles at some major loci affecting fiber traits when polyploids evolved, as a result of prior natural selection. By contrast, the D-subgenome (for which diploid forms do not produce spinnable fiber), may have come under selection at these primary fiber-determining loci only after polyploid formation, and therefore harbor greater allelic diversity among tetraploid forms.

9 GENOTYPE × ENVIRONMENT INTERACTIONS

Differential genotypic expression across environments, often referred to as Genotype × Environment interaction (G × E) is one of the central challenges facing plant and animal breeders. Many agriculturally important traits are endpoint measurements, reflecting the aggregate effects of large numbers of genes acting independently and in concert, throughout the life cycle of the organism—and external factors at any time during the life cycle may change the course of growth and development of an organism in ways that may or may not be predictable. The extent to which G × E affects a trait is an important factor in the degree of testing over years and locations that must be employed to satisfactorily quantify the performance of a crop genotype. Because testing is a major determinant in the time and cost of developing new crop varieties, G × E interactions and their consequences have received a considerable amount of attention from crop scientists (see Romagosa and Fox, 1993 for a recent review). While the specific environmental parameters contributing to G × E are often unknown, water availability is a particularly important factor in determining the performance of different crop genotypes.

The genetic control of cotton productivity and fiber quality, as reflected by a comprehensive genomewide QTL mapping, was markedly affected by specific differences in water management regimes (Saranga et al., 2001). There does appear to exist a basal set of QTLs that are relatively unaffected by environmental parameters and account for progress from selection in a wide range of environments. An especially important finding in regard to G × E interactions was that 33 QTLs were detected only in the water-limited treatment, while only 13 were specific to the well-watered treatment. This suggests that improvement of productivity and fiber quality under water stress may be even more complicated than improvement of this already complex trait under well-watered conditions.

The finding (both herein and by Tuinstra et al., 1997) that partly different sets of genetic loci account for productivity and quality under well-watered versus water-limited conditions indicates that genetic potential for productivity under arid conditions can be improved with little or no penalty under irrigated conditions. At face value, these results seem contradictory to the long-held notion that selection for stress tolerance will generally result in reduced productivity under favorable environments and a decrease in average overall production (Finley and Wilkinson, 1963; Rosielle and Hamblin, 1981). Our findings might be reconciled with the classical expectation (Finley and Wilkinson, 1963; Rosielle and Hamblin, 1981) in that simultaneous improvement of productivity (and/or quality) for both arid and irrigated conditions will reduce the expected rate of genetic gain, because of the need to manipulate larger numbers of genes and conduct more extensive field testing (Falconer, 1981). These difficulties may be partly ameliorated by efficiencies gained through identification and use of diagnostic DNA markers (cf. Paterson, 1997; Bastford and Cooper, 1998).

Genotype × Environment interactions affecting key attributes such as productivity and fiber quality present special challenges in the improvement of crops such as cotton, in which similarly large acreages are grown under irrigated and rain-fed conditions, respectively. While it is accepted that some genotypes are better suited to irrigation and others to rain-fed production, the study and manipulation of specific genes that confer adaptation to these very different environments has previously focused largely on simply inherited variants useful in disease or insect management. These new findings suggest that complex traits such as fiber quality may also be fine-tuned to arid conditions, presumably in conjunction with the development of genotypes that also contain genes conferring adaptations such as osmotic adjustment that help to maintain productivity under arid conditions (Saranga et al., 2001).

10 IMPLICATIONS FOR IMPROVEMENT OF PRODUCTIVITY AND QUALITY UNDER ARID CONDITIONS

The prospects for developing cotton genotypes that retain commercial levels of quality and productivity, but reassemble the sets of genes needed to confer adaptation to arid conditions, may depend heavily upon exploring crosses between diverse germplasm. The strategy of crossing two superior genotypes of different species to better exploit the genetic potential for productivity under arid conditions, was borne out by the finding that each of the two species contained different alleles and/or loci conferring adaptation to arid conditions. Specifically, we crossed GH cv. Siv'on with GB cv. F-177, each of which had the highest WUE among cultivars of their species grown in the test environment in Israel (Saranga et al., 1998). At loci related to plant productivity and physiological status, the GH allele is favorable at 58 loci, and the GB allele at 12 loci, showing that recombination of favorable alleles from each of these species may form novel genotypes that are better adapted to arid conditions than either of the parental species. Genetic mapping of lint quality traits (Paterson et al., 2003) in the same population indicated that improved productivity is not negatively associated with quality, and hence their concurrent improvement seems feasible.

The use of an interspecific cross in this work enabled us to further investigate the extent to which superior QTLs might be found in an apparently inferior parent (Tanksley and Nelson, 1995). The cotton species GH and GB are thought to be derived from a common polyploid ancestor that formed naturally perhaps 1 million years ago (Wendel, 1989), and has diverged into five modern polyploid species. Cultivated forms of the two species differ in that GH tends to have higher yield and earlier maturity, but GB has markedly superior fiber length, strength, and fineness. While most favorable QTLs for these traits were indeed derived from the expected parent, an appreciable number of exceptions support the notion that new interspecific gene combinations may be created that are superior for human purposes than either of the naturally occurring species. Although mainstream cotton breeders only occasionally use such crosses, introgression from GH has played a major role in the breeding of GB (Wang et al., 1995), and many of the problems such as segregation distortion and linkage drag associated with use of such crosses can be mitigated by the use of DNA markers to identify rare desirable genotypes (Jiang et al., 2000a; Chee et al., in preparation).

Some especially interesting examples of transgressive variation warrant further discussion. The first is our discovery (described earlier) of QTLs related to the same trait in homoeologous chromosomal regions, most notably QTLs related to OP and SC on chromosomes 6 and 25. Curiously, the favorable allele(s) on chromosome 6 were conferred by *GH*, while the favorable allele(s) on chromosome 25 were conferred by *GB*. In principle, by assembling interspecific hybrid genotypes that contain each favorable allele, both *GH* and *GB* might perhaps be improved relative to their native state. This exemplifies the unique opportunities to evolve favorable new traits that accrue as a result of polyploid formation. The "genomic exploration" of other accessions of these species, or other wild tetraploid cottons (*G. tomentosum, G. darwinii, G. mustelinum*) may yield still additional valuable alleles.

Analysis at the resolution of QTL mapping is a start toward gaining molecular-level understanding of the nature of plant response to water deficit, but several observations point to the need for finer-scale studies. For example, a chromosome 22 QTL associated with harvest index showed a favorable effect of the *GH* allele in one year, and a favorable effect of the *GB* allele in the next year. One possible explanation of this may be the presence of a gene cluster, with both structural and functional divergence among members of the cluster, and between *GH* and *GB*. More generally, associations among QTLs represent overlaps of 20 to 30 cM intervals containing many genes, and are a starting point for molecular dissection but by no means an endpoint.

The merger of genetics and physiology promises to yield better understanding, and more effective improvement, of plant response to arid conditions. While our data clearly implicate OP in adaptation to arid conditions, testing of further traits is needed to account for QTL alleles that have not yet been linked to their physiological basis. In particular, several QTLs were associated with increased harvest index specifically under water-limited conditions, suggesting that there exists a mechanism for reallocating resources to reproductive tissues under abiotic stress—but none of these QTLs were associated with physiological parameters that we measured. The relatively large number of QTLs associated with $\delta^{13}C$ emphasize the physiological and genetic complexity of WUE and suggests that further genetic dissection may be needed to identify physiological traits that contribute to plant g_s/Pn relationships. Near-isogenic lines being made for QTLs discovered herein will offer a powerful new tool useful toward identification of the underlying gene(s) by using fine-scale mapping approaches (Paterson et al., 1990). The availability of cotton bacterial artificial chromosome (BAC) libraries (Abbey et al., in preparation; Peterson et al., in preparation) and established transformation methods for cotton (Bayley et al., 1992), together with the possibility of using comparative approaches (Paterson et al., 1996) to exploit complete sequence data from botanical models such as *Arabidopsis*, may help to provide clues as to the identities of genes in the genomic regions containing QTLs. Clues as to the physiological roles of these genes may help to design appropriate probes for parallel high-throughput expression studies (Schena et al., 1995; DeRisi et al., 1997; Hieter and Boguski, 1997; Ruan et al., 1998) and/or mutation searches (Underhill et al., 1997) to winnow the genes in these regions down to a few high-probability candidates. The prevalence (Kramer, 1980; Boyer, 1982) and possible spread (Le Houerou, 1996) of arid lands worldwide impel further efforts to dissect the molecular and physiological basis of adaptations to arid conditions in the world's leading crops.

ACKNOWLEDGMENTS

We gratefully acknowledge the support of Research Grant #US-2506-94R from the United States–Israel Binational Agricultural Research and Development (BARD) Fund, together with support from the Texas and Georgia Agricultural Experiment Stations, The Hebrew University of Jerusalem, the USDA National Research Initiative, the U.S. National Science Foundation, the Georgia Cotton Commission, and the Israel Cotton Production and Marketing Board.

REFERENCES

Ackerson RC, Krieg DR. Stomatal and non-stomatal regulation of water use in cotton, corn and sorghum. Plant Physiol 1977; 60:850–853.

Agrama HAS, Moussa ME. Mapping QTLs in breeding for drought tolerance in maize (*Zea mays* L). Euphytica 1996; 91:89–97.

Araus JL, Bort J, Ceccarelli S, Grando S. Relationship between leaf structure and carbon isotope discrimination in field grown barley. Plant Physiol Biochem 1997; 35:533–541.

Bastford KE, Cooper M. Genotype × environment interactions and some considerations to their implications for wheat breeding in Australia. Aust J Agric Res 1998; 49:153–174.

Bayley C, Trolinder NL, Ray C, Morgan M, Quisenberry JE, Ow DW. Engineering 2,4-D resistance into cotton. Theor Appl Genet 1992; 83:645–649.

Blum A. Plant Breeding for Stress Environment. Boca Raton: CRC, 1988.

Boyer JS. Plant productivity and environment. Science 1982; 218:443–448.

Brubaker CL, Paterson AH, Wendel JF. Comparative genetic mapping of allotetraploid cotton and its diploid progenitors. Genome 1999; 42:184–203.

Cohen A. Water-use efficiency of interspecific F4 and F5 cotton lines. M.Sc. dissertation, The Hebrew University of Jerusalem, Rehovot, 2001 (in Hebrew with English abstract).

Condon AG, Hall AE. Adaptation to diverse environments: variation in water-use efficiency within crop species. In: Jackson LE, ed. Ecology in Agriculture. San Diego: Academic Press, 1997:79–116.

Condon AG, Richards RA, Farquhar GD. The effect of variation in soil water availability, vapor pressure deficit and nitrogen nutrition on carbon isotope discrimination in wheat. Aust J Agric Res 1992; 43:935–947.

Cornish K, Radin JW, Turcotte EL, Lu Z, Zeiger E. Enhanced photosynthesis and stomatal conductance of Pima cotton (*Gossypium barbadense* L) bred for increased yield. Plant Physiol 1991; 97:484–489.

Cowan IR. Economics of carbon fixation in higher plants. In: Givnish TJ, ed. On the Economy of Plant Form and Function. Cambridge: Cambridge University Press, 1986:133–170.

DeRisi JL, Iyer VR, Brown PO. Exploring the metabolic and genetic control of gene expression on a genomic scale. Science 1997; 278:680–686.

El Hafid R, Smith DH, Karrou M, Samir K. Physiological attributes associated with early-season drought resistance in spring durum wheat cultivars. Can J Plant Sci 1998; 78:227–237.

Ehleringer JR, Hall AE, Farquhar GD, eds. Stable Isotopes and Plant Carbon–Water Relations. San Diego, CA: Academic Press, 1993.

Falconer D. Introduction to Quantitative Genetics. 2d ed. London: Longman Press, 1981.

Farquhar GD, Ehleringer JR, Hubick KT. Carbon isotope discrimination in photo-synthesis. Annu Rev Plant Physiol Plant Mol Biol 1989; 40:503–537.

Farquhar GD, Lloyd J. Carbon and oxygen isotope effects in the exchange of carbon dioxide between plants and the atmosphere. In: Ehleringer JR, Hall AE, Farquhar GD, eds. Stable Isotopes and Plant Carbon–Water Relations. San Diego: Academic Press, 1993:47–70.

Farquhar GD, O'Leary MH, Berry JA. On the relationship between carbon isotope discrimination and intercellular carbon dioxide concentration in leaves. Aust J Plant Physiol 1982; 9:121–137.

Farquhar GD, Richards RA. Isotopic composition of plant carbon correlates with water-use efficiency of wheat genotypes. Aust J Plant Physiol 1984; 11: 539–552.

Finley K, Wilkinson G. The analysis of adaptation in a plant-breeding program. Aust J Agric Res 1963; 14:742–754.

Gerik TJ, Gannaway JR, El-Zik KM, Faver KM, Thaxton PM. Identifying high yield cotton varieties with carbon isotope analysis. Proceedings Beltwide Cotton Conference. Vol. 3. Memphis, TN: National Cotton Council of America, 1996:1297–1300.

Hall AE, Richards RA, Condon AG, Wright GC, Farquhar GD. Carbon isotope discrimination and plant breeding. Plant Breed Rev 1994; 12:81–113.

Henderson S, von Caemmerer S, Farquhar GD, Wade L, Hammer G. Correlation between carbon isotope discrimination and transpiration efficiency in lines of the C4 species *Sorghum bicolor* in the glasshouse and the field. Aust J Plant Physiol 1998; 25:23–35.

Hieter P, Boguski M. Functional genomics: it's all how you read it. Science 1997; 278:601–602.

Hutmacher RB, Krieg DR. Photosynthetic rate control in cotton—stomatal and nonstomatal factors (*Gossypium hirsutum*). Plant Physiol 1983; 73:658–661.

Ismail AM, Hall AE. Carbon isotope discrimination and gas exchange of cowpea accessions and hybrids. Crop Sci 1993; 33:788–793.

Ismail AM, Hall AE, Bray EA. Drought and pot size effects on transpiration efficiency and carbon isotope discrimination of cowpea accessions and hybrids. Aust J Plant Physiol 1994; 21:23–35.

Jiang C, Wright RJ, El-Zik KM, Paterson AH. Polyploid formation created unique avenues for response to selection in *Gossypium* (cotton). Proc Natl Acad Sci USA 1998; 95:4419–4424.

Jiang C, Chee P, Draye X, Morrell P, Smith C, Paterson A. Multi-locus interactions restrict gene flow in advanced-generation interspecific populations of polyploid *Gossypium* (cotton). Evolution 2000; 54:798–814.

Jiang C, Wright R, Woo S, Delmonte T, Paterson A. QTL analysis of leaf morphology in tetraploid *Gossypium* (cotton). Theor Appl Genet 2000; 100: 409–418.

Jordan WR. Water relations in cotton. In: Teare JD, Peet MM, eds. Crop Water Relations. New York: Wiley Interscience, 1982.

Kohel RJ. Influence of certain morphological characters on yield. Cotton Grow Rev 1974; 51:281–292.

Kramer PJ. Drought, stress, and the origin of adaptation. In: Turner NC, Kramer PJ, eds. Adaptation of Plants to Water and High Temperature Stress. New York: John Wiley and Sons, 1980:7–20.

Kumar A, Singh DP. Use of physiological indices as a screening technique for drought tolerance in oilseed *Brassica* species. Ann Bot 1998; 81:413–420.

Larsen RJ, Marx ML. An Introduction to Probability and its Applications. Englewood Cliffs: Prentice-Hall, 1985.

Le Houerou HN. Climate change, drought and desertification. J Arid Environ 1996; 34:133–185.

Lebreton C, Lazic-Jancic V, Steel A, Pekic S, Quarrie SA. Identification of QTL for drought responses in maize and their use in testing casual relationship between traits. J Exp Bot 1995; 46:853–865.

Lilley JM, Ludlow MM, McCouch SR, O'Toole JC. Locating QTL for osmotic adjustment and dehydration tolerance in rice. J Exp Bot 1996; 47:1427–1436.

Lin YR, Schertz KF, Paterson AH. Comparative mapping of QTLs affecting plant height and flowering time in the Poaceae, in reference to an interspecific *Sorghum* population. Genetics 1995; 141:391–411.

Lu Z, Chen J, Percy RG, Sharifi MR, Rundel PW, Zeiger E. Genetic variation in carbon isotope discrimination and its relation to stomatal conductance in Pima cotton (*Gossypium barbadense*). Aust J Plant Physiol 1996; 23:127–132.

Ludlow MM, Santamaria JM, Fukai S. Contribution of osmotic adjustment to grain yield in *Sorghum bicolor* (L) Moench under water-limited conditions. II. Water stress after anthesis. Aust J Agric Res 1990; 41:67–78.

Mansur LM, Lark KG, Kross H, Oliveira A. Interval mapping of quantitative trait loci for reproductive, morphological, and seed traits of soybean (*Glycine max* L). Theor Appl Genet 1993; 86:907–913.

Martin B, Nienhuis J, King G, Schaefer A. Restriction fragment length polymorphisms associated with water use efficiency in tomato. Science 1989; 243:1725–1728.

Masle JM, Shin JS, Farquhar GD. Analysis of restriction fragment length polymorphisms associated with variation of carbon isotope discrimination among ecotypes of *Arabidopsis thaliana*. In: Ehleringer JR, Hall AE, Farquhar GD, eds. Stable Isotopes and Plant Carbon–Water Relations. San Diego: Academic Press, 1993:371–386.

McCarty J, Jenkins J. Cotton germplasm: characteristics of 79 day-neutral primitive race accessions. Tech Bull-Miss Agric For Exp Stn Tech Bull 1992; 184:17.

Mian MAR, Bailey MA, Ashley DA, Wells R, Carter TA Jr, Parrot WA, Boerma HR. Molecular markers associated with water use efficiency and leaf ash in soybean. Crop Sci 1996; 36, 1252–1257.

Mian MAR, Ashley DA, Boerma HR. An additional QTL for water use efficiency in soybean. Crop Sci 1998; 38:390–393.

Morgan JM. Osmoregulation and water stress in higher plants. Annu Rev Plant Physiol 1984; 35:299–319.

Morgan JM. Growth and yield of wheat lines with differing osmoregulatory capacity at high soil water deficit in seasons of varying evaporative demand. Field Crops Res 1995; 40:143–152.

Morgan JM, Tan MK. Chromosomal location of a wheat osmoregulation gene using RFLP analysis. Aust J Plant Physiol 1996; 23:803–806.

Paterson A. Molecular Dissection of Complex Traits. Boca Raton: CRC, 1997.

Paterson AH, DeVerna J, Lanini B, Tanksley SD. Fine mapping of quantitative trait loci using selected overlapping recombinant chromosomes, from an interspecies cross of tomato. Genetics 1990; 124:735–742.

Paterson AH, Lander ES, Hewitt JD, Peterson S, Lincoln S, Tanksley SD. Resolution of quantitative traits into Mendelian factors by using a complete linkage map of restriction fragment length polymorphisms. Nature 1988; 335:721–726.

Paterson AH, Lin YR, Li Z, Schertz KF, Doebley JF, Pinson SRM, Liu SC, Stansel JW, Irvine JE. Convergent domestication of cereal crops by independent mutations at corresponding genetic loci. Science 1995; 269:1714–1718.

Paterson AH, Lan TH, Reischmann KP, Chang C, Lin YR, Liu SC, Burow MD, Kowalski SP, Katsar CS, Del Monte TA, Feldmann KA, Schertz KF, Wendel JF. Toward a unified map of higher plant chromosomes, transcending the monocot–dicot divergence. Nat Genet 1996; 14:380–382.

Paterson AH, Saranga Y, Menz M, Jiang C-X, Wright RJ. QTL analysis of genotype × environment interactions affecting cotton fiber quality. Theor Appl Genet 2003; 106:384–396.

Pettigrew WT, Meredith WR Jr. Leaf gas exchange parameters vary among cotton genotypes. Crop Sci 1994; 34:700–705.

Quarrie SA, Gulli M, Calestani C, Steed A. Location of gene regulating drought-induced abscisic acid production on the long arm of chromosome 5A of wheat. Theor Appl Genet 1994; 89:794–800.

Quisenberry JE, Jordan WR, Roark BA, Fryrear DW. Exotic cottons as genetic sources for drought resistance. Crop Sci 1982; 21:889–895.

Radin JW. When is stomatal control of water loss consistent with the thermal kinetic window concept? Proceedings Beltwide Cotton Conference. Memphis, TN: National Cotton Council of America, 1989:46–49.

Radin JW. Reconciling water-use efficiencies of cotton in field and laboratory. Crop Sci 1992; 32:13–18.

Radin JW, Kimball BA, Hendrix DL, Mauney JR. Photosynthesis of cotton plants exposed to elevated levels of carbon dioxide in the field. Photosynth Res 1987; 12:191–203.

Radin JW, Lu Z, Percy RG, Zeiger E. Genetic variability for stomatal conductance in Pima cotton and its relation to improvements of heat adaptation. Proc Nat Acad Sci U S A 1994; 91:7217–7221.

Reinisch AJ, Dong J-M, Brubaker C, Stelly D, Wendel JF, Paterson AH. A detailed RFLP map of cotton (*Gossypium hirsutum* × *Gossypium barbadense*): chromosome organization and evolution in a disomic polyploid genome. Genetics 1994; 138:829–847.

Ribaut JM, Jiang C, Gonzalez-de-Leon D, Edmeades GO, Hoisington DA. Identification of quantitative trait loci under drought conditions in tropical maize. 2. Yield components and marker-assisted selection strategies. Theor Appl Genet 1997; 94:887–896.

Richards RA, Condon AG. Challenges ahead in using carbon isotope discrimination in plant breeding programs. In: Ehleringer JR, Hall AE, Farquhar GD, eds. Stable Isotopes and Plant Carbon–Water Relations. San Diego: Academic Press, 1993:451–462.

Romagosa I, Fox P. Genotype × environment interaction and adaptation. In: Hayward MD, Bosemark N, Romagosa I, eds. Plant Breeding: Principles and Prospects. London: Chapman and Hall, 1993:373–390.

Rosenow DT, Quisenberry JE, Wendt CW, Clark LE. Drought tolerant sorghum and cotton germplasm. Agric Water Manag 1983; 7:207–222.

Rosielle A, Hamblin J. Theoretical aspects of selection for yield in stress and non-stress environments. Crop Sci 1981; 21:943–946.

Ruan Y, Gilmore J, Conner T. Towards *Arabidopsis* genome analysis: monitoring expression profiles of 1400 genes using cDNA microarrays. Plant J 1998; 15:821–833.

Saranga Y, Flash I, Yakir D. Variation in water-use efficiency and its relation to carbon isotope ratio in cotton. Crop Sci 1998; 38:782–787.

Saranga Y, Menz M, Jiang C-X, Wright RJ, Yakir D, Paterson AH. Genomic dissection of genotype × environment interactions conferring adaptation of cotton to arid conditions. Genome Res 2001; 11:988–995.

Saranga Y, Jiang C-X, Wright RJ, Yakir D, Paterson AH. Genetic dissection of cotton: physiological responses to arid conditions and their inter-relationships with productivity. Submitted.

Schaffer B, Whilley AW, Searle C, Nissen RJ. Leaf gas exchange, dry matter partitioning, and mineral element concentrations in mango as influenced by elevated atmospheric carbon dioxide and root restriction. J Am Soc Hortic Sci 1997; 122: 849–855.

Schena M, Shalon D, Davis RW, Brown PO. Quantitative monitoring of gene expression patterns with a complementary DNA microarray. Science 1995; 270:467–470.

Specht JE, Chase K, Macrander M, Graef GL, Chung J, Markwell JP, Germann M, Orf JH, Lark KG. Soybean response to water: a QTL analysis of drought tolerance. Crop Sci 2001; 41:493–509.

Tangpremsri T, Fukai S, Fischer KS. Growth and yield of sorghum lines extracted from a population for differences in osmotic adjustment. Aust J Agric Res 1995; 46:61–74.

Tanksley S, Nelson J. Advanced backcross QTL analysis: a method for the simultaneous discovery and transfer of valuable QTLs from unadapted germplasm into elite breeding lines. Theor Appl Genet 1995; 92:191–203.

Teulat B, Monneveux P, Wery J, Borries C, Souyris I, Charrier A, This D. Relationship between relative water content and growth parameters under water stress in barley: a QTL study. New Phytol 1998a; 137:99–107.

Teulat B, This D, Khairallah M, Borries C, Ragot C, Sourdille P, Leroy P, Monneveux P, Charrier A. Several QTLs involved in osmotic adjustment trait variation in barley (*Hordeum vulgare* L). Theor Appl Genet 1998b; 96:688–698.

Tuberosa R, Sanguineti MC, Landi P, Salvi S, Casarini E, Conti S. RFLP mapping of quantitative loci controlling abscisic acid concentration in leaves of drought-stressed maize (*Zea mays* L). Theor Appl Genet 1998; 97:744–755.

Tuinstra MR, Grote EM, Goldsbrough PB, Ejeta G. Identification of quantitative loci associated with pre-flowering drought tolerance in sorghum. Crop Sci 1996; 36:1337–1344.

Tuinstra MR, Grote EM, Goldsbrough PB, Ejeta G. Genetic analysis of post-flowering drought tolerance and components of grain development in *Sorghum bicolor* (L) Moench. Mol Breed 1997; 3:439–448.

Ulloa M, Cantrell RG, Percy RG, Zeiger E, Lu Z. QTL analysis of stomatal conductance and relationship to lint yield in an interspecific cotton. J Cotton Sci 2000; 4:10–18.

Underhill PA, Jin L, Lin AA, Mehdi SO, Jenkins T, Vollrath D, Davis RW, Cavalli-Sforza LL, Oefner PJ. Detection of numerous Y chromosome biallelic polymorphisms by denaturing high-performance liquid chromatography. Genome Res 1997; 7:996–1005.

Wang G, Dong J, Paterson AH. Genome composition of cultivated *Gossypium barbadense* reveals both historical and recent introgressions from *G. hirsutum*. Theor Appl Genet 1995; 91:1153–1161.

Wendel JF. New world tetraploid cottons contain old world cytoplasm. Proc Natl Acad Sci USA 1989; 86:4132–4136.

Whilley AW, Searle C, Schaffer B, Wolstenholme BN. Cool orchard temperatures or growing trees in containers can inhibit leaf gas exchange of avocado and mango. J Am Soc Hortic Sci 1999; 124:46–51.

Wright GC, Hubick KT, Farquhar GD, Nageswara Rao RC. Genetic and environmental variation in transpiration efficiency and its correlation with carbon isotope discrimination and specific leaf area in peanut. In: Ehleringer JR, Hall AE,

Farquhar GD, eds. Stable Isotopes and Plant Carbon–Water Relations. San Diego: Academic Press, 1993:245–267.

Wright RJ, Thaxton PM, El-Zik KM, Paterson AH. D-Subgenome bias of Xcm resistance genes in tetraploid *Gossypium* (cotton) suggests that polyploid formation has created novel avenues for evolution. Genetics 1998; 149:1987–1996.

Wright RJ, Thaxton P, Paterson AH, El-Zik KM. Molecular mapping of genes affecting pubescence of cotton. J Heredity 1999; 90:215–219.

Yakir D, De Niro MJ, Ephrath JE. Effect of water stress on oxygen, hydrogen and carbon isotope ratios in two species of cotton plants. Plant Cell Environ 1990; 13: 949–955.

Zhang J, Nguyen HT, Blum A. Genetic analysis of osmotic adjustment in crop plants. J Exp Bot 1999; 50:291–302.

14

Molecular Dissection of Abiotic Stress Tolerance in Sorghum and Rice

M. S. Pathan and Henry T. Nguyen
University of Missouri–Columbia
Columbia, Missouri, U.S.A.

Prasanta K. Subudhi
Louisiana State University
Baton Rouge, Louisiana, U.S.A.

Brigitte Courtois
CIRAP-Biotrop
Montpellier, France

1 INTRODUCTION

Drought, salinity, and flooding are the major abiotic stresses that limit crop production all over the world. Drought is, by far, the leading environmental stress in agriculture in the United States and the rest of the world. Most of the crop production areas of the world face frequent and short periods of water deficits almost every year and severe drought stress in some years. Drought at any stage of crop development affects growth and production, but drought during the flowering stage causes maximum crop damage. Drought is endemic in large areas of Asia and Africa, where most of the world's poor people live and where they fully depend on rainfall for crop production. The developed world is not exempt from the problem. In the United States alone, about 40% of crop losses occur yearly due to drought (Boyer, 1982). Efficient irrigation systems can mitigate crop loss due to drought and can also bring more areas under cultivation. However, the increasing cost of irrigation water is a great concern for profitable crop production, especially for poor farmers in subsistence agriculture. All over the world, most of the cereal crops are grown under rainfed conditions where irrigation is unavailable or, if available, too costly for marginal farmers. Under a water-limited environment, the genetic improvement of a crop for drought resistance is a sustainable and economically feasible solution to reduce the problem of drought (Blum, 1988).

Drought stress is a major constraint to sorghum [*Sorghum bicolor* (L.) Moench] production worldwide, although sorghum is considered as a highly drought tolerant cereal. Sorghum is mostly cultivated in rainfed environments. Drought at any stage of the life cycle of a sorghum crop affects its growth and production, but most severely at the postflowering stage. The situation is similar for rice grown in nonirrigated ecosystems. Plants possess several adaptive traits to endure periods of drought. Stay green is an important trait in sorghum for continued grain filling under postflowering drought stress. In rice, the ability of the root system to provide water for evapotranspirational demand from deep soil moisture and capacity for osmotic adjustment (OA) are considered major drought resistance traits. If lack of water (drought) is a problem for rice production, excess of water (flooding) can be a constraint, too. Unlike other crops, rice can tolerate submergence for a short period, but prolonged submergence significantly reduces growth and production. Submergence tolerance is a desirable trait for rainfed lowland rice, and stem elongation is desirable for deepwater rice. In rice and sorghum, salinity is also considered as an important abiotic stress limiting production. Flowering stage is the most sensitive stage for all kinds of abiotic stresses.

The genetic improvement of adaptation to drought and other abiotic stresses has been addressed through the conventional breeding approach by selecting for yield performance over locations and years. This approach remains slow because of the difficulty in finding optimal environments for evaluation. The recent development of molecular marker techniques offers new opportunities to develop abiotic stress tolerant crops through understanding of the tolerance components. Quantitative trait loci (QTLs)/genes related to drought tolerance have been identified in several crops, and progress from employing marker-assisted selection rather than phenotype-based progeny selection is anticipated.

This chapter focuses on examples of molecular dissection of abiotic stress tolerance, especially drought tolerance in sorghum, and drought, submergence, and salinity tolerance in rice. Because drought tolerance is a complex phenomenon, dissection into individual component makes it easier to understand its genetic bases and to apply this knowledge in marker-assisted breeding programs. This chapter also focuses on recent progress of new technologies such as functional genomics and proteomics in the area of abiotic stress tolerance in rice.

2 DROUGHT RESISTANCE IN SORGHUM

Sorghum is considered the most drought-resistant cereal and a model crop for evaluation of drought resistance mechanisms, but the genetic and physiological mechanisms involved in its expression are poorly understood (Subudhi and Nguyen, 2000). Drought resistance in sorghum is a complex trait affected by genotype by environment interaction. Preflowering and postflowering responses to drought stress are generally distinguished in sorghum (Rosenow, 1987, 1993). The preflowering drought response has been well characterized and observed when plants are under significant water stress before flowering, from panicle differentiation until flowering. Preflowering drought stress reduces panicle size, grain number and, finally, grain yield. Postflowering drought response occurs during grain development stages and is characterized by premature leaf senescence. Nonyellowing or stay green phenotype during grain filling is considered as a desirable trait in agriculture (Hauck et al., 1997). One of the possible ways to increase productivity is to increase or retain photosynthetic activity during grain filling. Stay green mutants have been detected and studied in *Festuca pratensis*, *Phaseolus vulgaris*, soybean, sorghum, pea, maize, and rice with the objective of increasing agronomic potential such as grain yield, biomass, resistance to abiotic stresses, and forage quality (Duncan et al., 1981; Thomas and Smart, 1993). Stay green ability delays premature leaf senescence under moisture stress during grain filling. Sorghum genotypes with stay green trait are able to fill grain normally under soil moisture stress and show increased resistance to charcoal rot and lodging (Rosenow and Clark, 1981; Rosenow, 1984). Premature leaf senescence leads to charcoal rot disease, stalk lodging, and yield reduction (Rosenow and Clark, 1995). In recent years, the stay green trait has been recognized as a major mechanism of postflowering drought resistance in grain sorghum (Rosenow et al., 1996). Expression of this trait has been reported in other crops including maize (Tollenaar and Daynard, 1978), oat (Helsel and Frey, 1978), and tomato (Akhter et al., 1999). Recently, stay green mutants were identified in rice and pasture grass, *F. pratensis* Huds, which leads to better understanding of senescence (Hauck et al., 1997; Thomas et al., 1999). The stay green trait has been successfully used in Australia to improve lodging resistance under terminal drought stress (Henzell et al., 1992).

Table 1 QTLs Related to Abiotic Stress Tolerance in Sorghum and Rice

References	Mapping population, number of lines, population type	Trait(s) studied	Total number of detected QTL, phenotypic variation (R^2), types of marker used
Sorghum: Traits related to drought tolerance			
Tuinstra et al. (1996)	TX7078 × B35, 98, RILs	Preflowering	Six, R^2 = 14–43%, 170 RFLP, RAPD
Crasta et al. (1999)	B35 × TX430, 96, RILs	Stay green	Seven, R^2 = 53%, RFLP
Tao et al. (2000)	QL39 × QL41, 196, RILs	Stay green	Five, R^2 = 10–15%, RFLP
Xu et al. (2000b)	B35 × TX7000, 98, RIL	Stay green	Four, R^2 = 30–46% average of five locations, RFLP
		Chlorophyll content	Three, R^2 = 21–32%, RFLP
Kebede et al. (2001)	SC56 × TX7000, 125, RILs	Stay green	Nine, R^2 = 10–26%, 170 RFLP
		Lodging tolerance	Three, R^2 = 15–19%
		Preflowering	Four, R^2 = 10–15%
Sanchez et al. (2002)	B35 × TX7000, 98, RILs	Stay green	Four, R^2 = 54%, 274 RFLP, SSR
Rice shoot related to drought tolerance			
Lilly et al. (1996)	CO39 × Moroberekan, 52, RILs	Osmotic adjustment	One, R^2 = 32%, 127 RFLP
		Dehydration tolerance	Four, R^2 = 27–36%
Quarrie et al. (1997)	IR20 × 63-83, 123, F2	Leaf size	Seven, 228 RFLP, AFLP
		ABA accumulation	Ten
Courtois et al. (2000)	IR64 × Azucena, 135, DHLs	Leaf rolling	Eleven, R^2 = 5–22%, 175 RFLP, AFLP, Isozyme
		Leaf drying	Ten, R^2 = 5–19%
		RWC	Eleven, R^2 = 6–19%
		Growth under stress	Ten, R^2 = 5–22%

Reference	Population	Trait	QTLs
Hemamalini et al. (2000)	IR64 × Azucena, 56, DHLs	Morphological and physiological traits	Thirty-five, R^2 = 11–29%, 175 RFLP, AFLP, Isozyme
Tripathy et al. (2000)	CT9993 × IR62266, 154, DHLs	Cell-membrane stability	Nine, R^2 = 13–42%, 315 RFLP, AFLP, SSR
Zhang et al. (2001)	CT9993 × IR62266, 154, DHLs	Osmotic adjustment	Five, R^2 = 8–13%, 315 RFLP, AFLP, SSR
Kamoshita et al. (2002a)	CT9993 × IR62266, 154, DHLs	Shoot biomass	Seven, R^2 = 9–32%, 315 RFLP, AFLP, SSR
Kamoshita et al. (2002b)	IR58821 × IR52561, 184, RILs	Shoot biomass	Two, R^2 = 13–14%, 399 RFLP, AFLP
Price et al. (2002b)	Bala × Azucena, 205, RILs	Leaf rolling	Five, R^2 = 5–20%, 135 RFLP, AFLP
		Leaf drying	Eleven, R^2 = 6–18%
		RWC	Eight, R^2 = 9–26%

Rice root related to drought tolerance

Reference	Population	Trait	QTLs
Champoux et al. (1995)	CO39 × Moroberekan, 203, RILs	Root thickness	Eighteen, R^2 = 13–33%, 127 RFLP
		Root shoot ratio	Sixteen, R^2 = 9–22%
		Root dry weight/tiller	Fourteen, R^2 = 11–18%
		Deep root weight	Eight, R^2 = 6–17%
Ray et al. (1996)	CO39 × Moroberekan, 203, RILs	Total root number	Nineteen, R^2 = 8–19%, 125 RFLP
		Root penetration index	Six, R^2 = 7–13%
		Tiller number	Ten, R^2 = 7–14%
		Penetrated root number	Four, R^2 = 6–8%
Price and Tomos (1997)	Azucena × Bala, 178, F2	Root length	Seventeen, R^2 = 5–38%, 71 RFLP
		Root cell length	One, R^2 = 10%
		Root thickness	Three, R^2 = 7–21%
Yadav et al. (1997)	IR64 × Azucena, 125, DHLs	Root morphology and Root distribution	Forty-three, R^2 = 4–21%, 175 RFLP, RAPD, isozyme

Table 1 Continued

References	Mapping population, number of lines, population type	Trait(s) studied	Total number of detected QTL, phenotypic variation (R^2), types of marker used
Ali et al. (2000)	IR58821 × IR52561, 166, RILs	Total root number	Two, R^2 = 9–12%, 399 RFLP, AFLP
		Penetrated root number	Seven, R^2 = 11–27%
		Root penetration index	Six, R^2 = 12–26%
		Penetrated root thickness	Eight, R^2 = 6–14%
		Penetrated root length	Five, R^2 = 6–13%
Price et al. (2000a)	Azucena × Bala, 205, RILs	Total roots	Three, R^2 = 5–10%, 135 RFLP, AFLP
		Penetrated root	Seven, R^2 = 5–17%
		Penetration ratio	Seven, R^2 = 7–18%
		Tiller number	One, R^2 = 12%
Zheng et al. (2000)	IR64 × Azucena, 109, DHLs	Root penetration index	Four, total R^2 = 35%, 175 RFLP, RAPD, Isozyme
		Penetrated root	Four, total R^2 = 35%
		Total root number	Two, total R^2 = 18%
		Penetrated root number	Two, total R^2 = 4%
Zhang et al. (2001)	CT9993 × IR62266, 154, DHLs	Root penetration index	Four, R^2 = 8–33%, 315 RFLP, AFLP, SSR
		Basal root thickness	Six, R^2 = 9–37%
		Penetrated root thickness	Eleven, R^2 = 9–31%
		Penetrated root length	One, R^2 = 17%
		Total root dry weight	Five, R^2 = 9–20%
		Penetrated root dry wt.	Three, R^2 = 11–17%
		Root pulling force	Six, R^2 = 9–20%
Kamoshita et al. (2002a)	CT9993 × IR62266, 154, DHLs	Root morphology	Thirty-seven, R^2 = 8–56%, 315 RFLP, AFLP, SSR

Reference	Cross, population	Trait	QTL information
Kamoshita et al. (2002b)	IR58821 × IR52561, 184, RILs	Root morphology	Twenty-nine, R^2 = 6–30%, 399 RFLP, AFLP
Submergence tolerance in rice			
Nandi et al. (1997)	IR74 × FR13A, 74, RILs	Submergence tolerance	Five (major one on chr. 9) and Four other, R^2 = 19–27%, 202 AFLP
Sripongpangkul et al. (2000)	IR74 × Jalamagna, 165, RILs	Submergence tolerance	Thirteen, R^2 = 11–36%, 144 RFLP, AFLP, and isozyme
		Leaf elongation	Three, R^2 = 9–14%
		Internodes elongation	Three R^2 = 8–37%
Xu et al. (2000a)	DX18-121 × M-202, 2950, F2	Submergence tolerance	One, *Sub1* gene
Siangliw et al. (2003)	KDML105 × RF13A, IR49830, IR67819F2; 467, BC	Submergence tolerance	One, *Sub1* gene
Toojinda et al. (2003)	IR49830 × CR6241, 65, DHLs FR13A × CT6241, 172, RILs Jao-Hom-Nin × KDML105, 188, F2	Submergence tolerance	Five traits, Integrated map with 298 RFLP, AFLP, and others, Ten, R^2 = 9–74%, Nineteen, R^2 = 2–77%, Sixteen, R^2 = 2–18%
Salt tolerance in rice			
Zhang et al. (1995)	M-20 × 77-170, 85, F2	Salt tolerance	One, 130 RFLP
Flowers et al. (2000)	Nona Bokra × Pokkali //IR4630 × IR10167, 150, RILs	Na, K uptake, and ion concentration	Sixteen, AFLP
Prasad et al. (2000)	IR64 × Azucena, 76, DHLs	Salt tolerant traits	Seven, R^2 = 13–20%, RFLP
Koyama et al. (2001)	IR4630 × IR15324, 118 RILs	Ion concentration, ratios and ion uptake (7 traits)	Eleven, R^2 = 6–20%, RFLP, AFLP, SSR

2.1 Preflowering Drought Stress

Tuinstra et al. (1996) detected six genomic regions associated with preflowering drought tolerance in recombinant inbred lines (RILs) derived from the cross TX7078 × B35. Five out of six alleles were derived from TX7078, a preflowering drought-tolerant variety. B35 is susceptible to preflowering drought. These regions were not detected under fully irrigated conditions, indicating that these QTLs were only expressed under drought stress. Kebede et al. (2001) identified four QTLs for preflowering drought stress on linkage groups C, E, F, and G in the RIL population developed from SC56 × TX7000. One major QTL for preflowering drought resistance, *pfrG*, was consistently detected in two environments and accounted for 15–37% of phenotypic variation. This region is very important because it harbors QTL for other traits as well (stay green, lodging resistance, flowering time, and plant height). Another preflowering drought-tolerant QTL, *pfr F*, accounting for 22–25% of the phenotypic variation, was consistently detected in two environments. This region also overlaps with a QTL for flowering time and a strong phenotypic correlation between these two traits has been noticed. Rosenow et al. (1996) indicated that this association was probably a result of the effect of preflowering drought stress on flowering time.

2.2 Postflowering Drought Stress

In the last decade, significant progress was made in the analysis of QTLs associated with drought resistance in sorghum. Several authors have identified different genomic regions associated with preflowering and postflowering drought resistance (Tuinstra et al., 1996; Crasta et al., 1999; Tao et al., 2000; Xu et al., 2000a; Kebede et al., 2001; Sanchez et al., 2002) (Table 1). QTLs for stay green ability have been mapped in five different populations. Sorghum genotype B35 is resistant to postflowering drought stress but susceptible to preflowering drought stress. The genotype QL41 is a stay green line developed from a cross between sorghum genotypes B35 and QL33. The genotype SC56, a well-characterized stay green line, was selected from Sudan. In contrast, genotype TX7000 is resistant to preflowering stress but susceptible to postflowering drought stress. A number of stay green QTLs have been detected by several authors using populations developed from the different genetic backgrounds. Direct comparisons of stay green QTLs detected in different populations proved difficult due to the use of different types of markers. But comparative study of stay green QTLs indicated that four stay green QTLs (*stg1*, *stg2*, *stg3*, and *stg4*) were consistently found in different genetic backgrounds (Subudhi et al., 2000). Two stay-green QTLs, *stg1* and *stg4*, were mapped in identical regions of B35 × TX430 (Crasta et al., 1999), B35 × TX7000 (Xu et al., 2000a; Sanchez et al., 2002), and SC56 × TX700 (Kebede et al., 2001). Stay green QTL *stg2* was also detected in both B35 × TX430 and

B35 × TX7000 populations. Among the four QTLs, *stg2* was considered to be the most important, explaining about 30% of phenotypic variation in both B35 × TX430 and B35 × TX7000 populations. Stay green QTL *stg3* was mapped in the identical genomic region of populations B35 × TX430 (Crasta et al., 1999), and SC56 × TX7000 (Kebede et al., 2001). Xu et al. (2000a) detected one QTL for chlorophyll content in the same genomic region as the stay green QTL *stg1*. This region showed synteny with a stay green QTL of maize on chromosome 8 (Beavis et al., 1994). This region also encoded for maize hsp70 (Davis et al., 1999), root penetration index and total root number, two drought resistance traits in rice (Ray et al., 1996). This result suggested that this genomic region in sorghum, maize, and rice harbored many important genes responsible for drought tolerance.

2.3 Lodging Tolerance

Sorghum stalk lodging is related to postflowering drought stress as it appears during grain filling (Rosenow, 1977). For the first time, Kebede et al. (2001) mapped three QTLs related to lodging tolerance in sorghum and found a positive correlation between lodging tolerance and stay green under postflowering drought stress. One lodging tolerance QTL, *Idg G*, overlaps with QTLs of stay-green ability, preflowering drought tolerance, flowering time, and plant height in linkage group G. This QTL might be the same QTL as the ones reported by Crasta et al. (1999) and Tuinstra et al. (1997) on linkage groups G and F, respectively. Another lodging tolerance QTL, *Idg F*, overlaps with a plant height QTL on linkage group F. This region is syntenic to plant height and stay green QTLs on maize chromosome 2 (Beavis et al., 1994; Veldloom and Lee, 1996). Lodging-tolerant sorghum varieties have been developed in Australia using the stay green trait (Henzell et al., 1992).

2.4 Fine Mapping of Stay-Green QTL

Four stay green QTLs (*stg1*, *stg2*, *stg3*, and *stg4*) are consistently detected in different genetic backgrounds (Subudhi et al., 2000). With the objective of fine mapping and transfer of stay green QTLs to elite nonstay green sorghum lines, several near-isogenic lines (NILs) have been developed through marker-assisted selection (MAS) approach at Texas Tech University, Lubbock, TX, U.S.A., in collaboration with Texas A and M University Agricultural Research and Extension Center, Lubbock, TX, U.S.A. These BC_4F_6 NILs are being evaluated in the field for stay green trait and drought tolerance.

3 DROUGHT RESISTANCE MECHANISMS IN RICE

Rice is a unique cereal crop as compared to others, as it is grown under a wide range of agro-climatic conditions from deep-water to dry land.

Growing environments are classified as irrigated, rainfed lowland, upland, and flood-prone or deep water. Asia produces 92% of the world's rice, and China and India together produce 58%. Any shortfall in the major rice-growing countries could be a disaster for food security (Hossain et al., 2000). Drought is a common phenomenon in all rice growing environments except irrigated areas. Water stress is the most severe limitation to rice productivity in the rainfed ecosystems (Widawsky and O'Toole, 1996). During drought, marginal farmers cannot provide supplemental irrigation and, therefore, completely depend on rainfall for rice cultivation. Drought may occur at any time from early seedling stage to grain filling. Fukai and Cooper (1995) classified drought stress as: early season drought stress that induces delays in transplanting and seedling establishment, intermittent stress that affects tillering and panicle initiation, and late stress that affects flowering and yield. It has been observed that, when rice genotypes are subjected to water stress, certain genotypes can survive and yield under water stress better than others.

The rice plant can use different mechanisms to cope with drought stress, namely drought escape and drought resistance (Levitt, 1980). Drought escape allows the plant to complete its life cycle before drought, during the period of maximum water supply, via short life cycle. Drought resistance is generally divided into drought avoidance and tolerance. Drought avoidance helps the plants to maintain a relatively high leaf water potential during water stress by extracting more water from the soil through a well-developed root system and/or by leaf rolling, reducing water loss through stomatal closure and thick leaf cuticle (O'Toole and Bland, 1987; Ludlow and Muchow, 1990). Drought tolerance allows plants to maintain turgor and cell volume at low leaf water potential, thus continue metabolic activity longer under water stress through osmotic adjustment, antioxidant capacity, and cell membrane stability (CMS). Recently, different authors reviewed the important traits related to drought resistance in rice (Fukai and Cooper, 1995; Nguyen et al., 1999; Zhang et al., 1999; Courtois et al., 2002 (personal communication), Price et al., 2002a).

4 SHOOT-RELATED TRAITS

4.1 Osmotic Adjustment

Plants respond to different abiotic stresses at morphological, anatomical, cellular, and biochemical levels. Price et al. (2002a) reviewed several shoot-related traits associated with drought resistance mechanism in rice. Significant genetic variation was reported for traits such as osmotic adjustment (OA), dehydration tolerance, epicuticular wax, membrane stability, leaf rolling, stomatal closure, water use efficiency, and photoinhibition resistance.

Among these traits, OA is a basic cellular response to water deficit that appears to be conserved across all plant species and an important drought tolerance mechanism in plants (reviewed by Zhu et al., 1997; Zhang et al., 1999). Osmotic adjustment is defined as a decrease in osmotic potential resulting from a net accumulation of compatible solutes in response to water deficit. It allows plants to maintain higher turgor to sustain normal physiological functions. Osmotically active and compatible inorganic and organic compounds, or both solutes, help the cell to lower its osmotic potential and attract water into the cell. Although OA is considered as an important mechanism for drought tolerance, it has not been used in rice breeding programs, mainly because of labor-intensive and time-consuming measurement methods and lack of convincing genetic relationship between OA and yield performance under drought conditions. Different methods have been used for measuring OA. Babu et al. (1999) compared four different methods of OA analysis using 12 rice genotypes. Morgan's (1992) method, although time- and labor-consuming, is comprehensive for measuring OA in a small sample size. However, this method is not suitable for screening a large number of genotypes. While dealing with a large sample size, the rehydration method (Blum and Sullivan 1986; Blum 1989) is considered as most suitable. QTLs for OA have been mapped in rice (Lilley et al., 1996; Zhang et al., 2001; Robin et al., 2003), barley, (Teulat et al., 1998, 2001), and wheat (Morgan and Tan, 1996). In rice, *indica* cultivars are known to have high OA capacity as compared to *japonica* cultivars. On the other hand, *japonica* cultivars have a well-developed root system compared to *indica* cultivars. In rice, Lilley et al. (1996) first detected QTLs for OA and dehydration tolerance using 52 RILs developed from CO39 × Moroberekan. One major QTL for OA detected on chromosome 8 between markers RG978 and RG1, explained 32% of phenotypic variation. They also detected five putative QTLs for dehydration tolerance on chromosomes 1, 3, 7, and 8. Two of the five QTLs for dehydration tolerance, between markers RG96 and RG910 (chromosome 3) and RG351 and CDO533 (chromosome 7), fell within the chromosomal regions which contained QTLs associated with vegetative stage leaf rolling under water stress in the field (Champoux et al., 1995) and total root number (Ray et al., 1996).

Zhang et al. (2001) used 154 doubled haploid lines (DHLs) developed from a cross between CT9993-5-10-1-M and IR62266-42-6-2. IR62266-42-6-2 has high OA capacity and low root penetration ability, while CT9993-5-10-1-M has low OA capacity with high root penetration ability. Five QTLs associated with OA were detected on chromosomes 1, 2, 3, 8, and 9, together explaining 32% of the phenotypic variation. Robin et al. (2003) used an advanced backcross population developed from the cross between IR62266-42-6-2 and IR60080-46A, detected 14 QTLs located on chromosomes 1, 2, 3, 4, 5, 7, 8, and 10, explained together 58% of the phenotypic variation. Most, but not all, of the favorable alleles were from the *indica* parent, IR62266. Both the *indica* parents, IR62266 and CO39, contributed all the favorable alleles for OA in two other populations. None of the root QTL overlaps with the OA QTL on chromosome 8 but there is a root depth QTL near RG1.

As of today, three rice populations are available for comparative analysis of QTLs controlling OA (Table 2A) and four other populations for comparative analysis of other leaf-related traits (Table 2B). On chromosome 8, Zhang et al. (2001) identified one QTL for OA flanked by the RFLP markers G2132 and R1394A, that was located in the same genomic region as the OA QTL detected by Lilley et al. (1996). On chromosome 8, Robin et al. (2003) detected two QTLs for OA (in the intervals RG1–RM80 and RM284–RM210) that were in good agreement with previous studies. On chromosome 1, one OA QTL (in the interval RG140-ME2_12) was found to be common in both populations (Zhang et al., 2001; Robin et al., 2003). Tripathy et al. (personal communication) detected 14 QTLs for different osmolytes. A major QTL for proline and total sugar overlapped with an OA QTL reported earlier in rice and other crops.

Morgan (1991) reported that OA has a monogenic control in wheat. However, several authors suggested that more than one gene might regulate OA in rice (Lilley et al., 1996; Zhang et al., 2001; Robin et al., 2003), sorghum (Basnayake et al., 1995), and barley (Teulat et al., 1998, 2001). The probability of detection of QTLs increases with the population size. Syntenic comparisons help to test the consistency of the QTLs for the same traits across species. One of the most important rice OA QTL was possibly syntenic with one from wheat and one from barley OA QTL (Fig. 1). Morgan and Tan (1996) detected a putative gene for osmoregulation on the short arm of wheat chromosome 7A. The rice–wheat syntenic relationships are not simple for this chromosome with markers from wheat chromosome 7A mapping on rice chromosomes 1, 3, 6, and 8. Only one marker, CDO595, located near the rice OA QTL, bridges rice chromosome 8 with wheat chromosome 7A, so the evidence of synteny for the QTL is not strong (http://www.gramene.org). Teulat et al. (1998, 2001) identified several QTLs for OA and relative water content (RWC) in barley. Out of these, one QTL for OA and RWC on chromosome 1 (Q7HB) was homoeologous with the OA QTL region of rice chromosome 8. A high correlation between OA and RWC has been found in barley (Teulat et al., 1998). Price et al. (2002b) detected several QTLs associated with drought avoidance traits, namely, leaf rolling, leaf drying, and RWC in rice using RILs developed from Bala×Azucena. One QTL for RWC was identified on chromosome 8 (linked with marker G1073), in the location of the major OA QTL detected by others (Lilley et al., 1996; Zhang et al., 2001; Robin et al., 2003). The results mentioned above suggested that this genomic region contained a gene or cluster of genes that conferred

Table 2A Common QTLs for Osmotic Adjustment, Dehydration Tolerance, Cell Membrane Stability, and Abscicic Acid Across Different Genetic Background in Rice[a]

Chr. #	CO39 × Moroberekan[b]	IR20 × 63–83[c]	CT9993 × IR62266[d]	CT9993 × IR62266[e]	IR62266 × IR60080[f]
1	**RG109 (DT)**	—	**CDO345 (CMS)**	**ME2_12-RG140 (OA)**	**RM84-RM220 (OA)** OSR2-RM259 (OA) RM265-RM315 (OA) OSR9A-RG171 (OA) **RG224-CDO1305 (OA)**
2	—	—	—	RM263-R3393	—
3	**RG96 (OA)**	—	RZ403 (CMS)	**EM17_1-C63 (OA)**	—
4	—	—	—	—	RG375-C335 (OA)
5	—	—	—	—	OSR35-RZ390 (OA) RM31-BCD738 (OA)
7	CDO533 (DT) RG128 (DT)	**C507 (ABA)**	—	—	**RM11-OSR22 (OA)** RZ989-CDO38 (OA)
8	**RG1 (OA)** **RG20 (DT)**	—	**G2132-R1394A (CMS)**	**G2132-R1394A (OA)**	**RG1-RM80 (OA)** **RM34-RM25 (OA)** RM284-RM210 (OA)
9	—	—	**RZ698-RM219 (CMS)**	**EM14_6-ME4_13 (OA)**	—
10	—	—	—	—	C809-R716 (OA)

Where dashes appear, data not available/no QTL; linked markers and trait given otherwise.
[a] Common QTLs appear in bold; OA, osmotic adjustment; DT, dehydration tolerance; CMS, cell membrane stability; ABA, abscicic acid.
[b] Source: Lilley et al. (1996).
[c] Source: Quarrie et al. (1997).
[d] Source: Tripathy et al. (2000).
[e] Source: Zhang et al. (2001).
[f] Source: Robin et al. (in press, 2003).

Table 2B Common QTLs for Leaf Drying, Leaf Rolling, and Relative Water Content Traits Across Different Genetic Background in Rice[a]

Chr. #	CO39 × Moroberekan[b]	IR20 × 63–83[c]	IR64 × Azucena[d]	Bala × Azucena[e]
1	RG462 (LR)	—	RZ730–RZ801 (LR) **RG810–RG331 (LR, LD, RWC)** **RG146 (LD)**	**C949 (LR, RWC)** **RZ14, R117 (LD, LR, RWC)** **C178 (LD)**
2	RG324 (LR) RG139 (LR) RG437 (LR) RG544 (LR)	—	—	—
3	—		RZ892–RG100 (LR) RZ574–RZ284 (LR) RG910–RG418A (LR)	C136 (LR)
4	RG214 (LR) RG190 (LR)	—	RG163–RZ590 (LR)	
5	—	—	**RZ67–RZ70 (LR, RWC)** —	**C624 (LR, RWC)** C43 (LR)
6	RZ516 (LR) RZ192 (LR)	—	—	**C76 (LR)**
7	CDO533 (LR)	—	**RG477–PGMS07 (LR)** PGMS07–CDO59 (LR)	G20, G338 (LD) R1440 (LR)
8	RG1 (LR) —	**G1073 (RWC)** —	RG757–C711 (LR) RZ12–RG667 (LR)	— —
9	RG662 (LR)			
10	RZ892 (LR) RZ561 (LR)			
11	CDO365 (LR) RZ53 (LR) (LR)			

Where dashes appear, data not available/no QTL; linked markers and trait given otherwise.

[a] Common QTLs appear in bold; LD, leaf drying; LR, leaf rolling; RWC, relative water content traits.

[b] *Source*: Champoux et al. (1995).

[c] *Source*: Quarrie et al. (1997).

[d] *Source*: Courtois et al. (2000).

[e] *Source*: Price et al. (2002b).

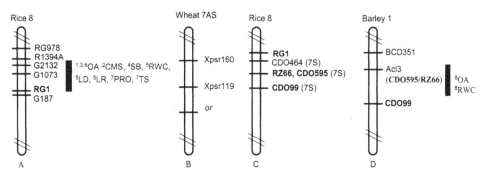

Figure 1 A common genomic region of rice chromosome 8 for drought tolerance related root and shoot traits and possibly, syntenic for osmotic adjustment between rice, wheat, and barley. Markers are assigned at the right side of the chromosome. (A) Zhang et al. (2001), (B) Morgan and Tan (1996), (C) Ahn et al. (1993), Causse et al. (1994), (D) Teulat et al. (1998, 2001). OA: osmotic adjustment, RWC: relative water content, LD: leaf drying, LR: leaf rolling, SB: shoot biomass, CMS: cell membrane stability, PRO: proline, TS: total sugar. Numbers with abbreviation refers to authors, as: [1]Lilley et al. (1996); [2]Tripathy et al. (2000); [3]Zhang et al. (2001); [4]Kamoshita et al. (2002a,b); [5]Price et al. (2002b); [6]Robin et al. (in press); [7]Tripathy et al. (unpublished data); [8]Teulat et al. (1998, 2001).

drought resistance in rice, wheat, and barley. This information is valuable for researchers targeting specific regions to produce near isogenic lines (NILs) and to validate the impact of that individual QTL on yield under drought stress.

On chromosome 3, one OA QTL was detected between markers RZ313 and RG224 in two rice populations (Zhang et al., 2001; Robin et al., 2003). In this region, Price et al. (1997) detected one QTL for stomatal behavior in an F2 population of rice. In maize, this region was also associated with stomatal conductance and leaf abscicic acid (ABA) concentration (Tuberosa et al., 1998). Zhang et al. (2001) detected one OA QTL on rice chromosome 9 between amplified fragment length polymorphism (AFLP) markers EM14_6 and ME4_13. Using the same population, Tripathy et al. (unpublished data) detected QTL for proline and total sugar accumulation, two important osmolytes, in this genomic region. In a different mapping population, Quarrie et al. (1997) detected one QTL for ABA concentration in the same region. In wheat, an ABA QTL was mapped in the long arm of chromosome 5A (Quarrie et al., 1994) that was homoeologous to rice chromosome 9. This is a conserved region for ABA accumulation in wheat and rice, cell membrane stability, and osmolyte accumulation under stress.

4.2 Cell Membrane Stability

In rice, Tripathy et al. (2000) detected nine QTLs for cell membrane stability (CMS) using 154 DHLs developed from the cross between CT9993 and IR62266. One QTL (marker interval G2132-R1394A) for CMS was mapped at the same locus as the OA QTL on chromosome 8 (Zhang et al., 2001).

4.3 Leaf Rolling, Leaf Drying, and Relative Water Content

Three rice populations were used to identify QTLs associated with drought avoidance mechanisms, such as leaf rolling, leaf drying, and RWC. A population of 203 RILs developed from a cross between CO39, a lowland indica cultivar, and Moroberekan, an upland japonica cultivar, was used to study root morphology and drought avoidance (Champoux et al., 1995). CO39 has a shallow, thin root system and is susceptible to drought, while Moroberekan has a deep, thick root system and is tolerant to drought. A total of 18 QTLs were associated with drought avoidance at seedling, early- and late-vegetative stages in the field. Of the 18 QTLs, 5 were common across the three different growth stages and 4 were common to at least two growth stages. In three separate trials, Courtois et al. (2000) evaluated about 100 DHLs developed from the cross between IR64 and Azucena in two different environments. They detected 11 genomic regions for leaf rolling, 10 for leaf drying and 11 for RWC. Among the 11 QTLs for leaf rolling, three QTLs, one each on chromosomes 1 (RG810–RG331), 5 (RZ67–RZ70), and 9 (RZ12–RG667), were found to be common in three trials. Four QTLs were also found common across two trials, one each on chromosomes 1 (RZ730–RZ801), 3 (RG910–RG418A), 4 (RG163–RZ590), and 9 (RG757–C711). One QTL for leaf drying on chromosome 4 (RG143–RG620) and one QTL for leaf RWC on chromosome 1 (RZ730–RZ801) were common in two trials. Two QTLs, one on chromosome 1 (RG810–RG331) and another on chromosome 5 (RZ67–RZ70) were common for leaf rolling and leaf drying in at least one trial. Two QTLs, one on chromosome 1 (RG810–RG331) and another on chromosome 6 (CDO544–RG653) were common for both leaf drying and leaf RWC in at least one trial. On chromosome 1, one QTL (RG810–RG331) was common for three different traits in at least one trial. Price et al. (2002b) used 100–176 RILs developed from the cross Bala × Azucena. Bala is a lowland indica cultivar and Azucena is an upland japonica cultivar. One of the parents, Azucena, is also a parent of the population used by Courtois et al. (2000). A total of 17 QTLs were detected for leaf rolling, leaf drying, and relative water content in two different years, in two different locations, one at International Rice Research Institute (IRRI), Philippines, and the other at West Africa Rice Development Association (WARDA), Ivory Cost (Côte d'Ivoire).

Although a large number of QTLs have been detected for various drought avoidance traits in different mapping populations, screening locations, and years, there is some strong evidence of consistent drought avoidance QTLs (Price et al., 2002b). Champoux et al. (1995) evaluated 203 RILs developed from CO39 and Moroberekan at IRRI for leaf rolling at three different growth stages. Courtois et al. (2000) used DHLs developed from IR64 and Azucena and tested about 100 lines at IRRI and India for leaf rolling, leaf drying, and leaf RWC. Price et al. (2002b) used between 110 and 176 RILs, developed from a cross between Bala and Azucena, and evaluated them at IRRI and WARDA for leaf rolling, leaf drying, and RWC. In all the cases, at least one experiment was conducted at IRRI. To bridge the different sets of markers, reference maps of Causse et al. (1994) and the comparative mapping website http://www.gramene.org were used. The region of chromosome 1 between markers RG331 and RZ14 that contains QTLs for leaf rolling and RWC was consistent across CO39 × Moroberkan and IR64 × Azucena populations, and also across locations and seasons. In the same region, a QTL for leaf drying was also detected in IR64 × Azucena population. This region also harbors QTLs for OA, root traits, and submergence tolerance across the different genetic backgrounds. The region of chromosome 3 between RZ519 and CDO795 carries QTLs for leaf rolling, drying, and RWC in all the popu-

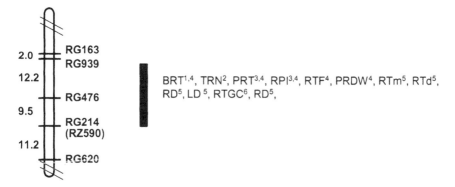

Figure 2 A common genomic region of rice chromosome 4 for different root traits related to drought tolerance. Solid bar on the left side of the chromosome indicates the location of QTL for different traits. Traits are abbreviated as: BRT, basal root thickness; PRT, penetrated root thickness; RPI, root penetration index; RTF, root pulling force; RTm, root thickness at 10–20 cm; RTd, root thickness at 20–30 cm; LD, leaf drying; RTGC, root thickness in glass chamber; TRN, total root number; RD, rooting depth. Numbers along with abbreviated traits refers to authors: [1]Champoux et al. (1995), [2]Ray et al. (1996), [3]Zheng et al. (2000), [4]Zhang et al. (2001), [5]Kamoshita et al. (2002a,b), [6]Price et al. (2002b).

lations. This region also contains QTLs for different root traits. Marker region between RZ590 and RG163 on chromosome 4 contains QTLs for both root and shoot related traits (Fig. 2). On chromosome 5, the region between marker RZ296 and RZ70 contains QTLs for leaf rolling and drying in both IR64 × Azucena and Bala × Azucena population, while none was detected in CO39 × Moroberekan population. Interestingly, this region harbors QTLs for root and submergence tolerance. On chromosome 7, a leaf rolling QTL linked with marker C39 in the Bala × Azucena population, is in the same region than a leaf rolling QTL at marker RG447 identified in the IR64 × Azucena population. The genomic region of chromosome 8 between markers G2132 and RG1 is the location of a QTL for OA and also for other traits identified in different populations (Fig. 1). In the CO39 × Moroberekan population, Champoux et al. (1995) detected QTL for leaf rolling at three different growth stages. In the IR64 × Azucena population, Courtois et al. (2000) detected a QTL for leaf drying just below the significance level. Price et al. (2002b) detected QTLs for leaf drying and RWC in the same region linked with the marker G1073. Marker G1073 is very close to RG1.

5 ROOT-RELATED TRAITS

The root system is an important organ that enables plants to anchorage into soil, and to extract water and minerals for growth, development, and production. Root development is fundamentally involved in plant response to stresses such as drought or mineral deficiency (Price et al., 2002a). Root tip is an important component in sensing and signaling environmental cues to the whole plant (Aiken and Smuker, 1996). A deep and thick root system allows the rice plant to absorb water from the deeper soil horizon and is considered an important drought resistance trait in upland rice (O'Toole 1982; Fukai and Cooper 1995). Under well-watered conditions, root system development has a positive effect on subsequent plant growth during progressive water stress. Root elongation to deeper soil resulted in improved water extraction (Azhiri-Sigari et al., 2000; Kamoshita et al., 2000, 2002a,b). Despite having fewer roots in deeper layers, rainfed lowland rice can extract water from below 15-cm soil depth in subsequent drought periods (Wade et al., 1999).

In rainfed lowland rice fields, water in deeper soil is often inaccessible because of the presence of compacted soil layers (hardpans). In the presence of soil hardpans, especially when water deficit occurs, root penetration ability through the compacted soil layers is considered as an important drought resistance trait (Nguyen et al. 1999; Ali et al., 2000; Zheng et al., 2000; Zhang et al., 2001).

QTLs for different root traits such as root thickness, root dry weight, deep root weight, root penetration index, penetrated root length, and root shoot ratio have been studied in several intra- and inter-subspecific crosses of rice (Champoux et al., 1995; Ray et al., 1996; Price and Tomos 1997; Yadav et al., 1997; Ali et al., 2000; Price et al., 2000, 2002b; Zheng et al., 2000; Zhang et al., 2001; Shen et al., 2001) (Table 1). To bridge the different sets of markers, reference maps of Causse et al. (1994) and comparative mapping website were used.

Champoux et al. (1995) used about 200 RILs developed from a cross between CO39 (an *indica* wetland adapted cultivar) and Moroberekan (a *japonica* dryland adapted cultivar) for QTL analysis in rice. The population was developed to study blast resistance in rice, and significant differences in root traits between the two parents of this population were observed. This was the first study to locate QTLs associated with root morphology and drought avoidance in the field and greenhouse. Most of the QTLs that were identified in this study were associated with root thickness, root/shoot ratio and root dry weight. There was a positive correlation between the field and greenhouse experiment. In support of this finding, 12 out of 14 chromosomal regions containing putative QTLs associated with field drought tolerance also contained QTLs associated with root morphology.

Root penetration ability is considered as an important trait for drought resistance for rainfed lowland and as well as upland rice ecosystems, but field evaluation for this trait is almost impossible for large-size mapping populations. Yu et al. (1995) designed a root penetration screening system that used wax–petrolatum barriers as a substitute for compacted soil layers. The wax petrolatum layers restrict vertical root growth with a barrier of consistent hardness. Using wax–petrolatum layers as a proxy for compacted soil layers, Ray et al. (1996) used the same RILs to locate QTLs associated with root penetration ability in rice. This was the first report of QTL analysis associated with root penetration ability through wax petrolatum layer simulating a compacted soil layer.

So far, five mapping populations have been used in six separate experiments to study root traits related to drought tolerance in rice (Table 3A). Out of these six experiments, four were conducted at Texas Tech University, Texas, U.S.A., under similar condition using a wax petrolatum layer as barrier to root penetration (Ray et al., 1996; Ali et al., 2000; Zheng et al., 2000; Zhang et al., 2001). Champoux et al. (1995) studied root thickness without any barrier, and Price et al. (2002b) used wax petrolatum layer (4:1) with modification. The same populations were also used for root morphology studies (Price and Tomos 1997; Yadav et al., 1997; Kamoshita et al., 2002a,b). Although these groups evaluated different mapping populations under different experimental conditions, a number of QTLs were found consistent over the experiments (Tables 3A and 3B). One QTL for root penetration index located between markers RG256 and PC32M10, on chromosome 2, was

Table 3A Common QTLs for Root Thickness, Root Penetration Index, Penetrated Root Number, Total Root Number Across Different Genetic Background in Rice[a]

Chr. #	CO39 × Moroberekan[b]	CO39 × Moroberekan[c]	IR64 × Azucena[d]	IR58821 × IR52561[e]	CT9993 × IR62266[f]	Bala × Azucena[g]
1	RG197, RG811 (RT)	—	RG730–RZ801 (RT)	PC15M11–PC3M3 (RT)	RG975–RG345 (RT)	—
	CDO920, RG140 (RT)			PC3M3–WG110 (RT)	EM19_11–EM13_1 (RT)	
		RG612 (RPN)	RG246 (RPN)	BCD134–RZ776 (RPN)		
		RG350, RG77 (TRN)				RG173, R117 (TRN)
		RG811, RG140 (TRN)	—	—	—	—
2	—	RG324–RG73 (RPI)	RZ123–RG520 (RPI, PRN)	RG256–PC32M10 (RPI)	—	G45, C601 (RPI)
	RG139, RG437 (RT)	—	Pall (RT)	RG256–PC32M10 (RT)		—
	—	—	RZ318-pall (PRN)			
				PC33M8–PC21M1 (PRN)	ME9_7–K706 (RT)	G45, C601 (PRN)
			CDO686 (PRN)	RG256–PC32M10 (PRN)		
			PC3M11–PC33M8	ME2_7–EMP2_7 RT		
3	RG104 (RT)	RG104–RG348 (RPI)	RG104–RG590 (RT)	EM19_4–EM13_1 (RPI)	—	e12m37.4, e13m36.16 (RPI)
	RG910 (RT)	RZ393 (PRN)	PC73M13–PC3M5 PRN	R1925–RG1356 (TRN)		el12m36.16 (PRN)

4. RG576 (RT)	RG745, RG103 (TRN) PC73M13–PC3M5 (RPI)	RG104 (PRN)	C746–RZ448 (PRN)	—	e12m37.4 (PRN)
RG788 (RT)	**RG620–RG476 (RPI)**	—	—	—	—
6 **RG214 (RT)**	**RG329 (RPI)**	**RG163–RZ590 (RT)**	PC33M5–PC38M9 (RT)	**RG939–RG476 (RT)** EM14_5–EM2_13 (RPI)	—
WAXY (RT)	—	—	—	R2549–RG716 (RT)	—
7 CDO533, RG528 (RT)	RG163 (TRN)	—	RZ978–RG351 (RT)	RG417–EM17_3 (RT)	—
—	—	RZ337B–CDO497 (TRN)	PC75M8–PC32M1 (TRN)	—	—
9 RG553 (RT) RZ12 (RT)	—	Amy3A–RZ228 (RT) **RZ892–BCD386 (RT)**	—	**RG553–EM14_6(RT)** ME9_6–K985 (RT)	—
10 RZ892 (RT)	—	**RZ892–BCD386 (RT)**	**RZ892–BCD386 (RPI)**	—	e12m37.2 (RPI)
11 CDO365 (RT)	CDO365, RG118 (RPI)	—	—	—	**C189 (RPI)**
12 RZ397	—	—	—	RG9	—

Where dashes appear, data not available/no QTL; linked markers and trait given otherwise.
Common QTLs appear in bold; RT, root thickness; RPI, root penetration index; PRN, penetrated root number; TRN, total root number.

[a] *Source:* Champoux et al. (1995).
[b] *Source:* Ray et al. (1996).
[c] *Source:* Zheng et al. (2000).
[d] *Source:* Ali et al. (2000).
[e] *Source:* Zhang et al. (2001).
[f] *Source:* Price et al. (2002).

Table 3B Common QTLs for Root Thickness, Root Weight, Shoot Weight, Deep Root Weight Below 30 cm, Maximum Root Length, Maximum Rooting Depth, and Tiller Number Across Different Genetic Backgrounds of Rice (from Kamoshita et al. 2002b)[a]

Chr. #	Bala × Azucena[b]	IR64/Azucena[c]	IR58822 × IR52561[d]	CT9993 × IR62266[e]
1	—	RG381–RZ730 (DRW, RD, TN)	PC31M10–PC34M6 (RD)	RG957 (DRW, RW, TN)
	—	RZ730–RZ802 (RT)	PC180M1–PC325 (RT)	ME1014–RZ909 (RT)
2	R683–RG171 (MRL)	RG171–RG157 (DRW, TN, RD)	PC41M2–PC173M5 (DRW, TN)	RG437–ME1018 (DRW, RW, TN, RD)
	—	—	—	RG437–C106 (RT)
	—	PALI-RZ58 (RT)	—	R3393–RG158 (RD)
	—	—	RG256–RG151 (DRW, RW, TN)	EM1813–ME97 (RD)
3	C643 (MRL)	RZ329–RZ892 (DRW)	PC73M7–PC20M12 (DRW, RW)	EM119–RG409 (DRW)
	C746 (MRL)	RZ519 (RD)	—	RZ474–ME82 (DRW)
	RG104A (RT)	—	—	—
4	—	—	PC150M11–PC11M4 (DRW, RW, RD)	RG939–RG214 (DRW, TN)
	RG260–RG190 (MRL)	RG190 (DRW, SW)	RG218–PC79M8 (RD)	—
	—	—	PC75M3–PC11M4 (RT)	RG939–RG214 (RT)
5	R3166–R2232 (DRW)	RZ390–RG403 (DRW, SW, RD)	—	ME513–RG403 (DRW, RW, TN)
7	—	RG711–RG351 (DRW, SW)	—	RG650–EM184 (RD)
	—	RG711 (DRW, SW, TN)	—	RG404–CDO38 (RD)
9	—	RZ206–RZ422 (DRW)	—	G103–ME58 (RD)
	—	AMY3ABC–RG667 (DRW, SW, TN, RD)	PC33M1–C570 (DRW, RW)	—
	—	—	PC32M8–C570 (RT)	—
11	—	—	G257–PC31M8 (DRW, RW)	G320–ME26 (DRW, RW, TN)
	RG2 (MRL)	—	—	CDO365–EM1819 (DRW, RW, TN)
	—	—	—	ME67–EM1819 (RT)

Where dashes appear, data not available/no QTL; linked markers and trait given otherwise.

[a] Rt, root thickness, RW, root weight, SW, shoot weight, DRW, deep root weight below 30 cm, MRL, maximum root length, RD, maximum rooting depth, TN, tiller number.

[b] *Source*: Price and Tomos (1997).

[c] *Source*: Yadav et al. (1997).

[d] *Source*: Kamoshita et al. (2002b).

[e] *Source*: Kamoshita et al. (2002a).

found in similar position in both IR58821 × IR52561 and Bala × Azucena populations. In the same genomic region, one QTL for root thickness was detected in IR58821 × IR52561 population. The same genomic region also harbors a common QTL for penetrated root number in IR64 × Azucena, IR58821 × IR52561, and Bala × Azucena populations. This genomic region carries QTL for traits related to root morphology, such as deep root weight and maximum rooting depth. On chromosome 3, Champoux et al. (1995) and Zheng et al. (2000) detected one QTL for root thickness between markers RG104 and RG590. In the same region, Ray et al. (1996) detected one QTL for penetrated root number. On the same chromosome, the genomic region RG910–RZ393 harbors QTLs for root thickness, penetrated and total root number as detected by Champoux et al. (1995), Ray et al. (1996), and Ali et al. (2000), respectively. On chromosome 4, the region between markers RG163 and RZ590 contains common QTLs for root thickness and root penetration index in four experiments. No common QTL for root traits was detected on chromosome 8, on which a major QTL for OA is located. On chromosome 9, one QTL for root thickness was found linked to marker RG553 in two populations (Champoux et al., 1995; Price et al., 2002a). A large region of chromosome 10 carries QTLs for root thickness and root penetration index in three different populations. From the above discussion, it is clear that there are certain points of concentration of QTLs for root or leaf related traits. For example, the RG1–RG978 region of chromosome 8 not only contains a major OA QTL, but also contains QTL for leaf RWC, rolling, drying, dehydration tolerance, and cell membrane stability (Fig. 1) and the RG939–RZ590 region of chromosome 4 also contains QTLs for different root traits (Fig. 2).

6 MAPPING OF SUBMERGENCE TOLERANCE IN RICE

Although rice is a semiaquatic plant, flooding and/or continuous submergence is a serious problem to rice plant growth, survival, and yield in South and Southeast Asia, and affects approximately 22 million ha in rainfed lowland and flood-prone areas (Khush, 1984). A few traditional rice varieties are adapted to these environments through two basic mechanisms, namely, submergence tolerance and elongation ability. Submergence tolerance is the capacity of rice plants to survive under completely submerged conditions for 10 days or more. Elongation ability is the ability of rice plants to elongate as water rises to keep the leaf canopy above the water surface (Sripongpangkul et al., 2000). Because of the complexity of submergence tolerance and elongation ability traits, little progress has so far been made to develop tolerant varieties using conventional breeding methods. During the last decade, molecular marker technologies have been successfully used to identify genes related to submergence tolerance and elongation ability and use them to facilitate the development of flood-tolerant modern rice varieties.

Submergence tolerance is a desirable trait for rainfed lowland genotypes. Adaptation to submergence is a complex process and several factors contribute to adverse effects of submergence. During flooding, growth and metabolism are affected by limited gas diffusion, light penetration, and siltation on leaves (Palada and Vergara 1972; Setter et al., 1988, 1995). There is a negative correlation between survival and elongation growth in completely submerged rice seedlings (Setter and Laureles 1996; Singh et al., 2001). These findings suggest that rainfed lowland rice varieties with less elongation ability will show increased submergence tolerance during flash flood. Metabolic changes occur differently in submergence tolerant and susceptible rice genotypes after desubmergence. Singh et al. (2001) reported that higher initial carbohydrate concentration (soluble sugar and starch) enable rice plants to sustain submergence stress and regenerate quickly during desubmergence. Kawano et al. (2002) reported that both tolerant and susceptible rice genotypes showed a decrease in ascorbate concentration. But 3 days after desubmergence, the tolerant genotypes showed a rapid recovery of total ascorbate and ascorbic acid, whereas the susceptible genotypes showed a slow recovery with an increase in malondialdehyde formation. Ascorbic acid is an important antioxidant in vivo for the recovery of submerged rice seedlings.

Several studies have been conducted for mapping QTLs for submergence tolerance, and leaf and internodes elongation (Table 1). Submergence tolerance is considered as a quantitative trait with a complex pattern of inheritance with the involvement of few to many genes (Suprihanto and Coffman 1981; Mohanty and Khush 1985; Haque et al., 1989). Zhang et al. (1995) and Xu and Mackill (1995) identified one major gene on chromosome 9 involved in submergence tolerance. Mishra et al. (1996) worked with three cross combinations and showed that a single dominant gene controls this trait. Using AFLP markers, Nandi et al. (1997) identified five QTLs for submergence tolerance, including one major QTL mapped on chromosome 9, in the same position as the *sub*1 gene reported by Xu and Mackill (1995). Xu et al. (2000b) constructed a high-density map of the *sub*1 gene region. Sripongpangkul et al. (2000) detected one gene with large phenotypic effect on submergence tolerance. This gene was mapped in the genomic region where the gene *sub*1 was positioned earlier in different genetic backgrounds (Xu and Mackill, 1995; Nandi et al., 1997). Siangliw et al. (2003) used three submergence-tolerant genotypes—FR13A, a submergence-tolerant landrace from India with low yielding capacity, a doubled haploid line (IR67819F2-CA-61) developed from the FR13A × CT6241-7-1-2-2 cross, and an improved genotype IR49830-7-1-2-2—as donor parents, and KDML105 as a recurrent parent. They were successful in transferring a genomic segment of chromosome 9, responsible for submergence tolerance to backcross lines, retaining all the desirable traits of KDML105. Toojinda et al. (2003) tested three different mapping populations (DHLs, RILs, and F_2) in different seasons to identify chromosomal regions responsible for submergence tolerance in rice. They mapped several QTLs related to submergence tolerance on different chromosomal regions. They identified one major QTL on chromosome 9 with the highest phenotypic effect, common across the three different mapping populations over the years. This QTL is also consistently detected in four other mapping populations, evaluated over different seasons and locations. A number of QTLs related to root and shoot traits are also identified in the same region. In addition to this submergence tolerance major QTL, several other QTLs with minor effects were detected on chromosomes 1, 5, 7, and 10.

Avoidance of complete submergence by stem elongation is a typical characteristic of deepwater rice, which can elongate its stem up to 25 cm a day to reach 5 m or more in height and produce grain despite several months of deepwater stress. Elongation ability under prolonged flooded condition is necessary in deepwater or floating water so that shoots reach the water surface and can photosynthesize (Ito et al., 1999). Rapid elongation in response to submergence is triggered by oxygen deficiency and is controlled by ethylene (Kende et al., 1998). Six genomic regions have been identified for leaf and stem elongation, but none of them was detected on chromosome 9 (Sripongpang-kul et al., 2000), where the major QTL for submergence tolerance was located. This result indicated that submergence tolerance and elongation ability were controlled by two different sets of genes.

Transgenic rice plants containing the YK1 gene (showing homology with maize HM1 gene) were resistant to rice blast disease. The transgene also conferred high tolerance to several abiotic stresses such as NaCl, UV-C, submergence, and hydrogen peroxide (Uchimiya et al., 2002). Recently, Straeten et al. (2002) isolated 1-aminocyclopropane-1-carboxylic acid (ACC) from IR36, a lowland rice variety that helped for rapid elongation. Their result suggested that ACC limited the survival of submerged lowland rice seedlings. The pyruvate decarboxylase gene, *pdc*1 also increased submergence tolerance in rice. It is clear from the above results that the genomic region on chromosome 9 is the best candidate for the development of submergence tolerant rice genotypes. Flooding or submergence is not an issue for sorghum.

7 MAPPING OF SALINITY TOLERANCE IN RICE

Salt stress or salinity is the most prevalent constraint in the arid and semiarid regions of the world that significantly reduces crop productivity. About 20%

of the world's cultivated land and nearly half of all irrigated lands are affected by salinity (Rhoades and Loveday, 1990). About half of the irrigated lands of California are severely affected by salinity. Salinity occurs mainly attributable to the presence of high water tables or use of insufficient irrigation water to leach salts out of the soil. The main sources of irrigation water are canals, rivers, and ground water. Irrigation water contains significant amounts of salts and causes considerable salinization worldwide (Ashraf, 1994). Salt stress disturbs ion homeostasis at both cellular and whole-plant levels, which leads to molecular damage, reduced growth, and even death of plants (Zhu, 2001). Different methods, such as soil reclamation, irrigation, and drainage, are used to reduce soil salinity but these methods are not economically viable. Reclamation of saline soil has been widely carried out by leaching of surface salts through infiltration. However, this system is not feasible in arid and semiarid areas where rainfall is low and irrigation waters are saline and inadequate for the purpose. Adding gypsum to the saline soil and irrigating with salt-free water can temporarily amend the soil. But these methods are expensive and short-term solutions. Salinity affected areas can be brought under successful cultivation through the introduction of salt-tolerant culti- vars. The development and utilization of high yielding salt-tolerant cultivars is an economically feasible and permanent solution to overcome salinity.

Levitt (1980) categorized salt resistance mechanisms as salt tolerance and salt avoidance. Salt tolerance is the capacity of a plant to grow well under saline condition by compatible solutes accumulation, ion exclusion, inclusion, and selectivity. During an osmotic stress, plants accumulate organic solutes or inorganic ions, which reduces the cell osmotic potential to a level that provides turgor potential to maintain growth. Organic solutes include amino acids, sugars, and polyols. Sugars alone contribute about 50% of the total osmotic potential under salt stress. Plants accumulate significant amount of proline under severe drought and salt stress. Zhu (2001) discussed three aspects of salt tolerance in plants, such as detoxification, homeostasis, and growth regulation. Plants produce different osmolytes and stress proteins under salt stress. Many of these osmolytes and stress proteins detoxify plants by scavenging reactive oxygen species (ROS), or by preventing them from damaging cellular structure. Late-embryogenesis-abundant (LEA)-type stress proteins such as RD29A are proposed to function in the detoxification or alleviation of damages. C-repeat/dehydration responsive element binding protein (CBF/DREB) transcription factors mediate some of the stress protein gene expression in response to salt stress. Another aspect of salt tolerance is re-establishment of ionic and osmotic homeostasis under stressful environ- ment. The SOS pathway mediates ionic homeostasis and sodium ion (Na^+) tolerance, and the mitogen-activated protein kinase (MAPK) mediates osmotic homeostasis stress (Zhu, 2001; Zhu et al., 1997). Total ion activity and their ratios in the external environment have considerable effects on plant growth. Ion exclusion restricts the uptake of salts or toxic ions into shoots. Salt includers absorb large amount of salts and store them in cytosols of plant cells, compartment salts into vacuoles, or store them in special glands on the leaf surface. Both K^+ and Ca^+ are required to maintain integrity of the cell membrane. A high K/Na ratio can be used as selection criteria in certain species but not all. As reviewed by Bohnert and Sheveleva (1998), two different sets of mechanisms are involved in plant adaptation to salt stress. One includes morphological, developmental, and physiological mechanisms, and the other biochemical mechanisms. The first one is more complex and requires functions of many gene products. In contrast, biochemical mechanisms are simple, typically involving the action of a few gene products.

In a conventional rice breeding program, varietal development for salt tolerance involves selection of progeny lines based on phenotype, such as growth and yield performances over times and locations. Because of the complexity of the traits, such breeding programs are slow, expensive, and laborious. According to Flowers and Yeo (1995), little progress has been made for the development of salt-tolerant rice varieties in the last decade.

Recently, molecular marker technologies started to be used to overcome the above problems and QTLs related to salt tolerance traits have been mapped across divergent genetic backgrounds. Unfortunately, the findings are not as consistent as for the other abiotic stress tolerance traits. Zhang et al. (1995) detected one QTL for salt tolerance on chromosome 7. Prasad et al. (2000) also detected one in the same location. Gong et al. (1999) reported a major QTL for salt tolerance on rice chromosome 1, but it is not in the same region as reported by Koyama et al. (2001). Prasad et al. (2000) have also detected another QTL on chromosome 6 involved in salt tolerance and dry biomass, which may be related to the QTL for dry mass reported by Koyama et al. (2001). Flowers et al. (2000) have reported that none of the markers in their study showed any association with similar traits in a closely related population. He concluded that direct knowledge of the genes involved was needed.

8 MARKER-ASSISTED SELECTION AND CANDIDATE GENE ANALYSIS

Backcrossing is an established tool in plant breeding used to accelerate the recovery of a recurrent parent genome. In conventional breeding, thousands of progenies are screened in the field to identify plants with desirable traits based on phenotype. A limited number of plants can be evaluated in the field per season. Space, time, and manpower are the major limitations of traditional plant breeding along with uncertainty of obtaining desirable plants.

With the advancement of molecular marker technology, selection is based on markers, as well as expression of trait, rather than on the trait alone. DNA markers are being used in breeding programs to follow the inheritance of those gene(s) that are difficult to evaluate in the field. Because the selection is based on DNA markers that are tightly linked with the gene/QTL, it is known as marker–assisted selection (MAS). MAS can enhance the efficiency of plant breeding. A prerequisite of this technique is the availability of a saturated map and markers tightly linked to the gene or QTL with significant phenotypic effect. Among PCR-based markers, simple sequence repeats (SSR) and sequence-tagged sites (STS) are suitable to evaluate a large number of progeny lines in the lab with a very small amount of DNA. The efficiency of MAS has been demonstrated for traits such as disease and insect resistance that are controlled by a few genes/QTLs. Few MAS studies have been conducted for abiotic stress tolerance, but the results of these experiments are promising. Shen et al. (2001) have successfully introgressed several QTLs for root depth in rice from Azucena, a deep-rooted variety, into IR64, a shallow-rooted elite line. Under field condition, introgressed lines performed better than nonintrogressed lines and the yield performance of some of the introgressed lines was higher than that of IR64 (Lafitte et al., in preparation). In collaboration with the International Rice Research Institute (IRRI), Philippines, and Texas Tech University, Lubbock, TX, U.S.A., major QTLs for root thickness and root penetration ability have been selected to introgress into IR62266, an indica line with high OA capacity but having thin and poor root penetration ability. BC_4 lines carrying specific introgressed segments from CT9993 are being evaluated in the greenhouse of Texas Tech University for root thickness and penetration ability (Chamarerk, unpublished data). Preliminary results show that introgressed lines with thick roots were obtained.

Stay green is an important trait for drought tolerance in grain sorghum, and four stay green QTLs (*stg1*, *stg2*, *stg3*, and *stg4*) were consistently detected in different genetic backgrounds. To transfer stay green QTLs to elite nonstay green sorghum lines, several near-isogenic lines (NILs) have been developed through backcrossing and MAS approach at Texas Tech University, Texas, U.S.A. These BC5- and BC6-derived NILs are being evaluated under field conditions in Texas, U.S.A., and Queensland, Australia.

Advanced backcross QTL analysis (AB-QTL), an alternate method of MAS developed by Tanksley and Nelson (1996), allows the simultaneous discovery and transfer of valuable QTL alleles from unadapted donor lines into established lines. In this method, QTL analysis is delayed until the BC_2 or BC_3 generation. Near-isogenic lines introgressed with QTLs can be obtained from the AB population in one or two additional generations. AB-QTL analysis explored unadapted and exotic germplasm for quantitative trait improvement in tomato, rice, and other crops (Xiao et al., 1998; Bernacchi et al., 1998; Moncada et al., 2001). The AB-QTL method was applied to transfer osmotic adjustment QTLs from IR62266, an indica donor parent with high OA capacity to IR60080-46A, a japonica elite line with low OA capacity. A population of 150 BC_3F_3 families was evaluated for OA capacity under greenhouse condition and 14 QTLs together explained 58% of phenotypic variation (Robin et al., 2003). In barley, AB-QTL has been applied to uncover beneficial QTL alleles for drought tolerance from wild progenitor (Forster et al., 2000).

The goal of QTL analysis is to reduce the genetic distance between marker and QTL, and finally to identify the gene underlying the QTL. A QTL region contains thousands of genes and, by fine mapping of the QTL region, this number may be reduced to hundreds of candidate genes (Wayne and McIntyre, 2002). Candidate genes are known genes that map to the same locations as QTLs. A significant proportion of the QTLs affecting a trait variation are composed of candidate gene associated with that trait (Rothchild and Soller, 1997). In the case of resistance genes, this approach utilizes degenerate oligonucleotide primer pairs from conserved motives of the sequences of known genes to map resistance gene analogs. The candidate gene approach is being used for studying quantitative, as well as qualitative, traits. This approach has been successfully applied to locate resistance genes in rice, wheat, *Arabidopsis thaliana* (Chen et al., 1998; Faris et al., 1999; Sillito et al., 2000), mostly for qualitative traits or traits controlled by a few genes. So far, little progress has been made in identifying candidate genes for quantitative traits. Robin et al. (2003) used this approach to identify candidate genes for OA, but a limited number of candidate genes showed polymorphic in their population. Most recently, Wayne and McIntyre (2002) developed a technique combining QTL mapping and microarray analysis to identify potential candidate genes for a phenotype of interest whose expression varies between parental lines. Using this approach, they identified 34 candidate genes for ovariole number in *Drosophila melanogaster*.

9 FUNCTIONAL GENOMICS AND PROTEOMICS

Genome analysis helps to identify and classify all the genes of a particular species, while functional genomics seeks to understand the precise function of these genes, including unique and redundant functions (Fernandes et al., 2002). The first step of rapid gene discovery is by large-scale partial sequencing of randomly selected cDNA clones, known as expressed sequence tag (ESTs). ESTs are now being used in mapping to identify candidate genes where ESTs/genes share similarity in structure, function, and pathway (Davis et al., 1999). Large numbers of ESTs are available from different species at the dbEST section of GenBank (http://www.ncbi.nlm.nih.gov/dbEST/dbEST_summary.html). As of today, about 200,000 ESTs of rice and 130,0000 ESTs of sorghum are deposited in GenBank.

Microarray is a new powerful technique to monitor the changes in gene expression as a result of environmental cues such as salinity, cold, drought, and heat stress. This tool allows expression profiling of thousands of genes in a single experiment to monitor transcript profiles in response to imposed stress. Significant progress has been made in transcriptome profiling through microarray analysis in *Arabidopsis*, rice, maize, and barley (Deyholos and Galbraith, 2001; Kawasaki et al., 2001; Zhu, 2001; Seki et al., 2001; Bray, 2002; Ozturk et al., 2002; Yu and Setter, 2003). Recently, microarray analysis has been successfully used to analyze genes involved in stress tolerance in plants. During stress, a plant perceives signals leading to changes in gene expression (Ozturk et al., 2002; Seki et al., 2001), inducing gene with adaptive function (Bray et al., 2000; 2002). Kawasaki et al. (2001) first investigated transcript regulation in response to salt stress in rice with 1728 cDNAs from salt stressed rice root libraries. They used Pokkali, a salt-tolerant rice genotype and IR29, a salt-sensitive one. Most of the transcripts in Pokkali displayed constant expression levels in all levels of salt stress. Transcripts for energy supply, transcription, transport facilitation, biogenesis, and DNA synthesis were unaffected at different time points. Many transcripts among those that remained unchanged in Pokkali showed little change in IR29 at an early time point, but were mostly downregulated within 6 h of stress. On the other hand, few transcripts that showed downregulation in Pokkali in the initial phase of stress returned to the original level of expression in later stages of stress. Between the two varieties, the salt-tolerant Pokkali can induce transcripts early to stimulate protein synthesis to overcome stress but the salt-susceptible IR29 delays the transcript processing.

Hazen et al. (personal communication) have investigated the expression profile for osmotic adjustment (an important component of drought tolerance) in two rice lines, IR62266-42-6-2, an *indica* genotype with high OA, and CT9993-5-10-1-M, a *japonica* genotype with low OA capacity, and their five DH lines with the highest and lowest OA capacity. This mapping population has been extensively studied for drought tolerance related traits (Zhang et al., 2001; Kamoshita et al., 2002a,b). The two parents and the DH lines were subjected to gradual water stress and leaf tissues were collected at control (95% leaf RWC), moderate (80% leaf RWC), and severe (65% leaf RWC) water stress. Rice genechip genome arrays containing 21,000 elements were used for array hybridization. More than 1200 transcripts showed greater than twofold changes in expression between the two parents. At moderate water stress, nearly tenfold more genes were induced in the high OA parent than in the low OA parent. A total of 117 genes were upregulated in all high OA lines and 14 of them were unique, not expressed in any of the low OA lines. Two genes, polyadenylate-binding protein and PP2A regulatory subunit, had a regulatory effect on high OA. Several genes were also identified for each of the OA QTL including a GTP-binding protein, two homeodomain transcription factors, a *trans*-acting transcriptional protein, a eukaryotic cap-binding protein, a protein phosphatase 2C, and an mRNA cleavage factor subunit-like protein. Two genes, GSH-dependent dehydroascorbate reductase and chloroplast superoxide dismutase, were expressed in all high OA lines. Wayne and McIntyre (2002) successfully integrated QTL analysis, fine mapping, and microarray techniques for an efficient, precise, and quantitative evaluation of genes underlying QTLs in *D. melanogaster*. In the future, this approach will be applied to rice, sorghum, and other agriculturally important crops. QTL analysis should be viewed as an integral part of functional genomics that can help to determine the relevance of allelic variation in grain yield (Tuberosa et al., 2002). Schadt et al. (2003) described comprehensive genetic screens of mouse, maize, and human transcriptomes by considering gene expression value as quantitative traits. They used mRNA transcript abundance as quantitative traits, mapped gene expression QTL for these traits, and demonstrated that such a combination of gene expression and genetic data had the potential to overcome the difficulties in dissecting complex traits.

Proteomics studies the function of all expressed proteins in a genome. Proteomics complements other functional genomics approaches, including microarray-based expression profiles, systematic phenotypic profiles at the cell and organism level, systematic genetics, and small-molecule-based arrays. There are five different approaches for proteomics research—mass spectrometry-based proteomics, array-based proteomics, structural proteomics, informatics, and clinical proteomics (reviewed by Tyers and Mann 2003). Cutler (2003) suggested to analyze the entire protein complement of a cell on a similar scale to mRNA, as mRNA levels do not always correlate with the protein. Protein arrays allow rapid integration of protein activity on a proteomic scale. These arrays may be based on either recombinant proteins or, conversely, reagents that specifically interact with proteins, including antibodies, peptides, and small molecules. All the techniques are in very initial stage, and are not yet performed in a coordinated fashion among the laboratories. A significant work has been carried out with array-based protein analysis with mouse and human. Proteomics is a new tool to analyze drought tolerance in plants. Salekdeh et al. (2002) examined drought responsive proteins of two rice genotypes, CT9993-5-10-1-M and IR62266-42-6-2. These two varieties were also used for molecular mapping and microarray analysis. They detected 16 drought responsive proteins using mass spectrometry. In a microarray analysis with these two genotypes and five extreme DH lines for OA, Hazen et al. (unpublished data) reported that out of 16 of the corresponding genes, 11 had complementary probes in rice GeneChip and two of the corresponding genes, GSH-dependent dehydroascorbate reductase and chloroplast superoxide dismutase, were induced in all the osmotically adjusting lines. Salekdeh et al. (2002) detected four novel drought-responsive mechanisms through this work, namely upregulation of S-like RNAse homologue, actin depolymerizing factor, rubisco activase, and downregulation of isoflavone reductase-like protein.

10 GENETIC ENGINEERING OF ABIOTIC STRESS TOLERANCE

Abiotic stress, particularly drought, salinity, and submergence is a complex phenomenon that is influenced by coordinated and differential expression of a network of genes. Abiotic stresses alter the metabolism, growth, and development of an organism. Under the influence of abiotic stress, plants accumulate various organic compounds of low molecular weight, collectively known as compatible solutes or osmolytes, synthesize late-embryogenesis-abundant (LEA) proteins, and activate several detoxification enzymes. The compounds produced serve as osmoprotectant and stabilize biomolecules under stress (Bajaj et al., 1999; Zhang et al., 2001; Datta, 2002). The recent development of DNA microarray technology allows the analysis of gene expression profile at the mRNA level, leading to the discovery of candidate genes. The discovery of novel genes, determination of their expression patterns in response to abiotic stress, and improved understanding of their roles in stress adaptation will provide the basis for effective engineering strategies leading to greater stress tolerance (Cushman and Bohnert, 2001). Transgenic approaches expand gene resources, because genes from both prokaryotes and eukaryotes can be potentially cloned and used to increase stress tolerance in rice (Cheng et al., 2002). A large number of stress tolerance genes have been isolated from different organisms but so far, few genes have been successfully incorporated into agriculturally important crops to increase abiotic stress tolerance. Most of the genetic engineering works were focused on *Arabidopsis* and tobacco, but few on cereal crops, especially rice. Increased abiotic tolerance of plants to stresses have been demonstrated by introducing specific genes, such as a late embryogenesis abundant (LEA group 3) protein (Hong et al., 1988; Xu et al., 1996), an abscisic acid responsive protein (RAB) (Chandler and Robertson 1994; Mundy and Chua, 1988), and several other proteins (Bajaj et al., 1999). Several stress-related genes have been cloned and transferred into rice to enhance osmolyte production and some transgenic lines showed tolerance to osmotic stress (Datta, 2002). Progress has been made in developing transgenic rice and the incorporated genes, gene products, and performance of transgenic rice plants are summarized in Table 4. More information on this topic is covered by Ho et al. (this volume).

11 CONCLUSION AND PERSPECTIVE

Significant progress has been made in the last two decades in QTL analysis related to drought tolerance in sorghum and rice. A number of QTLs for different abiotic stress related traits were detected that were consistent across different genetic backgrounds, locations, and seasons, and even across species. Thus QTL analysis followed by marker-assisted introgression of consistent QTLs is considered as an ideal strategy. In this regard, transfer of stay green in sorghum and root QTLs in rice to elite lines through marker-assisted back cross breeding is quite encouraging. We now have enough information about the prospective QTLs, the effect of which need to be validated in the target field environments using near-isogenic lines (NILs). A few reports are showing positive impact of these QTL in improving crop performance in field. Similarly, work on transgenic rice has shown great promise on the genetic improvement of drought and salt tolerance in rice. Future studies should focus on the evaluation of the transgenic plants in target field environments. The discovery of novel genes, determination of their expression patterns in response to abiotic stresses, and enhanced understanding of their roles in stress adaptation will provide the foundation for designing effective engineering strategies leading to abiotic stress tolerance.

So far, associating DNA sequence variation with variation of the organism phenotype to understand the complexity of quantitative traits such as abiotic stress tolerance is in progress. In the post-genomic era, the greatest challenge is to establish links between DNA sequence, transcript levels (functional genomics), proteins (proteomics), and metabolites (metabolomics). Progress has been made in gene expression profiling of rice and other plants. The availability of whole genome rice arrays will provide a new opportunity for profiling gene transcript levels during drought, salinity, and submergence stress. A combination of QTL mapping and microarray analysis has great potential to identify genes associated with abiotic stress tolerance traits. Rapid progress is now anticipated through comparative genomics investigations in diverse set of model organisms coupled with techniques such as high-throughput analysis of expressed sequence tags (EST), large-scale expression profiling, targeted mutagenesis, and mutant complementation. In addition, newly evolved tools such as proteomics and metabolic profiling will provide critical information to understand abiotic stress tolerance. Integration of these available tools and technologies, from QTL mapping to functional genomics and bioinformatics, will provide better understanding of the genetic

Table 4 Progress in Developing Transgenic Rice with Abiotic Stress Tolerance Genes

Gene	Gene product	Performance of transgenic plants	Reference
Genes encoding enzymes that synthesize osmoprotectants			
Adc	Arginine decarboxylase	Reduction in chlorophyll loss under drought stress	Capell et al. (1998)
CodA	Choline oxidase (glycine betaine synthesis)	Increased tolerance to salt and cold	Sakamoto et al. (1998)
P5cs	Pyrroline carboxylate synthetase (proline synthesis)	Increased biomass production under water and salinity stress	Zhu et al. (1998)
TLP-D34	Thaumatin-like protein (PR-5 member related to osmotin)	Enhanced fungal protection and osmotic adjustment	Datta et al. (2000)
GS2	Chloroplastic glutamine synthetase	Increased salt resistance and chilling tolerance	Hoshida et al. (2000)
pdc1	Pyruvate decarboxylase	Increased submergence tolerance	Quimio et al. (2000)
TPSP	Trehalose-6-phosphate synthase phosphatase	Increased photosynthetic capacity, drought, and salt tolerance	Garg et al. (2002)
Late embryogenesis abundant (LEA) related genes			
HVA1	Group 3 LEA protein gene	Increased drought and salinity tolerance	Rohila et al. (2002)

Gene	Product	Effect	Reference
HVA1	Group 3 LEA protein gene	Increased tolerance to drought and salt stress	Xu et al. (1996)
Regulatory genes			
OsCDPK7	transcription factor	Increase in cold and salt/drought tolerance	Saijo et al. (2000)
Adh1	Alcohol dehydrogenase	Submergence tolerance	Rahman et al. (2001)
ADC	Arginine decarboxylase overexpression	Polyamine accumulation and salt resistance	Roy and Wu (2001)
Naat	Nicotianamine aminotransferase activity	Fe efficiency	Takahashi et al. (1999)
GPAT	Glycerol-3-phosphate acyltransferase	Improvement in photosynthesis and growth under chilling	Ariizumi et al. (2002)
YK1		Resistance to rice blast and increased tolerance to several abiotic stresses such as, NaCl, UV-C, submergence, and hydrogen peroxide	Uchimiya et al. (2002)
Oxidative-stress related genes			
CAT	Catalase	Reduction in hydrogen peroxide under chilling stress	Matsumura et al. (2002)
Genes encoding for molecular chaperons			
Wx	Amylose synthesis	Increased amylose content at low temperature	Hirano and Sano (1998)

and molecular basis of abiotic stress tolerance toward the development of stress-tolerant plants useful for agriculture.

REFERENCES

Ahn S, Anderson JA, Sorrells ME, Tanksley SD. Homoeologous relationships of rice, wheat and maize chromosomes. Mol Gen Genet 1993; 241:483–490.

Akhter MS, Goldschmidt EE, John I, Rodoni S, Matile P, Grierson D. Altered patterns of senescence and ripening in gf, a stay green mutant of tomato (*Lycopersicon esculentum* Mill.). J Exp Bot 1999; 50:1115–1122.

Aiken RM, Smuker AJM. Root system regulation of whole plant growth. Annu Rev Phytopathol 1996; 34:325–346.

Ali ML, Pathan MS, Zhang J, Bai G, Sarkarung S, Nguyen HT. Mapping QTLs for root traits in a recombinant inbred population from two *indica* ecotypes in rice. Theor Appl Genet 2000; 101:756–766.

Ariizumi T, Kishitani S, Inatsugi R, Nishida I, Murata N, Toriyama K. An increase in unsaturation of fatty acid in phosphatidylglycerol from leaves improves the rates of photosynthesis and growth at low temperatures in transgenic rice seedlings. Plant Cell Physiol 2002; 43:751–758.

Ashraf M. Breeding for salinity tolerance in plants. Crit Rev Plant Sci 1994; 13:17–42.

Azhiri-Sigari T, Yamauchi A, Kamoshita A, Wade LJ. Genotypic variation in response of rainfed lowland rice to drought and rewatering. II. Root growth. Plant Prod Sci 2000; 3:180–188.

Babu RC, Pathan MS, Blum A, Nguyen HT. Comparison of measurement methods of osmotic adjustment in different rice cultivars. Crop Sci 1999; 39:150–158.

Bajaj S, Targolli J, Liu L-F, Ho T-H D, Wu R. Transgenic approaches to increase dehydration-stress tolerance in plants. Mol Breed 1999; 5:493–503.

Basnayake J, Cooper M, Ludlow MM, Henzell RG, Snell PJ. Inheritance of osmotic adjustment to water stress in three grain sorghum crosses. Theor Appl Genet 1995; 90:675–682.

Beavis WD, Smith OS, Grant D, Fincher R. Identification of quantitative trait loci using a small sample of topcrossed and F4 progeny from maize. Crop Sci 1994; 34:882–896.

Bernacchi D, Beck-Blum T, Eshed Y, Lopez J, Petiard V, Uhlig J, Zamir D, Tanksley SD. Advanced back-cross QTL analysis in tomato. 1. Identification of QTLs for traits of agronomic importance from *Lycopersicon hirsutum*. Theor Appl Genet 1998; 97:381–397.

Blum A. Plant Breeding for Stress Environments. Boca Raton, FL, USA: CRC Press, 1988.

Blum A. Osmotic adjustment and growth of barley cultivars under drought stress. Crop Sci 1989; 29:230–233.

Blum A, Sullivan CY. The comparative drought resistance of landraces of sorghum and millet from dry and humid regions. Ann Bot 1986; 57:835–846.

Bohnert HJ, Sheveleva E. Plant stress adaptations, making metabolism move. Curr Opin Plant Biol 1998; 1:267–274.

Boyer JS. Plant productivity and environment. Science 1982; 218:444–448.

Bray EA, Bailey-Serres J, Weretilnyk E. Response to abiotic stress. In: Buchanan BB, Grussiem W, Jones RL, eds. Biochemistry and Molecular Biology of Plants. Rockville, MD: American Society of Plant Physiologists, 2000:1158–1203.

Bray EA. Classification of genes differentially expressed during water-deficit stress in *Arabidopsis thaliana*: an analysis using microarray and differential expression data. Ann Bot 2002; 89:803–811.

Capell T, Escobar C, Liu H, Burtin D, Lepri O, Christou P. Overexpression of the oat arginine decarboxylase cDNA in transgenic rice (*Oryza sativa* L.) affects normal development patterns in vitro and results in putrescine accumulation in transgenic plants. Theor Appl Genet 1998; 97:246–254.

Causse M, Fulton TM, Cho YG, Ahn SN, Chunwongse J, Wu K, Xiao J, Yu Z, Ronald PC, Harrington SB, Second GA, McCouch SR, Tanksley SD. Saturated molecular map of the rice genome based on an interspecific backcross population. Genetics 1994; 138:1251–1274.

Champoux MC, Wang G, Sarkarung S, Mackill DJ, O'Toole JC, Huang N, McCouch SR. Locating genes associated with root morphology and drought avoidance in rice via linkage to molecular markers. Theor Appl Genet 1995; 90:969–981.

Chandler PM, Robertson M. Gene expression regulated by abscisic acid and its relation to stress tolerance. In: Somerville CR, Jones RL, eds. Annual Review of Plant Physiology and Plant Molecular Biology. Annual Reviews I. Palo Alto, CA, 1994:113–141.

Chen X, Line RF, Leung H. Resistance gene analogs associated with a barley locus for resistance to stripe rust. In: Heller SR, ed. International Conference on the Status of Plant and Animal Genome Research VI [abstr]. San Diego, California: Scherago International I, New York, 1998.

Cheng Z, Targolli J, Huang X, Wu R. Wheat LEA genes, PMA80 and PMA1959, enhance dehydration tolerance to transgenic rice (*Oryza sativa* L.). Mol Breed 2002; 10:71–82.

Courtois B, McLaren G, Sinha PK, Prasad K, Yadav R, Shen L. Mapping QTLs associated with drought avoidance in upland rice. Mol Breed 2000; 6:55–66.

Crasta OR, Xu WW, Rosenow DT, Mullet J, Nguyen HT. Mapping of post-flowering drought resistance traits in grain sorghum: association between QTLs influencing premature senescence and maturity. Mol Gen Genet 1999; 262:579–588.

Cushman JC, Bohnert HJ. Genomic approach to plant stress tolerance. Curr Opin Plant Biol 2001; 3:117–124.

Cutler P. Protein arrays: the current state-of-the-art. Proteomics 2003; 3:3–18.

Datta K, Koukolikova-Nicola Z, Baisakh N, Oliva N, Datta SK. *Agrobacterium*-mediated engineering for sheath blight resistance of *indica* rice cultivars from different ecosystems. Theor Appl Genet 2000; 100:832–839.

Datta SK. Recent developments in transgenics for abiotic stress tolerance in rice. JIRCAS (Japan) Work Rep 2002; 43–53.

Davis GL, McMullen MD, Baysdorfer C, et al. A maize map standard with sequenced with core markers, grass genome reference points and 932 expressed sequence tagged sites (ESTs) in a 1736-locus map. Genetics 1999; 152:1137–1172.

Deyholos M, Galbraith DW. High density microarrays for gene expression analysis. Cytometry 2001; 43:229–238.

Duncan RR, Bockholt AJ, Miller FR. Descriptive comparison of senescent and nonsenescent sorghum genotypes. Agron J 1981; 73:849–853.

Faris JD, Li W, Gill BS, Liu D, Chen P. Candidate gene analysis of quantitative disease resistance in wheat. Theor Appl Genet 1999; 98:219–225.

Fernandes J, Brendel V, Gai X, Lal S, Chandler VL, Elumalai RP, Galbraith DW, Pierson EA, Walbot V. Comparison of RNA expression profile based on maize expressed sequence tag frequency analysis and microarray hybridization. Plant Physiol 2002; 128:896–910.

Flowers TJ, Koyama ML, Flowers SA, Sudhakar C, Singh KP, Yeo AR. QTL: their place in engineering tolerance of rice to salinity. J Exp Bot 2000; 51:99–106.

Flowers TJ, Yeo J. Breeding for salinity resistance in crop plants: where next? Aust J Plant Physiol 1995; 22:875–884.

Forster BP, Ellis RP, Thomas WT, Newton AC, Tuberosa R, This D, el-Enein RA, Bahri MH, Ben Salem M. The development and application of molecular markers for abiotic stress tolerance in barley. J Exp Bot 2000; 51:18–27.

Fukai S, Cooper M. Development of drought resistant cultivators using physio-morphological traits in rice. Field Crops Res 1995; 40:67–86.

Garg AK, Kim J-K, Owens TG, Ranwala AP, Choi YD, Kochian LV, Wu RJ. Trehalose accumulation in rice plants confers high tolerance levels to different abiotic stresses. Proc Natl Acad Sci USA 2002; 99:15898–15903.

Gong JM, He P, Qian QA, Shen LS, Zhu LH, Chen SY. Identification of salt tolerant QTL in rice (*Oryza sativa* L.). Chin Sci Bull 1999; 44:68–71.

Hauck B, Gay AP, Macduff J, Griffiths CM, Thomas H. Leaf senescence in a non-yellowing mutant of *Festuca pratensis*: implication of the stay green mutation for photosynthesis, growth, and nitrogen nutrition. Plant Cell Environ 1997; 20:1007–1018.

Haque QA, Hille Ris Lambers D, Tepora NM, Dela Cruz QD. Inheritance of submergence tolerance in rice. Euphytica 1989; 41:247–251.

Helsel LDB, Frey KJ. Grain yield variations in oats associated with differences in leaf area duration among oat lines. Crop Sci 1978; 18:765–769.

Hemamalini GS, Shashidar HE, Hittalmani S. Molecular marker assisted tagging of morphological and physiological traits under two contrasting moisture regimes at peak vegetative stage in rice (*Oryza sativa* L.). Euphytica 2000; 112:69–78.

Henzell RG, Bregman RL, Fletcher DS, McCosker AN. Relationship between yield and non-senescence (stay green) in some grain sorghum hybrids grown under terminal drought stress. Proc 2nd Australian sorghum conference, February 4–6, 1992, Gatton, Queensland. Melbourne, Australia: Australian Institute of Agricultural Sciences, Melbourne, Australia, 1992:355–358. Publication no. 68.

Hirano HY, Sano Y. Enhancement of Wx gene expression and the accumulation of amylose in response to cool temperatures during seed development in rice. Plant Cell Physiol 1998; 39:807–812.

Hong B, Ukaes SJ, Ho T-H D. Cloning and characterization of a cDNA encoding an mRNA rapidly induced in barley aleurone layers. Plant Mol Biol 1988; 11:495–506.

Hossain M, Bennett J, Datta S, Leung H, Khush G. Biotechnology research in rice for Asia: priorities, focus and directions. In: Qaim M, ed. Agricultural Biotechnology in Developing Countries: Toward Optimizing the Benefits for the Poor. Boston, USA: Kluwer Academic Publishers, 2000:99–120.

Hoshida H, Tanaka T, Hibino T, Hayashi Y, Tanaka A, Takabe T. Enhanced tolerance to salt stress in transgenic rice that overexpresses chloroplast glutamine synthetase. Plant Mol Biol 2000; 43:103–111.

Ito O, Ella E, Kawano N. Physiological basis of submergence tolerance in rainfed lowland rice ecosystem. Field Crop Res 1999; 64:75–90.

Kamoshita A, Wade LJ, Yamauchi A. Genotypic variation in response of rainfed lowland rice to drought and rewatering. III. Water extraction during drought period. Plant Prod Sci 2000; 1:183–190.

Kamoshita A, Wade LJ, Ali ML, Pathan MS, Zhang J, Sarkarung S, Nguyen HT. Mapping QTLs for root morphology of a rice population adapted to rainfed lowland conditions. Theor Appl Genet 2002a; 104:880–893.

Kamoshita A, Zhang J, Siopongco J, Sarkarung S, Nguyen HT, Wade LJ. Effects of phenotyping environment on identification of QTL for rice root morphology under anaerobic conditions. Crop Sci 2002b; 42:255–265.

Kawano N, Ella E, Ito O, Yamauchi Y, Tanaka K. Metabolic changes in rice seedlings with different submergence tolerance after desubmergence. Environ Exp Bot 2002; 47:195–203.

Kawasaki S, Borchert C, Deyholos M, Wang H, Brazille S, Kawai K, Galbraith D, Bohnert HJ. Gene expression profiles during the initial phase of salt stress in rice. Plant Cell 2001; 13:889–905.

Kebede H, Subudhi PK, Rosenow DT, Nguyen HT. Quantitative trail loci influencing drought tolerance in grain sorghum (*Sorghum bicolor* L. Moench). Theor Appl Genet 2001; 103:266–276.

Kende H, van der Knaap E, Cho HT. Deep water rice: a model plant to study stem elongation. Plant Physiol 1998; 118:1105–1110.

Khush GS. Terminology for rice growing environments. Terminology for Rice Growing Environments. Manila, Philippines: IRRI, 1984:5–10.

Koyama ML, Levesley A, Koebner RMD, Flowers TJ, Yeo AR. Quantitative trait loci for component physiological traits determining salt tolerance in rice. Plant Physiol 2001; 125:406–422.

Levitt J. Response of Plants to Environmental Stresses. Vol. 2. Water, Radiation, Salt and Other Stresses. New York: Academic Press, 1980:93–128.

Lilley JM, Ludlow MM, McCouch SR, O'Toole JC. Locating QTL for osmotic adjustment and dehydration tolerance in rice. J Exp Bot 1996; 47:1427–1436.

Ludlow MM, Muchow RC. A critical evaluation of traits for improving crop yields in water-limited environments. Adv Agron 1990; 43:107–153.

Matsumura T, Tabayashi N, Kamagata Y, Souma Chihiro, Saruyama H. Wheat catalase expressed in transgenic rice can improve tolerance against low temperature stress. Physiol Plant 2002; 116:317–327.

Mishra SB, Senadhira D, Manigbas NL. Genetics of submergence tolerance in rice (*Oryza sativa* L.). Field Crop Res 1996; 46:177–181.

Mohanty HK, Khush GS. Diallel analysis of submergence tolerance in rice (*Oryza sativa* L.). Theor Appl Genet 1985; 70:467–473.

Moncada P, Martinez CP, Borrero J, Chatel M, Gauch H, Guimaraes E, Thome J, McCouch SR. QTL analysis for yield and yield components in an *Oryza sativa* × *O. rufipogon*. Theor Appl Genet 2001; 102:41–52.

Morgan JM. A gene controlling differences in osmoregulation in wheat. Aust J Plant Physiol 1991; 18:249–257.

Morgan JM. Osmotic components and properties associated with genotypic differences in osmoregulation in wheat. Aust J Plant Physiol 1992; 19:67–76.

Morgan JM, Tan MK. Chromosomal location of a wheat osmoregulation gene using RFLP analysis. Aust J Plant Physiol 1996; 23:803–806.

Mundy J, Chua NH. Abscisic acid and water stress induce the expression of a novel rice gene. EMBO J 1988; 7:2279–2286.

Nandi S, Subudhi PK, Senadhira D, Manigbas NL, Sen-Mandi S, Huang N. Mapping QTLs for submergence tolerance in rice by AFLP analysis and selective genotyping. Mol Gen Genet 1997; 255:1–8.

Nguyen HT, Babu RC, Blum A. Breeding for drought resistance in rice: physiology and molecular genetics considerations. Crop Sci 1999; 37:1426–1434.

O'Toole JC. Adaptation of rice to drought-prone environments. Drought Resistance in Crops with Emphasis on Rice. Manila, Philippines: The International Rice Research Institute, 1982:195–213.

O'Toole JC, Bland WL. Genotypic variation in crop plant root systems. Adv Agron 1987; 41:91–145.

Ozturk ZN, Talame V, Dyholos M, Michalowski CB, Galbraith DW, Gozukirmizi N, Tuberosa R, Bohnert HJ. Monitoring large-scale changes in transcript abundance in drought- and salt-stressed barley. Plant Mol Biol 2002; 48:551–573.

Palada MC, Vergara BS. Environmental effects on the resistance of rice seedlings to complete submergence. Crop Sci 1972; 12:209–212.

Prasad SR, Bagali PG, Hittalmani S, Shashidhar HE. Molecular mapping of quantitative trait loci associated with seedling tolerance to salt stress in rice (*Oryza sativa* L.). Curr Sci 2000; 78:162–164.

Price AH, Cairrns JE, Horton P, Jones HG, Griffiths H. Linking drought-resistance mechanisms to drought avoidance in upland rice using a QTL approach: progress and new opportunities to integrate stomatal and mesophyll response. J Exp Bot 2002a; 53:989–1004.

Price AH, Steele KA, Moore BJ, Barraclough PB, Clark LJ. A combined RFLP and AFLP linkage map of upland rice (*Oryza sativa* L.) used to identify QTLs for root penetration ability. Theor Appl Genet 2000; 100:49–56.

Price AH, Tomos AD. Genetic dissection of root growth in rice (*Oryza sativa* L.). II: mapping quantitative trait loci using molecular markers. Theor Appl Genet 1997; 95:143–152.

Price AH, Townend J, Jones MP, Audebert A, Courtois B. Mapping QTL associated with drought avoidance in upland rice grown in the Philippines and West Africa. Plant Mol Biol 2002b; 48:683–695.

Price AH, Young EM, Tomos AD. Quantitative trait loci associated with stomatal conductance, leaf rolling and heading date mapped in upland rice (*Oryza sativa* L.). New Phytol 1997; 137:83–91.

Quarrie SA, Gulli M, Calestani C, Steed A, Marmiroli N. Location of a gene regulating drought-induced abscisic acid production on the long arm of chromosome 5A of wheat. Theor Appl Genet 1994; 89:794–800.

Quarrie SA, Laurie DA, Zhu J, Lebreton C, Semikhodskii A, Steed A, Witsenboer H, Calestani C. QTL analysis to study the association between leaf size and abscisic acid accumulation in droughted rice leaves and comparison across cereals. Plant Mol Biol 1997; 35:155–165.

Quimio CA, Torrizo LB, Setter TL, Ellis M, Grover A, Abrigo EM, Oliva NP, Ella ES, Carpena AL, Ito O, Peacock WJ, Dennis E, Datta SK. Enhancement of submergence tolerance in transgenic rice overproducing pyruvate decarboxylase. J Plant Physiol 2000; 156:516–521.

Rahman M, Grover A, Peacock WJ, Dennis ES, Ellis MH. Effects of manipulation of pyruvate decarboxylase and alcohol dehydrogenease levels on the submergence tolerance in rice. Aust J Plant Physiol 2001; 28:1231–1241.

Ray JD, Yu L, McCouch SR, Champoux MC, Wang G, Nguyen HT. Mapping quantitative trait loci associated with root penetration ability in rice (*Oryza sativa* L.). Theor Appl Genet 1996; 92:627–636.

Rhoades JD, Loveday J. Salinity in irrigated agriculture. Agronomy 1990; 30:1089–1142.

Robin S, Pathan MS, Courtois B, Lafitte R, Scarandang C, Lanceras S, Amante M, Nguyen HT, Li Z. Mapping osmotic adjustment in an advanced backcross inbred population of rice. Theor Appl Genet, 2003. In press.

Rohila JS, Rajinder K, Wu RJ. Genetic improvement of Basmati rice for salt and drought tolerance by regulated expression of a barley Hva1 cDNA. Plant Sci 2002; 163:525–532.

Rosenow DT. Breeding for lodging resistance in sorghum. Proc 32nd Annual corn and sorghum research conference, December 6–8, Chicago, Illinois, USA, 1977:171–175.

Rosenow DT. Breeding for drought resistance under field conditions. Proc 18th Biennial Grain Sorghum Research. Feb. 28–Mar. 2, 1993. Lubbock, Texas, 1993:122–126.

Rosenow DT. Breeding for resistance to root and stalk rot in Texas. In: Mughogho LK, ed. Sorghum Root and Stalk Diseases: A Critical Review: Proc of the Consultative Discussion of Research Needs and Strategies for Control of Sorghum Root and Stalk Diseases (November 27–December 2, 1983, Bellagio, Italy). Patancheru, AP, India: ICRSAT, 1984:209–217.

Rosenow DT. Breeding sorghum for drought resistance. In: Menyonga JM, Bezune T, Yuodeowei A, eds. Proceedings of the International Drought Symposium. Ouagadougou, Burkina Faso: OAU/STRC-SAFGRAD Coordination Office, 1987:19–23.

Rosenow DT, Clark LE. Drought tolerance in sorghum. In: Loden HD, Wilkinson D, eds. Proceedings of the 36th Annual Corn and Sorghum Research Conference. 9–11 Dec 1981, Chicago, Ill., USA. Washington, DC: American Seed Trade Association, 1981:18–31.

Rosenow DT, Clark LE. Drought and lodging research for a quality crop. 5th Annual Corn and Sorghum Industry Research Conference, December 6–7, 1995, Illinois, USA, 1995:82–97.

Rosenow DT, Ejeta G, Clark LE, Gilbert ML, Henzell RG, Borell AK, Muchow RC. Breeding for Pre- and post-flowering drought stress resistance in sorghum. Proc International Conference on Genetic Improvement of Sorghum and Pearl Millet, September 23–27, 1996, Lubbock, Texas, USA, 1996:400–411.

Rothchild MF, Soller M. Candidate gene analysis to detect genes controlling traits of economic importance in domestic livestock. Probe 1997; 8:13–20.

Roy M, Wu R. Arginine decarboxylase transgene expression and analysis of environmental stress tolerance in transgenic rice. Plant Sci 2001; 160:869–875.

Saijo Y, Hata S, Kyozuka J, Shimamotok K, Izui K. Overexpression of a single Ca^{2+}-dependent protein kinase confers both cold and salt/drought tolerance on rice plants. Plant J 2000; 23:319–327.

Sakamoto A, Alia HH, Murata N. Metabolic engineering of rice leading to biosynthesis of glycine betaine and tolerance to salt and cold. Plant Mol Biol 1998; 38: 1011–1019.

Salekdeh GH, Siopongco J, Wade LJ, Ghareyazie B, Bennett J. Proteomic analysis of rice leaves during drought stress and recovery. Proteomics 2002; 2:1131–1145.

Sanchez A, Subudhi RK, Rosenow DT, Nguyen HT. Mapping QTLs associated with drought resistance in sorghum (*Sorghum bicolor* L. Moench). Plant Mol Biol 2002; 48:713–726.

Schadt EE, et al. Genetics of gene expression surveyed in maize, mouse and man. Nature 2003; 422:297–302.

Seki M, Narusaka M, Abe H, Kasuga M, Yamaguchi-Shinozaki K, Carninci P, Hayashizaki Y, Shinozaki K. Monitoring the expression pattern of 1300 *Arabidopsis* genes under drought and cold stress by using a full length cDNA microarray. Plant Cell 2001; 13:61–72.

Setter TL, Kupkanchanakul T, Water I, Greenway H. Evaluation of factors contributing to diurnal changes in O_2 concentrations in floodwater of deepwater rice fields. New Phytol 1988; 110:151–162.

Setter TL, Laureles EV. The beneficial effect of reduced elongation growth on submergence tolerance of rice. J Exp Bot 1996; 47:1551–1559.

Setter TL, Ramakrishnayya G, Ram MPC, Singh BB. Environmental characterization of flood water in eastern India: relevance to flooding tolerance of rice. J Indian Plant Physiol 1995; 38:34–40.

Shen L, Courtois B, McNally KL, Robin S, Li Z. Evaluation of near-isogenic lines of rice introgressed with QTLs for root depth through marker-aided selection. Theor Appl Genet 2001; 103:75–83.

Siangliw M, Toojinda T, Tragoonrung S, Vanavichit A. Thai Jasmine rice carrying QTLch9 (SubQTL) is submergence tolerant. Ann Bot 2003; 91:255–261.

Sillito D, Parkin IA, Mayerhofer R, Lydiate DJ, Good AG. *Arabidopsis thaliana*: a source of candidate disease resistance for *Brassica napus*. Genome 2000; 43:452–460.

Singh HP, Singh BB, Ram PC. Submergence tolerance of rainfed lowland rice: search for physiological marker traits. J Plant Physiol 2001; 158:883–889.

Sripongpangkul K, Posa GBT, Senadhira DW, Brar D, Huang N, Khush GS, Li ZK. Genes/QTLs affecting flood tolerance in rice. Theor Appl Genet 2000; 101:1074–1081.

Straeten DVD, Zhou Z, Prinsen E, Onckelen HAV, Montagu MCV. A comparative molecular–physiological study of submergence response in lowland and deepwater rice. Plant Physiol 2002; 125:955–968.

Subudhi PK, Rosenow DT, Nguyen HT. Quantitative trait loci for the stay-green trait in sorghum (*Sorghum bicolor* L. Moench): consistency across genetic backgrounds and environments. Theor Appl Genet 2000; 101:733–741.

Subudhi PK, Nguyen HT. Biotechnology—New Horizons. In: Wayne Smith C, Frederiksen RA, eds. Sorghum: Origin, History, Technology, and Production. New York, NY: John Wiley and Sons, 2000:349–397.

Suprihanto B, Coffman WR. Inheritance of submergence tolerance in rice (*Oryza sativa* L.). SABRAO J 1981; 13:98–108.

Takahashi M, Yamaguchi H, Nakanishi H, Shioiri T, Nishizawa NK, Mori S. Cloning two genes for nicotianamine aminotransferase, critical enzyme in iron acquisition (strategy II) in graminaceous plants. Plant Physiol 1999; 121:947–956.

Tanksley SD, Nelson JC. Advanced back-cross QTL analysis: a method for the simultaneous discovery and transfer of valuable QTLs from unadapted germplasm into elite breeding lines. Theor Appl Genet 1996; 92:191–203.

Tao YZ, Henzel RG, Jordan DR, Butler DG, Kelly AM, McIntyre CL. Identification of genomic regions associated with stay green in sorghum by testing RILs in multiple environments. Theor Appl Genet 2000; 100:1225–1232.

Teulat B, Borries C, This D. New QTLs identified for plant water status, water-soluble carbohydrate and osmotic adjustment in a barley population grown in a growth-chamber under two water regimes. Theor Appl Genet 2001; 103:161–170.

Teulat B, This D, Khairallah M, Borries C, Ragot C, Sourdille P, Leroy P, Monneveux, Charrier A. Several QTLs involved in osmotic adjustment trait variation in barley (*Hordeum vulgare* L.). Theor Appl Genet 1998; 96:688–698.

Thomas H, Morgan WG, Thomas MG, Ougham HJ. Expression of stay green character introgressed into *Lolium Temulentum* Ceres from a senescence mutant of *Festuca pratensis*. Theor Appl Genet 1999; 99:92–99.

Thomas H, Smart CM. Crops that stay green. Ann Appl Biol 1993; 123:193–233.

Tollenaar M, Daynard TB. Leaf senescence in short season maize hybrids. Can J Plant Sci 1978; 58:869–874.

Toojinda T, Siangliw M, Tragoonrung S, Vanavichit A. Molecular genetics of submergence tolerance in rice: QTL analysis of Key traits. Ann Bot 2003; 91:243–253.

Tripathy JN, Zhang J, Robin S, Nguyen TT, Nguyen HT. QTLs for cell-membrane stability mapped in rice (*Oryza sativa* L.) under drought stress. Theor Appl Genet 2000; 100:1197–1202.

Tuberosa R, Gill BS, Quarrie SA. Cereal genomics: ushering in a brave new world. Plant Mol Biol 2002; 48:445–449.

Tuberosa R, Sanguineti MC, Landi P, Salvi S, Casarini E, Conti S. RFLP mapping of quantitative trait loci controlling abscisic acid concentration in leaves of drought-stressed maize (*Zea mays* L.). Theor Appl Genet 1998; 97:744–755.

Tuinstra MR, Grote EM, Goldsbough PB, Ejeta G. Genetic analysis of post-flowering drought tolerance and components of grain development in *Sorghum bicolor* (L.) Moench. Mol Breed 1997; 3:439–448.

Tuinstra MR, Grote EM, Goldsbough PB, Ejeta G. Identification of quantitative trait loci associated with pre-flowering drought tolerance in sorghum. Crop Sci 1996; 36:1337–1344.

Tyers M, Mann M. From genomics to proteomics. Nature 2003; 422(6928):193–197.

Uchimiya H, et al. Transgenic rice plants conferring increased tolerance to rice blast and multiple environmental stresses. Mol Breed 2002; 9:25–31.

Veldloom RL, Lee M. Genetic mapping of quantitative trait loci in maize in stress and nonstress environments. II. Plant height and flowering. Crop Sci 1996; 36:1320–1327.

Wade LJ, Fukai S, Samson BK, Ali A, Mazid MA. Rainfed lowland rice: physiological environment and cultivar requirement. Field Crops Res 1999; 64:3–12.

Wayne ML, McIntyre LM. Combining mapping and arraying: an approach to candidate gene identification. Proc Natl Acad Sci USA 2002; 99:14903–14906.

Widawsky DA, O'Toole JC. Prioritizing the rice research agenda for Eastern India. In: Evenson RE, Herdt RW, Hossain M, eds. Rice Research in Asia: Progress and Priorities. Wallingford, UK: CAB international, 1996:109–129.

Xiao J, Li J, Grandillo S, Sang-Nang A, Yuan L, Tanksley SD, McCouch SR. Identification of quantitative trait loci alleles from a wild relative, *Oryza rufipogon*. Genetics 1998; 150:899–909.

Xu D, Duan X, Wang B, Hong B, Ho T-H D, Wu R. Expression of a late embryo-genesis abundant protein gene, HVA1, from barley confers tolerance to water deficit and salt stress in transgenic rice. Plant Physiol 1996; 110:249–257.

Xu K, Mackill DJ. A major locus for submergence tolerance mapped on rice chromosome 9. Mol Breed 1995; 2:219–224.

Xu W, Subudhi PK, Crasta OR, Rosenoe DT, Mullet JE, Nguyen HT. Molecular mapping of QTLs conferring stay green in grain sorghum (*Sorghum bicolor* L. Moench). Genome 2000a; 43:461–469.

Xu K, Xu X, Ronald PC, Mackill DJ. A high resolution linkage map of the vicinity of the rice submergence tolerance locus *Sub1*. Mol Gen Genet 2000b; 263:681–689.

Yadav R, Courtois B, Huang N, Mclaren G. Mapping genes controlling root morphology and root distribution in a doubled-haploid population of rice. Theor Appl Genet 1997; 94:619–632.

Yu L-X, Ray JD, O'Toole JC, Nguyen HT. Use of wax petrolatum layers for screening for rice root penetration. Crop Sci 1995; 35:684–687.

Yu L-X, Setter TL. Comparative transcriptional profiling of placenta and endosperm in developing maize kernels in response to water deficit. Plant Physiol 2003; 131:568–582.

Zhang GY, Guo Y, Chen SL, Chen SY. RFLP tagging of salt tolerance gene in rice. Plant Sci 1995; 110:227–234.

Zhang J, Nguyen HT, Blum A. Genetic analysis of osmotic adjustment in crop plants. J Exp Bot 1999; 50:291–302.

Zhang J, Zheng HG, Aart A, Pantuwan G, Nguyen TT, Tripathy JN, Sarial AK,

Robin S, Babu RC, Nguyen BD, Sarkarung S, Blum A, Nguyen HT. Locating genomic regions associated with components of drought resistance in rice: comparative mapping within and across species. Theor Appl Genet 2001; 103:19–29.

Zheng HG, Babu RC, Pathan MS, Ali ML, Huang N, Courtois B, Nguyen HT. Quantitative trait loci for root penetration ability and root thickness in rice: comparison of genetic backgrounds. Genome 2000; 43:53–61.

Zhu JK. Plant salt tolerance. Trends Plant Sci 2001; 6:66–71.

Zhu JK, Hasegawa PM, Bressan RA. Molecular aspects of osmotic stress in plants. Crit Rev Plant Sci 1997; 16:253–277.

Zhu B, Su J, Chang MC, Verma DPS, Fan YL, Wu Ray. Overexpression of a pyrroline-5-carboxylate synthetase gene and analysis of tolerance to water and salt stress in transgenic rice. Plant Sci 1998; 139:41–48.

15

Genetic Dissection of Drought Tolerance in Maize: A Case Study

Jean-Marcel Ribaut and David Hoisington
CIMMYT
Mexico City, Mexico

Marianne Bänziger
CIMMYT-Zimbabwe
Harare, Zimbabwe

Tim L. Setter
Cornell University
Ithaca, NY, U.S.A.

Gregory O. Edmeades
Pioneer Hi-Bred International, Inc.
Waimea, Hawaii, U.S.A.

1 INTRODUCTION

In developing countries, drought and low soil fertility are the main sources of reduced yields for cereal crops (Heisey and Edmeades, 1999; Edmeades et al., 2001). Drought strongly affects the production of maize, sorghum, rice, wheat, and pearl millet, and poses a serious threat to the food security of households, countries, and even entire subcontinents. Globally, sub-Saharan

Africa is the most severely affected region. FAO estimates that 44% of the land surface in sub-Saharan Africa is subject to a high risk of meteorological drought, affecting 21% of the maize area and reducing yields by 33%, on average. In the tropics where over 90% of all maize is grown under rainfed conditions, irrigation is often unavailable or not within the financial capacity of farmers. Droughts in 1992 and again in 2002 have reduced maize production in the parts of the southern African region by 50% or more.

In the future, the destructive impact of drought may grow as the specter of climate change becomes a reality. According to recent predictions by the Inter-Governmental Panel on Climate Change (IPCC), global mean temperatures are expected to rise 1.4 to 5.8°C over the course of this century. Intraregional and seasonal effects will even be greater. Even if the lower of these predicted increases occurs, this will have considerable effect on crop production systems. Most scenarios also foresee increased threats to crop systems stemming from erratic weather, resulting in more major storms and drought spells. Climate change may increase flooding in some regions while further intensifying the frequency and magnitude of droughts in central Asia, northern and southern Africa, the Middle East, the Mediterranean region, and Australia (CGIAR Annual Report, 2000). Furthermore, the impact of climate change on precipitation will pose a serious problem for water supplies in many areas of the world, a potential crisis that will be exacerbated by population growth and the continuing migration of rural populations to urban areas. Given the likelihood of reduced rainfall and the certainty of increased competition for water, the Consultative Group on International Agriculture Research (CGIAR) and CIMMYT (International Center for Maize and Wheat Improvement), in particular, have made the improvement of cereals for water-limited environments a top priority. During the past three decades, CIMMYT scientists have devoted considerable effort to improving drought tolerance in maize for the period during and after flowering (e.g., Beck et al., 1996). Extensive research has been conducted in the areas of breeding, physiology, agronomy, and most recently, biotechnology. Although to date major progress has been achieved through conventional breeding (for review, see Heisey and Edmeades, 1999; Bänziger et al., 2000a,b), this approach remains slow and time-consuming. Combining approaches should increase significantly the potential for genetic gain under water-limited conditions.

During the past two decades, molecular tools have advanced the identification, mapping, and isolation of genes in a wide range of crop species. The first DNA markers, restriction fragment length polymorphisms (RFLPs), have been widely used to construct linkage maps for several crop species, including maize (Helentjaris et al., 1986), tomato (Paterson et al., 1988), and rice (McCouch et al., 1988). During the 1990s, a new class of markers based on the polymerase chain reaction (PCR) revolutionized molecular marker assays; they are easy to use and suitable for automation, thereby allowing the large-scale screening of progeny. With the development of new PCR-based DNA markers such as simple sequence repeats (SSRs) (Powell et al., 1996), amplified fragment length polymorphisms (AFLPs) (Vos et al., 1995), allele specific marker-like molecular beacons (Bonnet et al., 1999), and most recently single nucleotide polymorphism (SNPs) (Gilles et al., 1999), marker technology today offers a palette of powerful tools to analyze the plant genome.

DNA markers have enabled the identification of genes and genomic regions associated with the expression of numerous qualitative and quantitative traits, and have made the manipulation of such genomic regions feasible through marker-assisted selection techniques. The extensive knowledge generated through the application of molecular markers has given scientists the ability to identify genes and pathways that control important biochemical and physiological parameters, and to better understand how they are regulated. Another application of DNA markers, that logically follows this "discovery phase," is to pyramid favorable alleles in targeted genetic backgrounds through marker-assisted selection (MAS) experiments. For simply inherited traits—those that have high heritability and are regulated by only a few genes—or for the manipulation of genomic region expressing a large percentage of the phenotypic variance, the use of molecular markers to accelerate germplasm improvement has been well documented (e.g., Ragot, 1995; Tanksley et al., 1996). On the other hand, when several genomic regions must be manipulated, the usefulness of MAS is less certain (Ribaut and Hoisington, 1998; Young, 1999; Dekkers and Hospital, 2002) with few success stories published (Johnson and Mumm, 1996; Hittalmani et al., 2000; Ragot et al., 2000; Shen et al., 2001; Ribaut et al., 2002). Unfortunately, most important agronomic traits, including yield, are complex and regulated by several genes. The dearth of MAS success stories related to germplasm improvement for polygenic traits has generated some uncertainty about the utility of DNA markers in practical breeding programs.

This chapter presents and discusses results from complementary areas of investigation that are helping scientists at CIMMYT better understand the phenotypic, genetic, and physiological responses of tropical maize under water-limited conditions (Fig. 1). Central to this pursuit is the identification of quantitative trait loci (QTL) involved in the expression of yield components, secondary traits of interest, and physiological parameters associated with improved drought tolerance. This identification has resulted from experiments conducted on various segregating populations, at different inbreeding levels, and in diverse environments. Recent experiments and ideas on how to efficiently bridge the gap between differential gene expression and plant phenotype by conducting profiling experiments on suitable germplasm are

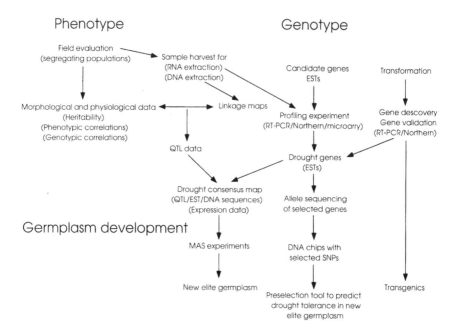

Figure 1 Multidisciplinary approaches conducted at CIMMYT to better understand the genetic basis of drought tolerance in tropical maize.

also presented. Finally, the chapter reports on MAS experiments to improve drought tolerance in target germplasm and reviews some innovative MAS strategies, including MAS schemes based on a drought consensus map.

2 THE CHALLENGE OF BREEDING FOR DROUGHT TOLERANCE

Drought is a climatic phenomenon encompassing water-limited conditions that reduce plant vegetative growth and can have severe effects on seed production. The term drought is loosely applied to wide variations in the timing and intensity of water-limited environment resulting in water deficit in plants. Except for seasonal patterns of rainfall, the occurrence of drought is largely unpredictable, making breeding for drought tolerance under rainfed conditions especially challenging. Because drought occurs unpredictably, drought tolerant selections should be competitive in performance when rainfall is adequate, while showing superiority under stress. Thus, selecting for improved performance under water-limited environment can be complicated by environmental variation and adaptation, and selection is slower than for other abiotic stresses. Multi-environment testing for drought tolerance in temperate areas is particularly inefficient because there is rarely a warm and reliably dry test site in target areas that provides a repeatable stress at a specific growth stage. When this does occur, trial managers often reject data because they are considered to be too variable. Good crop agronomy that minimizes water use by weeds or evaporation from bare soil surfaces can help considerably, but the effect of drought on grain yield can best be mitigated by irrigation. In the (sub)tropics, scientists have used seasonally rain-free dry periods for screening under drought conditions. Thus efficient screening for drought tolerance therefore calls for a rainfree environment where the timing and intensity of stress is modulated by irrigation, but where the effects of other factors such as disease, day length, temperature, and soil characteristics have as small an effect as possible on crop performance.

The optimal strategy for a plant type to produce the most seeds or kernels under drought may vary greatly depending on the nature of the stress. Following are some examples to illustrate how difficult it is to develop "a drought tolerant genotype" adapted to different water-limited environments, considering the different natural water regimes associated with drought conditions. Under drought produced by frequent but reduced rains during the plant cycle (a few millimeters of water per rainfall), one can imagine that the most appropriate phenotype might possess a dense, predominantly horizontal, basal root system. For environments where there are significant major rains at the beginning of the plant cycle, followed by no rain for several weeks, a phenotype with deep vertical roots would be the best adapted because it would be able to access water deep in the soil profile and therefore produce seeds. In environments characterized by short periods without rain during the entire crop cycle, plants that maintain assimilation and growth during stress periods should produce more seeds. On the contrary, if the drought periods are extended, plants that switch to a survival mode (e.g., "latente" phenotype; Castleberry et al., 1979) may survive the stress periods and produce some seeds at the end of the crop cycle. The optimal strategies under water-limited conditions depend also on the level of tolerance of the crop and the environmental conditions (e.g., type of soil, etc.). In wheat, for example, CIMMYT selected germplasm can produce about two tons per hectare, with only one heavy irrigation at planting time. Limited progress for grain yield under water-limited conditions may be obtained from spillover effects from the higher yield potential of improved germplasm when grown in place of local and low-yielding germplasm. Nevertheless, a breeding program focused at increasing the frequency of genes that increase productivity under water-limiting conditions is the best route for substantial and sustained gains. Such a program requires a good definition of major drought scenarios in the target environment and needs to be conducted under repeatable experimental conditions reflecting those scenarios.

Drought at any stage of plant development affects grain production, but maximum damage is inflicted on maize when it occurs around flowering. Farmers may respond to drought at the seedling stage by replanting their crop. When the lack of water occurs during grain filling, some yield may be salvaged because maize is relatively drought tolerant at this stage. Severe drought at flowering can cause extensive loss of grain, and the season has usually progressed too far for replanting the crop at that stage. CIMMYT scientists, therefore, have devoted considerable effort to improving drought tolerance in tropical maize for the period immediately before and during flowering.

3 TARGETED TRAITS

Germplasm-by-environment (G×E) effects under stress conditions can easily bias selection if based on yield components only, especially if there is a range of flowering dates in the germplasm under evaluation (Bänziger and Cooper, 2001). Most breeders therefore opt for a selection index that incorporates various yield components and secondary morphological traits in their breeding program to maximize gain in grain yield and stress tolerance while minimizing the possibility of selecting escapes. Proportional weights for each trait used in selection are calculated based on the phenotypic and genotypic covariance between that trait and grain yield (Bänziger and Lafitte, 1997). An extensive body of literature exists on the physiology and genetics of traits associated with tolerance to specific abiotic stresses (e.g., Shannon, 1998; de la Fuente-Martinez and Herrera-Estrella, 1999; Duncan and Carrow, 1999; Saini and Westgate, 2000; Edmeades et al., 2001). However, very few proposed secondary traits meet the criteria needed for successful incorporation into a drought breeding programs. A suitable secondary trait is one that is (i) genetically associated with grain yield under drought, (ii) highly heritable, (iii) inexpensive and quickly measured, (iv) stable over the measurement period, (v) observed at or before flowering, and (vi) not associated with yield loss under unstressed conditions (Edmeades et al., 1998).

A commonly used secondary trait for drought tolerance in maize is the asynchrony between silk emergence and pollen shedding. Under water-limited conditions, this asynchrony, termed the anthesis-silking interval (ASI), is highly correlated with grain yield; a short ASI is usually associated with good kernel set and high grain yield (Edmeades et al., 2000). Anthesis-silking interval has a heritability that on average is slightly greater than that for grain yield under stress (Bolaños and Edmeades, 1996). An advantage to using ASI in breeding is that it is easily measured in the field, on either a plant or row basis, and data are available at flowering. Thus it could be used to direct crossing in an accompanying well-watered nursery that was planted a few weeks later.

Stay-green is another trait of interest, because an active green canopy is essential to continued production under transient stress. In cereals, foliar senescence releases N required by the developing grain from the photosynthetic enzymes. Consequently, simultaneous selection for delayed senescence and enhanced sink size under drought may still result in accelerated senescence because the demands of the grain for N take precedence (Chapman and Edmeades, 1999). Greenness per se does not guarantee continued C assimilation (Thomas and Howarth, 2000) and may be detrimental if it indicates a weak grain sink and/or an inability to remobilize stem reserves (Blum, 1997). However, stay-green is generally associated with improved performance under drought in cereals because photosynthesis is maintained longer in stay-green types (Thomas and Smart, 1993). In sorghum, stay-green is considered the most important post-flowering secondary trait associated with grain production under water-limited conditions (Borrell, 2000).

In rice, scientists investigating drought tolerance have focused more on the structure and physiological characteristics of the roots, such as root penetration (Clark et al., 2002) and osmotic adjustment (OA) (Zhang et al., 2001). The rationale for selection based on root traits and/or OA is that a well-developed deep root system can promote water uptake from lower soil layers, while OA can help maintain the turgor of meristems when plants experience water deficits (Nguyen et al., 1997; Zhang et al., 1999). Successful MAS experiment manipulating QTLs for OA in rice has been reported (Shen et al., 2001).

Leaf temperature drops when water evaporates from leaf surfaces, a trait affected directly by stomatal conductance, which is affected by feedback mechanisms of other processes such as carbon fixation and vascular transport of water. Canopy temperature depression (CTD), the difference between the temperature of the air and the leaves of a field plot measured with an infrared thermometer, is another frequently used indicator of a genotype's ability to capture water. CIMMYT wheat breeders successfully used CTD measurements on small plots in their heat tolerance nurseries to identify the highest yielding entries (Reynolds et al., 2000a). When measured during grain filling under stress conditions, CTD correlated well (genetic correlation coefficients of 0.6–0.8) with final grain production in wheat (Reynolds et al., 2000b). Strong relationships between grain yield and canopy temperature have also been reported by Fischer et al. (1989) among full-sib progenies of a maize population grown under drought conditions, although these have been weaker in inbred progeny trials (Bolaños and Edmeades, 1996).

The nature and adaptive value of these secondary traits are quite diverse, as evidenced by the diversity of phenotypes (based on the allelic composition at the genes involved in their expression) when crops are grown under water-limited conditions. Some traits, such as stay-green, are of value to a wide range of crops, while others are specific to certain crops, such as ASI in maize. Another avenue of study has been the quantification of changes in key physiological parameters related to plant responses to water-limited conditions. Among the most informative physiological parameters are changes in cell water status (relative water content); compatible osmolyte accumulation; the accumulation of antioxidant (e.g., superoxide dismutase) to protect cell membranes from oxidative damage; the accumulation of protectants such as glycinebetaine or late embryogenesis abundant (LEA) proteins thought to stabilize plasma membrane structure; and changes in concentration of the stress hormone abscisic acid (ABA). Abscisic acid exhibits significant concentration changes under a range of abiotic stresses and is involved in regulating several key pathways such as cell growth, stomatal closure, root hydraulic conductivity, and ion transport.

Although most of the secondary morphological traits such as ASI, stay-green, and CTD can be routinely used in breeding programs, this is not the case for most physiological parameters, such as OA, ABA concentration, or the quantification of LEA proteins, because their measurement is too time consuming and/or too expensive and those parameters generally show low correlation with yield components. Nevertheless, plant improvement for drought environments requires more complete knowledge of the genetic basis of both the morphological traits and the physiological parameters involved in the plant responses under water-limited conditions. Genetic dissection of "drought physiological traits," at the structural and functional levels, would allow identification of the key pathways involved in plant responses, how they interact, and how differential expression of those key pathways can result in a broad spectrum of phenotypes that display different levels of drought tolerance.

4   GENETIC DISSECTION OF TARGETED TRAITS IN MAIZE

CIMMYT's work on drought tolerant maize spans three decades, with the tools of biotechnology being brought to bear on the challenge during the last 10 years. The focus of this approach has been the genetic dissection of drought tolerance at flowering and the identification of QTL(s) for yield components, secondary morphological traits of interest, and more recently, physiological parameters. To date, genetic dissection has been conducted in four different crosses, at different inbreeding levels [hybrids, F_2:F_3 families, and recombinant inbred lines (RILs)], under different water regimes [well-watered conditions (WW), intermediate stress (IS), and severe stress conditions (SS)] and in several environments (Kenya, Mexico, and Zimbabwe). This effort has generated a large data set of QTL summarized in Table 1.

Table 1 Segregating Populations Analyzed for Yield Components, Morphological Traits, and Physiological Parameters, Under Different Stress Regimes, in Different Locations, and at Different Inbreeding Levels

Populations	Trials	Traits (at flowering stage)
Ac7643×Ac7729/TZSRW (236 F2/3 families)	92A WW (TL) 94A IS, SS (TL)	GY, ENO, HKDW, KNO, MFLW, FFLW, ASI, LNO, ELW, ELL, YLW, YLL, EHT, PHT, EPO
Ac7643×Ac7729/TZSRW (236 RIL families)	96A IS, SS (TL) 96B WW (TL) 99A WW, IS, SS (TL) 01A IS, SS (TL)	GY, ENO, HKDW, KNO, MFLW, FFLW, ASI, LNO, EHT, PHT, EPO, RWC, OP, OA, RCT, YLCHL, ELCHL, ABAE, ABAS, ABAEL, EDW, EG, SDW, SG
Ac7643×CML247 (236 F2/3 families)	96A SS (TL)	GY, ENO, HKDW, KNO, MFLW, FFLW, ASI, LNO, EHT, PHT
K64R×H16 (280 F3 Topcross families)	99B IS, SS (ZW) 00A IS, SS (KY)	GY, ENO, MFLW, FFLW, ASI, PHT, SEN
K64R×H16 (170 F4 families)	00B SS (ZW)	GY, ENO, MFLW, FFLW, ASI, PHT, EHT, EPO, SEN
CML444×SC-Malawi (234 F3 families)	00B WW, IS, SS (ZW) 01A IS, SS (TL)	GY, GYDT, ENO, ENODT, HKDW, HKDWDT, KNO, KNODT, MFLW, FFLW, FFLWDT, ASI, ASIDT, EHT, PHT, EPO, SEN, TBNO

Populations	Trials	Traits (at plantlet stage)
Ac7643×Ac7729/TZSRW (140 RIL families)	98 (Laboratory test)	Root parameters under hydroponics
Ac7643×Ac7729/TZSRW (220 RIL families)	00 (Phytotron) Low temperature	Pigments, photosystem parameters, RWC, SHDW, RDW
Ac7643×Ac7729/TZSRW (220 RIL families)	01 (Phytotron) Drought, drought, and low temperature	Pigments, photosystem parameters, RWC, SDW, RDW

Note that the traits reported in the last column were not always measured in all trials per cross.
Location: KY: Kenya/TL: Tlaltizapan, Mexico/PR: Poza Rica, Mexico/ZW: Zimbabwe.
Stress regime: WW = Well watered; IS = intermediate water stress; SS = severe water stress; lowN = low nitrogen trial; HiN = high nitrogen trial.
Yield components: GY = grain yield; GYDT = grain yield for detasseled plant; ENO = ear number per plant; ENODT = ear number per plant for detasseled plant; HKDW = hundred kernel dry weight; HKDWDT = hundred kernel dry weight for detasseled plants; KNO = kernel number; KNODT = kernel number for detasseled plants.
Morphological traits: MFLW = male flowering; FFLW = female flowering; ASI = anthesis-silking interval; LNO = leaf number; ELW = ear leaf width; ELL = ear leaf length; YLW = young leaf width; YLL = young leaf length; EHT = ear height; PHT = plant height; EPO = ear position; SEN = senescence; TBNO = number of tassel branches; EDW = ear dry weight; EG = ear growth for 1 week; SDW = silk dry weight; SG = silk growth for 1 week; SHDW = shoot dry weight; RDW = root dry weight.
Physiological parameters: RWC = relative water content; OP = osmotic potential; OA = osmotic adjustment; YLCHL = chlorophyll content in a young leaf; ELCHL = chlorophyll content in the ear leaf; RCT = root conductivity; ABAE = abscisic acid content in the ear; ABAS = abscisic acid content in the silk; ABAEL = abscisic acid content in the ear leaf.

4.1 Experimental Design

Most of the experiments described in this chapter were conducted during consecutive dry winter seasons (mid-November to April) at the CIMMYT Experimental Station in Tlaltizapan (TL), Mexico (18°N, 940 masl). This location is consistently dry during the entire winter (November to May) season, with maximum temperatures around 35°C and no special disease/pest problem, thus allowing researchers to control stress severity by irrigation. Irrigations were applied by sprinkler for germination and by furrow irrigation at the plantlet stage and thereafter. The water supply was stopped at different stages of crop development, depending on the material's level of inbreeding and soil water holding capacity, to obtain the desired drought stress intensity before and during flowering and in some cases during grain filling. On average, irrigation was stopped 5, 3, and 2 weeks before the first genotypes started to flower for hybrids, F_3 families, and RIL, respectively. In experiments that were severely stressed during flowering (SS), irrigation was resumed at the end of the flowering stage, corresponding to the end of silk emergence for most plants, until harvest to allow grain filling of the pollinated embryos. For some experiments with intermediate stress intensity at flowering (IS), no further irrigation was applied up to harvest. For all trials, plots consisted of a single row 2.5 m in length (12 plants), with 20 cm between hills and 0.75 m between rows. Plots were overplanted and thinned to one plant per hill. The first two plants of each plot were considered to be border plants and were not used in measurements, because plants nearest to the alley show more vigor than plants under full competition inside the row. Experiments were planted in an alpha (0,1) lattice design with two replications for each water regime.

4.2 Yield Components and Secondary Traits

Many of the trials presented in Table 1 focused on measuring yield components and secondary traits. Yield components, in general, have been consistently evaluated across all field experiments, and grain yield (GY) has been quantified in all trials because at CIMMYT improvement in drought tolerance is associated with an increase in grain production under water-limited conditions. Some of the secondary traits have also been consistently measured across trials, such as the flowering traits, male and female flowering (MFLW and FFLW), allowing the calculation of ASI. Out of the stress trials (IS and SS) presented in Table 1, GY and ASI were evaluated in 18 environments. Other traits, such as leaf size (length and width), were only measured in one cycle (94A, IS, and SS). These latter traits were subsequently removed from the measurement procedure because analysis revealed only a number of small QTLs and these traits were only weakly correlated with yield components and other target traits such as ASI. During the past 10 years, CIMMYT scientists have established standard procedures for evaluating key morphological traits for QTL identification, including four yield components (grain yield (GY), ears per plant (ENO), 100 kernel dry weight (HKDW), and kernel number per ear (KNO)), male and female flowering, ASI, ear (EHT) and plant height (PHT) (including relative ear position, EPO), chlorophyll content (CHLO) (SPAD meter), senescence (SEN), and root capacitance (RCT) (measured using an electrical capacitance meter; van Beem et al., 1998). In addition to this set of yield components and secondary traits, a few traits are more cross-specific, e.g., tassel size.

4.2.1 Tassel Size

Tassel primary branch number (TBNO) has proven specific to only one of the four segregating population (CML444, few tassel branches, and SC-Malawi, large number of tassel branches) (Fig. 2). As difference in tassel branches, or tassel size, can indicate competition for resource allocation between tassel and ear/silk development, this trait was quantified in Tlaltizapan (2001A) under severe stress conditions in segregating F_3 families, and QTLs involved in the expression of this trait have been identified. Although the values were low, TBNO was significantly correlated with other traits of interest such as FFLW (0.33), ASI (0.31), GY (−0.19), and ENO (−0.36). Correlations show the presence of tassel branches to be associated with early female flowering, short ASI, and higher grain and ear production. In tropical maize, unlike temperate maize, the indirect pressure of selection to reduce tassel size by selecting for

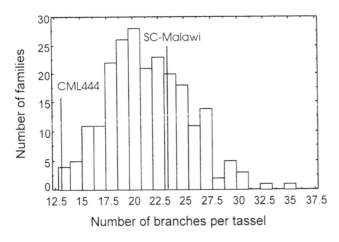

Figure 2 Distribution of tassel branch number per plant (average of 6 plants per plot) evaluated on a segregating population of 240 F3 families (CML444×Sc-Malawi) under drought conditions.

increased grain production has had relatively modest effects on tassel size. Consequently, most tropical inbreds still possess a relatively large tassel (12–20 branches), except for highland germplasm (with 1–10 tassel branches). When water-stress occurs at flowering, competition occurs between the developing tassel and ears. Because they are physically separated in maize, and assimilate moves to them from different parts of the canopy that vary in illuminance, silk emergence is impaired more than pollen shedding (Ribaut et al., 1996). This delay in silk emergence relative to anthesis results in ASI. In a recurrent selection study of Tuxpeño Sequía, tassel size fell with selection cycles as grain yield under stress improved, although tassels were not included in the direct selection strategy for this population (Bolaños and Edmeades, 1993). This suggests that tassel growth competes directly with ear growth under stress at flowering. To further investigate the impact of the absence of tassels on a set of selected traits, including silk emergence, each plot of a trial conducted in 2001 was planted twice (two neighbor rows of 2.5m length) and the tassels of all plants in the second row of each plot were removed about 2 weeks before pollen shedding when the tassel was still enveloped in the last leaves. Under severe stress conditions, only 25% of the nondetasseled plants produced silks, compared to 53% for the detasseled plants. The corresponding grain yield (GY) increase was more than 50% (479 kg/ha vs. 718 kg/ha). In addition, genetic correlations between some traits increased significantly using data obtained from detasseled plants. For example, the genetic correlation between ASI and GY was -0.36 in plants with tassels compared to -0.93 for detasseled plants, where ASI was calculated using the male flowering information registered on the nondetasseled plants.

4.2.2 Quantitative Trait Loci Identification

The identification of QTLs involved in the expression of the different morphological traits and physiological parameters was performed using Composite Interval Mapping (CIM) software (Zeng, 1994). Composite interval mapping uses a mixed model and likelihood methods; QTL identification is achieved by successively running three models and comparing the different outputs. Model 1 uses a single interval mapping (SIM) approach similar to Mapmaker-QTL (Lander and Botstein, 1989), which does not include any marker as a cofactor. In model 2, pre-selected unlinked markers, closely linked to significant QTL peaks identified through SIM (model 1), are used as cofactors to reduce the residual variation throughout the genome. In model 3, markers flanking the tested interval are used as cofactors to block the effects of possible QTL that are linked to the tested interval. Two different "window" sizes were successively tested, 30 and 20 centiMorgan (cM). At each significant QTL, allelic effects, additivity, and dominance were obtained directly from the output of the software. The presence of a QTL was declared significant if the LOD threshold value was >2.5 when analyzing data from a single environment or when running a joint analysis. Joint analysis of phenotypic data from different environments (different stress levels from the same location or different experiments across locations) was conducted to evaluate the environmental interaction on QTL identification (Q×E), thereby identifying the "stable QTL" across field evaluations (Jiang and Zeng, 1995). Multiple regression was performed for the phenotypic values of all markers closely linked to a QTL position (one per QTL), including both additive and dominance effects, to evaluate the total percentage of phenotypic variation accounted for by all of the significant QTLs.

Independent of cross and location, about 200 QTL profiles have been produced for secondary morphological traits using the 18 stressed field trials (IS and SS) listed in Table 1, generating a total of more than 1000 QTLs. Because all morphological traits studied were complex and regulated by several genes, with yield being among the most polygenic and complex traits, no major QTL (explaining more than 25% of the phenotypic variance) has been identified. Most of the individual QTL expressed 5–15% of the phenotypic variance; total phenotypic variance expressed by all of the significant QTLs combined was generally 20% to 50% and never surpassed 65%. The QTL results for flowering traits and yield components for the 92A and 94A field trials conducted on the Ac7643×Ac7729/TZSRW cross (referred as P1×P2) have been published (Ribaut et al., 1996, 1997). Below, we present and discuss QTL results for leaf senescence and root mass, two important traits not yet addressed in the molecular genetics literature on drought tolerance maize.

4.2.3 Leaf Senescence

The QTLs involved in the expression of leaf senescence evaluated under severe stress conditions in Zimbabwe in a segregating population of F_4 families derived from the H16×K64 cross are presented in Table 2. Taking advantage of the nature of the QTL identified for leaf senescence, we would like to elaborate on the different outputs obtained running the different models of CIM to demonstrate the danger of considering only SIM and the power of combining models. Using SIM, five significant QTLs were identified, with two QTLs located on chromosome 8, at 55 and 99 cM, respectively. By running model 2, one new significant QTL was identified on chromosomes 4 (116 cM). From the output of model 3, a new significant QTL appeared on chromosome 4 at 0 cM, while the QTLs identified at 55 cM with the previous models on chromosome 8 disappeared. Based on the results from models 2 and 3, one can conclude that two genomic regions on chromosomes 4 and 8 are involved in the expression of senescence in this segregating population. The two linked QTLs separated by 116 cM on chromosome 4 are in repulsion, as demon-

Table 2 QTL Identification of Senescence Evaluated in Zimbabwe Under Water-Limited Conditions at Flowering Time in a F4 Segregating Population (H16×K64R)

Chromosome	cM	LOD	Marker	R^2	Additivity	Dominance
Model I
2	152	2.03	umc4	6.62	−0.07	0.66
6	27	2.31	bmc1371	7.62	0.32	0.48
7	47	7.33	bmc1792	20.20	−0.41	1.09
8	55	2.77	umc152	7.61	0.52	0.24
8	99	2.59	umc2	7.86	0.46	0.15
				38.78		
Model II
2	151	2.71	umc4	6.50	−0.07	0.66
4	116	2.03	bnl5.71	4.86	−0.25	0.48
6	25	3.64	bmc1371	7.60	0.40	0.53
7	48	6.88	bmc1792	20.07	−0.43	0.92
8	54	2.87	umc152	7.47	0.51	0.31
8	97	3.43	umc2	7.70	0.49	0.02
				41.43		
Model III
2	152	3.07	umc4	6.62	−0.20	0.66
4	0	2.17	phi72	4.00	−0.28	0.43
	116	2.06	bnl5.71	4.86	0.06	0.64
6	24	3.21	bmc1371	7.46	0.40	0.47
7	49	5.61	bmc1792	19.73	−0.41	0.85
8	97	2.32	umc2	7.70	0.46	0.01
				46.65		

QTL results for the three different models of CIM are presented with their position by chromosome in centiMorgan (cM), the log likelihood ratio (LOD), the closest marker to the QTL peak, the percentage of the phenotypic variance accounted for (R^2), and the additive and dominance effect of individual QTL. For each model, the last line corresponds to the percentage of the phenotypic variance expressed when considering all the significant QTL acting together. A positive sign of the additivity indicates that H16 contribute to increase the value of the trait.

strated by the opposite sign of additivity (Table 2), meaning that the allele that favors senescence is coming from different parental lines at both QTLs. The two QTLs on chromosome 8 are separated by about 45 cM and are in phase, with the same sign of additivity at the two peak loci. Models 1 and 2 tend to overestimate the genetic effects of linked QTL in phase while underestimating their effect when they are in repulsion. Therefore the identification of the QTL on the extremity of the short arm of chromosome 4 (0 cM) was possible only by removing the genetic effects of the second QTL (at 116 cM). This was accomplished by using flanking markers of the tested interval as cofactors. When using flanking markers as cofactors on chromosome 8, the QTL at 55 cM was no longer significant. The difference in QTL results obtained under the three models illustrates how important it is to compare and compile their outputs to obtain the best possible QTL characterization, especially for linked QTLs. Model 3 identify the over- or undervaluation of genetic effect, due to the presence of linked QTLs in phase or in repulsion that might occur under models 1 and 2. However, because of the lower number of recombination events that are taken into account in model 3, that considers only the genetic information at markers within the tested interval plus the flanking windows (generally 30 or 20 cM), this model can slightly bias the evaluation of the genetic effects at the QTL (additivity and dominance). This bias increases when the window size and/or the number of segregating genotypes is reduced. Therefore a better QTL characterization can be obtained by looking at the output of model 2 for unlinked QTLs. We also recommend verifying that the QTLs identified with the largest LOD score, such as that identified on chromosome 7 for senescence, are present under the three different models. In conclusion, QTL results from all three models should be compared in each analysis.

The QTL at 47 cM for leaf senescence on chromosome 7, compared to QTLs identified for morphological traits under drought conditions, is remarkable because it was identified with a LOD score from 5.61 to 7.33 and on its own accounted for about 20% of the phenotypic variance looking at the outputs from the three models. The genetic effects of this large QTL when combined with the five other significant QTLs represents 46.7% of the phenotypic variance (model 3)—a high figure compared to that generally obtained with other target morphological traits. This particular cross was developed to conduct MAS at an early stage of recombination on a large segregating population (for details see Ribaut and Betrán, 1999). The two parental lines (H16 and K64R) are both drought tolerant but were selected because they presented good allelic complementarity when evaluated by hybrid performance under drought. This contrasts with the three other segregating populations that were developed by crossing tolerant and suscep- tible lines, and explains why H16 and K64 contributed equally favorable alleles that delayed senescence at three of the six significant QTLs identified with model 3 (three positive and three negative additivity signs).

4.2.4   Root Mass

Several studies demonstrated the role of root structure and the root capacity to uptake water in the drought tolerance process. Unfortunately, root characterization is not easy when working under field conditions. Root measurement such as pulling force is destructive, and root measurements (e.g. length, diameter or weight) cannot be conducted on a large scale in the field. Root mass evaluated using a capacitance meter (van Beem et al., 1998) has been considered an attractive alternative to evaluate root structure in our segregating populations, as it is a quick (few seconds per plant) and nondestructive method. Root mass was evaluated on six plants per plot a couple of days after irrigation making the end of the stress period had been applied. Based on data collected in a field trial conducted under severe stress conditions on RIL families in 1996 in Tlaltizapan, Mexico, root capacitance presented a heritability of 0.67, and several very significant QTLs have been identified (Table 3). Under model 3, five QTLs were identified; four of them with a LOD score higher than 4.00, with each accounting for at least 7% of the

Table 3 QTL Identification for Root Mass Evaluated Using a Portable Capacitance Meter Under Water-Limited Conditions in Tlaltizapan, Mexico, in a Segregating RIL Population (Ac7643×Ac7729/TZSRW), and Segregating F3 Families Derived from the Ac7643×CML247 Cross

Chromosome	cM	LOD	Marker	R^2	Additivity	Dominance
Ac7643×Ac7729/TZSRW (RILs) Tlaltizapan 1996						
1	159	6.57	umc119	10.65	2.69	
2	93	4.19	csu133	7.45	−2.43	
4	106	4.01	bnl5.71	8.15	2.54	
4	146	4.30	csu26	7.33	2.47	
9	95	2.52	umc95	2.96	1.69	
				31.80		
Ac7643×CML247 (F3 families) Tlaltizapan 2002						
1	141	6.08	csu61	10.23	0.38	0.04
3	0	6.00	csu32	9.95	0.30	0.09
4	66	2.51	csu84	4.47	−0.24	0.03
9	86	3.75	umc114	6.76	0.28	0.04
10	31	3.14	bnl7.49	3.73	0.28	0.07
				32.66		

Only the QTL output obtained by running model III is presented in the table, with the QTL position by chromosome in centiMorgan (cM), the log likelihood ratio (LOD), the closest marker to the QTL peak, the percentage of the phenotypic variance accounted for (R^2), the additive, and the dominance (only for the F_3 families) effect of individual QTL. The last line corresponds to the percentage of the phenotypic variance expressed when considering all the significant QTL acting together. A positive sign of the additivity indicates that Ac7643 contributes to increase the value of the target trait.

phenotypic variance, and the five QTLs taken together accounting for 31.8% of the total phenotypic variance. One of the QTLs, identified on chromosome 1, had a very significant LOD of 6.57. Heritability for root capacitance was not always as high as the one observed on the 1996 trial. A heritability of 0.23 was calculated based on field data collected during an experiment conducted under severe stress in 2002 in Tlaltizapan, Mexico, on segregating F_3 families derived from the cross Ac7643×CML247. Although the heritability was low, the quality of the QTL identified remained good. Two QTLs on chromosomes 1 and 3, identified with a LOD score of 6.0, each explained about 10% of the phenotypic variance. Two QTLs on chromosomes 1 and 9 appear to be common to the set of QTL identified on the RIL families in 1996. The total percentage of the phenotypic variance accounted for by the five QTLs together (32.7%) was as high as for the 1996 trial.

4.3 Physiological Parameters

To further explore the genetics of maize response under water-limited conditions, one can identify QTLs involved in the differential expression of the key physiological pathways underlying drought tolerance. Identifying such QTLs might advance our understanding of the physiological basis of target morphological traits. To achieve this objective, a RIL population was developed by single seed descent from F_3 families obtained by crossing Ac7643 (short ASI under drought) with Ac7729/TZSRW (long ASI under drought). The same morphological traits measured in the F_3 families were also evaluated in this RIL population (Table 1). RIL families are better suited than F_3 families for physiological measurements because they are genetically fixed; on the other hand, they are poor material for evaluation of yield components because they usually demonstrate high inbreeding depression, especially in tropical germplasm.

4.3.1 Tissue Harvest

Under water-limited conditions stress intensity increases over time. Two options might be considered for harvesting tissue across segregating populations grown in the field to quantify physiological parameters. It is important to realize that even if the two parental lines of a segregating population flower at the same time, a "flowering window" of 10 to 15 days can easily be observed among families derived from this cross, as male and female traits generally display transgressive segregation and are regulated by many genes. The first option is to harvest samples from all families on a given day, thus insuring that all plants will be at the same stress level in terms of days without water supply. However, due to differences in earliness the developmental stage at harvest time will differ among families. The second option is to harvest target tissue at a given morphological/physiological stage on a plant by plant basis, but in doing so the level of stress will vary among families, as early flowering families are subjected to less stress than late ones. Depending on the nature of the target tissues (the ear leaf, the ear, or the silk) both harvest options have been followed at CIMMYT. With regard to the ear leaf, as it is fully expanded at flowering time, tissue was harvested at set dates on all families. For ear and silk tissues that are in full expansion at flowering, the second approach targeting a given developmental stage was considered most suitable for a comparison across families. This approach represents a greater amount of work as it involves harvesting on a plant-by-plant basis.

4.3.2 Quantitative Trait Loci Identification

Methodology for QTL identification related to physiological parameters is the same as for yield components and secondary traits described above. The only difference is that we consider a LOD score of 2.0 (instead of 2.5 for F_2) to declare a QTL significant in a RIL population, as the probability to identify false positive is lower compared to a F_2 population (difference in degrees of freedom). The fact that most of the physiological parameters are not evaluated through simple direct measurement has some implication for QTL identification. Indeed, extraction or the combination of various variables decreases generally the accuracy in evaluating a physiological parameter, resulting often in a reduced heritability, as well as reducing the precision of QTL identification. Based on our experience, the heritability of physiological parameters evaluated indirectly is generally below 0.5 under field conditions, and the phenotypic variance accounted for by all of the significant QTLs rarely exceed 30%. Another complication of QTL identification for physiological parameters is that a phenotype cannot generally be visualized looking at the entire plant, and, therefore, several measurements must be taken at different times during the stress to understand the kinetics of the parameter. Data collection therefore can be time consuming, potentially expensive, and the interpretation of the phenotypic and genetic data quite challenging. Nevertheless, we are convinced that genetic dissection of the key physiological parameters is essential because it provides critical insights on how a certain plant phenotype induces drought tolerance. As one of our main objectives is to better understand under water-limited conditions (1) the source–sink relationship between the ear leaf and the ear, and (2) the physiological mechanisms involved with cell growth in the ear and the silk during the development of the female organ, we focused our QTL investigation on quantifying compounds/mechanisms significantly correlated with ear and silk growth at flowering.

4.3.3 Osmotic Adjustment

Many experimental results have been published about the capacity of a plant to regulate its osmolyte concentration under stress. Results are somewhat controversial, because depending on the crop OA has varied from relatively small adjustment in crops such as maize (Bolaños and Edmeades, 1991) to very significant adjustments in crops such as rice, wheat, and sorghum (for review, see Zhang et al., 1999). Osmotic adjustment is an important phenomenon because maintenance of turgor or water content in a cell under water-stressed conditions helps ensure cell growth, meristematic activity and, as the stress becomes more severe, compatible osmolytes may stabilize macromolecular structure and permit cell survival. Under our experimental conditions, OA has been measured in the ear leaves in the Ac7643×Ac7729/ TZSRW RIL population under three different stress environments (TL 96A SS, TL 99A IS, and SS). Leaf disks were punched on the ear leaf of five plants per plot and samples were collected in the morning (predawn), when temperatures were cool. The OA was calculated as the difference of the osmotic potential of the target tissue under well-watered condition minus the osmotic potential of the target tissue under stressed condition considering the RWC of a given genotype:

$$OA = \psi_{ns} - \psi_s^{100}$$

where ψ_{ns} is the osmotic potential of the target tissue under well-watered condition, and ψ_s^{100} is the osmotic potential of the target tissue under stress condition adjusted for the RWC of a given genotype.

Due to the manner in which leaf samples are processed to extract cell sap to quantify osmolytes, and because, in addition to the osmolite concentration, the calculation of OA takes into account the RWC under stress conditions, OA is typically a physiological parameter evaluated indirectly, with a large experimental error when measured under field conditions. In our trials, the heritability of this parameter was below 0.3, resulting in "poor" QTL identification, mainly due to poor phenotypic data. In the 96A trial, four QTLs were identified with a significant but reduced LOD score (between 2.00 and 2.50), except for the second QTL on chromosome 8 (Table 4A). In addition, three QTLs accounted for less than 4% of a phenotypic variance, rising to 6% for the most significant QTL (identified at 123 cM on chromosome 8). All the QTLs combined accounted for 16% of phenotypic variance. When working on a RIL population (fixed lines), it is important to remember that only additive effects are evaluated. A combined analysis on field data from three different environments (96, SS, and 99, IS and SS trials) identified six QTLs (Table 4B). From the combined analysis, three QTLs were identified with a LOD score around 2.50 and three with a LOD score slightly higher than 3.50. In reviewing the high Q×E interaction, which was significant for five of the six QTLs, it becomes apparent that there is little consistency in terms of QTL identification across environments, as confirmed also by the low percentage of the phenotypic variance obtained on the individual trials at the

Table 4A QTL Identification for Osmotic Adjustment Evaluated Under
Water-Limited Conditions in a 1996A Trial (SS) Conducted in Tlaltizapan,
Mexico, on a Segregating RIL Population (Ac7643×Ac7729/TZSRW)

Chromosome	cM	LOD	Marker	R^2	Additivity
5	186	2.21	umc51	3.75	0.53
7	117	2.27	umc80	2.41	0.59
8	23	2.33	bnl13.05	3.91	0.63
8	123	5.49	umc89	6.09	0.87
				15.93	

This table presents the QTL output obtained by running model III with the QTL position by
chromosome in centiMorgan (cM), the log likelihood ratio (LOD), the closest marker to the
QTL peak, the percentage of the phenotypic variance accounted for (R^2) and the additive
effect of individual QTL. The last line corresponds to the percentage of the phenotypic
variance expressed when considering all the significant QTL acting together. A positive sign
of the additivity indicates that Ac7643 contributes to increase the value of the target
parameter.

Table 4B QTL Identification for Osmotic Adjustment Evaluated Under Water-Limited
Conditions Combining the Phenotypic Data from Three Different Trials (SS 96A, IS 99,
and SS 99) Conducted in Tlaltizapan, Mexico, on a Segregating RIL Population
(Ac7643×Ac7729/TZSRW)

Chromosome	cM	LOD (combined)	Q×E (combined)	Markers	R^2 SS 99	R^2 IS 99	R^2 SS 96A	Additivity (combined)
3	25	3.57	1.40[a]	csu16	0.08	6.48	3.84	0.52
3	71	2.57	2.00[a]	umc10	1.43	0.71	0.34	−0.24
7	31	3.47	1.37[a]	umc116	0.16	4.03	4.29	0.50
7	94	3.86	3.34[a]	umc149	4.75	0.02	3.93	0.30
8	13	2.43	2.32[a]	bnl13.05	4.40	1.89	3.14	0.10
8	123	2.45	1.13	umc89	0.89	1.17	8.06	0.36
					10.92	13.85	19.57	

This table presents the QTL output obtained by running model III with the QTL position by chromosome
in centiMorgan (cM), the log likelihood ratio (LOD, considering the three trials together in a combined
analysis), the genetic interaction across environments (Q×E), the closest marker to the QTL peak, the
percentage of the phenotypic variance expressed in each individual trial (R^2), and the additivity
evaluated through the combined analysis. The last line corresponds to the percentage of the phenotypic
variance expressed per trial when considering all the significant QTL acting together. A positive sign of
the additivity indicates that Ac7643 contributes to increase the value of the target parameter.
[a] The genetic interaction across environments is significant.

loci with the highest LOD scores under the combined analysis. In 96A, OA was evaluated on the entire segregating population (about 220 families) under a single stress level, while in 99A, and considering resources available, OA was evaluated under three water regimes (SS, IS, and WW) but for only 130 genotypes randomly chosen. Although limited QTL stability across environments is expected for a parameter such as OA (nature of the measurement and low h), it is likely that the combined analysis taking into account 96 and 99 data underestimated these QTL as it was conducted only on the common 130 genotypes evaluated across all three stress trials.

4.3.4 Glucose in the Ear Leaf

A complex parameter such as OA is affected by several underlying parameters in its calculation. In contrast, there may be fewer physiological parameters underlying the glucose content measured in the ear leaf. It has been well demonstrated that sucrose maintains embryo growth in maize under water-limited conditions (Westgate and Boyer, 1985; Zinselmeier et al., 1995, 1999) and changes in sucrose invertase activities have been observed in specific maize tissues in plants grown under water-limited conditions (Kim et al., 2000). Therefore glucose concentration in the ear leaf might play an important role in the drought tolerance process, especially considering that this glucose might represent an enhanced osmolyte contribution due to invertase hydrolysis of sucrose to glucose and fructose. Table 5 presents the QTLs for glucose content in the ear leaf quantified at flowering time, under water-limited conditions, 2, 3, and 4 weeks after the last irrigation on a RIL population of 220 families derived from the P1×P2 cross. Glucose quantification was more reproducible across repetitions compared to OA, and heritabilities of 0.61, 0.52, and 0.54 were calculated for the three different harvest times. Four, five, and four significant QTLs were identified at 2, 3, and 4 weeks after the last irrigation, respectively. Two of them on chromosomes 5 (103–111 cM) and 8 (106–122 cM) were common across the three harvests. The QTLs on chromosomes 4 and 5 had a very significant LOD value (4.41 and 8.03, respectively) at 2 weeks, as it was the case at 3 weeks for the QTLs on chromosomes 5 and 7 (4.65 and 4.44). No very significant QTL was identified at 4 weeks. Considering the QTL results at the three different harvest times, it appears clearly that the level of significance for most QTL changes over time. The QTL on chromosome 5 had a LOD score of 5.62, 4.65, and 2.17 over time. The QTL on chromosome 7 was not identified significantly at 2 weeks (LOD of 1.61) but presented a very significant LOD at 3 weeks (4.44) that decreased to a LOD of 2.43 at 4 weeks. The significance of the QTL on chromosome 4 decreased from 3.80 at 2 weeks to 2.32 at 3 weeks to a final value of 1.35 at 4 weeks, and a new significant QTL appeared at 4 weeks on chromosome 2 expressing the largest percentage of the phenotypic at this harvest time. These

Table 5 QTLs Characterization for Glucose (μg/gDW) Content Quantified in Ear Leaf (Discs) Harvested at Flowering Time on Segregating RIL Families (Ac7643×Ac7729/TZSRW) 2, 3, and 4 Weeks After the Last Irrigation

Chromosome	cM	LOD	Marker	R^2	Additivity
2 weeks					
4	163	3.80	umc66	4.41	242.55
5	103	5.62	bnl6.22	8.03	−279.94
8	122	2.65	umc89	4.35	190.00
10	13	2.51	npi285	6.33	189.76
				23.34	
3 weeks					
4	162	2.32	umc66	2.65	157.37
5	111	4.65	bnl6.22	8.21	−236.24
7	144	4.44	umc91	6.03	218.99
8	117	2.17	umc89	3.97	160.99
8	156	2.66	umc30	1.90	−152.75
				26.37	
4 weeks					
2	182	2.01	umc150	5.78	−112.05
5	103	2.17	bnl6.22	4.61	−123.24
7	139	2.43	umc91	4.51	124.63
8	106	2.08	csu68	1.99	125.32
				15.55	

This table presents the QTL output obtained by running model III with the QTL position by chromosome in centiMorgan (cM), the log likelihood ratio (LOD), the closest marker to the QTL peak, the percentage of the phenotypic variance accounted for (R^2), and the additive effect of individual QTL. The last line corresponds to the percentage of the phenotypic variance expressed when considering all the significant QTL acting together. A positive sign of the additivity indicates that Ac7643 contributes to increase the value of the target parameter.

QTL results indicate that genetic control of glucose concentration in the ear leaf changes over time, concomitantly as the stress intensity increases. It can be expected, looking at the QTL results, that a profiling experiment conducted at 3 or 4 weeks after the last irrigation on ear leaf tissue will generate different results for changes in gene expression involved in the carbohydrate biosynthetic pathway.

4.3.5 The Challenge of Measuring Physiological Parameters Accurately on Large Population

There is a real challenge to evaluate with high accuracy physiological parameters under field conditions in large segregating populations. The nature of the parameter further challenges this effort as presented above. For most parameters, the field harvest step represents the most critical step for quality data. Based on our experience we decided to consider at least 200 RILs are needed for all physiological measurements conducted in the field, unless the cost of processing samples becomes prohibitive. Although it is challenging to evaluate with high accuracy physiological parameters on large segregating populations, and although in most of the cases it is time consuming and expensive compared to measurement of morphological traits, the information generated at the phenotypic (correlation, heritability) and QTL levels is extremely valuable. Indeed, this information provides an essential bridge between the data emerging from functional genomics and morphological plant responses, allowing the identification and characterization of major pathways related to drought tolerance.

For complex parameters such as OA, experiments conducted under controlled conditions (e.g., green/screen houses) will provide higher quality data compared to the field. Such type of experiments can provide useful and complementary information to better understand the physiological response of a plant under water-limited conditions. Working with plants in pots, one can simulate in the greenhouse stress conditions very close to those in the field, especially if working on small statured crops such as wheat or rice. For maize, with a rooting depth from 1 to 3 m depending on the genotype, working with pots is more challenging due to the stress imposed at the root level when working with mature plant. However, experiments conducted under controlled conditions should provide complementary information to field data, allowing better understanding of the genetic basis of physiological mechanisms involved in stress tolerance.

5 FUNCTIONAL GENOMICS

One of the weaknesses of the quantitative genetic approach is that it does not provide much information about mechanisms and pathways involved in drought tolerance, nor does it identify the multitude of genes involved in plant response. Although the number of genes identified in plant regulation under abiotic stress is increasing (Bohnert et al., 1995, 2000; http://stress-genomics.org), knowledge of their function, interaction, and, perhaps more importantly, the time frame they operate remains poor. Recent developments in functional genomics approaches should help overcome this problem and provide important information for evaluating the role of potential pathways, as they allow the simultaneous study of the expression of several thousand genes. To date, and partly because there are many constraints and uncertainties inherent in evaluating germplasm under controlled abiotic stress environments, few profiling experiments have reported changes in gene expression in maize grown under stress conditions in the field (Zinselmeier et al., 2002) or in the glasshouse (Setter and Flannigan, 2001; Yu and Setter, 2002; Zinselmeier et al., 2001).

5.1   Quantitative Trait Loci Information to Help Validating Candidate Gene

Microarrays or DNA chip experiments generate a large amount of data on each sample and the expression levels of large numbers of genes presenting under contrasting experimental conditions can be observed (e.g., Seki et al., 2001, 2002). The greatest challenge of functional genomics approach is the organization and interpretation of these data to identify the most informative changes in gene expression and to evaluate their associated phenotypes (tolerant vs. susceptible). In this regard, QTL data can help identify candidate genes responsible for changes in plant phenotype. Indeed, the correlation on a linkage map between a differentially expressed sequence tag (EST) cDNA and a QTL related to the same physiological parameter is a key step in the identification of candidate genes. Combining information at three different levels: phenotypic characterization of the plant, QTL data, and changes in gene expression, for selected physiological pathways and morphological traits, provides an integrated approach for better understanding response to a water-limited environment. These three sources of information are complementary because the QTL characterization, which defines the genetic basis of the key physiological parameters, represents the link between differential gene expression mechanisms and the morphological plant response.

The results presented in Fig. 3 illustrate how QTL information can help to bridge the gap between yield components, secondary traits of interest, physiological parameters, and, eventually, the differential expression of target ESTs. A very significant phenotypic correlation was observed between grain yield (GY) and ear number (ENO), ENO and ear weight (EW), EW and silk weight (SW), and SW and ABA content in the silk (ABAS). The measurements were taken at pollen shedding in the P1×P2 RIL population under water-stress field conditions. Although the correlation between ABAS and GY was relatively low, it was still significant. The correlation between ABAS vs. grain yield or silk weight was negative, indicating that under our experimental conditions tolerant genotypes accumulated less ABA compared to the susceptible ones at pollen shedding. Although quantified in different tissue, this result is in agreement with the fact that selection for increased tolerance to drought in maize was accompanied by a reduction in the concentration of leaf ABA (Mugo et al., 2000), suggesting that improved performance under stress is associated with reduced ABA production. As expected from the significant phenotypic correlations, common QTLs were

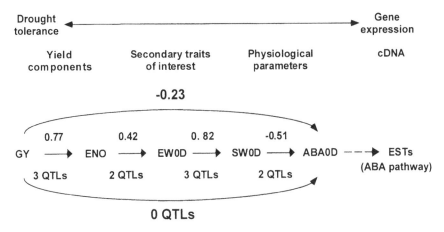

Figure 3 Phenotypic correlations and common QTLs between target phenotypic traits considering grain yield (GY), ear number per pant (ENO), ear weight at pollen shedding (EWOD), silk weight at pollen shedding (SWOD), and ABA content in the silk at pollen shedding (ABAOD).

identified for the different pair-wise comparisons. Three common QTLs were identified for GY and ENO, two for ENO and EW, three for EW and SW, and two for SW and ABAS. No common QTLs were identified between GY and ABAS. However, one can conclude looking at the results presented in Fig. 3 that ABA changes in the silk are related to grain production under water-limited conditions, as it is correlated at the phenotypic and the genetic levels (common QTLs) with silk and ear growth, which are themselves correlated with ear number per plant. The next step is to map ESTs associated with differential expression under our experimental conditions and involved in the ABA biosynthesis and see if they map to a QTL previously identified for ABA concentration. This "cascade" of measurements from grain yield to the quantification of the target physiological parameters should help in the validation and interpretation of the result obtained from profiling experiments.

5.2 Germ Plasm

Designing a profiling experiment includes the selection/identification of the germ plasm, the target tissue, the timeline, the experimental conditions, and a suitable number of repetitions. The choice of germplasm differs depending on the objective of the experiment, the nature of the target trait, and/or the target environment. Changes in gene expression quantified in contrasting lines provide useful information related to pathways involved in plant response under the target environments. It might also represent the most suitable contrasting material to understand the genetic bases of traits with non-transgressive segregation, where the tolerant line provides favorable alleles at all loci involved in the expression of the target trait.

Target traits or physiological parameters involved in the regulation of drought tolerance generally have a transgressive segregation, as demonstrated by the phenotypic distribution of segregating population and the QTL data presented in this chapter. Opposite allelic contribution at two QTL (opposite sign of additivity) involved in the expression of a given traits indicates that changes in the gene expression will be opposite at those two loci when looking at the two parental lines. In addition, due to the genetic complexity of drought tolerance and the reduced breeding effort dedicated to improvement of drought tolerance in maize during the last century, compared to "optimal" environments, it might be expected that a tolerant line does not have most of the favorable alleles for all the adaptive traits. As an example, chlorophyll content was evaluated (SPAD meter) in the ear leaf at flowering time and under water-limited conditions in the segregating RIL population derived from the $P1 \times P2$ cross (Fig. 4). Under stress conditions this trait was significantly and positively correlated with grain yield (0.31) and ear number per plant (0.45), confirming that a high level of chlorophyll content under water stress is associated with better grain production in this cross. However, P1, the drought tolerant line, had lower chlorophyll content than P2 (Fig. 4),

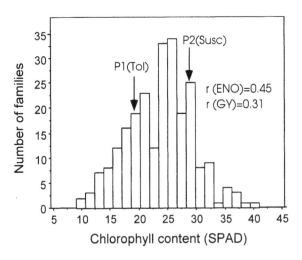

Figure 4 Chlorophyll content measured using a SPAD meter in a segregating population of 240 RIL families (Ac7643×Ac7729/TZSRW) under water-limited conditions.

indicating that P2 had more favorable alleles for chlorophyll content than P1. This situation is unique for this cross when looking at the other segregating populations developed at CIMMYT. Finally, working with contrasting lines with different genetic backgrounds, there is always a risk that changes in gene expression are due to the background and not because genes are associated with the expression of a target trait. This risk decreases as the number of lines involved in the profiling experiment increases. The use of segregating families (e.g., RILs, NILs), with the same genetic background, but contrasting phenotype, might represent very suitable germplasm to evaluate changes in gene expression under abiotic stress.

6 MARKER-ASSISTED SELECTION STRATEGIES

For approximately 15 years, genetic dissection of polygenic traits has been hailed as a promising application of DNA markers, resulting in extensive mapping experiments aimed toward the development of MAS (Lee, 1995). Few concrete MAS results have been published that would justify the initial enthusiasm (Mohan et al., 1997; Ribaut and Hoisington, 1998), although some clear success of using molecular markers for improving quantitative traits have been reported (Tanksley et al., 1996; Shen et al., 2001; Ribaut et al., 2002). The difficulty of manipulating quantitative traits is related to their genetic complexity, principally the number of genes involved in their expression and interactions between genes (epistasis). As several genes are involved in the expression of polygenic traits, they generally have smaller individual effects on the plant phenotype and are cross-dependent. This implies that several regions must be manipulated simultaneously to have a significant impact, and that the effect of individual regions is not easily identified. In addition, $Q \times E$ affects QTL stability across environments, creating a handicap for the efficient use of MAS (Beavis and Keim, 1996).

6.1 Backcross Marker-Assisted Selection

Until recently, a clear technical limitation of the BC-MAS scheme has been the restricted population sizes that can be handled, limiting the flexibility and power of selection. With the development of reliable PCR-based markers, a substantial increase in the segregating population size that can be screened is now feasible. A backcross MAS (BC-MAS) experiment, launched in 1994, aimed to improve the drought tolerance of CML247. CML 247 is an elite tropical inbred line with outstanding combining ability and good yield in hybrid combination under well-watered conditions, but with poor agronomic performance under drought conditions because it has a large ASI. The drought tolerant line Ac7643, characterized by a short ASI under water-limited conditions, was used as donor line. Five genomic regions involved mainly in the expression of a short ASI under severe stress conditions were selected for transfer from Ac7643 into CML247. MAS was conducted at two BC and two self-pollination cycles at selected loci, to identify donor alleles, and at nonselected loci to recover as many as possible of the recurrent parent alleles. After two BCs and two self-pollinations, the 70 best BC_2F_3 (i.e., S_2 lines) were identified and crossed with two CIMMYT testers of the opposite heterotic group, CML254 and CML274. These hybrids and the BC_2F_4 families (S_3 lines) derived from the selected BC_2F_3 plants were evaluated during the dry winter season in 1998, 1999, and 2001 in Tlaltizapan, Mexico, under several water regimes. Results show that under stress conditions that induce a yield reduction of at least 80%, the mean of the 70 selected genotypes performed better than the control when crossed with CML254 and CML274. In addition, the best genotype among the 70 selected ($BC_2F_3 \times$ testers) performed two to four times better than the control. This difference became less marked when the intensity of stress decreased, and for stress inducing less than 40% yield reduction, hybrids resulting from the MAS, or developed with the original version of CML247 performed the same. Although performance among the selected genotypes varied with stress intensity, few genotypes consistently performed significantly better than the controls across the six water-limited trials. No yield reduction was observed under well-watered conditions (Ribaut et al., 2002). This experiment demonstrates the efficiency/limitation of manipulating QTL in a given cross for polygenic trait improvement, as the MAS was successful, but only under the same type of environment as for the QTL identification.

The transfer of genomic segments from a donor to a recipient elite line through BC has time and cost constraints (Morris et al., 2003). However, the BC approach for polygenic trait improvement might be necessary when only a few donor lines have been characterized, as was the case when this project was initiated in 1994.

6.2 Single Large-Scale Marker-Assisted Selection

To date a large amount of information about elite line performance under drought has been generated by CIMMYT's maize programs resulting in the identification of drought tolerant germplasm with different genetic background. With this new germplasm, recurrent selection, and the pyramiding of favorable alleles by crossing two elite lines that perform well under the target environment conditions, became attractive a few years ago. Considering this situation a new MAS approach described as "single large-scale MAS" (SLS-MAS) has been implemented at CIMMYT (Ribaut and Betrán, 1999). This approach has been tested in collaboration with two national programs

(Kenya and Zimbabwe), with the goal of improving drought tolerance in tropical maize in Africa (the overall strategy is presented in Fig. 5). Parental lines, belonging to different heterotic groups, have been selected based on their line per se and hybrid performance under drought conditions. The two elite lines were crossed to develop an F_2 segregating population of 2000 plants, from which an F_3 segregating population of 4000 plants was generated (two F_3 plants from each F_2 plant). During the summer of 1998, a subset of 400 F_3 plants was crossed with two well-adapted African testers (CML202 and CML311) from two different heterotic groups to develop the top-cross populations. Leaf tissue was collected from this subset, from which 280 plants were selected, based on seed production, to construct the RFLP linkage map. Top crosses have been evaluated under IS and SS in Kenya and Zimbabwe by breeders from the national program. During the 1998 summer cycle, the large F_3 population was self-pollinated to develop $F_{3:4}$ families, and leaf tissue from the 4000 plants was collected. By combining phenotypic performance of testcrosses with the allelic segregation at the different loci used to construct the linkage map, four target genomic regions were identified focused mainly on QTL for grain yield, ear number per plant, and ASI. The SLS-MAS was conducted on the DNA extracted from the 4000 F_3 plants, and about 200 F_4 families were identified with fixed favorable alleles at the four target regions. Germplasm will be evaluated this coming 2002 drought cycle in Africa.

6.3 Consensus Map

Another alternative to cross-specific approach is MAS based on "universal drought QTLs" identified on a consensus map that incorporates (1) QTL information for yield components, morphological traits, and physiological parameters; (2) candidate genes and pathways; and (3) information provided by gene expression. The underlying rationale for this approach is that genes involved in drought response are most likely located at the same position in the maize genome, independent of the germplasm performance, and that phenotypic differences across germplasm are created by the nature/quality of the alleles at those genes. When a QTL is identified in a segregating population for a given trait, it means that the difference in the contribution of the two alleles coming from the two parental lines is significant at a specific genomic position. Allelic contribution has to be taken in the broad sense here, as contrasting phenotype might be due to difference in the gene, its promoter, to epistasis effects with other genes, etc. If in another segregating population the difference between the two alleles does not have a significant impact on the plant phenotype (e.g., they are both either good or bad), and although the gene(s) are still at the same location, no significant QTL will be identified. For this reason, the results of QTL identification for a given trait are germplasm dependent, making the extrapolation of this information questionable. However, one can postulate that if QTL identification is conducted on several segregating populations for a target trait, the same key QTLs should be identified in more than one cross, thereby clearly demonstrating the importance of these genomic regions for the expression of the target trait. This would be even more likely when the crosses share a significant part of their pedigree, as is the case with many private sector temperate maize breeding programs. Based on this assumption, our objective is to develop a consensus map to visualize and summarize the QTL information produced so far. Practically, the consensus map has been developed using about 40 anchor markers common for the four crosses. All of the QTL information available for each cross is being compiled on the consensus map (see Table 1) assigning a "weight" to each QTL. This weight will have two major components: (1) the nature of the trait and (2) the threshold value of the QTL identified under a specific environment (likelihood ratio, LR). The greatest weight will be given to yield components. For secondary traits, consideration will be given to the phenotypic and genetic correlation between the trait and yield components (mainly grain yield and ears per plant). The heritability of the trait might also be considered as a third component. In addition to our "in house QTL," QTL related to drought tolerance in maize identified by other groups for morphological traits (Agrama and Moussa, 1996; Austin and Lee, 1998; Frova et al., 1999; Sari-Gorla et al., 1999) or physiological parameters (Lebreton et al., 1995; Tuberosa et al., 1998; de Vienne et al., 1999) will be considered. Taking

---

Figure 5 Ongoing single large-scale marker-assisted selection (SLS-MAS) experiment aimed at fixing favorable alleles for drought tolerance in a F3 segregating population obtained by crossing two elite inbred lines adapted to southern African growing conditions. The two parental lines perform well under water-limited conditions and have been selected for their high level of allelic complementarity evaluated by the performance of crosses between those two lines under stress conditions (phase I). From this cross, a large segregating population of 3600 F3 plants was developed in Mexico. During the 1998B cycle at Tlaltizapan (TL 98B Mx), leaf tissue from a subset of this large population (280 genotypes) was harvested, and DNA extracted, to construct a genetic linkage maps. In the following cycle (1999A) the hybrids obtained by crossing those 280 F3 plants with two lines used at tester line in Africa, and belonging to two different heterotic groups, were evaluated in Kenya (Ky) and Zimbabwe (Zw) to collect phenotypic data that will allow the QTL identification (phase II). The F4 families presenting fixed favorable alleles at four QTLs were selected and derived to inbred lines through self pollination (phase III). At the end of this scheme hybrid vigor is recovered by crossing the germplasm obtained through SLS-MAS with the tester used in phase IIa for the QTL identification. ⊗ refers to self-pollination step.

Phase I

Elite lines outstanding for target traits, field evaluation (Ky and Zw)

Elite parental line selection
Genetic experimental design
(diallel/factorial crosses, fingerprinting)

Selected parental lines : K64R and H16

Phase II

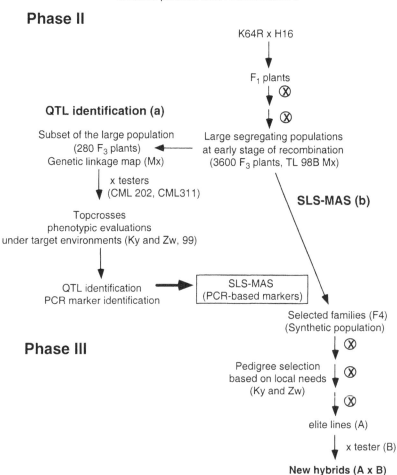

K64R x H16

F_1 plants

QTL identification (a)

Subset of the large population
(280 F_3 plants)
Genetic linkage map (Mx)

Large segregating populations
at early stage of recombination
(3600 F_3 plants, TL 98B Mx)

x testers
(CML 202, CML311)

SLS-MAS (b)

Topcrosses
phenotypic evaluations
under target environments (Ky and Zw, 99)

QTL identification
PCR marker identification

SLS-MAS
(PCR-based markers)

Selected families (F4)
(Synthetic population)

Phase III

Pedigree selection
based on local needs
(Ky and Zw)

elite lines (A)

x tester (B)

New hybrids (A x B)

advantage of the synteny across genomes, QTL identified under drought conditions in other crops might also be considered (for review, see Ribaut and Poland, 2000). With all the QTL information integrated into a single consensus map, outstanding regions involved in the expression of the same trait (different crosses or environments) or different target traits (same cross and/or different crosses or environments) should be identified. Indeed, a plant phenotype is the result of a differential expression of several physiological/ biochemical pathways. A short-ASI phenotype, for instance, might involve carbohydrate and hormone metabolism and translocation as well as water potential regulation and/or membrane stability. Therefore we should expect to see QTL for interrelated physiological and morphological traits identified at the same genomic location, given that changes in physiological pathways have an impact on the plant phenotype. If such "universal" drought genomic regions are identified, the objective should be to conduct MAS experiments on those regions for new crosses (good by good lines with different genetic background), without making a QTL identification in each target cross.

6.4 Target Regions of Interest

The success of the approach described above is highly dependent on the distribution along the genome of key genes conferring drought tolerance. If those genes are equally distributed along the genome, the notion of MAS without mapping QTL for a target cross will not be efficient because too many different regions, with reduced impact on the plant phenotype, will have to be considered. On the other hand, if gene clustering based on function occurs, this approach might be feasible. Although some results tend to confirm gene clustering of developmental functions based on QTL results (Khavkin and Coe, 1997) or gene location (Langridge et al., 2002), there are few data supporting either postulate at this time. However, looking at the large QTL data set we have produced, in conjunction with other published studies, some key regions for drought tolerance in maize can already be clearly identified. As an example, within a window of 15 cM on chromosome 2 (umc34-csu133), we identified on the P1×P2 RIL population and across six different stressed environments, QTLs for ASI, MFLW, EHT, EPO, RCT, EDW, and SDW at pollen shedding, EDW and SDW 7 days after pollen shedding, and ABA in the ear. At the same location a locus for sucrose invertase was identified by Prioul et al. (1999). Using two different genetic populations (Polj 17×F-2 and Os420×IABO78), Drs. Steve Quarrie and Roberto Tuberosa's research groups (Lebreton et al., 1995; Tuberosa et al., 1998) identified under water-limited conditions common QTLs for both leaf ABA content and root pulling resistance in this same genomic region. The fact that in three different populations, and at three different locations, QTL for ABA and root structure have been identified in addition to the QTL for the various other morphological traits demonstrates that several genes involved in drought tolerance are located within those 15 cM. Therefore this specific region on chromosome 2 might be suitable for gene discovery (map-based cloning) or germplasm improvement (MAS based on genomic regions identified on a drought consensus map).

7 CONCLUSIONS

Based on progress to date, it is clear that a multidisciplinary approach combining breeding, physiology, and biotechnology is required for an effective understanding of a plant's response to drought stress. Structural and functional genomics are two complementary and closely linked approaches that will permit the identification of key pathways involved in drought stress and further our understanding of how they interact. In addition, the identification of QTLs for target traits and physiological pathways serves as a valuable source of information in the validation phase of candidate gene identification when they map to the same genomic location, and vice versa. The QTL characterization effort initiated at CIMMYT several years ago will continue, with particular emphasis given to the genetic dissection of physiological drought-adaptive traits. At the same time, we will intensify our effort to quantify changes in gene expression in contrasting genotypes identified from the segregating populations developed for this project. Increasing our understanding of the genetic basis of the essential morphological traits and physiological parameters of drought tolerance in maize will lead to new MAS strategies to improve the tolerance of maize and other crops to water-limited conditions. A MAS strategy based on "universal drought QTLs" identified on a consensus map that incorporates genetic data from previous studies might offer an attractive option, if genomic regions involved in the drought tolerance process can be identified across germplasm and environments.

ACKNOWLEDGMENTS

The authors would like to thank Jeff Habben, and Chris Zinselmeier for their collaboration in some activities presented in this chapter. The assistance of David Poland, CIMMYT editor, in the writing of this chapter is gratefully acknowledged. Special thanks goes also to the Rockefeller Foundation that has supported most of our drought activities since January 2001 (grant 2001 FS 001).

REFERENCES

Agrama HAS, Moussa ME. Mapping QTLs in bredding for drought tolerance in maize (*Zea mays* L.). Euphytica 1996; 91:89–97.

Austin DF, Lee M. Detection of quantitative trait loci for grain yield and yield components in maize across generations in stress and nonstress environments. Crop Sci 1998; 38:1296–1308.

Bänziger M, Cooper ME. Breeding for low-input conditions and consequences for participatory plant breeding—examples from tropical maize and wheat. Euphytica 2001; 122:503–519.

Bänziger M, Lafitte HR. Efficiency of secondary traits for improving maize for low-nitrogen target environments. Crop Sci 1997; 37:1110–1117.

Bänziger M, Mugo S, Edmeades GO. Breeding for drought tolerance in tropical maize—conventional approaches and challenges to molecular approaches. In: Ribaut JM, Poland D, eds. Molecular Approaches for the Genetic Improvement of Cereals for Stable Production in Water Limited Environments. Proceedings of a Symposium, June 21–25, 1999, CIMMYT, El Batán, Mexico, 2002a:69–72.

Bänziger M, Pixley KV, Vivek B, Zambezi BT. Characterization of elite maize germplasm grown in eastern and southern Africa: Results of the 1999 regional trials conducted by CIMMYT and the Maize and Wheat Improvement Research Network for SADC (MWIRNET). Harare, Zimbabwe: CIMMYT, 2000b:1–44.

Beavis WD, Keim P. Identification of quantitative trait loci that are affected by environment. In: Kang MS, Gauch HG, eds. Genotype-by-environment interaction. Boca Raton, Florida: CRC Press, 1996:123–149.

Beck D, Betrán J, Bänziger M, Ribaut J-M, Willcox M, Vasal SK, Ortegà A. Progress in developing drought and low soil nitrogen tolerance in maize. In: Wilkinson DB, ed. Proceedings of the 51st Annual Corn & Sorghum Research Conference. Washington: ASTA, 1996:85–111.

Blum A. Constitutive traits affecting plant performance under stress. In: Edmeades GO, Bänziger M, Mickelson HR, Peña-Valdivia CB, eds. Developing Drought and Low-N Tolerant Maize. El Batan. Mexico: CIMMYT, 1997:131–135.

Bohnert HJ, Nelson DE, Jensen RG. Adaptations to environmental stresses. Plant Cell 1995; 7:1099–1111.

Bohnert H, Fischer R, Kawasaki S, Michalowski C, Wang H, Yale J, Zepeda G. Cataloging stress-inducible genes and pathways leading to stress tolerance. In: Ribaut J-M, Poland D, eds. Molecular Approaches for the Genetic Improvement of Cereals for Stable Production in Water-Limited Environments. A Strategic Planning Workshop Held at CIMMYT, El Batan, Mexico, 21–25 June 1999. Mexico, D.F., Mexico: CIMMYT, 2000:156–161.

Bolaños J, Edmeades GO. Value of selection for osmotic potential in tropical maize. Agron J 1991; 83:948–956.

Bolaños J, Edmeades GO. Eight cycles of selection for drought tolerance in lowland tropical maize: II. Responses in reproductive behavior. Field Crops Res 1993; 31:253–268.

Bolaños J, Edmeades GO. The importance of the anthesis-silking interval in breeding for drought tolerance in tropical maize. Field Crops Res 1996; 48:65–80.

Bonnet G, Tyagi S, Libchaber A, Kramer FR. Thermodynamic basis of the enhanced specificity of structured DNA probes. Proc Natl Acad Sci U S A 1999; 96:6171–6176.

Borrell AK, Tao Y, McIntyre CL. Physiological basis, QTL and MAS of the stay green drought resistance trait in grain sorghum. In: Ribaut J-M, Poland D, eds. Molecular Approaches for the Genetic Improvement of Cereals for Stable Production in Water-Limited Environments. A Strategic Planning Workshop Held at CIMMYT, El Batan, Mexico, 21–25 June 1999. Mexico, D.F., Mexico: CIMMYT, 2000:142–146.

Castleberry RM, Lerette RJ. Latente, a new type of drought tolerance? In: Loden H, Wilkinson D, eds. Proc 34th Annual Corn and Sorghum Res. Conf., Chicago, Dec 11–13, 1979. Washington: ASTA, 1979:46–56.

Chapman SC, Edmeades GO. Selection improves drought tolerance in tropical maize populations: II. Direct and correlated responses among secondary traits. Crop Sci 1999; 39:1315–1324.

Clark LJ, Cope RE, Whalley WR, Barraclough PB, Wade LJ. Root penetration of strong soil in rainfed lowland rice: Comparison of laboratory screens with field performance. Field Crops Res 2002; 76:189–198.

Cornic M. Leaf photosynthesis under drought stress. In: Baker NR, ed. Advances in Photosynthesis. Vol 5. Photosynthesis and the Environment. Kluwer Acad. Publishers, 1996:347–363.

De la Fuente-Martinez JM, Herrera-Estrella L. Advances in the understanding of aluminum toxicity and the development of aluminum-tolerant transgenic plants. Adv Agron 1999; 66:103–120.

De Vienne D, Leonardi A, Damerval C, Zivy M. Genetics of proteome variation for QTL characterization: application to drought-stress responses in maize. J Exp Bot 1999; 50:303–309.

Dekkers JCM, Hospital F. The use of molecular genetics in the improvement of agricultural populations. Nat Rev 2002; 3:22–32.

Duncan RR, Carrow RN. Turfgrass molecular genetic improvement for abiotic/edaphic stress resistance. Adv Agron 1999; 67:233–305.

Edmeades GO, Bolaños J, Bänziger M, Ribaut J-M, White JW, Reynolds MP, Lafitte HR. Improving crop yields under water deficits in the tropics. In: Chopra VL, Singh RB, Varma A, eds. Crop Productivity and Sustainability—Shaping the Future. Proceedings of Second International Crop Science Congress. New Delhi: Oxford and IBH, 1998:437–451.

Edmeades GO, Bolaños J, Elings A, Ribaut J-M, Bänziger M, Westgate ME. The role and regulation of the anthesis-silking interval in maize. In: Westgate ME, Boote KJ, eds. Physiology and Modeling Kernel Set in Maize. CSSA Special Publication No. 29. Madison, WI: CSSA, 2000:43–73.

Edmeades GO, Cooper M, Lafitte R, Zinselmeier C, Ribaut J-M, Habben JE, Löffler C, Bänziger M. Abiotic stresses and staple crops. In: Nosberger J, Geiger HH, Struik PC, eds. Crop Science: Progress and Prospects. Proceedings of the Third International Crops Science Congress, 17–21 August 2000. Wallingford, UK: CABI, 2001:137–154.

Fischer KS, Edmeades GO, Johnson EC. Selection for the improvement of maize yield under moisture-deficits. Field Crops Res 1989; 22:227–243.

Frova C, Krajewski P, Di Fonzo N, Villa M, Sari-Gorla M. Genetic analysis of drought tolerance in maize by molecular markers: I. Yield components. Theor Appl Genet 1999; 99:280–288.

Gilles PN, Wu DJ, Foster CB, Dillon PJ, Chanock SJ. Single nucleotide polymorphic discrimination by an electronic dot blot assay on semiconductor microchips. Nat Biotechnol 1999; 17:365–370.

Heisey PW, Edmeades GO. Part 1. Maize production in drought stressed environments: technical options and research resource allocation. In: CIMMYT eds.World Maize Facts and Trends 1997/98. Mexico D.F., Mexico: CIMMYT, 1999:1–36.

Helentjaris T, Slocum M, Wright S, Schaefer A, Nienhuis J. Construction of genetic linkage maps in maize and tomato using restriction fragment length polymorphisms. Theor Appl Genet 1986; 72:761–769.

Hittalmani S, Parco A, Mew TV, Zeigler RS, Huang N. Fine mapping and DNA marker-assisted pyramiding of the three major genes for blast resistance in rice. Theor Appl Genet 2000; 100:1121–1128.

Johnson GR, Mumm RH. Marker assisted maize breeding. Proc. 51st Annual Corn and Sorghum Res Conf, Chicago, IL. 11–12 Dec. 1996. Washington, DC: American Seed Trade Assoc, 1996:75–84.

Jiang C, Zeng ZB. Multiple trait analysis of genetic mapping for quantitative trait loci. Genetics 1995; 140:1111–1127.

Khavkin E, Coe E. Mapped genomic locations for developmental functions and QTLs reflect concerted groups in maize (*Zea mays* L.). Theor Appl Genet 1997; 95:343–352.

Kim JY, Mahe A, Brangeon J, Prioul JL. A maize vacuolar invertase, IVR2, is induced by water stress. Organ/tissue specificity and diurnal modulation of expression. Plant Physiol 2000; 124:71–84.

Lander ES, Botstein D. Mapping Mendelian factors underlying quantitative traits using RFLP linkage maps. Genetics 1989; 121:185–199.

Langridge P, Dong C, Whitford R, Sutton T, Wolters P, Rafalski A, Morgante M, Gumaelius L, Uhlmann N. Tingey S. Early meiotic genes and the PH2 region of wheat. Abstract W176, Proceeding of the Plant, Animal and Microbe genome, January 12–16, 2002, San Diego, USA, 2002:50.

Lebreton C, Lazic-Jancic V, Steed A, Pekic S, Quarrie SA. Identification of QTL for drought responses in maize and their use in testing causal relationships between traits. J Exp Bot 1995; 46:853–865.

Lee M. DNA markers and plant breeding programs. Adv Agron 1995; 55:265–344.

McCouch SR, Kochert G, Yu ZH, Wang ZY, Khush GS. Molecular mapping of rice chromosomes. Theor Appl Genet 1988; 76:815–829.

Mohan M, Nair S, Bhagwat A, Krishna TG, Yano M, Bhatia CR, Sasaki T. Genome mapping, molecular markers and marker-assisted selection in crop plants. Mol Breed 1997; 3:87–103.

Morris M, Dreher K, Ribaut J-M, Khairallah M. Money matters (II): costs of maize inbred line conversion schemes at CIMMYT using conventional and marker-assisted selection. Mol Breed 2003; 11:235–247.

Mugo SN, Bänziger M, Edmeades GO. Prospects of using ABA in selection for drought tolerance in cereal crops. In: Ribaut J-M, Poland D, eds. Molecular Approaches for the Genetic Improvement of Cereals for Stable Production in Water-Limited Environments. A Strategic Planning Workshop Held at CIMMYT, El Batan, Mexico, 21–25 June 1999. Mexico, D.F., Mexico: CIMMYT, 2000b:73–78.

Nguyen HT, Babu RC, Blum A. Breeding for drought resistance in rice: physiology and molecular genetics considerations. Crop Sci 1997; 37:1426–1434.

Paterson AH, Lander ES, Hewitt JD, Peterson S, Lincoln SE, Tanksley SD. Resolution of quantitative traits into Mendelian factors by using a complete linkage map of restriction fragment length polymorphisms. Nature 1988; 335:721–726.

Powell W, Machray GC, Provan J. Polymorphism revealed by simple sequence repeats. Trends Plant Sci 1996; 1:215–221.

Prioul J-L, Pelleschi S, Sene M, Thevenot C, Causse M, de Vienne D, Leonardi A. From QTLs for enzyme activity to candidate genes in maize. J Exp Bot 1999; 337:1281–1288.

Ragot M, Biasiolli M, Delbut MF, Dell'orco A, Malgarini L, Thevenin P, Vernoy J, Vivant J, Zimmermann R, Gay G. Marker-assisted backcrossing: a practical example. Techniques et utilisations des marqueurs moléculaires, (Les Colloques, no 72). Paris: INRA, 1995:45–56.

Ragot M, Gay G, Muller J-P, Duroway J. Efficient selection for adaptation to the environment through QTL mapping in manipulation in maize. In: Ribaut J-M, Poland D, eds. Molecular Approaches for the Genetic Improvement of Cereals for Stable Production in Water-Limited Environments. A Strategic Planning Workshop Held at CIMMYT, El Batan, Mexico, 21–25 June 1999. Mexico, D.F., Mexico: CIMMYT, 2000:128–130.

Reynolds MP, van Ginkel M, Ribaut J-M. Avenues for genetic modification of radiation use efficiency in wheat. J Exp Bot 2000a; 51:459–473.

Reynolds MP, Skovmand B, Trethowan R, Pfeiffer W. Evaluating a conceptual model for drought tolerance. In: Ribaut J-M, Poland D, eds. Molecular Approaches for the Genetic Improvement of Cereals for Stable Production in Water-Limited Environments. A Strategic Planning Workshop Held at CIMMYT, El Batan, Mexico, 21–25 June 1999. Mexico, D.F., Mexico: CIMMYT, 2000b:49–53.

Ribaut J-M, Hoisington D. Marker-assisted selection: new tools and strategies. Trends Plant Sci 1998; 3:236–239.

Ribaut J-M, Betrán FJ. Single large-scale marker-assisted selection (SLS-MAS). Mol Breed 1999; 5:531–541.

Ribaut J-M, Poland D, eds. Proceedings of Workshop on Molecular Approaches for the Genetic Improvement of Cereals for Stable Production in Water-Limited Environments. El Batan, Mexico: CIMMYT, 2000.

Ribaut J-M, Hoisington DA, Deutsch JA, Jiang C, González-de-León D. Identification of quantitative trait loci under drought conditions in tropical maize: 1. Flowering parameters and the anthesis-silking interval. Theor Appl Genet 1996; 92:905–914.

Ribaut J-M, Jiang C, González-de-León D, Edmeades GO, Hoisington DA. Identification of quantitative trait loci under drought conditions in tropical maize: 2. Yield components and marker-assisted selection strategies. Theor Appl Genet 1997; 94:887–896.

Ribaut J-M, Bänziger M, Betrán J, Jiang C, Edmeades GO, Dreher K, Hoisington D. Use of molecular markers in plant breeding: drought tolerance improvement in tropical maize. In: Kang Manjit S, ed. Quantitative Genetics, Genomics, and Plant Breeding, Chapter 7. Wallingford, UK: CABI Publishing, 2002:85–99.

Saini HS, Westgate ME. Reproductive development in grain crops during drought. Adv Agron 2000; 68:59–96.

Sari-Gorla M, Krajewski P, Di Fonzo N, Villa M, Frova C. Genetic analysis of drought tolerance in maize by molecular markers: II. Plant height and flowering. Theor Appl Genet 1999; 99:289–295.

Seki M, Narusaka M, Abe H, Kasuga M, Yamaguchi-Shinozaki K, Carninci P, Hayashizaki Y, Shinozaki K. Monitoring the expression pattern of 1300 Arabidopsis genes under drought and cold stresses by using a full-length cDNA microarray. Plant Cell 2001; 13:61–72.

Seki M, Narusaka M, Ishida J, Nanjo T, Fujita M, Oono Y, Kamiya A, Nakajima M, Enju A, Sakurai T, Satou M, Akiyama K, Taji T, Yamaguchi-Shinozaki K, Carninci P, Kawai J, Hayashizaki Y, Shinozaki K. Monitoring the expression profiles of 7000 Arabidopsis genes under drought, cold and high-salinity stresses using a full-length cDNA microarray. Plant J 2002; 31:279–292.

Setter T, Flannigan BA. Water deficit inhibits cell division and expression of transcripts involved in cell proliferation and endoreduplication in maize endosperm. J Exp Bot 2001; 52:1401–1408.

Shannon MC. Adaptation of plants to salinity. Adv Agron 1998; 60:75–120.

Shen L, Courtois B, McNally KL, Robin S, Li Z. Evaluation of near-isogenic lines of rice introgressed with QTLs for root depth through marker-assisted selection. Theor Appl Genet 2001; 103:75–83.

Tanksley SD, Grandillo S, Fulton TM, Zamir D, Eshed Y, Petiard V, Lopez J, Beck-Bunn T. Advanced backcross QTl analysis in a cross between an elite processing line of tomato and its wild relative *L. pimpinellifolium*. TAG 1996; 92:213–224.

Thomas H, Smart CM. Crops that stay-green. Ann Appl Biol 1993; 123:193–219.

Thomas H, Howarth CJ. Five ways to stay green. J Exp Bot 2000; 51:329–337.

Tuberosa R, Sanguineti MC, Landi P, Salvi S, Casarini Conti S. RFLP mapping of quantitative trait loci controlling abscisic acid concentration in leaves of drought-stressed maize (*Zea mays* L.). Theor Appl Genet 1998; 97:744–755.

Van Beem J, Smith ME, Zobel RW. Estimating root mass in maize using a portable capacitance meter. Agron J 1998; 90:566–570.

Vos P, Hogers R, Bleeker M, Reijas M, Van de Lee T, Hornes M, Frijters A, Pot J, Peleman J, Kuiper M, Zabeau M. AFLP: A new technique for DNA fingerprinting. Nucleic Acids Res 1995; 23:4407–4414.

Westgate ME, Boyer JS. Carbohydrate reserves and reproductive development at low leaf water potentials in maize. Crop Sci 1985; 25:762–769.

Yu LX, Setter TL. Comparative transcriptional profiling of developing maize kernels in response to water deficit. Paper presented at the Annual Meeting of the American Society of Plant Biologists, August 3–7, 2002, Denver, CO, Plant Biology, 2002.

Young ND. A cautiously optimistic vision for marker-assisted breeding. Mol Breed 1999; 5:505–510.

Zeng ZB. Precision mapping of quantitative trait loci. Genetics 1994; 136:1457–1468.

Zhang J, Nguyen HT, Blum A. Genetic analysis of osmotic adjustment in crop plants. J Exp Bot 1999; 50:291–302.

Zhang J, Zheng HG, Aarti A, Pantuwan G, Nguyen TT, Tripathy JN, Sarial AK, Robin S, Babu RC, Nguyen BD, Sarkarung S, Blum A, Nguyen HT. Locating genomic regions associated with components of drought resistance in rice: Comparative mapping within and across species. Theor Appl Genet 2001; 103:19–29.

Zinselmeier C, Lauer MJ, Boyer JS. Reversing drought-induced losses in grain yield: Sucrose maintains embryo growth in maize. Crop Sci 1995; 35:1390–1400.

Zinselmeier C, Jeong BR, Boyer JS. Starch and the control of kernel number in maize at low water potentials. Plant Physiol 1999; 121:25–35.

Zinselmeier C, Habben JE, Westgate ME, Boyer JS. Carbohydrate metabolism in setting and aborting maize ovaries. In: Westgate ME, Boote KJ, eds. Physiology and Modeling Kernel Set in Maize. CSSA Special Publication No. 29. Madison, WI: CSSA, 2000:1–13.

Zinselmeier C, Sun Y, Helentjaris, Beatty M, Yang S, Smith H, Habben J. The use of gene expression profiling to dissect the tress sensitivity of reproductive development in maize. Field Crop Res 2002; 75:111–121.

16

Physiology and Biotechnology Integration for Plant Breeding: Epilogue

Abraham Blum
The Volcani Center
Bet Dagan, Israel

Henry T. Nguyen
University of Missouri–Columbia
Columbia, Missouri, U.S.A.

In the breeder's perspective, plant and crop physiology were often seen as disciplines that served to explain change after the fact rather than to generate change. While crop physiology is acknowledged for having been able to explain yield modifications brought about by breeding (e.g., see chapter on "Wheat and Barley"), it was often blamed for not being effective enough in bringing about yield increases. For example, the contribution of the dwarfing genes to cereal yield improvement was well explained by an increase in harvest index and by faster and timely assimilate partitioning with hardly any increase in photosynthesis and total biomass production. Crop physiology had little to do with the progress made in breeding by using these genes. In fact, plant breeders and physiologists did not even realize initially that height genes had any value in this respect besides an expected improvement in lodging resistance. However, the ensuing physiological studies of the function of these genes did have an impact on the continuing breeding efforts. Plant physiology research outlined the pros and cons of the dwarfing genes in terms of adaptation to different environments, which helped breeders develop a strategy of deployment of these genes in different target environments. The ensuing studies of gibberellic acid responsiveness of the height genes led to the development of a phenotypic screening method for these genes at the seedling stage.

Beyond the above example, studies of the physiology of yield variation and plant response to the environment brought about the development of physiological selection criteria, methods, and protocols to accelerate and improve breeding programs. Table 1 summarizes just a few examples. Physiological knowledge also impacted the theories of plant breeding and selection and thus was important in improving breeding efficiency.

As the application of physiology to conventional breeding expanded during the 1990s, it has become clear that the physiological and biochemical techniques were too elaborate, slow, and often expensive for routine use in phenotypic selection in large breeding populations. At the same time, marker-assisted selection (MAS) emerged as a tool that allows selecting the genotype by probing the chromosomal location of the alleles rather than by measuring the phenotype. This was a timely solution, which came about by integrating molecular and physiological knowledge towards plant breeding.

Table 1 Use of Physiological Traits in the Genetic Improvement of Crops

Physiological trait improved	Crop	Physiological test/assay used
Plant water status under drought stress	Wheat	Use of canopy temperature as a selection index
Stomatal conductance under irrigated hot conditions	Several	Use of canopy temperature as a selection index
Water-use efficiency	Wheat	Carbon isotope discrimination and leaf porometry
Crop nitrogen-use efficiency	Maize	Monitoring leaf chlorophyll by leaf reflective properties
Effective radiation interception (reduced extinction coefficient)	Maize	Selection for erect leaves
Aluminum toxicity resistance	Wheat	Hematoxylin root staining
Legume N_2 fixation	Pulses	Leaf ureide assay
Delayed senescence	Various	Visual or instrumental methods
Stem reserve utilization for grain filling under stress	Wheat	Chemical desiccation of plants

Details and references are presented in the various chapters.

Plant breeders have always strived to achieve a state of knowledge to allow breeding by design rather than by trial and error or by manipulating probabilities. This has indeed been achieved for genetically simple traits such as certain plant morphological and phenological traits or resistance to certain diseases. However, for more complex traits involving the components of yield potential and yield stability, the present insufficient state of knowledge in physiology and genomics does not yet allow breeding by design.

While yield is the target of many breeding programs, it cannot be improved by design and it is normally improved by phenotypic selection for yield in given field environments. This approach has been successful in the past but its low efficiency and increasing cost is in the way of further progress. The problem of phenotypic selection for yield is that yield per se is not a genetic entity. Much has already been written on this dilemma. It is now being further underscored by the failed attempts to identify and select for QTLs assumed to be associated with "yield." The explanation of failure was given in the inconsistencies of the yield QTL mapping in different environments and different "genetic backgrounds." This reminds us of the failure of quantitative genetics to resolve the "inheritance of yield" in the 1950s to the 1960s and the ensuing compensation for that failure by the development of intricate biometrical methods to allow phenotypic selection for yield. Such biometrical methods have been recently revived by the sheer power of modern computing. However, a viable solution to a continued and sustained genetic improvement of yield lies in the understanding of its physiology and genomics. This cannot be done without the dissection of yield into its physiological and biochemical components and a true understanding of their molecular genetics. The prerequisite for the improvement of yield is the resolution of the functional genomics of the building blocks of yield at various levels of plant organization. The limited information available today in this respect has been presented in this book and it underlines the urgent need for more research.

What are some of the prospects of possible physiology–biotechnology integration to impact breeding by design for higher yield potential and better yield stability? Some discussions of this question can be found in this book. Here we shall highlight a few points of special consideration.

1 RAISING THE YIELD POTENTIAL

Solar radiation received by a unit ground area can produce much more plant dry matter than ever recorded in the field. Plant breeding did little to approach this potential. In some cases total production or biomass were increased by extending the duration of growth or by optimizing crop radiation interception (via early growth vigor and erect leaf posture) and by delaying leaf senescence. However, as discussed above, for most crops the increase of harvest index was the main route for grain yield improvement. This is a finite approach and further breeding progress relies on increasing total plant production (principally biomass).

One can recognize three approaches to the improvement of crop biomass as an avenue for increasing yield: (1) genes of wild relatives of crops that were shown to increase crop biomass when introgressed into modern cultivars, (2) heterosis, and (3) engineering photosystem biochemistry.

Genes from wild relatives of crops were demonstrated in a few case studies to increase biomass and sometimes yield when introgressed into cultivated genetic background. For example, a chromosome translocation containing the *Lr19* gene from *Agropyron elongatum* has been shown to be associated with a significant increase in biomass and yield when introduced into already high-yielding backgrounds (see "Wheat and Barley" chapter). However, for all cases the physiological/genetic route by which biomass has been improved was not explored.

This potential resource for improved crop biomass production is not fully utilized and it should by seriously explored with the modern tools available today. For example, Tanksley and Nelson (1996) proposed the "advanced backcross QTL analysis" as a tool for identifying yield-enhancing QTLs in exotic germplasm and as a technique to dissociate yield QTLs from any negative linkage drag carried over from the exotic parent. This tool should allow identifying within a relatively short period the exotic genetic resources, which carry yield-enhancing genes. However, because "yield genes" are elusive, the approach can be made much more efficient if it would address the QTLs linked to the physiological and biochemical components of the yield-enhancing factors coming from the wild species. This requires extensive physiological/genetic investigation.

Heterosis is a major but enigmatic genetic phenomenon that is responsible for high yield and historical progress in the genetic improvement of yield. After many years of utilizing and merchandizing heterosis, the phenomenon remains basically unexplained. Most attempted explanations are actually hypotheses. For some 50 years we have been mainly able to describe the manifestations of heterosis in different crops. These were seen in some morphological and phenological features, total biomass production, plant vigor, high rates of organ differentiation, higher stomatal conductance and assimilation rate, greater root depth, greater number of fruit (grain) per inflorescence—all leading to greater yield independently of harvest index. The genetic basis for these manifestations remains indistinct and at best speculative.

The expression of heterosis in different crops is surprisingly similar (although not perfectly uniform), leading to the conclusion that there must be a unified basis for the phenomenon at the molecular level. Some leads for pursuing possible explanations at the molecular level are proposed in the chapter on heterosis.

It must be quite obvious today that the solution of this enigma should pay a huge bonus in terms of advancing plant breeding as a whole. The integration of biotechnology and physiology is essential for solving the riddle of heterosis.

Photosynthesis is the foundation of plant production. There is a wide agreement across the different chapters of this book that indeed yield is source limited. Once canopy structure has been optimized, leaf nitrogen and chlorophyll content has been maximized, and sufficient sink structures have been differentiated, then single leaf photosynthesis becomes the main target to achieve further progress in plant production. The interaction between photosystem activity and stomatal conductance as an optimizing mechanism for leaf gas exchange has been utilized to develop the carbon isotope discrimination analysis and to achieve some yield improvement (see chapter on "Wheat and Barley"). This approach is apparently effective but limited. Once stomatal conductance is genetically optimized for the given environment (using carbon isotope discrimination or simpler and cheaper selection methods such as the leaf porometer), the limiting factor then becomes the capacity of the photosystem to process more efficiently a given rate of CO_2.

With the recent emergence of biotechnology, attention is turning towards the prospects of engineering a more efficient photosystem. This topic is discussed through several chapters of this book. Attempts to improve photosystem metabolism proceed by way of two biotechnological approaches: improving the efficiency of RUBISCO (as discussed in the chapter on rice) and attempting to engineer C_4 metabolism into the less efficient C_3 plants. For the second approach, conclusions to date are that for the primary enzymes of the C_4 pathway preliminary results are promising. However, a correct posttranslational regulation of the introduced, heterologous enzymes, fine-tuning of the levels of ancillary enzymes (such as adenylate kinase and pyrophosphatase), and metabolite transporters are still to be addressed. Whether the desirable genetic modification will be effectively expressed without modifying the C_3 plant to posses the Krantz anatomy remains under serious debate.

Here also, the synthesis of physiology and biotechnology research carries a great potential for progress, as, for example, in the case of temperature and photosynthesis. The photosystem presents a temperature response curve with a narrow maximum temperature that varies from cool- to warm-season crop plants. Because all plants grow under conditions of daily change in temperature, all plants are exposed to nonoptimal temperatures for photosynthesis during a large part of their life. It is fascinating to realize that the temperature homeostasis of the photosystem has rarely been approached as a means for improving total daily photosystem production. The only clue available to show that this is possible and crucial is the positive effect of heterosis on the temperature homeostasis of photosynthesis seen in sorghum hybrids (Blum, 1989). The molecular basis for temperature homeostasis of the photosystem and the consequence towards total leaf daily net assimilation is considered here as a potentially rewarding area of investigation.

2 STABILIZING YIELD AND THE REDUCTION OF INPUTS

The stabilization of yield under changing environments involves breeding varieties resistant to environmental stress. The reduction of inputs (such as fertilizers) to achieve greater income or to reduce environmental pollution involves breeding varieties that will not reduce yield excessively as inputs are reduced. Sometimes such varieties are defined as "efficient," meaning that they are capable of maintaining a high output/input ratio when input is reduced. The common denominator for progress in all of these directions is stress physiology and stress genomics.

Crop and stress physiology have made a significant contribution to the conventional breeding of stress-resistant varieties. This was possible due to plant breeding research integration with whole-plant physiology, crop physiology, and conventional genetics. The main approach has been the dissection of physiological traits at the crop and whole-plant level followed by physiological phenotypic selection programs. However, progress by this approach has become constrained due mainly to the technical limitation and cost of phenotypic selection for physiological and biochemical traits in large breeding populations. Marker-assisted selection is therefore a timely technological breakthrough in this respect, which is indeed expected to support and expedite conventional breeding for stress resistance.

Marker-assisted selection as a tool for selecting the desirable physiological phenotype is especially crucial in cases where phenotypic interactions in the selection nursery do not allow progress in selection through conventional phenotypic selection. An example can be seen in the attempt to select for improved drought resistance in field-grown populations that segregate for both root depth and osmotic adjustment (OA). Besides its control by genetic factors, OA is strongly affected by leaf water potential (LWP), whereas a reduction in LWP induces an increase in OA in plants that have the genetic capacity. Hence, phenotypic selection for OA should be performed only when all materials are water stressed to the same LWP. In materials segregating also for root depth under drought stress, the shallow-rooted segregates might express lower LWP than the deep-rooted ones. Hence, shallow-rooted plants may also express greater OA and deep-rooted ones may express smaller OA by the token of their LWP rather than by their OA genotype. Hence, phenotypic selection for both traits in the field is impossible. This is a true case for MAS, provided that all the well-known prerequisites for effective MAS are accounted for.

Further improvement of the molecular tools is crucial. However, where stress physiology is concerned, the weakest link in the molecular marker technology is achieving near-perfect phenotyping. Experience gained, as reflected in some of the chapters of this book, shows that phenotyping stress resistance is still often imperfect. The integration of physiology and biotechnology has a crucial role in perfecting MAS as a tool for improving stress tolerance.

Functional genomics approaches stress physiology from the basic end, namely, the gene and its immediate function. As it is seen today, functional genomics of stress resistance is addressing the primary metabolic expressions of the gene with some rudimentary probes of function at the whole-plant level. Impressive preliminary indications were achieved by transformation technology of model plants evidencing a potential role for certain genes in stress tolerance (see web link http://www.plantstress.com/admin/Files/abiotic-stress_gene.htm for a list of genes and their expressions). However, the assessment of function of these genes at the whole-plant level is lagging and it is often subjected to criticism. The assessment of function at the crop level is practically nonexistent. If function is viewed towards application in agriculture then functional genomics of stress resistance genes must be assessed in the perspective of the end user. In this respect, there is a large genotype–phenotype gap, because gene expression is manipulated and studied at the molecular level while function is implied towards the whole plant and its agronomic performance.

Stress tolerance and resource-use efficiency are rarely a function of one gene. This is why we find molecular biologists discuss the pyramidizing of genes as a practical solution to bridging the genotype–phenotype gap. However, this wish will not materialize into action if we do not widen and intensify our understanding of whole-plant stress physiology. In this context, it is not just an issue of pyramidizing several genes but mainly an issue of whole-plant interactions in affecting both resistance and productivity. It is sufficient to mention the well-recognized problem of certain negative associations between stress resistance and potential yield, which are sometimes defined as the "cost" of resistance. Hence, the expected bonus from research in functional plant genomics will not materialize without deciphering its significance in the physiological and agronomic domains.

If we consider extracting the essence of this book in one sentence it would be that the integration of physiology and biotechnology towards plant breeding should serve to eliminate the genotype–phenotype gap to the extent the functional genomics of plants will be able to impact crop production and its problems.

REFERENCES

Blum A. The temperature response of gas exchange in sorghum leaves and the effect of heterosis. J Exp Bot 1989; 40:453–460.

Tanksley SD, Nelson JC. Advanced backcross QTL analysis: a method for the simultaneous discovery and transfer of valuable QTLs from unadapted germplasm into elite breeding lines. TAG 1996; 92:191–203.

Index